DEPARTMENT OF THE ARMY TECHNICAL MANUAL

90-MM GUN TANK M47 PATTON TECHNICAL MANUAL

By DEPARTMENT OF THE ARMY • JANUARY 1952

©2013 Periscope Film LLC
All Rights Reserved
ISBN#978-1-937684-55-6
www.PeriscopeFilm.com

DISCLAIMER:

This document is a reproduction of a text first published by the Department of the Army, Washington DC. All source material contained herein has been approved for public release and unlimited distribution by an agency of the US Government. Any US Government markings in this reproduction that indicate limited distribution or classified material have been superseded by downgrading instructions promulgated by an agency of the US government after the original publication of the document No US government agency is associated with the publication of this reproduction. This manual is sold for historic research purposes only, as an entertainment. It contains obsolete information and is not intended to be used as part of an actual training program. No book can substitute for proper training by an authorized instructor.

©2013 Periscope Film LLC
All Rights Reserved
ISBN#978-1-937684-55-6
www.PeriscopeFilm.com

RESTRICTED
Security Information

DEPARTMENT OF THE ARMY TECHNICAL MANUAL
TM 9-718A

90-mm GUN TANK M47

DEPARTMENT OF THE ARMY • JANUARY 1952

United States Government Printing Office
Washington : 1952

RESTRICTED
Security Information

This manual is correct to 21 December 1951.

DEPARTMENT OF THE ARMY
WASHINGTON 25, D. C., *9 January 1952*

TM 9-718A is published for the information and guidance of all concerned.

[AG 470.8 (3 Jan 52)]

BY ORDER OF THE SECRETARY OF THE ARMY:

OFFICIAL:
WM. E. BERGIN
Major General, USA
The Adjutant General

J. LAWTON COLLINS
Chief of Staff, United States Army

DISTRIBUTION:
Active Army:
Tech Svc (1); Arm & Svc Bd (2); AFF (2); AA Comd (2); OS Maj Comd (10); Base Comd (2); MDW (3); Log Comd (5); A (20); CHQ (2); D (2); R 9 (2); Bn 9 (2); C 9 (2); FC (2); Sch (5) except 9 (50); Gen Dep (2); Dep 9 (10); PE (5), OSD (2); PG 9 (10); Ars 9 (10); Dist 9 (10); Mil Dist (3); One copy to each of the following T/O & E's: 5-15N, 5-16N, 5-215N, 5-217N, 7-11N, 17-25N, 17-26N, 17-27N, 17-35N, 17-36N, 17-37N, 17-51, 17-55.

NG: Same as Active Army except one copy to each unit.
ORC: Same as Active Army except one copy to each unit.
For explanation of distribution formula, see SR 310-90-1.

CONTENTS

		Paragraphs	Page
CHAPTER 1.	**INTRODUCTION**		
Section	I. General	1-2	1
	II. Description and data	3-5	13
CHAPTER 2.	**OPERATING INSTRUCTIONS**		
Section	I. Service upon receipt of matériel	6-9	21
	II. Controls and instruments	10-66	22
	III. Operation under usual conditions	67-91	77
	IV. Operation of matériel used in conjunction with major item	92-98	104
	V. Operation under unusual conditions	99-108	109
CHAPTER 3.	**ORGANIZATIONAL MAINTENANCE INSTRUCTIONS**		
Section	I. Parts, special tools, and equipment for organizational maintenance	109-112	117
	II. Lubrication and painting	113-119	132
	III. Preventive maintenance services	120-123	143
	IV. Trouble shooting	124-144	160
	V. Engine description and maintenance in vehicle	145-156	217
	VI. Power plant removal and installation	157-161	245
	VII. Engine removal and installation	162-164	263
	VIII. Engine maintenance with engine removed from vehicle	165-170	267
	IX. Fuel, air-intake, and exhaust systems	171-182	275
	X. Ignition system	183-188	304
	XI. Starting system	189-191	319
	XII. Hull electrical system	192-198	327
	XIII. Instrument and hull accessories panels, instruments, switches, and sending units	199-220	344
	XIV. Batteries and generating system	221-224	370
	XV. Horn and lighting system	225-233	381
	XVI. Transmission	234-239	391
	XVII. Oil coolers and lines	240-242	403
	XVIII. Final drives and universal joints	243-247	414
	XIX. Tracks and suspension	248-255	420
	XX. Driver's controls and linkage	256-259	462
	XXI. Hull	260-268	476
	XXII. Turret	269-280	488
	XXIII. Turret electrical system	281-299	499
	XXIV. Auxiliary generator and engine	300-315	517
	XXV. Fire extinguisher systems	316-320	539
	XXVI. Auxiliary equipment	321-329	543
	XXVII. Radio interference suppression	330-336	553
	XXVIII. 90-mm gun T119E1	337-340	555
	XXIX. Breech mechanism	341-346	561

CHAPTER 3. ORGANIZATIONAL MAINTENANCE INSTRUCTIONS—Continued

	Paragraphs	Page
XXX. Combination gun mount M78	347–348	573
XXXI. Firing linkage	349–354	574
XXXII. Recoil mechanism	355–359	578
XXXIII. Elevating mechanism	360–362	581
XXXIV. Machine gun mounts	363–366	582
XXXV. Sighting and fire control equipment	367–369	585
XXXVI. Maintenance under unusual conditions	370–376	588

CHAPTER 4. MATÉRIEL USED IN CONJUNCTION WITH MAJOR ITEM

Section			
	I. Ammunition for 90-mm gun T119E1	377–387	594
	II. Communication system	388–394	628

CHAPTER 5. SHIPMENT AND LIMITED STORAGE AND DESTRUCTION OF MATÉRIEL TO PREVENT ENEMY USE

Section			
	I. Shipment and limited storage	395–398	641
	II. Destruction of matériel to prevent enemy use	399–403	657

APPENDIX	REFERENCES	663
INDEX		669

CHAPTER 1

INTRODUCTION

Section I. GENERAL

1. Scope

a. These instructions are published for information and guidance of the personnel to whom the matériel is issued. They contain information on the operational and organizational maintenance of the matériel as well as descriptions of major units and their functions in relation to other components of the matériel.

b. The appendix contains a list of current references, including supply catelogs, technical manuals, and other available publications applicable to the matériel.

c. In general, the prescribed organizational maintenance responsibilities will apply as reflected in the allocation of tools and spare parts in the appropriate columns of the current ORD 7 supply catalog pertaining to this vehicle and in accordance with the extent of disassembly prescribed in this manual for the purpose of cleaning, lubricating, or replacing authorized spare parts. In all cases where the nature of repair, modification, or adjustment is beyond the scope or facilities of the using organization, the supporting ordnance maintenance unit should be informed in order that trained personnel with suitable tools and equipment may be provided or other proper instructions issued.

Note. The replacement of major components is normally an ordnance maintenance operation, but may be performed in an emergency by the using organization, provided authority for performing these replacements is obtained from the responsible commander. A replacement assembly, any tools needed for the operation which are not carried by the using organization, any necessary special instructions regarding associated accessories, etc., may be obtained from the supporting ordnance maintenance unit.

d. This first edition manual is published in advance of complete technical review. Any errors or omissions will be brought to the attention of the Chief of Ordnance, Washington 25, D. C., ATTN: ORDFM-Pub.

2. Forms, Records, and Reports

a. GENERAL. Forms, records, and reports are designed to serve necessary and useful purposes. Responsibility for the proper execution of these forms rests upon commanding officers of all units operating and maintaining vehicles. It is emphasized, however, that forms, records, and reports are merely aids. They are not a substitute for thorough practical work, physical inspection, and active supervision.

b. AUTHORIZED FORMS. The forms generally applicable to units operating and maintaining these vehicles are listed in the appendix. No forms other than those approved for the Department of the Army will be used. For a current and complete listing of all forms, refer to current SR 310-20-6.

c. ARTILLERY GUN BOOK.

(1) The Artillery Gun Books 28–F–67990 (OO Form 5825) is used to keep an accurate record of the weapon. The gun book is stored in gun book cover CO20-7228906 (M539) (fig. 85). The book is divided as follows: record of assignment; battery commander's daily gun record; and the inspector's record of examination.

Note. Record of assignment data must be removed and destroyed prior to entering combat.

These records are important for the following reasons:

(*a*) They inform the unit commander of the condition and serviceability of the weapons under his jurisdiction.

(*b*) They serve as the record of use and maintenance of the material and expedite effective maintenance.

(*c*) They serve as a source of technical data to the Ordnance Corps for the improvement of weapons, and furnish valuable design data for the development of new weapons.

(2) Instructions on how to make entries in the artillery gun book are contained therein. *It is essential that the gun book be kept complete and up to date, and that the gun book accompany the matériel at all times regardless of where it may be sent.* In order to facilitate proper maintenance of the guns and related matériel (i. e., mounts, recoil mechanism, and associated fire control equipment), and to avoid unnecessary duplication of repairs and maintenance, the following additional entries in the gun book are prescribed:

(*a*) A record of completed modification work orders. The record will show the date completed and bear the initial of the officer or mechanic responsible for completion of the modification.

(*b*) A record of the seasonal changes of lubricant in sufficient detail to prevent duplication and afford proper identification by the inspector.

(c) The estimated accuracy life of this weapon has not been determined at this time. The gun book contains information on the method of calculating full service rounds. The reference to OFSB 4–1 as the source of data for estimated accuracy of life in paragraph 6 of the gun book should be deleted.

(d) When a removable tube is replaced, an entry of the proof firing and bore diameter data shown on the star gage record which accompanies a new tube will be made in the gun book. Proof facilities will complete and forward star gage records with new tubes, attached in the same general manner as prescribed for gun books ((3) below). The tube serial number and all pertinent firing data for the removed tube which are contained in the gun book will be extracted onto a history of cannon report. History of gun reports will be extracted in letter form and forwarded through technical channels to the Chief of Ordnance, Washington 25, D. C., ATTN: ORDFM-Weapons Section.

(3) The following procedure is prescribed to insure that the gun book will always accompany the matériel whenever it is shipped or transferred from one organization to another:

(a) During transfer or shipment, the gun book will be kept in a waterproof envelope securely fastened to the matériel with waterproof tape.

(b) Under one of the wrappings of tape, one end of a small tab will be inserted reading "Gun Book Here."

(4) *Instructions for making gun book entries and the procedure for keeping the gun book with the weapon whenever it is shipped or transferred from one organization to another must be strictly followed.* Field maintenance units and depot maintenance shops will insist that the gun book accompany each weapon when it enters their shop for repairs or maintenance.

(5) When a gun book is completely filled, an additional new gun book, requisitioned through normal ordnance supply channels ((6) below), will be added to it by stapling the covers together so that the two books will remain together as a single unit.

(6) If a gun book is lost, it will be replaced at once and all available data will be entered in the new gun book. Additional copies of Artillery Gun Book (OO Form 5825, Federal Stock No. 28–F–67990) may be requisitioned through normal ordnance supply channels. A gun book which has become separated from the weapon to which it pertains and for which

3

efforts to locate the weapon have failed, will be forwarded immediately to the Chief of Ordnance, Washington 25, D. C., ATTN: ORDFM-Weapons Section.

(7) When the gun, including breech ring, is condemned, destroyed, turned in for salvage, or otherwise lost from service, the gun book will be forwarded with proper notation to the Chief of Ordnance, Washington 25, D.C., ATTN: ORDFM-Weapons Section. Information contained in the gun book which pertains to the mount, recoil mechanism, or other weapon components being retained in service will be extracted and inserted in the gun book pertaining to the replacement gun.

d. FIELD REPORTS OF ACCIDENTS. The reports necessary to comply with the requirements of the Army safety program are prescribed in detail in the SR 385-10-40 series. These reports are required whenever accidents involving injury to personnel or damage to matériel occur. Whenever an accident or malfunction involving the use of ammunition occurs, firing of the lot which malfunctions will be immediately discontinued. In addition to any applicable reports required above, details of the accident or malfunctions will be reported as prescribed in SR 385-310-1.

e. REPORT OF UNSATISFACTORY EQUIPMENT OR MATERIALS. Any suggestions for improvement in design and maintenance of equipment, safety and efficiency of operation, or pertaining to the application of prescribed petroleum fuels, lubricants, and/or preserving materials, will be reported through technical channels, as prescribed in SR 700-45-5, to the Chief of Ordnance, Washington 25, D. C., ATTN: ORDFM, using DA AGO Form 468, Unsatisfactory Equipment Report. Such suggestions are encouraged in order that other organizations may benefit.

Note. Do not report all failures that occur. Report only REPEATED or RECURRENT failures or malfunctions which indicate unsatisfactory design or material. However, reports will always be made in the event that exceptionally costly equipment is involved. See also SR 700-45-5 and the printed instructions on DA AGO Form 468.

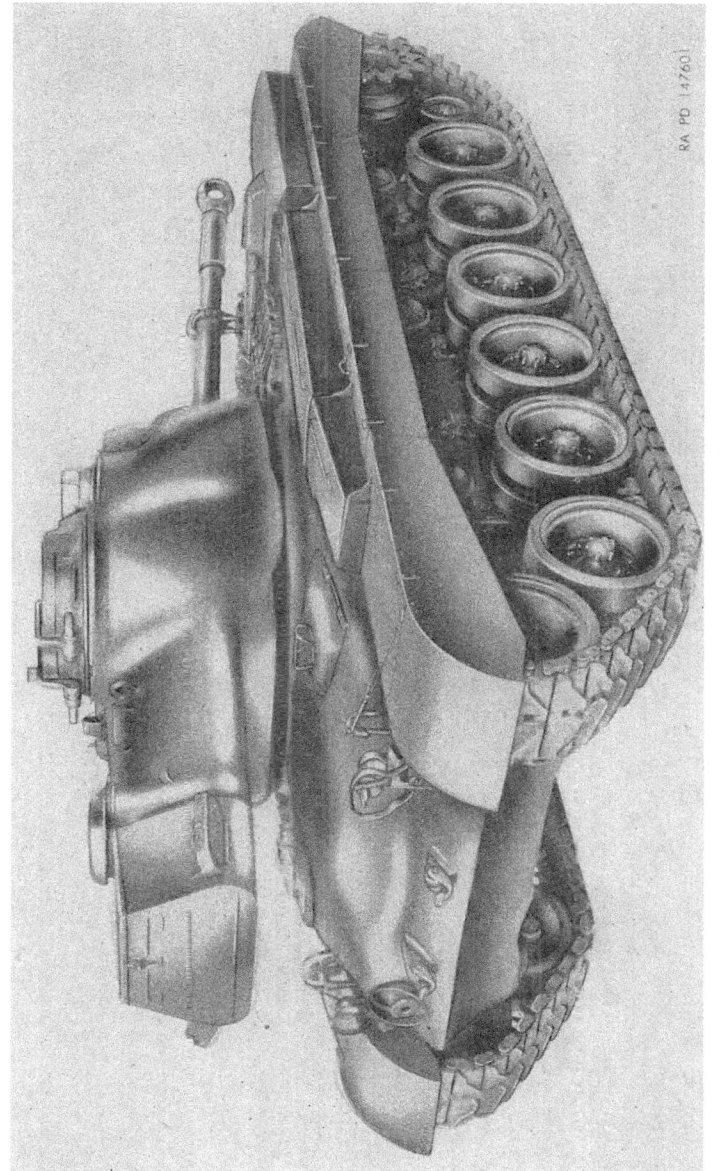

Figure 1. 90-mm gun tank M47—left front view.

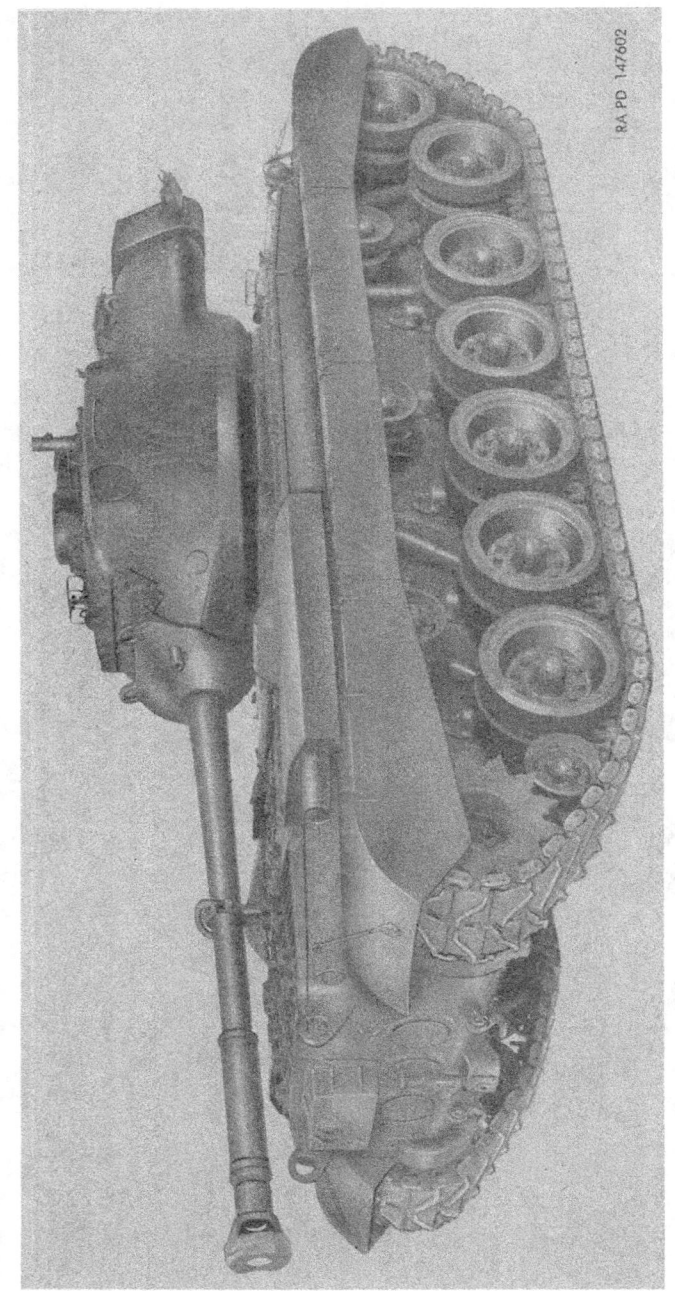

Figure 2. 90-mm gun tank M47—right rear view.

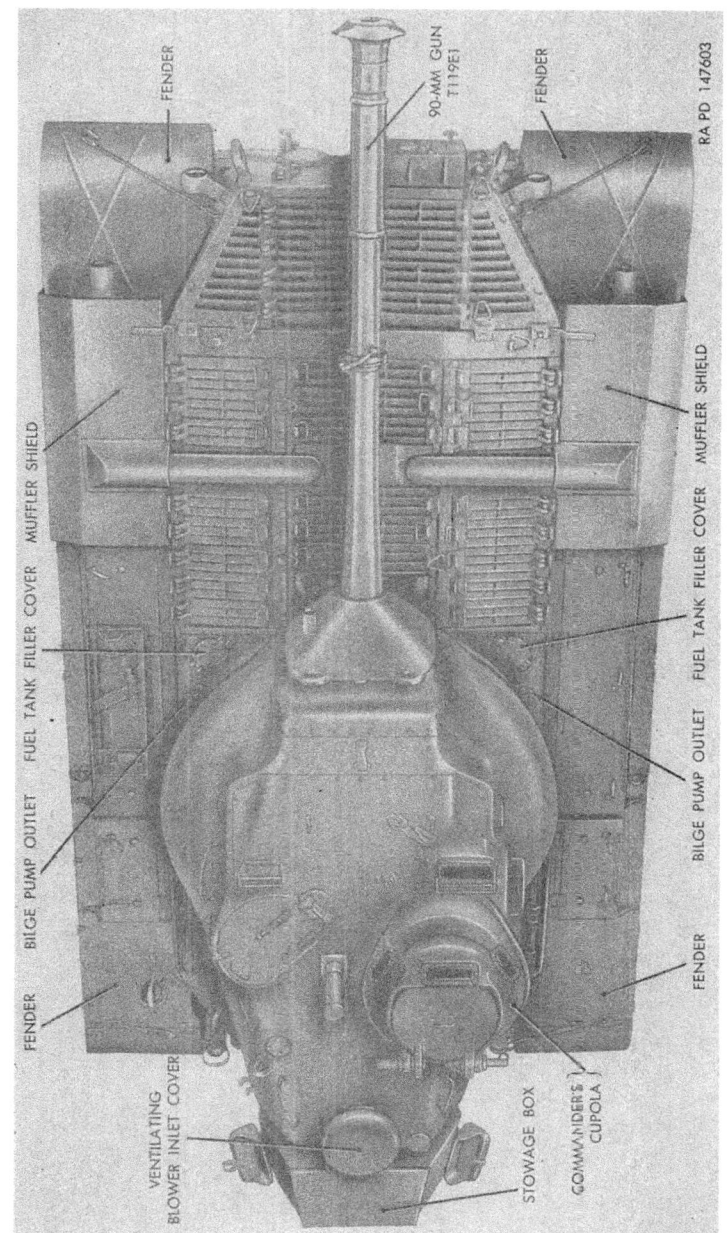

Figure 3. 90-mm gun tank M47—top view.

7

Figure 4. 90-mm gun tank M47—left side view.

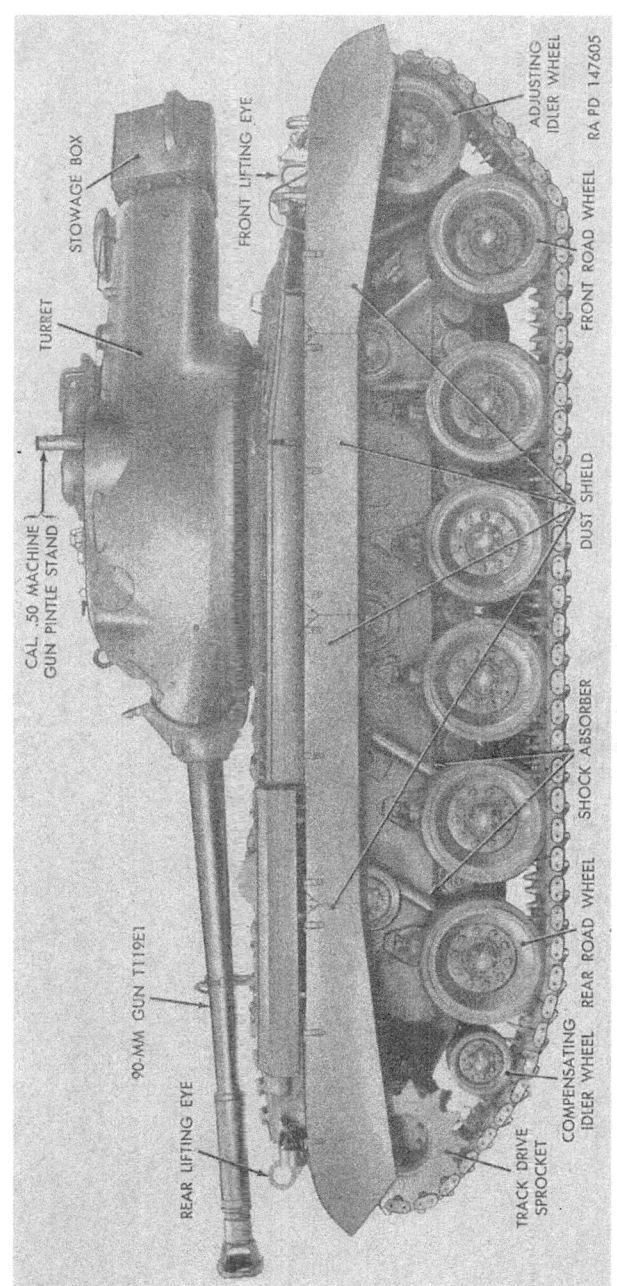

Figure 5. 90-mm gun tank M47—right side view.

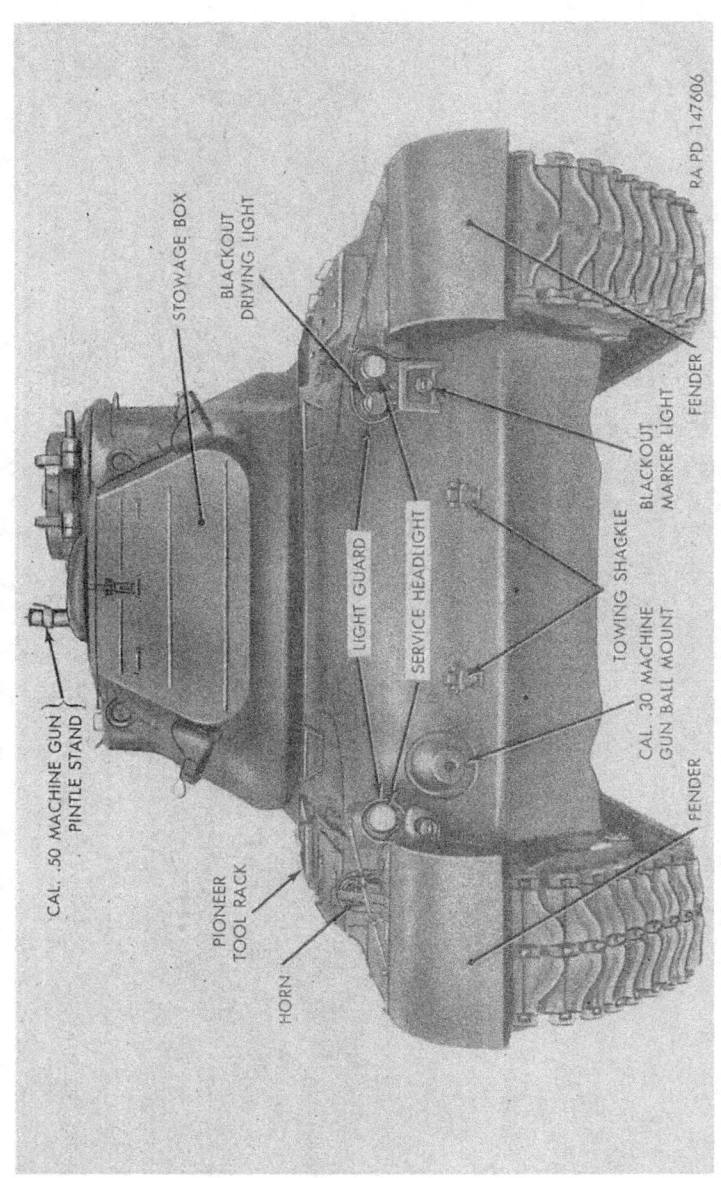

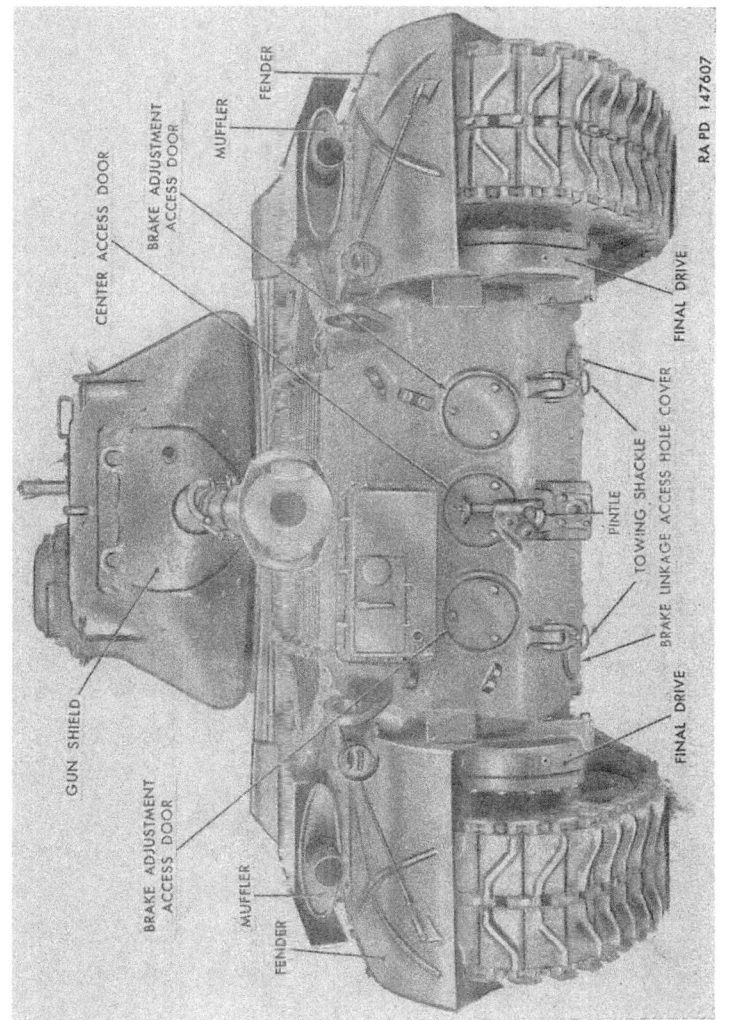

Figure 7. 90-mm gun tank M47—rear view.

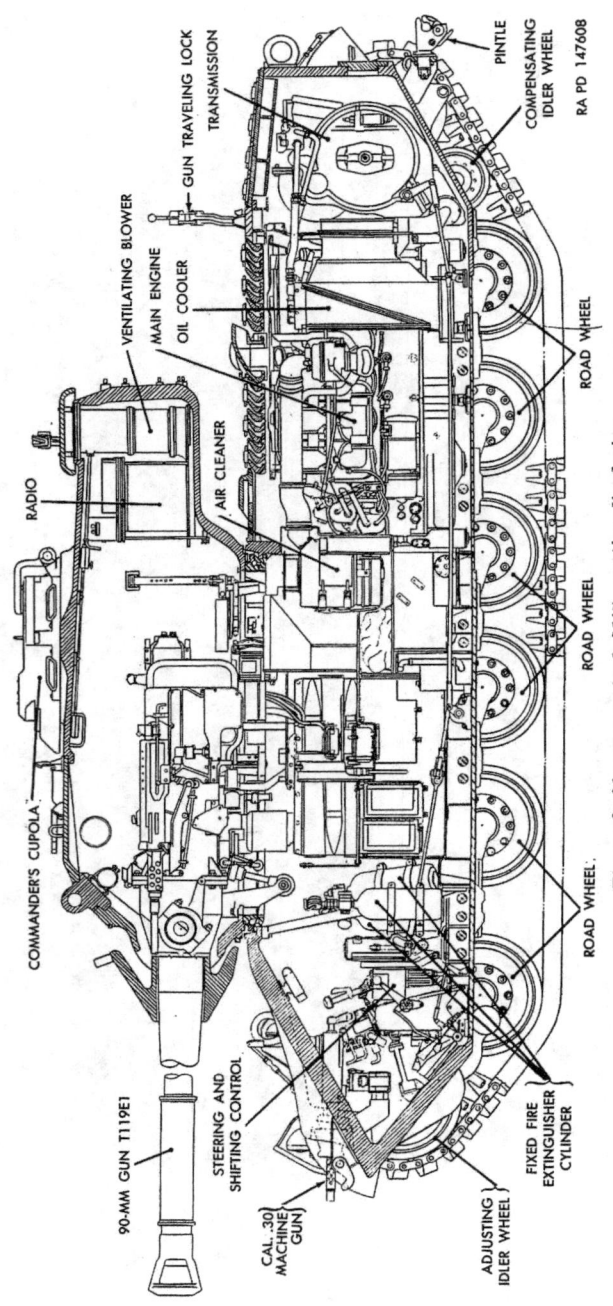

Figure 8. 90-mm gun tank M47—sectionalized view.

Section II. DESCRIPTION AND DATA

3. Description

a. GENERAL. The 90-mm gun tank M47 is a heavily-armored, full-track-laying, low-silhouette, combat vehicle, mounting a 90-mm gun T119E1 (figs. 1 through 8). One cal. .30 machine gun M1919A4 is installed in a flexible ball mount in the bow (fig. 13) and a cal. .50 machine gun M2, HB or a cal. .30 machine gun is mounted coaxially with the 90-mm gun in the combination gun mount (fig. 39). One cal. .50 machine gun M2, HB with a pintle mount is installed in a pintle stand on the turret roof (fig. 5). This vehicle carries a crew of five—vehicle commander, driver, assistant driver (also, cal. .30 gunner), loader, and gunner.

b. POWER TRAIN AND SUSPENSION. The M47 tank is powered by a Continental model AV–1790–5B, 12-cylinder, V-type, 4-cycle, air-cooled engine. The cylinders are individual replaceable units arranged in two banks of six each with a camshaft for each bank. The valve rocker arm assemblies are located in the head of each cylinder. Two mechanically driven fans located on top of the engine circulate air around the cylinders for cooling. Power is transmitted to the final drives and track sprockets through a cross-drive transmission which is a combined transmission, differential, and steering unit. The cross-drive transmission, model CD–850–4, incorporates a split-hydraulic-power path providing split-torque drive; variable-steering mechanism; and built-in-disk-type brakes. The operation of the transmission is controlled by means of a manual control box installed in the driving compartment and connected to the transmission through mechanical linkage. Both steering and drive-range shifting are controlled by a manual control lever on the manual control box. The vehicle is equipped with torsion-bar-type suspension and individually sprung wheels.

c. HULL. The hull of the M47 tank is made from armor plate and cast armor sections welded together and reinforced. Access hole covers and plates (fig. 285) are located in the bottom and rear of the hull. Escape hatch doors for the driver and assistant driver are in the bottom of the hull at the front (fig. 284). The hull is divided into two compartments—the crew compartment (driving and fighting) at the front and the engine compartment (engine and transmission) in the rear. These compartments are separated by a bulkhead that extends across the hull. Four drain valves are provided in the hull floor, one in crew compartment and three in engine compartment.

d. TURRET. The turret mounts the 90-mm gun T119E1 (fig. 4) in a combination gun mount M78 (fig. 350). The turret can be traversed 360° in either direction both manually and by means of a hydraulic-power-turret-traversing mechanism. The 90-mm gun T119E1 can be elevated or depressed through 24° (19° elevation and 5° depression) both manually and by means of a hydraulic-power-gun-elevating mechanism. All sighting and fire control instruments except driver's and assistant driver's periscope are installed in the turret.

e. AUXILIARY GENERATOR AND ENGINE (fig. 309). The M47 tank is equipped with an auxiliary generator driven by a two-cylinder, gasoline, air-cooled engine to furnish auxiliary electrical power for the vehicle's electrical equipment in addition to that supplied by the generator on the main engine. The auxiliary generator and engine can be operated while the main engine is not running to recharge batteries and to provide power for communications and other electrical equipment. Fuel for the auxiliary engine is supplied from the vehicle fuel tanks.

f. FIRE EXTINGUISHERS (fig. 324 and 327). The vehicle is equipped with a fixed fire extinguisher system which may be operated from the crew compartment, or by a remote control handle from outside the tank. A portable fire extinguisher is located in the crew compartment.

g. BILGE PUMPS (figs. 334 and 335). Two bilge pumps for deep water operations are installed in the vehicle, one in the engine compartment and the other in the crew compartment.

h. VENTILATING BLOWER (fig. 304). An electric-motor-driven-ventilating blower is installed in the rear of the turret to provide ventilation in the crew compartment.

i. CREW COMPARTMENT HEATER (fig. 328). A gasoline-type heater is installed in the driver's compartment to provide heat in the crew compartment when required. Fuel for the crew compartment heater is supplied from the vehicle fuel tanks.

4. Name, Caution, and Instruction Plates

a. VEHICLE NAME PLATE. The vehicle name plate is mounted on the hull wall at left side of crew compartment. Data on vehicle nomeclature, number in crew, manufacturer's name, vehicle publications, maximum allowable vehicle speeds, vehicle dimensions, engine cruising speed, and date of manufacture are furnished.

b. MAIN ENGINE NAME PLATE (fig. 137). The main engine name plate is located on left-front side of crankcase, behind starter. Data on engine model number, serial number, rated speed, compression ratio, oil pressure, spark advance, valve timing and clearance, and firing order are furnished.

c. TRANSMISSION NAME PLATE (fig. 230). The transmission name plate is located on top of transmission housing near the front. Data on model number, serial number, and recommended oil are furnished.

d. Main Generator Name Plate. The main generator name plate includes data on manufacturer's model number, manufacturer's drawing number, type, style, minimum revolutions per minute, and rated amperage and voltage. It is located on side of generator.

e. Starter Name Plate (figs. 177 and 178). The starter name plate, located on the starter housing, includes data on manufacturer's type, model number, rated voltage, and torque.

f. Auxiliary Engine Name Plate (fig. 309). The auxiliary engine name plate is located on top of the flywheel shroud. Data on manufacturer's model number, serial number, piston bore and stroke, specifications number, and operating instructions are furnished.

g. Oil Cooler Fan Name Plate (fig. 233). The oil cooler fan name plate is located on top of oil cooler fan shroud. It includes manufacturer's name and serial number.

h. Ignition Harness Name Plate (figs. 96 and 97). The ignition harness name plate is located on harness near No. 1 cylinders. It furnishes assembly number, specification number, manufacturer's order number, class, serial number, and type.

i. Magneto Name Plate (fig. 166). The magneto name plate is located on the magneto housing behind the cam oiler. Serial number, type, manufacturer's drawing number, and specification number are given on this plate.

j. Governor Name Plate. The governor name plate is located on the side of the governor housing and gives the manufacturer's name and part number.

k. Booster-and-filter-coils Name Plate (fig. 206). The booster-and-filter-coils name plate is located on the side of the booster-and-filter-coils housing. It furnishes the manufacturer's part number and serial number. A warning plate also is located on the front side of the housing.

l. Auxiliary Generator Name Plate. The auxiliary generator name plate, located on the side of the auxiliary generator, includes data on manufacturer's model number, manufacturer's drawing number, type, style, minimum revolutions per minute, and rated amperage and voltage.

m. Manual Control Box Name Plate. The manual control box name plate is located on the rear of the manual control box. It furnishes information on ordnance number, serial number, oil specification, and proper oil level.

n. Ventilating Blower Motor Name Plate. The ventilating blower motor name plate is located on the bottom of the ventilator blower motor. Data on model, horsepower, rated voltage and amperage, serial number, revolutions per minute, and manufacturer's number are furnished.

o. MASTER JUNCTION BOX NAME PLATE. The master junction box name plate is located on the bottom of the box, beneath one of the mounting straps. It includes manufacturer's data.

p. FIXED FIRE EXTINGUISHER INSTRUCTION PLATE (fig. 9). The fixed fire extinguisher instruction plate is located on rear of vertical support behind the driver's compartment. It supplies operating instructions for the fixed fire extinguisher system.

q. PORTABLE FIRE EXTINGUISHER INSTRUCTION PLATE (fig. 327). The portable fire extinguisher instruction plate, located on the portable fire extinguisher cylinder, furnishes data on type, weight, capacity, operating instructions, and recharging procedure.

r. HULL ACCESSORIES PANEL WARNING PLATE (fig. 190). The hull accessories panel warning plate, located on the hull accessories panel, contains a warning regarding the use of the master relay switch.

s. CREW COMPARTMENT HEATER INSTRUCTION PLATE (fig. 219). The crew compartment heater instruction plate, located on the hull top above the instrument panel, furnishes operating instructions for the crew compartment heater.

t. TURRET-TRAVERSING-AND-GUN-ELEVATING-SYSTEM-MOTOR NAME PLATE (fig. 32). The turret-traversing-and-gun-elevating-system-motor name plate is located on the side of the motor housing. Data on model, horsepower, rated voltage and amperage, manufacturer's number, serial number, and revolutions per minute, are furnished.

u. AZIMUTH INDICATOR NAME PLATE (fig. 58). The azimuth indicator name plate, located on the side of the housing, furnishes data on type, number and manufacturer.

v. COMBINATION GUN MOUNT NAME PLATE (fig. 350). The combination gun mount name plate, located on the right side of the gun mount, below the gun firing relay, supplies information regarding the manufacturer, serial number, date of manufacture, and publications.

w. GUNNER'S CONTROL NAME PLATE (fig. 24). The gunner's control name plate, located on the gunner's control handle assembly, furnishes information regarding voltage, manufacturer's number, and serial number.

x. PERISCOPE NAME PLATE (fig. 49). The periscope name plate, located on the front of the body assembly, furnishes the model and serial numbers of the periscope body.

y. ELEVATION QUADRANT NAME PLATE (fig. 57). The elevation quadrant name plate, located on the elevation quadrant body, furnishes the model and serial numbers.

5. Tabulated Data

a. GENERAL DATA.

Armament	One 90-mm gun T119E1 (combination gun mount M78); two cal. .50 machine guns M2, HB; one cal. .30 machine gun M1919A4
Crew	5
Engine	Continental V-12, air-cooled, model AB-1790-5B
Weight (combat loaded)	97,200 lb.

b. VEHICLE.

Dimensions:
- Length (gun in traveling position) --- 279$\frac{3}{16}$ in.
- Height (lowest operable) (to top of machine gun mount) --- 116$\frac{9}{16}$ in.
- Over-all length of hull --- 250$\frac{5}{16}$ in.
- Width --- 138$\frac{1}{4}$ in.
- Height (overall) w/machine gun --- 131 in.
- w/o machine gun --- 116$\frac{9}{16}$ in.
- Ground clearance --- 18$\frac{1}{2}$ in.
- Ground pressure --- 13.3 psi
- Electrical system --- 24 volt
- Number of batteries --- 4–12 volt (2 sets of 2)

Hull armor:
- Type --- homogeneous steel; front, armor plate; rear, cast armor
- Front:
  - Upper --- 4 in. at 60 deg.
  - Lower --- 3 to 3$\frac{1}{2}$ in. at 53 deg.
- Sides:
  - Front --- 3 in. at 0 deg.
  - Rear --- 2 in. at 0 deg.
- Rear --- 2 in. at 10 deg.
- Top --- $\frac{7}{8}$ in.
- Floor:
  - Front --- 1 in.
  - Rear --- $\frac{1}{2}$ in.

Turret armor:
- Type --- cast homogeneous
- Front --- 4 in. at 40 deg.
- Sides --- 2$\frac{1}{2}$ in. at 30 deg.
- Rear --- 3 in. at 3 deg.
- Top --- 1 in.

Capacities:
  Auxiliary engine crankcase _____ 3½ qt.
  Fuel tank, right _____ 100 gal.
  Fuel tank, left _____ 133 gal.
  Main engine carburetor air cleaner (each) _____ 3½ qt.
  Main engine crankcase (refill) _____ 16 gal.
  Transmission (aprx.) _____ 23 gal.
  c. ARMAMENT.

90-mm. gun T119E1:
  Weight of gun (complete) _____ 2,650 lb.
  Weight of gun tube _____ 1,750 lb.
  Length of bore (without muzzle brake) _____ 177.15 in.
  Type of breechblock _____ vertical sliding wedge
  Type of firing mechanism _____ percussion, inertia

Ammunition—refer to chapter 4 for complete data.

Combination gun mount M78:
  Gun shield armor thickness _____ 4½ in.
  Type of recoil mechanism _____ concentric hydrospring
  Length of recoil (normal) _____ 12 in.
  Length of recoil (max) _____ 14 in.
  Capacity of recoil mechanism (including replenisher
    assembly) _____ 5½ gal.
  Operation of firing linkage _____ manual or electrical

d. SIGHTING AND FIRE CONTROL.
  (1) *On-carriage equipment.*

Azimuth indicator T24:
  Dial graduations (100-mil) _____ 0–3,200, 3–3,200
  Dial graduations (1-mil) _____ 0–100
  Gunner's aid dial _____ 0–50 R and L

Ballistic drive T23E1 w/instrument light M30:
  Scale graduations:
    Mil scale _____ mils
    HE _____ 200 yd.
    AP _____ 200 yd.
    HVAP _____ 200 yd.
    Heat _____ 200 yd.

Elevation quadrant T21 w/instrument light T22:
  Scale graduations:
    Elevation (100-mil) _____ 200 to 0 to +60
    Elevation (1-mil) _____ 0–100
    Limit of elevation _____ 600 mils
    Limit of depression _____ 200 mils

Periscope M6, to M13, or M13B1:
  Magnification _____ 1X

Periscope T35:
    Magnification:
        Periscope _____ 1X
        Telescope _____ 6X
    Field of view (periscope):
        Horizontal_____ 33 deg. 18 min.
        Vertical_____ 7 deg. 9 min.
    Field of view (telescope)_____ 8 deg.
    Deflection graduation on reticle_____ L 20 to 0 to 20 R
Periscope mounts:
    Commander's, T177
    Gunner's, T176
Range finder T41:
    Magnification_____ 7.5X
    Field of view_____ 5 deg.
    Range scale graduations on reticle_____ 500 to 5,000 yd.
    Ammunition graduations on reticle__ cal. .50, HEAT T108, HE
                                                                         M71, AP T33, HVAP
                                                                         M304, HVAP T65, and
                                                                         HVAP T67
    Reticle movement (mils)_____ L 20 to 0 to 20R
    Diopter scale range_____ −3 to +3 diopters
    Interpupillary adjustment_____ 58 to 72 mm
Superelevation transmitter T13:
    Quadrant, gunner's, M1 w/e
    Setter, fuze, M27 or M14
   (2) *Off-carriage equipment.*
Binocular, M17A1 w/case, carrying M24
Light, aiming post, M14 (red and green filter)
Post, aiming, M1A1 or M1 w/cover M401
Table, graphical firing, M42
   e. COMMUNICATIONS SYSTEM.
Radio and interphone consisting of alternative combinations of:
    AN/GRC-3, 4, 5, 6, 7, or 8 with appropriate antennas; and
    AN/VIA-1.
   f. PERFORMANCE.
Maximum allowable speed_____ 12 (low range); 30 (high range);
                                      12 (reverse) mph
Maximum vehicle speed_____ 30 mph
Cruising range (aprx)_____ 80 miles
Gun elevating mechanism_____ hydraulic
Depression of gun (max)_____ (89 mils) 5 deg.
Elevation of gun (max)_____ (338 mils) 19 deg.
Maximum rate of power elevation of gun_____ 4 deg. per sec.
Fording depth (max)_____ 4 ft.

Fuel consumption_____ ⅓ mpg
Maximum allowable oil consumption (main engine)__ 2 gal. in 1½ hr.
Grade ascending ability (max)_____ 60 percent
Traverse of turret_____ (6,400 mils) 360 deg.
Minimum time required for power traverse of turret 360 deg.___ 10 sec.
Minimum turning circle (right or left)_____ pivot
Vertical obstacle vehicle will climb (max)_____ 36 in.
Width of ditch vehicle will cross (max)_____ 102 in.

g. DETAILED DATA REFERENCES. Additional detailed tabular data pertaining to individual components and systems are contained in the following paragraphs:

Ammunition (authorized rounds)_____ 382
Auxiliary generator and engine_____ 300
Crew compartment heater_____ 321
Engine _____ 145
Fuel system_____ 171
Ignition _____ 183
Starting system_____ 189
Tracks _____ 248
Transmission_____ 234

CHAPTER 2

OPERATING INSTRUCTIONS

Section I. SERVICE UPON RECEIPT OF MATÉRIEL

6. Purpose

a. When a new or reconditioned vehicle is first received by the using organization, it is necessary for the organizational mechanics to determine whether the vehicle has been properly prepared for service by the supplying organization and to be sure it is in condition to perform any mission to which it may be assigned when placed in service. For this purpose, inspect all assemblies, subassemblies, and accessories to be sure they are properly assembled, secure, clean, and correctly adjusted and/or lubricated. Check all tools and equipment to be sure every item is present, in good condition, clean, and properly mounted or stowed.

b. In addition, perform a "run-in" of at least 50 miles on all new or reconditioned vehicles, and a sufficient number of miles on used vehicles to completely check their operation, according to procedures in paragraph 9 herein.

c. Armament parts when received from storage coated with rust-preventive compound should be thoroughly cleaned with waste, wiping cloths, or a brush saturated with dry-cleaning solvent, or one part of grease-cleaning compound to four parts dry-cleaning solvent or volatile mineral spirits. After complete removal of the compound, lubricate as specified in the lubrication order (par. 113). Component parts of each weapon should be cleaned separately where practicable. Although like parts are interchangeable, the parts originally assembled work best together.

d. Whenever practicable, the vehicle crew will assist in the performance of these services.

7. Preliminary Service

Perform the Commander's "C" (quarterly or 750-mile) preventive-maintenance service, with the following variations:

a. Line out the other services on the work sheet (DA AGO Form 462) and write in "New (or Rebuilt) Vehicle Reception."

b. Perform item 19 before starting the road test. If a processing tag on the vehicle states that the engine contains preservative oil that is suitable for 500 miles of operation, and of the correct seasonal viscosity, check the level but do not change the oil; otherwise change it. Lubricate all points, regardless of interval. Check the levels of the lubricant in all gear cases. If the gear lubricant is known to be of the correct seasonal grade, do not change it; otherwise change it.

c. Perform items 12 and 27. Inspect breaker points; dressing should not be necessary.

8. Run-in Procedures

a. RUN-IN TEST. Refer to section III of this chapter for operating instructions. Continue the road test (items 1 through 9) for at least 50 miles, unless the vehicle has been driven to the using organization. In the latter case, make the road test only long enough to make the usual observations. Stop at least every 10 miles and make external observations around the vehicle; look particularly for overheated sprocket, idler, roadwheel, or support-roller hubs and leaks from their lubricant seals.

b. AFTER RUN-IN TEST. Upon completion of the run-in test, change the engine and transmission oil and place the vehicle in normal service. It will be due for its first regular preventive-maintenance service after 1 month or 250 additional miles.

9. Correction of Deficiencies

a. Ordinary deficiencies disclosed during the preliminary inspection and servicing or during the break-in period will be corrected in the usual way; that is, by the using organization or a higher maintenance echelon.

b. Serious deficiencies, which appear to involve unsatisfactory design or material, will be reported on DA AGO Form 468, Unsatisfactory Equipment Report. The commander of the using organization will submit the completed form to the Chief of Ordnance, Washington 25, D. C., ATTN: ORDFM, or chief of appropriate technical service or other than ordnance equipment.

Section II. CONTROLS AND INSTRUMENTS

10. General

This section describes, locates, illustrates, and furnishes the crew with sufficient information pertaining to the various controls and instruments provided for the proper operation of the vehicle.

11. Manual Control Lever

The vehicle is equipped with a manual control lever (fig. 10) for steering the vehicle and selecting the transmission gear ranges. The manual control box is located between the driver and the assistant driver. The manual control lever has four shift positions: neutral ("NEU"), low ("LO"), high ("HI"), and reverse ("REV"). Selecting the proper gear range is accomplished by moving the lever forward or to the rear. Steering is accomplished in any shift position by moving the lever to the left or the right. To move the lever in or out of neutral position, whether in steer or neutral steer position, it is necessary to actuate the hand grip handle. To move the lever from low to high or vice versa, neither the finger lift trigger nor the hand grip handle need be actuated. Detent cams and springs incorporated in the manual control box and in the transmission control valve, make it possible to definitely locate the low and high positions. To shift into reverse from either low or high, or vice versa, the finger lift trigger or the hand grip handle must be actuated.

Note. On some early models of tank M47, a dual manual control box is provided. However, the right hand portion of the dual box is not used.

12. Service Brake Pedal

A service brake pedal (fig. 9) is located in front of the driver. To apply the brakes, push the pedal forward. The pedal will return to the "OFF" position as soon as pressure is released

13. Hand Brake Lock Handle

The hand brake lock handle (fig. 10) is located to the right of the driver. To lock the brakes for parking, depress and hold the service brake pedal in applied position; then pull the handle back as far as it will go. To release the brakes, step on the pedal and the brake lock will automatically release.

Caution: Never apply the handle roughly. The handle and locking mechanism can be damaged.

14. Accelerator Pedal

An accelerator pedal (fig. 9), provided for the driver, is located on the floor in front of the driver's set. To accelerate the vehicle, apply pressure to the accelerator pedal.

15. Hand Throttle Lever

The hand throttle lever (fig. 9) is located to the right of the driver. To open the carburetor throttle valve plates and thereby increase engine speed, push the lever forward. To close the throttle valve plates, pull the lever back as far as it will go.

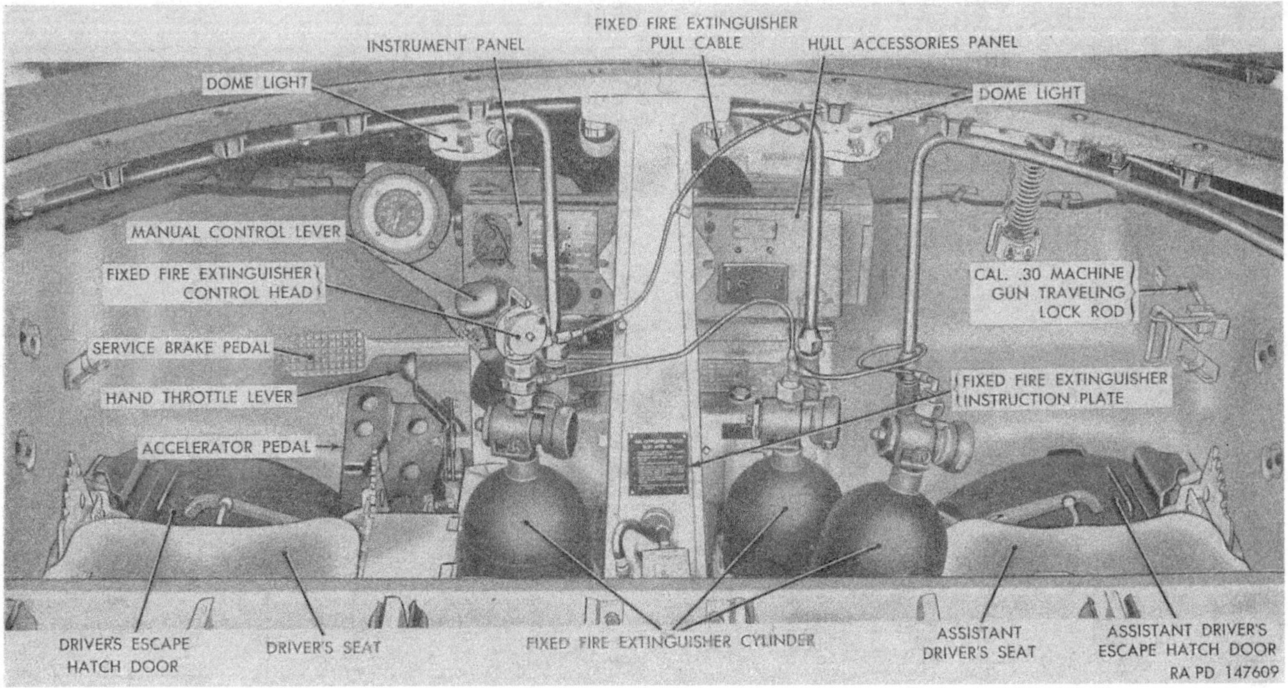

Figure 9. Driver's compartment.

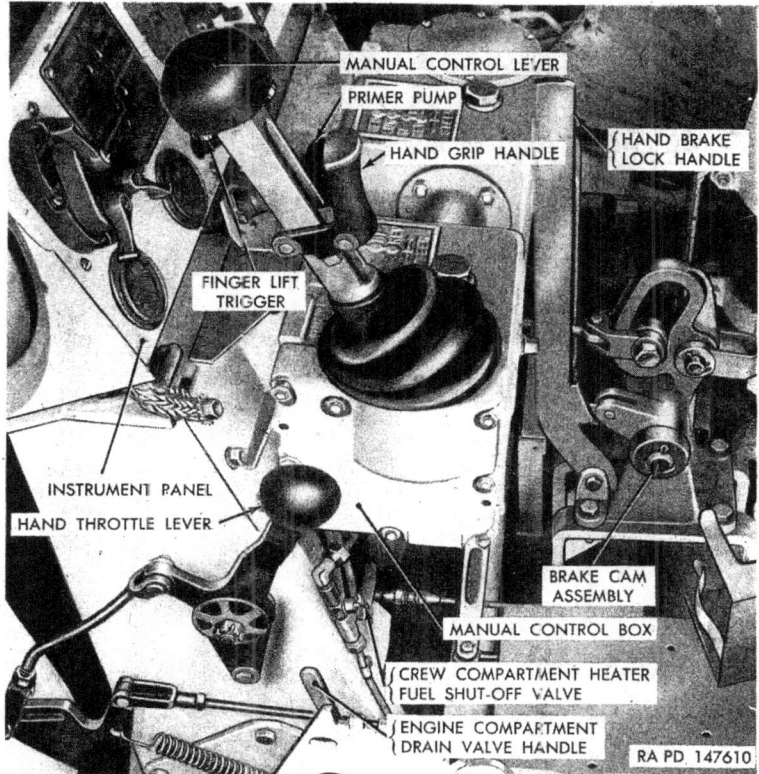

Figure 10. Driver's controls.

16. Primer Pump

The primer pump (fig. 158), is located on a bracket just below the instrument panel to the right of the driver's manual control lever. It is manually operated to force fuel into the engine cylinders to facilitate starting in cold weather.

17. Fuel-Tank-Shut-Off-Valve Handles

The two fuel-tank-shut-off-valve handles (fig. 11) are located underneath the horizontal section of the bulkhead below the rear center of the turret. They are accessible through a plate in the horizontal section of the bulkhead. The rear handle controls the left fuel tank supply; the front handle controls the right fuel tank supply. To shut off the fuel supply from either tank, push the handle down. To open, pull the handle up.

Figure 11. Fuel-tank-shut-off-valve handles.

18. Escape Hatch Doors

Escape hatch doors (figs. 12 and 13) are located in the hull floor in front of both the driver's and the assistant driver's seats. They are used as emergency exists from the tank. To operate an escape hatch door, pull up on the release lever and allow the door to drop to the ground.

19. Driver's Doors

Above each driver's seat is an armor-plate door (fig. 14). The doors are hinged so that they swing upward and outward (fig. 14). Each door is held in the closed position by operating the locking plunger lever (fig. 15) and seating the locking plunger securely in its boss. To open, move the locking plunger lever and push the door outward. It is held in the open position by a latch. To close the door, pull out on the latch trigger rod handle on the inner side of the door and carefully lower the door.

20. Driver's Seats

Form-fitting padded seats (figs. 12 and 13) with padded backs are provided for the driver and the assistance driver. A control lever on each seat is used for adjusting to any of five positions. The seat control levers are located below the front edges of the seats. To adjust seat height, pull the lever back, raise or lower the seat as desired, and release the lever to lock the seat in place.

Caution: Do not attempt to adjust the seat height unless sitting in seat. Weight of the body is required to control action.

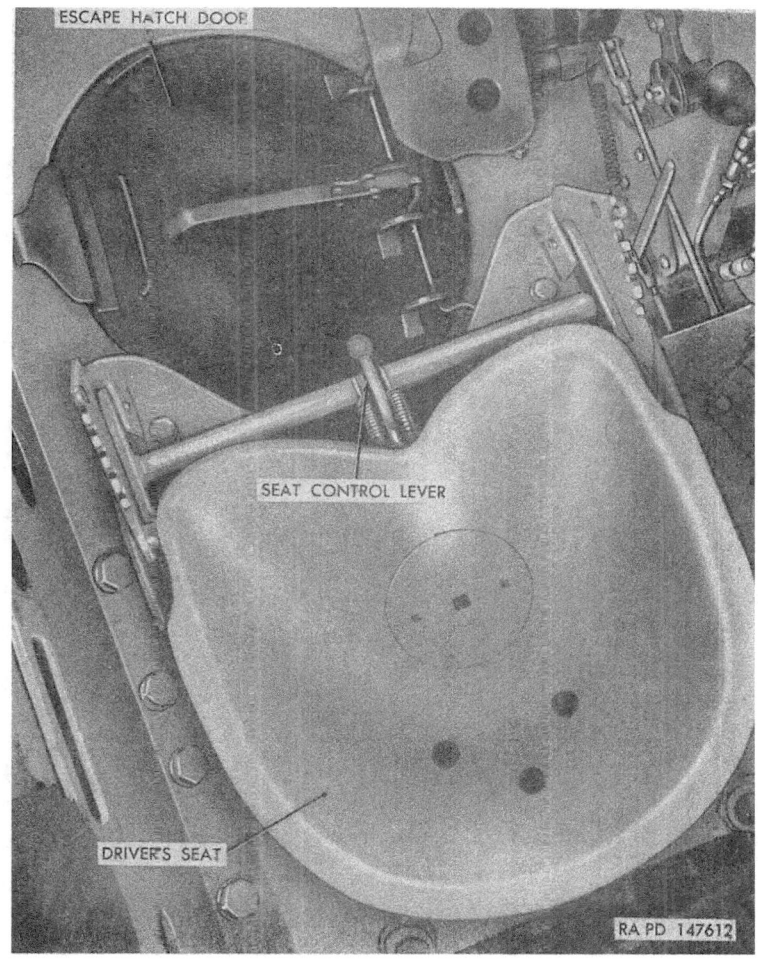

Figure 12. Driver's seat controls and escape hatch door.

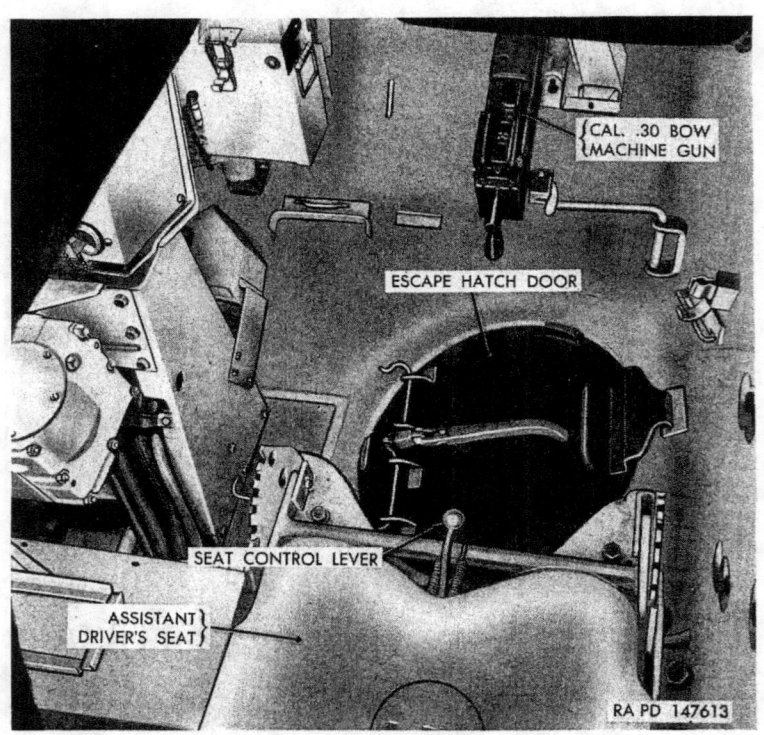
Figure 13. Assistant driver's controls and escape hatch door.

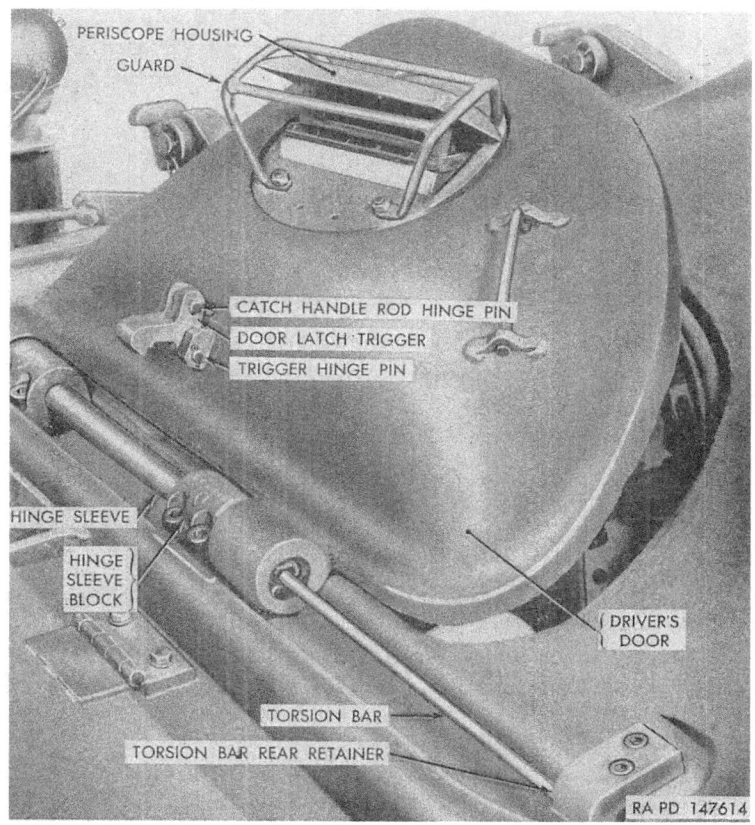

Figure 14. Driver's door partially open.

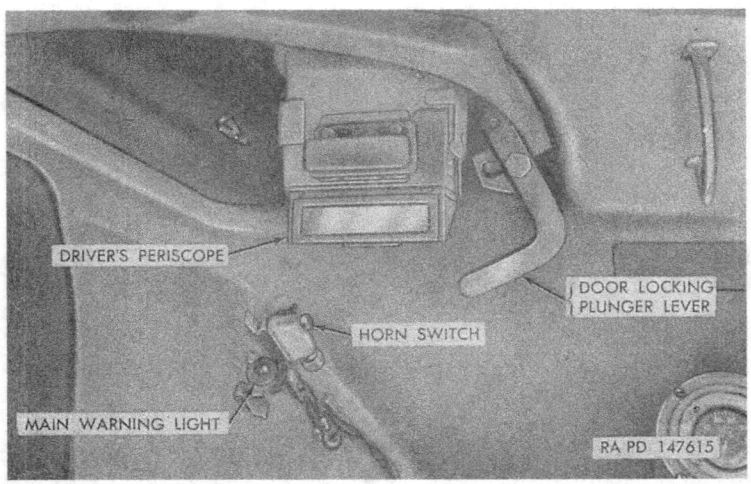

Figure 15. Driver's door locking plunger lever and periscope.

21. Drain Valves

a. GENERAL. Four poppet-type, spring-loaded, drain valves are provided in the hull floor to drain the vehicle of any accumulated water, oil, or fuel. Two individual handles are used to operate the valves.

Note. When the vehicle is to be parked for a period of time, long enough for water to accumulate, always leave the drain valves open.

b. CREW COMPARTMENT DRAIN VALVE. The handle for the crew compartment drain valve is located below the instrument panel (fig. 16). To open the valve, pull the handle up and back. To close the valve, push the handle forward.

c. ENGINE COMPARTMENT DRAIN VALVE. There are three engine compartment drain valves: left front, right front, and rear. One spring-loaded handle (fig. 17), located on the driver's right side below the throttle control lever, operates the three valves. To drain the engine compartment of fluid, pull back on the handle and engage it in the hold-open position by inserting the handle in the slot on the left of the handle support. To close the valve, release the handle from the hold-open slot.

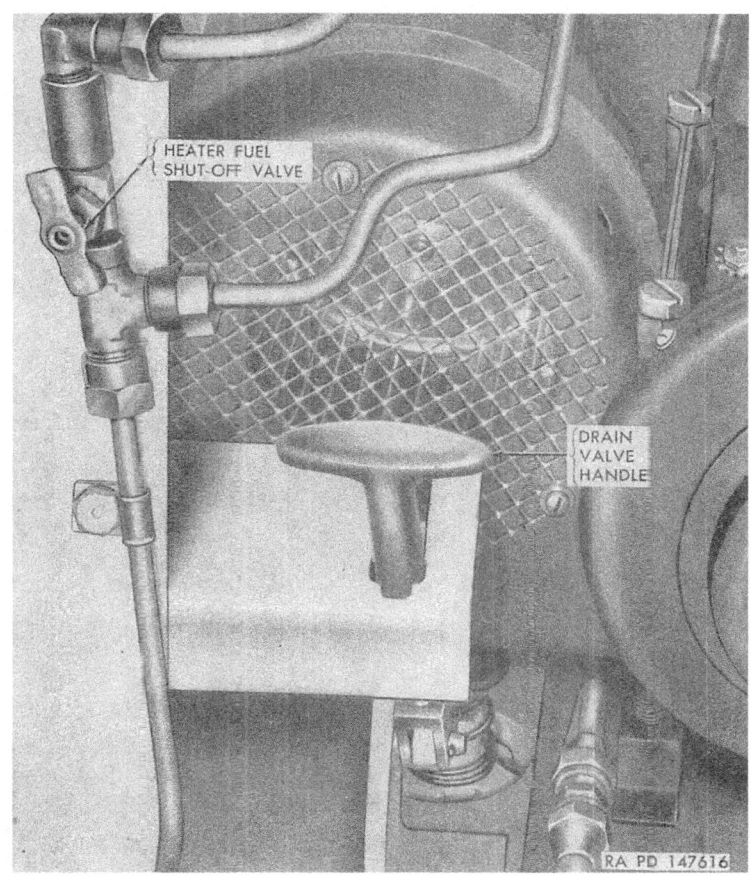

Figure 16. Drain valve handle—crew compartment.

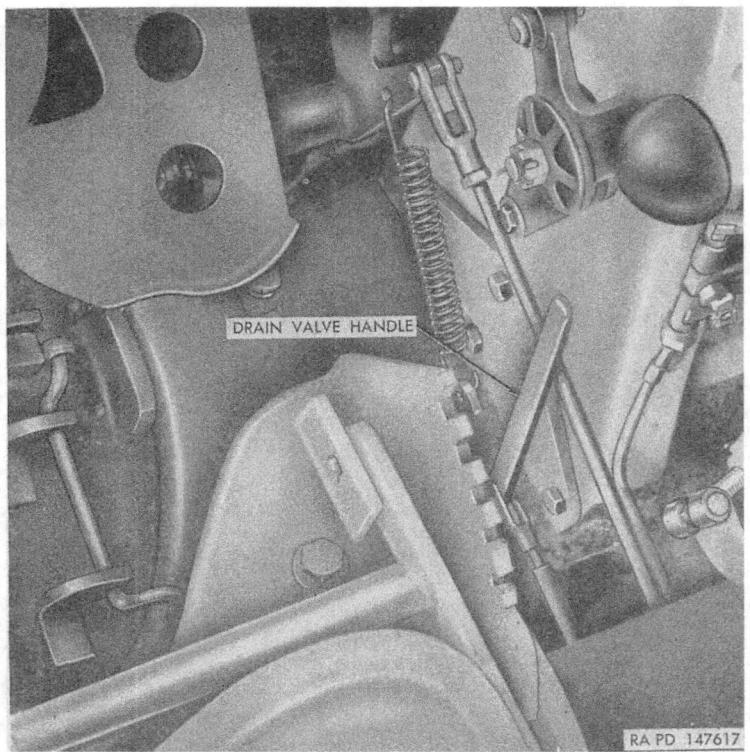

Figure 17. Drain valve handle—engine compartment.

22. Master Relay Switch

The master relay switch (fig. 18), marked "MASTER," is located on the lower-right side of the instrument panel. This switch controls the master relay in the master junction box which controls the entire vehicle electrical system by connecting or disconnecting both sets of storage batteries from all other electrical equipment.

Caution: Do not turn switch to "OFF" position if radio switch is on.

23. Main-Engine-Magneto Switch

a. GENERAL. The main-engine-magneto switch (fig. 18), marked "MAGNETOS," is located in the upper-right corner of the instrument panel. It is a four-position switch which controls the four main engine magnetos by making and breaking the magneto's ground connection.

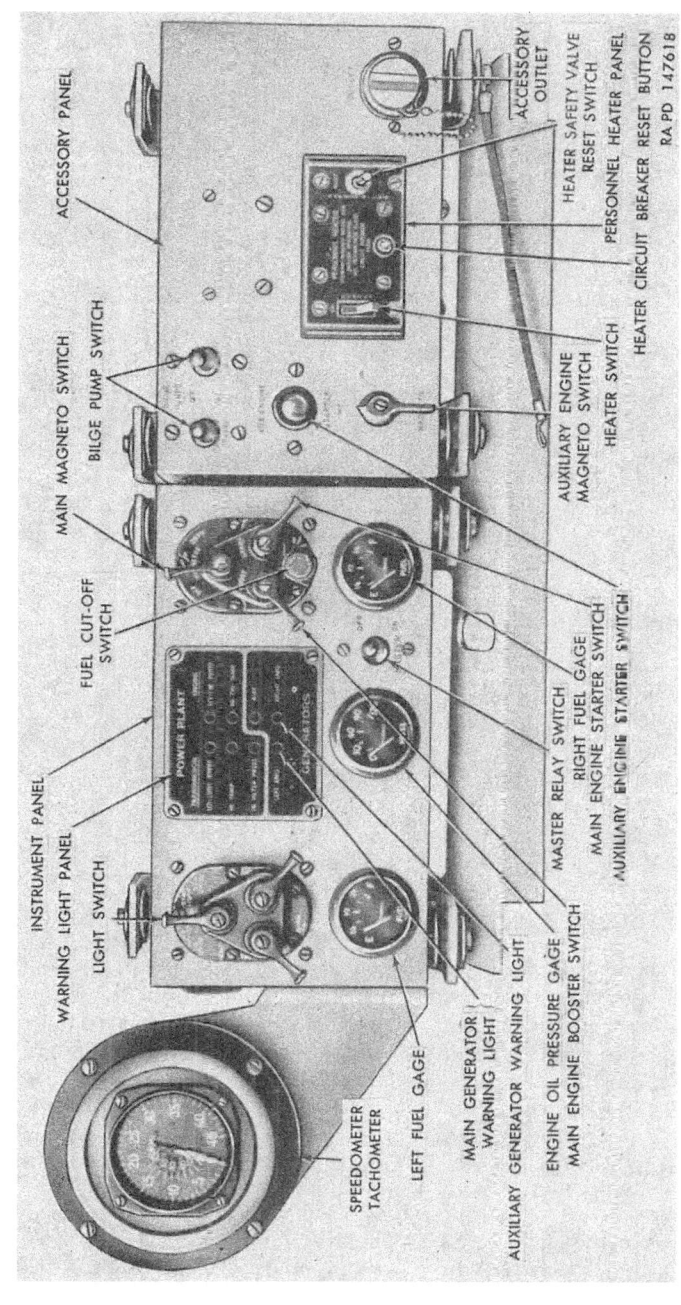

Figure 18. Instrument and hull accessory panels.

b. OPERATION.
  (1) *"OFF" position.* All magnetos are grounded and therefore inoperative.
  (2) *"A" position.* The two magnetos that supply current to the spark plugs on the accessory end of each cylinder are operative, while the other two are grounded.
  (3) *"F" position.* The two magnetos that supply current to the spark plugs on the flywheel end of each cylinder are operative, while the other two are grounded.
  (4) *"BOTH" position.* All four magnetos are operative.

24. Main-Engine-Starter Switch

The main-engine-starter switch (fig. 18), marked "START," is located in the upper-right corner of the instrument panel, on the same housing as the main-engine-magneto switch. It has two positions, "ON" and "OFF," selected by means of a spring-loaded lever. The switch is normally in the "OFF" position. To operate, turn the lever to the "ON" position and hold; when released, it will return automatically to the "OFF" position.

25. Main-Engine-Booster Switch

The main-engine-booster switch (fig. 18), marked "BOOST," is located in the upper-right corner of the instrument panel, on the same housing as the main-engine-magneto switch. It is used to energize the boosters to facilitate starting of the main engine. It has two positions, "ON" and "OFF," selected by means of a spring-loaded lever. The switch is normally in the "OFF" position. To operate, turn the lever to the "ON" position and hold; when released, it will return automatically to the "OFF" position.

26. Fuel-Cut-Off Switch

The fuel-cut-off switch (fig. 18), marked "FUEL OFF," is located in the upper-right corner of the instrument panel, in the lower-center of the main-engine-magneto switch housing. It is used to operate the degassers which shut off the flow of fuel to the idle circuits in the carburetors. To stop the engine, press the switch button in and hold until the engine stops.

27. Light Switch

a. GENERAL. The light switch is located in the upper-left corner of the instrument panel (fig. 18). The switch is a three-lever-type switch with main switch, auxiliary switch, and mechanical lock lever (fig. 19).

(1) *Main switch.* The main switch is located at the top of the light switch housing. On this vehicle, this switch is used only in "BO DRIVE," "BO MARKER," "OFF," and "SER DRIVE" positions. The "STOP LIGHT" position is not used.

(2) *Auxiliary switch.* The auxiliary switch is located on the lower-left corner of the light switch housing. It is not used in this vehicle.

(3) *Mechanical lock.* The mechanical lock lever is located on the lower-right of the light switch housing. The lever must be turned counterclockwise and held in this position to allow the main switch to move from "OFF" to "SER DRIVE" or from "BO MARKER" to "BO DRIVE" positions.

b. OPERATION.

(1) *Service head lights.* To turn the service head lights on, turn the mechanical lock lever counterclockwise and hold; then turn the main switch lever to "SER DRIVE" position.

(2) *Blackout driving light.* To turn the blackout driving light on, turn the mechanical lock lever counterclockwise and hold; then turn the main switch lever to "BO DRIVE" position.

(3) *Blackout marker lights.* To turn the blackout marker lights on, turn the main switch lever to "BO MARKER" position.

(4) *All lights off.* To turn all lights off, turn the main light switch to "OFF" position.

28. Dimmer Switch

The dimmer switch is mounted on a bracket in the left-front corner of the hull floor. It is used by the driver to turn either the service head lights or the blackout driving lights from "HI" to "LO" position or vice versa. To change the beam direction in either respect, apply pressure to the switch with the foot.

29. Dome Light Controls

Two dome lights (figs. 20 and 219) are installed in the driver's compartment and two in the turret. Each light is equipped with two lamps. The lenses are divided into red and white sections. Each dome light is controlled by a switch equipped with a safety plunger to prevent accidental turning on of the white light. To turn the red light on, move the switch lever toward the red lens. To turn the white light on, push in the plunger on the end of the switch lever; then move the lever toward the white lens. Turn off either red or white light by returning lever to its center position.

30. Horn Switch

A push-button-type-horn switch (fig. 21) for sounding the horn is located on the left hull wall in the driver's compartment.

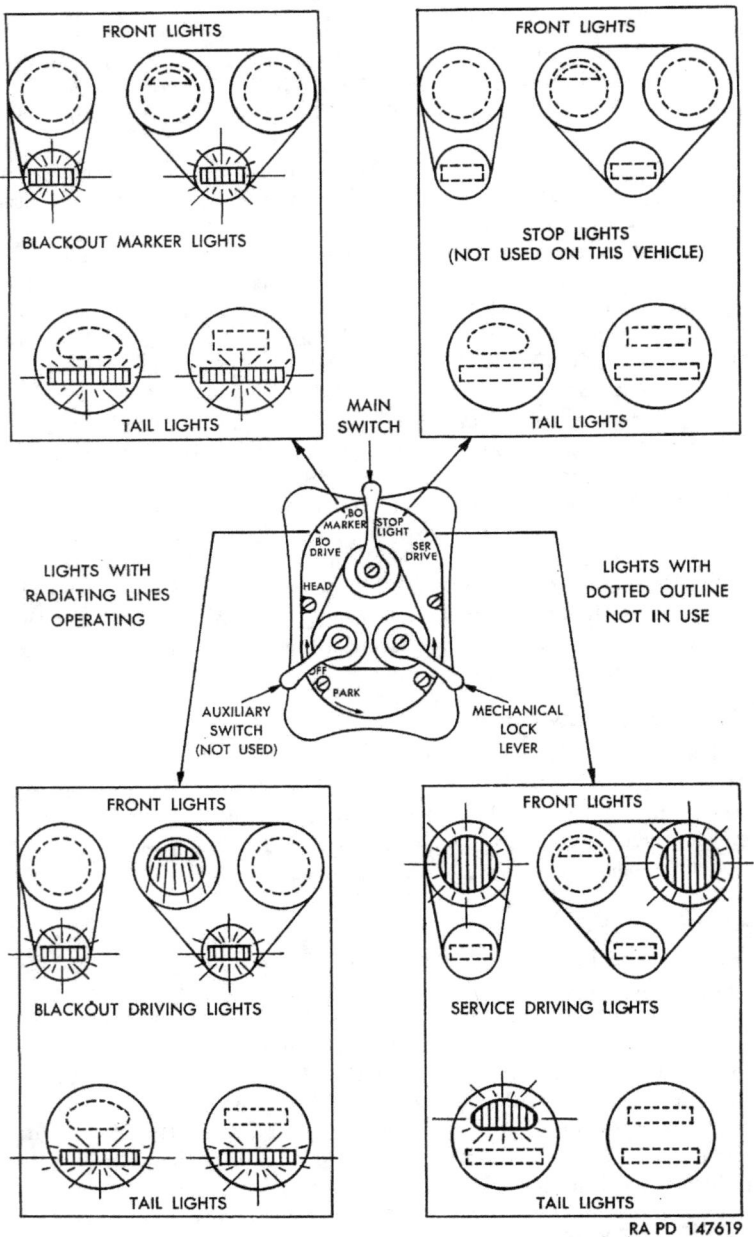

Figure 19. Driving lights chart.

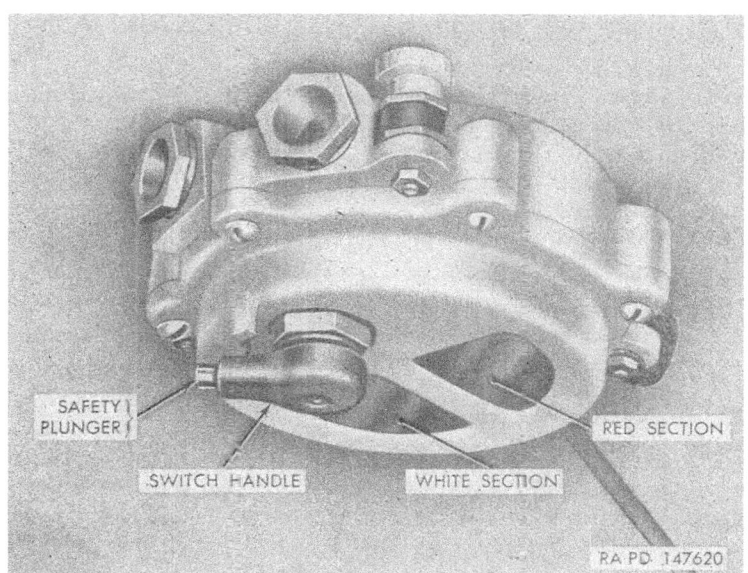

Figure 20. Dome-light controls.

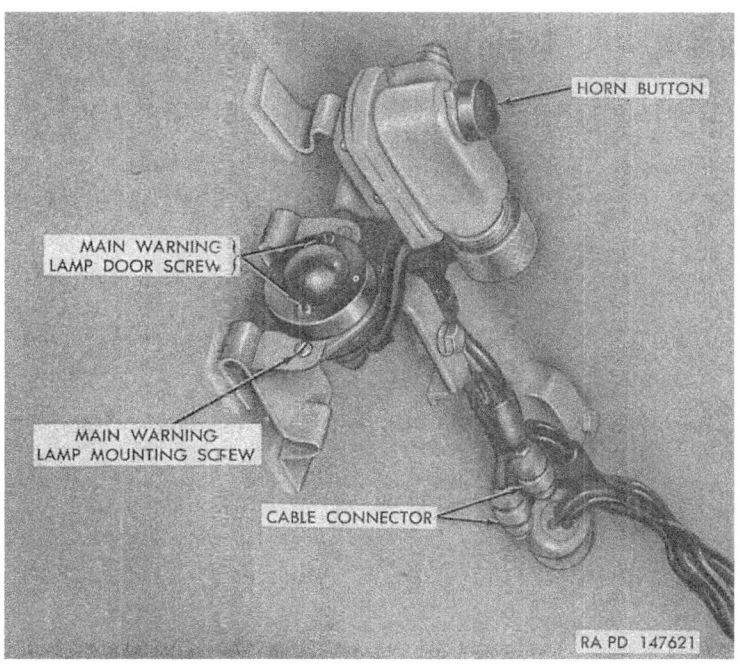

Figure 21. Horn switch and main warning light.

31. Speedometer-Tachometer

The speedometer-tachometer (fig. 18) is secured to the left side of the instrument panel. The inner dial indicates vehicle speed in miles per hour, with a range of 0 to 50. The outer dial indicates engine crankshaft speed in hundreds of revolutions per minute with a range of 0 to 45.

32. Fuel Gages

Two fuel gages (fig. 18) are mounted on the instrument panel. The gage in the lower-left corner of the panel indicates the amount of fuel in the left fuel tank. The gage in the lower-right corner of panel indicates the amount of fuel in the right fuel tank.

33. Engine Oil Pressure Gage

The engine oil pressure gage (fig. 18) is mounted in the lower-center section of the instrument panel. The gage indicates main engine oil pressure, with a range of 0 to 120 psi. Normal engine oil pressure is 60 to 70 psi.

34. Warning Signal Lights

a. GENERAL. Six warning signal lights are mounted in a panel in the upper-center section of the instrument panel (fig. 18). Two of these lights provide warning of trouble in the transmission, two of trouble in the main engine, and two of excessive reverse current in the generators. A main warning light (fig. 21), mounted on the left hull wall in the driver's compartment, is provided to attract the driver's attention when one of the engine or transmission warning lights is on.

b. NORMAL INDICATIONS.

(1) *Transmission warning signal lights.*

(*a*) *Transmission-low-lubrication-pressure-warning-s i g n a l light.* A red-jewel-warning light, marked "LO LUBE PRESS" is provided to warn the driver when the transmission oil pressure falls below 10 psi.

(*b*) *Transmission-high-oil-temperature-warning-signal light.* A red-jewel-warning-signal light, marked "HI TEMP" is provided to warn the driver when the transmission oil temperature rises above 280° F.

Note. The warning signal light marked "HI FILTER PRESS" is not used on this vehicle.

(2) *Engine warning signal lights.*

(*a*) *Engine-low-oil-pressure-warning-signal light.* A red-jewel-warning-signal light, marked "LO OIL PRESS" is provided to warn the driver when the main engine oil pressure drops below 11 psi.

(b) *Engine-high-oil-temperature-warning-signal light.* A red-jewel-warning-signal light, marked "HI OIL TEMP" is provided to warn the driver when the main engine oil temperature rises above 225° F.

(3) *Generator warning signal lights.*

(a) *Main-generator-warning-signal light.* A red-jewel-warning-signal light, marked "LEFT ENG" is provided to warn the driver when the main generator current relay (par. 194) has opened and indicates that the main generator is not charging the batteries.

(b) *Auxiliary-generator-warning-signal light.* A red-jewel-warning-signal light marked "RIGHT ENG" is provided to warn the driver when the auxiliary-generator-current relay (par. 194) has opened and indicates that the auxiliary generator is not charging the batteries.

Note. The warning signal light marked "AUX" is not used on this vehicle.

35. Gunner's Power Traversing and Elevating Control Handles

a. GENERAL. The gunner's power traversing and elevating control handles (fig. 22) are located directly in front of the gunner's seat. A firing button is located on each handle.

Note. Each handle provides independent control over traversing, elevating, and firing.

b. OPERATING TO TRAVERSE THE TURRET. To traverse the turret clockwise, turn the handle toward the right from neutral; to traverse the turret counterclockwise, turn the handle toward the left from neutral (fig. 23). The amount the handle is turned away from the neutral position regulates the rate at which the turret traverses.

c. OPERATING TO ELEVATE OR DEPRESS THE GUN. To elevate the gun, turn the top of the handle away from neutral toward the gunner (fig. 23). To depress the gun, turn the top of the handle away from the gunner. The amount the handle is turned away from the neutral position regulates the rate at which the gun is elevated or depressed.

Note. The gun can be elevated or depressed while the turret is being traversed.

36. Gunner's Manual Traversing Control Handle

The gunner's manual traversing control handle (fig. 23) is located above and to the right-front of the gunner's seat. To manually traverse the turret, squeeze the release lever on the handle; then rotate the handle in the desired direction while holding the lever in the re-

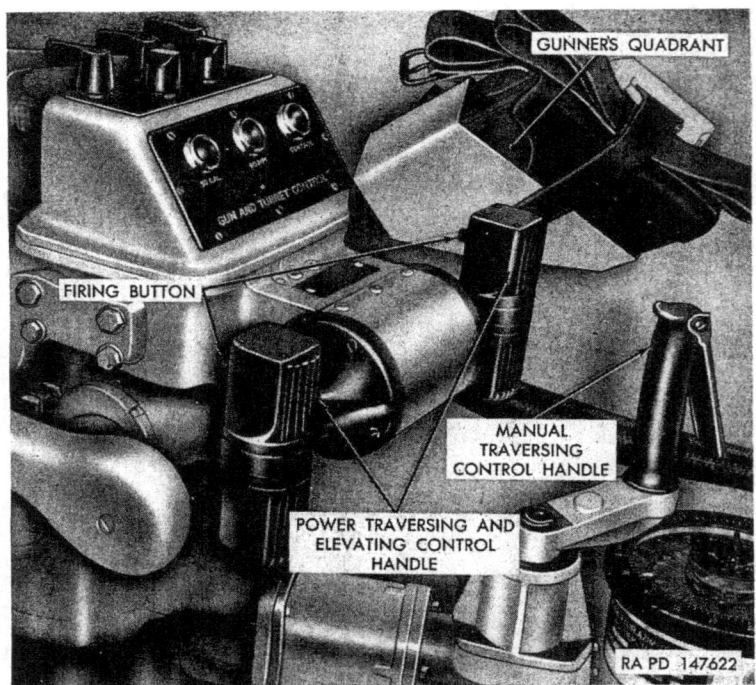

Figure 22. Gunner's traversing and elevating controls.

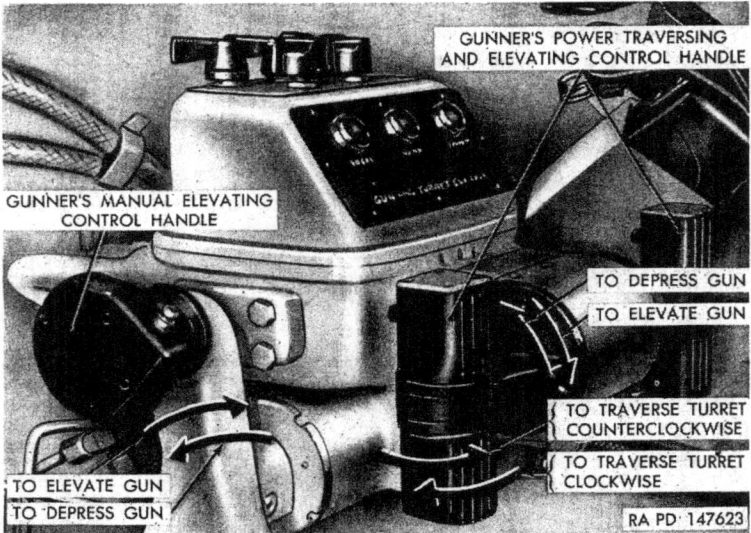

Figure 23. Operation of gunner's controls.

lease position. The turret rotates in the same direction as the handle. The rate at which the handle is rotated regulates the rate at which the turret traverses. A no-back mechanism automatically holds the turret in position and prevents turret drift when the vehicle is not on a horizontal plane.

37. Gunner's Manual Elevating Control Handle

The gunner's manual elevating control handle is located to the left of the gunner's-power traversing and elevating control handle (figs. 22 and 24). A gunner's firing button is located on the handle (fig. 25). To elevate the gun, rotate the handle clockwise; to depress the gun, rotate the handle counterclockwise. The rate at which the handle is rotated regulates the rate at which the gun is elevated or depressed.

38. Commander's Power Traversing and Elevating Control Handle

a. GENERAL. The commander's power traversing and elevating control handle (fig. 26) is located above and to the right of the commander's seat. A commander's firing trigger is located on the handle.

b. OVERRIDE OF GUNNER. If the commander so desires, he may take over control of power traversing, elevating, and firing from the gunner by squeezing the override lever against the control handle (fig. 26). This action will disengage the gunner's control handle and engage the commander's control handle.

c. OPERATING TO TRAVERSE THE TURRET. To traverse the turret clockwise, squeeze the override lever (*b* above) and turn the control handle toward the right from neutral (fig. 27). To traverse the turret counterclockwise, squeeze the override lever and turn the control handle toward the left from neutral (fig. 27). The amount the handle is turned away from the neutral position regulates the rate at which the turret traverses.

d. OPERATING TO ELEVATE OR DEPRESS THE GUN. To elevate the gun, squeeze the override lever (*b* above) and turn the top of the control handle away from neutral toward the rear of the turret (fig. 27). To depress the gun, squeeze the override lever and turn the top of the control handle away from neutral toward the front of the turret (fig. 27). The amount the handle is turned away from the neutral position regulates the rate at which the gun is elevated or depressed.

Note. The gun can be elevated or depressed while the turret is being traversed.

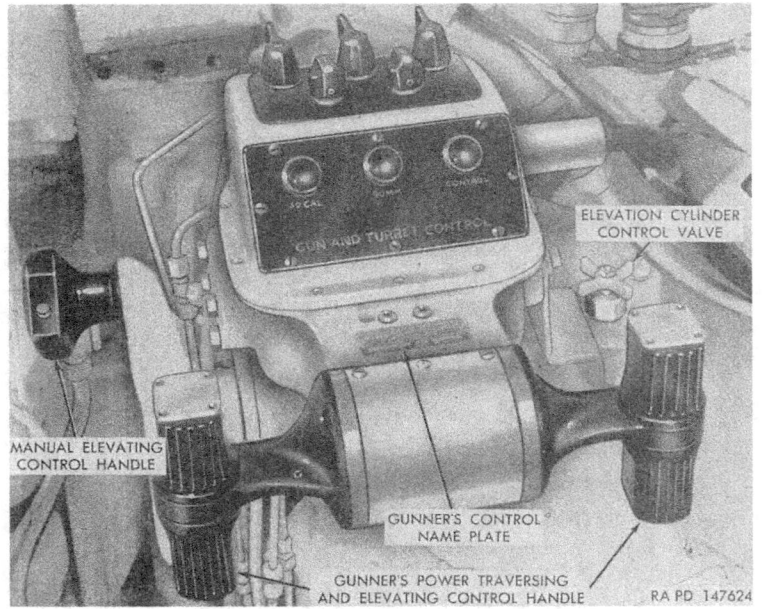

Figure 24. Gunner's controls.

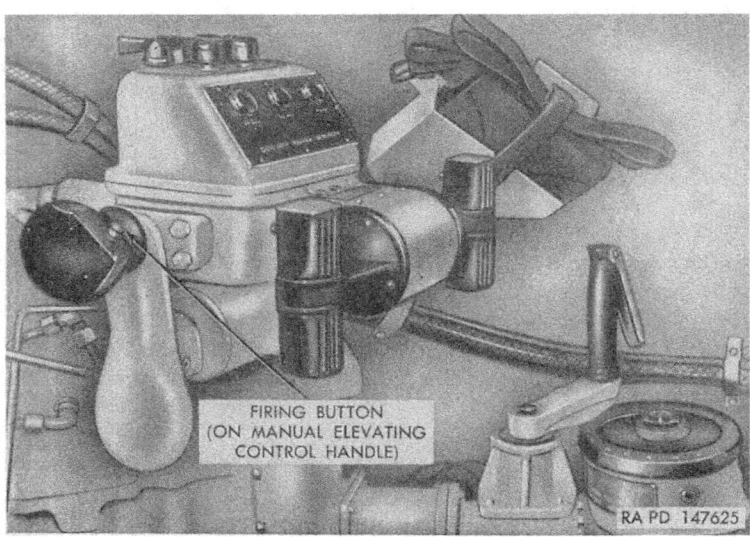

Figure 25. Gunner's controls and firing buttons.

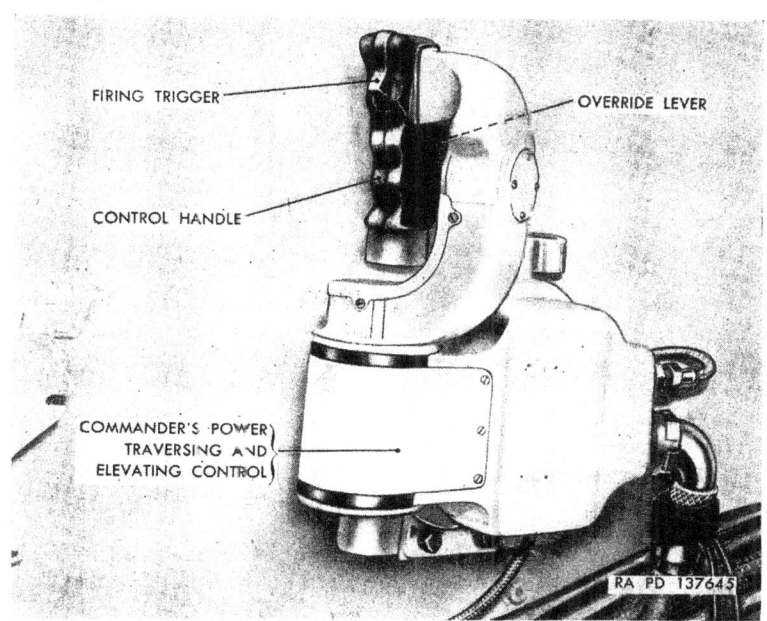

Figure 26. Commander's power traversing and elevating control.

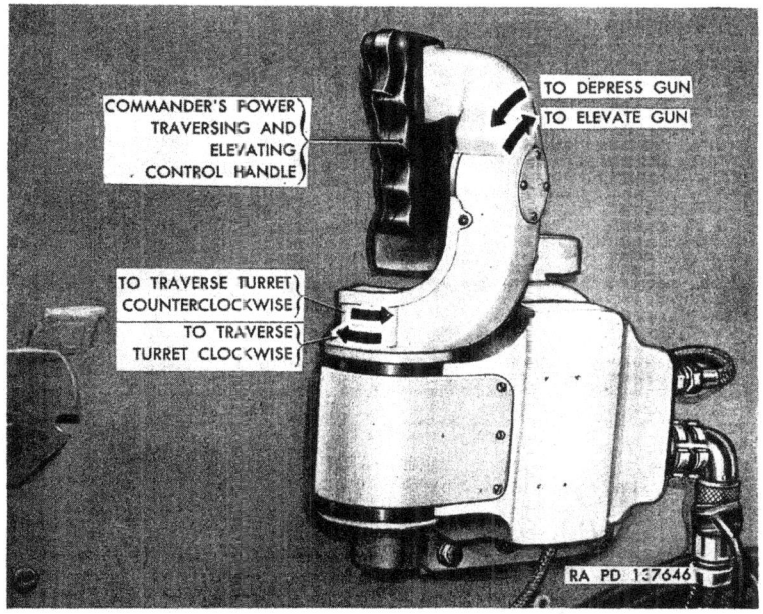

Figure 27. Operation of commander's controls.

39. Turret Control Switch, Firing Selector Switches, and Indicator Lights

a. GENERAL. Three selector switches (*b*, *c*, and *d* below) and their respective red-jewel-indicator lights are located on the gunner's control switch panel (fig. 28). Any of the three switches can be turned on by turning the switch knob to the left, toward the "ON" position on the gun-and-turret-control-name plate (fig. 27). To turn off any of the switches, turn the knob to the right, away from the "ON" position.

b. TURRET CONTROL SWITCH AND INDICATOR LIGHT. The turret control switch controls the power electric motor circuit and allows the turret to be power traversed and gun elevated (par. 35). The red-jewel-indicator light will be lighted when this switch is in the "ON" position.

c. 90-MM GUN FIRING SELECTOR SWITCH AND INDICATOR LIGHT. The 90-mm gun firing selector switch energizes the firing circuit for the 90-mm gun. When it is in the "ON" position, the red-jewel-indicator light located beneath the switch will be lighted.

d. CAL. .50 MACHINE GUN FIRING SELECTOR SWITCH AND INDICATOR LIGHT. The cal. .50 machine gun firing selector switch energizes the firing circuit of the coaxial machine gun. When it is in the "ON" position, the red-jewel-indicator light located beneath the switch will be lighted.

e. TRAVERSE ADJUSTING KNOB. A traverse adjusting knob, marked "TRAVERSE ADJ" is located on top of the gunner's power traversing and elevating control. To increase traversing speed, turn knob clockwise; to decrease speed, turn knob counterclockwise.

f. ELEVATION ADJUSTING KNOB. An elevation adjusting knob, marked "ELEVATION ADJ" is located on top of the gunner's power traversing and elevating control. To increase elevating speed, turn knob clockwise; to decrease speed, turn knob counterclockwise.

40. Loader's Traverse Safety Switch and Indicator Light

The loader's traverse safety switch (fig. 29) is located on the left-front wall of the turret. This switch acts as a safety device in the power-elevating-and-traversing circuit. To close the circuit, turn the safety switch to the "ON" position; to open the circuit, turn the safety switch to the "OFF" position. The red-jewel-indicator light located above the safety switch indicates that the turret can be traversed.

Caution: The loader must not try to obtain ammunition from under the turret floor when the power-unit-acts indicator light is on.

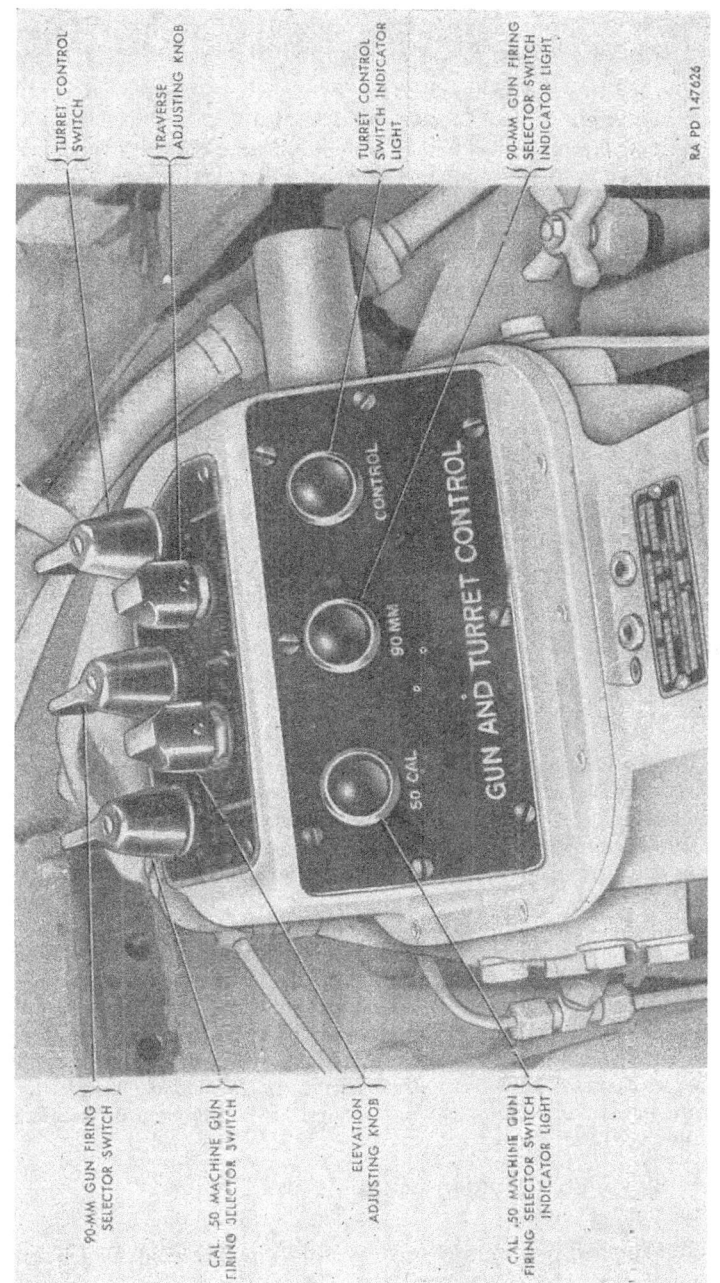

Figure 28. Gunner's control switch panel.

41. Loader's Firing Reset Safety Switch and Indicator Light

The loader's firing reset safety switch (fig. 30) is located on the left-rear wall of the turret. Switch opens and closes the 90-mm gun firing circuit, and, when set, enables the gun to be fired electrically. The switch is of the manual reset type. To reset, push in the switch button. The red-jewel-indicator light located to the right of the button indicates that the 90-mm gun is ready to be fired.

Caution: The loader must stand to the side and clear of the breech end of the gun (when this switch is set and its indicator light is lighted) to avoid being struck by the gun during recoil.

42. Turret Lock

The turret lock (fig. 31), located to the right of the gunner's seat, holds the turret stationary by means of a gear segment which engages the teeth of the turret ring. The turret lock should be in the locked position whenever the vehicle is in motion unless the turret is to be traversed or the gun elevated or depressed. To unlock the turret lock, grasp the handle of the lock, turn clockwise several turns until the gear segment disengages from the turret ring gear. To lock, turn the handle counterclockwise several turns until the gear segment engages the turret ring gear. If the teeth of the lock gear segment do not fully engage with the teeth of the turret ring, traverse the turret slightly until the teeth are in mesh.

43. Hand Pump

The hand pump (fig. 32) is located on the right side of the turret floor. This pump is used to charge the manual elevation system and is used only if the vehicle has been inactive for a period of time or if the system has been serviced. The pump is operated by means of a hand pump handle located to the left of the gunner's seat. To operate the pump, move the pump handle back and forth several times. An accumulator, which is a chamber filled with an inert gas, is located on the front of the turret floor to the right of the hydraulic pumps (fig. 32). The accumulator is used to keep the manual elevating system under hydraulic pressure. A small amount of oil from the power pack reservoir is occasionally pumped into the accumulator by the hand pump and stored under pressure until required during normal operation. The accumulator is also charged by oil from the traverse pump when the turret is accelerated or decelerated in power traverse.

44. Elevation Cylinder Control Valve

The elevation cylinder control valve (fig. 24) is located to the right of the gunner's power traversing and elevating control. The purpose of this valve is to supercharge the elevation cylinder when shifting from power to manual operation of the turret. To supercharge

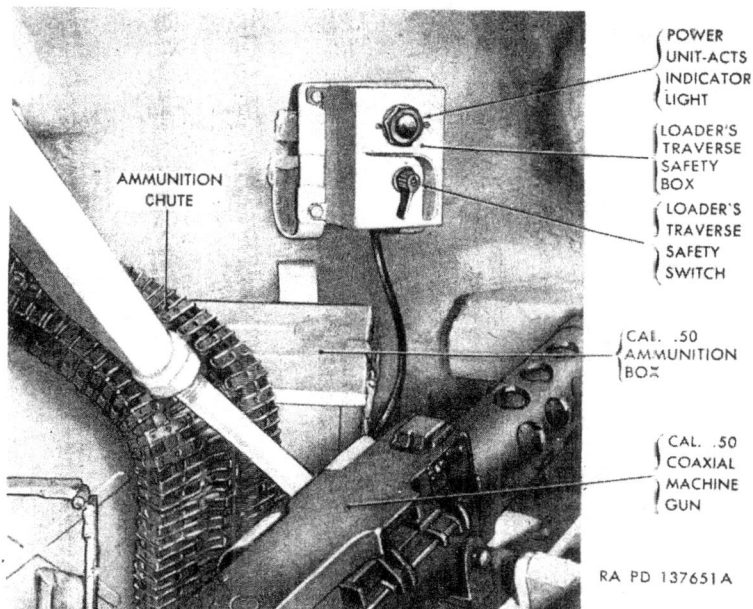

Figure 29. Loader's traverse safety switch.

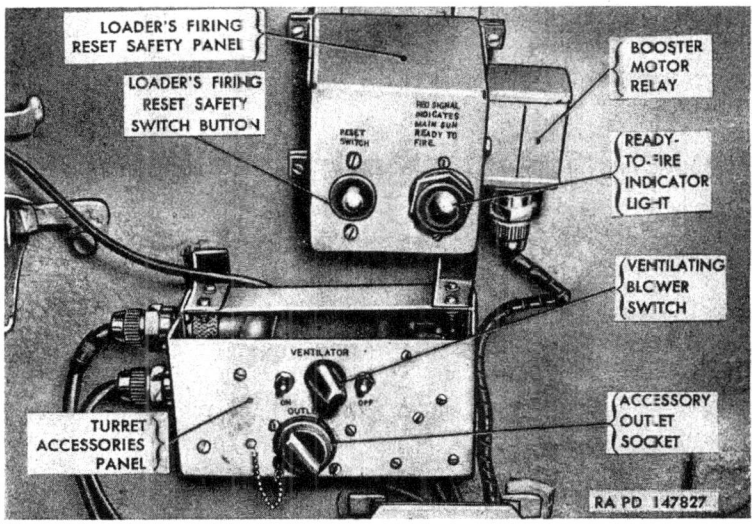

Figure 30. Loader's firing reset safety switch and turret accessories panel.

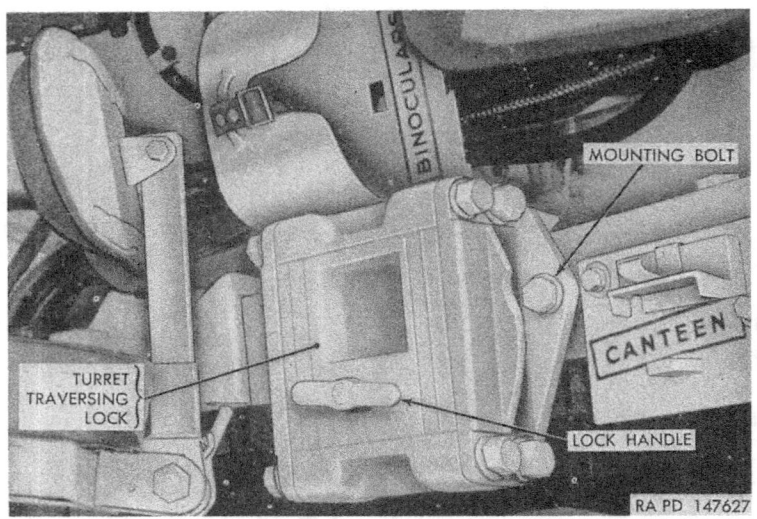

Figure 31. Turret lock.

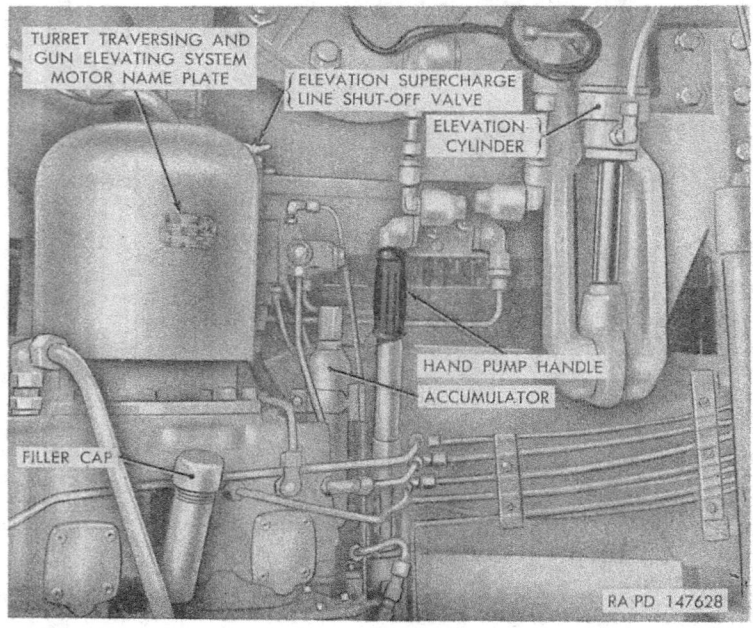

Figure 32. Hand pump.

the elevation cylinder, turn the elevation cylinder control valve handle counterclockwise to open the valve, hold for one second to allow passage of a few drops of oil; then turn the handle clockwise to close the valve.

Note. When draining the system, this valve should be open to hasten the action.

Caution: The valve must be closed when elevating or depressing the gun.

45. Oil-Return-Valve Handle

An oil-return valve (fig. 33) is located on the front of the turret floor to the right of the hydraulic pumps behind the accumulator. The purpose of this valve is to drain oil and relieve the supercharge pressure from the manual elevating system whenever repairs are to be made. To open this valve, turn the oil-return-valve handle counterclockwise; to close this valve, turn the handle clockwise.

Caution: The valve must be closed at all times except when draining the system or replacing or repairing an inoperative component of the manual hydraulic elevation system.

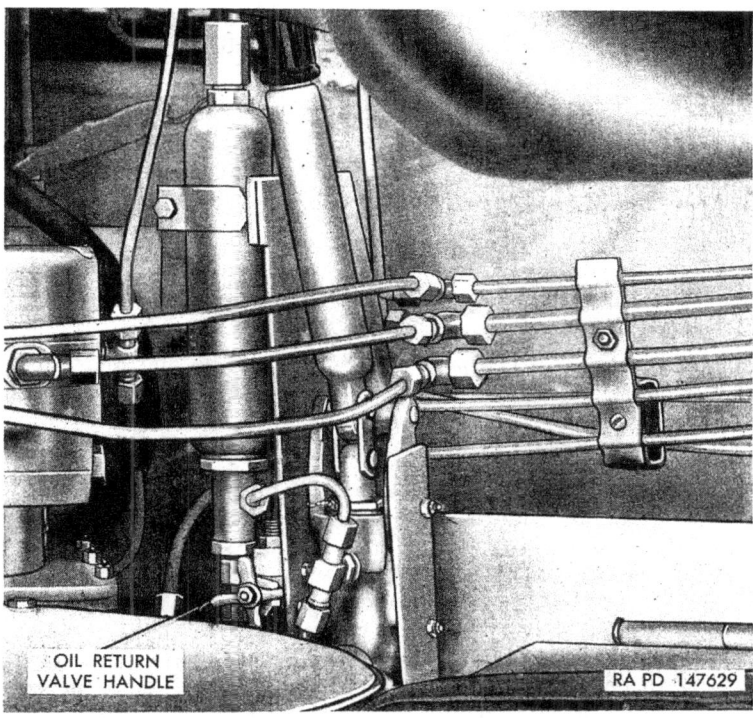

Figure 33. Oil-return-valve handle.

46. Elevation Supercharge Line Shut-Off Valve

The elevation supercharge line shut-off valve (figs. 32 and 301) is located on top of the booster actuated selector valve. The purpose of this valve is to shut off the oil supply to the elevation circuit of the oil gear power pack for maintenance of the elevating system. To shut off this oil supply, turn the valve clockwise.

47. Ammunition Ready Racks

Ready racks (fig. 34) for ammunition are provided in the turret basket. To open a ready rack, loosen the rack clamp from the retainer; then move the clamp out of the way.

48. Gunner's Seat Control Handle

The gunner's seat is mounted on the turret ring to the right of the gun. It has a detachable back and a seat control handle (fig. 35) for height adjustment.

Caution: Do not operate the handle unless seated in the seat. To raise the seat, pull the control handle toward the seat and at the same time, gradually remove body weight from the seat. To lower the seat, push the handle away from the seat and put body weight on the seat.

49. Commander's Seat Control Handle

The commander's pad-type seat is adjustable in height and can be folded out of the way when not in use. To adjust height, lift the seat control handle (fig. 36) to the left-rear of the seat, raise or lower the seat, and push the handle down.

50. Loader's Seat

The loader's seat (fig. 34) is mounted on the turret ring at the left-rear of the turret. The pad-type seat is nonadjustable and can be folded out of the way when not in use.

51. Commander's Cupola Controls

The commander's cupola consists of a stationary ring bolted to the turret roof, with a hinged door mounted on top. Five direct-view prisms or vision blocks are positioned around the stationary ring. The forward portion of the cupola mounts the commander's periscope T35. The door is opened by releasing the locking handle and pushing upward. It is held in the open position by the hold-open lock. To close the cupola, pull on the hold-open-lock-trigger-operating-rod eye, and carefully lower the door into its closed position. Lock the door in the closed position with the locking handle.

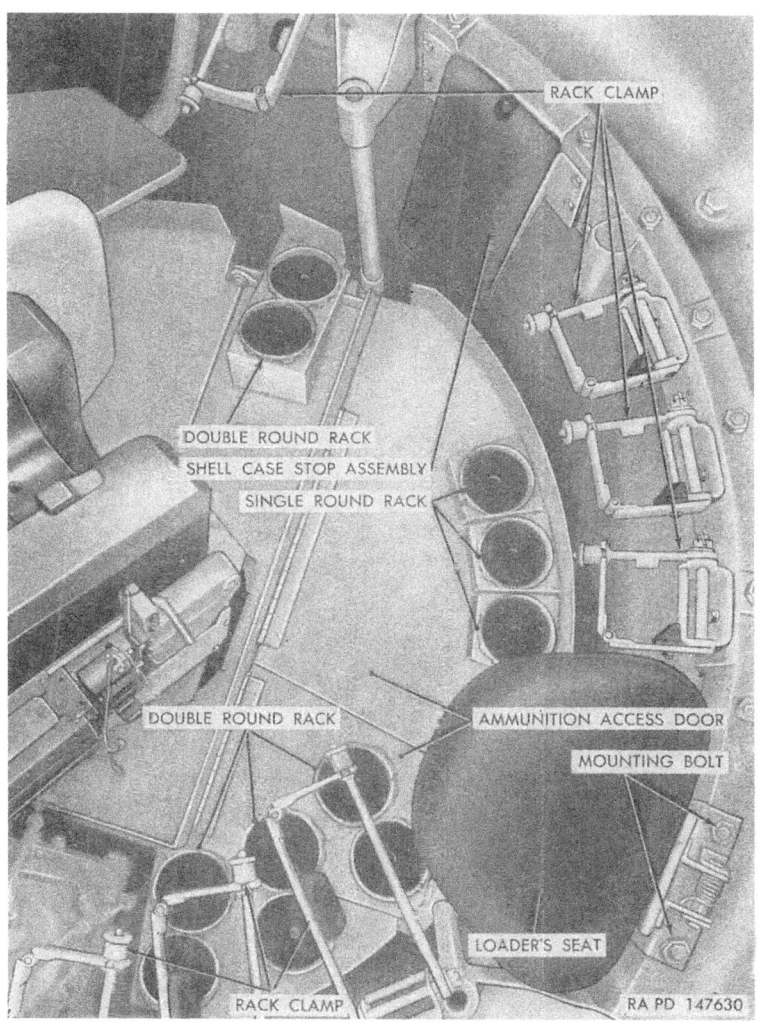

Figure 34. Ammunition ready racks.

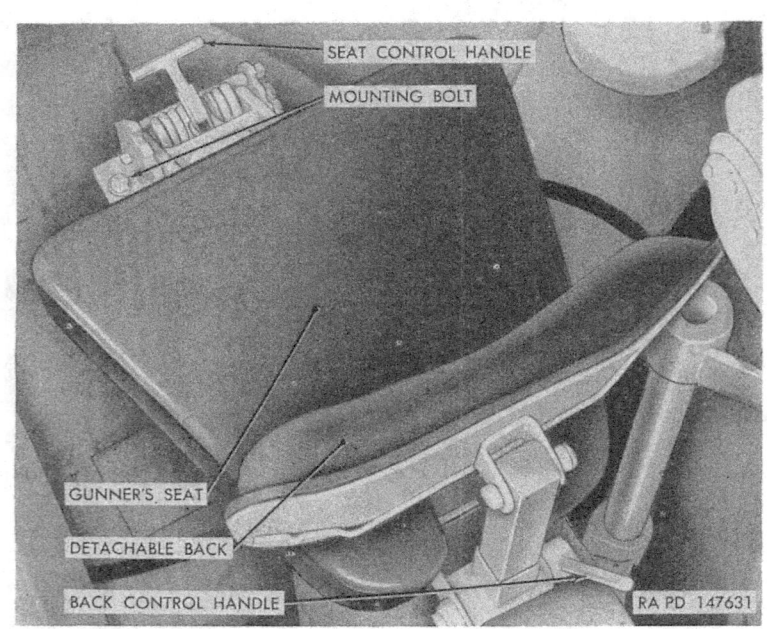

Figure 35. Gunner's seat and seat control handle.

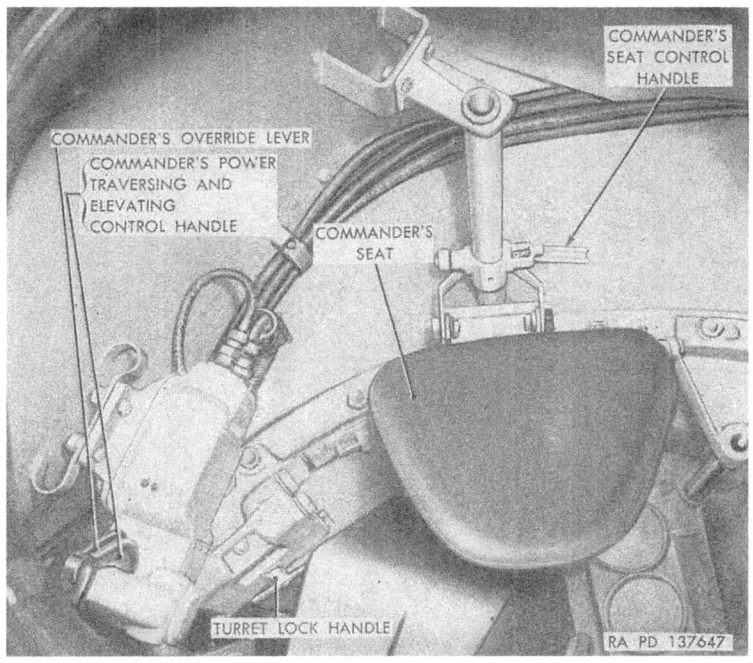

Figure 36. Commander's seat.

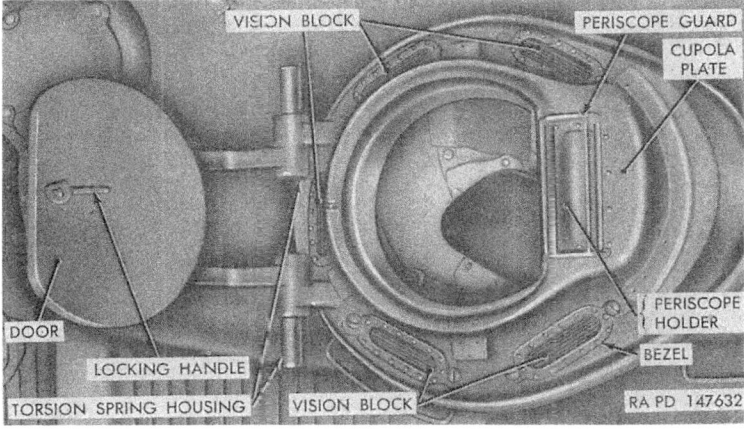

Figure 37. Commander's cupola controls.

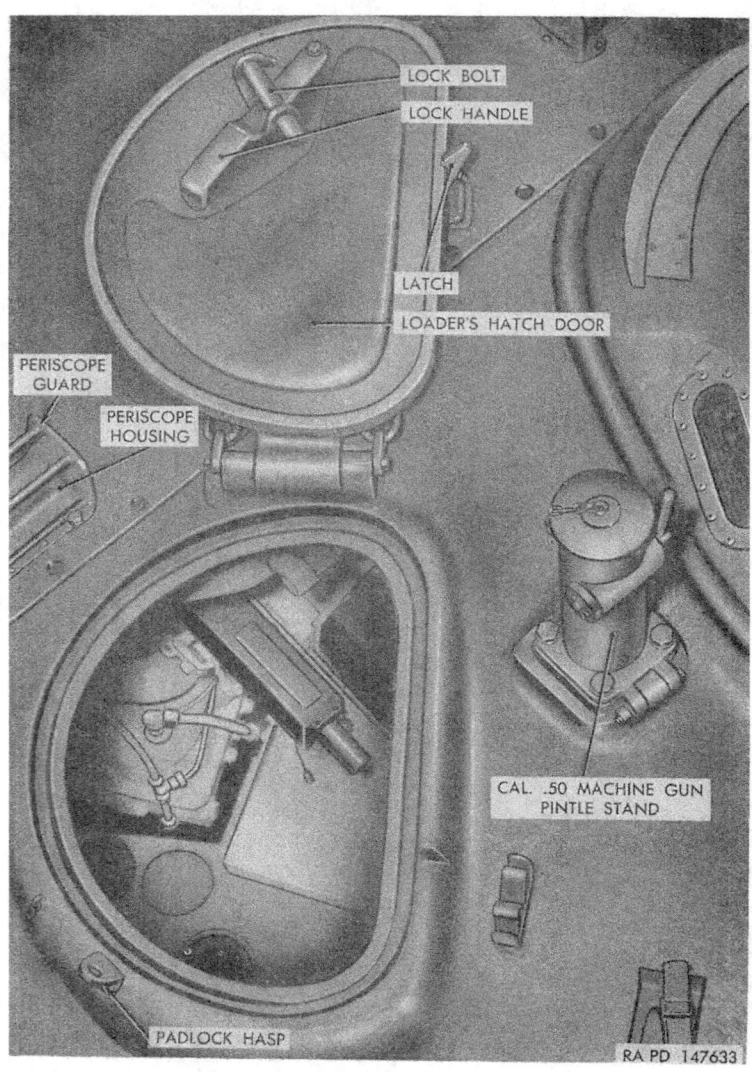

Figure 38. Loader's hatch-door controls.

52. Loader's Hatch Door Controls
(fig. 38)

The loader's hatch is an oval-shaped opening in the top-left side of the turret. A hinged, spring-controlled, armor-plated door covers the opening. A spring-loaded catch is provided to hold the door open. The door, when closed, is locked from inside the turret by a lock bolt which is seated by swinging the lock handle to its closed position.

53. Cal. .50 Coaxial Machine Gun Manual Controls

a. SAFETY LATCH. The cal. .50 coaxial machine gun safety latch (fig. 39) is a small "L"-shaped component of the trigger assembly and can be pivoted 90° from its safe position shown in figure 41.

b. FIRING TRIGGER. When the safety latch is pivoted forward 90°, the machine gun can be fired manually by pressing forward on the firing trigger (figs. 39 and 41) which forces the tip of the safety latch into contact with the solenoid plunger which releases the sear and permits firing of the machine gun.

c. ELEVATING AND TRAVERSING MECHANISM. For every slight adjustment of the coaxial machine gun, an elevating screw and a traversing screw (fig. 40) are provided on the cradle assembly just beneath the rear locking pin.

d. COCKING LEVER. For initially charging the coaxial machine gun and for cocking when some malfunction occurs, a machine gun cocking lever (fig. 40) is provided for the gun. This lever is connected to the bolt stud on the gun and is bracketed to the breech ring of the 90-mm. gun.

e. CRADLE LOCKING PINS. Front and rear locking pins (fig. 40) are provided on the coaxial machine gun cradle to secure the machine gun in operating position.

54. 90-mm Gun Manual Controls

a. SAFETY LEVER. The 90-mm. gun safety lever (fig. 41) locks the firing plunger and makes it inoperative when in "SAFE" position. As shown in figure 41, the gun is in the "SAFE" position. By pressing down on the lever and rotating it 180°, the gun is made ready for firing.

b. HAND FIRING HANDLE. When firing the 90-mm. gun manually, the hand firing handle (fig. 302) is used to actuate the firing linkage.

c. COCKING LEVER. The 90-mm. gun cocking lever (fig. 41) is actuated manually by pulling the lever to the rear. Normally it is automatically actuated as the breechblock is lowered by the camming action between the cylindrical lug on the right side of the cocking lever shaft and the bottom arm of the cocking lever.

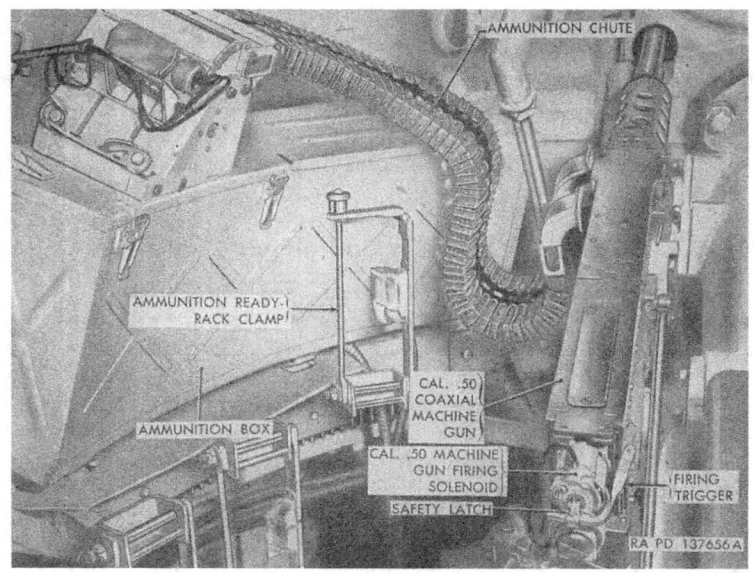

Figure 39. Coaxial machine gun—installed view.

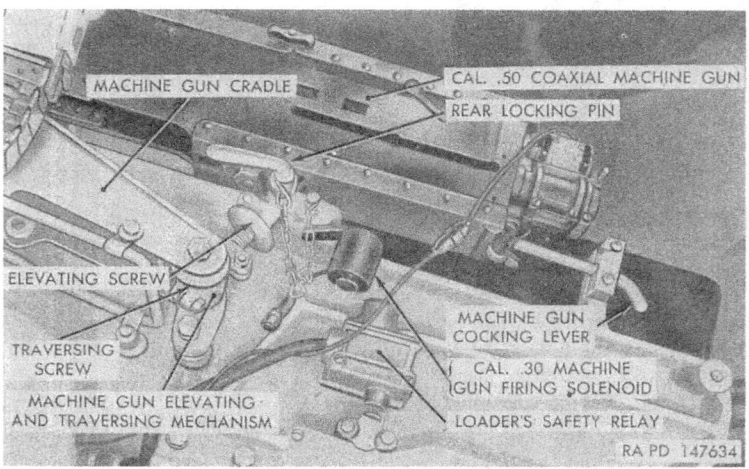

Figure 40. Coaxial machine gun elevating and traversing mechanism.

d. BREECH OPERATING HANDLE. The breech operating handle (fig. 41) is a long flat bar with a locking plunger at one end and a hub at the other end which fits over the right end of the operating shaft. To open the breech manually, press the locking plunger and pull the operating handle to the rear.

55. Gun Traveling Lock

The gun traveling lock (fig. 289) supports the gun barrel, near the muzzle end, when the vehicle is traveling. To open the lock, lift the lever from the top of the cap and unscrew the lever bolt from its bracket by turning the lever counterclockwise. Swing the cap and lever back. Lay the traveling lock flat on the top deck. Before the lock can be raised again into the vertical position, the stop pawl must first be lifted. To lock, swing the cap over the gun barrel and tighten by turning the lever clockwise. Rotate the lever into position over the cap and push the lever down.

56. Replenisher Indicator

The replenisher indicator assembly (fig. 42) is used to indicate the amount of oil in the concentric recoil mechanism. To determine whether there is too much or too little oil reserve in the replenisher, grasp the indicator tape with the fingers on both edges through the accessible part of the assembly. If both edges are serrated, the replenisher is empty and needs refilling; if one edge is smooth and the other serrated, the replenisher has the normal amount of reserve; if both edges are smooth, the replenisher has too much oil and should be bled from one of the drainage plugs at the left end of the replenisher.

Note. Do not read indicator while gun is still hot from firing.

57. Arrangement and Use of Sighting and Fire Control Equipment

a. GENERAL. The 90-mm. gun tank M47 is equipped with both direct fire and indirect fire, as well as miscellaneous, sighting and fire control equipment.

b. DIRECT FIRE EQUIPMENT.

(1) *Primary equipment.* Primary equipment are used for direct fire only and consists of a range finder T41 and superelevation transmitter T13 (par. 58).

> *Note.* First production models of 90-mm gun tank M47 are not equipped with range finder T41 and superelevation transmitter T13. In these vehicles secondary equipment is used.

(2) *Secondary equipment.* Secondary equipment can be used for direct fire. However, they may also be used for indirect fire. It consists of a ballistic drive T23E1 (par. 61) with instru-

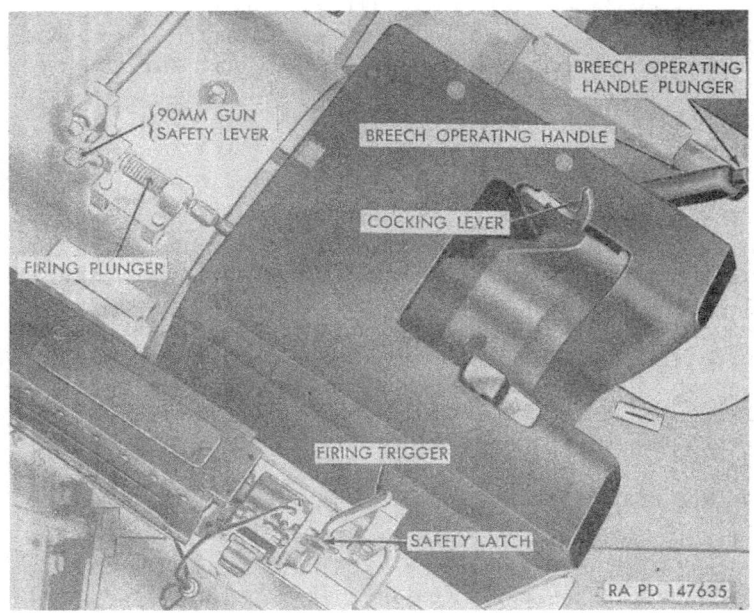

Figure 41. 90-mm. gun manual controls.

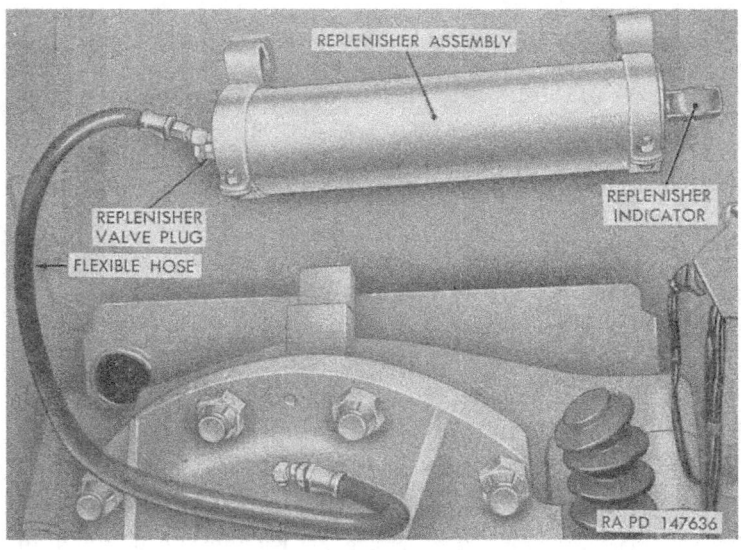

Figure 42. Replenisher assembly.

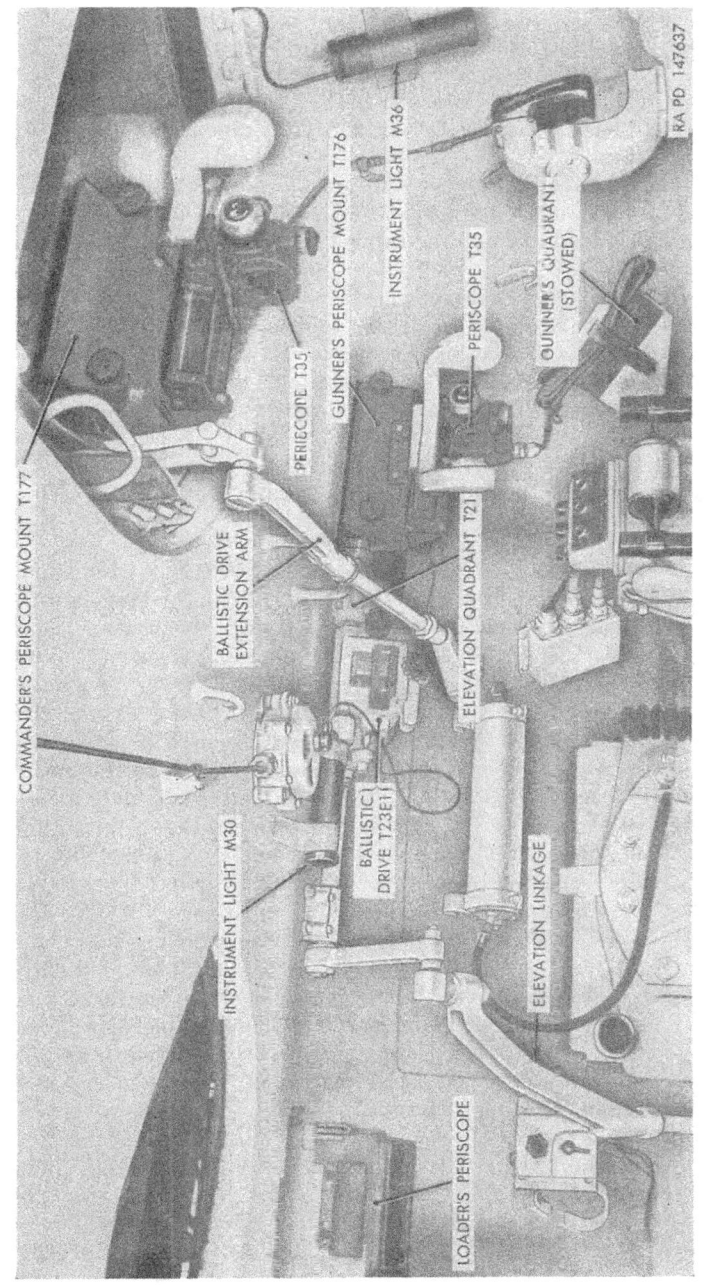

Figure 43. Sighting and fire control instruments—installed view.

ment light M30 (par. 65, gunner's periscope mount T176 (par. 60) and periscope T35 (par. 59) with instrument light M36 (par. 65); and commander's periscope mount T177 (par. 60) with periscope T35 (par. 59) with instrument light M36 (par. 65).

c. INDIRECT FIRE EQUIPMENT. Indirect fire control equipment consists of an elevation quadrant T21 (par. 63) with instrument light T22 (par. 65), and azimuth indicator T24 (par. 64) with instrument light D78454 (par. 65).

d. MISCELLANEOUS EQUIPMENT. Periscope M13B1, M13, or M6 (par. 62) are provided for the driver, assistant driver, and loader for observation of fire or terrain. Fuze setter M14 (par. 66) or M27 are provided for setting fuzes in the ammunition. Gunner's quadrant M1 is provided for use in adjusting the elevation quadrant micrometer (par. 369).

58. Range Finder T41 and Superelevation Transmitter T13

a. DESCRIPTION.
  (1) *Range finder T41.* The range finder T41 (fig. 44) is an autocollimated stereoscopic range finder having a magnification of 7.5 and a field of view of 5°. Two main bearings, right and left, which are bolted to the mounting plate on the roof of the tank turret, support the range finder on steel radial ball bearings. A parallel gun linkage between the gun trunnion and the range finder assures that the line of sight of the range finder reproduces the exact motion of the gun in elevation. In addition, the range finder is independently moved in elevation, by means of the rotation of a ballistic cam, through a "superelevation" angle determined by the ballistic correction and type of ammunition used, as set into the range finder by the gunner.
  (2) *Superelevation transmitter T13.* The superelevation transmitter T13 is connected by a cable to a receptacle at the rear of the range finder. As the gunner operates the range knob of the range finder T41 to maintain the target image in proper contact with the stereoscopic pattern on the reticle, the superelevation transmitter T13 automatically elevates or depresses the gun to keep it on target.

b. RETICLE PATTERN.
  (1) *General.* The reticle pattern (fig. 45) for the range finder combines the range scale, stereoscopic pattern, sighting reticle, and ammunition scale.
  (2) *Range scale.* The range scale (fig. 46) indicates the range at which the range finder is set and is graduated from 500 to 5,000 yards. The "B" at the left end of the scale indicates

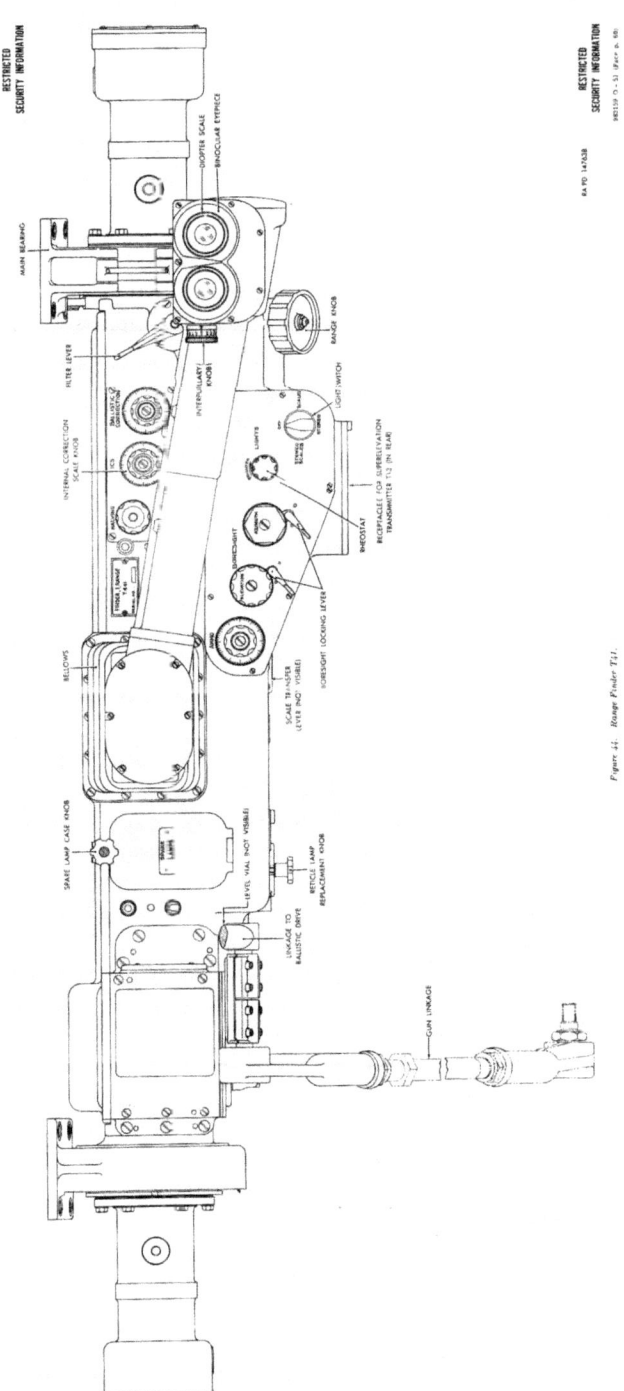

Figure 44. Range Finder T31.

This page intentionally left blank.

the boresighting range point when all superelevation is removed.
(3) *Stereoscopic pattern.* The stereoscopic pattern when viewed simultaneously in both eyepieces appears as a fused pattern forming a "V." The lower (apex) bar is the one used for ranging. When this apex bar is in stereoscopic contact with the target image, the correct range is indicated on the range scale of the reticle pattern.
(4) *Sighting reticle.* The sighting reticle is 40 mils across with each horizontal space and line representing a deflection of 5 mils. The top of the vertical bar is used for sighting when no lateral lead is required.
(5) *Ammunition scale.* The ammunition scale (fig. 46) indicates the ammunition for which the range finder is set and is graduated for cal. 50, HEAT T108, HE M71, AP T33, HVAP M304, HVAP T65, and HVAP T67 ammunition. The "B" at the right end of the scale indicates the boresighting setting.

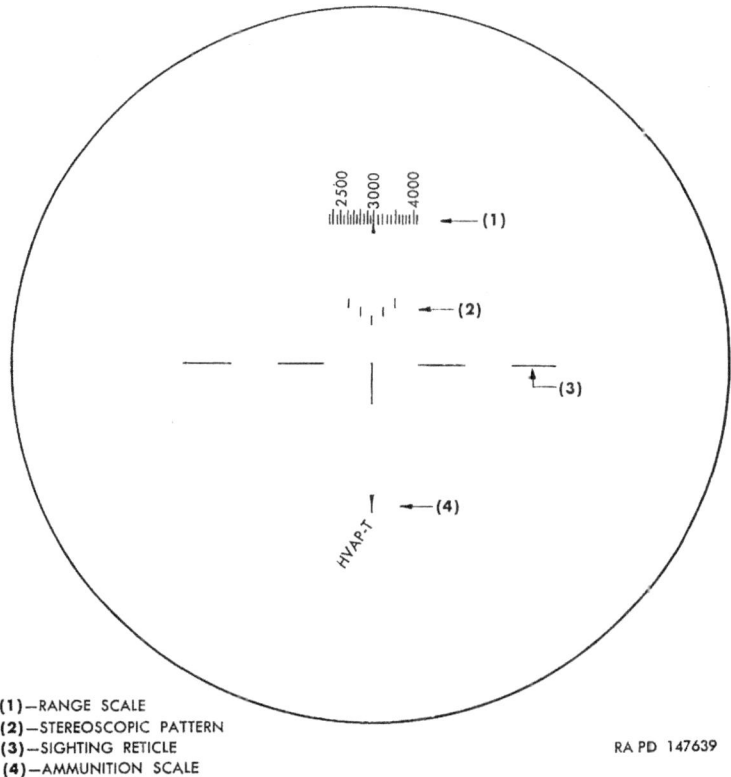

(1)—RANGE SCALE
(2)—STEREOSCOPIC PATTERN
(3)—SIGHTING RETICLE
(4)—AMMUNITION SCALE

RA PD 147639

Figure 45. Reticle pattern for range finder T41.

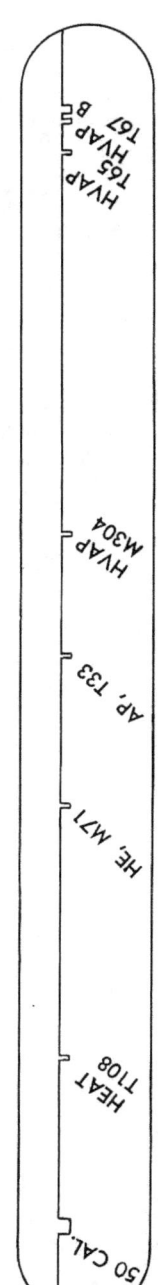

Figure 46. Range and ammunition scales for range finder T41.

c. CONTROLS (fig. 44).
   (1) *Diopter scales.* Each eyepiece has a diopter scale for focusing the eyepieces of the range finder for the eyes of the gunner. Each scale is graduated from minus 3 to plus 3 diopters so that an observer may record his corrections and facilitate resetting the scales once his correction is known.
   (2) *Interpupillary knob.* An interpupillary knob is provided to adjust the separation between the eyepieces to correspond to the observer's eye separation. The scale is graduated from 58 to 72 millimeters in 1-millimeter intervals so that the observer may record his eye separation and facilitate setting the interpupillary knob once his eye separation is known.
   (3) *Filter lever.* A filter lever is provided to insert filters into the optical system to improve the contrast between the target image and the reticle pattern.
   (4) *Scale transfer lever.* A scale transfer lever is provided to transfer the range and ammunition scales from the left to the right eyepieces for emergency operation. Under normal operation (*b* above) the scales should be in the left eyepiece.
   (5) *Range knob.* A range knob is used to bring the lower (apex) bar of the stereoscopic pattern on the reticle (fig. 45) into stereoscopic contact with the target, and as a result set in the correct range and indicate this range on the range scale of the reticle pattern.
   (6) *Ballistic correction knob.* The ballistic correction knob marked "BALLISTIC CORRECTION" sets in the required ballistic corrections. Obtain the ballistic correction for the particular muzzle velocity, air density, air temperature, and rear wind by referring to the ballistic correction plate adjacent to the range finder.
   (7) *Ammunition knob.* The ammunition knob, marked "AMMO," is provided to set in the desired ammunition correction as indicated on the ammunition scale of the reticle pattern. Graduations are provided on the knob to allow for selection of the proper ammunition correction should ammunition other than that marked on the ammunition scale of the reticle be used.
   (8) *Boresight knobs.* Two boresight knobs, marked "ELEVATION" and "AZIMUTH" allow for boresighting the range finder with the gun. Two locking levers lock the knobs in position when the range finder is properly boresighted.
   (9) *Halving knob.* A halving knob completes the boresighting operation by bringing the stereoscopic pattern into correct alinement.

(10) *Internal correction scale knob.* An internal correction scale knob, marked "ICS," introduces the observer's personal correction to enable the observer to obtain precise stereoscopic contact with the target.

(11) *Light switch.* A four-position rotary switch gives direct illumination to the reticle scales, to the stereoscopic pattern, or to both the scales and the pattern.

(12) *Light rheostat.* A light rheostat regulates the intensity of illumination of the scales and stereoscopic pattern. To increase the illumination, turn the rheostat knob clockwise.

d. ILLUMINATION. Three lamps illuminate the reticle, scales, and stereoscopic pattern. The lamps for the scales and the left stereoscopic reticle are enclosed in a housing which can be opened by means of a toggle type knob for lamp replacement. The lamp for the right stereoscopic reticle is enclosed in a housing on the right underside of the range finder.

59. Periscope T35

a. DESCRIPTION.

(1) *General.* The periscope T35 (figs. 47 and 48) is a monocular-type instrument having two built-in optical systems: with a 33-degree 18-minute vertical and 7-degree 9-minute horizontal field of view for wide, close-in vision of the terrain and a six-power system with an 8-degree field of view for sighting on distant targets. The periscope is composed of two separate parts: a head assembly (fig. 47) which is supported in the periscope mount by two clamping screws; and a body assembly (figs. 47 and 49) which is secured to the periscope mount by four screws.

(2) *Commander's periscope.* The commander's periscope T35 serves as a target-designating device. It is supported in the commander's periscope mount T177 (fig. 47). An elevation lever assembly couples the periscope to an arm and connector on the commander's periscope mount which in turn is connected to an extension arm on the ballistic drive. Thus the commander's periscope moves with the gunner's periscope and the gun as the gun is elevated or depressed.

(3) *Gunner's periscope.* The gunner's periscope T35 serves as a secondary sight for fire control. It is supported in the gunner's periscope mount T176 (fig. 48). An elevation lever assembly couples the periscope directly to the shaft of the ballistic drive. Thus the gunner's periscope moves with the commander's periscope and the gun as the gun is elevated or depressed.

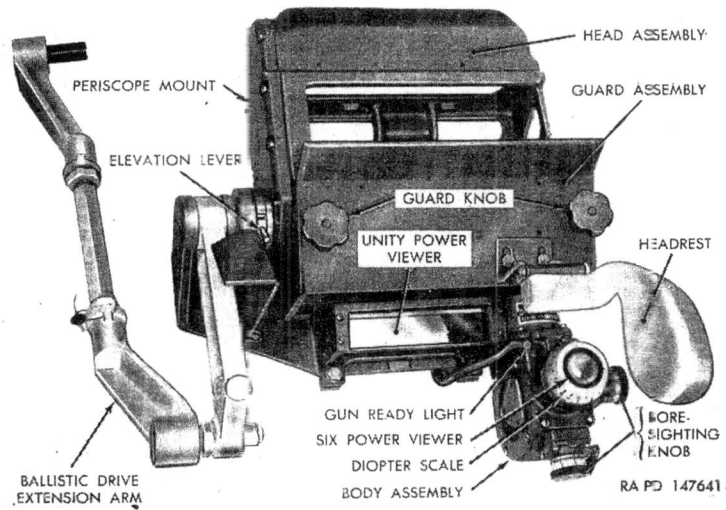

Figure 47. Periscope T35 installed in commander's periscope mount T177.

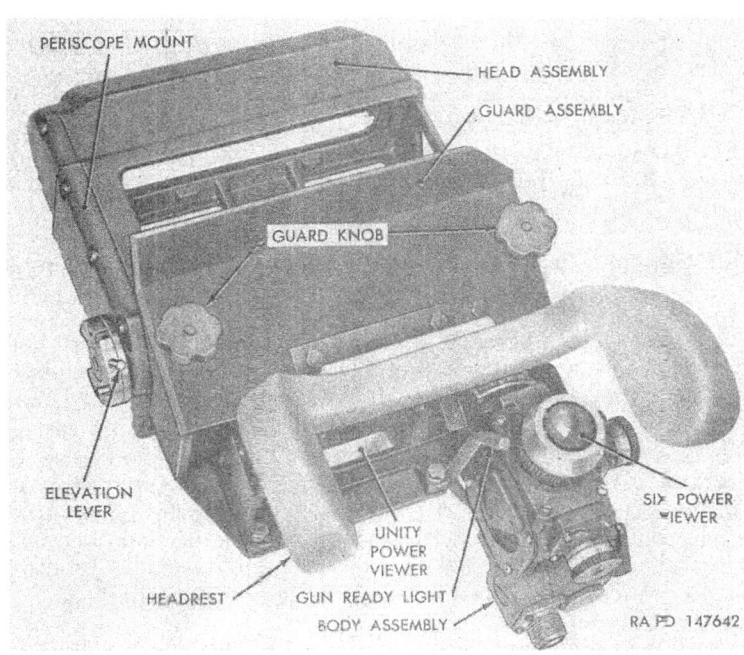

Figure 48. Periscope T35 installed in gunner's periscope mount T176.

b. RETICLE PATTERN. The reticle pattern (fig. 50) is 40 mils across with each horizontal space or line on the reticle pattern representing a deflection of five mils. The top of the vertical bar is used for sighting where no lateral lead is necessary. The unity power system has no reticle pattern.

c. CONTROLS.

(1) *Elevation lever.* An elevation lever assembly (figs. 47 and 48) is provided so that the line of sight of the periscope may be manually elevated 22° or depressed 18° from zero elevation. However, when the periscope is installed for secondary direct fire sighting, the maximum elevation is 20° while the maximum depression is 14°. The lever is spring loaded to allow it to return to the level line of sight or, if installed for secondary direct fire, to the same elevation as the gun.

(2) *Diopter scale.* A diopter scale (fig. 49) is provided on the six-power eyepiece for focusing the periscope to the eyes of the observer. The scale is graduated from plus 3 diopters to minus 3 diopters, so that an observer may read his correction and facilitate resetting the scale once his correction is known.

(3) *Boresighting knobs* (fig. 49). A vertical boresighting knob and a lateral boresighting knob are provided for making boresighting adjustments. Each knob is provided with a graduated scale for recording the boresight adjustment. A locking lever serves to hold each knob in correct adjustment.

d. ILLUMINATION. Reticles of periscope T35 are illuminated by means of a lamp instrument light M36 (par. 64*c*). A dovetail slot above the eyepiece of the telescope receives the lamp bracket.

60. Gunner's Periscope Mount T176 and Commander's Periscope Mount T177

a. DESCRIPTION. The gunner's periscope mount T176 (fig. 48) and the commander's periscope mount T177 (fig. 47) are used to support periscope T35. Both mounts are attached to the turret roof. A guard (figs. 47 and 48) is provided at the rear of each mount to protect the observer should the periscope head assembly be hit. Two knobs secure the guard to the mount. The gunner's mount has a full headrest; the commander's mount a half headrest. An arm and connector on the left side of the commander's mount connects the elevation lever assembly of the periscope T35 to the extension arm on the ballistic drive. An elevation level assembly on the commander's mount is used to level the mount.

b. GUN READY LIGHT (figs. 47 and 48). A cable from the loader's safety panel is connected to the receptacle at the base of the periscope

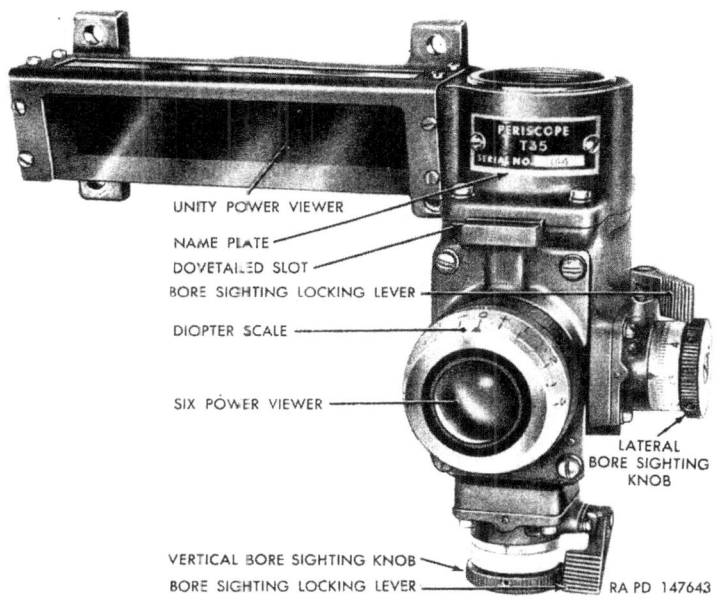

Figure 49. Periscope T35 (body assembly).

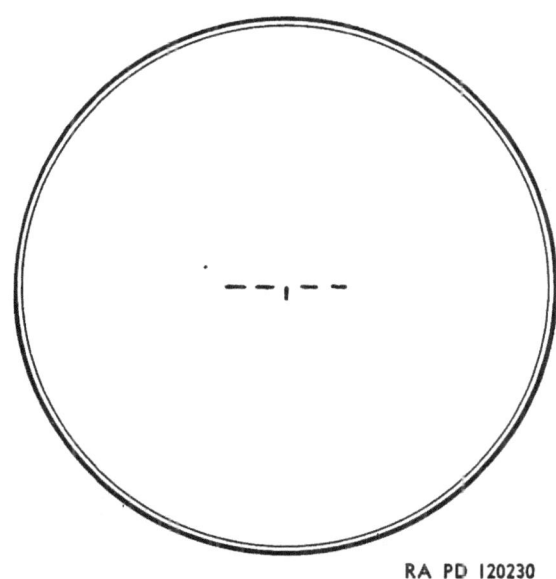

Figure 50. Reticle pattern for periscope T35.

mount T177, and at the rear of the periscope mount T176 to supply power to light a lamp when the loader operates the firing reset safety switch (par. 41). An illuminating tube carries the light to the gunner's or the commander's viewing position.

61. Ballistic Drive T23E1

a. DESCRIPTION. The ballistic drive T23E1 (figs. 51, 52, 53, and 54) is composed of two units: a ballistic unit and an arm assembly. The ballistic unit applies an elevation correction to the angle of sight depending on the range and type of ammunition used. The arm assembly (fig. 52) connects periscope T35 to the gun linkage by means of a connector.

Note. This connector is furnished with the ballistic drive. When range finder T41 is used stow this connector in the turret.

The ballistic drive is secured to the turret roof by means of three mounting brackets. A level vial is provided on the ballistic unit to determine that the ballistic drive is mounted in a level position.

b. CONTROLS (fig. 54).

(1) *Shutter assembly spring loaded lever.* A shutter assembly, operated by a spring loaded lever, allows the shutter to be moved up or down to select the desired range scale for the particular type of ammunition used. When the lever is released, the shutter assembly is locked in position in a groove.

(2) *Range knob.* A range knob on the ballistic unit turns in the desired range on the range scale. The range scales are graduated as shown in table I.

Table I. Range Graduations on Ballistic Unit Scales

	Marked	Graduated
Mil scale	MILS	Mils.
HE–M71	HE	200 yards.
AP–T33	AP	200 yards.
HVAP–T67	HVAP	200 yards.
HEAT–T108	HEAT	200 yards.

(3) *Light switch.* The light switch turns off or selects the source of power to illuminate the scales.

c. ILLUMINATION (fig. 54). Two built-in electric lamp assemblies are provided for illuminating the scales of the ballistic unit. The 24-volt supply from the vehicle power source is connected to the upper receptacle while the lower receptacle is provided for receiving the plug from instrument light M30 (par. 64*b*) should the 24-volt supply fail. When the light switch is in the lower position, the 24-volt supply is connected to the ballistic unit; in the upper position instrument light M30 is connected into the circuit and in the middle position no illumination is provided.

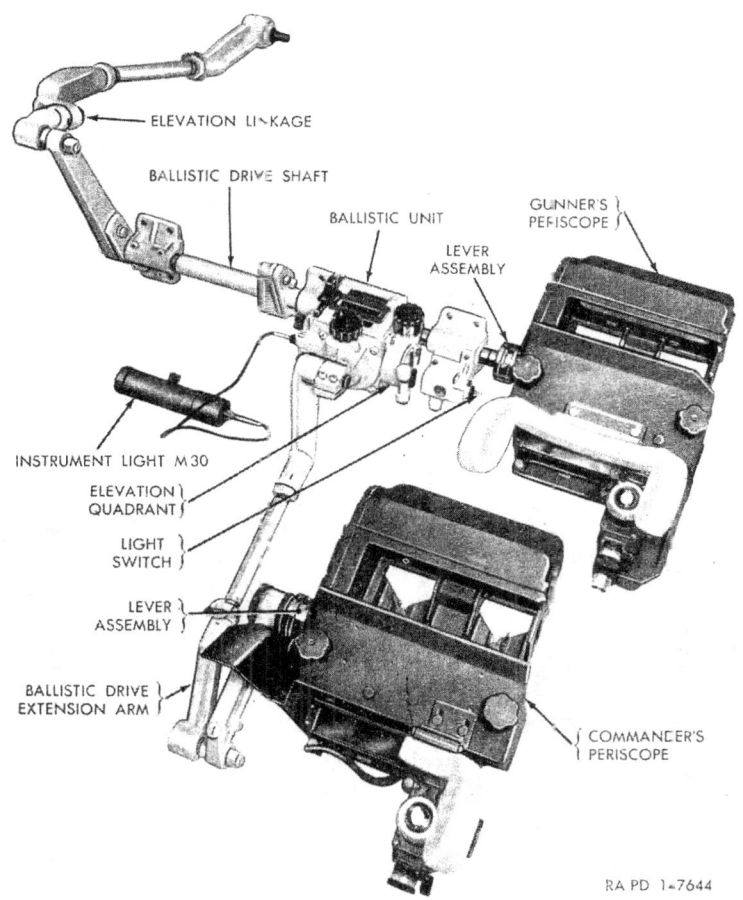

Figure 51. Arrangement of ballistic drive T23E1, gunner's periscope mount T176, and commander's periscope mount T177.

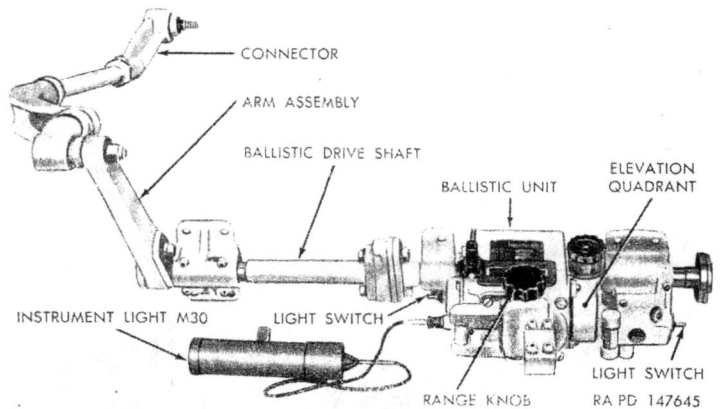

Figure 52. Ballistic drive T23E1, elevation quadrant T21 and instrument Light M30.

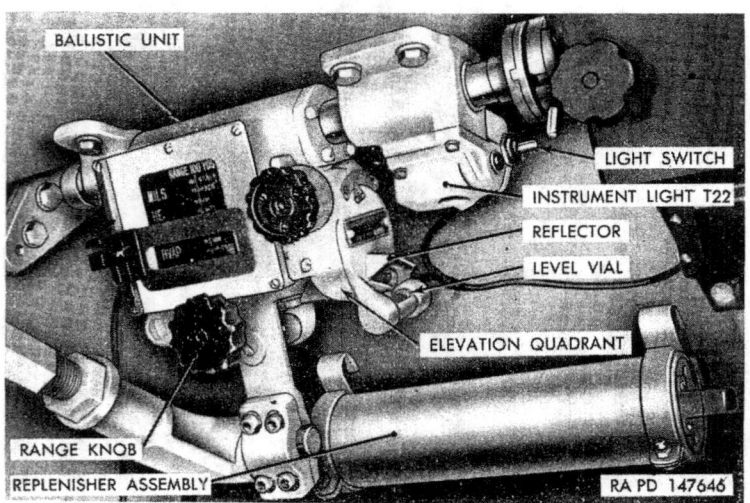

Figure 53. Ballistic drive T23E1—right side view.

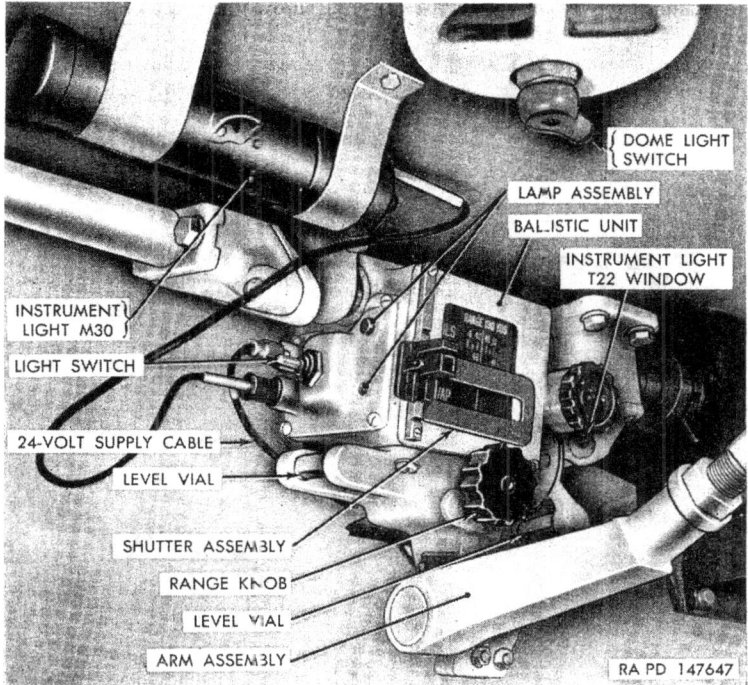

Figure 54. Ballistic drive T23E1—left side view.

62. Elevation Quadrant T21

a. DESCRIPTION. Elevation quadrant T21 (fig. 57), is used for setting elevation or depression angles for indirect fire operation of the gun. The elevation scale is graduated in 100-mil intervals from minus 200 to plus 600 mils and numbered every 200 mils from minus 2 to plus 6. A level vial mounted on the quadrant moves with the scale index when elevation or depression angles are set. The quadrant moves with the gun so that when the level bubble is centered by elevating or depressing the gun, the gun is aimed to the elevation set on the scale and micrometer. A reflector set at 45 degrees over the level bubble allows viewing the centering of the bubble more readily.

b. ELEVATING KNOB. An elevating knob is provided with an elevation micrometer graduated in 1-mil intervals from 0 to 100 mils and is numbered in every 10 mils from 0 to 90. The micrometer has two scales reading in opposite directions for putting in elevation or depression angles. For elevation angles, read the inner scale; for depression angles, read the outer scale. One complete revolution of the

71

elevating knob represents 100 mils and moves the scale index one complete division on the scales.

c. ILLUMINATION. The level vial, scale, and elevation micrometer of elevation quadrant T21 is illuminated by means of instrument light T22 (par. 64*d*).

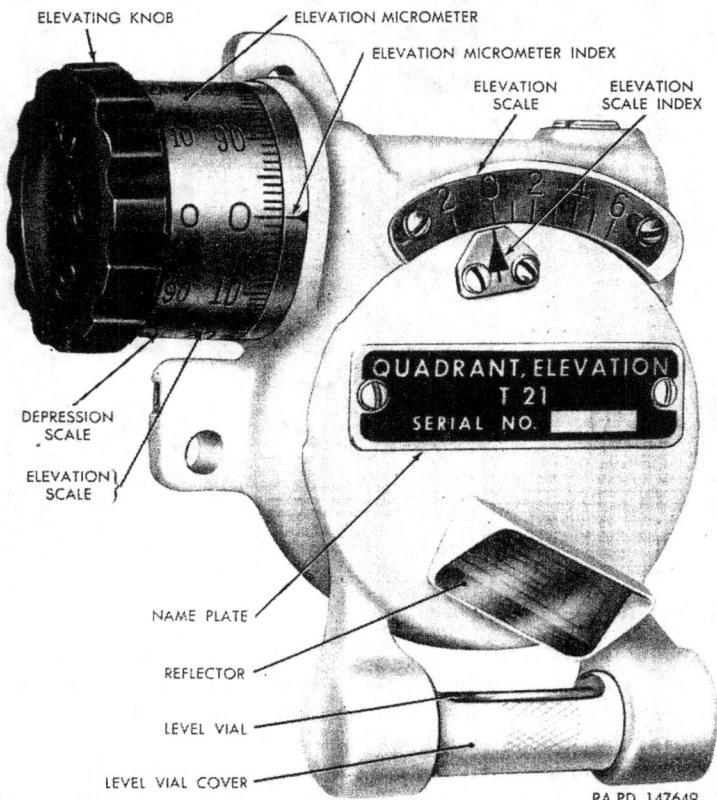

Figure 55. Elevation quadrant T21.

63. Azimuth Indicator T24

a. DESCRIPTION. The azimuth indicator T24 (fig. 58) is mounted in the right-forward section of the turret. It is used for indirect fire operation in conjunction with elevation quadrant T21. The azimuth indicator dial (fig. 59) contains three scales: a 100-mil (inner) scale, a 1-mil (center) scale, and a gunner's aid dial on the perimeter of the dial face. The inner scale is graduated in 100-mil intervals and the center scale in 1-mil intervals. If the azimuth angle required is 1,650 mils, the inner scale should read midway between 16 and 17 and the center scale at 50.

b. RESETTER KNOB. The resetter knob is used to adjust the top and middle pointers to zero azimuth when the sighting equipment and gun is directed on an aiming point.

c. ILLUMINATION. Instrument light assembly D78454 (par. 64a) illuminates the azimuth indicator scales for night operation.

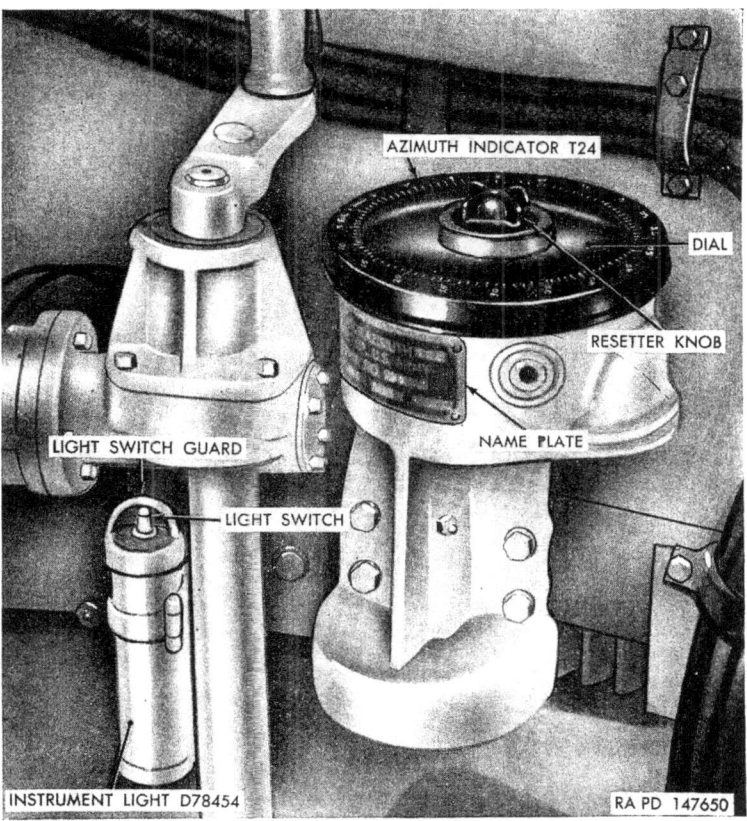

Figure 56. Azimuth indicator T24—installed view.

64. Instrument Lights

a. INSTRUMENT LIGHT D78454 (fig. 58). Instrument light D78454 is a battery assembly supplying power for the lamp in azimuth indicator T24 (par. 63c). A toggle switch on top of the instrument light is used to turn the light on or off. A guard protects the switch from being turned on or off accidentally.

b. INSTRUMENT LIGHT M30 (fig. 54). Instrument light M30 is a battery assembly to provide emergency power to the ballistic drive T23E1 in case the main power cable is out of order. A switch on the ballistic drive is used to turn off or to select the source of power.

c. INSTRUMENT LIGHT M36. An instrument light M36 is provided for each periscope T35. A dovetailed slot above the eyepiece of the

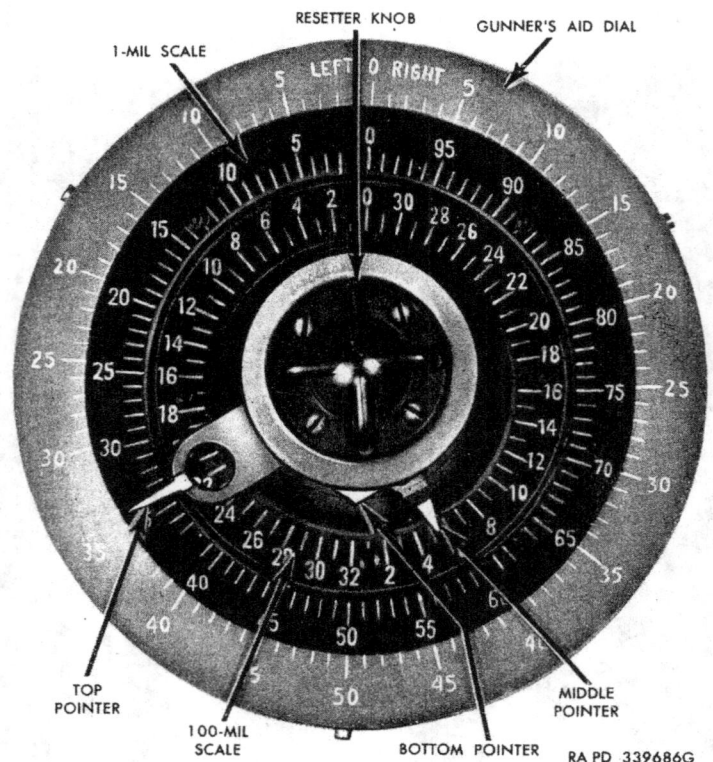

Figure 57. Azimuth indicator T24 dial.

telescope (fig. 49) is provided for attaching the lamp bracket of instrument light M36 for illumination of the six-power reticle (par. 59*d*). A rheostat knob is provided to turn the light off or to regulate the illumination.

d. INSTRUMENT LIGHT T22 (fig. 53). Instrument light T22 is used to illuminate the elevation quadrant T21 by directing light to the level vial, scale, and micrometer through a window (fig. 54). The instrument light has a twin receptacle in the rear. The upper receptacle is provided for receiving the plug from the 24-volt vehicle supply while the lower receptacle is provided for the plug from the instrument light M30 which is for emergency use only. The light switch (fig. 53) has three positions: down, 24-volt supply; up, instrument light M30; middle position, off.

65. Periscope M13, M13B1, and M6

Periscope M13 (fig. 55), periscope M13B1, or periscope M6 (fig. 56) is provided for the driver, assistant driver, and loader for observation of fire or terrain. The periscopes are unity-power instruments with a wide field of view and are mounted in the periscope housings located in each driver's door (fig. 14) and, in the turret roof forward of the loader's door (fig. 39). A clamping knob (fig. 56) and latch are provided for securing the periscope to its mount. Periscope M6 has a metal body; periscope M13 and M13B1 have an all plastic body.

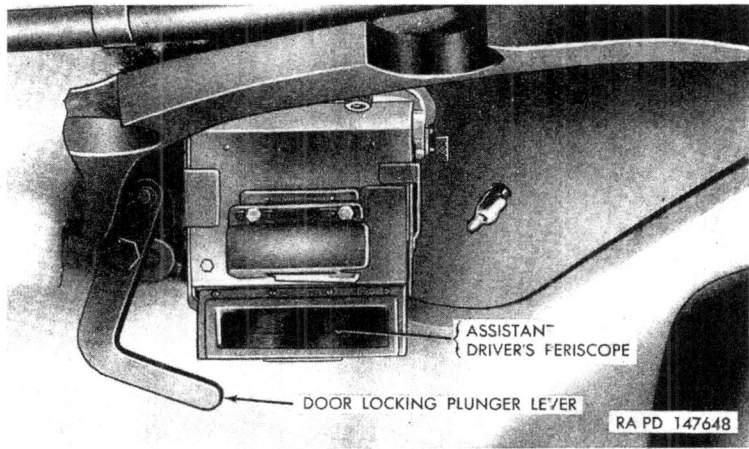

Figure 58. Assistant driver's periscope—installed view.

66. Fuze Setters

a. GENERAL. Fuze setters M14 and M27 are hand operated instruments used for setting mechanical time (MT), mechanical time and superquick (MTSQ), and time and superquick (TSQ) fuzes.

b. FUZE SETTER. Fuze setter M14 (fig. 60) is a flat-handled wrench having a circular tapered hole at one end to fit over the fuze and a key which protrudes through the hole to engage the slot in the fuze; This fuze setter has no scale. The required time setting is obtained by reference to an index and scale engraved on the fuze.

c. FUZE SETTER M27. Fuze setter M27 is a flat-handled wrench-type fuze setter similar to fuze setter M14 except that it is built stronger.

Figure 59. Periscope M6.

Figure 60. Fuze setter M14.

Section III. OPERATION UNDER USUAL CONDITIONS

67. General

This section contains instructions for the mechanical steps necessary to operate the medium tank M47 under conditions of moderate temperatures and humidity. It also includes operation of the turret. For operation under unusual conditions, refer to section V, this chapter.

68. Starting the Main Engine

a. GENERAL.
  (1) Before a new or reconditioned vehicle is placed in service, be sure that the services described in paragraphs 6 through 9 have been performed.
  (2) Before attempting to start the main engine, the driver must become familiar with the purpose and location of the various instruments and controls as outlined in section II, this chapter.

Caution: If there is a warning tag attached in driver's compartment, correct deficiency indicated on tag before starting engine.

  (3) Perform the first five before-operations services in table III (par. 122), before attempting to start engine.

b. STARTING ENGINE.

Caution: Be sure radio switch is off before starting engine. After engine is started, radio can be turned on. Do not race engine while radios are in operation.

Note. When battery charge is low (par. 224) or when starting a cold engine when temperature is below 32° F. it is advisable to first start the auxiliary generator (par. 93). The starting current for starting the main engine will then be furnished by the auxiliary engine generator.

(1) Move the manual control lever to the neutral (forward) position (fig. 61).

Note. This lever must be in neutral position before the engine starter can be operated. Placing the manual control lever in neutral position causes a microswitch (neutral-position switch) in series with the starter circuit to be closed.

(2) Set hand brake lock (par. 13).
(3) Turn on master relay switch (fig. 18).
(4) Check the quantity of fuel in each tank by noting the readings on the two fuel gages (fig. 18).

Note. Magneto switch must be in the "A," "F," or "BOTH" position before fuel gage will indicate.

(5) If fuel-tank-shut-off valves are closed, open the fuel-tank-shut-off valve for the fuel tank to be used (par. 17). If both the tanks are full, open both valves.
(6) Advance (open) the hand throttle lever approximately 1 inch beyond free-play travel (par. 15).
(7) Turn the main-engine-magneto switch (fig. 18) to the "BOTH" position.
(8) If the atmospheric temperature is below 32° F., it may be necessary to use the hand primer pump (par. 16). Care must be exercised in the use of the primer pump to avoid flooding the engine. Normally, from 3 to 9 strokes are required, depending upon the temperature; the lower the temperature, the more strokes necessary.

Note. The driver must learn from personal experience with his vehicle how best to use the primer pump effectively.

After the engine starts, an occasional stroke of the primer pump may be required to keep the engine running.

(9) Turn the main-engine-starter switch (fig. 18) to the "ON" position. At the same time turn the main-engine-booster switch (fig. 18) to the "ON" position.

Caution: Do not hold the starter switch on longer than 30 seconds at a time. So doing will seriously deplete the battery charge.

When the engine starts, release the starter switch.

Note. If the starter fails to turn the engine, release the starter switch immediately and investigate the cause (par. 126).

As soon as engine is running at 500 to 550 rpm, as indicated on the tachometer (fig. 18), release the booster switch. Depress the accelerator pedal (fig. 9) two or three times to assist in keeping the engine running.

(10) Adjust hand throttle so that engine will run between 1,000 and 1,100 rpm during warm-up period.

c. CHECKING ENGINE OPERATION.

(1) As soon as the engine is running, check the oil pressure reading on the engine-oil-pressure gage (fig. 18). Proper pressure is a minimum of 60 psi and a maximum of 70 psi. Stop engine and investigate the cause if the engine-low-oil-pressure-warning-signal light does not go out within 10 seconds after engine has started (par. 126). Observe the oil pressure frequently during operation. Shut off auxiliary engine (par. 93) unless an unusual amount of current is required for such purposes as traversing the turret or operating the bilge pumps or if the battery charge is low.

(2) When the engine runs smoothly without misfiring, idle the engine by moving the hand throttle lever until the tachometer indicates 650 rpm.

Caution: Do not idle engine at less than 650 rpm at any time. Test the magnetos by moving the main engine magneto switch to the "F" position and then to the "A" position, and comparing tachometer readings at these positions with the reading (650 rpm) when the switch is at the "BOTH" position. If a drop of more than 150 rpm is indicated when running engine with the magneto switch at either the "F" or "A" positions, investigate the cause (par. 126)

Caution: Do not run the engine for more than 1 minute with the magneto switch in either the "A" or the "F" positions.

(3) Depress accelerator pedal until engine is running at 1,000 rpm. Check the main-generator-warning-signal light. It should not be lit. If the generator-warning-signal light is on, check as outlined in paragraph 132.

69. Placing Vehicle in Motion

Caution: Before placing the vehicle in motion, be sure that the gun traveling lock (fig. 4) is in the traveling position, and the turret lock (fig. 31) is engaged, to prevent injury to personnel and damage to equipment. Tie antenna back securely when radio is not in use. Keep antenna vertical and not touching anything when radio is in use.

a. Start the main engine and check engine operation, as outlined in paragraph 68. Do not operate the vehicle until the engine is running smoothly without misfiring. Check instrument panel (fig. 18) to be sure no warning signal light is on. Release the hand brake lock handle in the manner outlined in paragraph 13. Move the manual control lever to the shift positions desired (*b* below). Depress the accelerator pedal slowly. As the pedal is depressed causing the engine speed to increase, the tank will begin to move. Depress the accelerator pedal until the desired vehicle speed is obtained. When turning the vehicle, maintain or increase the engine speed as required.

Note: Do not pump accelerator pedal while driving. This can cause a flooded engine.

Caution: Do not operate the vehicle with the hand throttle lever except in emergency.

b. The manual control lever has four shift positions: neutral ("NEU"), low ("LOW"), high ("HI"), and reverse ("REV").

Note: Charts showing the position of the lever for each gear range selection are cast on top of the manual control box (fig. 61).

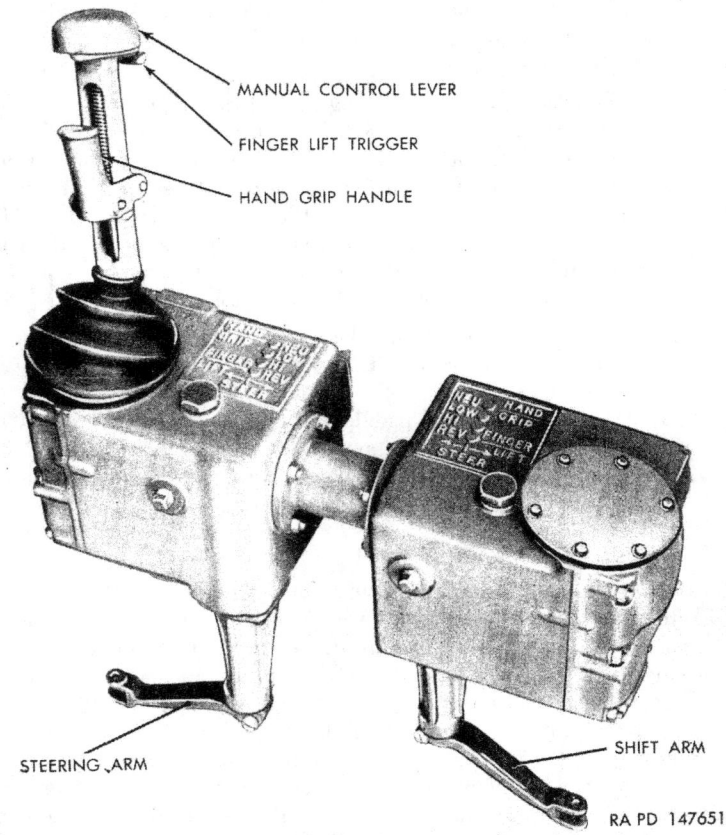

Figure 61. Steering and shifting control

To move the manual control lever in or out of neutral shift position, whether in steer or in neutral steer, it is necessary to actuate the hand grip handle (fig. 31). To move the manual control lever from low- to high-shift position and vice versa, neither the finger lift trigger nor the hand grip handle need be actuated. Detent cams and springs, incorporated in the manual control box and in the transmission control valve, make it possible to definitely locate the low- and high-shift positions. To shift into the reverse-shift position from either the low or high positions, or vice versa, the finger lift trigger or the hand grip handle must be actuated.

c. For normal driving on hard-surface roads, shift directly into the high-shift position before starting. The low-shift position is for ascending and descending steep grades, or for soft or very rough terrain. If the vehicle is set in motion with the manual control lever in the low-shift position, the lever can be moved to the high-shift when the vehicle has reached a speed of from 6 to 12 mph. When it is necessary to shift from high to low, perform the shift operation at approximately 11 mph. Depress the accelerator pedal simultaneously with the high- to low-shift operation.

d. To reverse the vehicle, move the manual control lever to the reverse position. The lever can be moved directly to reverse from any of the other three positions.

Caution: Bring the vehicle to a complete stop before shifting into the reverse-shift position. Completely stop vehicle before shifting from the reverse-shift position into one of the forward positions. When tactical situation permits, station an observer in front of the vehicle to direct the driver while backing.

70. Driving the Vehicle

a. PRELIMINARY INSTRUCTIONS. Observe all instruments and gages frequently during operation, paying particular attention to the engine-oil-pressure gage (fig. 18) and high-oil-temperature-warning-signal light (fig. 18). Normal engine oil pressure is 60 to 70 psi. Do not operate the vehicle until the engine is running smoothly without misfiring. Stop the engine immediately and investigate the cause if any engine or transmission warning signal lights come on (pars. 126 and 134).

b. STEERING. Steering of the vehicle is accomplished by movement of the manual control lever to the right or left. When released, the manual control lever automatically will return to the neutral-steer position. When the vehicle is moving in the forward direction (either in low- or high-shift position), moving the manual control lever to the right will turn the vehicle to the right; movement to the left will turn the vehicle to the left. The turn radius, at a constant vehicle speed, is decreased as manual pressure on the control level is increased, due to reduction of slippage in the friction elements of the steering clutch in the cross-dirve transmission. When the vehicle is in reverse, the rear of the vehicle will turn in the direction opposite to the direction in which the manual control lever is moved. That is, to move the rear of the vehicle to the right, move the manual control lever gradually to the left; to move the rear of the vehicle to the left, move the manual control lever gradually to the right.

Note. Do not apply a jerking motion to the steering controls.

Steering with the manual control lever in the neutral-shift position causes the vehicle to pivot in place with the tracks turning in opposite directions. By using a combination of neutral-steer and low-steer, the vehicle can be made very maneuverable.

71. Stopping the Vehicle

To stop the vehicle, release foot pressure from the accelerator pedal and apply pressure to the service brake pedal.

Caution: Do not move manual control lever to neutral shift position until vehicle is brought to a complete stop. Some steering may be necessary to control direction of the vehicle during braking and stopping operation. The parking brakes are set by maintaining pressure on the service brake pedal and pulling back on the parking brake lock handle as far as possible.

72. Stopping Main Engine

After stopping the vehicle, set the hand throttle lever (fig. 9) to run the engine 650 rpm. Idle the engine at this speed for 5 minutes to assure a gradual and uniform cooling of valves and other engine parts. To stop the engine, push in the fuel-cut-off switch on the instrument panel (fig. 18), and hold in this position until the engine stops. Turn the main engine magneto switch to "OFF" position after the engine has stopped. If the vehicle is to be stopped for any length of time, and no accessories, radio, or lights are to be used during the halt, turn off the master relay switch.

73. Driving Precautions

a. GENERAL. Until the driver (even one experienced on other types of track-laying vehicles) becomes thoroughly familiar with this vehicle, every precaution must be taken not to overdrive or allow the vehicle to get out of control.

b. HARD PAVEMENT. Great care in driving must be observed at all times when operating on hard surfaces or pavement. Normally, the vehicle may be driven safely at recommended speeds (par. 5), providing the driver has the proper skill and experience in operating this type of vehicle. Careless handling can result in loss of steering control and consequent serious injury to personnel, damage to property, and damage to the vehicle.

c. DITCHES AND OBSTRUCTIONS. When crossing a ditch, shell hole, or trench, release the accelerator pedal momentarily until the vehicle reaches the bottom, then bring the engine up to full power as the vehicle starts to climb. Shift down to low-shift position, if necessary. When going over an obstruction, release the accelerator pedal and allow the vehicle to settle down over the obstruction before applying full power. In going over a high obstruction, it may be necessary to use full power at the start but, upon reaching the crest of the obstruction, release the accelerator until the vehicle is fully over the top of the obstacle.

Caution: Use care not to damage the gun (if in traveling position or rear fenders when climbing out of ditches or shell holes, or when pulling away from obstructions.

d. STARTING VEHICLE ON AN UPGRADE. If the vehicle is headed up an incline or hill, it will roll backwards when the parking brakes are released, unless increased engine power is applied to provide sufficient power through the transmission to move the vehicle forward.

Caution: Do not attempt to hold the vehicle stationary for any length of time by this method. The transmission will overheat.

e. DESCENDING STEEP GRADES. To negotiate steep down-grades, the transmission should be shifted into low range. If this will not sufficiently brake the vehicle, it should be slowed by intermittent application of the brakes. If the down-grade is very steep (50 to 60 percent), the vehicle can be braked very effectively by placing the transmission in reverse-shift at the start of the grade. Increased braking effect then is obtained by accelerating the engine. Usually, low range will brake the vehicle sufficiently so that no application of the brakes is necessary. At bottom of descent, shift transmission into desired forward-speed range and proceed.

74. Towing the Vehicle

a. GENERAL. The vehicle is equipped with four towing shackles and two towing cables. Two shackles are mounted on the front of the vehicle (fig. 6) and two on the rear (fig. 7). The two 10-foot towing cables are coiled on outside rear of the vehicle.

b. TOWING TO START ENGINE. Under normal conditions, and on level terrain, an engine can be started by towing. Connect the towing bar or cable. Station a driver in the towed vehicle and an observer at the rear of the towing vehicle. Set the manual control lever in the high-shift position (fig. 61). Turn the main-engine-magneto switch (fig. 18) to "BOTH" position. Set the hand throttle level (fig. 9) forward approximately 1 inch beyond free-play travel. Hold main-engine-booster switch (fig. 18) in "ON" position while being towed, until engine starts to run smoothly at about 500 rpm; then release switch lever. Signal the observer that the engine has started. The observer then will signal drivers of both vehicles when to apply brakes so that the towed vehicle will not ram the towing vehicle. The engine can be started in this manner due to a freewheeling device in the transmission which locks the torque converter input and output shafts together when the speed of the output shaft exceeds that of the input shaft.

  c. Towing a Disabled Vehicle.

Caution: The towing cable must not be connected by any means other than by the cable eyes, as doubling the cable causes short bends which break the wire strands and leaves the cable weak and dangerous to handle.

A driver must be in control of the towed vehicle. The manual control lever (fig. 61) must be in neutral-shift position before starting to tow. Make changes of direction by a series of short turns, as this helps keep the vehicle in line. Keep the vehicles in line as far as possible, especially before entering and crossing wet or muddy terrain. Maximum speed when towing a disabled vehicles must never exceed 12 mph.

75. Preparation of Matériel for Firing

  a. Preliminary Procedure.
   (1) Remove the muzzle cover and the combination breech and empty shell case cover from the 90-mm gun. Install the combination breech and empty shell case cover in place to catch the fired cartridge cases.
   (2) Install the empty cartridge case on the coaxially mounted cal. .50 machine gun.
   (3) Unlock the turret (par. 42).
   (4) Unlock the gun traveling lock and latch it in the unlocked position (par. 55).
   (5) Install the sighting and fire control equipment (par. 84).
  b. Inspection Before Firing.
   (1) *Wipe bore and chamber.* Assemble wiper ring to cleaning staff. Insert dry jute burlap through loop of wiper ring and run it through the bore until all surfaces are dry.
   (2) *Clean gun tube.* Clean the gun tube and coat with grease at shield bore as specified in lubrication order (fig. 87).

(3) *Check functioning of breech mechanism.* Open and close the breech several times, using a dummy round or an empty shell. Operation should be smooth, without binding. If it is not, adjust the closing spring cap (par. 346*b*).

(4) *Check functioning of percussion mechanism.* Cock the percussion mechanism by pulling the hand cocking lever handle to the rear and manually "dry fire" the gun several times. Check the clearance between the firing plunger and the trigger plunger (fig. 349). When the percussion mechanism is cocked, the clearance should be one-thirty-second of an inch. If it is more or less than one-thirty-second of an inch, push the gun safety lever (fig. 349) into "SAFE" position, loosen the jam nut, adjust the firing plunger cap (fig. 349) until the desired clearance is obtained, and tighten the jam nut.

(5) *Check functioning of firing linkage.* Examine firing linkage for general condition. Check for broken, missing, or deformed parts. If linkage does not operate properly, notify ordnance maintenance personnel.

(6) *Check replenisher oil supply.* Refer to paragraph 56.

(7) *Check functioning of elevating and traversing mechanism.* Check oil level in oil reservoir. If low, see lubrication order (fig. 87). Traverse the turret and elevate the gun (par. 76) through the full limits. Operation should be smooth, without binding. If binding occurs after lubrication, notify ordnance maintenance personnel.

(8) *Check functioning of commander's control handle.* Operate commander's control handle (par. 38) to make sure that it cuts out the gunner's control handle.

76. Traversing Turret and Elevating Gun

a. GENERAL. The turret may be traversed through any angle in either direction, at any rate from 0 to a maximum of 6 rpm. The gun may be elevated to 19 degrees above the horizontal or depressed to 5 degrees below the horizontal, at any rate from 0 to a maximum of 4 degrees per second. Both traversing and elevating may be controlled by either the commander or the gunner.

b. PRECAUTIONS. Before traversing the turret or elevating the gun, release the gun traveling lock (par. 55) and the turret lock (par. 42). The driver and the assistant driver must lower their seats and keep their heads down. Check to see that the gun will clear any obstructions on hull top.

Caution: The traversing and elevating controls must never be used unless personnel in or on the vehicle are aware of the operator's intentions.

c. MANUAL TRAVERSING. The turret may be traversed manually by use of the gunner's manual traversing control handle (par. 36) when conditions as outlined in *b* above are met.

d. MANUAL ELEVATING. The gun may be elevated or depressed manually by the use of the gunner's manual elevating control handle (par. 37) when conditions outlined in *b* above are met.

e. HYDRAULIC POWER OPERATION. To operate the hydraulic power traversing and elevating system, first be sure that the gunner's power traversing and elevating control handle is in the neutral position. Observe precautions outlined in *b* above. Turn on the master relay switch (fig. 18).

Note. Allow approximately 3 seconds for the shift mechanism to automatically operate before attempting to elevate the gun or traverse the turret.

If the vehicle main engine is not running, start the auxiliary engine (par. 93) to avoid excessive drain on the batteries while operating the turret. Set the turret control switch to the "ON" position (fig. 28). Set the loader's traverse safety switch (fig. 29) to the "ON" position. Operate the gunner's power traversing and elevating control handle (par. 35). The commander can assume control of the gun and turret by operating his override control (par. 38). When the turret power mechanism is not in use, be sure the turret control switch (fig. 28) is in the "OFF" position. The turret lock should be engaged (par. 42). A no-back mechanism automatically holds the turret in position when the hydraulic mechanism is stopped and prevents turret drift when the vehicle is not on a horizontal plane.

77. Operation of the Breech Mechanism

a. GENERAL. The breech mechanism is designed for automatic opening during counterrecoil and automatic closing by spring action upon insertion of a round.

b. OPEN THE BREECH. Manual opening of the breech, which is necessary when beginning a firing period, is accomplished by depressing the plunger at the top of the breech operating handle (fig. 41), then pulling the operating handle to the rear and downward.

Caution: The operating handle should always be swung up into its locked position on the breech as soon as the breechblock is locked in its open position. This leaves the handle immobile. Failure to do this may cause injury to personnel or damage to the mechanism when closing spring compression is released as the breech closes.

c. CLOSING THE BREECH. Manual closing of the breech, which is necessary at the end of the firing period, is accomplished by forcing the extractors forward with the back end of a cartridge case or a block of wood.

Caution: Do not use the hands to trip the extractors.

78. Loading the 90-mm Gun

Open the breech (par. 77) and return the breech operating handle (fig. 41) to its latched position. Push a round into the chamber as far as it will go. The breech will close automatically as the flange of the cartridge case trips the extractors. Exercise caution to avoid striking the fuze in any manner. If round fails to chamber completely, refer to paragraph 143 for proper procedure.

79. Firing the 90-mm Gun and Coaxial Machine Gun

a. ELECTRICAL FIRING OF 90-MM GUN.
  (1) Be sure that the turret is unlocked (par. 42).
  (2) Turn the 90-mm gun firing selector switch to the "ON" position (fig. 28).
  (3) After round has been taken from hull compartment, turn the loader's traverse safety switch to the "ON" position (fig. 29).
  (4) Press the loader's firing safety reset switch button (fig. 30) to close the 90-mm gun firing circuit.

  Caution: All personnel must stand clear of the breech end of the gun to avoid being struck by the gun during recoil.

  (5) Push the 90-mm gun safety lever out of "SAFE" position to unlock the firing plunger (fig. 41).
  (6) To fire the gun electrically, squeeze the firing button on one of the gunner's control handles or press down the commander's override lever (fig. 26) and squeeze the commander's firing trigger.

b. MANUAL FIRING OF 90-MM GUN.
  (1) Unlock the turret (par. 42).
  (2) Push the 90-mm gun safety lever out of "SAFE" position (fig. 41).

  Caution: All personnel must stand clear of the breech end of the gun to avoid being struck by the gun during recoil.

  (3) Grasp the hand firing handle (fig. 302) and pull it toward the breech end of the gun, thus activating the trigger plunger and gun firing components.

c. ELECTRICAL FIRING OF COAXIAL MACHINE GUN.
  (1) Turn the cal. .50 machine gun firing selector switch (fig. 28) to the "ON" position.
  (2) To fire the machine gun, squeeze the firing button on one of the gunner's control handles or press down the commander's override lever and squeeze the commander's firing trigger.

d. MANUAL FIRING OF COAXIAL MACHINE GUN.
  (1) Rotate the L-shaped safety latch (figs. 59 and 41) forward 90 degrees.

(2) Press the trigger lever forward until the tip of the safety latch contacts the solenoid plunger thereby firing the machine gun. Hold down the trigger lever for as long as necessary to achieve the desired burst.

Caution: Be sure to rotate the safety latch to its safe position whenever the machine gun is not being manually fired, in order to prevent accidental firing by contact with the trigger lever.

80. Extracting the Fired Cartridge Case

The breech is automatically opened during counterrecoil by the breech operating mechanism; thus, the firing cartridge case is extracted automatically from the chamber and ejected out of the rear of the gun. If the extractors fail to extract the fired cartridge case, refer to paragraph 143 for proper procedure.

81. Immediate Action in Case of Failure to Fire

a. Failure to fire may be caused by defective ammunition or by a malfunction of the weapon or mount.

b. If the gun fails to fire (no report is heard), recock the percussion mechanism by pulling the cocking lever (fig. 41) to the rear and make another attempt to fire. If the gun still does not fire, recock the percussion mechanism once more and make a third attempt to fire.

c. If the gun does not fire on the third attempt, proceed as in paragraph 143.

82. Unloading an Unfired Round

Prepare to catch the base of the round to prevent it from slipping out and dropping to the floor. Remove the round and return it to its rack.

83. Preparation of Armament for Traveling

a. Turn the loader's traverse safety switch (fig. 29) to the "OFF" position.

b. Turn the 90-mm gun firing selector switch (fig. 28) to the "OFF" position.

c. Push the 90-mm gun safety lever (fig. 41) to "SAFE" position.

d. Remove sighting and fire control equipment and stow in the place provided (par. 90).

e. Disengage the gun traveling lock from the latch on the vehicle. Swing the lock up and hold it up while the gun tube is being depressed into position in the lock bracket. Lower the cap (fig. 289) over the gun tube and rotate the lock lever so that the lever bolt is screwed into the bracket.

f. Engage the turret traversing lock (par. 42).

g. Remove the empty cartridge bag from the coaxially mounted cal. .50 machine gun. Install the muzzle cover on the 90-mm gun. Install the combination breech and empty shell case cover on the breech ring.

84. Preparation for Operation of Sighting and Fire Control Equipment

a. RANGE FINDER T41 (fig. 44).

(1) Look through the binocular eyepieces of the range finder. Turn the diopter scale on each eyepiece individually while having the other blacked out, until the image is seen with maximum sharpness.

(2) Look through the binocular eyespieces of the range finder and adjust the interpupillary knob until the distance between the eyepieces corresponds to the observer's eye separation.

(3) Turn the light switch to "STEREO-SCALES" position and adjust the light rheostat until the desired illumination of the reticle and scales is obtained.

(4) Move the filter lever to the left to insert filters in the optical system to improve the contrast between the target image and the scale.

(5) Make certain the scale transfer lever is positioned away from the observer to project the scales into the left eyepiece.

(6) Select the type of ammunition to be used and turn the ammunition knob so that the desired ammunition is indicated on the ammunition scale of the reticle pattern in the left eyepiece.

(7) Determine the ballistic corrections from the ballistic correction plate adjacent to the range finder and set this value on the ballistic correction knob.

b. PERISCOPE T35.

(1) Remove the two periscope T35 head assemblies from the stowage box (fig. 62).

(2) Remove the guard assemblies from the periscope mounts by loosening the two guard knobs (fig. 63).

(3) Loosen the two clamping screws (fig. 64) on the periscope mounts.

(4) With the prism side facing forward, insert a head assembly up into the commander's periscope mount until the brackets on the head assembly are positioned under the clamping screws (fig. 64). Holding the head assembly in position on

89

Figure 62. Stowage box for periscope T35 head assemblies.

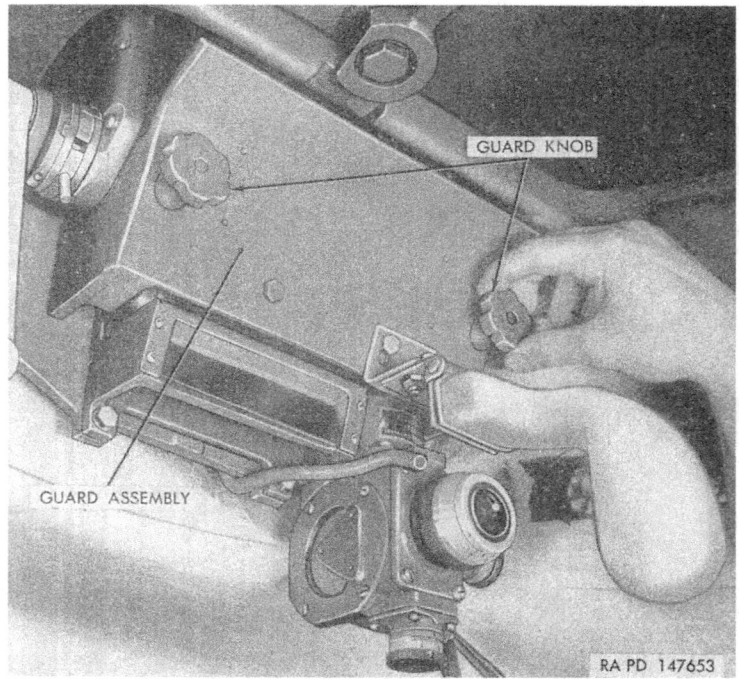

Figure 63. Removing guard on commander's periscope mount T177.

the locating pins, tighten the clamping screws. Repeat this procedure to install the second head assembly into the gunner's periscope mount.

(5) Replace the guard assemblies over the periscope heads. Tighten the two knobs locking the guards in position.

(6) Make certain the cable from the loader's safety panel is connected to the receptacles on the gunner's and the commander's mounts.

(7) Connect the shaft of the ballistic drive to the lever assembly on the gunner's periscope (fig. 51).

(8) Couple the level assembly on the commander's periscope to the arm on the commander's periscope mount. Make certain the arm of the commander's periscope mount is coupled to the extension arm of the ballistic drive.

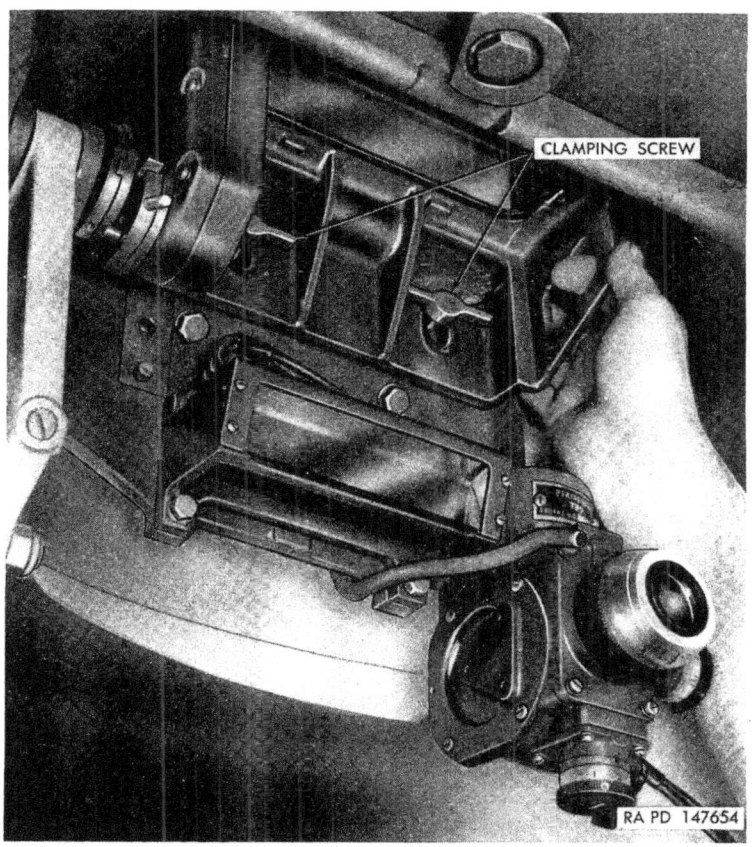

Figure 64. Installing periscope T35 head assembly.

(9) If night operation is required, install the lamp bracket of instrument lights M36 in the dovetailed slot above the six-power eyepiece on the periscopes (fig. 65).

(10) Look through the six-power eyepiece of the periscopes. Turn the diopeter scale on the eyepiece until the image is seen with maximum sharpness.

(11) If preparing for night operation, illuminate the reticle by rotating the rheostate knob of the instrument lights M36 until the desired illumination is obtained.

c. BALLISTIC DRIVE T23E1.

(1) Connect the cable from the 24-volt vehicle power supply to the upper receptacle on the ballistic unit.

(2) Connect the cable from instrument light M30 to the lower receptacle.

(3) Set the light switch to its lower position to illuminate the scales from the 24-volt supply from the vehicle or, for emergency operation, to its upper position.

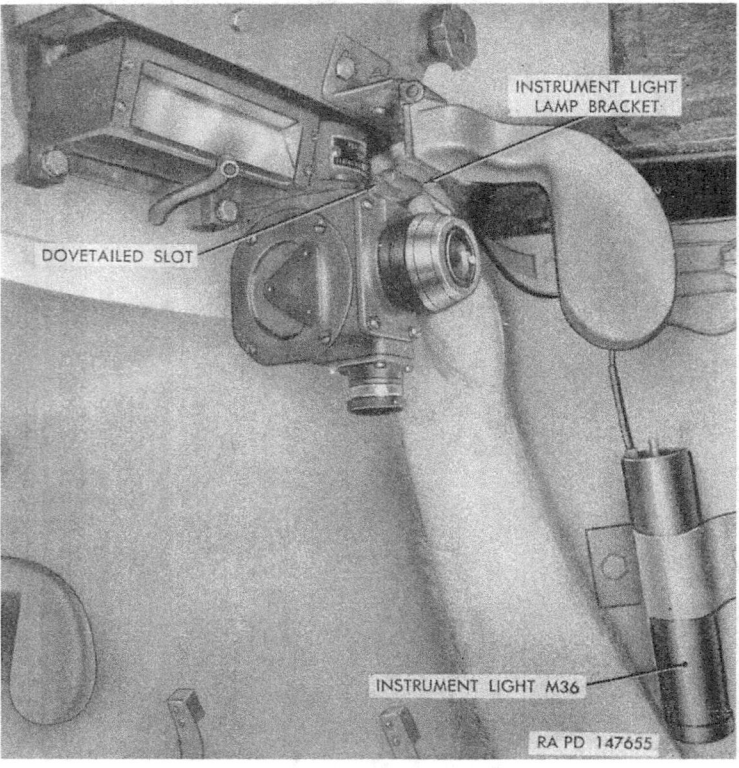

Figure 65. Installing lamp bracket on periscope T35.

(4) Slide the shutter assembly (fig. 68) so that the pertinent ammunition scale is exposed. To move the shutter, squeeze the clamp on the left side of the shutter assembly until the detent is freed from the notch; then slide the shutter assembly up or down as required. To lock in place, release the clamp.

d. ELEVATION QUADRANT T21.
  (1) Insert the plug from the instrument light T22 into the receptacle on the elevation quadrant T21.
  (2) Set the light switch on the instrument light to its lower position to illuminate the quadrant from the 24 volt supply from the vehicle or, for emergency operation, to its upper position.
  (3) Rotate the level vial cover on the elevation quadrant (fig. 57) to expose the graduations on the vial.

e. AZIMUTH INDICATOR T24.
  (1) Insert the plug from the instrument light D78454 into the receptacle on the side of the azimuth indicator T24.
  (2) Snap on the toggle switch of the instrument light to illuminate the scales of the indicator.

f. PERISCOPE M13, M13B1, or M6 (fig. 66).
  (1) Remove the periscopes M13, M13B1, or M6 from their stowage places. The driver's and assistant driver's periscopes are stowed in the stowage box mounted in the driver's compartment (fig. 327), the loader's periscope is stowed in the turret stowage box.
  (2) Install the periscopes in the periscope holders located in the driver's and assistant driver's door and in the turret roof forward of the loader's door. To install, loosen the knob on the front of the periscope and open the latch on the front of the periscope holder. Insert the periscope in the holder and secure in place by tightening the knob and moving the latch to the locking position.

g. FUZE SETTERS M14 or M27. Remove fuze setter M14 or M27 from the gun spare parts roll on the right side of turret bustle.

85. Direct Fire Operation

a. PRIMARY SIGHTING AND FIRE CONTROL SYSTEM.
  (1) Look through the eyepieces of the range finder T41 and elevate or traverse the gun to position the target directly under the apex of the V on the stereoscopic reticle pattern of the range finder. If necessary, adjust the internal correction scale knob to obtain precise stereoscopic contact with the target.
  (2) Turn the range knob (fig. 44) of the range finder until the apex bar of the "V" is in stereoscopic contact with the target.

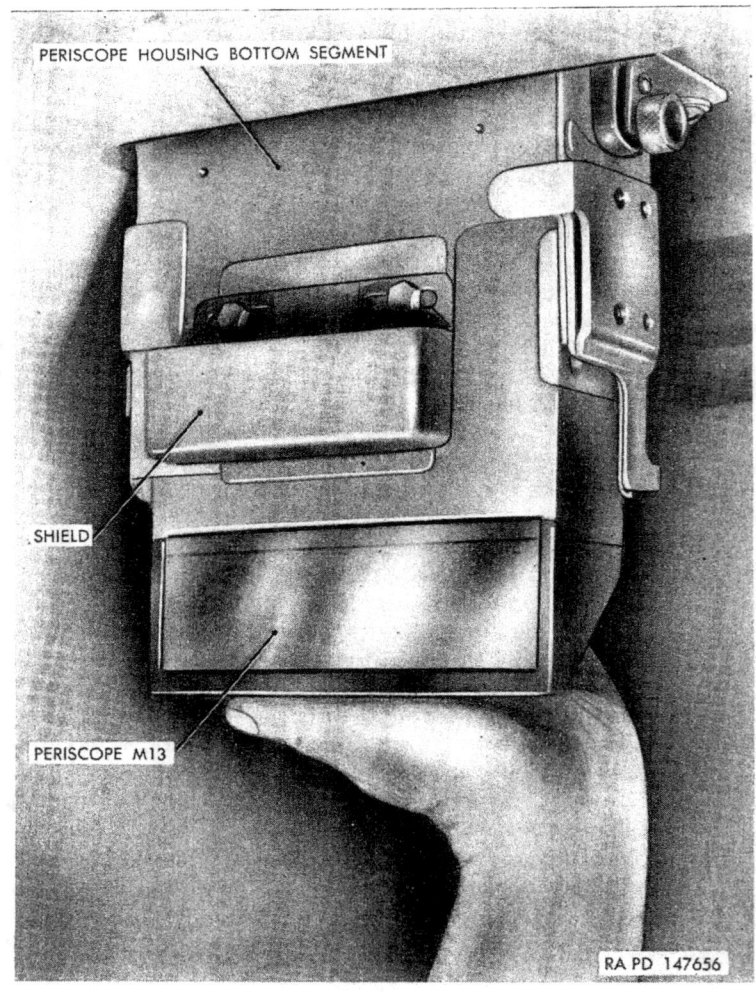

Figure 66. Installing periscope M13.

(3) Manipulate the gun azimuth and elevation controls to position the target exactly on top of the vertical bar of the sighting reticle (fig. 45) of the range finder. Insert any lead as required. The gun is now aimed for firing.

b. SECONDARY SIGHTING AND FIRE CONTROL SYSTEM.
  (1) Look through the eyepiece of the six power system of periscope T35.
  (2) Elevate and traverse the gun until target appears at the top of the vertical bar of the reticle pattern.

(3) Rotate the range knob of the ballistic unit until the estimated range appears under the index.

(4) Insert any lead required. The gun is now aimed to be fired.

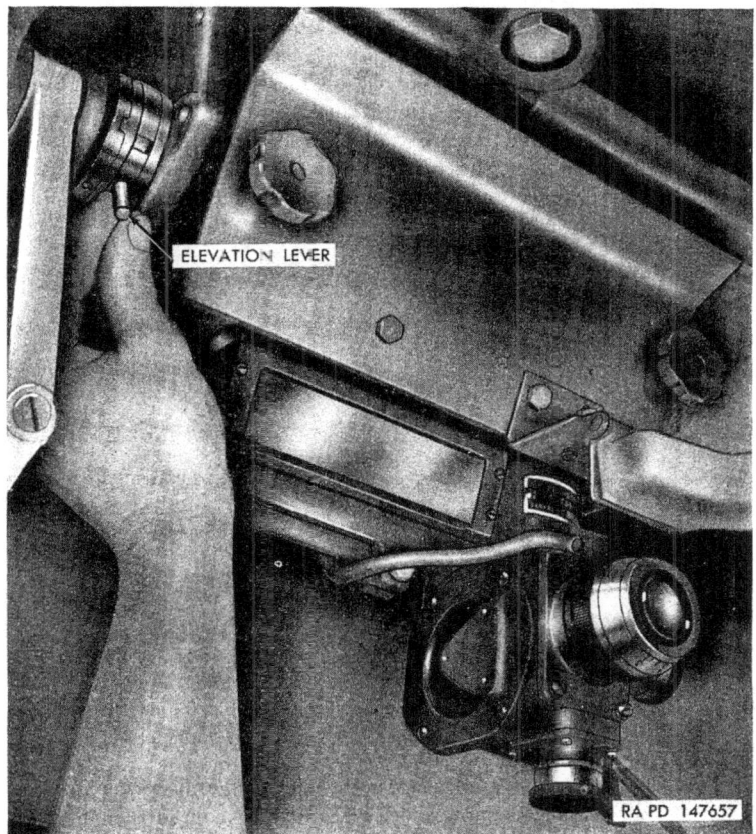

Figure 67. Operating lever on commander's periscope mount T177.

86. Indirect Fire Operation

a. LAYING GUN IN AZIMUTH.

(1) Traverse the turret and elevate the gun until the aiming point is exactly on top of the vertical bar of the reticle pattern of periscope T35. Turn the resetter knob of azimuth indicator T24 (fig. 59) until the top pointer coincides with the middle pointer. Press down on the knob and turn both the top and middle pointers until they read zero on both the 1-mil (inner) scale and 100-mil (outer) scale.

95

(2) Traverse the turret until the middle and top pointers read the required azimuth angle on their respective scales. If the azimuth angle required is 550 mils, the middle pointer should read midway between the 5 and 6 graduations (500 and 600 mils) on the 100-mil scale and the top pointer should read 50 on the 1-mil scale.

(3) To make a deflection correction subsequent to laying gun in azimuth, first rotate the gunner's aid dial on the azimuth indicator until the zero graduation is opposite the top pointer. Then traverse the turret until the top pointer reads the desired deflection correction on the gunner's aid dial.

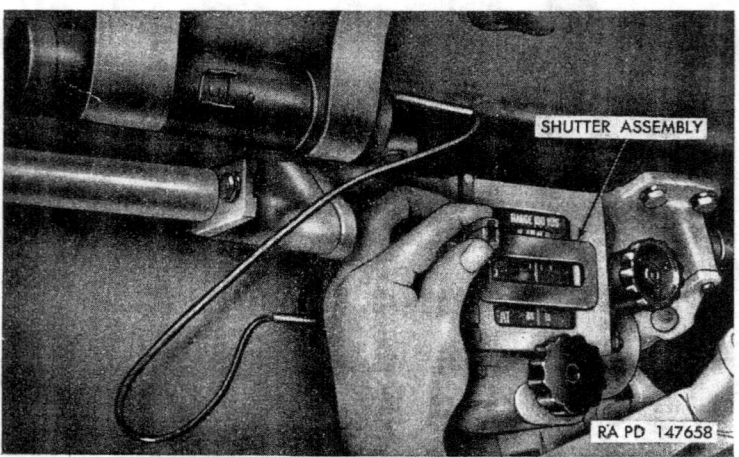

Figure 68. Sliding shutter assembly on ballistic drive T23E1.

b. LAYING GUN IN ELEVATION.
(1) For elevation angles, turn the elevating knob of elevation quadrant T21 counterclockwise until the sum of the elevation scale to the right of zero and inner (elevation scale) micrometer readings equals the required angle. Elevate the gun until the bubble in the level vial is centered.
(2) For depression angles, turn the elevating knob clockwise until the sum of the elevation scale to the left of zero and outer (depression scale) of the elevation micrometer readings equals the required depression angle. Depress the gun **until** the bubble in the level vial is centered.

87. Fuze Setting

Remove the safety wire from the fuze and engage the key on this wrench-type fuze setter (fig. 60) in the notch in the setting ring on the fuze. Rotate the setting ring in a clockwise direction until the desired time setting is opposite the index on the fixed ring on the fuze.

Caution: Rotate the fuze setter only in a clockwise direction. Incorrect settings and loosening of the fuze from the shell may result from failure to observe this precaution.

Figure 69. Turning resetter knob of azimuth indicator T24.

88. Observation of Fire

When a round has been fired, note the position of the shell burst in relation to the target on the reticle of the six-power eyepiece of periscope T35. Each horizontal line or space on the reticle represents a deflection of 5 mils. Apply this correction to the gunner's aid dial of the azimuth indicator (par. 86a(3)).

89. Observation of Terrain

   a. PERISCOPE M13B1, M13, OR M6. The periscope M13, M13B1, or M6 is used by the driver, assistant driver, or loader for observation from the interior of the vehicle. To operate any one of these periscopes, grasp the sides of the periscope and elevate or depress the holder and periscope until the desired panorama is brought into the field of view.

   b. PERISCOPE T35. The unity-power eyepiece of periscope T35 is used by the gunner or commander for observation from the interior of the vehicle. To vary the vertical direction of sight of the periscope T35, grasp the elevation lever (figs. 47 and 48) and manually elevate or depress the periscope.

   c. COMMANDER'S DIRECT-VIEW PRISMS. Five direct-view prisms or vision blocks are positioned around the commander's cupola (fig. 37) for observation of terrain around the vehicle.

90. Preparation for Traveling of Fire Control Equipment

   a. RANGE FINDER T41. Turn the light switch to "OFF" position.
   b. PERISCOPE T35.

   (1) Rotate the rheostat knob of instrument lights M36 to turn lights for the telescope reticle off.
   (2) Remove the instrument light lamp bracket from the dovetail slot of the periscope.
   (3) Uncouple the lever assembly on the gunner's periscope from the shaft of the ballistic drive (fig. 51).
   (4) Uncouple the lever assembly on the commander's periscope from the arm on the commander's periscope mount.
   (5) Remove the guard assemblies from each periscope mount by loosening the two guard knobs (fig. 47).
   (6) Supporting the periscope head assemblies, loosen the two clamping screws securing the assemblies to the mount, and remove assembly.
   (7) Place both head assemblies into stowage box for T35 head assemblies (fig. 62).
   (8) Replace guard assemblies on the periscope mounts and tighten the two knobs locking the guard in position.
   (9) Stow the two periscope T35 head assemblies in the stowage box (fig. 62).

   c. BALLISTIC DRIVE T23E1.

   (1) Set the light switch to its center position to turn lights off.
   (2) Disconnect the cables from the 24-volt vehicle power supply and instrument light M30 from the receptacles on the ballistic unit.

d. ELEVATION QUADRANT T21.
   (1) Set the light switch on the instrument light to its center position to turn lights off.
   (2) Remove the plug from the instrument light from the receptacle on the elevation quadrant.
  e. AZIMUTH INDICATOR T24.
   (1) Snap the toggle switch of instrument light D78454 to turn lights off.
   (2) Remove the plug from the instrument light from the receptacle on the side of the azimuth indicator.
  f. PERISCOPE M13, M13B1, OR M6.
   (1) Support periscope M13, M13B1, or M6 in periscope holder and loosen the clamping knob in front of periscope. Open the latch on the front of the periscope holder and pull the periscope down until it is free of the holder.
   (2) Stow the driver's and assistant driver's periscopes in the stowage box mounted in the driver's compartment (fig. 327).
   (3) Stow the loader's periscope in the turret stowage box.

  g. FUZE SETTER M14 AND M27. Stow fuze setter M14 or M27 in the gun spare parts roll on the right side of turret bustle.

91. Boresighting

  a. PURPOSE. The purpose of boresighting is to properly adjust the sighting equipment in relation to the axis of the gun bore to obtain accuracy of fire. Either the distant aiming point method or the testing target method can be used for boresighting.
  b. EMPLACEMENT. Place the tank on ground as level as possible since no means are provided for cross leveling the weapon trunnions.
  c. INSTALLATION OF BORE SIGHTS.
   (1) Open the breechblock of the gun and insert the breech bore sight (fig. 72) into the chamber. If the breech bore sight is not available, remove the percussion mechanism (par. 342) and with the breechblock closed, use the firing pin hole as a peep sight.
   (2) Attach the muzzle bore sight (fig. 72), stretching the linen cord (or string) tightly across the score marks on the muzzle and hold in place by the strap or rubber bands.
  d. AIMING POINT METHOD.
   (1) *General.* The aiming point must be a sharp and distinct object preferably in excess of the greatest range of employment and never less than the average range of employment, or at approximately 1,500 yards if neither of these ranges is known.
   (2) *90-mm gun.* Look through the center hole of the breech bore sight or firing pin hole and elevate and traverse the gun

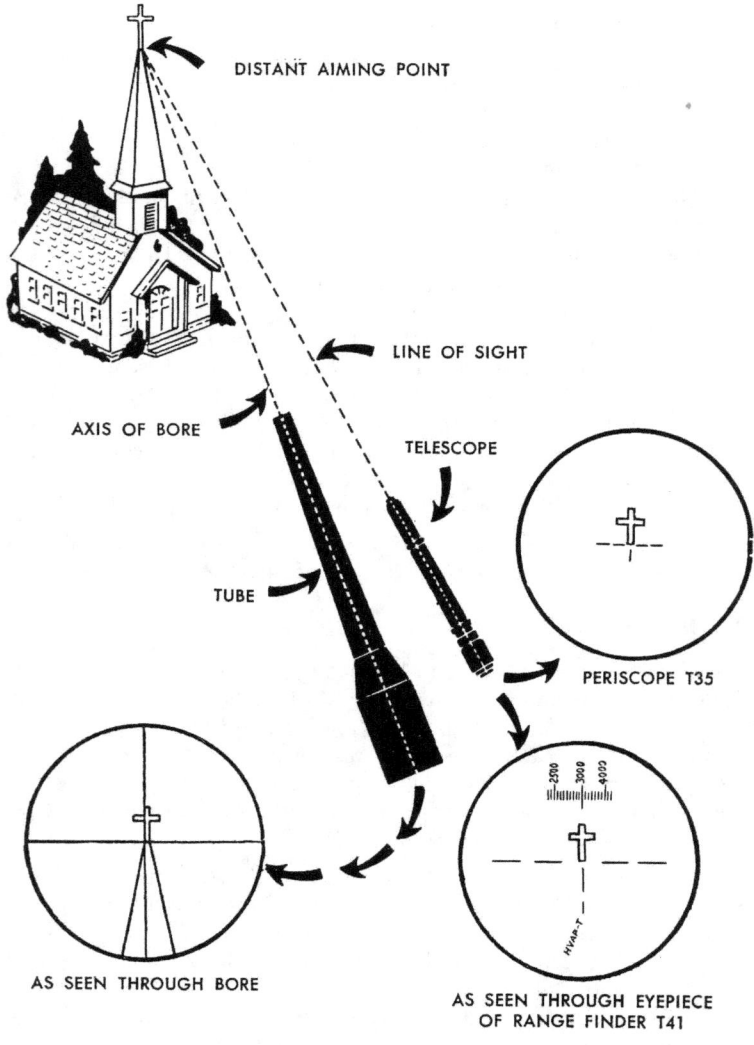

Figure 70. Aiming point method of boresighting.

(par. 76) until the intersection of the cross strings of the muzzle bore sight are lined accurately on the aiming point (fig. 70).

(3) *Coaxial machine gun.* Remove the back plate and bolt assembly of the coaxially mounted cal. .50 machine gun, look

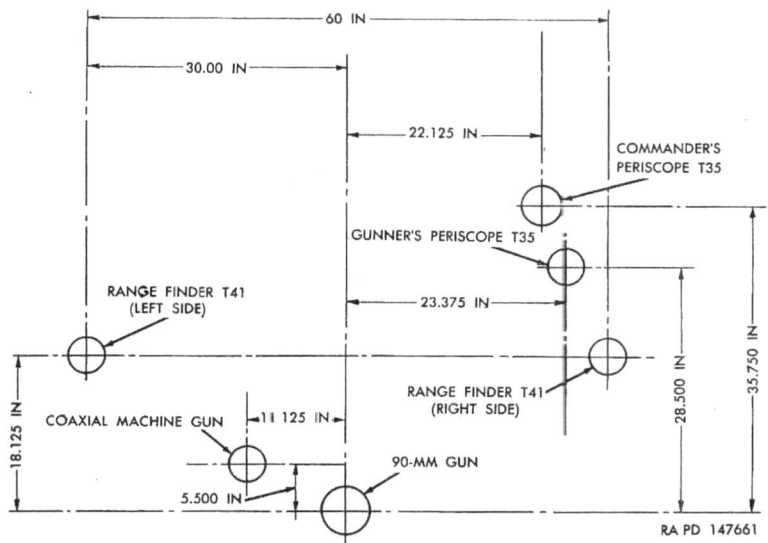

Figure 71. Testing target for boresighting.

through the machine gun barrel to see if the aiming point falls in the center of the barrel. If not, elevate or traverse the machine gun by means of the elevating screw or traversing screw on the elevating and traversing adjusting mechanism (fig. 40) until the center of the barrel is alined on the aiming point. Replace back plate and bolt assembly.

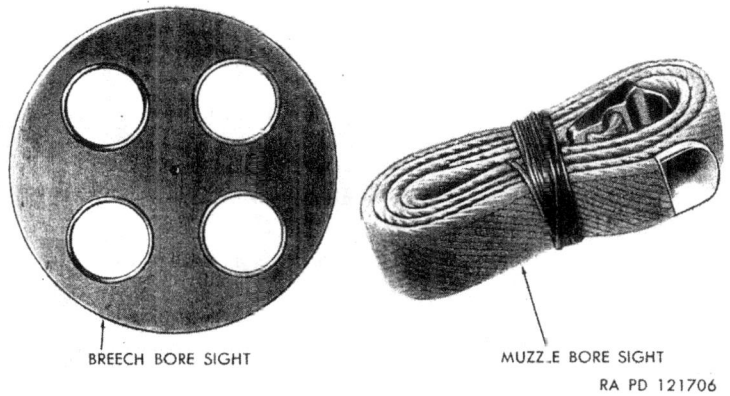

Figure 72. Bore sights—90-mm gun T119E1.

101

(4) *Range finder T41.*
  (*a*) Adjust the diopter and interpupillary settings on eyepieces (par. 84*a* (1) and (2)).
  (*b*) The range and ammunition scales should appear in the left eyepiece, if not, position the scale transfer lever away from the observer.
  (*c*) Turn the light switch to "STEREO-SCALES" position and adjust light rheostat until the desired illumination of the reticle is obtained. The stereoscopic pattern should be approximately 4.5 mils above the sighting reticle. Use the halving knob if adjustment of the stereoscopic reticle is necessary to fuze the left and right stereoscopic patterns.
  (*d*) Rotate the range and ammunition knobs until the range scale indicates the known range of the aiming point and the ammunition scale indicates "B" on the reticle pattern.
  (*e*) With both eyes open obtain precise stereoscopic contact on the aiming point using the internal correction system knob; this introduces the observer's personal correction. This setting may vary according to the observer. However, the normal setting is "25."
  (*f*) Rotate the range knob and ammunition knob until "B" at the end of the scales are indicated on the reticle pattern.
  (*g*) Set the ballistic correction knob to zero.

Note. When the range and ammunition scales are both set at "B" any position of the ballistic correction knob will have no effect on superelevation.

  (*h*) Sighting through the left eyepiece, aline the top of the vertical bar of the sighting reticle on the aiming point using the azimuth and elevation bore sight knobs (fig. 44).
  (*i*) Clamp the bore sight knobs and slip the scales to the normal setting at "3."

(5) *Periscope T35.*
  (*a*) The observer should set in his diopter setting on the eyepiece (par. 84*b*(10)).
  (*b*) Turn on and obtain the desired illumination on the reticle by rotating the rheostat knob on instrument light M36.
  (*c*) Make certain that the range knob on the ballistic drive is turned counterclockwise until it stops, indicating that all superelevation has been removed.

    Note. Also be certain that the range finder T41 range and ammunition scales are set at "B."

  (*d*) Sighting through the eyepiece, aline the top of the vertical bar of the reticle (fig. 50) with the aiming point, using the lateral and vertical boresighting knobs (figs. 49 and 73).
  (*e*) Clamp the boresighting knobs and slip the scales to "3" indicating normal.

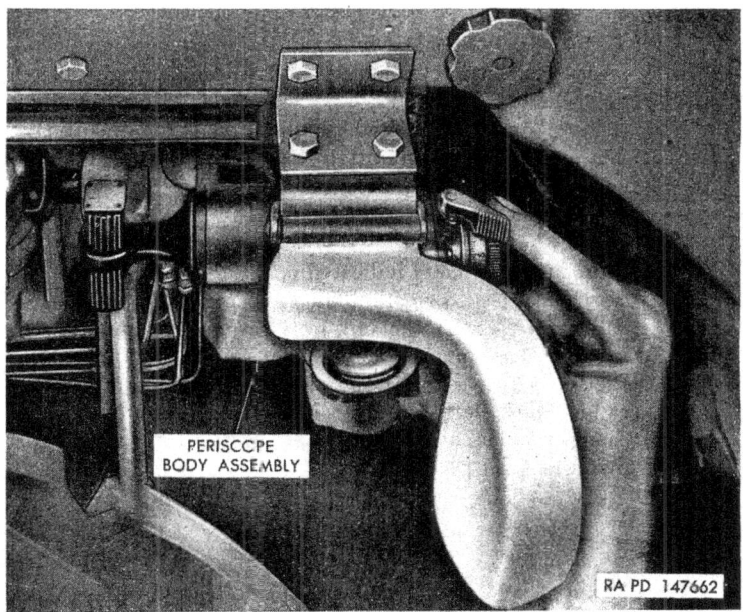

Figure 73. Adjusting azimuth boresighting knob on periscope T35.

c. TESTING TARGET METHOD.
  (1) *General.* If a testing target is not available, construct one on cardboard or some other suitable material using the dimensions shown in figure 71. Mark the top of the target "TOP" to avoid accidentally positioning the target in an inverted position. Then place the target at a distance of 80 to 120 feet from the muzzle of the weapon so that the surface is perpendicular to the line of sight and the horizontal lines of the target are on the same level as the gun trunnions.
  (2) *90-mm gun.* Look through the center hole of the breech bore sight or firing pin hole and elevate and traverse the gun (par. 76) until the cross lines of the muzzle bore sight are superimposed on 90-mm gun cross lines on the target.
  (3) *Coaxial machine gun.* Remove the back plate and bolt assembly of the coaxially mounted cal. .50 machine gun. Look through the barrel of the gun to see if the machine gun cross lines on the target appear in the center of the machine gun barrel. If not, elevate or traverse the machine gun by means of the elevating and traversing mechanism (fig. 40), until the cross lines on the target appear in the center of the barrel. Replace back plate and bolt assembly.

103

(4) *Range finder T41.*
   (a) Adjust the diopter and interpupillary settings on the eyepieces (par. 84a(1) and (2)). Turn on the light switch to "STEREO-SCALES" position and adjust the light rheostat until the desired illumination is obtained.
   (b) The range and ammunition scales should appear in the left eyepiece, if not, position the scale transfer lever away from the observer.
   (c) Rotate the range and ammunition knobs until "B" at the end of the scales are indicated on the reticle pattern.
   (d) Set the ballistic correction knob to zero.
   (e) Sighting through the left eyepiece and by moving the range finder azimuth and elevation bore sight knobs, aline the top of the vertical bar of the sighting reticle on the range finder (left side) cross lines of the target.
   (f) Clamp the bore sight knobs and slip the scales to the normal setting at "3."

   Note. There is no need to bore sight the right side of the range finder on a close target.

(5) *Periscope T35.*
   (a) Be sure that the range knob on the ballistic drive is turned counterclockwise until it stops, indicating that all superelevation has been removed from the ballistic drive. The range finder ammunition and range scales must be set at "B."
   (b) Sighting through the commander's periscope eyepiece, aline the top of the vertical bar of the reticle on the cross lines for the commander's periscope on the target by moving the horizontal and vertical bore sight knobs.
   (c) Clamp the bore sight knobs and slip the scales to "3."
   (d) Sighting through the gunner's periscope eyepiece, aline the top of the center vertical bar of the reticle, on the cross lines for the gunner's periscope, on the target, by moving the horizontal and vertical bore sight knobs.
   (e) Clamp the bore sight knobs and slip the scales to "3."

Section IV. OPERATION OF MATÉRIEL USED IN CONJUNCTION WITH MAJOR ITEM

92. General

This section includes the description of, and operating instructions for such auxiliary items as the auxiliary generator and engine, fire extinguisher, ventilating blower, crew compartment heater, bilge pumps, and slave battery receptacle. Ammunition and communications equipment are covered in chapter 4.

93. Auxiliary Generator and Engine

a. GENERAL. The auxiliary generator and engine (fig. 308) form one integral unit located to the rear of the right fuel tank (fig. 309). This unit is used to provide auxiliary power, controlled by the master junction box, for one or more of the following purposes:

  (1) To charge the batteries.
  (2) To use electrical equipment while the main engine is not running.
  (3) To provide additional current when the main generator is overloaded.

b. STARTING.

  (1) *General.* The auxiliary engine can be started electrically or manually, as outlined in (2) and (3) below. Normally, start the engine electrically. However, in extreme-cold weather operation, or if the batteries are discharged to the point they will not furnish the current for electric starting, start the engine manually. Before attempting to start the auxiliary engine by either method, perform the before-operation services outlined in paragraph 122.

   Caution: The master relay switch (fig. 18) must be in the "ON" position at all times when the auxiliary engine is running, or damage to master junction box components or radio equipment may result.

  (2) *Electric starting.* The auxiliary engine is started by connecting the battery into the auxiliary generator which thus acts as a motor to turn the engine. Therefore, to start the auxiliary engine, turn on the master relay switch (fig. 18). Move the auxiliary engine magneto switch to "ON" position (fig. 18). The auxiliary engine fuel shut off valve will open automatically unless the batteries are completely discharged. Check the quantity of fuel in the tanks (par 32). Set fuel tank shut off valves (par 17). Press in the auxiliary engine starter switch button (fig. 18), and hold button until engine starts.

   Caution: Do not hold button longer than 30 seconds at a time.

   If engine does not start readily, priming may be required. Prime engine by operating fuel pump manual priming lever (N, fig. 310) 20 to 30 strokes, as necessary, before attempting to start engine.

  (3) *Manual starting.* Turn on the master relay switch (fig. 18). Move the auxiliary-engine magneto switch to "ON" position (fig. 18). Pull open the five right grille doors (fig. 101). If the batteries are completely discharged, the auxiliary engine fuel shut off valve (fig. 323) will not automatically open

when the magneto switch is turned to "ON" position. Push the fuel shut off valve handle down and carefully turn the handle clockwise until one click is heard. If no click can be heard, the valve had been opened electrically when the magneto switch was turned to "ON" position. Grasp the manual starting handle, located on a bracket on the cylinder-head cover (fig. 74), and pull the handle upward with a quick movement. Allow the spring action to return the handle to its original position.

Note. In cold weather, choking is accomplished by pulling up on choke control handle, located on the same bracket as the manual starting handle.

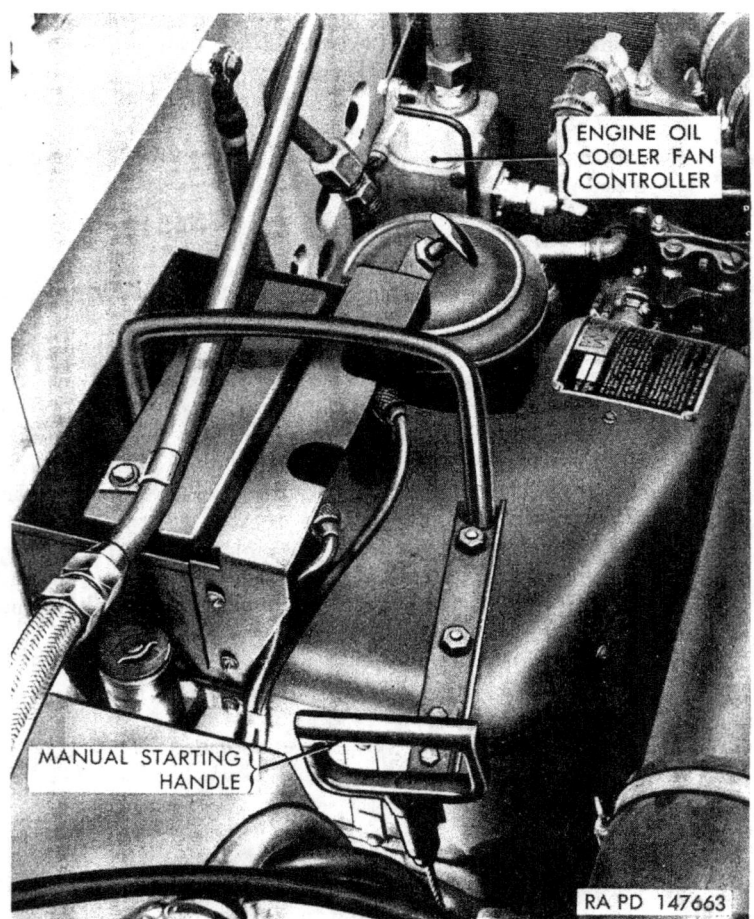

Figure 74. Auxiliary engine manual starting controls.

If the engine does not start on the first attempt, operate fuel pump manual priming lever (N, fig. 310) 20 to 30 strokes; then continue cranking until engine starts. When the engine starts, return the starting and choke control handles to their original positions and close the engine compartment grille doors.

(4) *Inspection after engine is started.* Check auxiliary generator warning signal light ("RIGHT ENG") (fig. 18) to see that it is not on.

  c. STOPPING. Move the auxiliary engine magneto switch (fig. 18) to "OFF" position. The auxiliary engine fuel shut off valve will close automatically. If the vehicle is not to be operated, and there is no immediate need for electric current to operate any accessories, turn off the master relay switch (fig. 18) as soon as the auxiliary engine stops.

94. Fire Extinguishers

  a. FIXED FIRE EXTINGUISHERS (fig. 324). Three fixed fire extinguisher cylinders are located between the driver's and the assistant driver's positions in the crew compartment. They are connected by lines to three discharge nozzles located throughout the vehicle. To operate from inside, pull the safety pin from the fixed fire extinguisher control on the left cylinder and pull down on the pull handle (fig. 324). To operate from outside the vehicle, pull out the remote control handle located on top of the hull behind the assistant driver's door (fig. 75) and let go.

  b. PORTABLE FIRE EXTINGUISHERS. Two portable fire extinguishers are provided. One is located in the driver's compartment (fig. 327)

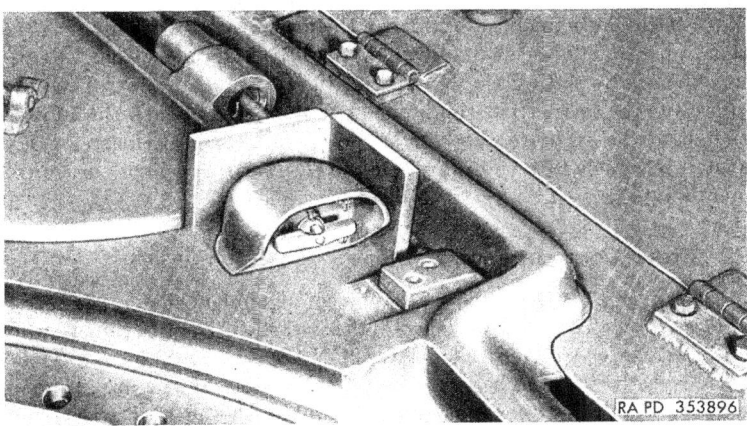

Figure 75. Remote control handle for fixed fire extinguisher system.

and the other in a stowage box located in the turret bulge. Remove the portable fire extinguisher from its mounting bracket, direct the discharge horn toward the base of the fire, and pull the trigger. Move the cone slowly from side to side to cover the burning area.

Caution: Handle the extinguisher with great care. Do not jar or drop.

95. Bilge Pumps

Two centrifugal-type bilge pumps (figs. 334 and 335) are installed in the hull to pump out any accumulated water, fuel, or oil. The compound-wound motor-type pumps are waterproof and capable of operating while immersed in water. Strainers are provided to prevent debris from entering the pump base and impeller area. A tube from the side of the pump impeller housing at the top of the hull acts as a discharge. The bilge pump motors are controlled by toggle-type switches located in the upper left hand corner of the hull accessories panel (fig. 18). The switch marked "FRONT" controls the crew compartment bilge pump motor and the switch marked "REAR" controls the engine compartment bilge pump motor. To operate either bilge pump motor, turn on the master relay switch (fig. 18) and then push the proper bilge pump switch to the "ON" position.

96. Ventilating Blower

A blower, located on the ceiling of the bustle compartment in the rear of the turret, provides ventilation for crew comfort (fig. 304). The ventilating blower inlet cover is located in the turret roof just forward of the stowage box (fig. 3). The electrically-operated ventilating blower is controlled by the ventilating-blower switch located on the turret accessories panel on the left-rear wall of the turret (fig. 30). To operate the blower, turn on the master relay switch (fig. 18); then push the blower switch to the left to the "ON" position. Regulate the air by setting the ventilating blower damper control handle (fig. 304) in desired position. To stop the blower, push the blower switch to the right to the "OFF" position. Turn off master relay switch if vehicle is to be stopped and no electrical accessories or equipment are to be used.

97. Crew Compartment Heater

a. GENERAL. A gasoline-type crew compartment heater (fig. 328) is located on the hull floor below and forward of the instrument and hull accessories panels. The heater "ON-OFF" switch, heater safety valve reset switch, and heater circuit breaker reset button are grouped together in the heater control box mounted on the hull accessories panel (fig. 18). The heater thermostat, which has high, medium, and low positions with a range of 70° F. between high and low positions, is installed in crew compartment on column (fig. 76) between driver and assistant driver.

b. STARTING HEATER. To start heater, turn thermostat knob (fig. 76) to "HIGH" position and snap heater switch (fig. 18) to "ON" position. Warm air should be felt at heat outlet within 3 minutes. As soon as heater starts, adjust thermostat knob to desired position.

c. HEATER FAILS TO START. If heater fails to start, safety valve (fig. 332) will close to stop flow of fuel to heater and circuit breaker will break the electrical circuit. Snap heater switch (fig. 18) to "OFF" position and push in circuit breaker reset button (fig. 18). Hold safety valve reset lever (fig. 18) in "RESET" position for 30 seconds. Lever will return to original position when released. Snap heater switch to "ON" position. If heater does not start after second try, correct cause of failure or replace heater (par. 322).

d. STOPPING HEATER. To stop the heater, snap the heater switch (fig. 18) to the "OFF" position. Fire in the heater will go out in a few seconds but the blowers will continue to operate for 2 or 3 minutes to cool and purge the heater of any unburned gases. When the heater has cooled, the blowers will stop. If the heater switch is turned to the "ON" position during this period, the heater will not ignite until the purge period is over.

98. Slave Battery Receptacle

A slave battery receptacle (fig. 76) is installed in the crew compartment on the column to the rear of and between the drivers. This receptacle is connected to the vehicle batteries through the master relay. The receptable is used to connect an external battery-charging source or additional power to operate any of the vehicle accessories or electrical equipment. In addition, if the vehicle batteries are discharged to the point where they cannot close the master relay when the master relay switch is turned on, an external battery can be connected into the slave battery receptacle to close the master relay. Extension cable 17-C-568 (fig. 78) is used to make these external connections to the slave battery receptacle.

Section V. OPERATION UNDER UNUSUAL CONDITIONS

99. General Conditions

a. In addition to the operating procedures described for usual conditions, special instructions of a technical nature for operating and servicing this vehicle under unusual conditions are contained or referred to herein. In addition to the normal preventive maintenance service, special care in cleaning and lubrication must be observed where extremes of temperature, humidity, and terrain conditions are present

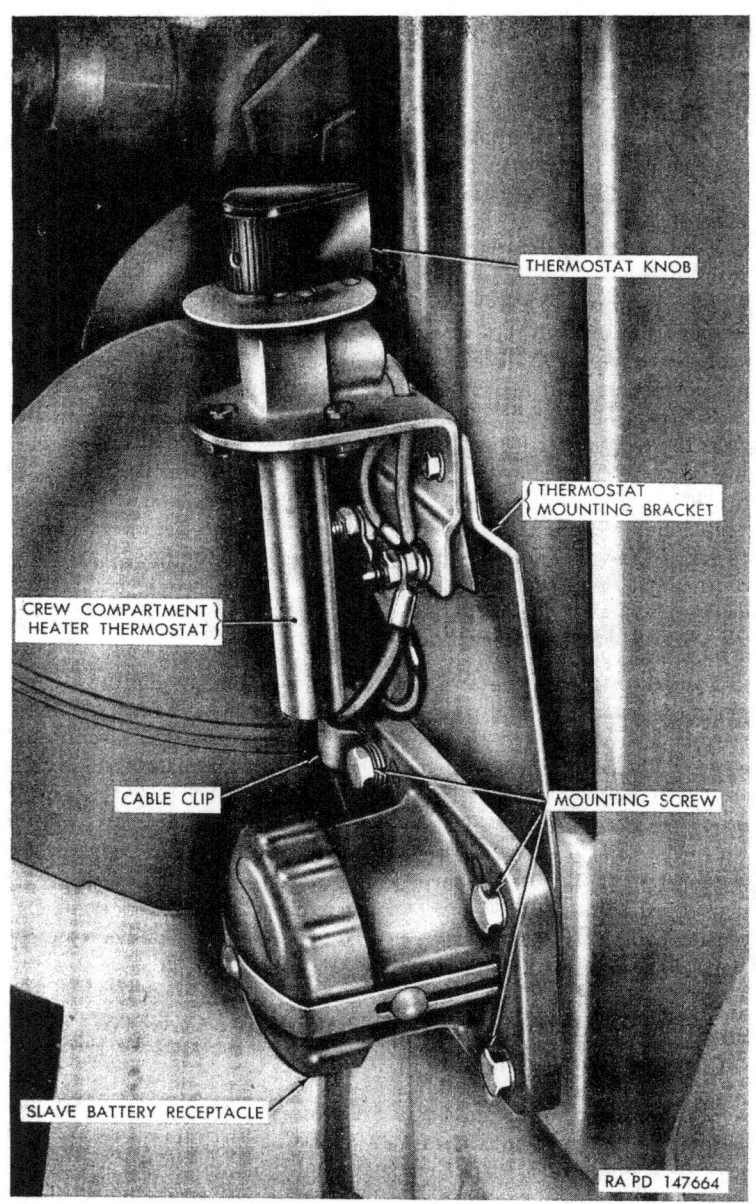

Figure 76. Crew compartment heater thermostat and slave battery receptacle.

or anticipated. Proper cleaning, lubrication, and storage and handling of fuels and lubricants not only insure proper operation and functioning, but also guard against excessive wear of the working parts and deterioration of the matériel.

b. TM 21-301 contains very important instructions on driver selection, training, and supervision and TM 21-306 prescribes special driving instructions for operating full-track and tank-like vehicles under unusual conditions.

Caution: It is imperative that the approved practices and precautions be followed. A detailed study of these TM's is essential for use of this matériel under unusual conditions.

c. Refer to paragraph 115 for lubrication under unusual conditions, to tables III and IV, paragraph 122 for preventive maintenance checks, and to chapter 3, section XXXVI for maintenance procedures.

d. When chronic failure of matériel results from subjection to extreme conditions, report of the conditions should be made on DA AGO Form 468 (par. 2).

100. Extreme-Cold Weather Conditions

a. GENERAL PROBLEMS.

(1) Extensive preparation is necessary of matériel scheduled for operation in extreme-cold weather. Generally, extreme cold will cause lubricants to thicken or congeal, freeze batteries or prevent them from furnishing sufficient current for cold-weather starting, crack insulation, and cause electrical short circuits, prevent fuels from vaporizing and properly combining with air to form a combustible mixture for starting, and will cause the various construction materials to become hard, brittle, and easily damaged or broken.

(2) TM 9-2855 also describes the method of correcting specific gravity readings for batteries exposed to extreme cold.

(3) Armament scheduled for extreme-cold operation must be checked for proper lubrication. Fire control and sighting instruments should not be transferred suddenly from cold to warm temperatures or vice versa. Condensation induced by this action may cause clouding of optics and rusting of internal parts. Strains may be set up in parts.

(4) For description of operations in extreme cold, refer to FM 70-15 as well as to TM 9-2855.

Caution: It is imperative that the approved practices and precautions be followed. TM 9-2855 contains general information which is specifically applicable to the vehicle as well as to all other vehicles. It must be considered an essential part of this manual, not merely an explanatory supplement to it.

b. WINTERIZATION EQUIPMENT. Special equipment is provided for the vehicle when protection against extreme-cold weather (0° to −65° F.) is required. This equipment is issued as specific kits. Each kit contains a technical bulletin which provides information on description, installation instructions, and methods of use. TM 9-2855 contains general information on winterization equipment and processing.

c. FUELS AND LUBRICANTS (STORAGE, HANDLING, AND USE).

(1) The operation of equipment at arctic temperatures will depend to a great extent upon the condition of the fuels and lubricants used in the equipment.

(2) The manner in which the fuels and lubricants are stored, handled, and used, greatly affects the service the fuels and lubricants will give.

(3) In arctic operations, contamination with moisture is the source of many difficulties. Moisture can be the result of snow getting into the product, condensation due to "breathing" of a partially filled container, or moisture condensed from warm air in a partially filled container when a product is brought outdoors from room temperatures. Other impurities will also contaminate fuels and lubricants so that their usefulness is impaired.

(4) Immediate effects of careless storage and handling or improper use of these materials are not always apparent, but any deviation from proper procedures may cause trouble at the least expected time.

(5) Refer to TM 9-2855 for detailed instructions.

101. Extreme-Cold Weather Operation

a. GENERAL.

(1) The driver or operators must always be on the alert for indications of the effect of cold weather on the vehicle.

(2) The driver or operators must be very cautious when placing the vehicle in motion after a shut down. Congealed lubricants may cause failure of parts. Tracks frozen to the ground must be considered. After warming up the engine thoroughly, place transmission in first gear and drive vehicle slowly about 100 yards, being careful not to stall the engine. This should heat gears and tracks to a point where normal operation can be expected.

(3) Constantly note instrument readings. If instrument reading consistently deviates from normal, stop the vehicle and investigate the cause. Normal engine oil temperature is from 140° to 160° F. If the engine-high-oil-temperature-warning-signal light (fig. 18) is on, stop the engine and investigate the cause.

(4) Refer to TM 21-306 for special instructions on driving hazards in snow, ice, and unusual terrain encountered under extreme-cold conditions.

b. AT-HALT OR PARKING.
   (1) When halted for short shut-down periods the vehicle should be parked in a sheltered spot out of the wind. If no shelter is available, it will be helpful to park so that the vehicle does not face into the wind. For long shut-down periods, if high dry ground is not available, effort should be made to prepare a footing of planks or brush. Chock in place is necessary.
   (2) When preparing a vehicle for shut-down period, place control levers in the neutral position to prevent possible freezing in an engaged position. Freezing may occur when water is present due to condensation.
   (3) Clean all parts of the vehicle of snow, ice, and mud as soon as possible after operation. Refer to table III, paragraph 122 for detailed after-operation procedures. If the winterfront and side covers are not installed be sure to protect all parts of the engine and engine accessories against entrance of loose, drifting snow during the halt. Snow flurries penetrating the engine compartment may enter the crankcase filler vent, etc. Cover and shield the vehicle but keep the ends of the canvas paulins off the ground to prevent them from freezing to the ground.
   (4) If no power plant heating device is present the battery should be removed and stored in a warm place.
   (5) Refuel immediately in order to reduce condensation in the fuel tanks. Prior to refueling, open fuel tank drains and drain off any accumulated water.
   (6) When the vehicle is equipped with a power plant heater as provided by the arctic winterization kit, start the heater and check to be sure that it is operating effectively. This heater should avoid the necessity of removing the battery to warm storage, and is designed to operate unattended during overnight stops. Instructions for operation of winterization equipment will be found in pamphlet packed with the kit.

   Note. This heater is used only while the vehicle is halted.

c. ARMAMENT. Weapons should be covered when not in use. Breech and firing mechanisms must be extremely clean and lightly lubricated. Keep checking recoil mechanism and handwheels for sluggishness. When cleaning, do not dilute rifle-bore cleaner or add and antifreeze. Store cleaning solutions in a warm place, if practical. Shake rifle-bore cleaner well before using.

d. SIGHTING AND FIRE CONTROL MATÉRIEL. Avoid breathing on oculars and causing condensation which might freeze. Cover ends of periscopes when not in use to prevent accumulation of snow or ice. Avoid sudden temperature changes from warm room to low temperatures outdoors.

102. Extreme-Hot Weather Operation

   a. GENERAL. Continuous operation of the vehicle at high speeds or long hard pulls in lower gear positions on steep grades or in soft terrain may cause the vehicle to register overheating. Avoid the continuous use of low gear ratios whenever possible. Continuously watch the temperature and halt the vehicle for a cooling-off period whenever necessary and the tactical situation permits. Keep ventilating blower on during operations. Make frequent inspections and servicing of cooling fans, oil filter, and air cleaner. If the engine-oil-temperature-warning-signal light is on, stop engine immediately and investigate the cause. Check radiator cores for any obstructions which would prevent free circulation of air.

   b. AT HALT OR PARKING.
   (1) Do not park the vehicles in the sun for long periods of time. When practicable, park vehicle under cover to protect it from sun, sand, and dust.
   (2) Cover inactive vehicles with paulins if no other suitable shelter is available. Where entire vehicle cannot be covered, protect periscopes, etc., against etching by sand, and protect engine compartment against entry of sand.
   (3) Vehicles inactive for long periods in hot humid weather are subjected to rapid rusting and accumulation of fungi growth. Make frequent inspections and clean and lubricate to prevent excessive deterioration.

   c. SIGHTING AND FIRE CONTROL MATÉRIEL. This matériel should be shielded as much as possible from the direct rays of the sun.

103. Operation on Unusual Terrain

   a. MUD. Select transmission low range to move vehicle steadily without digging in. If the vehicle becomes mired, arrange to be towed out of the mud instead of digging in. When a drop to below freezing temperatures is anticipated, make sure that the vehicle is parked on solid ground or footing to prevent the tracks from being frozen in the mud, and that accumulations of mud have been removed from track and wheel contacting surfaces.

   b. SNOW. It may be possible for the vehicle to ride heavily crusted snow with occasional breakthrough. To climb back onto the crust reduce engine speed, and shift into transmission low range to achieve very low track speed for forward movement without slippage. Avoid grades. Where grades must be taken, drive as nearly straight up and down as possible to equalize track load. Avoid sharp turns. For soft or fine snow select the transmission low range shift position which gives good traction.

c. ICE. Skidding is the general hazard encountered on ice. Select the proper gear ratio to move the vehicle steadily, without imposing undue strain on engine. When skidding occurs, decelerate the engine and proceed with caution.

d. SAND. The main objective when driving in sand is to avoid spinning the tracks. Reduce speed and use a gear low enough to move the vehicle steadily. Do not allow the engine to labor.

104. Fording Operation

a. GENERAL. In fording, vehicles may be subjected to water varying in depth from only a few inches to depths sufficient to completely submerge the vehicle. Factors to be considered are spray-splashing precautions, normal fording capabilities, deep-water fording using fording kits, and accidental complete submersion. Optical instruments should not be submerged, except those periscopes used for driving.

b. NORMAL FORDING. Fording of bodies of water up to maximum vehicle fording depth of 48 inches is based on the standard vehicle with waterproofing protection provided for critical units when manufactured, but without deep-water fording kit. Observe the following precautions:

(1) Make sure that battery cell vent caps are snug.
(2) Do not exceed the known ording limits of the vehicle.
(3) The engine must be operating at maximum efficiency before attempting to ford. Start bilge pump.
(4) Shift transmission into lowest speed position. Enter the water slowly. Should the engine stall while submerged, it may be started in the usual manner.
(5) All normal fording should be at speeds of from 3 to 4 mph to avoid forming a "bow wave." Check bilge pump operation.
(6) If accidental complete submersion occurs, the vehicle will be salvaged, temporary preservation applied as outlined in paragraph 375 and then sent to the ordnance maintenance unit as soon as possible for necessary permanent maintenance.

c. DEEP-WATER FORDING. Refer to TM 9-2853 for general information, descriptions, and methods of use of deep-water fording kits, and for general procedures for the operation of vehicles so equipped.

d. AFTER-FORDING OPERATION. Immediately after a vehicle emerges from the water, open all drain valves in hull. Also, at the earliest opportunity, check the engine oil level and check for presence of water in the crankcase. Heat generated by driving will evaporate or force out most water which has entered at various points. Also, any *small* amount of water which has entered the crankcase either through leakage or due to condensation will usually be dissipated by the ventilating system. Refer to paragraph 375 for maintenance operations after fording. Optical instruments that have been wetted should be quickly wiped dry and examined for indication of leakage into the instrument. Such instruments should be turned in for reconditioning at the first opportunity. Shut off bilge pumps.

105. Lubrication Under Unusual Conditions

Refer to paragraph 115.

106. Lubrication for Continued Operation Below 0° F.

Refer to paragraph 116.

107. Lubrication After Fording Operations

Refer to paragraph 117.

108. Lubrication After Operation Under Dusty and Sandy Conditions

Refer to paragraph 118.

CHAPTER 3

ORGANIZATIONAL MAINTENANCE INSTRUCTIONS

Section I. PARTS, SPECIAL TOOLS, AND EQUIPMENT FOR ORGANIZATIONAL MAINTENANCE

109. General

Tools, equipment, and spare parts are issued to the using organization for maintaining the matériel. Tools and equipment should not be used for purposes other than prescribed and, when not in use, should be properly stored in the chest and/or roll provided for them.

110. Parts

Spare parts are supplied to the using organization for replacement of those parts most likely to become worn, broken, or otherwise unserviceable, providing such operations are within the scope of organization maintenance functions. Spare parts, tools, and equipment supplied for the 90-mm gun tank M47 are listed in Department of the Army Supply Catalog ORD 7 SNL G–262 which is the authority for requisitioning replacements.

111. Common Tools and Equipment

Standard and commonly used tools and equipment having general application to this matériel are listed for issue by the ORD 7 SNL G–262 catalog and by T/A and T/O & E.

112. Special Tools and Equipment

Certain tools and equipment specially designed for organizational maintenance, repair, and general use with the matériel are listed in table II for information only. This list is not to be used for requisitioning replacements.

Table II. Special Tools and Equipment for Organization and Maintenance

Item	Identifying number	References Fig.	References Par.	Use
FOR TANK 90-MM GUN, M47				
ADAPTER, puller, thd 1 in–8NC–2 female and ⅝ in–18NF–2 male, lgh 2⅛ in.	41–A–18–242	79	250	Removing front road wheel arm shackle pin. Use w/PULLER 41–P–2957–33.
ADAPTER, puller, thd 1 in–18NF–2, lgh 1 11/16 in.	41–A–18–248	79	252	Removing idler wheel arm and support spindle assembly. Use w/PULLER 41–P–2957–33.
ADAPTER, remover and replacer, thd ⅝ in–18NF–2 female and ⅞ in–20NF–2 male, lgh 2⅝ in.	41–A–18–400	79	253	Removing or installing compensating idler wheel torsion bar. Use w/ REMOVER and REPLACER 41–R–2378–950.
ADAPTER, road wheel arm, spindle removal.	41–A–18–775	78, 265	250	Removing road wheel arm. Use w/ PULLER 41–P–2957–33.
CABLE, extn, rubber covered, 2 conductor, stranded, w/female plugs at both ends AWG No 1, lgh 20 ft.	17–C–568	78	98	To connect outside power source to slave battery receptacle.
FIXTURE, track connecting and link pulling L and RH. Composed of: 1 FIXTURE, track connecting and link pulling, LH. Including: 1 BAR, lever, track connecting fixture.	41–F–2995–200 41–F–2995–275 41–B–257–10	83, 249, 250 83 83	249	Connecting and disconnecting tracks.

118

Description	Part Number			Purpose
1 FIXTURE, track connecting and link pulling, RH. Including:	41-F-2995-375			
1 BAR, lever, track connecting fixture.	41-B-257-10		83	
GAGE, pressure oil, w/hose, size of dial 3½ in, graduated 0 to 200 lb.	45-G-438-500	80, 224	235	Testing transmission oil pressures.
GAGE, thkns.	7083769	80, 119, 120	155	Checking and adjusting main engine valve clearance.
HANDLE, remover and replacer, diam of ablt 1 in, lgth 13¾ in.	41-H-1396-510	79, 264, 267, 268	250, 252	Use w/REMOVER and REPLACERS.
KIT, extn cable and fuel lines, engine test stand.	41-K-86-850			Testing power plant outside of tank prior to installation.
LIFTER, aux tension idler, turn buckle type.	41-L-1380-100	81		Raising compensating idler wheel.
LIFTER, road wheel, front.	41-L-1390-5	82, 254	250	Raising front road wheels.
LIFTER, road wheel, intermediate and front.	41-L-1390-100	82, 253	250	Raising intermediate and rear road wheels.
LIGHT, timing, magneto, type B-1	41-L-1439	82	185	Timing engine.
PULLER, crankcase section removing	41-P-2906-280	77, 136	163, 222	Removing main engine generator adapter.
PULLER, slide hammer type, bogie gudgeon.	41-P-2957-33	82, 265	250, 252	Removing wheel arms and spindles.
REMOVER, lead seal from valve body stud, cutter type with T hdl.	41-R-2371-760			
REMOVER and REPLACER, bearing, OD 4⅛ in, lgh 1¾ in.	41-R-2373-460	79	250	Removing or installing front road wheel arm shackle bearing.
REMOVER and REPLACER, bearing cup (inner), thd ¼ in-12NF-2, OD 4.540 in, thkns 1 ¹¹⁄₃₂ in.	41-R-2374-630	79, 275	252, 254	Removing or installing compensating idler wheel and track support roller inner bearing cups. Use w/SCREW 41-S-1047-315.

Table II. Special Tools and Equipment for Organization and Maintenance—Continued

Item	Identifying number	Fig.	Par.	Use
REMOVER and REPLACER, bearing cup (inner), thd 1¼ in–12NF–2, OD 5.890 in, thkns 1 ²⁷⁄₃₂.	41–R–2374–635	79	250	Removing or installing adjusting idler wheel and road wheel hub inner bearing cups. Use w/SCREW 41–S–1047–315.
REMOVER and REPLACER, bearing cup (outer), thd 1¼ in–12NF–2.	41–R–2374–655	79, 275	250, 254	Removing or installing road wheel, idler wheels, and track support roller hub outer bearing cups. Use w/ SCREW 41–S–1047–315.
REMOVER and REPLACER, plug (camshaft drive quill and oil transfer), T hdl, thd ⅝ in–18NF–3.	41–R–2378–575			
REMOVER and REPLACER, torsion bar.	41–R–2378–950	78, 272	253	Removing or installing torsion bars.
REPLACER, bushing and oil seal (starter drive gear hub accessory case).	41–R–2388–715	81, 179	190	Installing starter drive gear hub oil seal and bushing.
REPLACER, oil seal (aux tension idler spindle), OD 5 in, lgh 8³⁄₁₆ in.	41–R–2391–420	79	252, 254	Installing compensating idler wheel and track support roller hub oil seals.
REPLACER, oil seal, compensating arm spindle.	41–R–2392–50	78	250	Installing adjusting idler wheel and road wheel arm support oil seals.
REPLACER, oil, compensating arm spindle, track wheel.	41–R–2392–65	78, 267	250	Installing adjusting idler wheel and road wheel arm spindle oil seals.
REPLACER, oil seal (front and compensating wheel), OD–6–¼ in., lgh ¹⁵⁄₁₆ in.	41–R–2392–630	79, 264	250	Installing adjusting idler wheel and road wheel hub oil seals.
REPLACER, oil seal (magneto drive driven shaft gear), OD 2.120 in., lgh 1⁵⁄₁₆ in.	41–R–2392–995	78		Installing magneto drive driven shaft gear oil seal.

SCREW, puller, thd ½ in–20NF–2 for 5¾ in., lgh over-all 6⅜ in.	41–S–1044–125	80	----------	Removing transmission end.
SCREW, remover and replacer (brg cup), thd 1¼ in-12NF–2, lgh 7½ in.	41–S–1047–315	79	250, 254	Use with remover and replacers.
SLING, differential, track connecting and bogie spring replacement.	41–S–3830–30	83, 242, 244	245, 246	Removing or installing drive sprocket hubs and final drives.
SLING, final drive	41–S–3832–34	79, 244	246	Removing or installing final drives.
SLING, lifting engine and transmission	41–S–3832–46	83–129	158, 160, 161	Removing or installing power plant.
SLING, lifting (transmission carrier case off output shaft).	41–S–3832–165	81, 134	163, 164	Removing or installing transmission.
STAND, engine and transmission transport	41–S–4941–50	84	159	Support power plant or engine only while out of vehicle.
WRENCH, cooling flange nut, sgle-hd box, dble-hex, size of opng $\frac{9}{16}$ in., lgh 7⅜ in.	41–W–869–410	80	----------	Removing or installing transmission fan controller.
WRENCH, crowfoot, sgle open end, T hdl, size of hex nut opng 0.763 in., lgh 11⅜ in.	41–W–871–70	80, 168, 169	186, 187	Removing or installing spark plug cables.
WRENCH, engine oil presending unit	7950049	----------	217	Removing engine oil pressure sending unit.
WRENCH, engine turning (sgle-hd plug w/exter splines), lgh $3\frac{1}{16}$ in.	41–W–906–125	81	164	Turning engine at accessory end.
WRENCH, input driving pinion turning (two key plug, ¾ in. sq-drive, lgh 8 in.).	41–W–1536–380	77, 117	155, 184	Turning engine at transmission input shaft (par. 185).
WRENCH, roller and track wheel brg nut (adj face spanner, cap c to c of pins 4 in.).	41–W–3242–300	77, 261	250, 254	Removing, installing, and/or adjusting wheel bearings.
WRENCH, spindle nut (hook spanner, diam of circle ⅝-in. lgh 14½ in.).	41–W–3252–375	77, 266	250, 252	Removing or installing road wheel spindle nut.
WRENCH, spark plug holding sgle opng end, crowfoot, size of hex nut opng 0.828 in., lgh 10¾ in.	41–W–3304–800	80, 169, 170	186	Removing spark plugs.

Table II. Special Tools and Equipment for Organization and Maintenance—Continued

Item	Identifying number	References Fig.	References Par.	Use
WRENCH, spark plug inserting thd opng ⅝ in.–24NEF–2, lgh 5¾ in.	41–W–3306–500	80, 170	186	Removing or installing spark plugs.
WRENCH, track slack adjusting (hook spanner, diam of circle 7⅜ in., lgh 13½ in.	41–W–3250–875	77, 252	249	Adjusting track tension.
Wrench, transmission case rear brg (socket (detachable), ¾ in. sq-drive, 6 point, size of opng 2¾₂ in. (2.284 in.)).	41–W–3031–940	226	236	Loosening or tightening transmission low and reverse gear band adjusting screw lock nut.
WRENCH, generator and starter nut, sgle-hd box, 90 deg offset hdl, dble-hex, size of opng 9/16 in. lgh 16 in.	41–W–1496–625	80, 312	222, 304	Removing or installing generator or starter.
WRENCH, tubr, dble end, hex, size of hex opngs 1.6555 x 1.915 in. lgh 6½ in.	41–W–3737–33	80, 108	150, 151	Removing or installing oil pressure control valve or oil filter bypass valves.
FOR GUN 90-MM. T–119E1				
BRUSH, bore 90-mm, M19	6181980	85	339	Used with staff-section to clean gun bore.
COVER, canvas, brush, bore, M518	5606252	85		Used to cover bore brush.
COVER, gun book M539	7228906	85	2	Used to store arty gun book form and publications.
COVER, muzzle, w/muzzle brake, assy	7727373	85		Used to cover muzzle of gun.
EYEBOLT, breechblock removing, threaded one end, diam of bolt ⅞ in, thread ⅞–11NC–2, lgh over-all 16 in.	41–E–3150	85	342, 346	To remove and install breechblock.

Item	Stock No.		Page	Purpose
FORM, govt, War Dept, arty gun book OO No. 5825.	28-F-67990	85	2, 114	To keep record of condition of gun.
HEAD, rammer, diam 3.52 in, lgh 10⅛ in	41-H-1826-150	85	143	Used with staff-section to remove unfired round.
RING, wiper	6181986	85	339	Used with staff-section to clean gun bore.
ROD, push, assembling and disassembling shaft, diam ½ in., lgh 6 in., point ⅛ in.	5206983	85	342	Removing breechblock operating shaft.
ROLL, gun spare parts, M13	6507349	85		To carry spare parts.
STAFF-SECTION, intermediate (48¼ in. lg).	6197240		339	Used with bore brush to clean gun bore.
TOOL, breechblock removing	7237831		342, 346	To remove and install breechblock.
TOOL, ramming and extracting	7328638		143	To ram and to extract stuck round.
WRENCH, fuze, M16	41-W-1496-115	85	384 (table IX)	To install and remove CP fuze and booster.
WRENCH, fuze, M18	41-W-1496-135	85	384 (table IX)	To install, set, and remove Sq-del fuzes. To install and remove VT fuzes.
WRENCH, spanner, face, pintype, c to c of pins 2 in., diam of pin ¼ in. lgh 6¾ in.	41-W-3247-720		342	Removing breechblock closing cap spring.
WRENCH, spanner pin, circle diam 6½ in., pin diam ½ in., lgh over-all 14¼ in.	7237270	85	338	To remove and install evacuator chamber (pars. 340 and 346).
WRENCH, tubr, sglt-end, pronged, w/pin hdl, OD 1¹⁄₃₂ in., No. of prongs 2.	41-W-3735-510	85	343	To remove and install firing plunger bushing.

123

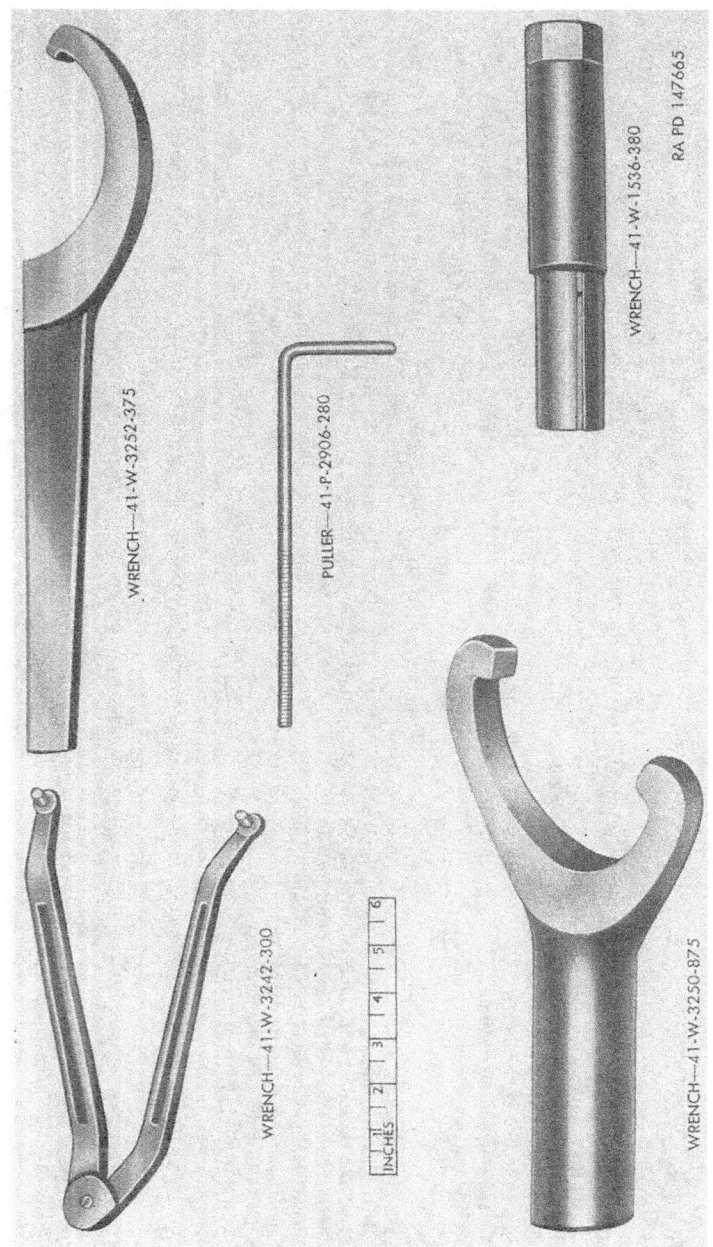

Figure 77. Special tools and equipment.

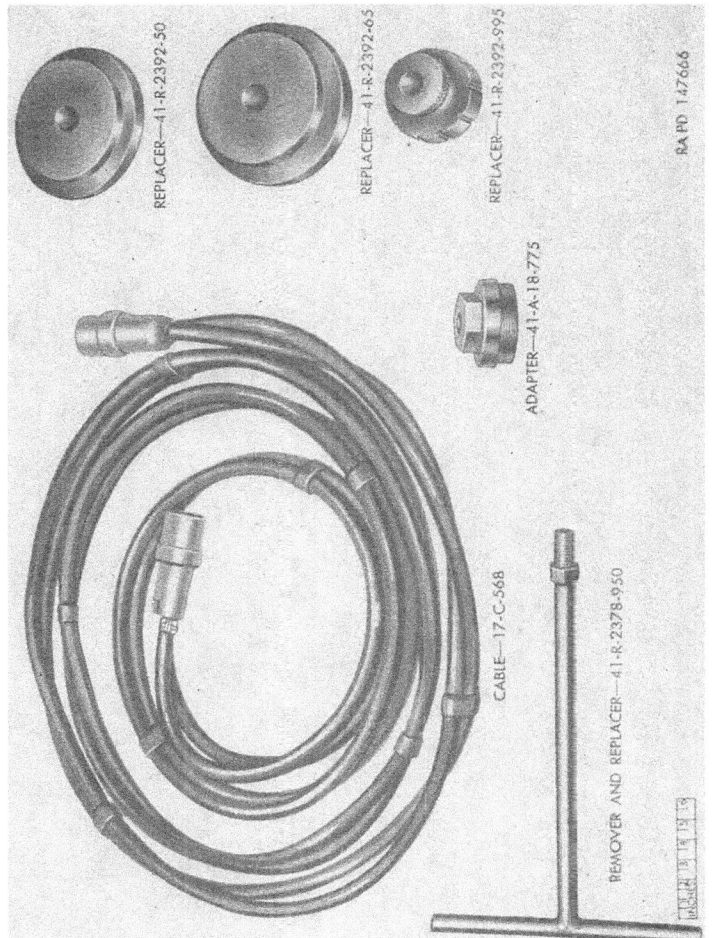

Figure 78. Special tools and equipment.

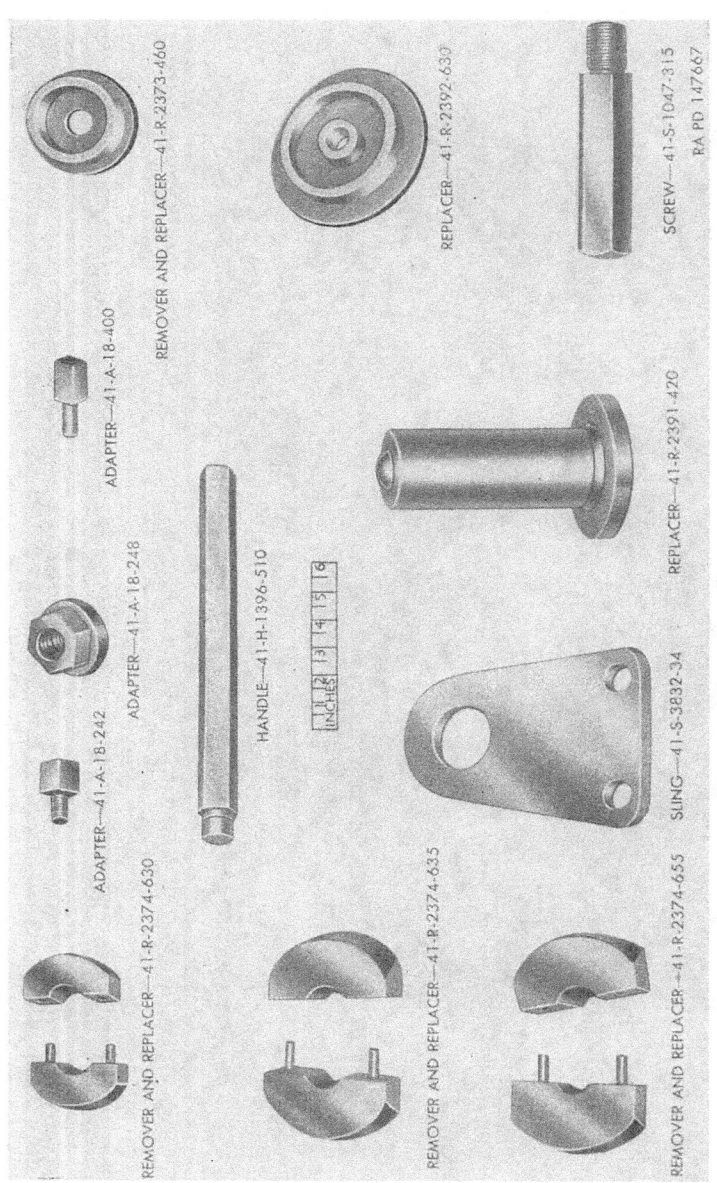

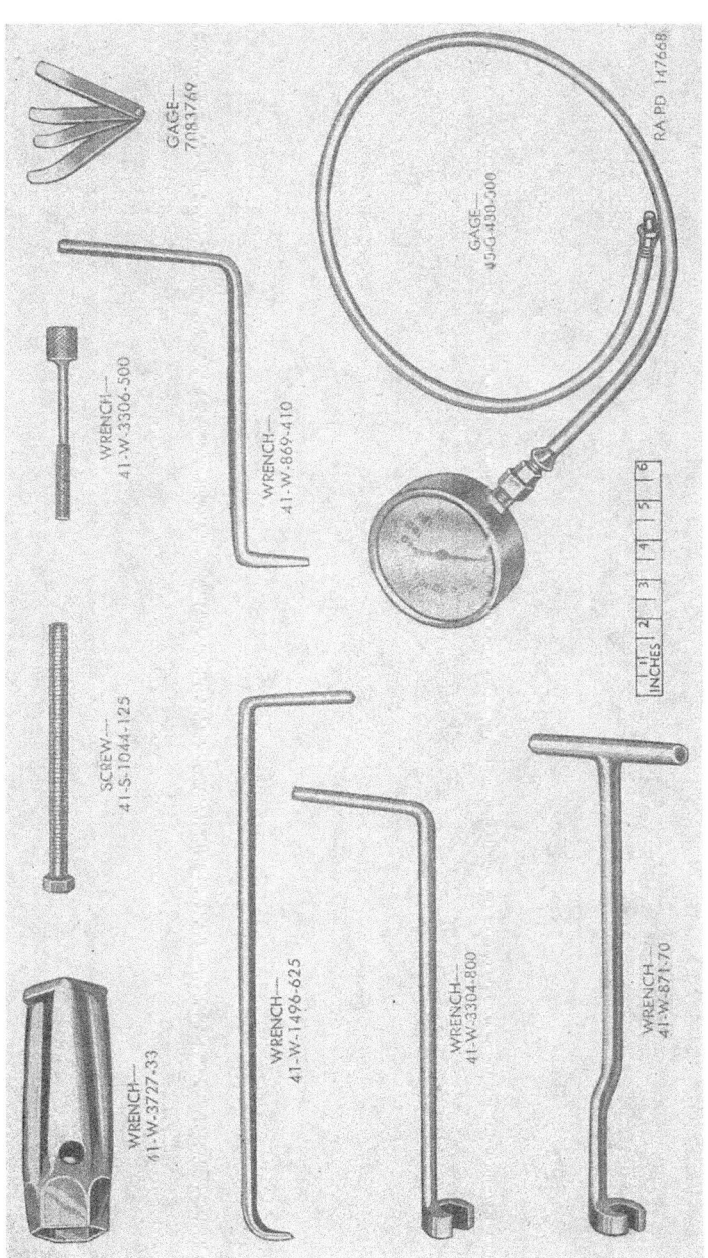

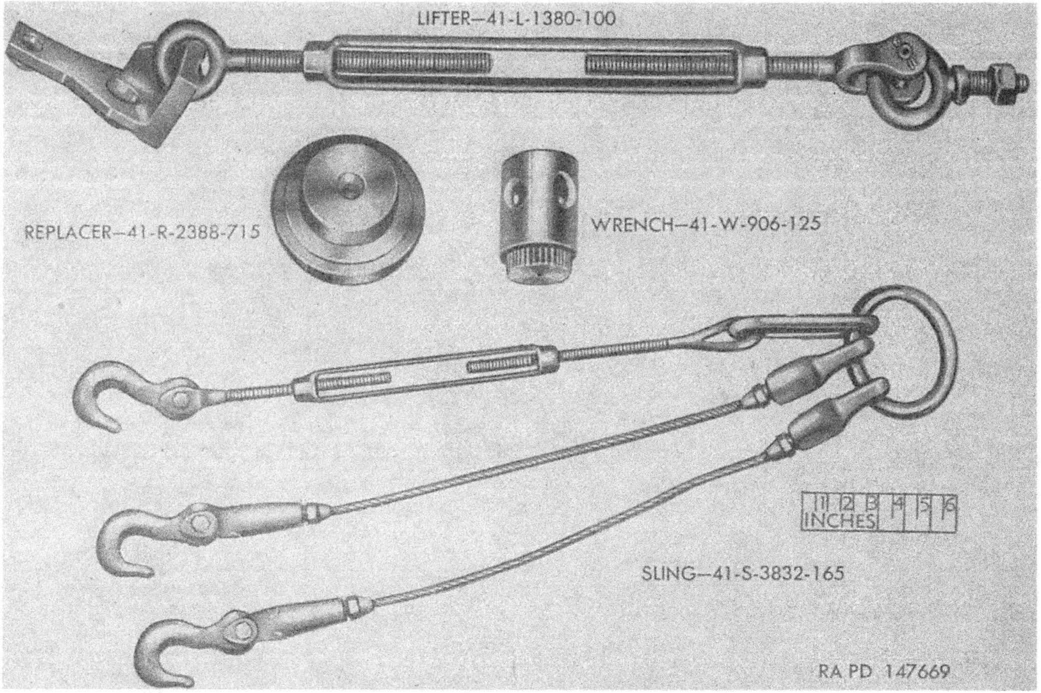

Figure 81. Special tools and equipment.

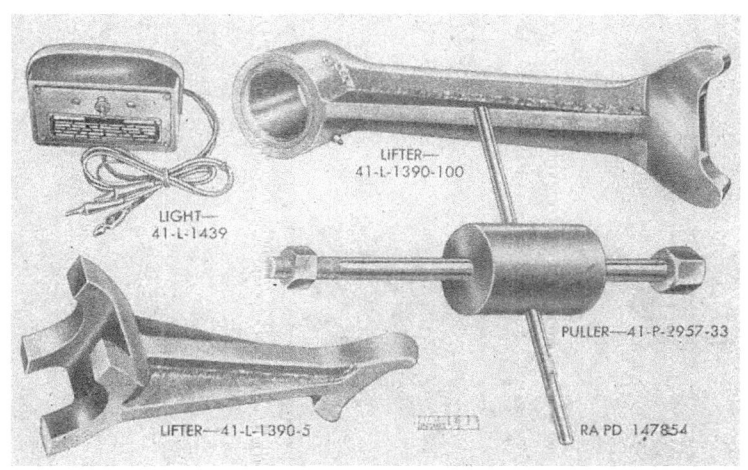

Figure 82 Special tools and equipment.

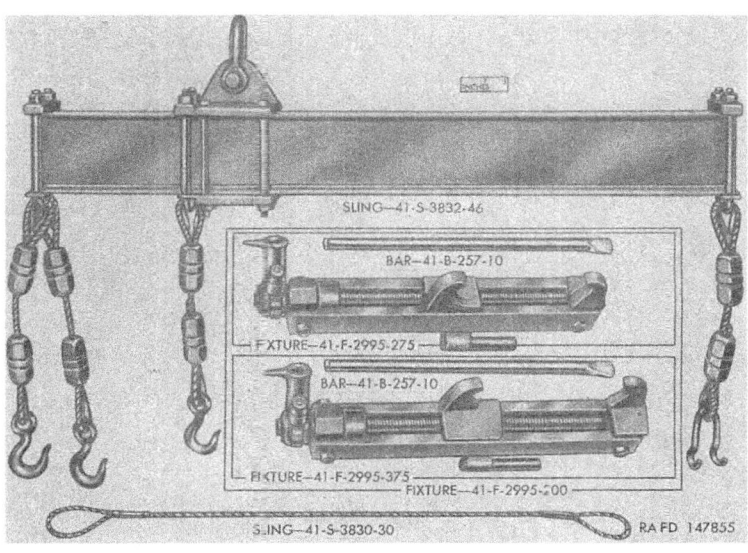

Figure 83. Special tools and equipment.

Figure 84. Special tools and equipment.

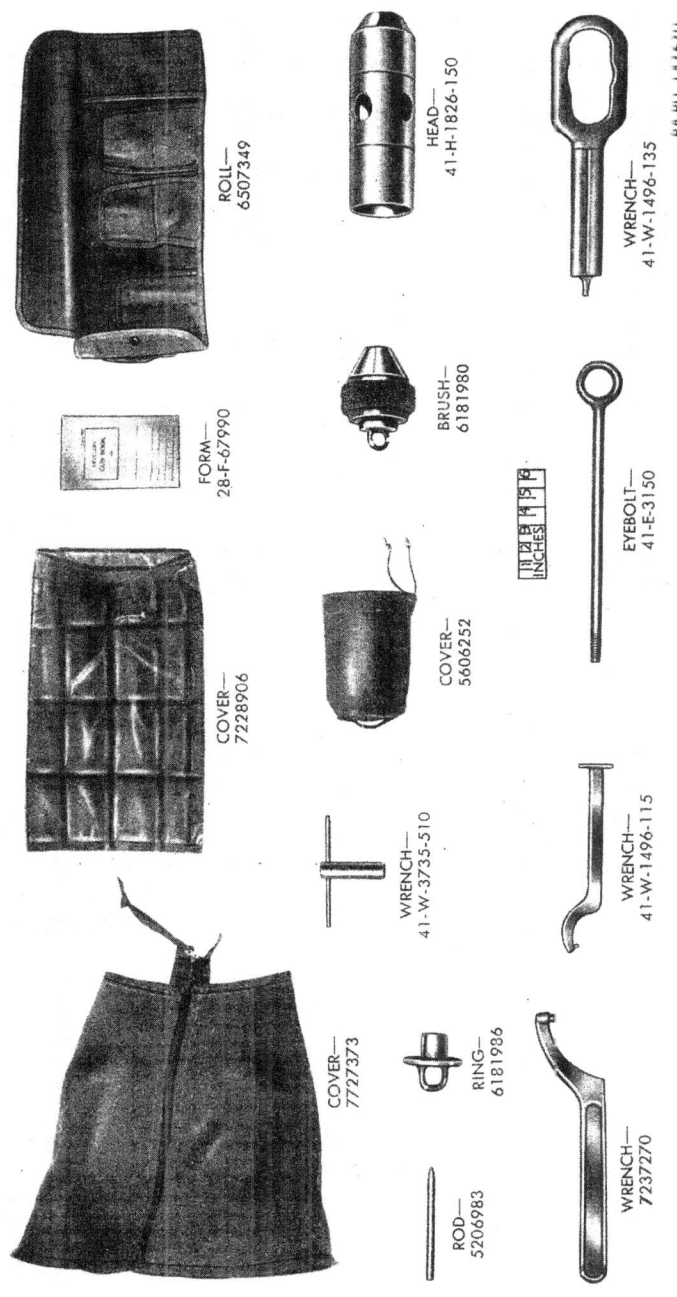

Figure 85. Special tools and equipment.

131

Section II. LUBRICATION AND PAINTING

113. Lubrication Order

Improvised lubrication order (figs. 86 and 87) prescribes cleaning and lubricating procedures as to locations, intervals, and proper materials for this vehicle.

Note. The lubrication instructions illustrated in figures 86 and 87 furnish advance information on lubrication for this vehicle and are to be used until such time that an LO is distributed.

When issued, the LO is to be carried with the vehicle at all times. Lubrication which is to be performed by ordnance maintenance personnel is listed on the lubrication order in the NOTES.

114. General Lubrication Instructions

a. USUAL CONDITIONS. Service intervals specified on the lubrication order are for normal operation and where moderate temperature, humidity, and atmospheric conditions prevail.

b. LUBRICATION EQUIPMENT. Each vehicle is supplied with lubrication equipment adequate for its maintenance. Clean this equipment both before and after use. Operate the lubricating guns carefully and in such a manner as to insure a proper distribution of the lubricant.

c. POINTS OF APPLICATION.
  (1) Lubricating fittings, grease cups, oilers, and oilholes are shown in figures 88 through 93 and are referenced to the lubrication order. Wipe these devices and the surrounding surfaces clean before and after lubricant is applied.
  (2) A ¾-inch red circle should be painted around each lubricating fitting and oilhole. Lubrication fittings on sighting and fire control instruments should not be used.

d. REPORTS AND RECORDS.
  (1) Report unsatisfactory performance of prescribed petroleum fuels, lubricants, or preserving materials, using DA AGO Form 468.
  (2) Maintain a record of lubrication of the vehicle on DA AGO Form 462.
  (3) Maintain a record of changes in grade of lubricant and recoil oil for the weapon in OO Form 5825, Artillery Gun Book.

115. Lubrication Under Unusual Conditions

a. UNUSUAL CONDITIONS. Reduce service intervals specified on the lubrication order, i. e., lubricate more frequently, to compensate for abnormal or extreme conditions, such as high or low temperatures, prolonged periods of high speed operation, continued operation in sand or dust, immersion in water, or exposure to moisture. Any one of these operations or conditions may cause contamination and quickly destroy the protective qualities of the lubricants. Intervals may be extended during inactive periods commensurate with adequate preservation.

b. CHANGING GRADE OF LUBRICANTS. Lubricants are prescribed in the "Key" in accordance with three temperature ranges; above $+32°$ F., $+40°$ to $-10°$ F., and from $0°$ to $-65°$ F. Change the grade of lubricants whenever weather forecast data indicate that air temperatures will be consistently in the next higher or lower temperature range or when sluggish starting caused by lubricant thickening occurs. No change in grade will be made when a temporary rise in temperature is encountered.

c. MAINTAINING PROPER LUBRICANT LEVELS. Lubricant levels must be observed closely and necessary steps taken to replenish in order to maintain proper levels at all times.

116. Lubrication for Continued Operation Below 0° F.

Refer to TM 9–2855 for instructions on necessary special preliminary lubrication of the vehicle, and to TB 9–2855–1 for installation and instructions on winterization kit.

117. Lubrication After Fording Operations

a. After any prolonged fording operation, in water 12 inches or over, lubricate all track and suspension points to cleanse bearings of water or grit as well as any other points required in accordance with paragraph 122 for maintenance operations after fording.

b. If the vehicle has been in deep water for a considerable length of time or was submerged beyond its fording capabilities precautions must be taken as soon as practicable to avoid damage to the engine and other vehicle components as follows:

(1) Perform a complete lubrication service (figs. 86 and 87).

(2) Inspect engine crankcase oil. If water or sludge is found, drain the oil and flush the engine with preservative engine oil PE–30. Before putting in new oil, drain the oil filter and install a new filter element (par. 149).

Note. If preservative engine oil is not available, engine lubricating oil OE–30 may be used.

(3) Operation in bodies of salt water enhances the rapid growth of rust and corrosion, especially on unpainted surfaces. It is most important to remove all traces of salt water and salt deposits from every part of the vehicle. For assemblies which have to be disassembled, dried and relubricated, perform these operations as soon as the situation permits. Wheel bearings must be disassembled and repacked after each sub-

LUBRICATION ORDER

TANK, 90-mm GUN, M47

References: TM 9-718A, ORD 7 SNL G-262

Intervals are based on normal operation. Reduce to compensate for abnormal operation and severe conditions or contaminated lubricants. During inactive periods, intervals may be extended commensurate with adequate preservation. Relubricate after washing or fording.

Clean fittings before lubricating. Clean parts with THINNER, paint, volatile mineral spirits (TPM) or SOLVENT, dry cleaning (SD). Dry before lubricating. (For exceptions, see notes 5 and 7.) Lubricate dotted arrow points on both sides of the equipment.

Lubricant ● Interval   Interval ● Lubricant

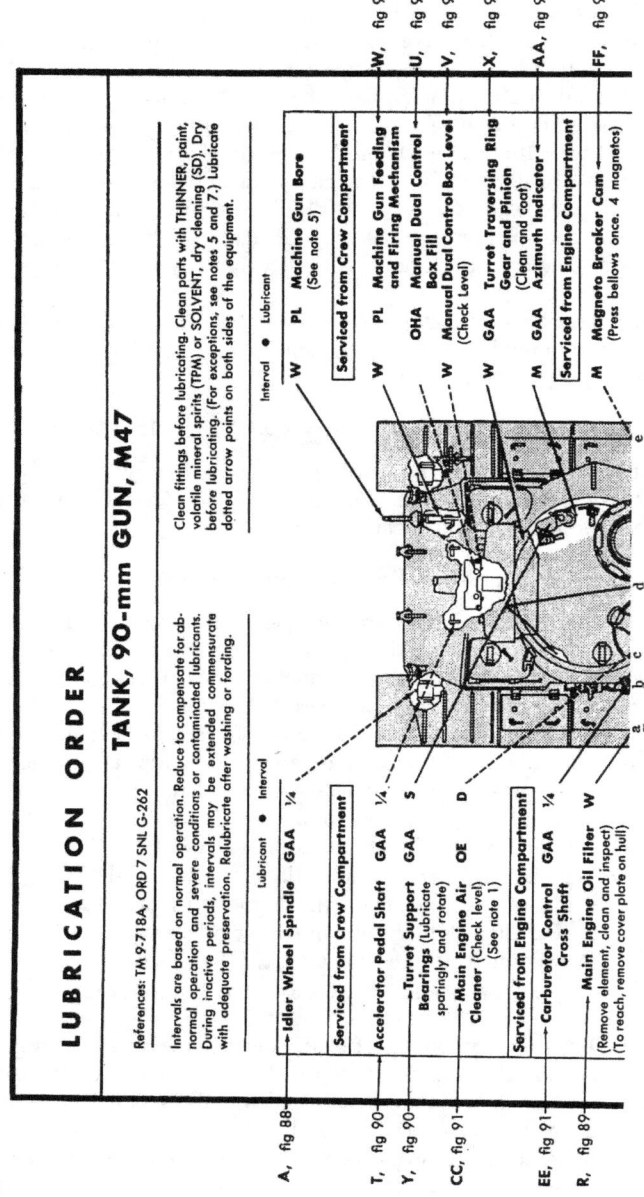

A, fig 88 → Idler Wheel Spindle — GAA — ½

Serviced from Crew Compartment

T, fig 90 → Accelerator Pedal Shaft — GAA — ½
Y, fig 90 → Turret Support Bearings (Lubricate sparingly and rotate) — GAA — S
CC, fig 91 → Main Engine Air Cleaner (Check level) (See note 1) — OE — D

Serviced from Engine Compartment

EE, fig 91 → Carburetor Control Cross Shaft — GAA — ¼
R, fig 89 → Main Engine Oil Filter (Remove element, clean and inspect) (To reach, remove cover plate on hull) — W

W, fig 90 → PL — Machine Gun Bore (See note 5)

Serviced from Crew Compartment

U, fig 90 → PL — Machine Gun Feeding and Firing Mechanism — W
V, fig 90 → OHA — Manual Dual Control Box Fill
X, fig 90 → Manual Dual Control Box Level (Check Level) — W
AA, fig 90 → GAA — Turret Traversing Ring Gear and Pinion (Clean and coat) — W
          → GAA — Azimuth Indicator — M

Serviced from Engine Compartment

FF, fig 91 → Magneto Breaker Cam (Press bellows once. 4 magnetos) — M

134

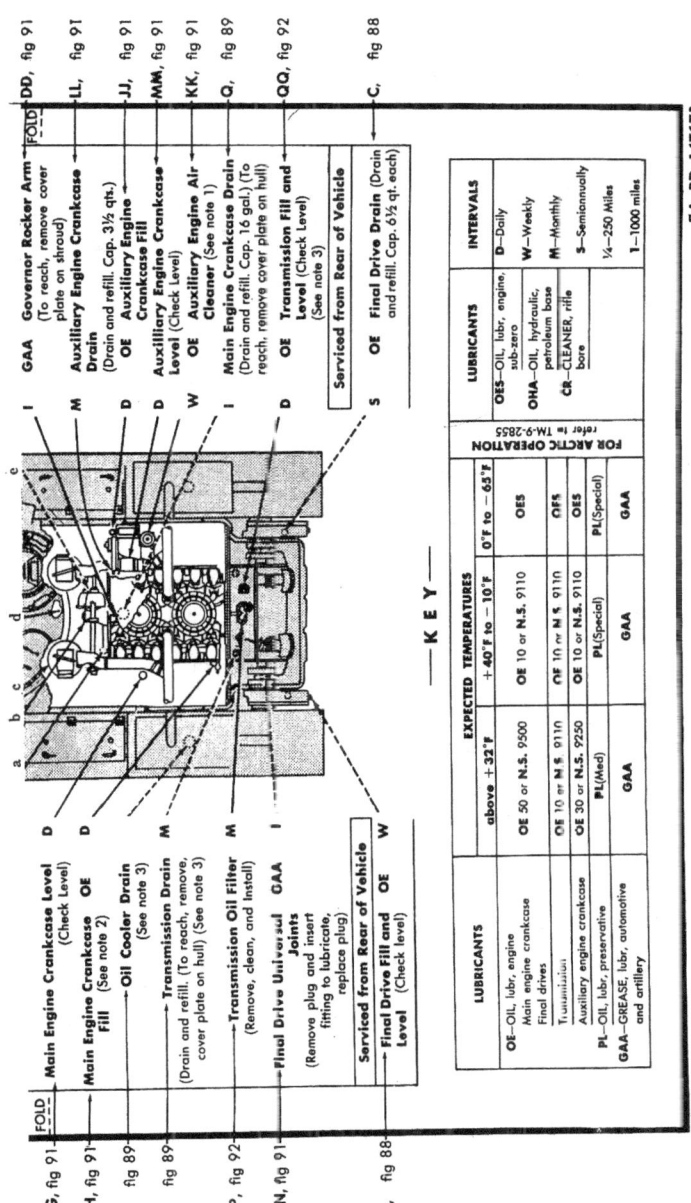

Figure 86. *Front side of lubrication order—90-mm gun tank M47.*

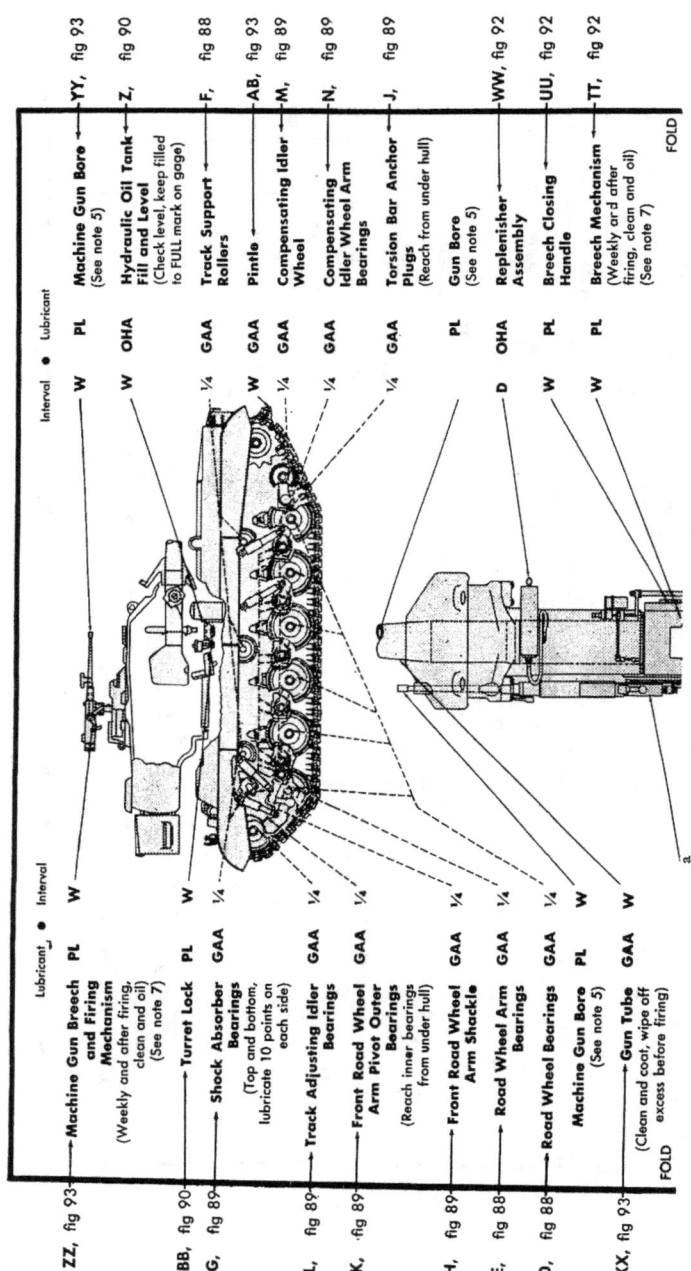

Lubricant	Interval		Lubricant	Interval	
ZZ, fig 93 → Machine Gun Breech and Firing Mechanism (Weekly and after firing, clean and oil) (See note 7)	PL	W	YY, fig 93 → Machine Gun Bore (See note 5)	PL	W
BB, fig 90 → Turret Lock	PL	W	Z, fig 90 → Hydraulic Oil Tank Fill and Level (Check level, keep filled to FULL mark on gage)	OHA	W
G, fig 89 → Shock Absorber Bearings (Top and bottom, lubricate 10 points on each side)	GAA	¼	F, fig 88 → Track Support Rollers	GAA	¼
L, fig 89 → Track Adjusting Idler Bearings	GAA	¼	AB, fig 93 → Pintle	GAA	¼
K, fig 89 → Front Road Wheel Arm Pivot Outer Bearings (Reach inner bearings from under hull)	GAA	¼	M, fig 89 → Compensating Idler Wheel	GAA	¼
H, fig 89 → Front Road Wheel Arm Shackle	GAA	¼	N, fig 89 → Compensating Idler Wheel Arm Bearings	GAA	¼
E, fig 88 → Road Wheel Arm Bearings	GAA	¼	J, fig 89 → Torsion Bar Anchor Plugs (Reach from under hull)	GAA	¼
D, fig 88 → Road Wheel Bearings	GAA	¼	WW, fig 92 → Gun Bore (See note 5)	PL	D
			UU, fig 92 → Replenisher Assembly	OHA	D
XX, fig 93 → Gun Tube (Clean and coat, wipe off excess before firing)	PL	W	TT, fig 92 → Breech Closing Handle	PL	W
			→ Breech Mechanism (Weekly and after firing, clean and oil) (See note 7)	PL	W

FOLD

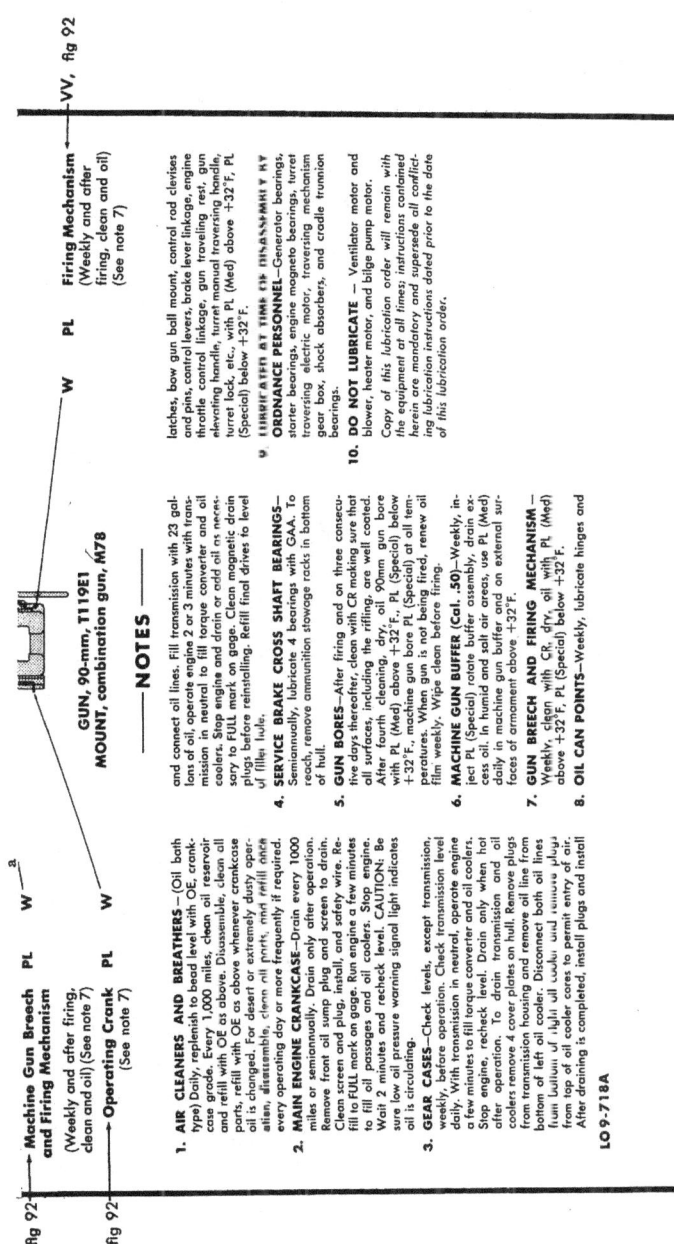

Figure 87. Back side of lubrication order—90-mm gun tank M47.

Figure 88. Localized lubrication points (points A through F).

Figure 89. Localized lubrication points (points G through S).

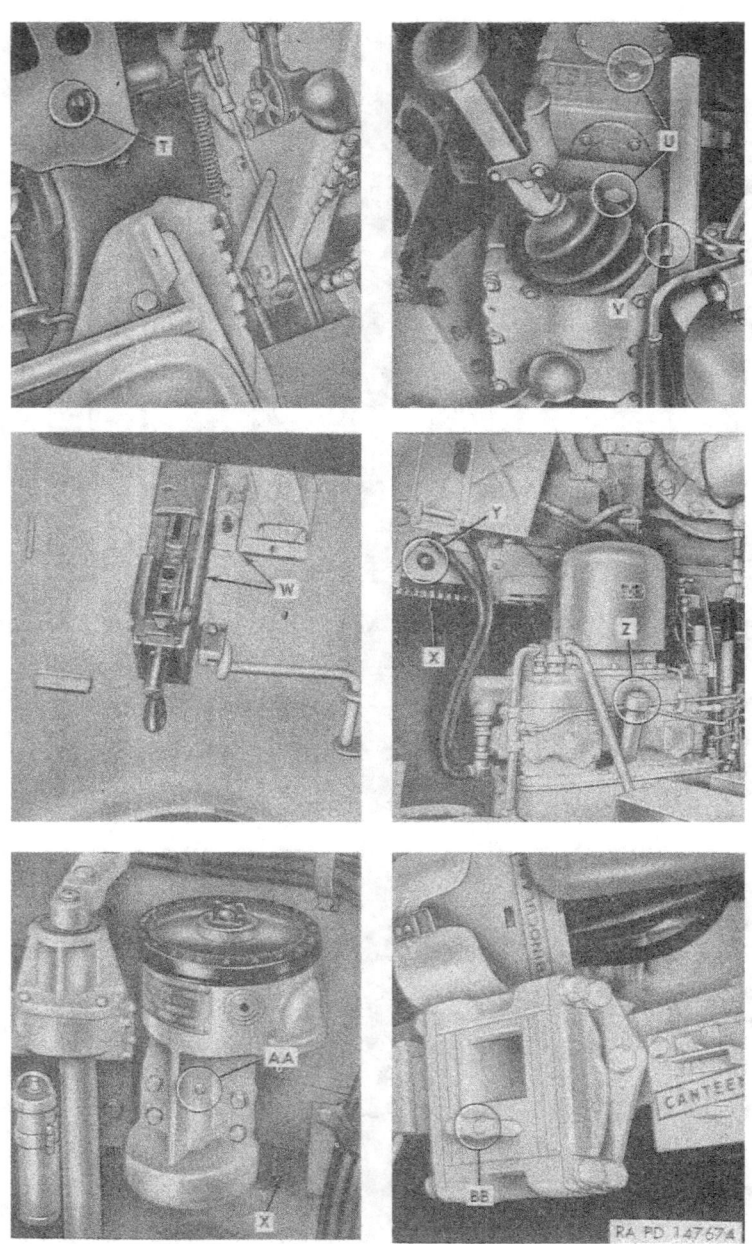

Figure 90. Localized lubrication points (points T through RB).

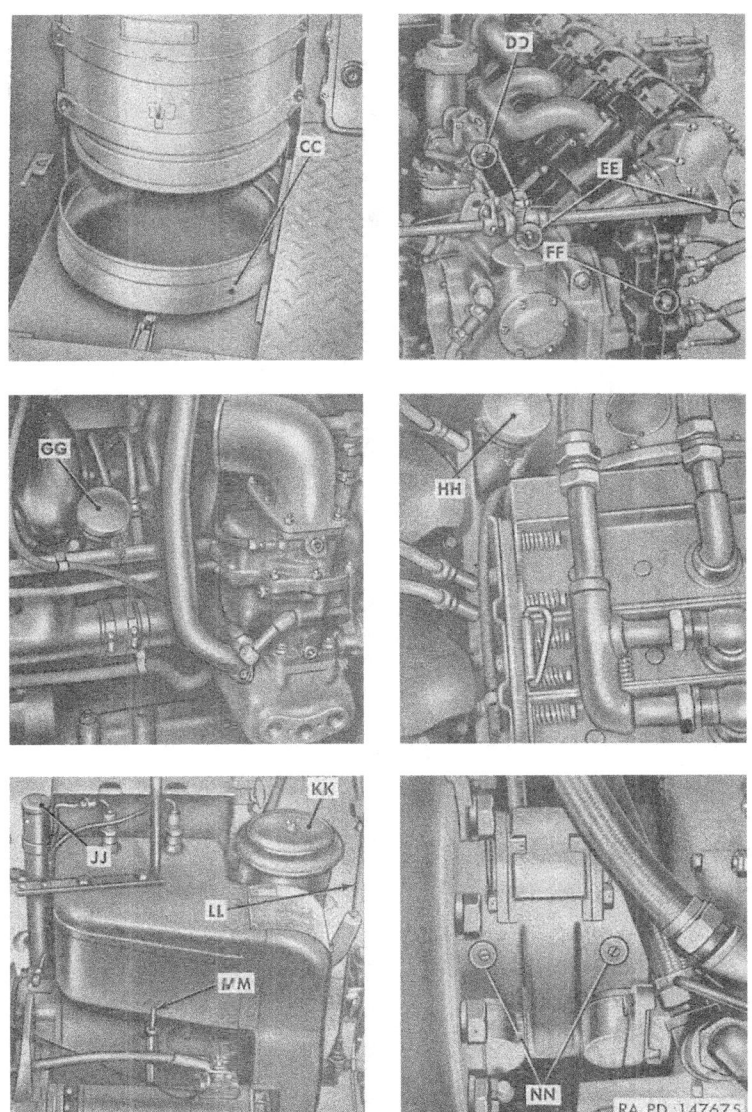

Figure 91. Localized lubrication points (points CC through NN).

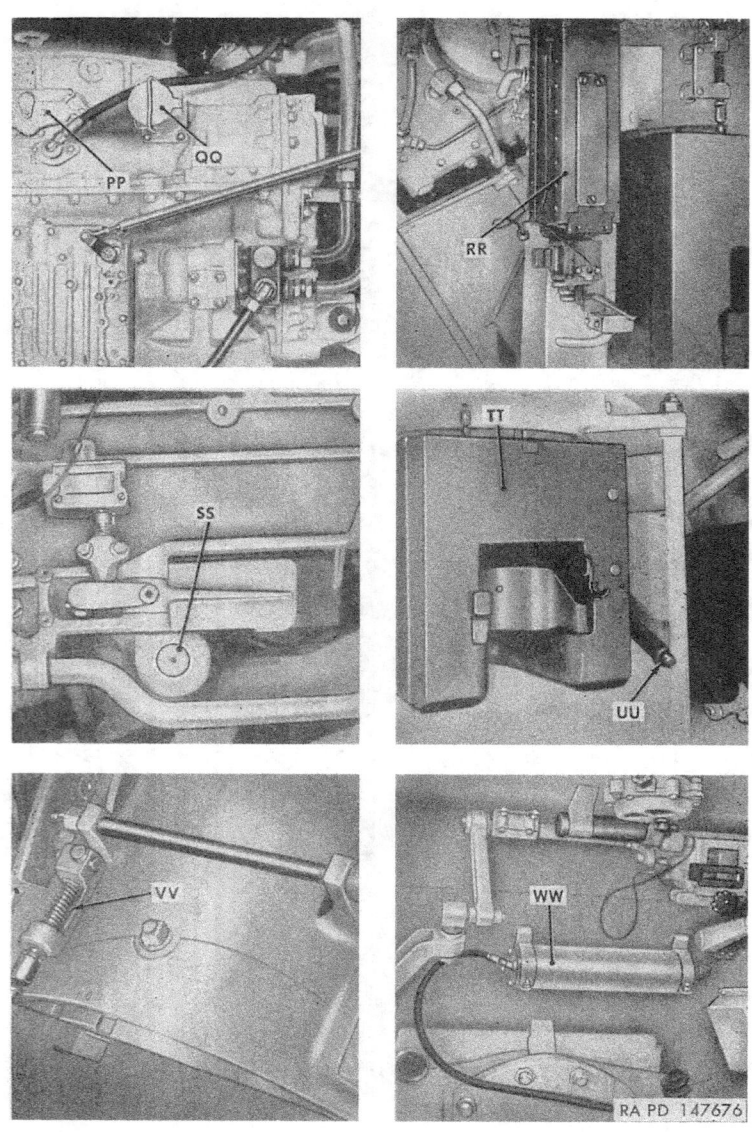

Figure 92. Localized lubrication points (points PP through WW).

Figure 93. Localized lubrication points (points XX through AB).

mersion. Regardless of the temporary measures taken, the vehicle must be delivered as soon as practicable to the ordnance maintenance unit.

118. Lubrication After Operation Under Dusty and Sandy Conditions

After operation under dusty or sandy conditions, clean and inspect all points of lubrication for fouled lubricants and relubricate as necessary.

Note. A lubricant which is fouled by dust and sand makes an abrasive mixture that causes rapid wear of parts.

119. Painting

Instructions for the preparation of the matériel for painting, methods of painting, and materials to be used are contained in TM 9–2851. Instructions for camouflage painting are contained in FM 5–20B.

Section III. PREVENTIVE MAINTENANCE SERVICES

120. General

a. RESPONSIBILITY AND INTERVALS. Preventive maintenance services are the responsibility of the using organization. These services consist generally of before-firing, during-firing, after-firing, before traveling (for weapon), of before-operation, during-operation, at the-halt, after-operation (for vehicle), and weekly services (for both) performed by the operator, crew or leader and the scheduled services to be preformed at designated intervals by organization mechanic or maintenance crews. Intervals are based on normal operations. Reduce intervals for abnormal operations or severe conditions. Intervals during inactive periods may be extended accordingly.

b. Definition of Terms. The general inspection of each item applies also to any supporting member or connection, and is generally a check to see whether the item is in good condition, correctly assembled, secure, and not excessively worn.

(1) Inspection for "good condition" is usually an external visual inspection to determine whether the unit is damaged beyond safe or serviceable limits. The term "good condition" is explained further by the following: not bent or twisted, not chafed or burned, not broken or cracked, not bare or frayed, not dented or collapsed, not torn or cut, not deteriorated.

(2) The inspection of a unit to see that it is "correctly assembled" is usually an external visual inspection to see whether it is in its normal assembled position in the vehicle.

(3) Inspection of a unit to determine if it is "secure" is usually an external visual examination or a check by hand, wrench, or pry-bar for looseness. Such an inspection must include any brackets, lock washers, lock nuts, locking wire, or cotter pins used.

(4) By "excessively worn" is meant worn beyond serviceable limits or to a point likely to result in failure if the unit is not replaced before the next scheduled inspection.

121. Cleaning

a. General. Any special cleaning instructions required for specific mechanisms or parts are contained in the pertinent section. General cleaning instructions are as follows:

(1) Use volatile mineral spirits or dry-cleaning solvent to clean or wash grease or oil from all parts, except those exposed to powder fouling during firing. These solvents will not readily dissolve the corrosive salts from powder and primer compositions.

(2) Use rifle-bore cleaner to clean all armament parts which have been exposed to powder fouling during firing.

Note. Rifle-bore cleaner is not a lubricant. Parts which require lubrication will be wiped dry and oiled.

(3) A solution of one part grease-cleaning compound to four parts of volatile mineral spirits or dry-cleaning solvent may be used for dissolving grease and oil from engine blocks, hull, and other parts. After cleaning, use cold water to rinse off any solution which remains.

(4) After the parts are cleaned, rinse and dry them thoroughly. Apply a light grade of oil to all polished metal surfaces (other than optical instruments) to prevent rusting.

(5) Before installing new parts, remove any preservative materials, such as rust-preventive compound, protective grease, etc.; prepare parts as required (oil seals, etc.); and for those parts requiring lubrication, apply the lubricant prescribed in the lubrication order.

b. GENERAL PRECAUTIONS IN CLEANING.

(1) Volatile mineral spirits and dry-cleaning solvent are inflammable and should not be used near an open flame. Fire extinguishers should be provided when these materials are used. Use only in well ventilated places.

(2) These cleaners evaporate quickly and have a drying effect on the skin. If used without gloves, they may cause cracks in the skin and, in the case of some individuals, a mild irritation or inflammation.

(3) Avoid getting petroleum products, such as volatile mineral spirits, dry-cleaning solvent, engine fuels, or lubricants on rubber parts, as they will deteriorate the rubber.

(4) The use of Diesel fuel oil, gasoline, or benzene (benzol) for cleaning is prohibited.

122. Preventive Maintenance by Operator, Leader or Crew

a. PURPOSE. To insure mechanical efficiency, it is necessary that the vehicle be systematically inspected at intervals every day it is operated and also weekly, so defects may be discovered and corrected before they result in serious damage or failure. Certain scheduled maintenance services will be performed at these designated intervals. Any defects or unsatisfactory operating characteristics beyond the scope of the operator or crew to correct must be reported at the earliest opportunity to the designated individual in authority.

b. SERVICES.

(1) *Vehicle.* Operator and leader preventive maintenance services are listed in table III. Every organization must thoroughly school its personnel in performing the maintenance procedures for this vehicle as set forth in this manual.

(2) *Armament.* Armament preventive maintenance services performed by the crew are described in table IV.

(a) Rust, dirt, grit, gummed oil, and water cause rapid deterioration of all parts of the weapon. Particular care should be taken to keep all bearing surfaces and exposed unpainted parts clean and properly lubricated. Wiping cloths, riflebore cleaner, dry-cleaning solvent, and lubricants are furnished for this purpose. Remove all traces of rust from surfaces with crocus cloth which is the coarsest abrasive to be used by organizational personnel for this purpose. Take care not to change the shape or dimensions of any part.

Table III. Operator's and Leader's "A" Preventive-maintenance Services

\multicolumn{4}{c}{Intervals}		
\multicolumn{4}{c}{Operator's}	Leader's "A" (weekly)	Procedure
Before-operation	During-operation	At halt	After-operation		

Before-operation	During-operation	At halt	After-operation	Leader's "A" (weekly)	Procedure
					USUAL CONDITIONS
X	----	X	X	X	*Fuel and oil.* Check fuel and oil levels. Look for leaks in engine compartment. Check spare containers for contents.
				X	Remove the drain cock from the fuel line filter and drain water and sediment (fig. 130).
X	----	----	----	X	*Tracks.* Inspect tracks visually.
			X	X	Inspect tracks for correct tension (par. 249).
X	----	X	X	X	*Leaks, general.* Look under vehicle for indications of fuel, engine-oil, gear-oil, or grease leaks.
X	----	----	----	----	*Vehicle tools and equipment.* Check Department of the Army Supply Catalog ORD 7 SNL G-262 to see that all tools and equipment issued with and carried on the vehicle are present and in serviceable condition. Inspect publications, lubrication order, Standard Form 91, and WD AGO Form 614. Examine the seals on the fire extinguisher control head (fig. 324), remote control handle (fig. 75), and safety valves (fig. 324). If the seal wires are broken or the safety seal is missing, replace the cylinders with fully charged ones (par. 317).
				X	See that fire extinguishers are charged and sealed. Remove and weigh fire extinguisher cylinders. Fixed-fire-extinguisher cylinders weigh 37 pounds (without connecting heads) when full, and 27 pounds when empty. If weight is less than 36 pounds, replace cylinder (par. 317). While cylinders are removed, check operation of remote control handle (par. 94). Clean and adjust discharge nozzles. Replace damaged lines and nozzles (par. 319).
X	----	----	X	X	Operate lights, horn (if tactical situation permits), windshield wipers. Visually inspect vision devices, hull, armor, towing connections, traveling locks, hatches, doors, paulins, tools, etc.
X	X	----	----	----	*Instruments.* Observe for normal readings during warm-up (par. 68) and during operation of vehicle. **Caution:** If any warning signal lights come on shut off engine immediately and investigate cause.
		X	----	----	*General operation.* Be alert for any unusual noises or improper operation of steering, brakes, or shifting.
		X	X	X	*Operating faults.* Investigate and correct or report any faults noted during operation.

Table III. Operator's and Leader's "A" Preventive-maintenance Services—Con.

Before-operation	During-operation	At halt	After-operation	Leader's "A" (weekly)	Procedure
		X	X	X	*Springs and suspensions.* Look at bumper springs, road wheel arms, shock absorbers, and torsion bars to see if they have been damaged.
			X	X	*Lubricate.* Lubricate daily or weekly items specified on lubrication order (par. 113).
			X	X	*Clean.* Clean glass, vision devices, and inside of vehicle. Wipe off exterior of vehicle.
				X	Wash vehicle. Clean engine and engine compartment.
				X	*Battery.* Clean. Check water level. Inspect terminals for corrosion, tightness, and coating of grease.
				X	*Assemblies.* Inspect assemblies such as carburetor, generator, heater, ventilating blower, magnetos, starter, fuel and bilge pumps, for looseness of mountings or connections.
				X	*Electrical wiring.* Inspect visually electrical wiring, conduits, and shielding.

UNUSUAL CONDITIONS

Preventive maintenance services for usual conditions will apply, with emphasis on servicing to combat the effect of unusual conditions of extreme cold, extreme heat, unusual terrain, and fording. The services described below are those required to assure best results with the winterization kit as well as the special services that may be required under unusual conditions.

Extreme Cold

(Pars. 370 through 373 and TM 9-2355)

Before-operation	During-operation	At halt	After-operation	Leader's "A" (weekly)	Procedure
		X	X		*Fuel system.* Refuel and add denatured alcohol as required.
			X		Drain fuel tank and clean sump to remove condensation and sludge; refuel tank.
			X		*Lubricants.* Check and, if necessary, change lubricants and special oils to conform with the lubrication order (par. 113).
			X		Check gear cases for collections of sludge and water and clean out if necessary and refill.
					Note. It is necessary to have lubricant warm and fluid for draining and refilling. Control lever.
		X	X		Position manual control lever in neutral position.

147

Table III. Operator's and Leader's "A" Preventive-maintenance Services—Con.

Before-operation	During-operation	At halt	After-operation	Leader's "A" (weekly)	Procedure
					Intervals — Operator's
X		X			*Tracks.* Check for tracks frozen to ground.
X				X	*Battery.* Check for proper charge and electrolyte level.
			X		Remove battery and store in warm place, if vehicle is not equipped with power plant heater.
X			X	X	*Clean.* Clean snow, ice, and mud from all parts of vehicle.
	X				*Operating observations.* Check for the feel of stiffness of lubricant in the final drives and suspension com- components. This will be indicated by unusual power demand when putting vehicle in motion. Listen for signs of malfunctions and inspect immediately to determine causes.
					Winterization equipment.
X				X	Check personnel heater for proper operation.
			X	X	Fill power plant heater fuel tank and check unit for proper functioning.
			X	X	Check all winterization equipment for secure installation and proper functioning.
					Extreme Heat (par. 374)
X		X		X	*Cooling and fuel systems.* Check oil cooler cores; engine and oil cooler fans, engine baffles, and top deck grille. Check air cleaner, fuel and oil filters, and clean as often as necessary to keep them in good condition.
X				X	*Battery.* Check for electrolyte level.
				X	Check for proper charge.
			X		Remove battery and store in cool place, if necessary to park for extended periods.
X		X		X	*Ventilating blower.* Check for proper operation.
					Unusual Terrain (par. 376)
			X		*Lubrication.* Check for fouled lubricants and lubricate as necessary (par. 115).
X				X	*Fuel system.* Check air cleaners, fuel and oil filters, and clean as often as necessary to keep them in good condition. **Warning:** Under extremely dusty conditions or blowing sand, it will be necessary to service the air cleaner several times daily during operation to prevent entry of dust or sand into the engine. Failure to do this may wear out engine parts in a short time.

Table III. Operator's and Leader's "A" Preventive-maintenance Services—Con.

Intervals					Procedure
Before-operation	During-operation	At halt	After-operation	Leader's "A" (weekly)	
			X	X	*Clean.* Clean all parts of vehicle of snow, ice, mud, dust, and sand, especially all suspension components. After cleaning, lubricate suspension components as prescribed in the lubrication order (par. 113).
				X	Check for any sand-blasted surfaces and paint as required.
X			X		*Vents.* Check engine vents and other exposed vents and keep them covered with cloth to prevent entry of dust, sand, or drifting snow.
					Fording Operations (par. 375)
X					*Fording limits.* Check vehicle fording limits. See paragraph 104 for operation precautions.
X					*Battery.* Check vent caps for tightness to prevent entrance of water.
			X		Check for seepage of water into battery. Check charge as soon as practicable and add electrolyte and charge if necessary.
			X		*Drain hull.* Immediately after fording, open hull drain valves to remove accumulated water.
			X		*Clean.* Remove water and sludge from all parts of the vehicle. If fording through salt water, wash with fresh water.
			X		*Engine and transmission.* Check for evidence of water or grit and replace oil if necessary. If engine oil must be replaced, use engine conditioning oil to flush before adding new oil.
			X		*Fuel system.* Check air cleaner, oil and fuel filters, and clean or replace if necessary. Drain any accumulation of water or sludge from fuel tank.
					Lubrication. Lubricate as specified in paragraph 117.

Table IV. Crew Preventive-Maintenance Services

	Intervals				Procedure
Before-firing	During-firing	After-firing	Weekly	Before-traveling	
					USUAL CONDITIONS
					Bore and chamber.
		X		X	Remove rounds of ammunition from chambers, if present.
X					Wipe dry.
		X			Clean thoroughly (par. 339) and lubricate (par. 113).
			X		Examine for powder fouling, corrosion or other damage.
				X	Cover muzzle.
					Breech mechanism.
X			X		Test operation for proper functioning.
			X		Examine for corrosion or other damage.
				X	Cover breech.
					Firing controls.
X			X		Test operation for proper functioning.
					Elevating and traversing mechanism.
X			X		Check for smooth operation throughout entire range (par. 76).
					Recoil mechanism.
X	X	X	X		Look for excessive oil leakage.
	X				Test for smooth operation, length of recoil, and complete return to battery without shock.
X			X		Check and reestablish oil reserve (pars. 56 and 358).
					Recoil slides.
			X		Look for rust or damage.
					Sighting and fire control matériel.
		X	X		Wipe clean. **Caution:** Optical glass is easily scratched. Use only a camel's hair brush, to remove dust. Clean only with lens tissue. Breathing on the glass before wiping is permissible. If the glass is greasy or oily, apply lens soap only, and wipe off with lens tissue.
		X			Cover.
				X	Stow.
		X	X	X	*Covers.* Examine for proper installation and condition of canvas (breech, muzzle, instruments, etc.).
					General.
		X		X	Inspect gun for general condition.
			X	X	Inspect mount for general cleanliness and condition of paint.
				X	Look for loose bolts or parts.
X				X	Secure or release traveling locks and latches, as applicable.

Table IV. Crew Preventive-Maintenance Services—Continued

Before-firing	During-firing	After-firing	Weekly	Before-traveling	Procedure
					UNUSUAL CONDITIONS
					Preventive maintenance services for usual conditions will apply, with emphasis on servicing to combat the effect of unusual conditions of extreme cold and extreme heat. The services described below are those required to assure best results under unusual conditions.
					Extreme Cold (pars. 101 and 372 and TB ORD 193)
		X	X		*Bore and chamber.* Do not dilute rifle-bore cleaner. For first cleaning after firing, clean while bore and chamber are still warm from firing, if practicable.
X			X		*Clean.* Clean snow, ice, and mud from all parts of the weapons and mounts.
		X		X	*Cover.* Make sure covers are secure and serviceable.
X		X	X	X	*Lubrication.* Lubricate sparingly at more frequent intervals in accordance with the lubrication order (par. 113).
		X	X		*Sweating.* Dry any condensed moisture from all parts and lubricate in accordance with the lubrication order (par. 113).
					Extreme Heat
	X				*Rate of fire.* **Warning:** Do not allow tube to become overheated due to prolonged high rate of fire or allow unfired round to remain in overheated chamber because of danger of "cook-offs" (par. 143).
		X		X	*Cover.* Make sure covers are secure and serviceable.
X		X	X		*Lubrication.* Lubricate at more frequent intervals in accordance with the lubrication order (par. 113).
X				X	In dusty and sandy terrain, during operation, leave recoil slides and exposed points dry to prevent sand and lubricant from forming an abrasive paste.

(*b*) Retouch painted surfaces as required to cover glossy paint or nicks, scratches, and worn spots which expose bare metal (par. 119).

(*c*) Tighten loose parts, as necessary.

(*d*) Each time a weapon is disassembled for cleaning or repair, carefully inspect all parts for cracks, excessive wear, rust, and like defects which might cause malfunction of the gun. See section IV, this chapter, on trouble shooting for information and certain parts which when worn, damaged, or improperly adjusted causes definite malfunctions. Thoroughly clean and properly lubricate all parts before assembly.

(e) At least every 6 months a check up will be made to see that all modification work orders have been applied. A list of current modification work orders is published in SR 310–20–4. If a modification has not been applied, promptly notify the local armament officer. No alteration or modification, which will affect the moving parts, will be made by organizational personnel, except as authorized by official publications.

(f) When the matériel is not in use, the proper covers must be used.

(g) When canvas or other type cover is used during periods of inactivity, moisture may condense on metal surfaces. To prevent rusting, the covers must be removed at least weekly, and all surfaces thoroughly dried and unpainted surfaces coated with the prescribed lubricant. In cold weather, apply lubricant sparingly (par. 101).

123. Preventive Maintenance by Organizational Maintenance Mechanics

a. INTERVALS. The indicated frequency of the prescribed preventive maintenance services is considered a minimum requirement for normal operation of the vehicle. Under unusual operating conditions, such as extreme temperatures, dust or sand, or extremely wet terrain, it may be necessary to perform certain maintenance services more frequently.

b. CREW OR OPERATOR PARTICIPATION. The operators and crew should accompany their vehicles and assist the mechanics while periodic organizational preventive maintenance services are performed. Ordinarily, the operator should present the vehicle for a scheduled preventive maintenance service in a reasonably clean condition.

c. SPECIAL SERVICES. These are indicated by the item numbers in the columns which show the interval at which the services are to be performed, and show that the parts or assemblies are to receive certain mandatory services. For example, an item number in one or both columns opposite a *Tighten* procedure means that the actual tightening of the object must be performed. The special services are as follows:

(1) *Adjust.* Make all necessary adjustments in accordance with the pertinent section of this manual, technical bulletins, or other current directives.

(2) *Clean.* Clean the unit as outlined in paragraph 121 to remove old lubricant, dirt, and other foreign material.

(3) *Special lubrication.* This applies either to lubrication operations that do not appear on the vehicle lubrication order or to items that do appear but which should be performed in connection with the maintenance operations if parts have to be disassembled for inspection or service.

(4) *Serve.* This usually consists of performing special operations, such as replenishing battery water, draining and refilling units with oil, and changing or cleaning oil filter, air cleaner, or fuel filter elements.

(5) *Tighten.* All tightening operations should be performed with sufficient wrench torque (force on the wrench handle) to tighten the unit according to good mechanical practice. Use a torque-indicating wrench where specified. Do not overtighten, as this may strip threads or cause distortion. Tighten will always be understood to include the correct installation of lock washer, lock nuts, jam nuts, locking wire, or cotter pins.

d. SPECIAL CONDITIONS. When conditions make it difficult to perform the complete preventive maintenance procedures at one time, they can sometimes be handled in sections. Plan to complete all operations within the week if possible. All available time at halts and in bivouac areas must be utilized, if necessary, to assure that maintenance operations are completed. When limited by the tactical situation, items with special services in the columns should be given first consideration.

e. WORK SHEET. The numbers of the preventive maintenance procedures that follow are identical with those outlined on DA AGO Form 462, Work Sheet for Full-Track and Tanklike Wheeled Vehicles—Preventive Maintenance Service and Technical Inspection. Certain items on the work sheet that do not apply to this vehicle are not included in the procedures in this manual. In general, the sequence of items on the work sheet is followed, but in some instances there is deviation for conservation of the mechanic's time and effort.

f. PROCEDURES. Table V lists the services to be performed by the organizational mechanic or maintenance crew at the designated intervals. Each page of the table has two columns at its left edge corresponding to Commander's "C" (quarterly or 750-mile) and Commander's "B" (monthly or 250-mile) maintenance respectively. Very often it will be found that a particular procedure does not apply to both scheduled maintenances. In order to determine which procedure to follow, look down the column corresponding to the maintenance procedure and, wherever an item number appears perform the operations indicated opposite the number.

Table V. *Commander's "B" and "C" Preventive-Maintenance Services*

Intervals		Procedure
Commander's "C" (quarterly or 750 miles)	Commander's "B" (monthly or 250 miles)	
		INSPECTION AND ROAD TEST
		Note. When the tactical situation does not permit a full road test, perform only those items that require little or no movement of the vehicle. When a road test can be made, it should be approximately three miles, but not over five miles.
		BEFORE OPERATION: *Oil, fuel, track, leaks, general visual inspection of vehicle and equipment.* Perform the *before-operation* service (table III, par. 122).
1	1	*Dash instruments, gages, warning signal lights—engine oil pressure and temperature, transmission oil pressure and temperature, generator charging speedometer-tachometer fuel.* Immediately after starting the engine, before the generator regulator has reduced the charging rate, notice if the generator warning signal light goes out. Check the oil pressure on the engine oil pressure gage. Observe all the dash instruments for normal readings (pars. 31, 32, and 33). Determine if all warning signal lights are working (par. 34).
2	2	*Horn, windshield wipers, heater, ventilating blower.* If the tactical situation permits, sound horn (par. 30) to see if signal is normal. Test windshield wipers for satisfactory operation. At the same time, test the operation of defroster, crew-compartment heater (par. 97) and ventilating blower (par. 96).
3	3	*Lamps—head, tail, dome, stop, blackout.* While the vehicle is stopped, if the tactical situation permits, test the operation of all lights, external and internal, and light switches (pars. 27, 28, and 29). Notice if the head lights appear to be correctly aimed. Note condition of lights and safety reflectors.
4	4	*Engine—idle, acceleration, power, noise, smoke, governed speed.* In warming up the engine, observe if it starts easily and if the operation of primer pump and hand throttle lever is satisfactory. Notice if idling speed is correct (650 rpm). Note any unusual noises or excessive vibration, at idle or higher speed. When operating the vehicle, notice if it has normal power and acceleration in each transmission range, or any tendency to stall. Listen for any unusual noises when the engine is under load. Look for excessive black or blue smoke issuing from the exhaust. Speed up the vehicle on a level stretch and see if the engine will reach, but not exceed, the governed speed of 2,800 rpm under load.
5	5	*Suspension—track, wheels, idlers, sprockets, shock absorbers, springs, torsion bars, arms.* Inspect these items. In the track, look particularly for dead track blocks and bottomed wedges; notice if tension appears to be satisfactory (par. 249). Inspect road wheels, sprockets, idlers, and support rollers, paying particular attention to lubricant leaks from the bearing seals and rubber tires separated from rims. Inspect bumper springs visually, noting if they have taken a permanent set. Pry up on each road wheel to detect broken torsion bars.

Table V. Commander's "B" and "C" Preventive-Maintenance Services—Con.

Commander's "C" (quarterly or 750 miles)	Commander's "B" (monthly or 250 miles)	Procedure
6	6	*Brakes—service, parking, braking effect, steering action.* Inspect these items. Test service brake pedal for specified free travel and floorboard clearance (par. 259c). Move manual control lever through its entire range of positions, and observe if the steering response is satisfactory and without wander when manual control lever is centered, chatter, noise, or any unusual behavior. Accelerate vehicle to a moderate speed and apply service brakes and see if they stop the vehicle smoothly and effectively, without side pull. When vehicle is on an incline, apply parking brake and observe if it locks securely and holds the vehicle safely.
8	8	*Transmission, lever action, vibration, noise, control, shifting action.* Inspect transmission for clogged breather line. Shift through all forward and reverse speed ranges of transmission noticing if it shifts smoothly without excessive vibration or unusual noise and if vehicle response is satisfactory. Note any tendency of the manual control lever to jump out of position. Examine all control levers, shafts, and linkage for freedom of action and correct adjustment (par. 257).
8		Tighten all assembly and mounting bolts of these units.
9	9	*Unusual noises—universal joints, final drives, road wheels, idlers, shock absorbers, fans, tracks, drive sprockets, support rollers.* Listen, both from inside and from outside the vehicle, for unusual noises emanating from these components, or others, that would indicate lack of lubrication, maladjustment, or damage. Be alert for unusual noises throughout the road test.

AFTER ROAD TEST

11	11	*Temperatures—wheel hubs, drive sprocket hubs, shock absorbers, transmission, final drives.* Immediately after the road test, feel these components cautiously. Feel the transmission also if there is doubt that the warning signal light is functioning properly. If shock absorbers are operating satisfactorily they should be warmer than the hull. An overheated road wheel, drive sprocket, idler, or support roller hub indicates a maladjusted, dry, or damaged bearing. An overheated gear case indicates internal maladjustment, damage, or inadequate lubrication.
12	12	*Auxiliary generator—power, smoke, governed speed, generator output, wiring.* Inspect engine and wiring; pay particular attention to the carburetor, fuel pump, generator, fuel and oil lines, and choke linkage. Operate the engine (par. 93), noticing if it has normal power and generally performs satisfactorily. See if excessive black or blue smoke issues from the exhaust. Determine if the generator output is satisfactory by the method described in paragraph 132b and c. If need is indicated by engine sound or low generator output, use a tachometer to see if engine governed speed is within specified limits (par. 172a (3)).

Table V. Commander's "B" and "C" Preventive-Maintenance Services—Con.

Commander's "C" (quarterly or 750 miles)	Commander's "B" (monthly or 250 miles)	Procedure
12	12	Service carburetor air cleaner in accordance with lubrication order (par. 113). Clean fuel-filter element and sediment bowl.
12	----	Clean spark plugs and adjust gap (par. 311d). Dress magneto breaker points and adjust gap (par. 310c). On every second quarterly or 750-miles inspection and service, examine generator brushes and commutator (par. 304). Inspect mounting brackets.
13	13	*Leaks—engine oil, fuel, transmission, final drive, road and idler wheels, grease seals, primer lines.* Make general observations inside and outside the vehicle for fuel, lubricant leaks from seals, gaskets, line and hose connections, oil coolers, tanks, or other sources.
14	14	*Hull—paint, fenders, hatches, grille doors, drain plugs and valves, seats, bilge pumps.* Inspect these items, gun traveling lock, all other attachments on deck, and optical devices (including spares) in the hull. *Caution:* Optical glass is easily scratched. Use only a camel's hair brush to remove dust. Clean only with lens tissue. Breathing on the glass before wiping is permissible. If the glass is greasy or oily, apply lens soap only, and wipe off with lens tissue. Notice particularly any bare spots in the interior or exterior paint, which might permit rust or reflection, and check legibility of markings. Observe if hatch covers, doors, and inspection-hole covers lock and seal properly. Look at grille doors for clogging. Examine hull drain valves (par. 21) and plugs for proper functioning. Test operation of bilege pumps (par. 95); see if strainers are clean. Test seat-adjusting devices to see if they are functioning properly (pars. 20, 48, and 49).
15	15	*Turret—basket, hatches, ventilating blower, locks, seats, pads.* Inspect these items including attachments and optical devices on turret. Test operation of turret lock (par. 42). Make observations similar to those made in item 14 on paint, hatches and locks, ventilating blower, seats, and optical devices. Notice particularly if crash pads are in serviceable condition. Test operation of azimuth indicator (par. 63).
16	16	*Armament—mounts, firing controls, elevating and traversing mechanism, recoil mechanism.* Inspect these items; include breech mechanism and spare parts. With vehicle tilted sidewise about 10°, test the operation of manual and power traversing mechanisms through their entire range (par. 76). *Caution:* Be sure driver's doors are closed and that gun has clearance to traverse. With gun pointed forward or rearward, test the operation of manual and power elevating mechanisms through their entire range (par. 76). Test the operation of manual and electric firing controls (par. 79), noticing particularly if safety devices are functioning. If optical glass is cleaned, observe the following:

Table V. Commander's "B" and "C" Preventive-Maintenance Services—Con.

Commander's "C" (quarterly or 750 miles)	Commander's "B" (monthly or 250 miles)	Procedure
		Caution: Optical glass is easily scratched. Use only a camel's hair brush to remove dust. Clean only with lens tissue. Breathing on the glass before wiping is permissible. If the glass is greasy or oily, apply lens soap only, and wipe off with lens tissue.
16	----	Tighten all gun-mount assembly and mounting bolts.
17	17	*Stowage boxes—ammunition stowage, racks.* Inspect these items, observing particularly if ammunition is securely and safely stowed, or if ammunition racks and stowage boxes are empty, observe if they are ready for use.
18	18	*Towing shackles, lifting eyes, pintle, towing cables.* Inspect these items. Observe if pintle hook locks satisfactorily.
19	19	*Lubrication—lubricate vehicle in accordance with LO (par. 113).* Lubricate vehicle and armament in accordance with lubrication orders (par. 113). Exclude those items that are to be disassembled later and lubricated before reassembly.
20	20	*Oil consumption.* Check whether or not the engine has been consuming an excessive amount of oil. Maximum allowable oil consumption for an engine operating at 2,800 rpm is 2 gallons in 1½ hours (par. 126).
		MAINTENANCE OPERATIONS
21	21	*Battery—specific gravity.* Make a hydrometer test of the electrolyte in each cell (par. 224) and record specific gravity in space provided on DA AGO Form 462.
22	22	*Battery—voltage.* Perform a high-rate discharge test according to instructions accompanying the test instrument. Record voltage of each cell in space provided on DA AGO Form 462.
22	----	After battery tests, clean top of batteries, coat terminals lightly with grease, and repaint carrier, if corroded. Look to see if battery requires water. *Note.* If distilled or approved water is not available, clean water, preferably rainwater, may be used.
23	----	*Engine compression.* Test compression in each cylinder, with one spark plug out of each cylinder and engine warm. Insert the compression gage in a spark plug hole, and with throttle wide open, revolve the engine at cranking speed until the maximum compression is indicated. Normal compression should be from 90 to 110 psi. Variation of 10 to 15 psi between cylinders is not ordinarily detectable in performance. Record readings in space provided on DA AGO Form 432. Repeat this process for each cylinder.
24	----	*Fuel line filter, bilge-pump strainers, oil filters, air cleaners.* Clean fuel line filter (par. 176), bilge-pump strainers (par. 329), engine oil filter (par. 149), and transmission oil filter (par. 238). Clean and service air cleaners (par. 175). If fuel filter shows signs of contaminated fuel, drain water and sediment from fuel tanks (par. 177), using suitable containers to catch the drainings.

Table V. *Commander's "B" and "C" Preventive-Maintenance Services*—Con.

Commander's "C" (quarterly or 750 miles)	Commander's "B" (monthly or 250 miles)	Procedure
27	----	*Spark plugs, booster and filter coil, wiring, timing, magnetos.* Remove and inspect spark plugs (par. 186). Inspect magneto covers, cams, and breaker points. Test booster coil, and capacitor with ohmmeter (par. 129) or with high-tension ignition-circuit tester according to instructions accompanying test instrument. Observe if timing is correct (par. 184). Test generator regulator with low-voltage circuit tester, following instructions accompanying test instrument.
27	----	Clean spark plugs and adjust gap (0.011–0.014 in.). Dress magneto breaker points and adjust gap (0.018 in.). If points are badly pitted, replace both points and capacitor (condenser) (par. 185).
28	----	*Valve mechanism—clearance.* Inspect valve mechanism, looking particularly for weak or broken valve springs. Gage valve clearance (par. 155). Inspect valve rocker covers.
28	----	Adjust valve clearance to 0.007 inch for intake valve rockers and 0.020 inch for exhaust valve rockers (par. 155).
29	----	*Carburetor, fuel pumps, carburetor control linkages.* Inspect these items, noticing particularly if carburetor shafts and linkage operate freely and are not excessively worn. Observe if the throttle valve opens fully when the accelerator is fully depressed. Note particularly if the throttles are synchronized. Make an engine vacuum test as follows: with the engine compartment grille doors open, attach a vacuum gage to the intake manifold. Run the engine at 650 rpm. The vacuum gage should read between 11 to 14 inches (of mercury), and the pointer should be steady. A needle fluctuating between readings of 10 and 15 may indicate a leak in the intake manifold or intake manifold gaskets. If carburetors are out of balance, one cylinder bank can read high vacuum and the other low. Momentarily accelerate engine with full throttle. The gage indicator should drop to approximately two inches as the throttle is opened quickly and recoil to at least 24 inches as the throttle is closed. If gage fails to recoil to at least 24 inches, these may be the reasons: diluted oil; poor piston ring sealing; or an abnormal restriction in the exhaust system, carburetor, or air cleaner. Repeat test on the manifold for the other bank of six cylinders. *Note.* The above readings apply at sea level. There will be approximately a 1-inch drop for each 1,000 feet of altitude. Adjust idle mixture (par. 172). Test fuel-pump pressure (par. 174).
30	----	*Exhaust pipes and mufflers.* Inspect; listen for excessive or unusual noises and look for exhaust leaks usually indicated by carbon streaks.
30	----	Tighten mountings (par. 170).
31	----	*Manifolds.* Inspect these items looking particularly for signs of leakage at the manifold gaskets.

Table V. Commander's "B" and "C" Preventive-Maintenance Services—Con.

Commander's "C" (quarterly or 750 miles)	Commander's "B" (monthly or 250 miles)	Procedure
32	----	*Brakes—adjustment, linkage.* Inspect these items observing particularly cross shafts, linkage, and hand brake lock handle.
32	----	Adjust service brakes and linkage (par. 259), if need was indicated in the road test. Tighten assembly and mounting bolts.
33	----	*Engine fans, shrouds, and cylinder air deflectors.* Inspect these items looking particularly for interference and looseness.
34	----	*Track tension.* Test track tension (par. 249).
34	----	Adjust track tension (par. 249), if it is not within permissible limits. Using a torque-indicating wrench tighten all track wedge nuts to 120–140 lb-ft. Tighten all assembly and mounting bolts of the suspension components. Whenever the track has been removed for replacement or repair, examine road wheels, drive sprockets, idlers, and support rollers for end play and bearing looseness or damage. Spin each wheel and listen for a damaged bearing. Pull outward and push inward on the wheel to determine if end play is excessive. Test for bearing looseness with a pry bar. Test for end play and bearing looseness or damage should be performed at least every third quarter. If the tracks have not been removed within that time, these tests can be performed without removing the tracks by the use of road-wheel lifters, jacks, and blocks. Adjust or replace bearings as necessary (pars. 250 and 251).
35	----	*Electrical controls, wiring, slip ring box, suppressors.* Inspect all exposed electrical control, junction, and terminal boxes and connecting wiring, cables, and conduits. Inspect all visible radio-noise-suppression ground straps, suppressors, and shields and radio mountings, radio controls, headsets, microphones, antenna mast and insulators.
35	----	Clean these items with a dry, soft cloth. Test the operation of the slip ring box (par. 296). If objectionable radio noise from the vehicle has been reported, make tests (par. 142) to determine the source. Clean contacting surfaces, tighten bonds, internal-external-toothed lock washers, and mountings of units bonded together. Replace noise-suppression units as required. If these procedures do not eliminate the trouble, the radio operator will report the condition to the designated authority.
36	36	*Oil coolers (engine and transmission).* Examine the oil coolers (figs. 130 and 131), including their cores and connecting lines, to see that they are in good condition, secure, and do not leak. Also check to see that the core air passages are not clogged with dirt and trash. Tighten all mounting bolts and hose connections securely.
36	----	Service oil coolers according to instructions on lubrication order (par. 113).

Table V. Commander's "B" and "C" Preventive-Maintenance Services—Con.

Commander's "C" (quarterly or 750 miles)	Commander's "B" (monthly or 250 miles)	Procedure
37	----	*Power plant removal (when required).* Remove power plant (par. 159), only if inspections made in items 4, 8, 13, 20, and 23, above indicates a definite need. Examine generator brushes and commutator (par. 222). Clean engine compartment and exterior of power plant, with volatile mineral spirits or dry-cleaning solvent, using care to keep these solvents away from electrical wiring and equipment. Hot water and soap, which is not harmful to insulation, should be used when available. Remove and clean fuel tank filters (par. 176). Inspect engine mountings (fig. 222) and mounting plates (fig. 133) to see that they are in good condition and secure.
38	38	*Modifications and final road test.* Check WD AGO Form 478 to determine whether all modification work orders have been completed. A list of current modification work orders is contained in SR 310-20-4. Enter any modifications or major unit assembly replacements during this service on WD AGO Form 478. Make a final road test, rechecking items 1 through 9. Confine this road test to the minimum distance necessary to make satisfactory observations.

Note. Correct or report all deficiencies found during final road test.

UNUSUAL CONDITIONS

Maintenance operations and road tests as prescribed under usual conditions will apply equally under unusual conditions for operations for all for all occasions except in extreme-cold weather. Intervals are necessarily shortened in extreme-cold weather servicing and maintenance. Vehicles subjected to salt-water immersion or complete submersion are evacuated to ordnance maintenance unit as soon as possible after the exposure.

Section IV. TROUBLE SHOOTING

124. Scope

a. This section contains trouble shooting information and tests for locating and correcting some of the troubles which may develop in the vehicle. Trouble shooting is a systematic isolation of defective components by means of an analysis of vehicle trouble symptoms; testing to determine the defective components, and applying the remedies. Each symptom of trouble given for an individual unit or system is followed by a list of probable causes of the trouble and suggested procedures to be followed.

b. This manual cannot cover all possible troubles and deficiencies that may occur under the many conditions of operation. If a specific trouble, test, and remedy therefor are not covered herein, proceed to isolate the system in which the trouble occurs and then locate the defective component. Use all the senses to observe and to locate troubles. Do not neglect use of any test instruments such as a 24-volt circuit tester, ohmmeter, voltmeter, ammeter, test lamp, hydrometer, and pressure and vacuum gages that are available (par. 48). Standard automotive theories and principles of operation apply in trouble shooting the vehicle. Question vehicle crew to obtain maximum number of observed symptoms. The greater the number of symptoms of troubles that can be evaluated, the easier will be the isolation of the defect.

125. Ohmmeter Method of Electrical Testing

a. GENERAL. The ohmmeter method of testing electrical circuits and devices is simply the use of continuity tests to determine whether the circuit or device being tested has a continuous electrical path through the cables and units connected between the two test points. An ohmmeter indicates, on a calibrated scale, the resistance of the circuit being tested. It is equipped with a source of power (battery or hand generator), usually installed inside the case which houses the meter.

Caution: Never attempt to make ohmmeter tests until all sources of power connected to the equipment to be tested are disconnected. The ohmmeter will be ruined if this procedure is not followed. All electrical circuits have some resistance. Some resistances are so low and others so high that they cannot be read conveniently with an ordinary ohmmeter. An ohmmeter with a full-scale reading of about 10 ohms is desirable for measuring low resistances. Higher range ohmmeters are better suited for testing insulation leaks. If the normal resistance of the circuit to be tested is known, select an ohmmeter full-scale range higher than the normal resistance.

b. CONTINUITY TESTS. The two test prods or clips from the ohmmeter are placed on the end points of the circuit being tested, or to the two terminals at which the ends of the component to be tested terminate. If a reading is obtained, the circuit has no breaks or opens.

c. RESISTANCE READING. A resistance reading is taken to determine the electrical condition of a coil, resistor, capacitor, or complete circuit. When the correct resistance of a unit or circuit is known, a resistance test will indicate, after interpretation of the readings, the fault in the circuit (fig. 94).

d. TERMINAL AND GROUND DESIGNATIONS. The following system of terminal designations is used in the tabular matter for electrical tests: "R5" indicates "receptacle 5;" "R2–D" indicates "receptacle 2, terminal D;" "P5" indicates "plug 5;" "P2–D" indicates "Plug 2, terminal

161

COMPONENTS AND CIRCUITS	CONDITION	OHMMETER READING
 (resistor)	NORMAL WINDING OR RESISTANCE	NORMAL
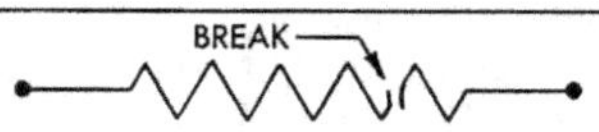 (resistor with BREAK)	OPEN	INFINITE
 (resistor with poor connection)	POOR CONNECTION	HIGHER THAN NORMAL
 (resistor with partial short)	PARTIAL SHORT	LOWER THAN NORMAL
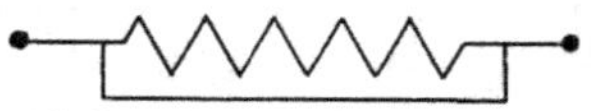 (resistor with complete short)	COMPLETE SHORT	ZERO
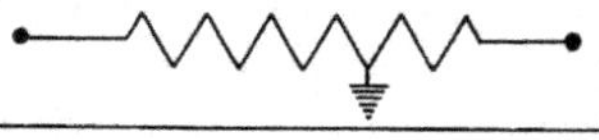 (resistor with ground)	GROUNDED WINDING	NORMAL FROM TERMINAL TO TERMINAL. LOWER FROM EITHER TERMINAL TO GROUND
 (capacitor)	NORMAL CAPACITOR	INFINITE
 (leaky capacitor)	LEAKY CAPACITOR	FROM A FEW OHMS UPWARD
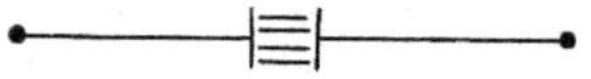 (shorted capacitor)	SHORTED CAPACITOR	ZERO
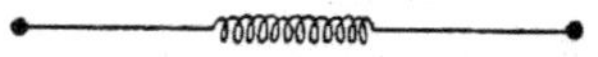 (coil)	NORMAL COIL	NORMAL
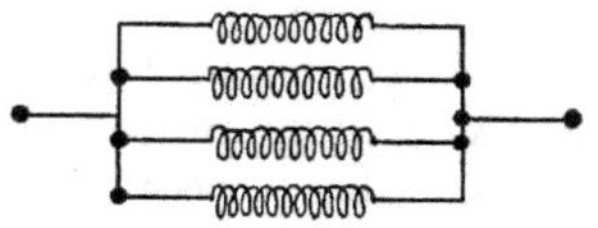 (four parallel coils)	FOUR PARALLEL EQUAL-VALUED COILS	¼ READING OF ONE COIL
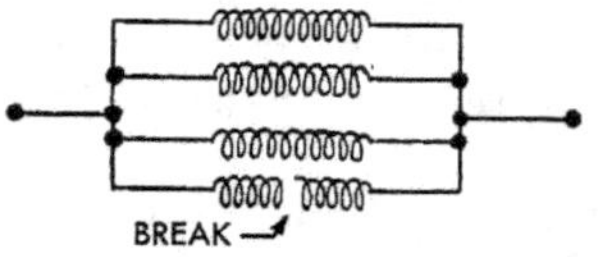 (four parallel coils, one broken)	FOUR PARALLEL EQUAL-VALUED COILS, ONE OPEN	⅓ READING OF ONE COIL

Figure 94. Trouble shooting with an ohmmeter.

D." See figure 195 for location of connector receptacles. Ground (gd) means any metal part of vehicle which has a good electrical connection with the unit or circuit being tested.

   e. INFINITE AND ZERO OHMMETER READINGS.

   (1) *Infinite reading.* An infinite reading is an open circuit reading. There will be no movement of the ohmmeter hand when an open circuit (infinite resistance) exists. Infinity position on the ohmmeter scale is usually marked by the symbol "∞" or abbreviation "INF."

   (2) *Zero reading.* A zero reading indicates a closed circuit with no resistance in the circuit being tested.

126. Main Engine

   a. GENERAL.

   (1) Ignition, starting, fuel, lubrication, and oil-cooling systems are all regarded as part of the engine, but are treated separately in this section to simplify procedures. Trouble in any one of the systems will be reflected in engine performance.

   (2) When trouble shooting the engine, or one of the systems mentioned in (1) above, it probably will be necessary to remove the top deck (par. 158*b*), or portions thereof, to get at components of the engine or the various systems. The units on the accessory end of the engine, such as oil pressure regulator, magnetos, fuel pumps, etc., may be made accessible by removing air cleaners (par. 175) and removing bulkhead covers in back of air cleaners. To gain access to main engine generator, filter by-pass valve, fuel tank drains, and engine oil drain, remove access hole covers on bottom of vehicle hull (fig. 285).

   b. ENGINE FAILS TO CRANK WHEN STARTER SWITCH IS HELD ON.

   (1) *Master relay switch not turned on.* Turn on switch (fig. 18).

   (2) *Manual control lever not in neutral-shift position.* Move manual control lever to neutral position (fig. 10).

   (3) *Fault in starting system.* Refer to paragraph 130.

   (4) *Fault in battery system.* Refer to paragraphs 132 and 224.

   (5) *Mechanical seizure of engine parts.* Notify ordnance maintenance personnel.

   (6) *Incorrect oil viscosity.* Drain and refill (par. 148) with correct grade of oil, as specified on lubrication order (fig. 86).

   c. ENGINE CRANKS BUT FAILS TO START.

   (1) *Preliminary instructions.* Make certain magneto switch is at "BOTH" position, magneto booster switch is on, fuel-tank-shut-off valves are open and there is fuel in the fuel tanks. Follow the procedure below for whichever conditions are applicable.

(a) *Engine is flooded.* If engine is flooded due to overmanipulation of the primer pump or accelerator pedal, turn off magneto switch, push the accelerator pedal down to hold the throttle wide open, and crank the engine intermittently for several seconds to exhaust the surplus fuel. Do not use primer pump.

(b) *Engine is extremely cold.* If engine is extremely cold, additional use of the primer pump is required. The primer pump should operate with a resistance of approximately 10 psi. If the handle goes in easily, it indicates broken primer pump lines, defective primer pump, or clogged fuel lines to the primer pump. If resistance is high, it indicates defective primer pump or clogged fuel lines from the primer pump. Make repairs as required (pars. 180 and 181).

(c) *Engine cranks slowly.* Check battery charge (par. 224). If battery charge is low, replace battery, or run auxiliary engine until batteries are charged. If improper engine oil is being used, drain and replace with proper oil for prevailing temperature, as prescribed on the lubrication order (fig. 86). If cranking speed is still slow, follow procedures under symptom outlined in paragraph 130.

(d) *Engine still fails to start.* If engine still fails to start after the above procedures have been followed, observe the procedures outlined below until the trouble has been found and corrected.

(2) *Fuel not reaching carburetor.* Disconnect the carburetor inlet flexible fuel line at either carburetor (figs. 130 and 131).

Caution: Use a suitable container to catch the fuel. Make sure magneto switch (fig. 18) is off. Crank engine a few revolutions with the starter. If a free flow of fuel is discharged from the pump, fuel is being delivered to the carburetor. If no fuel is discharged, fuel is not reaching the carburetor. Proceed as directed under symptom outlined in paragraph 128.

(3) *Current not reaching spark plugs.*

(a) *General.* Ignition trouble usually can be disregarded as a cause for the engine not starting as each cylinder bank has dual ignition and received its ignition voltage from two separate magnetos, with separate ground wires from magnetos to magneto switch. It is unlikely that the four magnetos for the two cylinder banks would fail at the same time. If the magneto drive gears are damaged, however, complete ignition failure will occur.

(b) *Procedure.* Disconnect the spark plug cable from the flywheel end of one cylinder on each bank. (Flywheel-end spark plugs are fed from bottom magnetos and boosters.) Turn the main magneto switch (fig. 18) to the position marked "BOTH." Hold the loose end of the spark plug cable approximately three-sixteenths of an inch from the cylinder block, hold the magneto booster switch (fig. 18) in the "ON" position, and crank the engine with the starter.

Caution: Do not touch the metal end of the wire when making this test. Observe if a spark jumps the gap. Perform this same test for both spark plug cables. If the spark jumps the gap on both cables, ignition may be dismissed from further consideration. If no spark is observed at either cable, remove the accessory drive shaft opening cover (fig. 130), at the magneto drive gears, and again crank the engine. Observe if the magneto drive gears are stripped. Report damaged gears to ordnance maintenance personnel.

d. ENGINE STARTS BUT FAILS TO KEEP RUNNING. Engine may start and run on fuel remaining in carburetor bowls, and stop when the fuel in bowls is exhausted. Operate primer pump several times and start engine again. If engine starts each time the primer pump is used, but will not continue running, it indicates that fuel is not reaching the carburetors. Proceed as directed under symptom outlined in paragraphs *h* (7) below and 128 *b*.

e. ENGINE DOES NOT DEVELOP FULL POWER.

(1) *Preliminary instructions.* The many different factors that may cause loss of power make it advisable to perform the quarterly preventive-maintenance service (par. 123) before proceeding further. If this service recently has been performed and the trouble was not eliminated, proceed as follows, omitting consideration of those factors known to be right.

(2) *Procedure.*

(*a*) *Check for cylinders missing at idle speed.* Remove the cable from number 1 spark plug (par 186), start the engine, and hold the wire three-sixteenths of an inch from the cylinder head.

Caution: Do not touch the metal end of the wire when making this test. Observe if the spark jumps the gap regularly without missing. Reconnect the wire and repeat this test at each spark plug. Follow whichever of the following conditions that applies.

1. *Satisfactory spark from all wires.* If a satisfactory spark is obtained from all wires, proceed with *3* below.

165

2. *Unsatisfactory spark.* If an unsatisfactory spark is obtained from one or more wires (a spark that will not jump a 3/16-inch gap, an irregular spark, or no spark), the magneto or spark plug cables for the side on which the missing occurs are at fault. Refer to paragraph 129.
3. *Inspect spark plugs and test compression.* Refer to paragraph 123, items 12 and 23, table V.

(b) *Check ignition timing.* Check ignition and time magneto if required (par. 184).

(c) *Check fuel mixture.* If the above inspections and tests do not locate the cause for lack of power, the trouble may be in a lean fuel mixture. Proceed as directed under symptom in paragraph 128c.

(d) *Additional possible causes of loss of engine power.*
1. *Brakes dragging.* Service brakes adjusted too tightly will cause sluggish vehicle action. If the vehicle tends to lead in either direction, it may be an indication of improper brake adjustment. Adjust brakes if required (par. 259).
2. *Improper engine valve timing.* No appreciable change in engine valve timing will occur during operation. Therefore, incorrect valve timing can occur only from improper assembly at the time of previous repairs. Check valve timing (par. 155).

f. EXCESSIVE OIL CONSUMPTION.

Note: Normal oil consumption for a new engine under usual operating conditions is five quarts per hour.

(1) *Improper grade of oil in engine.* Be sure oil being used is of correct viscosity for prevailing atmospheric temperatures as specified on lubrication order (fig. 86), and that oil changes are being made at specified intervals.

(2) *External leaks.* If excessive oil is seen on the floor of the engine compartment, an external leak is indicated. Examine for leaks at points accessible from the engine compartment, and make necessary repairs. If the leak is at the oil pan gasket, removal of the power plant from the vehicle will be necessary (par. 159).

(3) *Worn internal components.* If external leaks are not present, and proper grade of engine oil is being used, it may be assumed that internal components of the engine, piston rings, cylinder walls, bearings, etc., are worn or damaged. Notify ordnance maintenance personnel.

g. LOW OIL PRESSURE. Check engine oil level (par. 148b) and replenish as specified in lubrication order (fig. 86). If the oil has become overdiluted, it should be drained and replaced (par. 148).

If the oil is known to be satisfactory and oil level is correct, follow procedure under symptom outlined in paragraph 127a.

h. ENGINE OVERHEATS. Proper engine temperature is indicated by engine-high-oil-temperature-warning-signal light (fig. 18).

(1) *Improper grade or level of oil in engine.* Check grade and level of oil for prevailing atmospheric temperatures (par. 148). Drain and refill with proper grade, if required. Add oil as required.

(2) *Engine cooling fans operating improperly.* Run engine and observe action of two cooling fans on top of engine (fig. 96). Replace fans as required (par. 154).

(3) *Oil coolers clogged or fans inoperative.* With engine running, insert oil coolers, fans, and fan drives (figs. 130, 131, and 132). Clean or replace as required (par. 242).

(4) *Defective engine oil pump.* Check oil pressure. If improper, proceed as outlined under symptom in paragraph 127d.

(5) *Vehicle improperly operated.* Operate vehicle and engine as directed in chapter 2, section III.

(6) *Engine shrouds improperly installed.* The shrouds and hinged cover plates (figs. 96 and 170) direct the flow of air around engine. Check installation of shrouds.

(7) *Air induction system obstructed.* Check the air cleaner inlets to make sure they are free from obstructions. Disconnect air cleaner pipes (fig. 147) and inspect the pipes and carburetor intake elbows. Remove any materials which prevent a free flow of air.

127. Engine Lubrication System

a. LOW OIL PRESSURE. Low oil pressure will be indicated on the oil pressure gage and by the low-oil-pressure-warning-signal light (fig. 18).

Caution: Do not operate engine if the oil pressure is low; serious damage will result.

(1) *Improper lubrication oil.* Dilution of engine oil by overpriming, or as a result of too infrequent oil changes, will cause low oil pressure. It also can be caused by using a lighter viscosity oil than specified on lubrication order for prevailing atmospheric temperatures (figs. 86 and 87). Oil that is too heavy to flow in cold weather will also cause the oil pressure to be low when engine is started. Inspect engine oil while engine is cold and again when hot. Drain and refill if necessary (par. 148).

(2) *Defective oil pump.* If oil pump is believed to be defective, it can be checked only by removing pump and replacing it

with one known to be good. Notify ordnance maintenance personnel.
- (3) *Worn or burned-out connecting rod bearings.* Worn or burned-out connecting rod bearings will cause a low pressure reading on the gage. Defective bearings generally can be determined by an excessive rattling or knocking noise in the engine. If this condition is recognized, stop engine immediately and notify ordnance maintenance personnel.
- (4) *Defective oil pressure gage or low-oil-pressure-warning-signal-light switch.* Do not operate engine if either the gage or the warning signal light, or both, indicate low oil pressure. To check the low-oil-pressure-warning-signal-light switch, refer to paragraph 131. If defective, replace with a new one (par. 218). If oil pressure gage shows an incorrect reading after making any necessary corrections as outlined in (1), (2), or (3), above, check gage (par. 131). If defective, replace the gage (par. 205). If gage and warning-signal-light switch are correct, and warning signal light does not light when oil pressure is low, replace lamp in warning signal light (par. 206).
- (5) *Foreign material on oil-pressure-control-valve seat.* Remove and inspect oil-pressure-control valve to determine if the valve has foreign material on its seat. Clean or replace the valve as necessary (par. 150).

b. Excessive Oil Pressure.
- (1) *Incorrect grade of engine oil.* Excessive oil pressure may result from use of improper oil for prevailing atmospheric temperatures. Drain and refill (par. 148) with grade specified on lubrication order (fig. 86).
- (2) *Improperly adjusted or defective oil-pressure-control valve.* Adjust or replace valve (par. 150).
- (3) *Defective oil pressure gage.* Replace oil pressure gage (par. 205) if engine oil grade is correct for prevailing atmospheric temperatures and excessive oil pressure reading continues.

c. Excessive Oil Consumption.
- (1) *Oil leaks.* Examine engine compartment floor for evidence of oil leaks. Correct the cause of any leaks discovered or notify ordnance maintenance personnel.
- (2) *Incorrect grade of engine oil.* Oil having too low viscosity will be consumed rapidly. Drain and refill (par. 148) with grade specified on lubrication order (fig. 86) for prevailing atmospheric temperatures.
- (3) *Worn or scored pistons and cylinders.* Excessive oil consumption will result if pistons, rings, or cylinders are worn. This condition is indicated by a smokey exhaust, loss of power, and fouled spark plugs. If this cause is suspected,

perform a compression test for each cylinder (par. 123, item 23). If compression pressure is low or uneven between cylinders, notify ordnance maintenance personnel.

d. No Oil Pressure.

Caution: Stop engine immediately if no oil pressure is indicated within 10 seconds after starting engine.

  (1) *No oil in engine.* Check oil level (par. 148). Fill with proper grade of engine oil (par. 148), as specified on lubrication order (fig. 86), if level is incorrect.

  (2) *Clogged oil tubes or defective oil pump.* If the engine oil has been checked and found to be adequate and of correct viscosity, it can be assumed the trouble lies in a defective pump or clogged lines. Notify ordnance maintenance personnel.

e. High Oil Temperature. If the engine oil temperature, as indicated by high-oil-temperature-warning-signal light, is excessive, follow the procedures outlined in paragraph 126*h*.

128. Fuel System

a. General. Most fuel system and carburetion troubles will effect engine performance. Therefore, the symptoms listed below are established during main engine trouble-shooting procedures (par. 126).

b. Fuel Not Reaching Carburetor.

  (1) *Check fuel-tank-shut-off valves and fuel in tanks.* Be sure fuel-tank-shut-off valves are open (par. 17), and that there is fuel in the fuel tanks (par. 32).

  (2) *Check fuel shut-off switch.* Check the fuel shut-off (degasser) switch for a short (par. 131). If shorted, the carburetor degassers will operate to shut-off the flow of fuel to the carburetors. If defective, replace main engine magneto switch (par. 204).

  (3) *Check fuel lines from fuel pumps to carburetor.* Check the fuel-pump-to-carburetor lines to insure that they are free from obstructions. Replace fuel lines if necessary (par. 179).

  (4) *Check fuel lines from tank to fuel pumps (compressed air available).* Disconnect fuel line from inlet side of fuel pumps (fig. 144). Place suitable container at tube end to catch the fuel. Traverse turret, as required, to reach and open filler covers at rear of vehicle (fig. 154). Open fuel filler cap and insert compressed air hose, using clean rags to close opening around hose. Open compressed air valve slowly and check to see if fuel flows from open end of tube at fuel pump. Proceed according to applicable step below.

   (*a*) *Fuel flows from open end of tube at fuel pump.* If fuel flows from fuel line when compressed air is applied to fuel tank filler neck, check fuel cut-off (degasser) switch for a short (*b*(2) above) and fuel lines from fuel pumps to carburetor for obstruction (*b*(3) above). If these units are in good condition, the fuel pumps are defective and are the cause of fuel not reaching the carburetor.

>*Note.* Fuel pumps are interconnected so either pump can deliver fuel to both carburetors.

Replace fuel pumps (par. 174).
- (*b*) *Restricted tube or no fuel flows from open end of tube at fuel pump.* If fuel does not flow from fuel line when compressed air is applied, the fuel tank filter, fuel lines, or fuel line filter are damaged or clogged, or the fuel-tank-shut-off valves did not open when the controls are set at "OPEN" position. Replace or clean as required (pars. 176, 178, and 179).
- (5) *Check fuel lines from tank to fuel pump (compressed air not available).* Remove floor plate at rear of fighting compartment and set fuel-tank-shut-off valve to "CLOSE" position (fig. 11). Reach through opening and disconnect fuel line to main engine at the valve outlet. Place a container at opening to catch fuel and open one shut-off valve. If there is a flow of fuel, either the fuel pumps, fuel-line filter, or the fuel line from the shut-off valves to the pumps is defective or clogged. Replace the filter, fuel line, or the pumps (pars. 174, 176, and 179), as necessary.
- (6) *Fuel leaks.* Examine for leaking or broken fuel lines (par. 179). Replace as required.

c. FUEL MIXTURE TOO LEAN.

- (1) *Adjust idle screws on carburetors.* Adjust idle screws (par. 172).

  >*Note.* Idle screw adjustment will have little effect on engine performance above 650 rpm.

- (2) *Test fuel pump pressure.* Check the fuel pump pressure (par. 174). If the pressure is less than $4\frac{1}{2}$ psi, replace the fuel pump (par. 174). If the pressure still is below $4\frac{1}{2}$ psi after a new pump is installed, it indicates the fuel lines or fuel filters are restricted. Replace as required (pars. 176 and 179).
- (3) *Check for air leaks.* Check the carburetor and hot spot manifold flange gaskets for air leaks (fig. 143). Tighten mounting screws or replace gaskets, if required (pars. 172 and 173). If all gaskets are satisfactory and mounting screws are tightened, continued lean fuel mixture indicates the carburetors are at fault. Replace one or both carburetors, as required.

d. CARBURETOR FLOODS.
  (1) *Overmanipulation of accelerator pedal or primer pump.* Overmanipulation of accelerator pedal or primer pump when engine is not running is the most common cause for carburetor flooding and will prevent the engine from starting until the excess fuel is removed. Follow procedure outlined in paragraph 128. Flooding from other causes will cause the engine to run unevenly and to have a tendency to stall when idling. A strong odor of gasoline usually is present and black smoke is emitted from the exhaust.
  (2) *Check fuel pump pressure.* Check fuel pump pressure (par. 174). If pressure is greater than specified limit of 6 psi, check outlet pressure of each pump and replace the defective pump.
  (3) *Check carburetor float lever.* Check carburetor float lever (par 172). If fuel level is too high, it indicates the float mechanism is adjusted improperly or the needle valve seat is faulty. Replace carburetor. If carburetor still floods after all the above checks have been made, the carburetor is at fault and must be replaced.

e. ENGINE SLOWS DOWN BUT DOES NOT STOP WHEN THE FUEL CUT-OFF (DEGASSER) SWITCH IS PRESSED. This will indicate that some but not all of the degassers are operating effectively. To isolate which carburetor degassers are defective, disconnect cables at right degassers (K, of fig. 131). If engine stops, the left carburetor degassers are working. Repeat the operation for the right carburetor degassers, by disconnecting cable at left degassers (B, of fig. 130). If found to be defective, notify ordnance maintenance personnel.

f. OPERATION OF FUEL CUT-OFF (DEGASSERS) SWITCH HAS NO EFFECT ON ENGINE. This indicates that the degasser switch or instrument circuit breaker is defective, since it is not likely that all four degassers will become defective at the same time. Check switch and circuit breaker (par. 131) and replace as necessary (pars. 204 and 214).

129. Ignition System

a. GENERAL.
  (1) Engine performance affected by ignition system troubles is treated in paragraph 126. The following procedures provide test and analysis methods of specific ignition troubles.
  (2) Each cylinder bank is provided with dual ignition and receives its ignition voltages and currents from two separate magnetos. The 2 magnetos on the right side deliver current to the 12 spark plugs in the 6 cylinders of the right cylinder bank, and the 2 magnetos on the left side deliver current to

the 12 spark plugs of the left cylinder bank. The lower magneto on each side furnishes current to the spark plugs at the flywheel ends of the six cylinders on their respective sides, and the upper magneto furnishes current to spark plugs at the accessory ends of the six cylinders. A separate ground wire is provided from each magneto to the magneto switch. The two lower magnetos are assisted by booster coils. These magnetos will provide a spark while the engine is being cranked as long as the booster switch is on. The two upper magnetos are not aided by booster coils and will provide a spark only when engine speed is above 300 rpm. Above 300 rpm all spark plugs can be checked without use of the booster coils.

b. No Spark or Unsatisfactory Spark from Spark Plug Cables (Engine Cranked).

    (1) *Preliminary instructions.* Open spark plug hinged cover plates (fig. 170). Loosen cable nuts from spark plugs on flywheel ends of all cylinders on both right and left cylinder banks (par. 186). Turn magneto switch (fig. 18) to "BOTH" position. Hold booster and starter switches on "ON" position. Hold each wire three-sixteenths of an inch from cylinder block and observe if spark occurs.

    (2) *Procedure.*

        (*a*) *No spark from spark plug cables at one cylinder bank.* If no spark is obtained from all flywheel-end spark plug cables at one cylinder bank, remove ground wire from lower magneto serving those cables. Start the engine. If a spark now is obtained, it indicates the cable leading from magneto to magneto switch is faulty. Make corrections or replacement as required. If ignition is not corrected by removal of ground wire, it indicates that the magneto or ignition harness to spark plugs is at fault. Adjust or replace breaker points or replace magneto (par. 185), or replace ignition wiring harness (par. 187), whichever is required.

        (*b*) *No spark from cables at both cylinder banks.* If no spark is obtained from all flywheel-end spark plug cables of both cylinder banks, the most probable causes are a faulty magneto switch or booster switch, disconnected magneto switch and booster switch cables or cable connectors, or damaged magneto drive gears. Check switches and cables (*f* below). Check connectors in engine wiring junction box. Replace defective units as required. If these units are in good condition, remove accessory drive gear cover (k, of fig. 130) and examine magneto drive gears. If damaged, notify ordnance maintenance personnel.

c. No Spark or Unsatisfactory Spark from Spark Plug Cables (Engine Running).

(1) *Preliminary instructions.* Open spark plug hinged cover plates (fig. 170). Loosen spark plug cable nuts and remove cables from spark plugs on accessory ends of all cylinders on both banks. Turn magneto switch to "BOTH" position. Start engine. Hold each wire three-sixteenths of an inch from cylinder block and observe spark.

Caution: Do not touch metal end of wire making this test.

(2) *Procedure.*

(a) *No spark from cables at one cylinder bank.* If no spark is obtained from all accessory-end spark plug cables of one cylinder bank, remove ground wire from upper magneto serving these cables. Start the engine. If a spark is now obtained, it indicates that the ground wire extending from the magneto to the magneto switch is grounded, or the magneto switch is faulty. Make corrections or replacements as required. If trouble is not corrected by removal of the ground wire, it indicates that the magneto or the ignition harness to the spark plug is at fault. Adjust or replace breaker points (par. 185), replace magneto (par. 185), or replace ignition harness (par. 187), whichever is required.

(b) *No spark from cables at both cylinder banks.* If no spark is obtained from all accessory-end spark plug cables at both cylinder banks, the most probable cause is damaged magneto drive gears. Remove accessory drive shaft opening cover (k, of fig. 130) and examine gears. If damaged, notify ordnance maintenance personnel.

d. Irregular or Unsatisfactory Spark from Spark Plug Cables (Engine Cranked or Running). If an irregular or unsatisfactory spark (spark that will not jump $3/16$-inch gap) is obtained from all six spark plug cables at either the accessory end or flywheel end of the cylinder heads in one bank, the magneto serving those plugs is at fault. Examine magneto breaker points for pits or other conditions which would make them unserviceable (par. 185). Replace defective points. Check and correct breaker point adjustment if necessary. If a satisfactory spark still cannot be obtained, replace magneto (par. 185).

e. No Spark from Some Spark Plug Cables and a Satisfactory Spark from Others (Engine Cranked or Running). If a satisfactory spark is obtained from some cables, and no spark or an unsatisfactory spark is obtained from related others at either the flywheel or accessory end of a cylinder head, faulty spark plug cable or magneto distributor cap is indicated. Replace ignition harness (par. 187). If this does not correct the trouble, replace the magneto (par. 185).

Component	Circuit No.	Test From—	Test To—	Resistance (ohms)	Circuit (figs. 161 and 195)	Remarks
Booster, left	12A	P-9B	gnd	2.9	From P9-B through booster coil, through filter coil, through left lower magneto primary to ground.	0 ohms indicate a grounded cable between plug and booster coil. A reading of 1 ohm indicates a shorted booster coil.
Booster, right	12	P8-A	gnd	2.9	From P8-A through booster coil, through filter coil, through right lower magneto primary to ground.	0 ohms indicates a grounded cable between plug and booster cable. A reading of 1 ohm indicates a shorted booster coil.
Capacitors, filter						See magnet tests below.
Coils, filter						See magneto tests below.
Magneto, left lower	13A	P9-C	gnd	0.9	From P9-C through filter coil through magneto primary to ground.	
Magneto, left upper	13	P9-A	gnd	0.9	From P9-A through filter coil through magneto primary to ground.	0 ohms indicates grounded cable or shorted filter capacitor(s). Infinity indicates broken filter coil or damaged magneto.
Magneto, right lower	11A	P8-C	gnd	0.9	From P8-C through filter coil through magneto primary to ground.	
Magneto, right upper		P8-B	gnd	0.9	From P8-B through filter coil through magneto primary to ground.	
Switches, magneto, booster, and starter.						Refer to paragraph 131.

f. RESISTANCE AND CONTINUITY TESTS. Refer to paragraph 125 for methods of testing with an ohmmeter.

Caution: Disconnect connector No. 11 from instrument panel before making tests, otherwise ohmmeter will be ruined.

130. Starting System

a. ENGINE WILL NOT CRANK WHEN STARTER SWITCH IS TURNED ON.
   (1) *Check neutral position switch.* Be sure manual control lever (fig. 10) is in neutral position.
   (2) *Check batteries and connections.* Perform battery preventive maintenance service table III (par. 122).
   (3) *Check master relay switch, master relay, and master relay rectifier.*
      (*a*) Be sure master relay switch (fig. 18) is on.
      (*b*) Turn magneto switch to "A," "F," or "BOTH" position.
      (*c*) Notice oil pressure and fuel gages. If they give readings, there is power to the starter switch. If not, test master relay switch (par. 131). If faulty, replace switch (par. 207). If not faulty, examine master relay rectifier (fig. 195). If faulty, replace rectifier (par. 213).
   (4) *Check starter relay circuit.* A click in the starter relay (located in master junction box) should be heard when starter switch is turned on. If no click is heard, use a jumper cable and contact one end to battery terminal in master junction box and the other against terminal to which cable (14) is connected. Proceed according to applicable step below.
      (*a*) *If relay now clicks and engine cranks.* If relay clicks when jumper cable is contacted and the engine cranks, the starter switch, neutral position switch (in manual control box), or cable from starter switch to starter relay is at fault. Replace starter switch (par. 204), neutral position switch (par. 191), or cable, whichever is required.
      (*b*) *If relay does not click.* If relay does not click when jumper cable is contacted, replace master junction box (par. 194).
   (5) *Check for hydrostatic lock or seized engine.* Remove spark plugs (par. 186) and again attempt to crank engine with starter. Proceed according to applicable step below.
      (*a*) *If engine can now be cranked.* If the engine can be cranked after removing the spark plugs, the engine was locked hydrostatically. Notify ordnance maintenance personnel.
      (*b*) *If engine cannot be cranked.* If the engine cannot be cranked with the spark plugs removed, it indicates a moving part has seized or that the starter, battery, or starter

circuit is at fault. Remove the starter (par. 190) and connect it to a fully charged battery. If the starter does not run, it must be replaced. If it does run, examine cables from battery to starter. If faulty, replace cables. If cables are not faulty, the engine has seized. Notify ordnance maintenance personnel.

b. STARTER SPINS BUT DOES NOT CRANK ENGINE. If the starter spins but does not crank the engine, replace the starter (par. 190).

c. ENGINE CRANKS SLOWLY.

(1) *Check grade of engine oil.* Make sure engine oil viscosity is correct for prevailing atmospheric temperatures as specified on lubrication order (fig. 86).

(2) *Battery charge low.* Perform preventive maintenance on battery, items 21 and 22, table V (par. 123).

d. RESISTANCE AND CONTINUITY TESTS. Refer to paragraph 125 for methods of testing with an ohmmeter.

Caution: Do not attempt to make any ohmmeter tests from connector No. 11 until a piece of rolled tape or other insulating material is pushed into terminal "H" of the plugs or receptacle of connector No. 11 to prevent accidental contact with an ohmmeter test prod when performing tests. The ohmmeter will be ruined immediately if one ohmmeter test prod contacts terminal "H" of connector No. 11 and other test prod contacts ground.

131. Instrument Panel Components, Switches, and Sending Units

See paragraph 125 for methods of testing with an ohmmeter. On all tests in this chapter in which a circuit breaker is included in the circuit being tested, resistance valves given assume that the circuit breaker is closed. If no reading is obtained on the ohmmeter in such a case, a continuity test across the circuit breaker should be made from the rear of the panel.

Caution: Make sure the ohmmeter test rod does not touch terminal "H" of connector plug No. 11 or ohmmeter will be ruined. Also be sure connector plug No. 11 is disconnected from its receptacle.

Components	Circuit No.	Test F????—	Test T????—	Resistance (ohms)	Circuit (figs. 173 and 195)	Remarks
Coil, starter relay	14	P11–E	gnd	60	P11–E through starter relay coil to ground.	Infinity indicates damaged cable or starter relay coil.
Switch, neutral position	14	P10–L	P10–M	0 or inf	From P10–L through switch to P10–M.	Zero reading should be obtained with manual control lever in neutral position, infinity with lever in any other position.
Switch, starter						
Switch, master relay						Refer to paragraph 131.
Rectifier, master relay						
Relay, master						Refer to paragraph 132.

177

Component	Circuit No.	Test From—	Test To—	Resistance (ohms)	Circuit	Remarks
Clutch, oil cooler fan, circuit.	476	P8–F	gnd	inf	P8–F through two thermoswitches, through two clutch coils to ground (figs. 193 and 194).	A reading less than infinite when both cooler fans are inoperative indicates defective thermoswitch or grounded cable. *Note.* Engine and transmission oil must be hot to perform this test. See paragraph 220a for closing, temperatures of thermoswitches.
Degassers	54	P9–J	gnd	3.3	P9–J through four parallel degasser coils to ground (figs. 193 and 194).	A reading of 4.3 ohms indicates 1 open coil; 6.5 ohms, 2 open coils; 13 ohms, 3 open coils.
Gage, engine oil pressure (variable voltage coil).	36, 27 481, 10	R9–H	R11–C	330–350	R9–H through variable voltage coil, through main engine magneto switch, through circuit breaker to R11–C (figs. 193 and 194).	Main engine magneto switch in "A," "F," or "BOTH" position.
(Fixed voltage coil)	36	R9–H	gnd	25–30	R9–H through fixed voltage coil to ground (figs. 193 and 194).	Disconnect cable 27 on rear of gage before making test.
Gage, left fuel: Variable voltage coil.	30, 27 481, 10	R10–K	R11–C	330–350	R10–K through variable voltage coil, through main engine magneto switch, through circuit breaker to R11–C (figs. 193 and 194).	Main engine magneto switch in "A," "F," or "BOTH" position.
Fixed voltage coil	30	R10–K	gnd	25–30	R10–K through fixed voltage coil to ground (figs. 193 and 194).	Disconnect cable 27 from rear of gage before making test.

		R10-J	R11-C	330-350	R10-J through variable voltage coil, through main engine magneto switch, through circuit breaker to R11-C (figs. 193 and 194).	Main engine magneto switch in "A," "F," or "BOTH" position.
Gage, right fuel: Variable voltage coil.	31, 27, 481, 10					
Fixed voltage coil	31	R10-J	gnd	25-30	R10-J through fixed voltage coil to ground (figs. 193 and 194).	Disconnect cable 27 from rear of gage before making test.
Sending unit, engine oil pressure.	36	P9-H	gnd	See Remarks	P9-H through sending unit to ground (figs. 193 and 194).	0 psi—0 ohms, 60 psi—15 ohms, 120 psi—30 ohms.
Sending unit, left fuel tank.	30	P10-K	gnd	See Remarks	P10-K through sending unit to ground (figs. 193 and 194).	Empty tank—0 ohms. Half-full tank—15 ohms. Full tank—30 ohms.
Sending unit, right fuel tank.	31	P10-J	gnd	See Remarks	P10-J through sending unit to ground (figs. 193 and 194).	Empty tank—0 ohms. Half-full tank—15 ohms. Full tank—30 ohms.
Sending unit, speedometer.	431, 432, 433	P12-A	P12-B	90	P12-A through three speedometer coils to P12-B (figs. 193 and 194).	Any reading higher than 90 ohms indicates one or more open speedometer sending unit coils.
Sending unit, tachometer.	427, 428, 429	P6-A	P6-B	90	P6-A through tree tachometer coils to P6-B (figs. 193 and 194).	Any reading higher than 90 ohms indicates one or more open tachometer sending unit coils.
Speedometer	431, 432, 433	R12-A	R12-B	90	R12-A through three speedometer coils to R12-B (figs. 193 and 194).	Any reading higher than 90 ohms indicates one or more open speedometer coils.
Switch, booster: Coil, left booster	12A, 481, 10	R9-B	R-11C	0 or inf	R9-B through booster switch, through instrument circuit breaker to R11-C (fig. 191).	Zero if booster switch is "ON;" infinite if booster switch is "OFF."

Component	Circuit No.	Test From—	Test To—	Resistance (ohms)	Circuit	Remarks
Switch, booster—Con. Coil, right booster.	12, 481, 10	R8-A	R11-C	0 or inf	R8-A through booster switch, through instrument circuit breaker to R11-C (fig. 191).	Zero if booster switch is "ON," infinite if booster switch is "OFF."
Switch, engine "HI OIL TEMP" warning light.	35	P9-F	gnd	inf	P9-F through switch to ground (figs. 193 and 194).	Switch is normally open. Closes at 225° F.
Switch, engine "LO OIL PRESS" warning signal light.	34	P9-E	gnd	0 (engine not running)	P9-E through switch to ground (figs. 193 and 194).	Switch is normally closed. Opens at 11 to 13 psi.
Switch, fuel cut-off	54, 481, 10	R9-J	R11-C	0 or inf	R9-J through fuel cut-off switch, through instrument circuit breaker to R11-C (figs. 193 and 194).	Zero if fuel cutoff switch button is depressed; infinite if not depressed. Magneto switch in "A," "F," or "BOTH" position.
Switch lights: Head lights	16, 15, 10	R10-H	R11-C	0 or inf	R10-H through light switch, through horn and main lights circuit breaker to R11-C. (Figs. 211 and 212)	Zero to "SER DRIVE" position; infinite in all others.
Blackout marker lights.	20, 24, 15, 10	R10-C	R11-C	0 or inf	R10-C through light switch, through horn and main lights circuit breaker to R11-C.	Zero in "BO MARKER" position; infinite in all others.
Blackout driving lights.	19, 15, 10	R10-F	R11-C	0 or inf	R10-F through light switch, through horn and main lights circuit breaker to R11-C.	Zero in "BO DRIVE" position; infinite in all others.
Service tail light	21, 15, 10	R10-D	R11-C	0 or inf	R10-D through light switch, through horn and main lights circuit breaker to R11-C.	Zero in "STOP LIGHT" position; infinite in all others.

(Figs. 161 and 191)

Switch, main engine magneto:					
Lower left magneto ground.	13A	R9-C	gnd	0 or inf	R9-C through magneto switch to ground.	Zero in "OFF" and "A" positions; infinite in all others.
Lower right magneto ground.	11A	R8-C	gnd	0 or inf	R8-C through magneto switch to ground.	Zero in "OFF" and "A" positions; infinite in all others.
Upper left magneto ground.	13	R9-A	gnd	0 or inf	R8-A through magneto switch to ground.	Zero in "OFF" and "F" positions; infinite in all others.
Upper right magneto ground.	11	R8-B	gnd	0 or inf	R8-B through magneto switch to ground.	Zero in "OFF" and "F" positions; infinite in all others.
Switch, master relay	459	See circuit column.	See circuit column.	0 or inf	Disconnect bell-type connectors at rear of master relay switch and connect one ohmmeter test prod to each side of switch (fig. 191).	Zero if master relay switch is in "ON" position; infinite if master relay switch is in "OFF" position.
Switch, starter	14, 481, 10	R10-L	R11-C	0 or inf	R10-L through starter switch through instrument circuit breaker to R11-C (figs. 173 and 191).	Zero if starter switch is on; infinite if starter switch is off. Magneto switch in "A," "F," or "BOTH" position.
Switch, transmission "HI TEMP" warning-signal light.	327	P9-N	gnd	inf	P9-N through switch to ground (figs. 193 and 194).	Switch is normally closed. Opens at 280° F.
Switch, transmission "LO LUBE PRESS" warning-signal light.	339	P9-D	gnd	0	P9-D through switch to ground (figs. 193 and 194).	Switch is normally closed. Opens at 10 psi.
Tachometer	427, 428, 429	R6-A	R6-B	90	R6-A through three tachometer coils to R6-B (figs. 193 and 194).	Any reading higher than 90 ohms indicates one or more open tachometer coils.

181

132. Generating and Charging System and Master Junction Box

a. TESTING MAIN GENERATOR. Start main engine (par. 68). Set engine speed at 1,200 rpm. Operate turret at maximum traversing speed (par. 76).

Caution: Observe all precautions listed in paragraph 76 before attempting to traverse turret. Also make sure that radio equipment and all other electrical equipment switches are in "OFF" positions. While turret is operating, turn off master relay switch (fig. 18). Make sure that engine continues to turn at 1,200 rpm or slightly higher as observed on tachometer. Proceed according to applicable step below.

(1) *If turret continues to rotate at maximum speed.* If turret continues to rotate at maximum speed when master relay switch is turned off, generator is operating satisfactorily at approximately 50 percent of its total rated output. This test also shows that main generator line switch, reverse current relay, and pilot relay in master junction box (par. 192) are operating correctly.

(2) *If turret slows down.* If turret turns much slower when master relay switch is turned off, increase engine speed. If engine speed must be increased beyond 1,500 rpm to cause turret to rotate at maximum speed, replace main generator (par. 222).

(3) *If turret stops rotating.* If turret stops rotating when master relay switch is turned off, either generator is defective or main generator line switch, reverse current relay, or pilot relay in master junction box (par. 194) are defective. Refer to *c* below for testing master junction box components. Replace main generator (par. 222), or master junction box (par. 194), as required.

Caution: Turn on master relay switch (fig. 18) before turning off traversing motor switch.

b. TESTING AUXILIARY GENERATOR. Start auxiliary engine (par. 93). Operate turret at maximum traversing speed (par. 76).

Caution: Observe all precautions listed in paragraph 76 before attempting to traverse turret. Also make sure that radio equipment and all other electrical equipment switches are in "OFF" positions. With auxiliary engine operating, main engine not operating, turret rotating at maximum speed, turn off master relay switch (fig. 18). Proceed according to applicable step below.

(1) *If turret continues to rotate at maximum speed.* If turret continues to rotate at maximum speed when master relay switch is turned off, generator is operating satisfactorily. This test also shows that main generator line switch, reverse

current relay, and pilot relay in master junction box (par. 192) are operating correctly.

(2) *If turret slows down.* If turret turns much slower when master relay switch is turned off, either auxiliary generator or auxiliary engine is defective. Check auxiliary engine operation (par. 140). If engine is operating properly, replace auxiliary generator (par. 304).

(3) *If turret stops rotating.* If turret stops rotating when master relay switch is turned off, either auxiliary generator, auxiliary engine, auxiliary generator line switch, reverse current relay, or pilot relay in master junction box is defective. Check auxiliary engine operation (par. 140). If engine is operating properly, check master junction box (*e* below). If both auxiliary engine and master junction box are functioning properly, replace auxiliary generator (par. 304).

c. MASTER JUNCTION BOX.

(1) *Voltage measurements.*

(a) *Preliminary instructions.*

1. Start main and auxiliary engines and allow to run for at least 5 minutes to stabilize component temperatures.
2. Make all measurement from crew compartment side of master junction box by removing master junction box cover (par. 194). Be sure that no dirt, metallic particles, or other foreign substances enter the box. Be careful not to damage any of the items or dislodge spring contacts on polarized relays (fig. 184).
3. The voltmeter recommended is low voltage circuit tester FSN 17–T–5675. If this voltmeter is not available, use most accurate instrument available.
4. Make all electrical tests in order indicated below.
5. Refer to figure 95 or to chart inside cover of master junction box for voltage test points.

(b) *Procedure for voltage tests.*

1. *Ground strap.* Master junction box is provided with ground strap at rear of box (fig. 182). If faulty, voltage will rise too high.
2. *Auxiliary generator ballast lamp.* Voltage at "M" and none at "L" (fig. 95) indicates a failed filament. Replace only with Mazda 89 lamp (ORD 7539459).
3. *Main generator ballast lamp.* Voltage at "O" and none at "N" (fig. 95) indicates a failed filament. Replace only with Mazda 89 lamp (ORD 7539459).
4. *Auxiliary generator output.* With auxiliary generator in operation, voltage less than 3 volts when measured at "B" indicates that cables (61) and (62) between master

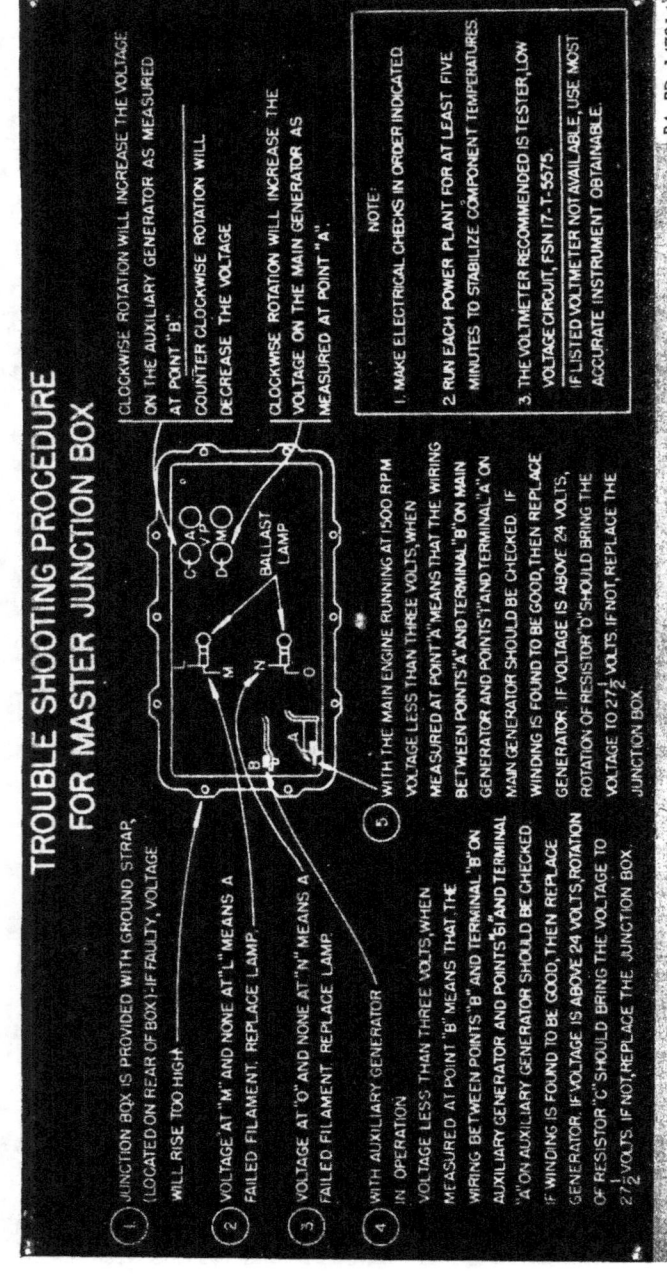

Figure 95. Master junction box trouble-shooting procedure.

junction box and auxiliary generator should be checked. If cables are found to be in good condition, replace auxiliary generator (par. 304). If voltage at "B" is above 24 volts but below 27½ volts, clockwise rotation of resistor "C" should bring voltage up to 27½ volts. If it does not, replace master junction box (par. 194).

5. *Main generator output.* With main engine running at 1,500 rpm, voltage less than 3 volts when measured at "A" indicates that cables (1) and (2) between master junction box and main generator should be checked. If cables are found to be in good condition, replace main generator (par. 222). If voltage at "A" is above 24 volts but below 27½ volts, clockwise rotation of resistor "D" should bring voltage up to 27½ volts. If it does not, replace master junction box (par. 194).

(2) *Resistance and continuity tests.* See paragraph 125 for methods of testing with an ohmmeter.

Note. After completing voltage measurements in (1) above, turn off main and auxiliary engines and turn master relay switch (fig. 18) to "OFF" position.

To proceed with tests below, remove master junction box cover (par. 194).

Caution: Terminals (81) and (459B) (fig. 187) are live terminals. Every precaution must be taken not to touch these three terminals with an ohmmeter test prod; to do so may immediately ruin the meter.

133. Horn and Lighting System

a. GENERAL. The electrical circuits of the horn and lighting system may be tested with reference to the wiring diagram, figure 211. The horn is connected through the horn switch into a circuit breaker and through to the instrument panel positive line. Both front and rear lights are connected through the main lights switch (fig. 18) into the same circuit breaker and through to the instrument panel positive line. The hull dome lights are connected through a switch in the dome-light housing into the dome lights circuit breaker and through to the instrument panel positive line.

b. NEITHER THE HORN, THE FRONT, NOR REAR LIGHTS OPERATE. Make sure master relay switch (fig. 18) is on. Then check horn-and-main-lights circuit breaker (fig. 211).

c. NONE OF THE HULL DOME LIGHTS OPERATE. Make sure master relay switch is on. Then check dome-lights circuit breaker (fig. 211).

d. SOME BUT NOT ALL OF FRONT OR REAR LIGHTS OPERATE. Replace bulb or lamp unit (pars. 228 to 232). If this does not correct the

Component	Circuit No.	Test		Resistance (ohms)	Circuit (figs. 185, 186, and 187)	Remarks
		From—	To—			
Relay, heater	479	479	gnd	60	From terminal (479) through relay coil to ground.	Zero indicates a grounded cable between terminal (479) and relay, or a shorted relay coil.
Relay, master	459	459	gnd	60	From terminal (459) through relay coil to ground.	Zero indicates a grounded cable between terminal (459) and relay or shorted relay coil.

Components	Circuit No.	Test		Resistance (ohms)	Circuit (fig. 187)	Remarks
		From—	To—			
Relay, safety	65	(65) (fig. 187)	gnd	300	From terminal (65) through relay coil to ground.	Zero indicates a grounded cable between terminal (65) and relay or shorted relay coil.
Relay, starter	14	(14) (fig. 187)	gnd	60	From terminal (14) through relay coil to ground.	Zero indicates a grounded cable between terminal (14) and relay or shorted relay coil.

trouble, check main light switch (par. 131), electrical wiring, or dimmer switch (*f* below).

e. HORN DOES NOT OPERATE BUT FRONT AND REAR LIGHTS DO. Connect a jumper cable across the horn switch. If horn operates, replace switch (par. 227). If horn still does not operate, replace horn (par. 226).

f. RESISTANCE AND CONTINUITY TESTS. See paragraph 125 for methods of making ohmmeter tests.

Caution: Do not touch ohmmeter test prod to terminal "H" of plug No. 11 or ohmmeter may be immediately ruined.

134. Transmission and Final Drives

a. PRELIMINARY INSTRUCTIONS.

(1) *Inspect for oil leaks.* Visually inspect all gasket joints, transmission valve body, and plugs for signs of escaping oil. Tighten all mounting bolts where oil leakage is noticed. If mounting bolts are tight and oil leaks continue, replace gaskets or notify ordnance maintenance personnel.

(2) *Inspect for water in transmission and final drives.* Remove drain plugs (figs. 222 and 223) and note if water flows from openings. If there is water, drain and refill (par. 235), as specified on the lubrication order (fig. 86). If no water is evident, replace drain plugs and add oil as required.

(3) *Check oil levels and operate vehicle.* Check oil levels in transmission (par. 235) and final drives (par. 246). Then operate vehicle in all speed ranges (par. 11) under various load and terrain conditions. Note any unusual conditions, such as power loss, bands slipping, overheating, oil foaming, noises, and the speed ranges and steering positions in which the conditions occur.

(4) *Check transmission oil pressures.* Transmission oil pressure tests are covered in paragraph 235. If replacement of the valve body is necessary, notify ordnance maintenance personnel.

b. VEHICLE WILL STEER OR PIVOT IN PLACE BUT WILL NOT DRIVE IN ANY GEAR. Check the oil pressure in torque converter oil feed line (fig. 225). If the presure is not within allowable range (par. 235), notify ordnance maintenance personnel.

c. VEHICLE WILL NOT DRIVE IN LOW GEAR. Adjust low range band (par. 236). If this does not correct trouble, check oil pressure in low range servo oil line and torque converter oil feed line (fig. 225). If oil pressures are not within allowable ranges (par. 235), notify ordnance maintenance personnel.

d. VEHICLE WILL NOT DRIVE IN HIGH GEAR. Check oil pressure in high range clutch oil line and torque converter oil feed line (fig. 225).

Component	Circuit No.	Test From—	Test To—	Resistance (ohms)	Circuit (figs. 195, 211 and 212)	Remarks
Horn	25	P10-A	gnd	5.3	P10-A through horn switch, through horn to ground.	Horn button must be pressed to get a reading.
Lights, blackout driving.	19, 514, 515.	P10-F	gnd	2	P10-F through two parallel lamps to ground (dimmer switch on "HI" position).	A reading of 4 ohms indicates one burned out lamp or open cable.
				2	P10-F through two parallel lamps to ground (dimmer switch on "LO" position).	
Lights, blackout marker.	20, 24	P10-C	gnd	3	P10-C through four parallel lamps to ground.	A reading higher than 3 ohms indicates one or more burnt-out marker lamps or damaged cables (20) or (24).
Lights, hull dome	38	P10-N	gnd	See remarks	P10-N through each dome lamp in turn to ground.	To get reading, switch each dome lamp on in turn. Reading for each one should be 10 ohms.

188

Component	Circuit No.	Test From—	Test To—	Resistance (ohms)	Circuit (figs. 211 and 195)	Remarks
Lights, service head_	16, 17, 18	P10–H	gnd	0.8	P10–H through two parallel head lamps to ground (dimmer switch on "HI" position).	A reading of 1.6 ohms indicates one burned out lamp or open cable.
				0.8	P10–H through two parallel head lamps to ground (dimmer switch on "LO" position).	
Light, service tail___	21	P10–D	gnd	7.3	P10–D through light to ground___	
Switch, dimmer, blackout section.	19, 514, 515.	R13–C	R13–A	0	R13–C through dimmer switch to R13–A.	Dimmer switch in "LO" position.
		R13–C	R13–B	0	R13–C through dimmer switch to R13–B.	Dimmer switch in "HI" position.
Switch, dimmer, head light section.	16, 17, 18	R13–G	R13–E	0	R13–G through dimmer switch on R13–E.	Dimmer switch in "LO" position.
		R13–G	R13–F	0	R13–G through dimmer switch to I13 F.	Dimmer switch in "HI" position.

If pressures are not within allowable ranges (par. 235), notify ordnance maintenance personnel.

e. VEHICLE WILL NOT DRIVE IN REVERSE GEAR. Adjust reverse range band (par. 236). If this does not correct trouble, check oil pressure in reverse servo oil line and torque converter oil feed line (fig. 225). If oil pressures are not within allowable ranges (par. 235), notify ordnance maintenance personnel.

f. TRANSMISSION LOW-OIL-PRESSURE-WARNING-SIGNAL LIGHT COMES ON. Check transmission oil level (par. 235). If oil level is correct, the main oil pressure is too low. Check the oil pressure in main line (fig. 225). If the pressure is too low (par. 235), remove and clean the oil filter (par. 238). If trouble still persists, notify ordnance maintenance personnel.

Note. Do not omit the possibility that the warning-signal-light switch is defective (par. 131).

g. TRANSMISSION-OIL-TEMPERATURE-WARNING-SIGNAL LIGHT COMES ON. Check the oil cooler cores and lines (fig. 239) to see that there are no restrictions to prevent flow of air. Another possible cause is slippage of low or reverse range bands. Adjust bands (par. 236). Check the service brakes (par. 259) to see that they are not adjusted too tightly. Adjust, if necessary (par. 259). If trouble still persists, notify ordnance maintenance personnel.

Note. Do not omit the possibility that the warning-signal-light switch is defective (par. 131).

135. Oil Coolers and Lines

a. PRELIMINARY INSTRUCTIONS. Troubles in the oil coolers and lines will be reflected in high engine and/or transmission oil temperatures, low oil pressures, or both. These symptoms can also be caused by improper operation of the vehicle, low or diluted oil, defective sending units, and faulty ignition. Do not disregard these possible causes in any investigation. Always check for oil leaks in the engine compartment.

b. BOTH ENGINE AND TRANSMISSION TEMPERATURES HIGH; OIL PRESSURE NORMAL. If the engine and transmission oil temperatures are high and the oil pressures in these units are normal, the oil is not being cooled adequately by the oil coolers. Check the oil cooler cores and fan shrouds to make sure the flow of air is free. If the temperatures remain high, replace the oil cooler fans (par. 241).

c. ENGINE TEMPERATURE HIGH; OIL PRESSURE LOW. High engine temperature combined with low oil pressure for the same unit indicates that the oil cooler lines for the engine may be restricted. Inspect the lines in question and remove the restrictions or replace the lines (par. 242). If the temperature continues high, check the engine cooling fans for proper operation and free flow of air. If the fans are operating satisfactorily, replace the engine oil coolers (par. 242).

d. TRANSMISSION TEMPERATURE HIGH; OIL PRESSURE LOW. High transmission temperature and low transmission oil pressure may be caused by restrictions in the oil cooler cores or lines. In addition, the transmission oil filter may be clogged. If necessary, replace the filter (par. 238). If this does not correct the trouble, proceed with a check of the oil coolers and lines (par. 242). Remove any obstructions and replace defective units (par. 242).

136. Tracks and Suspension

a. VEHICLE LEADS TO ONE SIDE.
   (1) *Unequal track tension.* Unequal track tension will cause the vehicle to lead to the side having the tighter track tension. Adjust both tracks (par. 249).
   (2) *Worn or distorted drive sprockets or tracks.* Worn or distorted sprockets or track parts will cause vehicle to lead to one side. Replace worn parts as required (pars. 244 and 249).
   (3) *Crowned road.* A crowned road will tend to cause the vehicle to pull to the low side of the road. Do not mistake this for the above causes. It is a normal tendency.

b. THROWN TRACK.
   (1) *Improper driving.* If track is thrown as a result of improper handling of vehicle, further instruction and review of driving methods is required (sec. III, ch. 2).
   (2) *Excessively loose or worn track.* Adjust track tension or replace track (par. 249).
   (3) *Compensating idler wheel ineffective.* If the compensating idler wheel torsion bar is broken, the compensating idler wheel will be ineffective. To test for broken torsion bar, attempt to raise compensating idler wheel with a crowbar. If the wheel can be lifted, replace the torsion bar (par. 253).

c. VEHICLE SAGS TO ONE SIDE.
   (1) *Broken torsion bar.* If vehicle sags to one side, it indicates a torsion bar is broken on the low side of the vehicle. Attempt to lift each road wheel with a crowbar. If any wheel can be lifted in this manner, the wheel's torsion bar is broken. Replace as required (par. 253).
   (2) *Excessively loose or worn track.* Adjust track tension or replace track (par. 249).

d. EXCESSIVELY HARD RIDING.
   (1) *Leaking shock absorber.* If shock absorber loses too much oil, hard riding will result. Test the shock absorber by the temperature method (par. 255).

(2) *Broken torsion bar.* Test by attempting to lift road wheel with crowbar. If any wheel can be lifted in this manner, replace torsion bar (par. 253).

137. Driver's Controls and Linkage

a. VEHICLE WILL NOT STEER. Check steering control linkage (fig. 276) to make sure it is operative. Movement of the manual control lever (fig. 10) in the driver's compartment must be reflected in movement of the steering control lever on the transmission valve body (F, fig. 221). Replace defective linkage, as necessary (par. 257). If this does not correct trouble, check transmission oil pressures (par. 235).

b. VEHICLE WILL STEER ONLY IN ONE DIRECTION. Adjust steering linkage at transmission valve body (par. 257). If this does not correct trouble, proceed as outlined in *f* below.

c. MANUAL CONTROL LEVER WILL NOT SHIFT OUT OF NEUTRAL OR REVERSE. Manual control-lever-control rod is out of adjustment. Adjust control rod (par. 257).

d. MANUAL CONTROL LEVER HAS PLAY, OR BINDS IN OPERATION. If operation of manual control lever indicates conditions of looseness or play, adjust steering and shifting linkage to eliminate looseness (par. 257). Particularly check for loose bellcranks. Replace linkage connections or parts, if necessary (par. 257). If binding is felt, inspect steering and shifting control linkage on top of transmission for bent rods. Straighten or replace linkage as necessary (par. 257). Adjust linkage (par. 257).

e. PRESSING SERVICE-BRAKE PEDAL WILL NOT EFFECTIVELY STOP VEHICLE. Check service-brake linkage and adjust as necessary to eliminate all looseness and play (par. 259). Adjust service brakes (par. 259). If this does not remedy trouble, notify ordnance maintenance personnel.

f. BRAKES DRAGGING AT ONE OR BOTH SIDES OF TRANSMISSION. Check to make sure hand brake (fig. 10) is released. Check service brake adjustment (par. 259). Adjust brake linkage or service brakes as necessary (par. 259).

g. ENGINE DOES NOT RESPOND PROPERLY TO THROTTLE CONTROL. If the engine does not respond properly to operation of the accelerator pedal or hand throttle control, check for loose or broken throttle control linkage (fig. 278). Adjust linkage or replace defective parts, as necessary (par. 258).

h. PRIMER PUMP HANDLE GOES IN TOO EASILY. The primer pump (fig. 158) should operate with a resistance of approximately 10 psi. If the handle goes in easily, it indicates broken primer pump lines, defective primer pump, or clogged filter or fuel lines to the primer pump. Make repairs as required (pars. 180 and 181).

138. Turret Traversing and Elevating Systems

a. PRELIMINARY CHECK. Before investigating hydraulic traversing, disengage turret traversing lock (fig. 31), gun traveling lock (fig. 4), and traverse turret manually (par. 76) to be sure it will traverse properly in both directions.

b. TURRET DOES NOT TRAVERSE FREELY UNDER MANUAL OPERATION. Proceed as outline in (1) through (5) below if turret fails to traverse freely.

(1) *Main driven pinion impeded by foreign matter on turret ring gear.* Inspect for foreign matter and clean, if required.

(2) *Improper lubrication.* Lubricate turret race bearings in accordance with lubrication order (fig. 86).

(3) *Turret lock pawl in contact with ring gear.* Adjust turret lock clearance if pawl prevents turret from traversing with lock in unlocked position (par. 272).

(4) *Turret control switch in "ON" position.* Turret control switch (fig. 28) must be in "OFF" position when traversing turret manually, or manual drive is ineffective. Turn switch to "OFF" position.

(5) *Damaged hand traversing mechanism.* Replace mechanism, if damaged (par. 271).

(6) *Damaged turret race bearings or ring gear.* If damaged bearings or ring gear prevent traversing turret, notify ordnance maintenance personnel.

c. HYDRAULIC PUMP OR ELECTRIC MOTOR FAILS TO OPERATE. After checking manual traversing, attempt to traverse turret using hydraulic power (par. 76). If hydraulic pump or electric motor fails to operate, notify ordnance maintenance personnel.

d. HYDRAULIC PUMP OPERATES BUT TURRET CANNOT BE TURNED IN EITHER DIRECTION.

(1) *Turret lock engaged.* Be sure turret lock is in disengaged position (par. 42).

(2) *Turret lock pawl in contact with ring gear.* Adjust turret lock clearance (par. 272) if pawl prevents turret from traversing with lock in unlocked position.

(3) *Low oil level in reservoir.* Check oil level (par. 270). If necessary, refill as specified on lubrication order (fig. 87).

(4) *Main drive pinion impeded by foreign matter on turret ring gear.* Inspect for foreign matter. Clean if necessary.

(5) *Defective hydraulic system.* If turret still does not traverse, hydraulic system must be repaired. Notify ordnance maintenance personnel.

e. TURRET-TRAVERSING SPEED LOW OR ERRATIC.
  (1) *Battery charge low.* Check batteries (par. 224) and replace or recharge as required.
  (2) *Poor electrical connections.* Check all wiring to be sure good mechanical and electrical connections are made. Clean and tighten connections.
  (3) *Low oil level in reservoir.* Check oil level (par. 270) and, if necessary, refill as specified in lubrication order (fig. 87).
  (4) *Foreign matter in traversing mechanism or turret ring gear.* Inspect for foreign matter that may be lodged in traversing mechanism and turret ring gear. Clean as required.
  (5) *Defective electric motor or hydraulic system.* Notify ordnance maintenance personnel.

f. TURRET CREEPS IN ONE DIRECTION WHILE TANK IS IN HORIZONTAL POSITION. Gunner's control handle not returned to neutral position. Return control handle to neutral position (par. 35).

g. TURRET CREEPS EXCESSIVELY WHEN VEHICLE IS NOT IN HORIZONTAL POSITION. Turret no-back mechanism is defective, notify ordnance maintenance personnel.

h. SLUGGISH OR UNSTEADY TURRET OPERATION.
  (1) *Battery low.* Recharge battery (par. 224).
  (2) *Ring gear and output gear bind.* Inspect, clean, or lubricate as necessary.
  (3) *Poor electrical connections.* Check all wiring for shorts and be sure good mechanical and electrical connections are made. Clean and tighten connections or replace wiring.
  (4) *Low oil level in reservoir.* Refill to proper level (par. 270) to prevent hydraulic pump from sucking air. Refer to lubrication order (fig. 86).
  (5) *Defective hydraulic system.* Notify ordnance maintenance personnel.

i. ABNORMAL NOISE IN HYDRAULIC PUMP OR HYDRAULIC MOTOR. Notify ordnance maintenance personnel.

j. COMMANDER'S CONTROL INOPERATIVE. Refer to paragraph 139.

k. GUNNER'S CONTROL INOPERATIVE.
  (1) *Defective hydraulic traversing control.* Notify ordnance maintenance personnel.
  (2) *Defective electrical control system.* Refer to paragraph 139.

l. HAND ELEVATING MECHANISM INOPERATIVE OR WEAK.

Failure of the hand elevating mechanism to function is usually due to low pressure in the accumulator. See that the oil return valve (fig. 33) is closed. Open the elevation supercharge-line-shut-off valve (fig. 32) for 1 second; then close it. If the hand elevating mechanism still does not function, operate the hand pump (fig. 32) a few strokes.

m. POWER ELEVATING MECHANISM INOPERATIVE OR SLOW.
  (1) Check that turret control switch (fig. 28) is in the "ON" position and oil return valve (fig. 33) and elevation-supercharge-line-shut-off valve are closed (fig. 32).
  (2) If the elevating mechanism is still inoperative, a defective hydraulic pump or elevation cylinder is probable. Notify ordnance maintenance personnel.

139. Turret Electrical System

a. GENERAL.
  (1) *Description.* The turret electrical system (fig. 298) consists of the turret traversing system, gun elevating system, firing control system, cal .50 ammunition feed control system, and the turret accessories system. In trouble shooting the turret electrical system, the trouble should first be tracked down into one of the above systems, and then the faulty component can be isolated. The theory of operation of each of these systems and the components involved are described in (2) through (6) below. Some of the common troubles in the turret electrical system are analyzed in the remaining portions of this paragraph. This manual cannot possibly cover all troubles that may arise. To analyze any troubles not mentioned here, first collect all possible symptoms of trouble, then, with the aid of circuit theory, isolate the system in which the trouble lies. Then, using the turret wiring diagram (fig. 298) and methods of electrical testing (par. 125), isolate the faulty component.
  (2) *Turret traversing system.* Electrically, the turret is traversed by the electric motor (fig. 296) which actuates one section of a double hydraulic pump. The motor can be controlled either by the gunner's control handle (fig. 23) or the commander's control handle (fig. 26) through cables (600), (601), and (602) (fig. 300). The electrical wiring from these two handles passes through the override motor control box (fig. 300). Within this box are two override relays which cut out the gunner's control when the commander's override lever is actuated. However, neither the gunner's nor the commander's control is operative unless the loader's traverse safety switch (fig. 29) and gunner's turret control switch (fig. 28) are in the "ON" position. The safety switch operates a relay in the override motor control box which connects the electric motor positive feed cable (100) to the slip ring positive feed cable (100).
  (3) *Gun elevating system.* Electrically the 90-mm gun is elevated or depressed by the same electric motor (fig. 296)

that is used for traversing the turret. This motor actuates the other section of the double hydraulic pump. The motor can be controlled either by the gunner's control handle (fig. 22) or the commander's control handle (fig. 26) through cables (608), (609), and (610). The electrical wiring from these two handles passes through the override motor control box (fig. 300). Within this box are two override relays which cut out the gunner's control when the commander's override lever is actuated. However, neither the gunner's nor the commander's control is operative unless the loader's traverse safety switch (fig. 29) and gunner's turret control switch (fig. 28) are in the "ON" position. The safety switch operates a relay in the override motor control box which connects the electric motor positive feed cable (100) to the slip ring positive feed cable (100). An elevation limit switch (fig. 301) opens up control cable (601) when the gun elevation reaches 19 degrees and the depression limit switch opens up control cable (602) when the gun depression reaches 5 degrees.

(4) *Firing control system.* The firing mechanism on the 90-mm gun consists of an impulse relay and a firing solenoid (fig. 302). This firing mechanism is connected by cable (113), through the loader's safety relay (fig. 40), through the turret filter box (fig. 336), into the gunner's control handle (fig. 24) where it passes through the 90-mm gun firing selector switch into the gunner's firing buttons (fig. 25). The loader's safety relay is actuated by the loader's firing reset safety (fig. 30). The machine gun firing solenoid (fig. 39) is connected by cable (103) through the filter box into the gunner's control handle where it passes through the machine gun firing selector switch into the gunner's firing buttons. The gunner's three firing buttons (figs. 22 and 25) are wired together and connected by a single cable (103 and 113) to the firing trigger on the commander's control handle (fig. 26).

(5) *Cal. .50 ammunition feed control system* (fig. 298). The booster motor (fig. 303), which feeds cal. .50 rounds into the ammunition chute, is connected by cable (642) through the booster motor relay secured to the side of the loader's firing reset safety panel (fig. 30), through a circuit breaker in the turret accessories panel (fig. 30) into positive cable (159). The relay control solenoid is connected by cable (641) through the ammunition chute micro switch (fig. 303) in to the turret filter box (fig. 336) where it is connected to the machine gun firing control cable (103). Thus, every time the machine gun is fired, the booster motor is actuated, unless the ammunition chute micro switch is open (which

occurs when pressure on the switch is increased due to a jammed machine gun).

(6) *Turret accessories system.* The turret accessories system includes a ventilating blower, two dome lights, a ballistic box feed, and an accessory outlet socket. The ventilating blower (fig. 304) is connected by cable (159) through the ventilating blower switch (fig. 30), through a circuit breaker in the turret accessories panel to positive feed cable (100). The dome lights are connected by cable (138) through a circuit breaker in the turret accessories panel into ventilating blower positive feed cable (159). The ballistic drive (fig. 43) is connected by cable (465) through a circuit breaker in the turret accessories panel to feed cable (159). The positive terminal of the accessory outlet socket (fig. 30) connected by cable (137) through the booster motor circuit breaker in the turret accessories panel to feed cable (159).

b. TURRET CANNOT BE TRAVERSED EITHER BY GUNNER'S OR BY COMMANDER'S CONTROL HANDLE.

(1) *Control switches not set.* Make sure that master relay switch (fig. 18), turret control switch (fig. 28), and loader's traverse safety switch (fig. 29) are all in the "ON" position.

(2) *No positive feed to electric motor.*

(*a*) *Preliminary check.* Disconnect cable (100) from electric motor. Using a voltmeter or test lamp, check presence of voltage at connector plug. If voltage is present, trouble is not in positive feed circuit. Proceed with (3) below. If voltage is not present, proceed with (*b*) below.

(*b*) *Check voltage at override motor control box output.* Disconnect cable 100 (leading from motor to override motor control box) at connector No. 101 (fig. 298). Using a voltmeter or test lamp, check presence of voltage at connector receptacle. If voltage is present, replace cable (100). If voltage is not present, check control box ground strap. Proceed with (*c*) below.

(*c*) *Check voltage at override motor control box input.* Disconnect connector No. 104 (fig. 298). Using a voltmeter or test lamp, check presence of voltage at connector plug. If voltage is not present, check cable (100) to slip ring box and check slip ring box. Replace cable (100) or slip ring box (par. 296), if faulty. If voltage is present, turn off master relay switch (fig. 18) and inspect loader's traverse safety switch (fig. 29). If switch is faulty, replace (par. 288). If switch is not faulty, inspect wiring from switch to override motor control box. If wiring is faulty, replace. If wiring is not faulty, replace override motor control box par. 282).

(3) *No control feed to electric motor.* Connect cable (100) to electric motor. Check override relay and override relay solenoids (*q* below). If faulty, replace override motor control box (par. 282). If not faulty, check cables (600), (601), (602), (608), (609), and (610) from override motor control box to electric motor. If cable is damaged, replace. If cable is not damaged, electric motor is faulty. Notify ordnance maintenance personnel.

c. TURRET CANNOT BE TRAVERSED BY GUNNER'S HANDLE, BUT CAN BE TRAVERSED BY COMMANDER'S HANDLE. Check override motor control box ground strap. Check override relay and override relay solenoids (*q* below) for damage. Actuate commander's override lever and inspect operation of relays. If relays are faulty, replace override motor control box (par. 282). If relays operate properly, inspect cable leading from control box to gunner's control mechanism. If cable is damaged, replace. If cable is not damaged, gunner's control mechanism is at fault. Replace gunner's control mechanism (par. 284).

d. TURRET CANNOT BE TRAVERSED BY COMMANDER'S HANDLE, BUT CAN BE TRAVERSED BY GUNNER'S HANDLE. Check override motor control box ground strap. Check override relay solenoids (fig. 299) for damage. Actuate commander's override lever and inspect operation of relays. If relays do not operate replace override motor control box (par. 282). If relays operate properly, inspect cable leading from control box to commander's control handle. If cable is damaged, replace. If cable is not damaged, commander's control handle is at fault. Replace commander's control handle assembly (par. 283).

e. 90-mm GUN CANNOT BE ELEVATED OR DEPRESSED.
  (1) *Control switches not set.* Refer to *b* (1) above.
  (2) *No positive feed to electric motor.* Refer to *b* (2) above.
  (3) *No control feed to electric motor.* Refer to *b* (3) above.
  (4) *Defective limit switch unit.* Perform visual inspection on elevation and depression limit switches (fig. 301). Check particularly the switch control rod and positions of adjusting screws on this rod. If defective, notify ordnance maintenance personnel.

f. 90-mm GUN ELEVATES TOO FAR OR NOT FAR ENOUGH. Adjust elevation limit switch (par. 286). If gun still elevates too far or not far enough, replace switch (par. 286).

g. 90-mm GUN DEPRESSES TOO FAR OR NOT FAR ENOUGH. Adjust depression limit switch (par. 286). If gun still depresses too far or not far enough, replace switch (par. 286).

h. GUNS CANNOT BE FIRED EITHER BY COMMANDER'S OR BY GUNNER'S CONTROL HANDLE.

Warning: Make sure both guns in combination gun mount and that they are cleared of ammunition before making the following tests.

(1) *Control switches not set.* Make sure either 90-mm gun firing selector switch or machine gun firing selector switch (fig. 28) is in the "ON" position. Make sure loader's firing reset safety switch button (fig. 30) has been depressed. Make sure master relay switch (fig. 18) is in the "ON" position.

(2) *No positive feed to firing control circuits.*
  (a) Check override motor control box ground strap. Replace if necessary.
  (b) Disconnect connector No. 104 (fig. 298) on override motor control box and check voltage at connector plug, using voltmeter or test lamp. If voltage is not present, check wiring to slip ring box. If wiring is damaged, examine slip ring box for damage and replace (par. 296), if necessary.
  (c) Replace connector plug No. 104 and disconnect connector No. 102 (fig. 298). Using voltmeter or test lamp, check voltage at pin "H" of receptacle No. 102. If voltage is not present, replace override motor control box (par. 282).
  (d) Replace connector plug No. 102 and disconnect connector No. 109 on gunner's control handle (fig. 298). Using voltmeter, or test lamp, check voltage at pin "H" of connector plug No. 109. If voltage is not present, replace cable connecting override motor control box to gunner's control handle. If voltage is present, replace connector plug No. 109 and proceed to (3) below.

(3) *No current flow to firing mechanism.*
  Note. To get a reading in tests (a) to (d) below, make sure machine gun firing selector switch (fig. 28) is in the "ON" position and a firing button is depressed.
  (a) Disconnect connector No. 110 (fig. 298) on gunner's control handle. Using voltmeter or test lamp, check voltage at pin "D" of receptacle No. 110. If voltage is not present when a firing button is depressed, replace gunner's control handle assembly (par. 284).
  (b) Replace connector No. 110 and disconnect connector No. 112 (fig. 298) on filter box. Using a voltmeter or test lamp, measure voltage at pin "D" of No. 112. If voltage is not present when a firing button is depressed, replace cable connecting gunner's control handle to filter box.
  (c) Replace connector plug No. 112 and disconnect connector No. 113 (fig. 298) on turret filter box (fig. 336). Using a voltmeter or test lamp, check voltage at pin "A" of receptacle 113. If voltage is not present, replace turret filter box (par. 335).
  (d) Check cable from turret filter box to firing mechanisms. If defective, replace.

i. Guns Cannot Be Fired by Commander's Control but Can Be Fired by Gunner's Control.

Warning: Make sure both guns in combination gun mount are cleared of ammunition before making the following tests.
- (1) *Control switches not set.* Make sure commander's override lever is actuated when firing trigger is pulled.
- (2) *No positive feed to commander's handle.*
    - (*a*) Disconnect connector No. 106 (fig. 298) on override motor control box. Using voltmeter or test lamp, check voltage at pin "G" of receptacle 106.

        Note. Master relay switch (fig. 18) must be in the "ON" position to make this test.

        If voltage is not present, replace override motor control box (par. 282).
    - (*b*) Replace connector plug No. 106 and disconnect connector No. 107 (fig. 298) on commander's control handle assembly. Using a voltmeter or test lamp, check voltage at pin "G" of plug No. 107.

        Note. Master relay switch (fig. 18) must be in "ON" position to get a reading.

        If voltage is not present, replace cable from override motor control box to commander's control handle assembly. If voltage is present, replace plug No. 107 and proceed to (3) below.
- (3) *No current flow to firing mechanisms.*

    Note. To get a reading in tests (*a*) and (*b*) below, commander's override lever and commander's firing trigger (fig. 26) must be depressed.

    - (*a*) Disconnect connector No. 108 (fig. 298) on commander's control handle. Using a voltmeter or test lamp, check voltage at receptacle No. 108. If voltage is not present when override lever and firing trigger are actuated, replace commander's control handle assembly (par. 283). If voltage is present, perform tests described in *d* above.
    - (*b*) Replace connector plug No. 108 and disconnect connector No. 110 (fig. 298) on gunner's control handle assembly. Using a voltmeter or test lamp, measure voltage at pin "A" of plug No. 110. If voltage is not present when commander's override lever and firing trigger are depressed, replace cable from commander's control handle assembly to gunner's control handle assembly. If voltage is present, replace gunner's control handle assembly (par. 284).

j. Guns Cannot Be Fired by Gunner's Control but Can Be Fired by Commander's Control. The fault is probably defective override relay. Check override relay and override relay solenoids (*q* below). If faulty, replace override motor control box (par. 282).

k. 90-mm Gun Cannot Be Fired Either by Gunner's or Commander's Control but Machine Gun Can.

Warning: Clear 90-mm gun of ammunition before making following tests.

  (1) *Control switches not set.* Make sure 90-mm gun firing selector switch (fig. 28) is in the "ON" position. Make sure loader's firing reset safety switch button (fig. 30) is depressed. Make sure breech is properly closed.

  (2) *Check firing circuit.*

    (*a*) If ready-to-fire-indicator light (fig. 30) does not light, remove firing safety panel (par. 287) and visually inspect reset switch. If defective, replace switch (par. 287). If not defective, inspect cable from firing safety panel to turret filter box (fig. 336). If damaged, replace.

    (*b*) Disconnect connector No. 110 (fig. 298). Using voltmeter or test lamp, check voltage at pin "F" of receptacle No. 110.

      Note. Depress a firing button to get a reading.

    If voltage is not present, replace gunner's control handle assembly (par. 284).

    (*c*) Reconnect connector plug No. 110 and disconnect connector No. 112 (fig. 298) on turret filter box. Using a voltmeter or test lamp, check voltage at pin "F" of plug No. 112.

      Note. Depress a firing button to get a reading.

    If voltage is not present, replace wiring from gunner's control handle assembly to filter box.

    (*d*) Reconnect connector plug No. 112 and disconnect connector No. 113 (fig. 298) on turret filter box. Using voltmeter or test lamp, measure voltage at pin "C" of receptacle No. 113.

      Note. Depress a firing button to get a reading.

    If voltage is not present, replace turret filter box (par 335).

    (*e*) Reconnect connector No. 113 and disconnect connector on loader's safety relay (fig. 40). Using voltmeter or test lamp, check voltage at pin "A" of loader's safety relay connector plug.

      Note. Depress a firing button to get a reading.

    If voltage is not present, replace cable from filter box to loader's safety relay.

    (*f*) Using ohmmeter, measure resistance from pin "A" to pin "B" of receptacle on loader's safety relay. If an infinite or zero reading is obtained, replace loader's safety relay (par. 288). If reading is neither infinite nor zero, fault lies in firing mechanism. Replace 90-mm gun firing mechanism (fig. 302).

l. MACHINE GUN CANNOT BE FIRED EITHER BY GUNNER'S OR BY COMMANDER'S CONTROL, BUT 90-MM GUN CAN.

Warning: Make sure machine gun is cleared of ammunition before making following tests.

    (1) *Control switches not set.* Make sure machine gun firing selector switch (fig. 28) is in the "ON" position.

    (2) *Check firing circuit.*

        (*a*) Disconnect connector No. 110 (fig. 298) on gunner's control handle assembly. Using voltmeter or test lamp, check voltage pin "D" of receptacle No. 110.

            Note. Depress a firing button to get a reading.

        If voltage is not present, replace gunner's control handle assembly (par. 284).

        (*b*) If voltage is present, reconnect plug No. 110 and disconnect connector No. 112 (fig. 298) on filter box. Using voltmeter or test lamp, check voltage at pin "D" of receptacle No. 112.

            Note. Depress a firing button to get a reading.

        If voltage is not present, replace cable from gunner's control handle assembly to turret filter box.

        (*c*) Reconnect plug No. 112 and disconnect connector No. 113 (fig. 298) on turret filter box. Using voltmeter or test lamp, check voltage at pin "A" of receptacle No. 113.

            Note. Depress a firing button to get a reading.

        If voltage is not present, replace turret filter box (par. 335).

        (*d*) Reconnect plug No. 113 and disconnect connector at machine gun firing solenoid. Using a voltmeter or test lamp, check voltage at plug of cal. .50 or cal. .30 solenoid (fig. 298).

            Note. Depress a firing button to get a reading.

        If voltage is not present, replace cable (103) from turret filter box to the solenoid.

        (*e*) Using ohmmeter, check resistance of cal. .50 or cal. .30 solenoid by connecting one lead of ohmmeter to receptacle on solenoid receptacle contact and the other to ground. If a zero or infinite reading is obtained, replace solenoid (par. 291).

m. CAL. .50 AMMUNITION DOES NOT FEED INTO AMMUNITION CHUTE.

Warning: Clear machine gun and ammunition chute of ammunition before making the following tests.

    (1) *Defective booster motor circuit.*

        (*a*) Disconnect connector on booster motor relay (fig. 30). Using voltmeter or test lamp, check voltage at pin "B" of connecting plug. If voltage is not present, replace booster motor circuit breaker (circuit 642) in turret accessories panel (par. 297).

(b) Using voltmeter or test lamp, check voltage at pin "A" of plug on booster motor relay (fig. 301).

Note. Depress a firing button to get a reading.

If voltage is not present, booster motor control circuit is defective. Proceed to (2) below.

(c) Reconnect plug on booster motor relay. Disconnect connector on cable (642) leading into booster motor (fig. 303). Using voltmeter or test lamp, measure voltage at socket contact of connector.

Note. Depress a firing button to get a reading.

If voltage is not present, replace booster motor relay (par. 293). If voltage is present, replace booster motor (par. 292).

(2) *Defective booster motor control circuit.*

(a) Remove ammunition chute micro switch cover (fig. 303) and inspect switch. Contacts should be normally closed. If switch is defective, replace switch (par. 294).

(b) Reinstall switch cover and disconnect connector No. 111 (fig. 298) on turret filter box. Using voltmeter or test lamp, check voltage at pin "A" of receptacle No. 111.

Note. Depress a firing button to get a reading.

If voltage is present, replace cable (641) from turret filter box to ammunition chute micro switch.

(c) If voltage is not present, reconnect plug No. 111 and disconnect connector No. 112 (fig. 298) on turret filter box. Using voltmeter or test lamp, check voltage at pin "D" of plug No. 112.

Note. Depress a firing button to get a reading.

If voltage is present, replace turret filter box (par. 335).

(d) If voltage is not present, reconnect plug No. 112 and disconnect connector No. 110 (fig. 298) on gunner's control handle assembly. Using a voltmeter or test lamp, check voltage at pin "D" of receptacle No. 110.

Note. Depress a firing button to get a reading.

If voltage is present, replace cable from gunner's control handle assembly to turret filter box. If voltage is not present, replace gunner's control handle assembly (par. 284).

n. VENTILATING BLOWER DOES NOT OPERATE.

(1) Disconnect connector at ventilating blower (fig. 304). Turn ventilator blower switch (fig. 30) to "ON" position. Using a voltmeter or test lamp, check voltage at connector plug. If voltage is present, replace ventilating blower (par. 298).

(2) Inspect cable from turret accessories panel to ventilating blower. If damaged, replace cable.

(3) Turn of master relay switch (fig. 18). Remove turret accessories panel (par. 297). Using ohmmeter, test ventilator blower switch (figs. 30 and 298). Reading should be zero when switch is in "ON" position and infinite when switch is in "OFF" position. If faulty, replace switch (par. 297).

(4) Using ohmmeter, test continuity of ventilating blower circuit breaker ((d) fig. 298) in turret accessories panel. Reading should be zero. If circuit breaker is defective, replace circuit breaker (par. 297).

o. TURRET DOME LIGHTS DO NOT OPERATE.

(1) *Both dome lamps inoperative.* If both dome lights are inoperative, the trouble is probably in the dome lights circuit breaker ((c) fig. 298) in the turret accessories panel. Using ohmmeter, check resistance of circuit breaker.

Caution: Be sure master relay switch is in the "OFF" position or ohmmeter will be damaged.

Reading should be zero. If it is not zero, replace circuit breaker (par. 297).

(2) *Both bulbs in one dome lamp inoperative.* If one dome lamp will not operate on either red or white lens, trouble is probably in dome light switch. Replace dome light (par. 232).

(3) *One bulb in one dome lamp inoperative.* If only one bulb does not operate, replace bulb (par. 232).

p. BALLISTIC BOX.

(1) Disconnect power cable connector at ballistic drive (fig. 54). Turn on master relay switch (fig. 18). Using voltmeter or test lamp, check voltage at connector plug. If voltage is present, ballistic drive lamp or switch is faulty. Notify ordnance maintenance personnel.

(2) Replace plug on ballistic drive. Turn off master relay switch (fig. 18). Remove turret accessories panel (par. 297). Using voltmeter, check continuity of ballistic drive circuit breaker (fig. 298) in turret accessories panel. Reading should be zero. If it is not, replace circuit breaker (par. 297).

q. Turret electrical system resistance and continuity tests.

Caution: Make sure master relay switch (fig. 18) is in the "OFF" position before readings are taken.

Refer to figure 298 for location of connectors mentioned below.

Component	Circuit No.	Test From—	Test To—	Resistance (ohms)	Circuit (fig. 299)	Remarks
Circuit breaker, override	111	R104	R102–H	0	R104 through circuit breaker to R102–H.	
Circuit breaker, traverse safety	625	R104	R103–C	0	R104 through circuit breaker to R103–C.	
Circuit breaker, turret motor	100	See remarks	See remarks	0		Remove override motor control box cover and measure across circuit breaker.
Relay, override, section I	111, 111A	R102–G	R106–G	0	R102–G through switch to R106–G.	
Relay, override, section II	601	R102–F	R105–B	0	R102–F through pole 1 of switch to R105–B.	
	602	R102–E	R105–C	0	R102–E through pole 2 of switch to R105–C.	
	609	R102–D	R105–E	0	R102–D through pole 3 of switch to R105–E.	
	010	R102–C	R105–F	0	R102–C through pole 4 of switch to R105–F.	
Relay, traverse safety	654A, 645B	R103–B	gnd	65	R103–B through traverse safety relay coil to ground.	
Resistors	600, 608	R105–A	R105–D	10	R105–A through two resistors to R105–D.	
Solenoids, override relay	623	R106–B	gnd	260	R106–B through two parallel solenoids to ground.	A reading of 520 ohms indicates one damaged solenoid.

140. Auxiliary Engine

a. GENERAL. Auxiliary engine trouble shooting is, in general, similar to trouble shooting of vehicle engine (par. 126). Consideration, however, must be given to the fact that only one magneto and one carburetor are used on the auxiliary engine and that paragraph references cited in paragraph 126 refer to vehicle engine.

b. ENGINE FAILS TO START.

(1) *Defective magneto switch or starter switch.* If engine fails to start when auxiliary magneto switch (fig. 18) is turned on and starter button fig. 18) is pressed in, attempt to start engine manually (par. 93). If engine still does not start, inspect magneto switch and wiring (par. 141). Replace as required (par. 210). If engine starts when cranked manually, check starter switch and wiring (par. 141). Replace as required (par. 209).

(2) *Worn or fouled spark plugs.* Badly worn or cracked spark plugs, or plugs fouled by carbon or lead will prevent engine starting. Remove and clean plugs (par. 311), or replace if required.

(3) *Magneto breaker points out of adjustment.* Adjust magneto breaker points (par. 310) to 0.012-inch gap as measured with feeler gage.

c. ENGINE STARTS BUT FAILS TO KEEP RUNNING.

(1) *Fuel shut-off valve closed.* If batteries are low or completely discharged, fuel shut-off valve (fig. 323) will not open electrically. Open manually before starting engine (par. 93).

(2) *Clogged fuel lines or filter.* Remove glass bowl and clean (par. 307). Clean fuel lines.

(3) *Damaged fuel pump.* Disconnect fuel line from carburetor (fig. 310). Crank engine with manual starter handle and check fuel flow through line. If flow is inadequate, replace fuel pump (par. 308).

d. ENGINE FAILS TO RUN AT FULL SPEED.

(1) *Governor adjusted incorrectly.* Check to see if governor spring (fig. 310) is hooked to tenth hole (third from top) in lever. Adjust if required.

(2) *Cylinders missing fire.* If engine runs unevenly because of cylinders missing fire, check both spark plugs (par. 311), spark plug cables, and magneto breaker point gap (par. 310). If trouble cannot be corrected, notify ordnance maintenance personnel.

141. Hull Accessories Panel Components and Auxiliary Equipment

See paragraph 125 for methods of testing with an ohmmeter. On all tests in this chapter in which a circuit breaker is included in the circuit being tested, resistance values given assume that circuit breaker is closed. If no reading is obtained on the ohmmeter, a continuity test across the circuit breaker should be made from the rear of the panel.

Caution: Make sure ohmmeter test prod does not touch plugs No. 3 or 5 or ohmmeter will be ruined.

Refer to figure 195 for connector numbers.

142. Radio Interference Suppression System

a. PRELIMINARY INSTRUCTIONS.
  (1) When radio interference resulting from operation of the vehicle is reported or experienced, the vehicle should be moved to a place which is comparatively free of high-tension lines, other vehicles, or machinery which could be a source of interference.

  > *Note.* If vehicle to be checked for insufficient radio interference suppression is not radio-equipped, keep another vehicle which is radio-equipped within 5 feet of the vehicle to be tested.

  (2) With engine operating but vehicle electrical equipment not operating, turn on radio. Notice type of noise evident under this condition so that, when checking equipment of vehicle, presence of new noise or interference can be detected.

  (3) If noise level without any vehicle equipment in operation is too high because of atmospheric or other outside causes, and if tactical situation will permit, delay further checking until such time as a moderate noise level prevails.

  (4) Disconnect hull and turret radio terminal box capacitors. If radio interference does not increase when these capacitors are disconnected, replace them. If interference does increase, leave terminal box capacitors disconnected for remainder of tests until defective equipment has been located, replaced, and tested.

  (5) Examine all shielded conduits and cables to make sure that couplings are tight and conduits and cables are clamped or bonded to hull at least every 2 feet.

b. RADIO INTERFERENCE WITH VEHICLE NOT IN MOTION, BUT WITH ENGINE RUNNING.
  (1) With engine running at 1,100 rpm, turn magneto switch (fig. 18) to "A" and then to "F." If interference is eliminated with either one of the two sets of magnetos off, interference may be attributed to those components of the ignition

207

Component	Circuit No.	Test From—	Test To—	Resistance (ohms)	Circuit	Remarks
Breaker, auxiliary engine starter and magneto circuit.	421	R7-B	R3	0	R10-B through magneto switch through circuit breaker to R7 (fig. 192).	Turn auxiliary engine switch to "ON" position.
Breaker, front bilge pump circuit.	452	R1	R3	0	R1 through front bilge pump switch, through circuit breaker to R3 (fig. 192).	Turn front bilge pump switch to "ON" position.
Breaker, rear bilge pump circuit.	451	R2	R3	0	R2 through rear bilge pump switch through circuit breaker to R3 (fig. 192).	Turn rear bilge pump switch to "ON" position.
Outlet, accessory	37	See circuit.	See circuit.	0	One test prod on center terminal of outlet; other prod at R3 (fig. 192).	Infinite reading indicates open circuit breaker.
Pump, bilge (front)	452	P1	gnd	0.9–1.2	P1 through motor winding to ground (fig. 181).	
Pump, bilge (rear)	451	P2	gnd	0.9–1.2	P2 through motor winding to ground (fig. 181).	
Switch, auxiliary engine magneto.	421	R7-B	R3	0 or inf	R7-B through fuel valve section of switch through circuit breaker to R3 (fig. 192).	Zero when magneto switch is on; infinite when magneto switch is off.
	422	R7-A	gnd	0 or inf	R7-A through magneto section of switch to ground (fig. 192).	
Switch, auxiliary engine starter.	65	R7-C	R7-B	0 or inf	R7-C through switch to R7-B (fig. 192).	Zero when starter switch depressed; infinite when not depressed.

Switch, bilge pump (front).	452	R1	R3	0 or inf	R1 through switch, through circuit breaker to R3 (fig. 192).	Zero when switch is on; infinite when switch is off.
Switch, bilge pump (rear).	451	R2	R3	0 or inf	R2 through switch, through circuit breaker to R3 (fig. 192).	Do.
Switch, heater in "ON" or "OFF" position.	------	R4–A	R5	0 or inf	R4–A through switch to R5 (fig. 192).	Zero when heater switch is on; infinite when heater switch is off.
Switch, heater reset	401, 402	R4–A	R4–F	0 of inf	R4–A through first section of switch to R4–F (fig. 192).	Zero when switch is in "RESET" position; infinite in "OFF" position.
	411, 402	R4–B	R4–F	0 or inf	R4–B through second section of switch to R4–F (fig. 192).	Infinite when switch is in "RESET" position; zero in "OFF" position.
	410, 411	R4–C	R4–B	0 or inf	R4–C through third section of switch to R4–B (fig. 192).	Do.

system served by that set of magnetos. Tighten all magneto conduit coupling connectors, and see that magneto ground wire is tightly secured at magneto ground terminal. Adjust breaker points (par. 184), replace magneto (par. 185), or replace defective spark plugs (par. 186), as required.

(2) If interference is in the form of an irregular clicking noise that continues a few seconds after magnetos are turned off, it may be attributed to the generator regulator units in the master junction box. Inspect master junction box (fig. 182) and generator (fig. 206) ground straps to be sure that good electrical connections are obtained. If this does not remove the interference, replace master junction box (par. 194).

(3) If interference is in the form of a whining noise that varies in pitch with engine speed and continues, but at a lowering pitch a few seconds after the magnetos are turned off, it may be attributed to the generating system. Check generator ground cable (fig. 206) by using a jumper wire to make connection between the generator frame and the generator mounting bracket. If a good ground connection corrects the interference, clean and tighten connections at each end of the ground cable.

(4) If interference does not cease when a good generator-to-bracket ground is established, replace the booster-and-filter coil assembly (par. 188).

(5) If interference still persists, the generator positive brush capacitors are at fault, or the brushes and commutator are in need of maintenance or adjustment. Replace brushes or generator (par. 222).

c. RADIO INTERFERENCE WITH AUXILIARY GENERATOR OPERATING.

(1) Examine auxiliary generator ground cable (63) for good electrical connection between auxiliary generator and hull. If interference is in the form of a regular clicking noise occurring at the same frequency as the exhaust reports, and stopping when the auxiliary engine magneto switch (fig. 18) is turned off, the auxiliary engine spark plug suppressor is at fault. Replace plug (par. 311).

(2) If the interference is in the form of a whining noise which continues at a lowering pitch a few seconds after the magneto switch is turned off, it may be attributed to the generating system. Check auxiliary generator ground cable (63) connection by using a jumper wire. If a good ground connection corrects the interference, clean and tighten connections at each end of the ground cable.

(3) If interference still persists after a good ground connection is made, replace magneto capacitor (par. 310).

(4) If interference still persists after magneto capacitor is replaced, replace auxiliary generator or brushes (par. 304).
(5) If interference is in the form of an irregular clicking noise which continues for a few seconds at a slower rate after auxiliary engine magneto switch (fig. 18) is turned off, it may be attributed to generator regulator units in master junction box. Inspect master junction box (fig. 182) and generator (fig. 308) ground straps to be sure that good electrical grounds are obtained. If this does not remove the interference, replace master junction box (par. 194).

d. Radio Intereference While Turret Is Being Traversed.
(1) Check turret ground cable (fig. 306) to make sure it is securely mounted and makes good electrical contact both at slip ring box connection and at mounting legs.
(2) Tighten connectors at all junction points of shielded cables and see that conduits are bonded to hull at least every 2 feet.
(3) Examine turret filter box. Disconnect each filter box capacitor in turn. If removing any capacitor makes no change in noise level, replace turret filter box (par. 335).
(4) Examine turret override motor control box capacitors. Disconnect each capacitor in turn. If removing either capacitor makes no change in noise level, replace override motor control box (par. 282).
(5) If interference still persists, turret traversing motor is at fault. Notify ordnance maintenance personnel.

e. Radio Interference With Ventilating Blower Operating.
(1) If noise is present when ventilating blower is operating, but stops as soon as blower is turned off, inspect shielding on blower cables and tighten cable connectors
(2) If interference persists, replace ventilating blower assembly (par. 298).

f. Radio Interference With Crew Compartment Heater Operating.
(1) If noise is present when crew compartment heater is operating, but stops as soon as heater is turned off, check heater ground straps.
(2) If interference persists, replace combustion air blower (par. 324) or ventilating air blower (par. 323), as required.

g. Excessive Radio Interference With Vehicle in Motion. If irregular crackling noise caused by motion of vehicle is believed to be excessive, examine all ground straps and toothed lock washers to make sure they are secure and making good electrical connection.

143. 90-mm Gun T119E1

  a. GENERAL. Although the malfunctions, described below, are rarely encountered when authorized and properly maintained ammunition is fired in properly maintained and operated weapons, it is important that all personnel concerned understand the nature of each kind of malfunction as well as the proper preventive and corrective procedure in order to avoid injury to personnel or damage to matériel.

  b. DEFINITIONS.

  (1) *Misfire.* A misfire is a *complete failure to fire* which may be due to faulty firing parts or a faulty element in the propelling charge explosive train. A misfire in itself is not dangerous but since it cannot be immediately distinguished from a delay in the functioning of the firing parts or from a hangfire, it should be considered as a possible delayed firing until such possibility has been eliminated. Such delay in the functioning of the firing parts, for example, could result from the presence of foreign matter such as grit, sand, frost, ice, or improper or excessive oil or grease, which might create initially a partial mechanical restraint which, after some indeterminate delay, is overcome as a result of the continued force applied by the spring, and the firing pin then driven into the primer in the normal manner. In this connection, no round should be left in a hot weapon any longer than the circumstances require to the possibility of a cook-off.

  (2) *Hangfire.* A hangfire is a delay in the functioning of a propelling charge explosive train at the time of firing. The amount of the delay is unpredictable but in most cases will fall within the range of a split second to several minutes. Thus, a hangfire cannot be distinguished immediately from a misfire and therein lies the principal danger—that of assuming that a failure of the weapon to fire immediately upon actuation of the firing linkage is a misfire whereas in fact it may prove to be a hangfire. It is for this reason that the time intervals prescribed in *d* below should be observed before opening the breech after a failure to fire. This time interval, based on experience and considerations of safety, has been established to minimize the danger associated with a hangfire and to prevent the occurrence of a cook-off.

  Caution: During the prescribed time interval, the weapon will be kept trained on the target and all personnel will stand clear of the muzzle and the path of recoil.

  (3) *Cook-off.* A cook-off is a functioning of any or all of the explosive components of a round chambered in a very hot weapon due to heat from the weapon. The primer and pro-

pelling charge, in that order, are in general more likely to cook off than the projectile of the fuze. If the primer or the propelling charge should cook off, the projectile *may* be propelled (fired) from the weapon with normal velocity even though no attempt was made to fire the primer by actuating the firing linkage. In such a case, although there may be uncertainty as to whether or when the round will fire, the precautions to be observed are the same as those prescribed for a hangfire. *However, should the bursting charge explosive train cook off, injury to personnel and destruction of the weapon may result.* To prevent a cook-off, a round of ammunition which has been loaded into a very hot weapon should be fired or removed within the time prescribed below, to prevent heating to the point where a cook-off may occur.

Caution: In case of a round, chambered in a very hot weapon which can neither be fired nor removed, all personnel will stand clear of the weapon until such time as the weapon and chambered round are cool, to avoid the danger from a possible cook-off.

c. FAILURE OF ROUND TO CHAMBER. If a round is impelled with sufficient force but fails to chamber, it is malformed. Proceed as follows:
  (1) While another crew member holds the breech operating handle down to prevent breechblock from accidental rising, install the extracting and ramming tool 7238638 so that the fulcrum slides in the breechblock guides in the breech ring, and the ramming plates contact the base of the cartridge.

  Caution: Do not contact primer.
  (2) Slowly pivot the tool to seat the round in the chamber.
  (3) With the round seated in the chamber, remove the tool and release the breechblock.

d. FAILURE TO FIRE. Because of the possibility of a hangfire (*b* above), the following procedure will be observed after any failure to fire:
  (1) Keep the weapon trained on the target and all personnel clear of the muzzle and path of recoil.
  (2) Recock the percussion mechanism by pressing down and releasing the cocking lever on the right side of the breechblock (fig. 41) and make two additional attempts to fire as quickly as possible.
  (3) After the last attempt to fire has been made, wait 2 minutes before opening the breech and attempting to remove the round. Personnel not required for this operation will be cleared from the tank.

  Warning: If the gun tube is very hot from continuous firing, only one attempt will be made to remove the defective round. This will be done by opening the breech slowly while another member of the crew catches the round as it is extracted. If the round cannot be removed, the breech will be closed and all personnel will evacuate the tank until such time as the gun is cool. If a cook-off does not occur by the time the tube is cool the defective round will then be removed and disposed of; under no circumstances will a round be fired which has been allowed to remain in a hot tube.

(4) Open the breech very slowly while another member of the crew catches the round as it is extracted.
(5) If fused ammunition is being used, set the fuse on Safe.
(6) Examine the primer for indent.
(7) If the primer has a normal indent, the round is defective and should be turned over to ordnance personnel. Insert another round and resume firing.
(8) If the primer has a light indent or no indent at all, failure to fire was caused by a malfunction of the weapon or mount. Proceed as follows:
   (a) See if gun is in battery.
   (b) If gun is not in battery, refer to paragraph 144.
   (c) If gun is in battery, failure to fire may be caused by improper functioning of the percussion mechanism, sear, or cocking mechanism parts. Proceed as follows:
      1. Remove the percussion mechanism, sear, and cocking mechanism parts (par. 342b).
      2. Disassemble the trigger plunger group (par. 343a) and the percussion mechanism (par. 343b).
      3. Wash parts with rifle bore cleaner, drying-cleaning solvent, or volatile mineral spirits. Remove any burs with oilstone or crocus cloth. Check firing pin for broken or damaged point. Replace if necessary. Check firing pin retracting spring (fig. 347) for wear or damage and replace if necessary. If any of the component parts of the percussion mechanism, other than the aforementioned pin and spring, are damaged, replace the percussion mechanism. If the firing spring, the sear spring, the cocking lever shaft spring, or the trigger plunger spring (fig. 346) is weak or broken, it should be replaced. Damage to the sear will necessitate notifying ordnance maintenance personnel. Lubricate as specified on the lubrication order.
      4. Assemble the percussion mechanism (par. 345a) and the trigger plunger group (par. 345b).
      5. Install the percussion mechanism, sear, and cocking mechanism parts (par. 346a).

6. If gun fails to fire, notify ordnance maintenance personnel.

e. PREMATURE FIRING. If the gun fires when the breech is closed but before the firing linkage is actuated, the cause may be that the sear holds the firing pin guide insecurely or that the jar of the breechblock closing the breech releases the firing pin due to worn cocking mechanism parts, sear, or sear spring. Perform *1* through *5* as in (*c*) above.

f. FAILURE OF BREECHBLOCK TO CLOSE.
  (1) *No round in chamber.*
    (a) If breechblock does not rise at all when extractors are tripped, it is seized. Notify ordnance maintenance personnel.
    (b) If breechblock rises but does not close completely when extractors are tripped, the cause may be a weak or improperly adjusted closing spring, removable obstruction (nicks, burs, and rough spots), or improper lubricant on operating surfaces of breechblock or in breech recess. Adjust closing spring if necessary (par. 346*b*). If closing spring is broken, notify ordnance maintenance personnel. Remove, maintain, and install the breechblock (pars. 342*a*, 344, and 346*b*).
  (2) *A round in chamber.*
    (a) If breechblock does not rise when the extractors are tripped by the base of the round, the breechblock may be seized. Notify ordnance maintenance personnel.
    (b) If breechblock rises but does not close completely, the cause may be a weak or incorrectly adjusted closing spring, removable obstructions (nicks, burs, and rough spots), or improper lubricant on operating surfaces of breechblock or in breech recess. Release the breech operating handle (fig. 342) and very slowly press down on the operating handle lever until the extractors lock the breechblock in the open position, while another crew member grasps the round as it comes out of the chamber and removes it. With the round removed, proceed as in (1) (*b*) above.

g. FAILURE OF BREECHBLOCK TO OPEN.
  (1) *No case in chamber.*
    (a) If breechblock cannot be manually opened at all, it is seized. Notify ordnance maintenance personnel.
    (b) If breechblock is closed at end of counterrecoil, with gun in battery, the extractor plunger springs may be weak or broken. Inspect and replace the springs if necessary (pars. 343 and 345).

(2) *Case in chamber.*
  (*a*) If the breechblock is completely closed at end of counterrecoil, with gun in battery, the cause may be a broken operating crank, broken operating cam, or broken or weak operating cam spring. Notify ordnance maintenance personnel.
  (*b*) If the breechblock is completely closed or only partly open, with gun not in battery, the cause may be weak or broken recoil spring or an excessive amount of oil in the recoil mechanism. Check replenisher indicator (par. 358*c*). If it shows too much oil, bleed to proper amount (par. 358*c*). If malfunction recurs, notify ordnance maintenance personnel.

h. FAILURE TO EXTRACT CARTRIDGE CASE.
  (1) If the extractors fail to extract the fired cartridge case from the chamber when the breech is opened, and if the tactical situation permits, insert the rammer staff into the muzzle end of the bore and tap the bottom of the inside of the case lightly until it is loosened and can be pushed out of the chamber.
  (2) If the tactical situation does not permit extracting the case in this manner, remove the case from the breech end, using extracting and ramming tool 7238638. Exercise caution in handling the empty case as it may still be hot.
  (3) While another crew member holds the breech operating handle (fig. 342) down to prevent accidental rising of the breechblock, install the extracting and ramming tool so that the tips of the fork fit in between the breech face and the rim of the cartridge case and the plug on the fork contacts the front side of the breech recess.
  (4) Pivot the tool up about its plug to extract the case. With the case extracted, remove the tool and release the breechblock.
  (5) If the extractors are broken, notify ordnance maintenance personnel.

i. UNLOADING A STUCK ROUND. When a round is stuck in the gun and it is either impossible or inadvisable to fire it out, it will be removed, except in combat, under the direct supervision of an officer. Ordinarily it is impossible to remove a stuck round with the ramming and extracting tool 7238638. Instead, the bore cleaning staff, with rammer head 41–H–1826–150 (fig. 85) attached, must be employed. Insert the rammer head in the muzzle of the gun and push it gently down the bore until it is seated on the ogive of the projectile. With the breech open, prepare to receive the round as it is pushed from the chamber. Exerting a steady pressure, shove the round clear so that it may be removed. If the weight of several men against the staff does not suffice, apply leverage by means of a 2 x 4-inch board, or other suitable object, connected to the tank by a rope at one end.

Caution: Under no circumstances should the staff be used to hammer against the projectile.

Keep all parts of the body as clear as possible from the muzzle or breech during the operation. If this procedure fails to remove the round, notify ordnance maintenance personnel.

144. Combination Gun Mount M78

a. Gun Returns to Battery with too Great a Shock. Check the oil level in replenisher (par. 358c) and add recoil oil if needed. If level is correct, or trouble continues, check to see that buffer regulator is fully closed (par. 356b).

b. Gun Overrecoils. Check oil level in replenisher (par. 358) and add oil if needed. If trouble continues with oil at correct level, overrecoil may be due to weak or broken recoil spring. Notify ordnance maintenance personnel.

c. Gun Underrecoils. Too much oil in recoil mechanism will cause underrecoil. Check oil level in replenisher and if too high, bleed off excess oil (par. 358c).

d. Gun Does Not Return to Battery. Check replenisher indicator (par. 358c) and if oil level is too high, bleed off excess oil (par. 358c). Under extremely cold conditions, trouble may be due to increased viscosity of the recoil oil. Open buffer regulator to allow gun to return to position (par. 356b).

Note. As firing continues, recoil oil will quickly become warm and buffer regulator should be closed again.

If malfunction continues, notify ordnance maintenance personnel.

e. Erratic Recoil Action. If recoil action is too violent, incomplete or erratic in any way, the most likely cause is either too much or too little oil in the recoil mechanism. Check this first by reading the tape of the replenisher oil indicator (par. 358c). If oil level is correct and corrections given above do not eliminate the erratic action, notify ordnance maintenance personnel.

Section V. ENGINE DESCRIPTION AND MAINTENANCE IN VEHICLE

145. Description and Data

a. Description. The medium tank M47 is powered by a 12-cylinder, V-type, 4-cycle, air-cooled, Continental engine model AV–1790–5B (figs. 96 and 97). The engine has overhead valves with a single overhead camshaft for each bank of six cylinders. The camshaft actuates the valves by means of rockers located in each cylinder head. The cylinders are individual replaceable units. Two conventional, double-venturi, downdraft carburetors are used, one for each bank of cylinders. Fuel is supplied to the carburetors by two interconnected fuel pumps mounted on the lower left side of the accessory case. Four magnetos are used, two for each bank of six cylinders, to supply dual ignition for each cylinder. An ignition harness connects the 2 magnetos to the 12 spark plugs on each side of the engine. The ignition harness and magnetos are shielded to prevent radio interference. Two mechanically driven fans (fig. 96), located on top of the engine, are provided to circulate air around the cylinders for air cooling. The engine, transmission, and oil coolers are considered as one unit (power plant) for removal and installation (ch. 3, sec. VI).

b. ENGINE NOMENCLATURE (fig. 98). The "front" or accessory end of the engine will be referred to as the accessory end. The "rear" or flywheel end of the engine will be referred to as the flywheel end. The term "left" and "right," as used with reference to the engine, are as viewed from the accessory end and looking toward the flywheel end.

Note. With the engine installed, the right bank of cylinders is on the left side of the vehicle and the left bank of cylinders is on the right side of the vehicle.

Cylinders are numbered beginning from the accessory end and are designated 1R, 2R, etc., for cylinders on the right bank and 1L, 2L, etc., for cylinders on the left bank. Viewing the engine from the accessory end, the crankshaft rotates in a clockwise direction and the two camshafts in a counterclockwise direction. The cooling fans rotate in a clockwise direction as viewed from above the engine.

c. TABULATED DATA.

Name and type	Continental, 12-cylinder, air-cooled, **V**-type
Model	AV-1790-5B
Over-all dimensions (including flywheel assembly):	
Length	73.69 in
Width	61.00 in
Height	40.88 in
Horsepower (gross)	810 at 2,800 rpm
Torque (gross)	1,560 lb-ft at 2,400 rpm
Cylinder bore	5.75 in
Piston stroke	5.75 in
Piston displacement	1,791.75 cu in
Compression ratio	6.5:1
Maximum governed speed (no load)	2,950 rpm
Maximum governed speed (full load)	2,800 rpm
Idling speed	650 rpm
Warm-up speed	1,000 to 1,100 rpm
Dry weight (complete with flywheel and transmission adapter)	2,505 lb

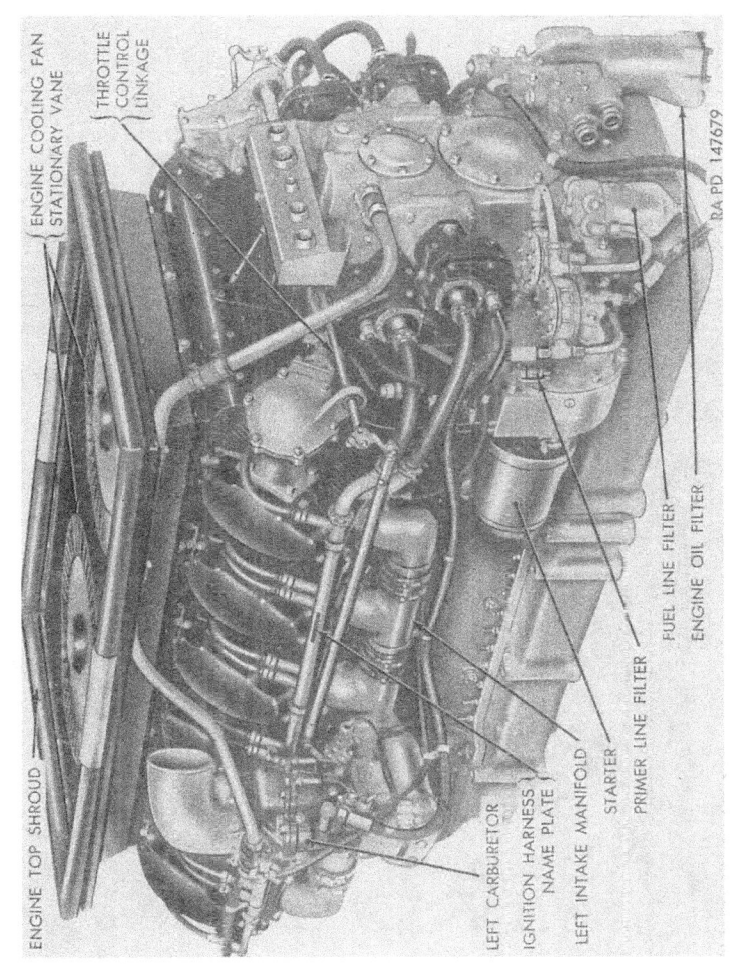

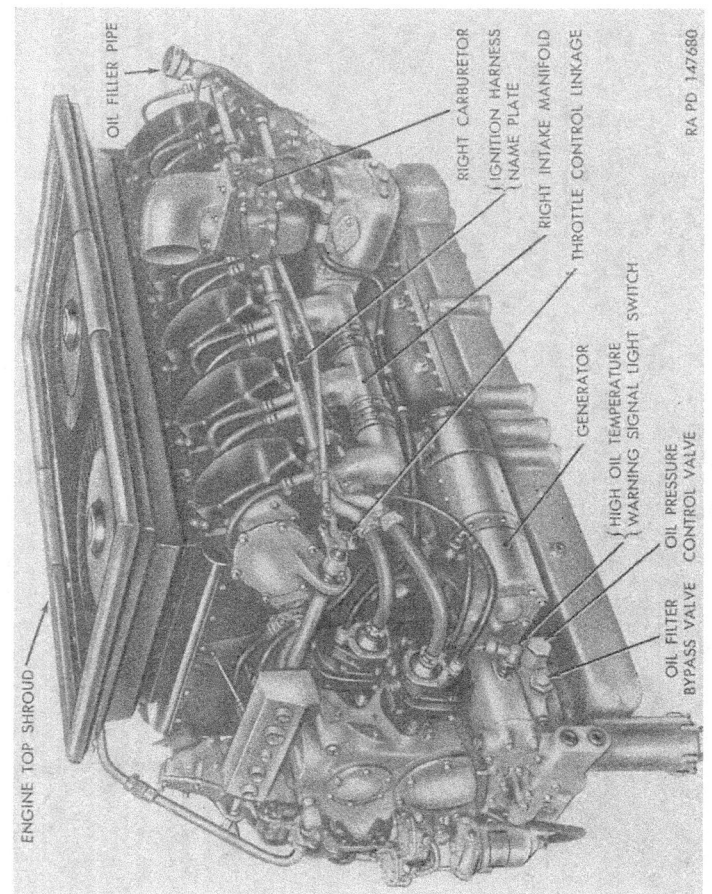

Figure 97. Right front view of engine.

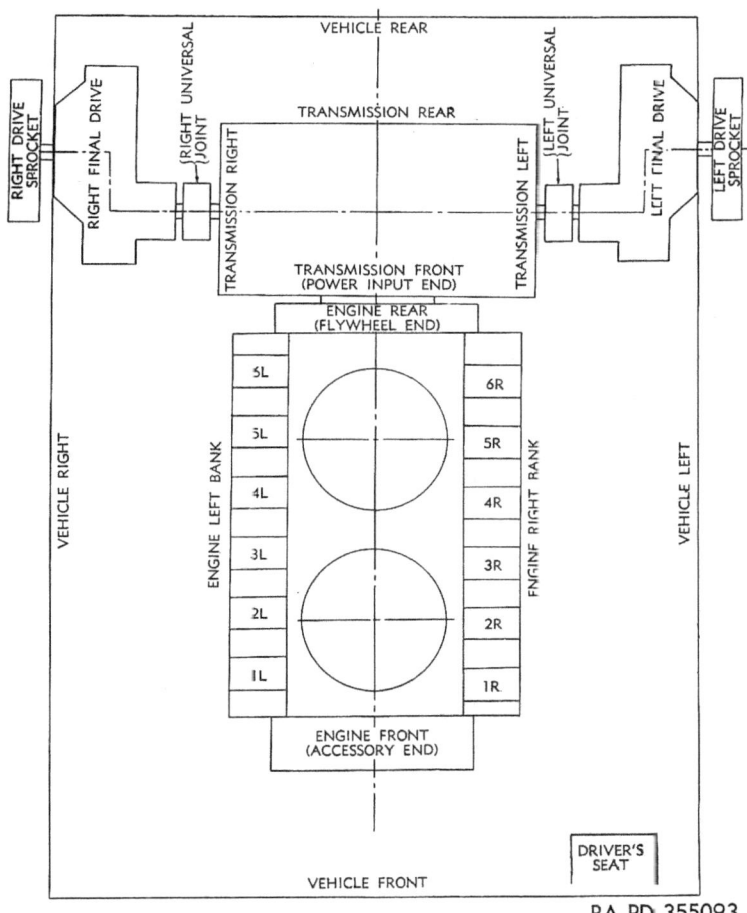

Figure 98. Power plant position identification diagram.

Cylinder arrangement_____ individual cylinders in a 90° upright **V**
Cylinder cooling_____ air supplied by 2 integral fans
Crankshaft rotation (viewed from accessory end)_____ clockwise
Camshaft rotation (viewed from accessory end)___ counterclockwise
Cooling fan rotation (viewed from above engine)_____ clockwise
Firing order_____ 1R, 2L, 5R, 4L, 3R, 1L, 6R, 5L, 2R, 3L, 4R, 6L
Valve timing setting_____ intake valve closes 50° after bottom center with 0.100-in clearance
Ignition timing_____ automatic advance, set 10° before top center

Accessory drive ratios:

Cooling fans	1.78 crankshaft speed
Tachometer drive	0.5 crankshaft speed
Generator drive	2.56 crankshaft speed
Magnetos	0.5 crankshaft speed
Power take-off	1.00 crankshaft speed
Governor	1.27 crankshaft speed
Starter	1.5 crankshaft speed
Fuel pump	0.58 crankshaft speed

146. Engine Tune Up

a. IGNITION. Check spark plugs and make any necessary replacements (par. 186). Check magneto breaker points and spark intensity (par. 129).

b. BATTERY CONNECTIONS See that all battery connections and ground connections to hull are clean and tight.

c. COMPRESSION. Check cylinder compressions (par. 123, item 23).

d. VALVES. Check valve clearances and make necessary adjustments (par. 155*b*).

e. CARBURETOR. Adjust idle mixture adjusting screws to get smoothest idling (par. 172*a*(2)). Check and adjust engine idle speed (par. 172*a*(3)).

147. Operations Performed With Engine in Vehicle

a. AIR CLEANERS. Service or replace (par. 175).

b. BOOSTER AND FILTER COILS. Replace (par. 188).

c. CARBURETORS. Adjust or replace (par. 172).

d. COOLING FAN ROTOR. Replace (par. 154).

e. CYLINDER AIR DEFLECTORS. Replace (par. 166).

f. ENGINE-LOW-OIL-PRESSURE-W A R N I N G-SIGNAL-LIGHT SWITCH. Replace (par. 218).

g. ENGINE LUBRICATION SYSTEM. Service (par. 148*b*, *c*, and *d*).

h. ENGINE OIL COOLER BYPASS VALVE. Replace (par. 151).

i. ENGINE OIL FILTER BYPASS VALVE. Replace (par. 151).

j. ENGINE OIL PRESSURE CONTROL VALVE. Replace (par. 150).

k. ENGINE OIL PRESSURE GAGE SENDING UNIT. Replace (par. 217).

l. ENGINE HIGH - OIL - TEMPERATURE - WARNING - SIGNAL - LIGHT SWITCH. Replace (par. 218).

m. ENGINE TOP SHROUD. Replace (par. 152).

n. ENGINE WIRING JUNCTION BOX. Replace (par. 195).

o. EXHAUST MANIFOLD. Replace (par. 153).

p. IGNITION HARNESS. Replace (par. 187).

q. MAGNETOS. Adjust or replace (par. 185).

r. OIL COOLERS. Replace (par. 242).

s. PRIMER LINES. Replace (par. 181).
t. SPARK PLUGS. Replace (par. 186).
u. TACHOMETER SENDING UNIT. Replace (par. 217).
v. VALVES. Check timing and adjust (par. 155).
w. VALVE ROCKERS. Replace (par. 156).

148. Engine Lubrication System

a. DESCRIPTION (fig. 99).

(1) Two positive displacement-gear-type oil pumps, scavenge and main, are inclosed in a single housing mounted to a machined mounting pad on the lower side of the accessory case. An inlet to the oil cooler is located on the discharge side of the main oil pump. Oil from the main oil pump is circulated externally to the oil coolers. It reenters the accessory case from the oil coolers. Adapters are provided for the external oil connections. During normal operation, oil passes from the main oil pump through the engine and transmission oil coolers and returns to the engine through the engine oil filter. Three valves, the oil cooler bypass valve, the oil pressure control valve, and the oil filter bypass valve control the flow of oil. Oil from the pressure inlet flows to the oil filter chamber through drilled passages in the accessory case.

(2) An oil cooler bypass valve (fig. 131) is located at the outlet of the main oil pump (in the upper portion of the filter housing). The purpose of this spring-balanced valve is to permit the oil to bypass the oil coolers. When the pressure differential exceeds 50 psi the valve opens and the oil passes directly to the return line.

(3) The oil filter bypass valve, located on the lower right side of the accessory case (fig. 131), is a balanced valve similar to the oil cooler bypass valve. If the oil filter becomes clogged, the bypass valve opens (50 psi) and bypasses the oil around the filter.

(4) An oil pressure control valve (fig. 131), placed on the oil inlet passage of the oil filter chamber, is set for the desired maximum engine oil pressure. If the pressure exceeds the specified amount, the valve opens and allows oil to return to the oil pan. Excess oil is passed directly to the oil pan.

(5) Drilled passages lead from the oil filter through the accessory case to provide lubrication for all bushings in the accessory case. A drilled passage at the right side of the accessory case connects to the oil gallery running the full length of the crankcase. A drilled passage extends from each main bearing web into the oil gallery. These passages are connected to annular grooves in the center of the main

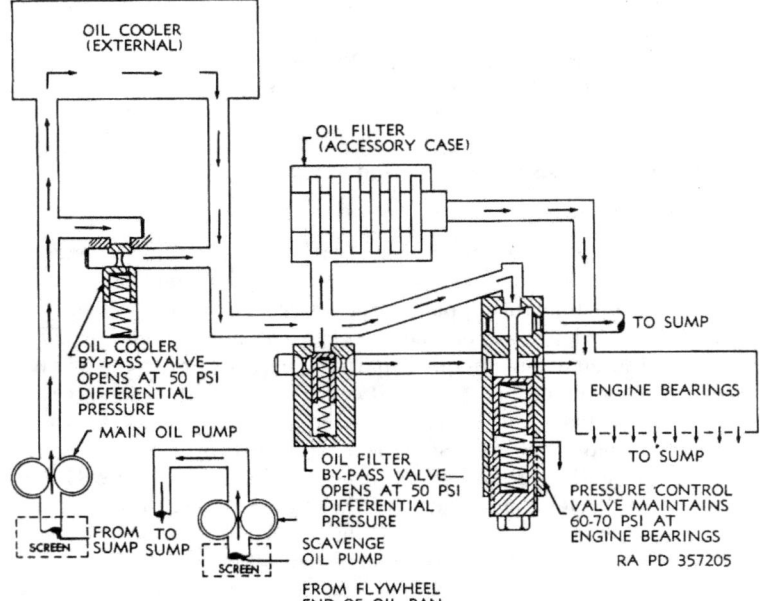

Figure 99. Oil flow diagram.

bearing caps and supports which coincide with holes in the main bearings. Steel tubes in the crankshaft direct the flow of pressure oil from the annular groove in the main bearings through the hollow main journals and hollow crankpins to provide full-pressure lubrication to the connecting rod bearings. Oil thrown off from the connecting rods lubricates the cylinder bores, piston pins, pistons, and rings.

(6) Oil for lubrication of both camshafts is picked up from supply lines in the accessory case. Several connecting passages transfer the supply to the camshaft housings and to a groove in the camshaft front bearing which feeds it under pressure to the hollow camshafts. Each bearing area on the camshaft is drilled to lubricate the bearing and also allow flow of oil into grooves in the rocker shaft support brackets. Drilled passages in the brackets conduct the flow to the rocker shafts. The shafts are drilled for lubrication of the rocker bushings. Throw-off oil from the camshafts and rocker rollers lubricates the valve stems. In addition, a drilled oil passage in the cylinder half of the camshaft bearing lubricates the exhaust valve stems.

(7) Surplus oil from the rocker boxes is returned to the crankcase through external oil drain lines (fig. 100). Surplus oil from

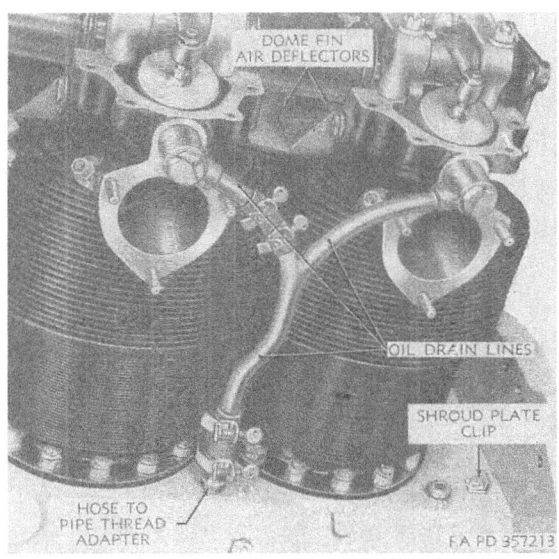

Figure 100. Rocker box oil drain lines.

the cam drive gears is drained back through the drive quill cover.

b. ENGINE OIL LEVEL CHECK. To check the engine oil, open the engine compartment 3 front left grille doors (fig. 101) to reach the engine oil level indicator gage (fig. 102) which is between right cylinders 1 and 2. Remove the oil gage, wipe it dry with a clean cloth, and then fully insert the gage into its housing. Remove the gage and read the oil level. Add oil (*d* below), as specified on the lubrication order (fig. 86), to bring the oil up to the "FULL" mark on the gage.

c. DRAIN ENGINE LUBRICATION SYSTEM. Drain oil only when warm. From underneath the vehicle, remove the engine oil drain access hole cover from the hull floor (figs. 103 and 235). Clean the oil pan around the drain plug. Position a clean container under drain plug and remove the locking wire, plug, and oil screen. Wash sediment from screen and plug in volatile mineral spirits or dry-cleaning solvent and, after oil has completely drained, reinstall the screen and plug. Secure plug with locking wire. Install access hole cover.

Note. If the engine is to be left dry after the engine oil has been drained, place a tag in a conspicuous place in driver's compartment (on master relay switch) indicating that engine oil is drained and that the engine must not be started.

d. FILL ENGINE LUBRICATION SYSTEM. Raise the engine compartment three rear left grille doors (fig. 101) to reach the engine oil filler cap (fig. 104). Pour engine oil, as specified on lubrication order (fig.

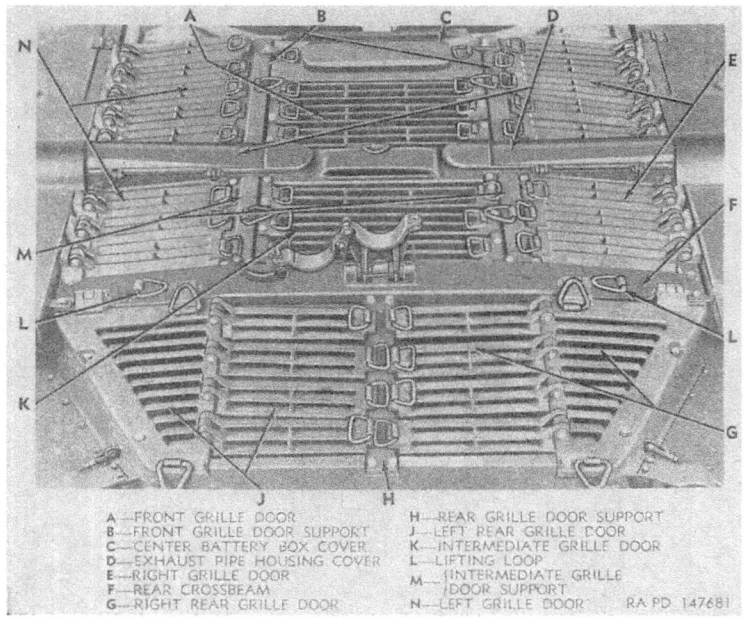

Figure 101. Top deck—installed view.

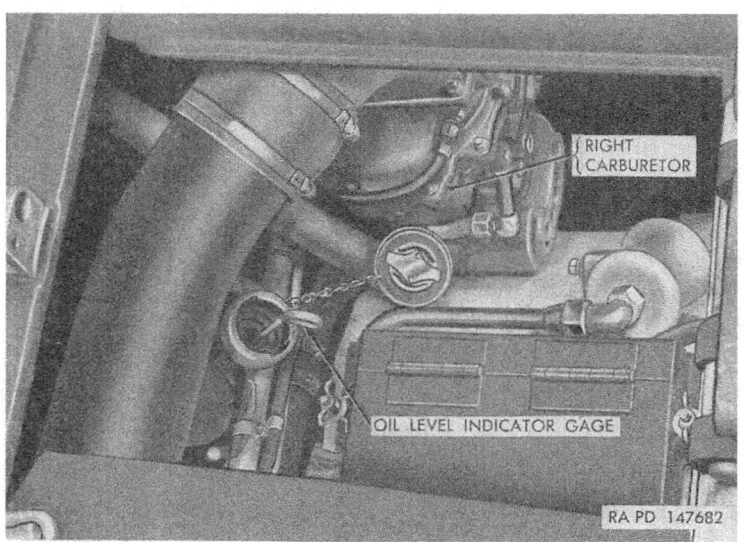

Figure 102. Oil level indicator gage.

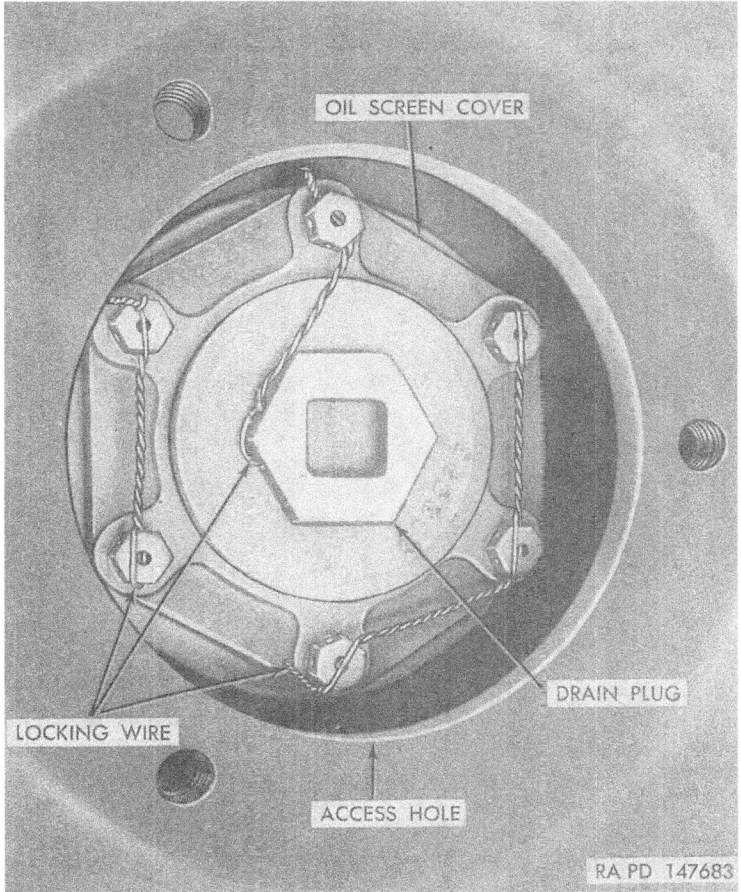

Figure 103. Main engine oil drain.

86) into the oil filler pipe. Check the oil level (*b* above). Install and lock the oil filler cap. Close grille doors.

149. Engine Oil Filter

a. REMOVAL. Remove the oil filter access hole cover from bottom of hull (fig. 285). Remove locking wire from six oil filter nuts and drain plug (fig. 105). Remove plug and drain oil from filter housing. Remove jam nuts, nuts, and washers from six studs and remove oil filter element.

Note. Pull element straight out of housing. Be careful to keep it centered in the filter housing.

227

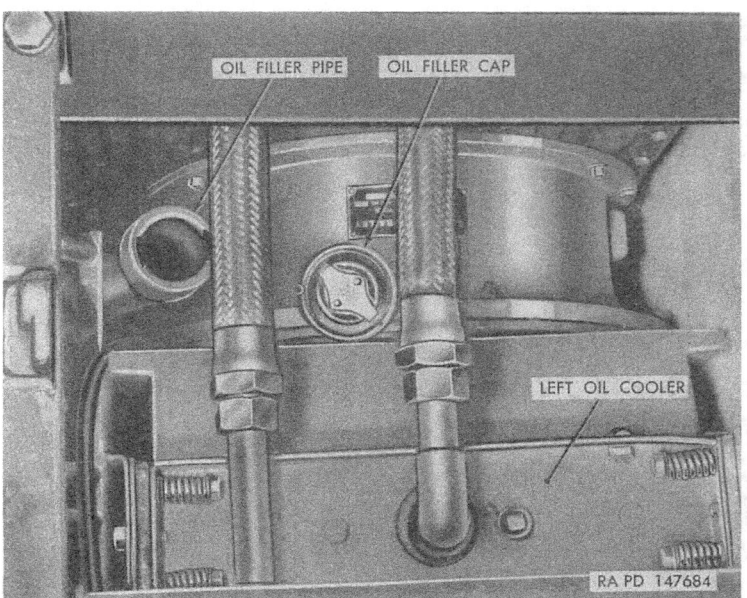

Figure 104. Oil filler pipe.

b. Cleaning. Wash the filter element in volatile mineral spirits or dry-cleaning solvent. Remove all foreign matter adhering to the filter. Disassemble oil filter element (fig. 106) and immerse the disks in a bath of clean volatile mineral spirits or dry-cleaning solvent, then dry thoroughly with compressed air. Clean all metallic particles from the magnetic drain plug. Assemble filter element with alternate disks and disk spacers.

c. Installation. Insert the filter element back into housing in the accessory case using the same precautions outlined for removal (*a* above). When installing the filter element, do not attempt to force it into place. If binding is noted, withdraw and inspect pilot for evidence of misalinement. A properly alined filter element will seat without forcing. When the filter element has seated, install new gasket. Secure with six nuts and washers and lock with jam nuts. Install drain plug. Lock drain plug and nuts with locking wire (fig. 105). Install oil filter access hole cover (fig. 285).

150. Oil Pressure Control Valve

a. Removal and Disassembly (figs. 107 and 108). Removal can be made with the power plant installed in the vehicle by gaining access though the bulkhead. Access can be obtained by removing the left

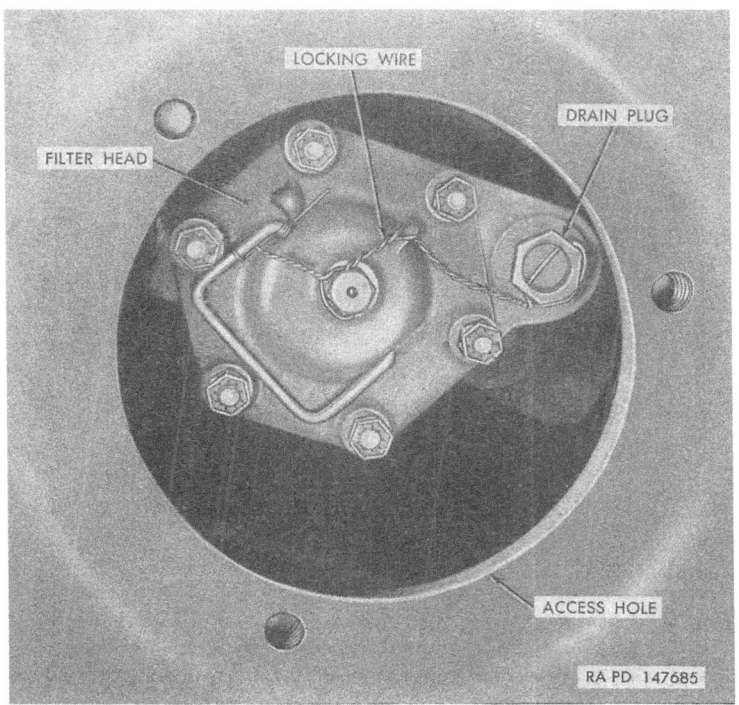

Figure 105. Oil filter element—installed view.

air cleaner (par. 175) and bulkhead inspection plate. Break locking wire and remove the valve assembly and annular gasket from the accessory case by using wrench 41-W-3727-33 (fig. 80). To disassemble, remove the housing cap and gasket from the outer end of control valve housing. Remove the control valve spring and spring seat from housing.

Note. A special washer may have been inserted in the valve housing cap to strengthen the spring.

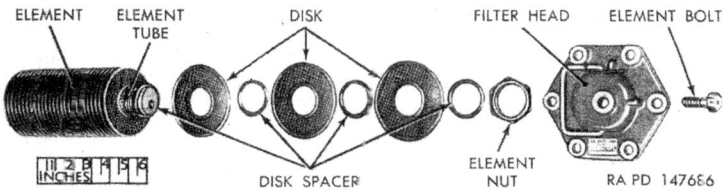

Figure 106. Oil filter element—exploded view.

229

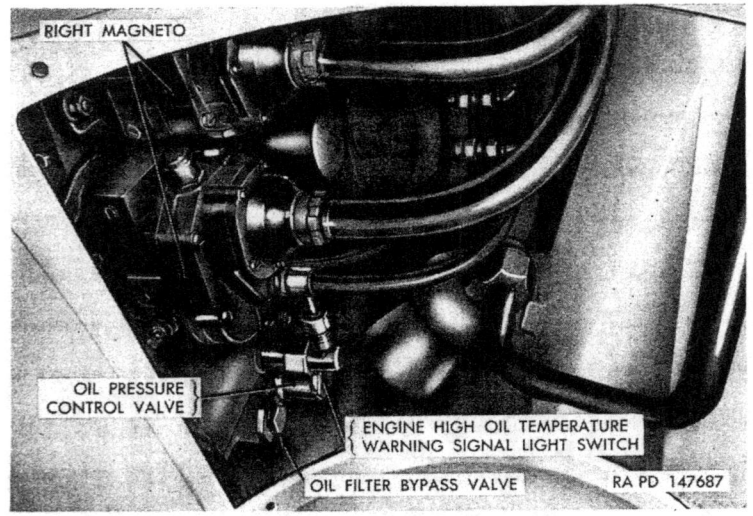

Figure 107. Oil pressure control valve—left bulkhead inspection plate removed.

Unscrew the control valve seat from the inner end of the control valve housing and remove the control valve.

b. CLEANING AND INSPECTION. Immerse all parts in volatile mineral spirits or dry-cleaning solvent and dry throughly with compressed air. Inspect all parts to see that they are without flaws, abrasions, or fatigue.

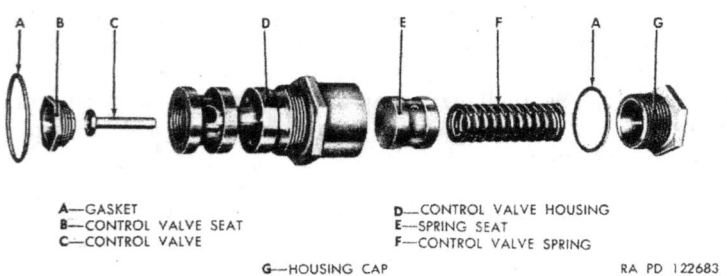

A—GASKET
B—CONTROL VALVE SEAT
C—CONTROL VALVE
D—CONTROL VALVE HOUSING
E—SPRING SEAT
F—CONTROL VALVE SPRING
G—HOUSING CAP

Figure 108. Oil pressure control valve—exploded view.

c. ASSEMBLY AND INSTALLATION. Insert control valve in inner end of housing and install control valve seat. Install spring seat, control valve spring, new gasket, and housing cap into outer end of valve housing. Install control valve housing into accessory case. Tighten and lock with locking wire. Install bulkhead inspection plate and air cleaner (par. 175).

151. Oil Filter Bypass and Oil Cooler Bypass Valves

a. GENERAL. The oil filter bypass and the oil cooler bypass valves (figs. 110 and 131) are virtually identical in construction, differing only in the valve housing length dimension. Procedures for removal, disassembly, and installation are identical for both valves.

b. ADJUSTMENT. The oil filter bypass and the oil cooler bypass valves require no adjustment, but all parts should be centrally seated and drawn tight for proper operation (figs. 110 and 131).

c. REMOVAL AND DISASSEMBLY. Replacement can be made with the power plant installed in vehicle by removing the right air cleaner (par. 175) and inspection plate. To remove the valve assembly from the accessory case, remove the locking wire and unscrew the valve assembly with wrench 41–W–3727–33 (fig. 109). To disassemble valve,

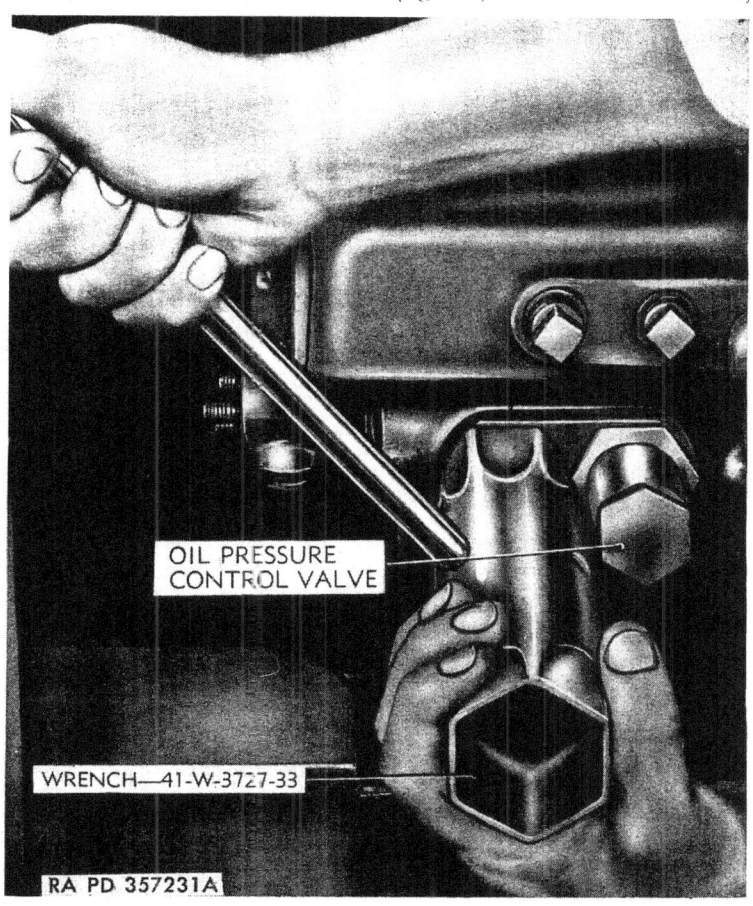

Figure 109. Removing oil filter bypass valve.

remove bypass housing cap and gasket from the bypass housing (fig. 110). Compress the bypass spring and remove the cotter pin from the valve stem. Remove the bypass spring and bypass valve.

Note. There will be some slight oil spillage upon removal of oil filter bypass valve from accessory case.

d. CLEANING AND INSPECTION. Immerse all parts in volatile mineral spirits or dry-cleaning solvent and dry thoroughly with compressed air. Inspect all parts to see that they are without flaws, abrasions, or fatigue. Replace faulty or damaged parts. Make certain the bypass valve operates freely within the housing.

Note. The valves are serviced only as complete units.

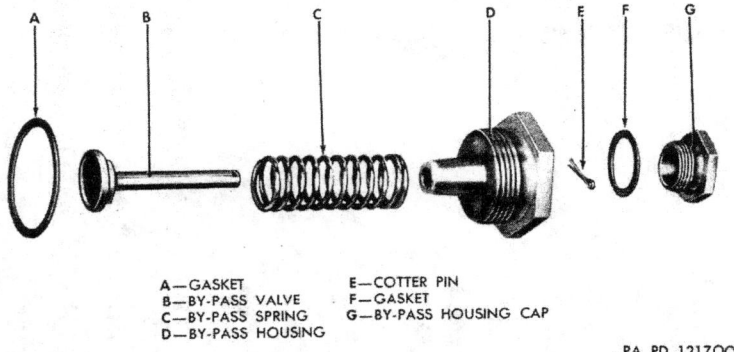

A—GASKET
B—BY-PASS VALVE
C—BY-PASS SPRING
D—BY-PASS HOUSING
E—COTTER PIN
F—GASKET
G—BY-PASS HOUSING CAP

RA PD 121700

Figure 110. Oil filter bypass valve—exploded view.

e. ASSEMBLY AND INSTALLATION. Position the spring over the bypass housing. Insert the bypass valve through spring into the bypass housing and compress the spring until a cotter pin can be inserted through valve stem. Install housing cap and new cap gasket. Install the valve assembly and new gasket into the accessory case. Tighten and lock with locking wire. Install the left air cleaner (par. 175) and inspection plate.

152. Engine Top Shroud

a. REMOVAL (fig. 111).

(1) Remove the front and intermediate sections of the top deck (par. 158*c*(1)).

(2) Remove hotspot tubes (par. 173).

(3) Remove 12 cotter pins, slotted nuts, flat washers and rubber washers securing the engine top shroud to top of cylinders.

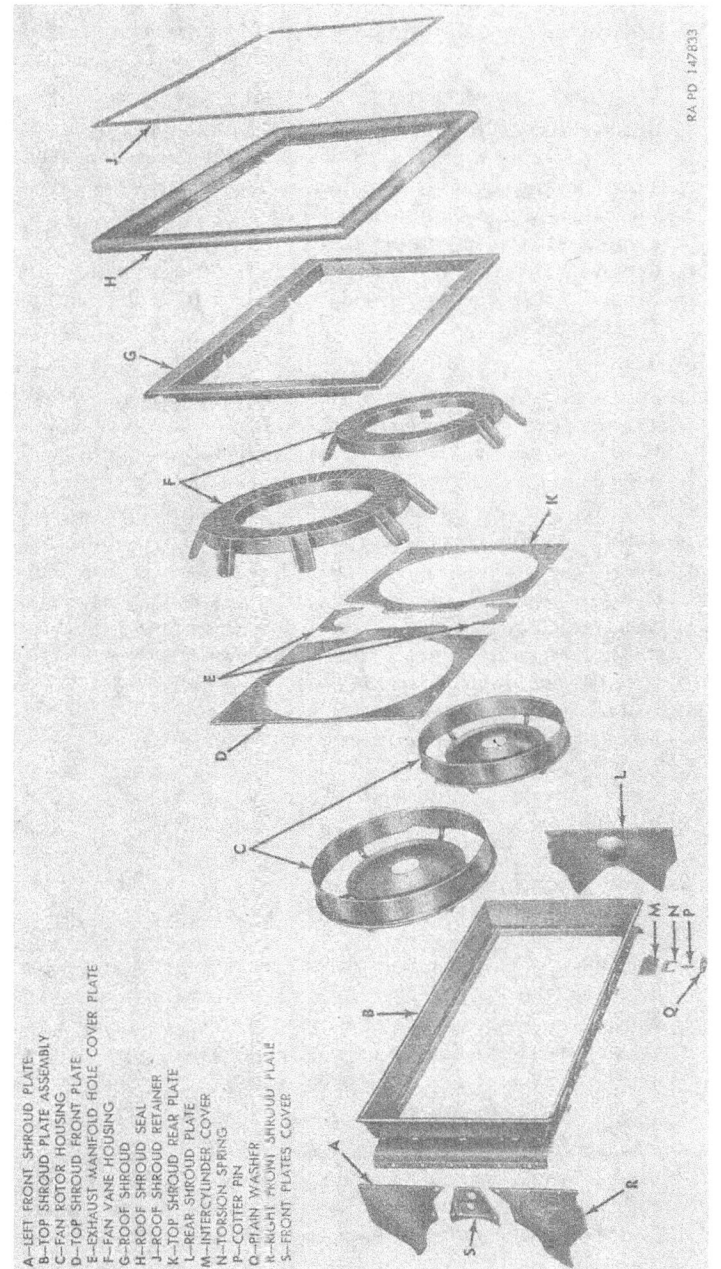

A—LEFT FRONT SHROUD PLATE
B—TOP SHROUD PLATE ASSEMBLY
C—FAN ROTOR HOUSING
D—TOP SHROUD FRONT PLATE
E—EXHAUST MANIFOLD HOLE COVER PLATE
F—FAN VANE HOUSING
G—ROOF SHROUD
H—ROOF SHROUD SEAL
J—ROOF SHROUD RETAINER
K—TOP SHROUD REAR PLATE
L—REAR SHROUD PLATE
M—INTERCYLINDER COVER
N—TORSION SPRING
P—COTTER PIN
Q—PLAIN WASHER
R—RIGHT FRONT SHROUD PLATE
S—FRONT PLATES COVER

Figure 111. Engine top shroud—exploded view.

(4) Remove the front shroud plates (fig. 131) by removing eight cap screws and two primer line clips.

Note. It is not necessary to remove the front plate cover.

(5) Remove the two fans (par. 154a(2), (3), and (4)).
(6) Remove locking wire from the six slotted nuts securing each rotor housing (fig. 114) to upper portion of fan drive housing. Remove the six nuts and washers.
(7) Remove two top shroud plates.
(8) Remove the two exhaust manifold hole cover plates.
(9) Remove the engine top shroud section by lifting it from the power plant.

b. INSTALLATION.

(1) Position the top shroud section on top of the cylinders. Make sure the fabric-type washers are installed on the studs on top of cylinders. Install top shroud section by installing 12 rubber washers, flat washers, slotted nuts, and cotter pins.
(2) Position the two exhaust manifold hole cover plates and install with four cap screws (two on each side).
(3) Secure the 2 top shroud plates to top shroud by installing 11 cap screws.
(4) Secure each rotor housing to the fan drive housing by installing six slotted nuts, washers, and locking wire.
(5) Install the two fans (par. 154).
(6) Install the two fan vanes (par. 154).
(7) Install the front shroud plates.
(8) Install hotspot tubes (par. 173).
(9) Install front and intermediate sections of top deck (par. 161b(2)).

153. Exhaust Manifold

a. REMOVAL (fig. 112).

(1) Remove engine top shroud plates, fan vanes, fan, and rotor housings (par. 154a).
(2) Remove self-locking nuts and plain washers securing each exhaust manifold flange to the cylinder exhaust port. Remove the exhaust manifold from the bank of cylinders.

Note. Each exhaust manifold is removed in the same manner.

b. CLEANING AND INSPECTION. Wash all parts in volatile mineral spirits or dry-cleaning solvent, cleaning the inside of the manifold sections with a stiff-bristle brush. Wipe and dry with compressed air. Inspect all mating flanges to see that they are without flaws, nicks, or burs, and that all welds are fused and secure. Inspect all areas for excessive rust or burning. Replace faulty or damaged parts.

c. INSTALLATION (figs. 112 and 113).

Note. The exhaust manifolds are interchangeable. Each manifold is made up of five separate sections; center, two intermediate, and two end elbow. The sections are provided with slip joints and may be pulled apart.

(1) Assemble the manifold by inserting the intermediate sections into the center manifold section and the end elbow sections into the intermediate sections (fig. 113).

(2) Replace all gaskets. Position the manifold flanges over the exhaust port studs. Secure each manifold flange to the cylinder exhaust port with four plain washers and self-locking nuts.

(3) Install the engine top shroud plates, fan vanes, fans, and rotor housing (par. 154*b*).

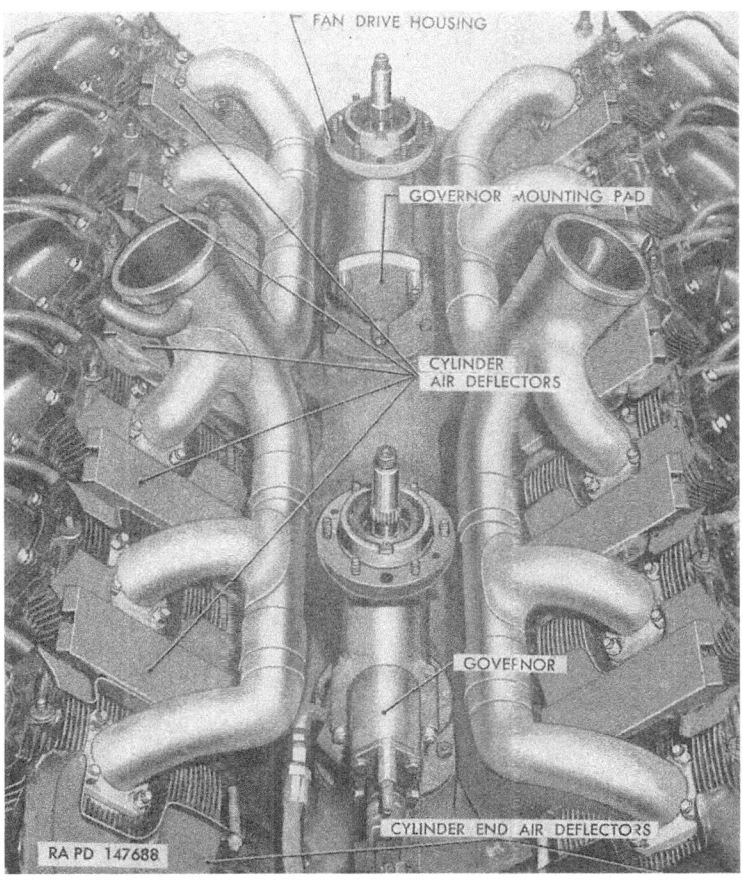

Figure 112. Exhaust manifolds—installed view.

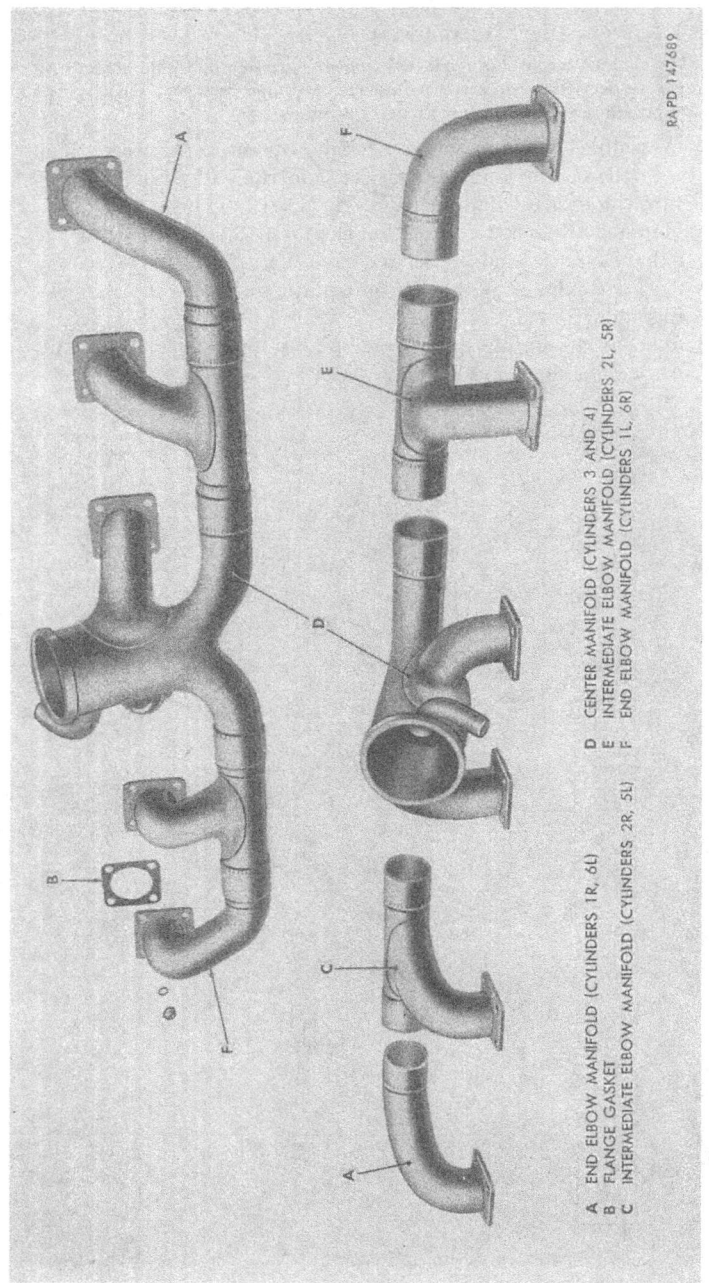

A END ELBOW MANIFOLD (CYLINDERS 1R, 6L)
B FLANGE GASKET
C INTERMEDIATE ELBOW MANIFOLD (CYLINDERS 2R, 5L)
D CENTER MANIFOLD (CYLINDERS 3 AND 4)
E INTERMEDIATE ELBOW MANIFOLD (CYLINDERS 2L, 5R)
F END ELBOW MANIFOLD (CYLINDERS 1L, 6R)

Figure 113. Exhaust manifolds—assembled and exploded view.

154. Engine Cooling Fan Rotors

a. REMOVAL.

(1) Remove front and intermediate sections of the top deck assembly (par. 158c(1)).

Note. When individual fan units require maintenance, it is necessary to remove only the affected section (front or intermediate).

Procedure is identical for both fans.

(2) Remove five cap screws securing stationary fan vane (fig. 96) to engine top shroud plates (fig. 114).

(3) Remove three machine screws and washers from top of shaft cover and lift off cover.

(4) Remove cotter pin and slotted nut from drive shaft. Tap drive shaft and remove fans from the splined shaft by lifting straight up.

(5) Remove outer row of 12 bolts connecting fan rotor to adapter and remove rotor.

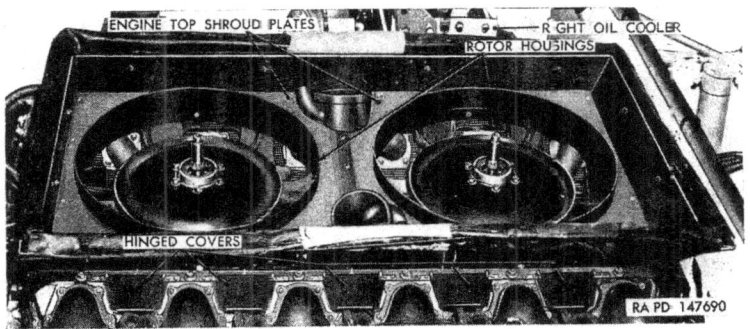

Figure 114. Engine top shroud with fans removed.

b. INSTALLATION.

(1) Position rotor on adapter and install 12 bolts, nuts, and cotter pins connecting rotor to adapter.

(2) Position fan (with thin edges of blades up) on fan drive shaft and secure with slotted nut and cotter pin. Install shaft cover, securing with three washers and machine screws. Install stationary fan vane with five cap screws.

(3) Install front and intermediate sections of top deck (par. 161*b*(2)).

155. Valve Timing and Adjustment

a. VALVE TIMING.

(1) *Remove front and intermediate sections of top deck.* Refer to paragraph 158c(1). Remove air-cleaner-to-carburetor

pipes (fig. 147). Open the engine compartment right rear grille doors (fig. 101).

(2) *Remove valve rocker covers from No. 1L and 1R cylinders.* Remove 8 cap screws, lock washers, and plain washers from the valve rocker cover of No. 1 right and No. 1 left cylinders (fig. 98). Remove the two cap screws and lock washers which extend through the camshaft drive gear housing (fig. 115) into the valve rocker covers and remove covers. Remove the engine-timing inspection plug (fig. 229), near transmission oil filter cap, to gain access to engine-timing pointer and flywheel markings (fig. 116).

(3) *Install tool for turning engine.* Remove three cap screws, lock washers, and plain washers from center access door on the rear of the tank hull (fig. 285). Remove the six cap screws securing the power-take-off cover (fig. 220) on the rear of the transmission and remove the cover. Install wrench 41–W–1536–380 (fig. 117) for turning the engine. This will be used to turn engine for valve timing.

(4) *Check valve timing.* Turn engine until the No. 1R cylinder intake rocker roller (cam follower) is directly opposite the toe (high point) of the cam. Adjust intake valve rocker clearance to 0.100 inch, using feeler gag 7083769 (fig. 80), by first loosening lock nut and then turning adjusting screw as required (fig. 121). Tighten lock nut. Turn the engine in normal direction of rotation to the point where the No. 1R cylinder intake valve rocker roller (cam follower) has just allowed the intake valve to close. This point may be determined precisely by attempting to rotate the intake valve rocker roller by hand as the camshaft is rotating. At the first instant the roller is free (rocker roller will turn with a small amount of friction) the engine-timing pointer (fig. 116) should be adjacent to the flywheel-timing mark "V-ENG, 1R-IN, CL 0.100 CLR" fig. 118). If pointer is adjacent to the timing mark, timing is correct. Reset intake valve rocker clearance to 0.007 inch. If the pointer is not adjacent to timing mark, valve timing is incorrect. Check valve timing for left bank of cylinders in the same manner, using the timing mark "V-ENG, 1L-IN, CL 0.100 CLR" on the flywheel. If either bank is incorrectly timed, notify ordnance maintenance personnel.

(5) *Install valve rocker covers.* Position valve rocker covers on No. 1R and 1L cylinders and secure each with eight cap screws, lock washers, and plain washers.

Note. Ignition harness conduit clip is secured by one of the valve rocker cover cap screws.

Install the two cap screws and lock washers each side through camshaft drive gear housing (fig. 115). Bend up tabs on lock washers to lock cap screws. Install air-cleaner-to-carburetor pipes (fig. 147). Install engine timing inspection plug (fig. 229).

(6) *Remove engine turning wrench.* Remove engine-turning wrench and install power-take-off cover, securing it with six cap screws. Install center access door on rear of tank hull, securing with three cap screws, lock washers, and plain washers.

(7) *Install front and intermediate sections of top deck.* Refer to paragraph 161*b*(2). Close engine compartment right rear grille doors.

b. VALVE ADJUSTMENT.
(1) *Remove front and intermediate sections of top deck.* Refer to paragraph 158*c*(1).
(2) *Remove valve rocker covers.* Remove the two spark plug access opening covers (fig. 139) from the engine rear shroud, near No. 6 cylinder in each bank. Remove the air-cleaner-

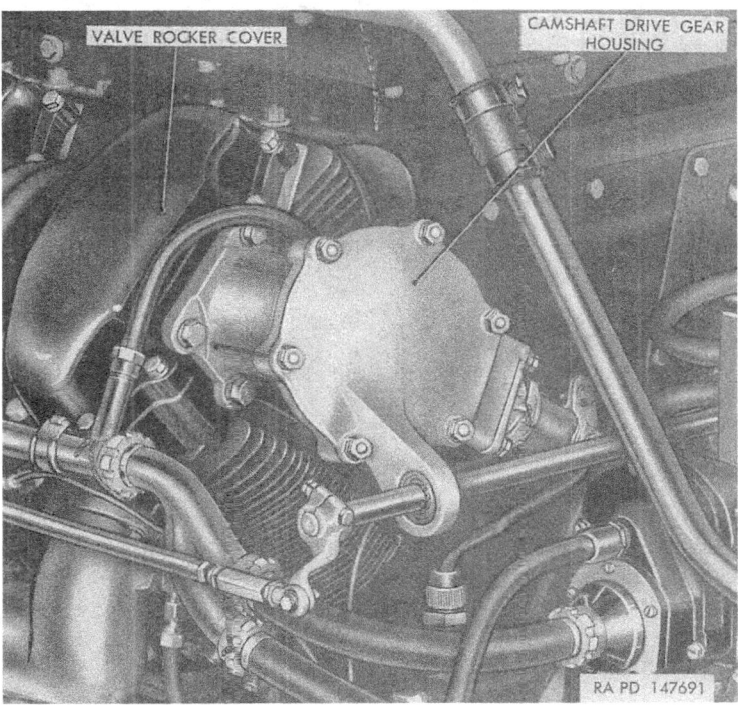

Figure 115. Valve rocker cover.

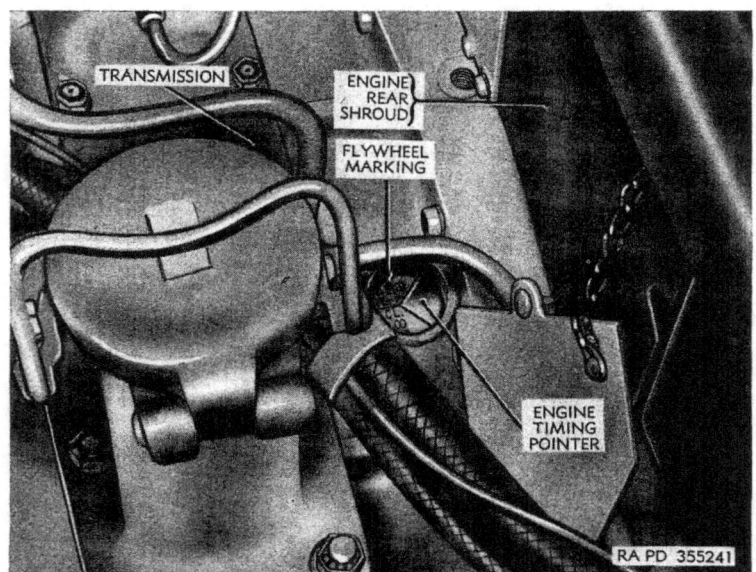

Figure 116. Engine timing pointer and flywheel markings.

Figure 117. Turning engine using wrench 41-W-1536-380.

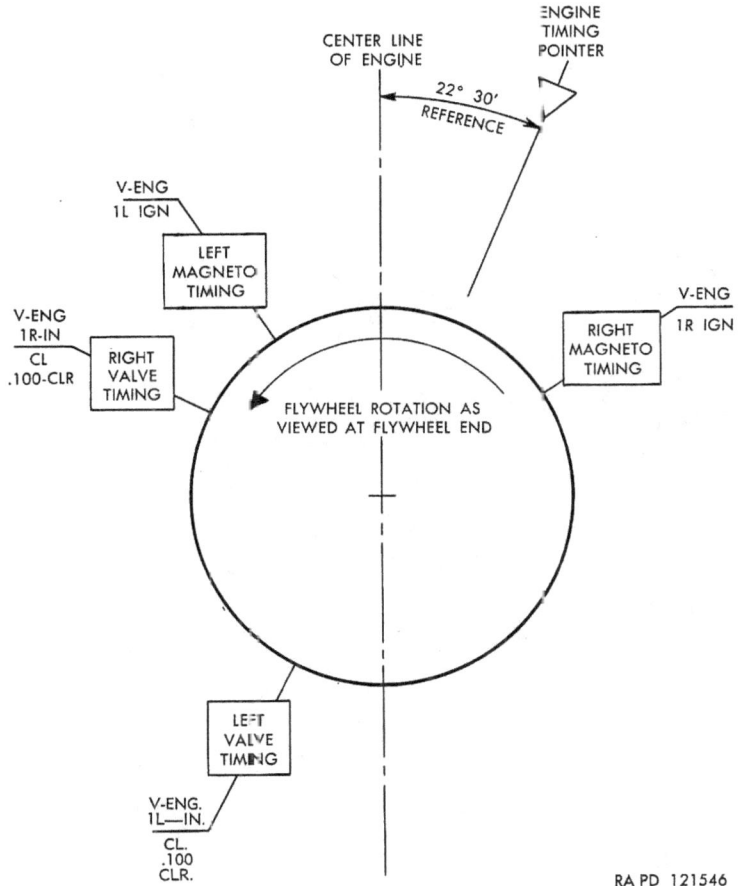

Figure 118. Engine flywheel markings.

to-carburetor pipes (fig. 147), and exhaust pipes (fig. 123). Remove the two cap screws which extend into the front of the two No. 1 cylinder valve rocker covers (through the camshaft drive gear housings) (fig. 115). Working through the holes in rear shroud, remove the two cap screws which extend into the rear of the No. 6 cylinders. Remove the 8 cap screws, lock washers, and plain washers which secure each valve rocker cover and remove the covers from the 12 cylinders.

(3) *Install engine-turning wrench.* Refer to *a*(3) above.
(4) *Check and adjust valve clearances.* Check intake and exhaust valve clearances and adjust to 0.007-inch clearance for intake valve rockers and 0.020-inch clearance for exhaust valve rock-

ers. Adjustments of all valves will be made with the rocker rollers (cam followers) on the base circles of the cams, opposite the cam lobes (high points of cams). Insert the feeler gage 7083769 between the tip of the valve stem and the adjusting screw pad ends (figs. 119 and 120). If adjustment is required, loosen the adjusting screw lock nut and turn the adjusting screw until the proper clearance is obtained. When adjusted, hold the adjusting screw in position and tighten the lock nut.

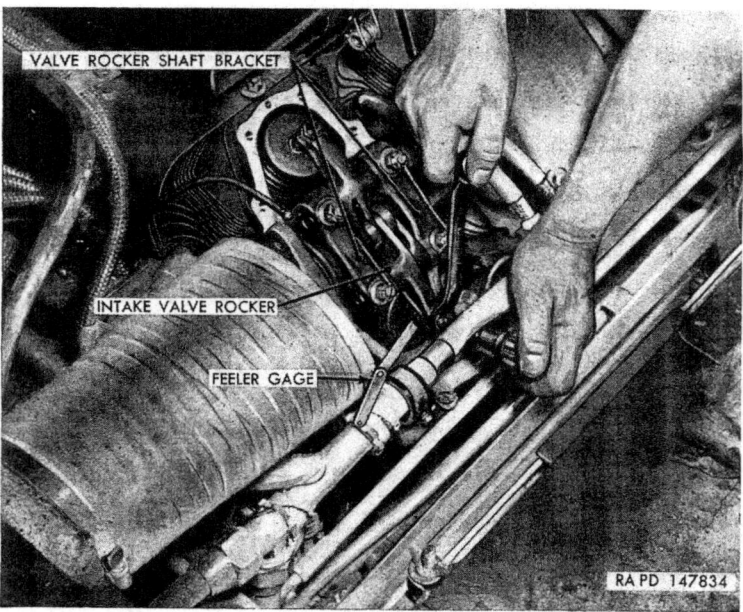

Figure 119. Adjusting intake valve with feeler gage 7083769.

(5) *Remove engine-turning wrench.* Remove engine-turning wrench and install power-take-off cover (fig. 220), and center access door (fig. 285).

(6) *Install valve rocker covers, exhaust and air-cleaner-to-carburetor pipes.* Position valve rocker covers on cylinders and secure each with eight cap screws, lock washers, and plain washers. Install the two cap screws through each of the camshaft drive gear housings (fig. 115) into the No. 1 cylinder valve rocker covers. Working through the space provided by the removal of the spark plug access opening covers, install the cap screws in the rear of each of the No. 6 cylinder valve rocker covers. Install the exhaust pipes (par. 182*d*) and air-cleaner-to-carburetor pipes (par. 160*b*(11)).

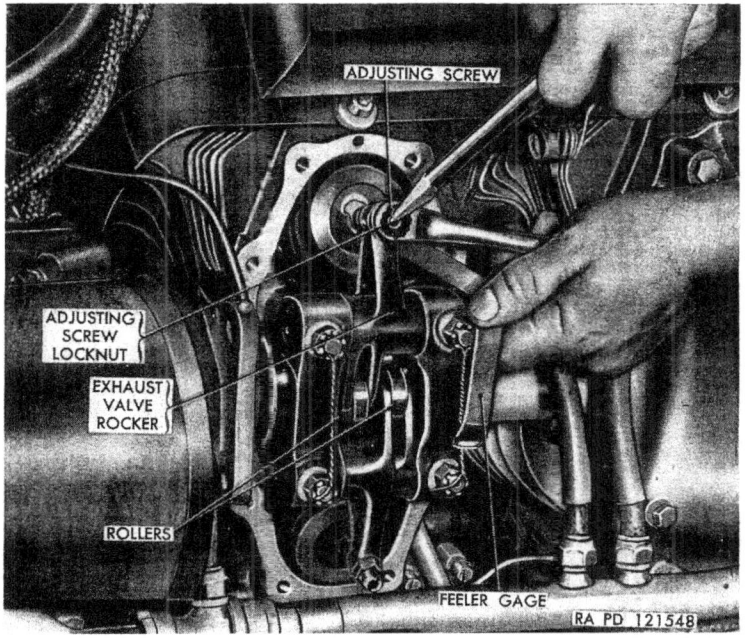

Figure 120. Adjusting exhaust valve with feeler gage 7083769.

(7) *Install front and intermediate sections of top deck.* Refer to paragraph 151b(2).

156. Valve Rocker Replacement and Adjustment

a. REMOVAL.

(1) *Remove front and intermediate sections of top deck.* Refer to paragraph 158c(1). Remove air-cleaner-to-carburetor pipes (par. 159d(13)), exhaust pipes (par. 182), and valve rocker covers (par. 155b(2)).

(2) *Remove valve rockers.* Remove the locking wire, nuts, and washers which secure the valve rocker shaft bracket (fig. 119).

> *Note.* Back off each rocker shaft bracket nut, in rotation, a few turns at a time, until all are loose. The rocker shaft brackets and cylinders are numbered. The brackets must be installed in the exact positions on the cylinders from which they were removed.

Lift off the rocker shaft brackets and rocker assemblies. Separate the rocker shaft brackets and the rocker assemblies. Repeat procedure for each cylinder.

b. CLEANING AND INSPECTION.
   (1) *Inspect.* Inspect the assembly for cracks, scratches, and scoring. See that the rocker rollers turn freely. Check to see if the adjusting screws turn freely. If the adjusting screws do not turn freely, clean up damaged threads.
   (2) *Clean.* Use volatile mineral spirits or dry-cleaning solvent to clean all parts.
   (3) *Repair.* If adjusting screws with damaged threads can not be repaired, install new screws.

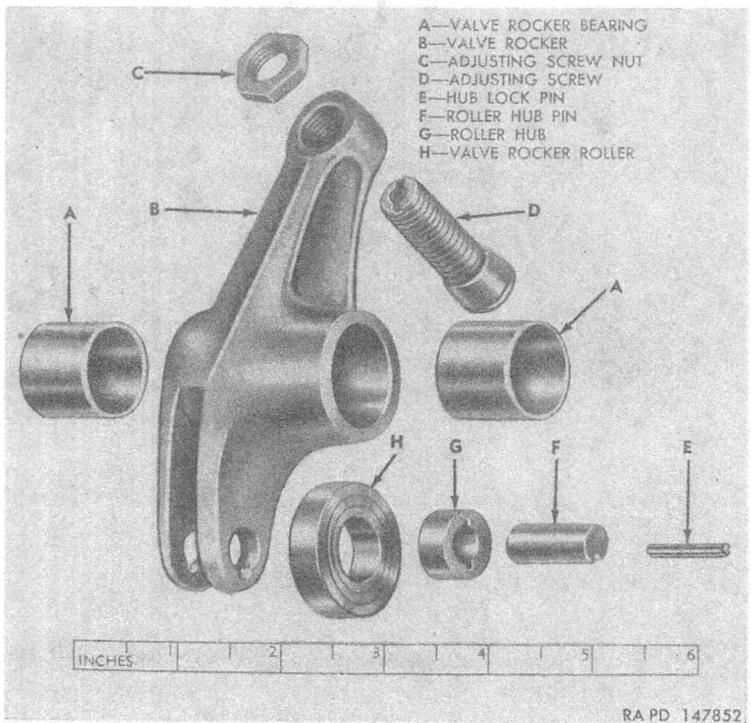

Figure 121. Valve rocker—exploded view.

c. INSTALLATION.
   (1) *Install valve rockers.* Assemble the rocker assemblies to the rocker shaft brackets and position over studs.
   Note. Make certain brackets are installed as marked (*a*(2) above). Secure the brackets with the four washers and nuts turning each nut, in rotation, a turn at a time. When the nuts are seated tighten to 175 to 200 lb-in torque, and lock wire each traverse set of nuts (fig. 119). Repeat procedure for each cylinder.

(2) *Adjust valves.* Refer to paragraph 155*b*(4).
(3) *Install valve rocker covers, exhaust pipes, and air-cleaner-to-carburetor pipes.* Refer to paragraph 155*b*(6).
(4) *Install front and intermediate sections of top deck.* Refer to paragraph 151*b*(2).

Section VI. POWER PLANT REMOVAL AND INSTALLATION

157. Coordination With Ordnance Maintenance Unit

Replacement of the power plant with a new or rebuilt unit is normally an ordnance maintenance operation, but may be performed in an emergency by the using organization, provided approval for performing this replacement is obtained from the supporting ordnance officer. A replacement power plant, any tools needed for the operation which are not carried by the using organization, and any necessary special instructions regarding accessories, etc., may be obtained from the supporting ordnance maintenance unit.

158. Top Deck Removal

a. GENERAL. The top deck can be removed as a complete unit or in individual sections. There are 24 hinged engine compartment grille doors (12 on each side) which can be lifted individually when access is needed to a certain section of the power plant or the covering part of the deck can be removed as a section. Of these, 16 (8 on each side) give access to the engine, oil coolers, batteries, and auxiliary engine. The other eight cover the center portion of the transmission. Access to the intermediate engine compartment section is provided by six removable grille doors.

b. TOP DECK REMOVAL AS A COMPLETE UNIT.
(1) *Traverse gun and turret.* Release gun traveling lock (par. 55). Elevate 90-mm gun sufficiently to clear obstructions (par. 76). Release the turret lock (par. 42). Using gunner's-manual-traversing-control handle (par. 36), traverse gun and turret 90° to the left so as to present the smooth side of the turret to the rear.
(2) *Remove muffler shields.* Remove shields covering mufflers, right and left.
(3) *Open engine compartment right and left grille doors.* Open the eight doors on each side (fig. 101). Removal of these doors is not necessary in order to remove the power plant.
(4) *Remove battery box cover.* Loosen two hold-down clamp cap screws (one on each side) securing center battery box cover to engine compartment front grille door supports (fig. 101), and turn clamps to clear cover. Lift off center battery box cover.

(5) *Remove center batteries and battery box.* Disconnect cables to the two center batteries (par. 224).

Caution: Tape live ends of the battery cables.

Remove two battery clamp hook nuts and remove battery clamp hooks. Remove battery clamp. Remove two batteries and then lift out battery box.

(6) *Remove engine compartment intermediate grille doors.* To give access to exhaust pipe clamps, remove the intermediate grille door adjacent to the exhaust pipe housing cover (fig. 101).

(7) *Remove exhaust pipe.* Loosen the two nuts on both the right and left exhaust pipe housing cover clamps. Release the clamps and swing both exhaust pipe housing covers back (fig. 123). Release the exhaust pipe clamps at both ends (fig. 122). Remove the exhaust pipes.

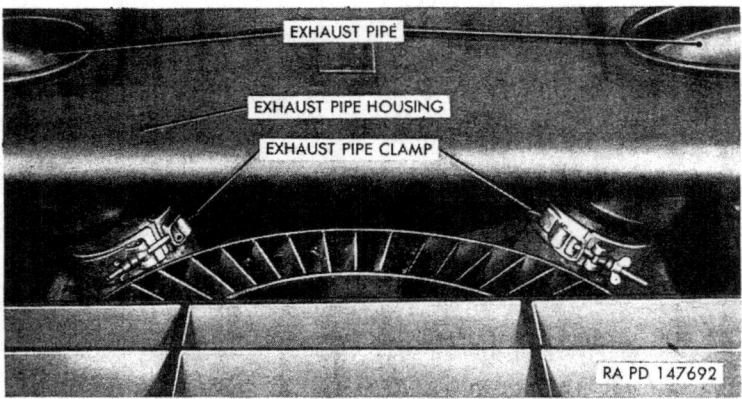

Figure 122. Grille doors removed for access to exhaust pipe clamps.

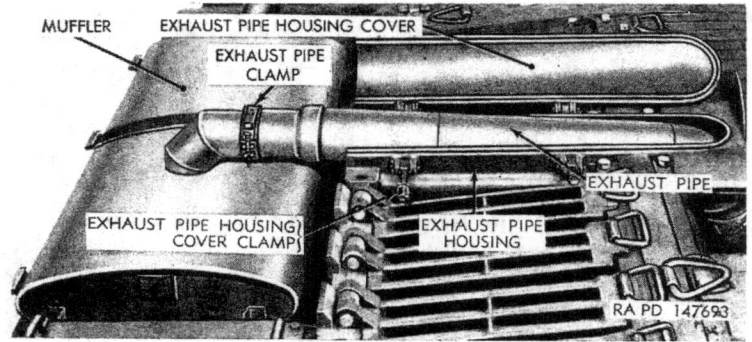

Figure 123. Right exhaust pipe—cover removed.

Caution: Place clean rags in exhaust manifold openings to prevent dirt or foreign matter from getting into cylinders.

(8) *Remove top deck assembly hold-down cap screws.* Remove two cap screws securing engine compartment rear grille door support to hull (fig. 101). Remove six cap screws (three on each side) securing engine compartment (right and left) rear grille door assemblies to hull. Remove four cap screws holding engine compartment rear crossbeam to hull. Remove four cap screws holding exhaust pipe housing to hull. Remove four cap screws holding front grille door supports to hull.

Figure 124. Grille doors open for removal of top deck.

(9) *Remove top deck assembly.* Install lifting sling 41-S-3832-46. Secure two hooks to lifting loops on engine compartment rear crossbeam (fig. 101) and two loops on engine compartment front grille door supports. Remove engine compartment top deck assembly by lifting it carefully from the tank (fig. 125).

c. Top Deck Removal in Individual Sections (fig. 101).

(1) *Removal of engine compartment front and intermediate sections.* Remove four cap screws and lock washers (two on each side) holding engine compartment intermediate grille door supports to rear crossbeam (fig. 101). Remove four cap screws and lock washers (two on each side) holding exhaust pipe housing to hull (fig. 123). Remove exhaust pipes (b (7) above). Remove four cap screws and lock washers (two on each side) holding engine compartment front grille door supports to hull. Remove center battery box cover, batteries, and battery box (b (4) and (5) above). Open the eight left and right intermediate grille doors on each side.

247

Figure 125. Removing or installing top deck with sling 41-S-3832-46.

Install lifting sling 41-S-3832-46 and remove the engine compartment front and intermediate sections as a unit from the vehicle.

(2) *Removal of engine compartment front section.* Open the five grille doors on each side. Remove center battery box cover and center batteries (*b* (4) and (5) above). Remove four cap screws and lock washers (two on each side) holding engine compartment front grille door supports to exhaust pipe housing. Remove four cap screws (two on each side) holding engine compartment front grille door supports to front portion of hull. Remove six cap screws (three on each side) securing engine compartment intermediate grille doors to grille door supports. Remove the three grille doors. Remove the grille door supports.

(3) *Removal of engine compartment intermediate section.* Open the three grille doors on each side. Remove four cap screws and lock washers (two on each side) holding engine compartment intermediate grille door supports to exhaust pipe housing. Remove four cap screws and lock washers (two on each side) holding grille door supports to engine compartment rear crossbeam. Remove six cap screws and lock washers (three on each side) securing grille doors to grille door support. Remove the three grille doors. Remove the two grille door supports.

(4) *Removal of engine compartment rear section.*

Note. Before the engine compartment rear section can be removed as a unit, the intermediate section must be first removed as outlined in (3) above.

Remove four cap screws and lock washers (two on each side) securing rear crossbeam to hull. Remove six cap screws and lock washers (three on each side) holding engine compartment rear grille doors to hull. Remove two cap screws and lock washers securing rear grille door support to hull. Install lifting sling 41–S–3832–46 (two hooks on loops on rear crossbeam and two hooks on loops on rear grille door assembly). Raise rear grille door assembly and remove from vehicle.

Note. The rear grille door assembly can be swung forward on four hinges to permit accessibility to the engine compartment rearward of the crossbeam.

159. Power Plant Removal

a. GENERAL. The power plant consists of the main engine, crossdrive transmission, and oil coolers (figs. 130, 131, and 132). The power plant is removed and installed as one unit.

b. PRELIMINARY INSTRUCTIONS. Check to be sure that the master relay switch (fig. 18) is in the "OFF" position. Close both right and left fuel tank shut-off valves (fig. 11). Manually traverse the turret

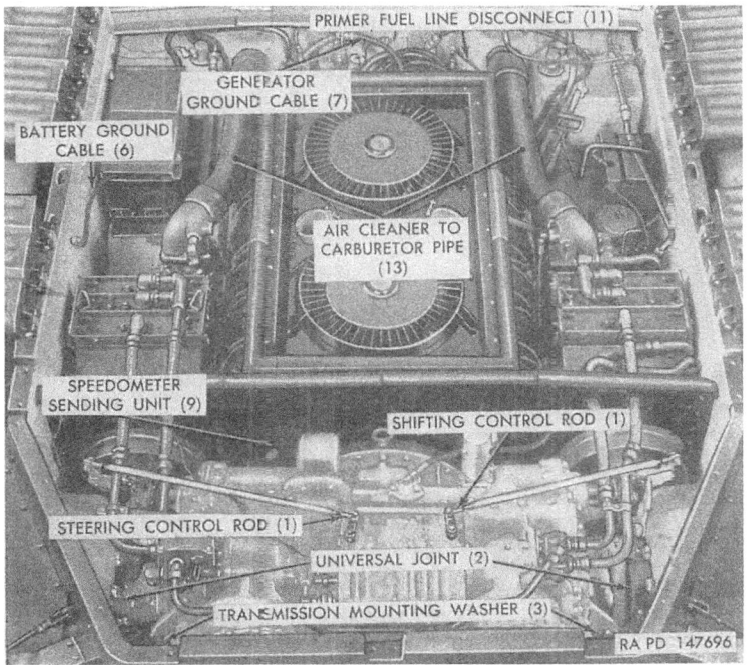

Figure 126. Power plant disconnect points.

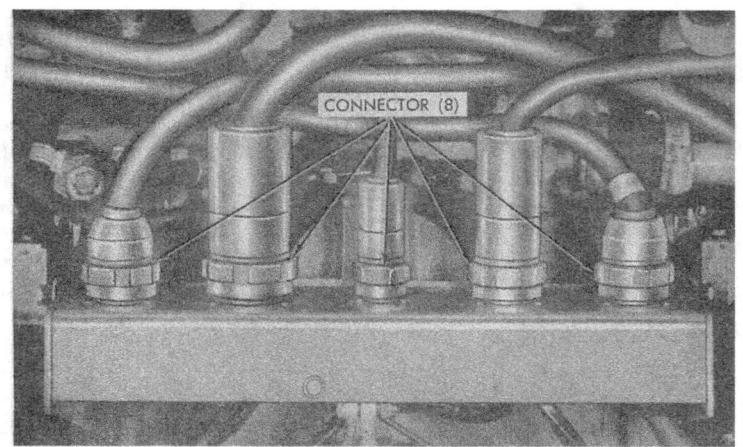
Accessory end of engine

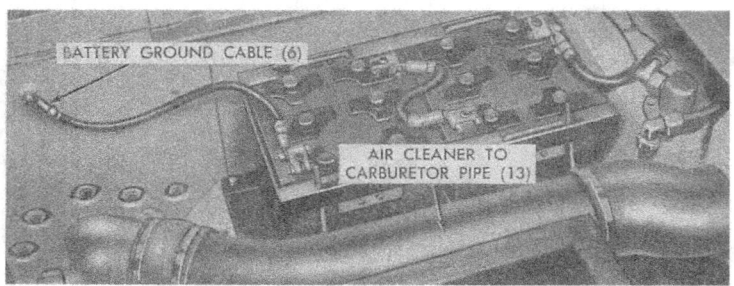
Left side of engine compartment

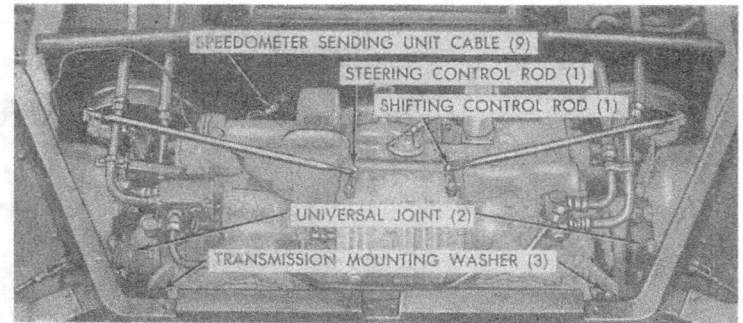
Rear of engine compartment

Figure 127. Power plant disconnect points.

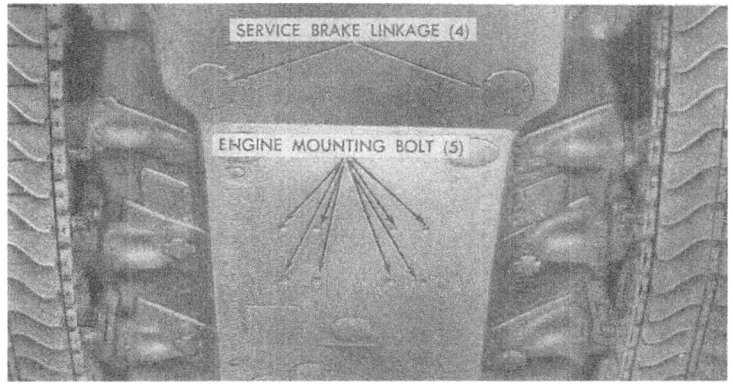
Under side of hull

Bulkhead center deck cover

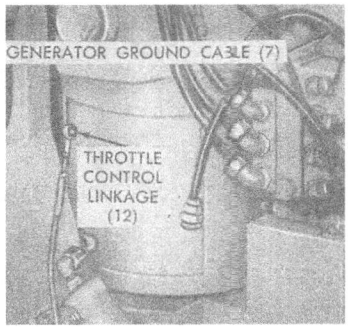
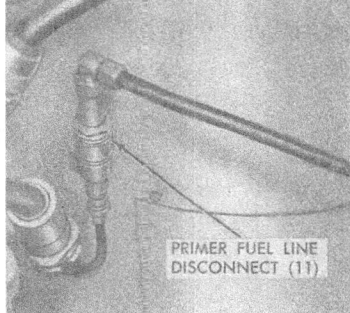
Front of engine compartment   RA PD 147698

Figure 123. Power plant disconnect points.

(par. 36), as necessary, to obtain maximum clearance above the front portion of the engine compartment.

c. DISCONNECT POINTS ILLUSTRATIONS. All disconnections made from above and below the engine compartment are illustrated in figure 126, 127, and 128. The key numbers on the illustrations coincide with step numbers under *d* below.

d. PROCEDURES.

(1) *Disconnect steering and shifting control rods.* Disconnect steering control rod at bellcrank (left side of transmission) (fig. 230) and swing clear. Disconnect shifting control rod at bellcrank on right side of transmission (fig. 229) and swing clear.

(2) *Disconnect universal joints.* Remove locking wire; remove four bolts (two from each bearing cap) from each universal joint; pry universal joints away from transmission flanges. Disconnect opposite universal joint in the same manner.

> *Note.* Moving the tank slightly, as required, will rotate the universal joints and make the bolts accessible. This step should be performed before step (*b*) above. The tank can be moved under its own power. If conditions prevent operation of engine, the tank can be pulled or towed. If brakes seize or other trouble prevents rotation of universal joints, it is possible to remove bolts inaccessible from the top by working through the space provided by the brake linkage access hole covers (fig. 285) in hull bottom.

Looping a piece of wire around the universal joint bearing cap and securing the opposite ends on a bolt inserted in one of the rear engine compartment hull tapped holes will aid in keeping the universal joints from binding when removing and installing power plant.

(3) *Remove transmission mounting washers.* Remove nuts from transmission mounts at each side of transmission at rear end of engine compartment. Remove the lock and flat washers. Put the protector nuts back on the mount bolt as guides to protect the holes in the mounting pads during removal.

(4) *Disconnect service brake linkage.* Remove two brake linkage access hole covers (fig. 285) beneath the service brake linkage near bottom of hull. Disconnect service brake linkage by removing clips and driving out pin which connects linkage rods to transmission brake control lever. Disconnect opposite brake linkage in the same manner.

(5) *Remove cap screws from engine mountings.* Working from beneath tank, remove eight cap screws (two from each mounting) which secure the four engine mountings to engine compartment floor.

(6) *Disconnect battery ground cable.* Disconnect the battery ground cable from hull at left and tape.

(7) *Disconnect generator cable.* Disconnect generator ground cable by separating connector on bracket at left top of master junction box.

(8) *Disconnect electrical cables at engine wiring junction box.* Disconnect all five cables from the engine wiring junction box by separating connectors.

(9) *Disconnect speedometer sending unit cable.* Disconnect speedometer sending unit cable at coupling at left top of transmission.

(10) *Disconnect main fuel line.* From the crew compartment, disconnect the fuel lines at the main engine fuel line disconnect.

(11) *Disconnect primer fuel line.* Disconnect primer fuel line at quick disconnect point in engine compartment.

(12) *Disconnect throttle control linkage.* Remove lock nut and disconnect throttle control linkage (fig. 97) at cross-shaft lever. Thread lock nut back on bolt, after disconnect is made, to prevent loss.

(13) *Remove air-cleaner-to-carburetor pipes.* Loosen hose clamps at both ends of each air-cleaner-to-carburetor pipe (fig. 147) and remove pipes.

Caution: Place clean rags in intake elbows on top of each carburetor to prevent dirt or foreign matter from getting into carburetors.

(14) *Remove power plant.* Check to be sure all disconnects have been accomplished. Install sling 41-S-3832-46.

Note. Be sure to balance power plant weight on lifting hook by moving sling lifting eye along rail.

Carefully lift power plant out of tank (fig. 129). Check to see that accessories and parts are clear as the power plant is being moved upward. Proceed carefully, using a series of short lifts. As lifting progresses, swing unit back to clear protruding air cleaner pipes. If no transport stand is available, put blocks under right and left sides of transmission and under front engine mountings.

Note. It may be necessary to adjust sling so that transmission end of power plant is raised slightly higher than engine end. Set power plant on transport stand 41-S-4941-50.

Caution: Be careful. Power plant dry weight is approximately 6,000 pounds. Injury can result to personnel upon careless handling. Damage can result to power plant from bumping against portions of hull or from a sudden drop on transport stand or blocking.

Figure 129. Removing power plant with sling 41-S-3832-46.

160. Power Plant Installation

a. PRELIMINARY INSTRUCTIONS. Prior to installation, check to be sure that all lines, cables, and parts are installed on the power plant that they were connected to and removed with it from the vehicle. If maintenance work was done on engine, be sure that all adjustments affected are made and are correct. If any components of the power plant were removed and installed, check all connections and related parts to be sure that the installations were accomplished correctly. Check the engine compartment (fig. 133) to make sure it is clear and that the proper shims are positioned over the engine-mounting bolt holes. Install guide nuts on the two transmission mounting bolts.

b. PROCEDURE (figs. 126, 127, and 128).

(1) *Install engine and transmission sling and lift power plant into tank.* Install sling 41-S-3832-46 on power plant (fig. 129). Lift and ease the power plant into the engine compartment by a series of short drops, making sure each time that clearance is provided on all sides. Guides for front engine mountings are provided (fig. 133).

(2) *Install cap screws in engine mountings.* With tension on engine and transmission sling, aline mounting holes and insert cap screws. Make sure that all shims are in place over mounting plates (fig. 133). Completely lower power plant and tighten the eight cap screws (two in each mounting).

Note. The power plant may have to be raised slightly to aline mounting holes and shims.

(3) *Connect service brake linkage.* Working from beneath the tank through the brake linkage access holes, connect the service brake linkage rods to transmission brake control levers (at transmission) by inserting pins and clips.

> *Note.* Insert pin so that clip is on outside (away from transmission). Alinement should be such that pins slip in easily.

Install the two service brake linkage access hole covers on hull (fig. 285).

(4) *Install transmission mounting washers.* Remove the two transmission mounting nuts. Install lock and flat washers. Install the two transmission mounting nuts and tighten.

(5) *Connect steering and shifting control rods.* Connect the steering and shifting control rods to bellcrank at the left and right sides of the transmission, respectively (figs. 229 and 230).

> *Note.* Check to see that rods are not bent and binding on top of the transmission. Pins should slip in without driving or controls will not operate properly.

(6) *Connect throttle control linkage.* Remove lock nut from bolt in throttle control linkage rod end and insert through rod end and throttle cross shaft lever. Install lock nut and tighten.

(7) *Connect main fuel line.* Connect fuel line at main engine fuel line (fig. 128) disconnect under floor in crew compartment.

(8) *Connect primer fuel line.* Connect primer fuel line at disconnect in front of engine compartment (fig. 128).

(9) *Connect electrical cables.* Connect all five connectors at engine wiring junction box. Connect generator cable connector at left top of master junction box. Secure battery ground cable to hull.

(10) *Connect speedometer drive.* Connect speedometer sending unit cable (fig. 127) (top of transmission left side).

(11) *Install air-cleaner-to-carburetor pipes.* Install the two air-cleaner-to-carburetor pipes (fig. 127). Tighten clamps at each end.

> **Caution:** Remove rags from elbow on top of carburetor before installing air cleaner pipes.

(12) *Connect universal joints.* Aline the universal joint and the transmission drive flange by rotating the drive flange.

> *Note.* Installing two bolts in the drive flange avoids slippage of the pry bar when rotating the flange. If universal joints are not alined for easy assembly, slip flange out of final drive and replace in desired position. Flange in transmission can then be rotated to match.

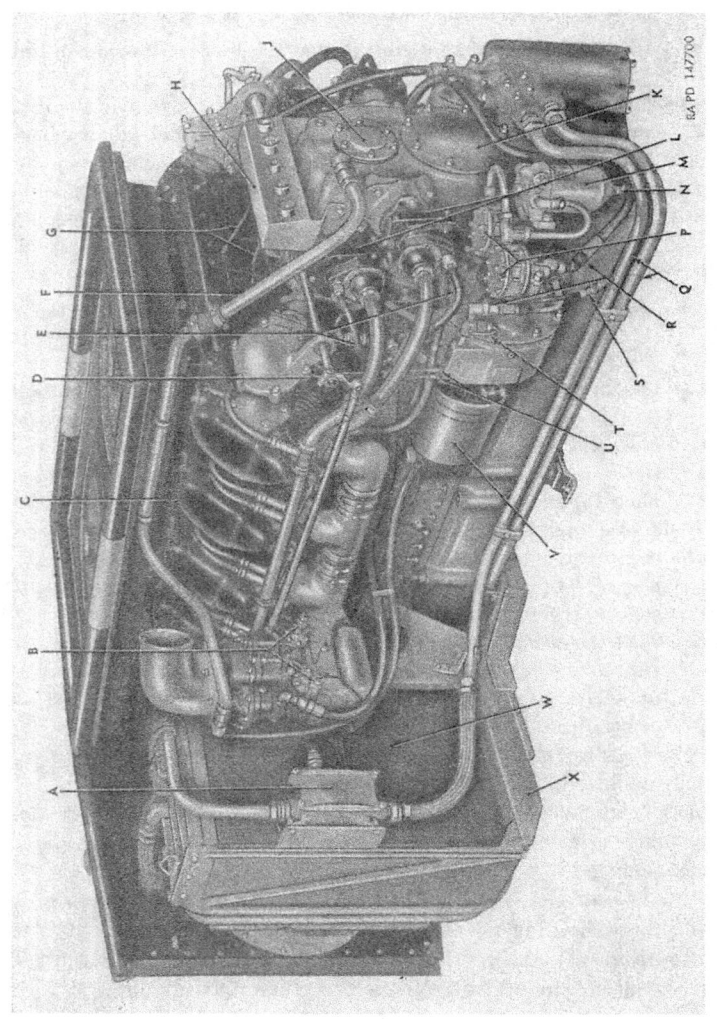

Figure 130. Power plant—left front view.

256

A—ENGINE OIL COOLER FAN CONTROLLER
B—DEGASSER CABLE CONNECTORS
C—LEFT CYLINDER BANK
D—THROTTLE LINKAGE
E—MAGNETO TO BOOSTER CABLE
F—ACCESSORY CASE VENT LINE
G—GOVERNOR CONTROL LINKAGE
H—ENGINE WIRING JUNCTION BOX
J—ACCESSORY DRIVEN SHAFT FRONT SUPPORT BEARING COVER
K—ACCESSORY DRIVE SHAFT OPENING COVER
L—LEFT MAGNETO
M—FUEL LINE FILTER
N—FUEL LINE FILTER DRAIN COCK
P—FUEL PUMP
Q—ENGINE OIL COOLER LINE
R—FUEL PUMP TO CARBURETOR FUEL LINE
S—PRIMER FUEL LINE
T—PRIMER LINE FILTER
U—PRIMER FUEL LINE TO CYLINDER
V—STARTER
W—RIGHT OIL COOLER
X—OIL COOLER CRADLE

Figure 130—Continued.

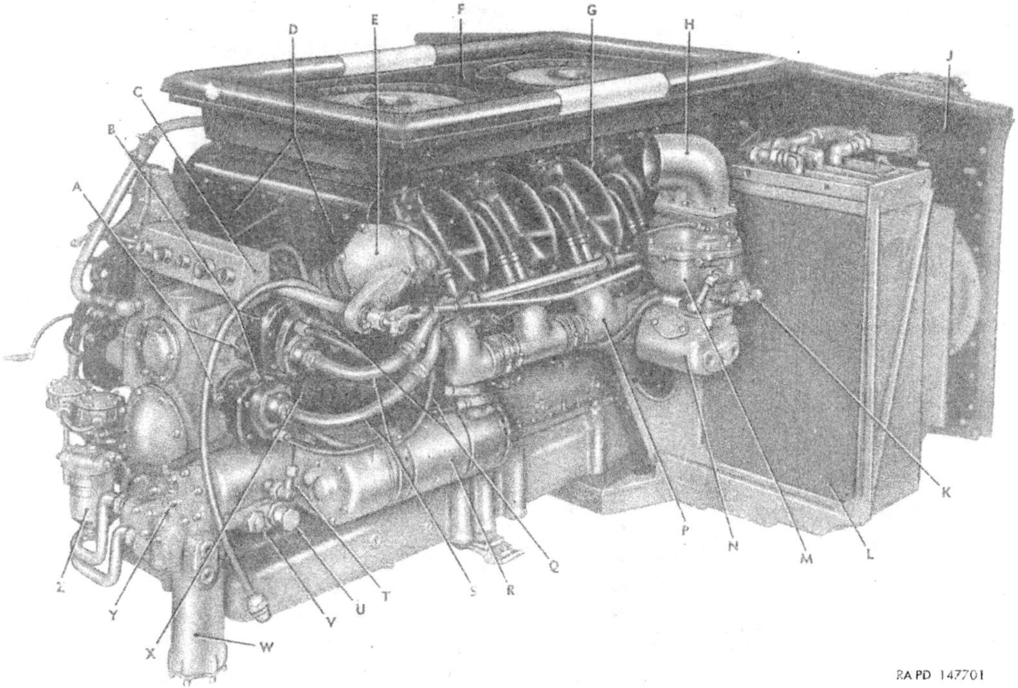

Figure 131. Power plant—right front view.

258

A—MAGNETO MOUNTING STUD AND NUT
B—RIGHT MAGNETO
C—ENGINE WIRING JUNCTION BOX
D—TOP SHROUD FRONT SHROUD PLATES
E—CAMSHAFT DRIVE GEAR HOUSING COVER
F—ENGINE TOP SHROUD
G—RIGHT CYLINDER BANK
H—CARBURETOR INTAKE ELBOW
J—ENGINE REAR SHROUD
K—DEGASSER
L—LEFT OIL COOLER (TRANSMISSION)
M—RIGHT CARBURETOR
N—HOT SPOT MANIFOLD
P—INTAKE MANIFOLD
Q—MAGNETO VENT LINE
R—MAIN GENERATOR
S—IGNITION HARNESS
T—ENGINE HIGH OIL TEMPERATURE WARNING SIGNAL LIGHT SWITCH
U—OIL PRESSURE CONTROL VALVE
V—OIL FILTER BYPASS VALVE
W—ENGINE OIL FILTER
X—TACHOMETER SENDING UNIT
Y—OIL COOLER BYPASS VALVE
Z—FUEL LINE FILTER

Figure 131—Continued.

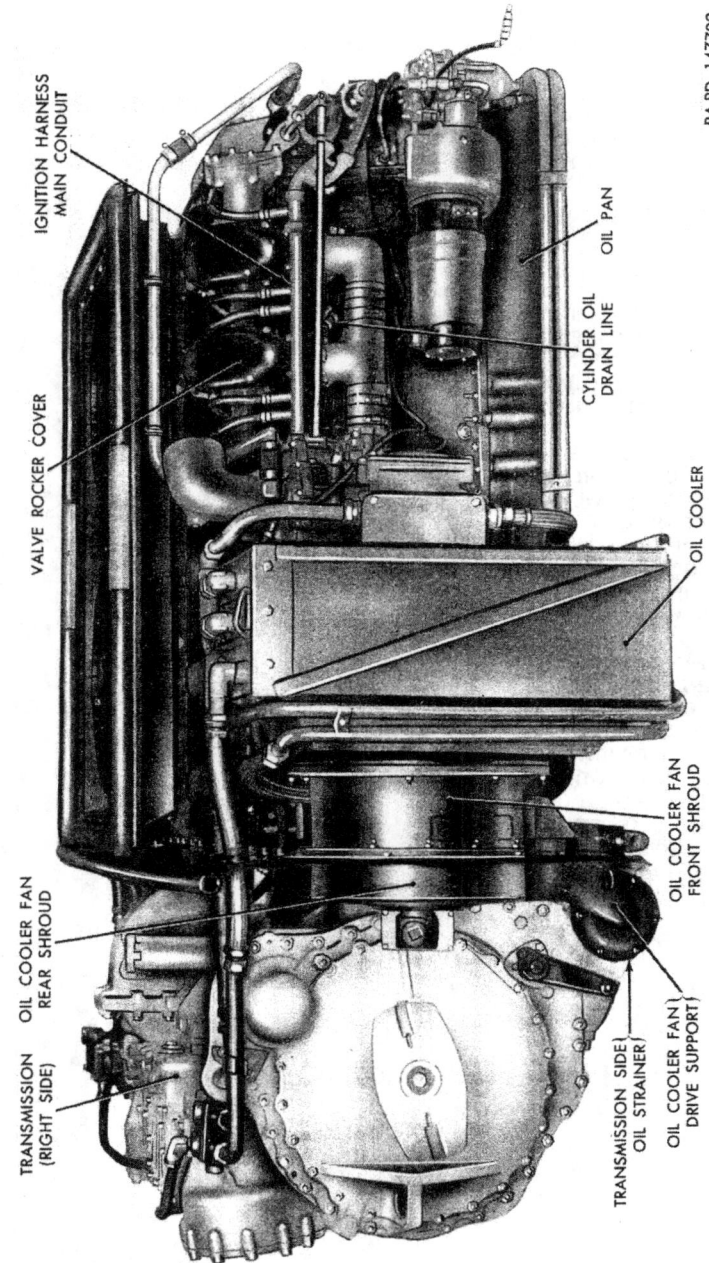

Figure 132. Power plant—left side view.

Lifting the universal joints will aid in alining bolt holes. Insert four bolts in each universal joint by installing two in each bearing cap. Tighten bolts and secure with locking wire.

(13) *Check engine operation.* Install exhaust pipes and tighten exhaust pipe clamps (fig. 122). Start engine (par. 68) and check operation to be sure it is in good running condition before installing top deck assembly (par. 161a). Check for oil leaks.

(14) *Record of replacement.* Record the replacement on DA AGO Form No. 478, MWO and Major Unit Assembly Replacement Record and Organizational Equipment File.

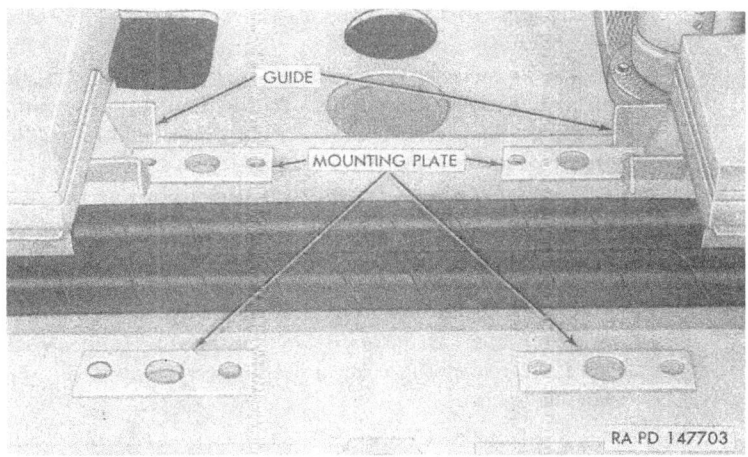

Figure 133. Engine mounting plates in engine compartment.

161. Top Deck Installation

(fig. 101)

a. INSTALLATION AS A COMPLETE UNIT.

(1) *Install lifting sling and lift top deck assembly.* Install lifting sling 41-S-2832-46. Secure hooks in two loops on engine compartment rear crossbeam and in two loops on engine compartment front grille door supports. Raise the top deck assembly (fig. 125) and carefully lower it into position on the hull.

(2) *Install top deck assembly hold-down cap screws.*

Note. Do not tighten hold-down cap screws until all have been started.

Install two cap screws securing engine compartment rear grille door support to hull (fig. 101). Install six cap screws (three on each side) securing engine compartment (right and left) rear grille door assemblies to hull. Install exhaust pipes and tighten exhaust pipe clamps (fig. 122). Swing the exhaust pipe housing covers (fig. 123) into position and tighten nuts (two on each side) on the exhaust pipe housing cover clamps.

Caution: Remove rags from exhaust manifolds before installing exhaust pipes. Install four cap screws securing engine compartment rear crossbeam to hull. Install four cap screws securing exhaust pipe housing to hull. Install four cap screws securing front grille door supports to hull (fig. 101). Tighten all hold-down cap screws securing top deck assembly to hull.

(3) *Install engine compartment intermediate grille door.* Install intermediate grille door adjacent to the exhaust pipe housing. Tighten two cap screws (one on each side) holding grille door to front door supports.

(4) *Install center batteries.* Install the battery box and the two batteries (par. 224). Install battery clamp, clamp hooks, and nuts. Connect cables to the two center batteries.

(5) *Install center battery cover.* Install center compartment battery box cover (fig. 101) by positioning it over the two dowel pins. Swing the two hold-down clamps into position and tighten the two battery cover hold-down clamp bolts (one on each side).

(6) *Close engine compartment doors.* Close the 16 engine compartment right and left grille doors (fig. 101).

(7) *Traverse turret.* Traverse the gun and turret, and secure the gun in the gun traveling lock (par. 55).

b. INSTALLATION AS INDIVIDUAL SECTIONS.

(1) *Installation of engine compartment rear section.* Install lifting sling 41–S–3832–46. Position engine compartment rear section carefully on top of engine compartment. Secure the rear crossbeam to hull by installing two cap screws and lock washers on each side. Secure the grille door assembly to the hull by installing three cap screws and lock washers on each side. Secure the rear grille door support to hull by installing two cap screws and lock washers.

Note. Do not tighten cap screws until all have been started.

(2) *Installation of engine-compartment front and intermediate sections.* Install lifting sling 41–S–3832–46 and raise engine compartment front and intermediate sections. Lower care-

fully into position. Install four cap screws and lock washers holding intermediate grille door supports to rear crossbeam.

Note. Do not tighten cap screws until all have been started.

Install four cap screws and lock washers holding exhaust pipe housing to hull (fig. 123). Install exhaust pipes and tighten exhaust pipe clamps (fig. 122). Install four cap screws and lock washers holding engine compartment front grille door supports to hull. Install center batteries and center battery box cover (par. 224). Close the eight grille doors on each side.

(3) *Installation of engine compartment front section.* Position the two front grille door supports (fig. 101) on the front portion of the hull and on the exhaust pipe housing and secure each with four cap screws and lock washers. Install the three intermediate grille doors and secure each to the intermediate grille door supports with one cap screw and lock washer on each side. Install center battery box, batteries, and center battery box cover (par. 224). Close the five grille doors on each side.

(4) *Installation of engine compartment intermediate section.* Position the two intermediate grille door supports on the rear crossbeam and on the exhaust pipe housing and secure each with four cap screws and lock washers. Install the three intermediate grille doors and secure each to the grille door supports with one cap screw and lock washer on each side. Close the three grille doors on each side.

Section VII. ENGINE REMOVAL AND INSTALLATION

162. Coordination With Ordnance Maintenance Unit

Replacement of the engine with a new or rebuilt engine is normally an ordnance maintenance operation, but may be performed in an emergency by the using organization, provided approval for performing this replacement is obtained from the supporting ordnance officer. A replacement engine, any tools needed for the operation which are not carried by the using organization, and any necessary special instructions regarding accessories, etc., may be obtained from the supporting maintenance unit.

163. Removal

a. REMOVE TOP DECK. Refer to paragraph 158*b*.

b. REMOVE POWER PLANT. Refer to paragraph 159.

c. REMOVE OIL COOLER FANS. Refer to paragraph 241*b.*

d. REMOVE OIL COOLERS AND OIL COOLER CRADLE. Refer to paragraph 242.

e. DISCONNECT ENGINE FROM TRANSMISSION. Install lifting sling 41-S-3832-165 on transmission (fig. 134). Make sure the engine is blocked securely. Remove the input-shaft access plug in center-rear of the transmission (fig. 135). Remove snap ring using snap ring pliers 41-P-1992-35, and pull input shaft to the rear with puller 41-P-2906-280 (fig. 136) to disengage engine and transmission drive connection. Remove transmission-to-engine mounting bolts (fig. 229). After removing the securing bolts, take up the weight of the transmission with the sling. Remove the transmission from the engine by working it straight off the dowels. Remove "O" ring packing from between engine and transmission flanges.

Caution: Use care in separating engine and transmission to prevent damage to mating parts.

164. Installation

a. GENERAL. Remove all cables, connections, and parts which are not furnished with the replacement engine from the engine being replaced. Install these items on the replacement engine before installing the power plant into the vehicle.

b. SUPPORT TRANSMISSION. Install lifting sling 41-S-3832-165 on transmission (fig. 134) and take up its weight.

c. CONNECT ENGINE TO TRANSMISSION. The accessory drive shaft opening cover (fig. 130) must be removed in order to turn the engine crankshaft to mate driving and driven gears and shafts. Use engine-turning wrench 41-W-906-125 (fig. 81) to turn. Coat mating surfaces with liquid-type gasket cement. Install "O" ring packing between engine and transmission flanges. Position engine to transmission flanges. Rotate engine crankshaft so as to mate transmission accessory gears and input drive shaft. Install transmission-to-engine mounting bolts (fig. 229)

d. INSTALL OIL COOLERS AND OIL COOLER CRADLES. Refer to paragraph 242.

e. INSTALL OIL COOLER FANS. Refer to paragraph 241.

f. INSTALL POWER PLANT. Refer to paragraph 160.

g. INSTALL TOP DECK. Refer to paragraph 161*a.*

h. RECORD OF REPLACEMENT. Record the replacement on DA AGO Form No. 478.

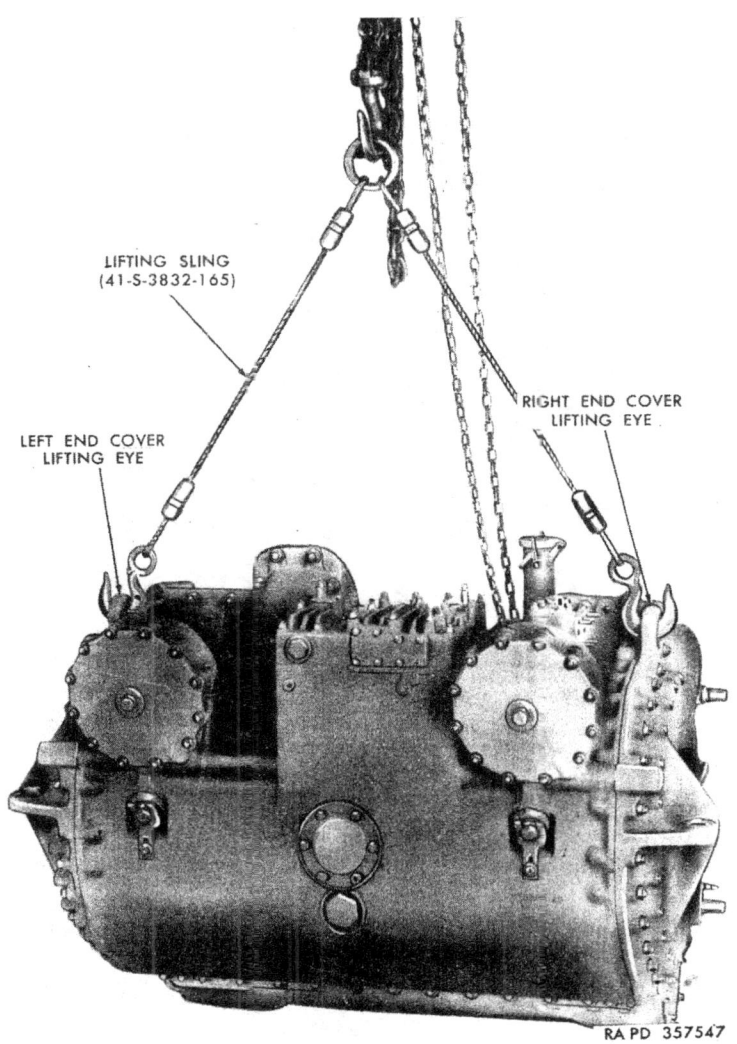

Figure 134. Lifting transmission with sling.

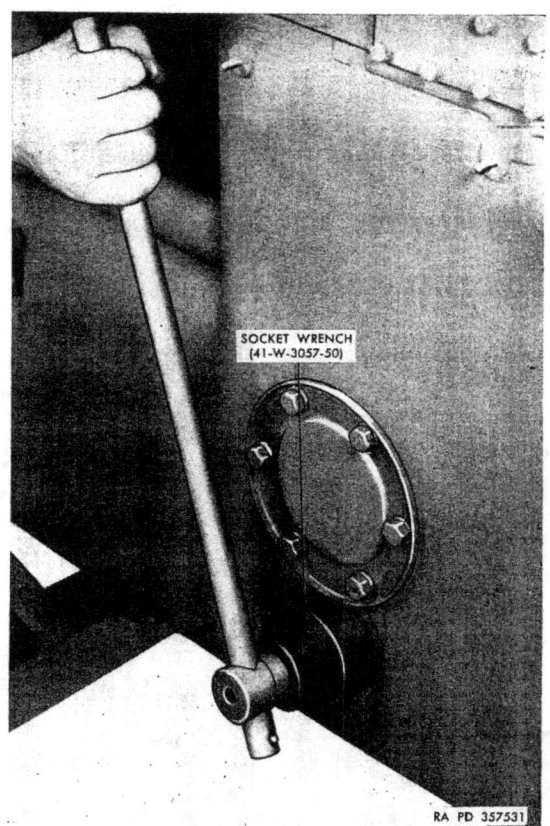

Figure 135. Removing input shaft access plug.

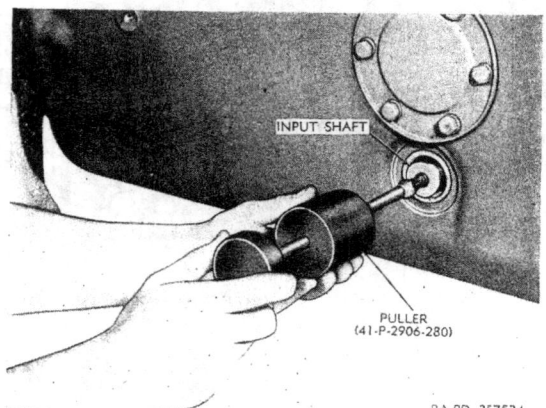

Figure 136. Pulling input shaft.

Section VIII. ENGINE MAINTENANCE WITH ENGINE REMOVED FROM VEHICLE

165. Operations With Engine Removed From Vehicle

a. ENGINE. Replace (pars. 163 and 164).
b. ENGINE MOUNTING. Replace (par. 170).
c. ENGINE REAR SHROUD. Repair or replace (par. 169).
d. FUEL FILTERS. Service or replace (par. 176).
e. FUEL TANKS. Service or replace (par. 177).
f. FUEL PUMPS. Replace (par. 174).
g. GENERATOR. Replace (par. 222).
h. GENERATOR BRUSHES. Replace (par. 222).
i. INTAKE MANIFOLDS. Replace (par. 167).
j. OIL COOLER CRADLE. Replace (par. 242).
k. OIL COOLER FANS. Replace (par. 241).
l. STARTER. Replace (par. 190).
m. STARTER BRUSHES. Replace (par. 190).
n. TRANSMISSION. Replace (par. 239).

166. Cylinder Air Deflectors

a. REMOVAL (fig. 112).
 (1) *Remove top deck.* Refer to paragraph 158*b.*
 (2) *Remove power plant.* Refer to paragraph 159.
 (3) *Remove engine top shroud.* Refer to paragraph 152*a.*
 (4) *Remove exhaust manifolds.* Refer to paragraph 153.
 (5) *Remove cylinder air deflectors.* Remove the four springs (one at each end of the cylinder banks) securing the end deflectors to the deflector clamps. Remove the cotter pins, slotted nuts, spacers, and bolts holding the end deflectors to adjacent intercylinder air deflectors. Remove the end deflectors.

 Note. In order to remove end deflectors from Nos. 1 right and 6 left, it is necessary to remove the dome fin deflectors (fig. 100) secured to these end deflectors with screws.

 Loosen the five nuts on the deflector clamps, located on the outer side of each cylinder bank. Remove the deflector clamp hooks from the cylinder deflectors. Remove the cylinder air deflectors from the cylinders.

b. INSTALLATION (fig. 112).
 (1) *Install air deflectors.* Position the ten cylinder air deflectors and secure them with the deflector-clamp hooks. Tighten the deflector-clamp nuts. Position the four end deflectors and secure them to the adjacent cylinder deflectors with spacers, bolts, slotted nuts, and cotter pins. Install the dome-fin de-

flectors on Nos. 1 right and 6 left end deflectors (fig. 100). Connect the deflector-clamp springs on the end deflectors.

(2) *Install exhaust manifolds.* Refer to paragraph 153.
(3) *Install engine top shroud.* Refer to paragraph 152*b*.
(4) *Install power plant.* Refer to paragraph 160.
(5) *Install top deck.* Refer to paragraph 161*a*.

167. Intake Manifolds

a. REMOVAL (figs. 137 and 138).
 (1) Remove top deck (par. 158*b*).
 (2) Remove power plant (par. 159).
 (3) Remove carburetor (par. 172*b*).
 (4) Remove hot spot tubes and hot spot manifold (par. 173).
 (5) Disconnect the magneto vent lines (fig. 137) at the accessory case end of both intake manifolds.

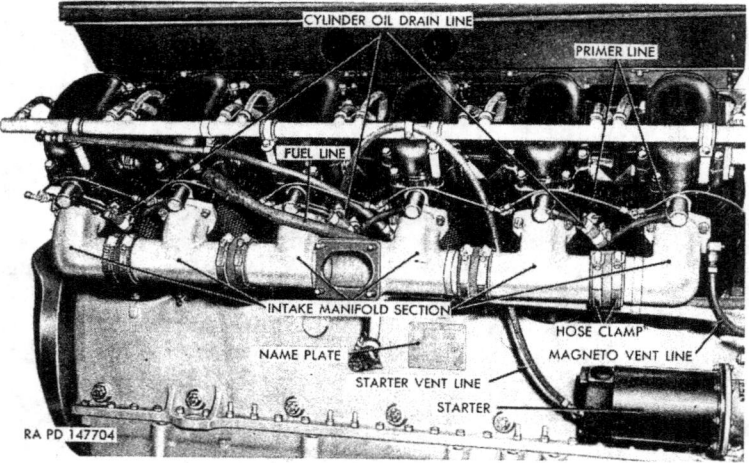

Figure 137. Left intake manifold—installed view.

 (6) To remove right manifold, disconnect mounting clip securing the carburetor inlet fuel line to manifold. Remove mounting clip securing oil filter breather line to manifold.
 (7) To remove left manifold, disconnect clip securing carburetor inlet fuel line to manifold.
 (8) Loosen the eight hose clamps (fig. 137). Remove the three jam nuts, nuts, and washers which secure each manifold section to the cylinder. Pull the manifold section free from hose and remove. Remove all hoses, hose clamps, and strip off gasket.

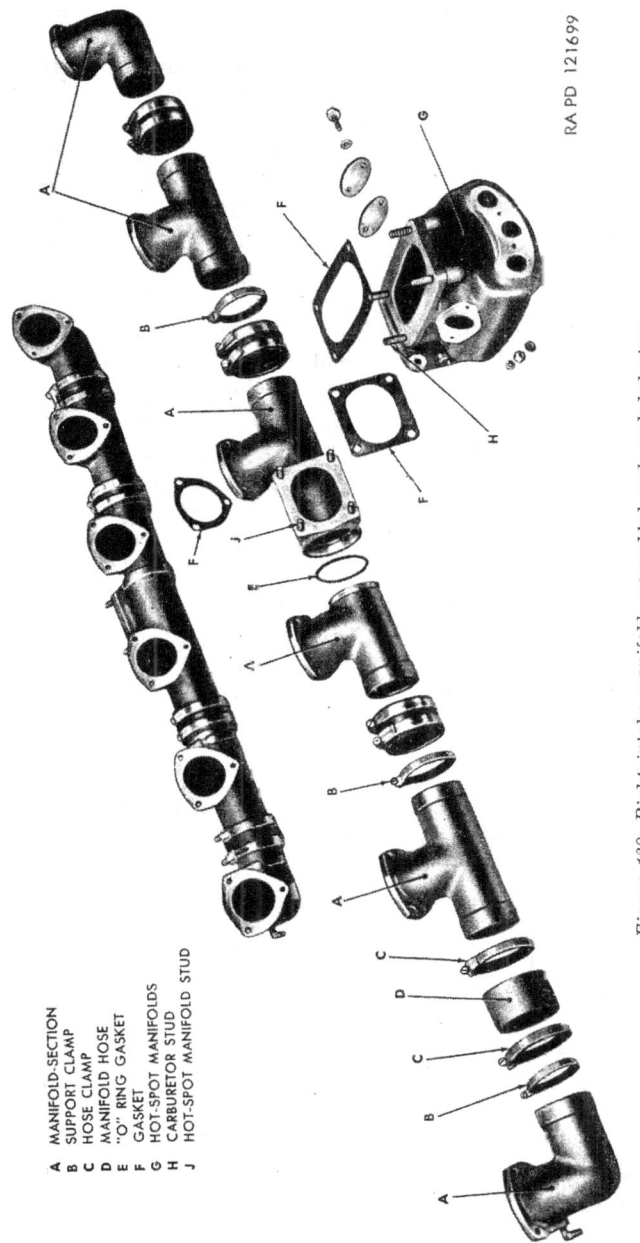

Figure 138. Right intake manifold—assembled and exploded view.

b. CLEANING AND INSPECTION. Clean the intake manifold section by imersing in volatile mineral spirits or dry-cleaning solvent and scrubbing with a stiff-bristle brush. Wipe, and dry with compressed air. Inspect the machined surface of each manifold section to see that it is flat, without warpage, nicks, or burs. Dress nicked or warped surfaces on a surface plate, or replace the manifold section. Inspect all castings for cracks and flaws, replacing unserviceable parts. Hose connections should be cleared with carbon tetrachloride. Replace all unserviceable hose clamps. Any hoses showing signs of deterioration should be replaced.

Note. The intake manifolds, consisting of seven similar castings, are interchangeable.

(1) Replace all gaskets and intake manifold hose (fig. 138). The intake manifolds can be assembled individually and installed as a unit to the engine or installed as individual sections directly to the engine. Install new "O" ring gaskets on the left center sections (fig. 138). Position the right center sections over the "O" ring gaskets. Install manifold hoses and secure with hose clamps. Install the intermediate sections and secure with hose clamps. Install the elbow sections and secure with hose clamps. Install the manifolds to the engine and secure each section with three washers, nuts, and jam nuts.

Note. If installing manifold in individual sections, start at one end and work toward the other end. Be sure that all hose clamps are tightened. Make sure that pipe plugs are in tapped holes in flywheel end sections.

(2) Connect the magneto ventilating lines (fig. 137).
(3) Install hot manifold and hot spot tubes (par. 173).
(4) Install carburetors (par. 172*c*).
(5) Install clip securing carburetor inlet fuel line to right manifold. Install clip securing oil filler breather line to right manifold.
(6) Install clip securing carburetor inlet fuel line to left manifold.
(7) Install power plant (par. 160).
(8) Install top deck assembly (par. 161*a*).

168. External Oil Lines

a. OIL COOLER LINES.
(1) *Removal.* Refer to paragraph 242.
(2) *Installation.* Refer to paragraph 242.
b. ROCKER BOX OIL DRAIN LINES (fig. 100).

(1) *General.* Six external "Y" shaped lines drain excess oil from the valve rocker boxes to the crankcase, each serving two adjacent rocker boxes.

(2) *Removal.*
   (*a*) Remove top deck (par. 158*b*).
   (*b*) Remove power plant (par. 159).
   (*c*) Remove carburetors (par. 172).
   (*d*) Remove oil coolers (par. 242).
   (*e*) Loosen clamps on hose adapters at crankcase, between each pair of cylinders. Remove bolts holding lines to rocker boxes. Pull lines and hose connections from hose adapters at crankcase. Strip off gaskets.

(3) *Installation.*
   (*a*) Position drain lines on hose adapters at crankcase and tighten clamps. Position lines on rocker boxes, using new gaskets, and install bolts securing drain lines to rocker boxes.
   (*b*) Install oil coolers (par. 242).
   (*c*) Install carburetors (par. 172).
   (*d*) Install power plant (par. 160).
   (*e*) Install top deck (par. 161*a*).

169. Engine Rear Shroud

a. REMOVAL (fig. 139).
   (1) Remove top deck (par. 158*b*).
   (2) Remove power plant (par. 159).
   (3) Remove right and left side seals.
   (4) Remove shroud center butt strap.
   (5) Disconnect top plates from upper plates, right and left, and remove top plates.
   (6) Disconnect transmission breather line at transmission (fig. 229) and remove clip securing line to transmission oil filler. Remove transmission breather line grommet from upper right plate and pull breather line through plate.
   (7) Remove right and left oil cooler fan cross-drive access covers.
   (8) Remove lower center plate.
   (9) Remove right and left oil cooler fan housing bracket seal retainers and seals.
   (10) Remove bolts securing lower right and left plates to oil cooler fan front shroud, and remove lower plates.
   (11) Remove bolts securing upper right and left plates to oil cooler fan front shroud, and remove upper plates.

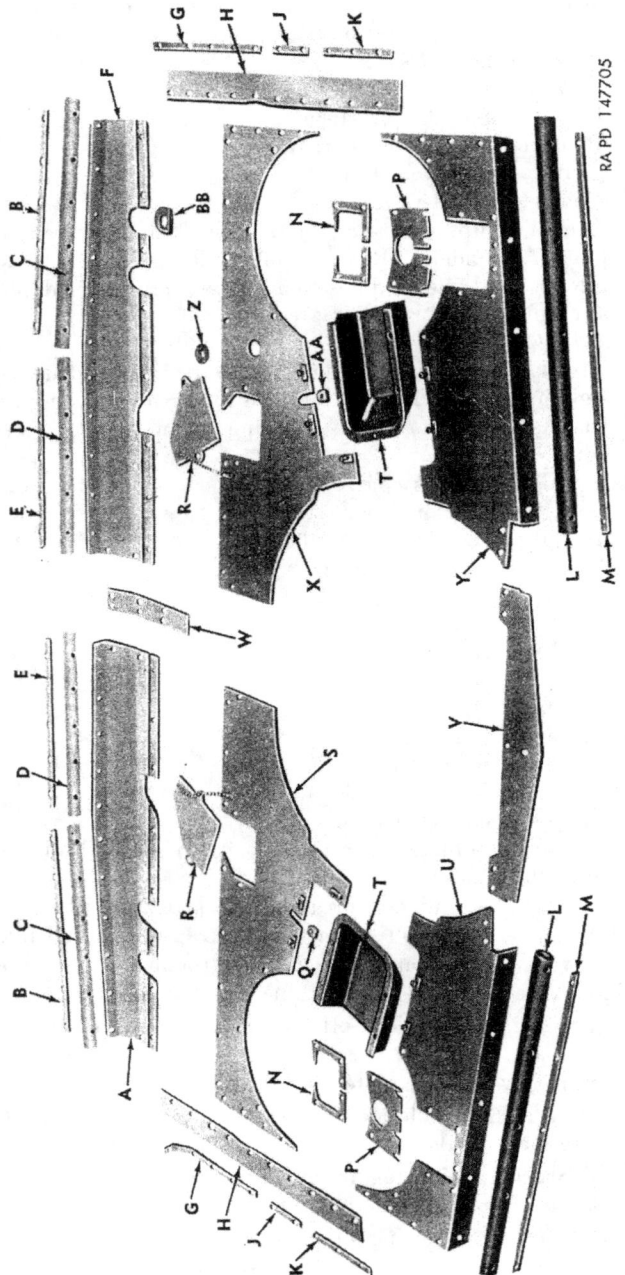

Figure 139. Engine rear shroud—exploded view.

A—TOP LEFT PLATE
B—TOP PLATE OUTER SEAL RETAINER
C—TOP PLATE OUTER SEAL
D—TOP PLATE INNER SEAL
E—TOP PLATE INNER SEAL RETAINER
F—TOP RIGHT PLATE
G—SIDE SEAL UPPER RETAINER
H—SIDE SEAL
J—SIDE SEAL CENTER RETAINER
K—SIDE SEAL LOWER RETAINER
L—LOWER PLATE SEAL
M—LOWER PLATE SEAL RETAINER
N—OIL COOLER FAN HOUSING BRACKET SEAL RETAINER
P—OIL COOLER FAN HOUSING BRACKET SEAL
Q—CABLE GROMMET—0.375-INCH HOLE
R—SPARK PLUG ACCESS OPENING COVER
S—UPPER LEFT PLATE
T—OIL COOLER FAN CROSS DRIVE ACCESS COVER
U—LOWER LEFT PLATE
V—LOWER CENTER PLATE
W—SHROUD CENTER BUTT STRAP
X—UPPER RIGHT PLATE
Y—LOWER RIGHT PLATE
Z—TRANSMISSION BREATHER LINE GROMMET
AA—CABLE GROMMET—0.160-INCH HOLE
BB—TRANSMISSION OIL COOLER LINE GROMMET

Figure 139.—Continued.

b. INSTALLATION (fig. 139).

(1) Install upper right plate on oil cooler fan front shroud, securing it with three bolts ($\frac{1}{4}$–28 x $1\frac{1}{8}$), six plain washers (one under each bolt and one under each nut), and three safety nuts. Install upper left plate in the same manner.

(2) Install lower right plate on oil cooler fan front shroud securing it with two bolts ($\frac{1}{4}$–28 x $1\frac{1}{8}$), four plain washers, and two safety nuts. Install lower left plate in same manner.

(3) Install right and left oil cooler fan housing bracket seals and seal retainers, securing each with three machine screws (No. 10–32NF x $\frac{1}{2}$), lock washers, and nuts.

(4) Install lower center plate securing with four bolts ($\frac{1}{4}$–28 x $\frac{7}{8}$), eight plain washers, and four safety nuts.

(5) Install right and left oil cooler fan across drive access covers, securing each with five bolts ($\frac{1}{4}$–28 x $\frac{5}{8}$) and five lock washers.

(6) Pass transmission breather line through upper right plate and install grommet in plate. Connect transmission breather line to fitting on transmission (fig. 229) and clip line to transmission oil filler.

(7) Attach top right and left plates to upper right and left plates with 16 bolts ($\frac{1}{4}$–28 x $\frac{3}{4}$) and lock washers.

(8) Install shroud center butt strap securing with six bolts ($\frac{1}{4}$–28 x $\frac{3}{4}$), three plain washers (used on right side), and six lock washers.

(9) Install right side seal and retainers securing it to upper right plate with four bolts ($\frac{1}{4}$–28 x $\frac{5}{8}$), lock washers, and hex nuts. Secure to lower right plate with three bolts ($\frac{1}{4}$–28 x $\frac{5}{8}$), lock washers, and hex nuts. Secure center section with two bolts ($\frac{1}{4}$–28 x $\frac{3}{4}$), lock washers, and hex nuts. Install bolt ($\frac{1}{4}$–28 x $\frac{3}{4}$) at outer end of top right seal. Install left side seal and retainers in the same manner.

(10) Install power plant (par. 160).

(11) Install top deck (par. 161*a*).

c. INSPECTION AND REPAIR.

(1) Straighten any bent plates or other parts and replace any which cannot be repaired.

(2) Top plate seals and lower plate shields are tubular with the retainers inside the shield and are tapped at proper intervals to receive the holding bolts. Note that the longer sections of the top plate seals are the outer sections and the shorter ones are the inner sections.

170. Engine Mounting

a. REMOVAL.
  (1) Remove top deck assembly (par. 158*b*).
  (2) Remove power plant (par. 159.)
  (3) Set power plant on stand (or use blocking) so mountings are clear of any obstructions.
  (4) Remove the cotter pin, castle nut, and plain washer, securing each mounting (fig. 140) to side of oil pan. To disassemble engine mounting, remove jam nut at bottom and unscrew stud from mounting cushion.

b. INSTALLATION.
  (1) Insert mounting stud through hole on side of oil pan. Position the dowel pin in the hole provided in the bottom of the oil pan. Install and secure each with plain washer, castle nut, and cotter pin.
  (2) Install power plant (par. 160).
  (3) Install top deck assembly (par. 161*a*).

Section IX. FUEL, AIR-INTAKE, AND EXHAUST SYSTEMS

171. Description and Data

a. DESCRIPTION.
  (1) *Fuel, air-intake, and exhaust systems.*
    (*a*) *Fuel system.* The fuel system is composed of two fuel tanks, each with its fuel-tank-shut-off valve and fuel tank filter, two fuel pumps, a fuel line filter, two carburetors, four degassers, a governor, a primer pump, a primer line filter, and the necessary connecting fuel lines (fig. 157).
    (*b*) *Air-intake system.* The air-intake system includes two air cleaners and the air cleaner pipes which connect air cleaners to the carburetors.
    (*c*) *Exhaust system.* The exhaust system is composed of two mufflers and the connecting exhaust pipes from the exhaust manifolds.
  (2) *Fuel tanks.* Two fuel tanks (figs. 152 and 153) are provided and are installed in the front left and front right corners of the engine compartment. These tanks are filled from the top deck of the vehicle, through fillers (fig. 154) located just to the rear of the turret, on the left and right. The filler caps are protected by fuel tank filler covers which are provided with locks. These locks are to prevent enemy troops from opening covers and dropping hand grenades.
  (3) *Fuel filters.* Three filters are provided in the fuel system, one mounted in each fuel tank at the point where the fuel tank outlet lines are connected (fig. 156) and the third

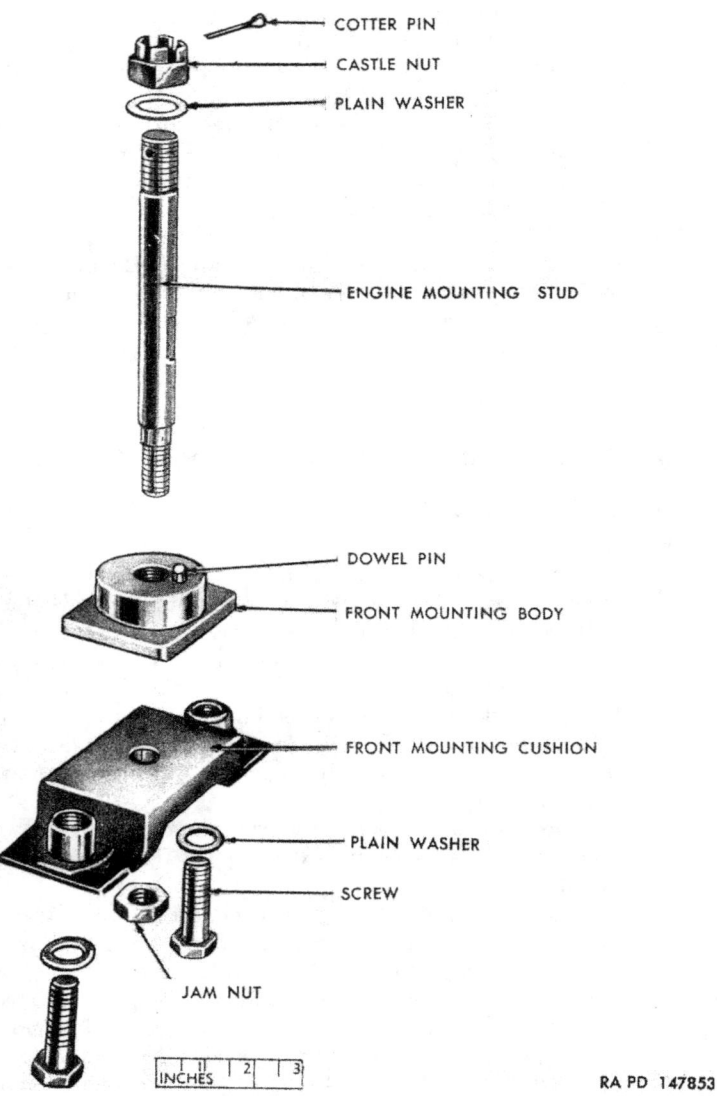

Figure 140. Engine mounting—exploded view.

mounted on the engine in front of the two fuel pumps (fig. 130). The filters contain disk-type elements (figs. 149 and 150).

(4) *Fuel-tank-shut-off valves.* Two tank-shut-fuel off valves are provided, one connected to each fuel tank outlet line. They are located between the fuel tanks (fig. 156). Access to the control handles for these two valves is obtained by traversing turret so 90-mm gun is pointing ahead and approximately 30° to the left, lifting ammunition access door (fig. 34) in turret floor, and removing fuel-shut-off-valve-cover plate (fig. 146).

(5) *Fuel pumps.* The fuel pumps are of the mechanical type, actuated by engine-driven eccentric cams. The two fuel pumps (figs. 130 and 144) are mounted on the accessory end of the engine and operate alternately. They are joined by an interconnecting line on the outlet side. Either pump will supply adequate fuel for both carburetors in the event of damage or failure of one of the pumps.

(6) *Carburetors.* Two carburetors, (fig. 141), one on each side of the engine, provide fuel for the engine. The carburetor on the left side of the engine provides fuel for the left bank and the carburetor on the right side provides fuel for the right bank of cylinders. The carburetor is a double-venturi, down-draft unit. It has two floats connected by one lever and operating one needle valve and seat. Vent lines from the ignition harness, crankcase oil breather, starter, accessory case, and transmission lead to the intake sides of the carburetor.

(7) *Degassers.* Each carburetor venturi tube has a **degasser** (fig. 141) incorporated in the idle system. The degasser shuts off the fuel supply when high-manifold vacuum is present during deceleration periods. It is controlled automatically by manifold vacuum. In addition, it has an electric solenoid, controlled by the fuel cut-off (degasser) switch (fig. 18) for positive shut-off of the fuel supply to stop the engine before the ignition system is turned off.

(8) *Governor.* The engine governor is the mechanical-hydraulic type. The governor comprises three basic systems. The first system is a fly-ball-and-race-type mechanical governor, which actuates the second, or pilot valve, which in turn controls the third, or oil pressure system. The oil pressure is supplied from the main oil gallery of the engine, and produces the amplified energy required to actuate the governor-to-carburetor control linkage. The pilot valve moves back and forth over oil passage orifices, as governed by the fly-ball

system, thereby allowing oil pressure to be increased or decreased in the hydraulic system, controlling the maximum engine revolutions per minute between 2,800 rpm full-load speed and 2,950 rpm no-load speed. No maintenance is covered in this manual as none is authorized the using organization.

(9) *Primer pump.* The primer pump (figs. 158 and 159) is a cylinder-and-piston type pump which is operated manually. It has primer fuel line connections with a fuel line filter to draw fuel from fuel tanks and inject it into the intake manifolds.

(10) *Air cleaners.* Two oil-bath-type air cleaners (fig. 146) are provided, one for each carburetor. They are mounted in the crew compartment, and have provision for drawing intake air from either the crew compartment or engine compartment. Air normally is drawn from the engine compartment. Operation of the air cleaner inlet duct control handle will cause air to be drawn from crew compartment.

b. TABULATED DATA.

Carburetor:
    Make _____ Bendix-Stromberg
    Model _____ NA–Y5G3
    Type _____ double-venturi; down-draft
Fuel pump:
    Make _____ AC Spark Plug Div
    Model _____ BF
Hydraulic governor:
    Make _____ Novi Equipment Company
    Part No _____ 52440A

172. Carburetors

a. ADJUSTMENTS.

(1) *Carburetor float level check.* Open engine compartment side grille doors. With the vehicle on level ground, start the engine and run it at idling speed. Stop the engine. Remove the locking wire from the float chamber fuel level checking plug in the carburetor (fig. 141) and remove the plug. If the fuel is not visible at the bottom of the threads, the float level is too low. If the fuel flows freely from the hole, the float level is too high. Install and tighten the plug, and secure with locking wire. Repeat the test on the remaining three float bowls of the two carburetors. If fuel level is too high or too low, replace the carburetor (*b* and *c* below).

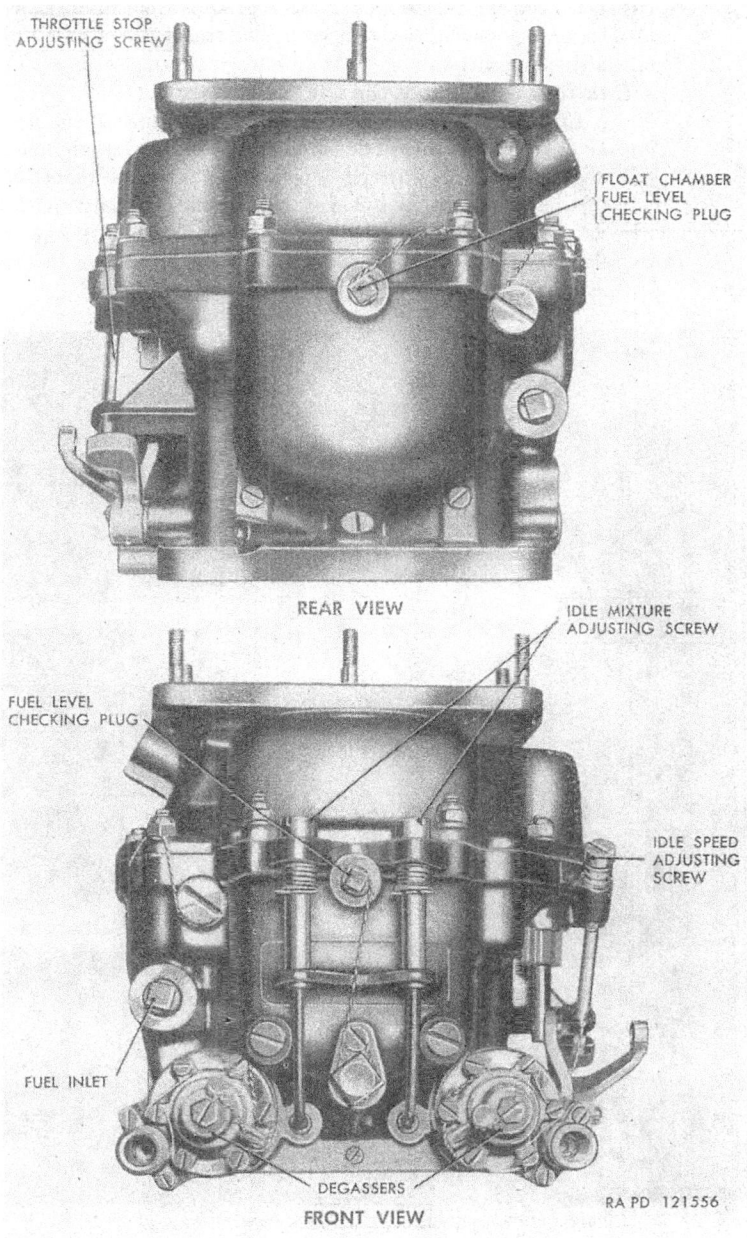

Figure 141. Carburetor—front and rear views.

(2) *Carburetor idle mixture.* Adjust the carburetor idle mixture with the engine stopped. Turn each of the idle mixture adjusting screws (fig. 142) on both carburetors clockwise until they seat lightly.

Caution: Do not turn the screws down tight.

Back off each screw one-quarter turn. Start and run the engine until it is warmed up. Return the hand throttle lever to the closed position. If the engine does not idle smoothly, vary the adjustment by turning the screws slightly in the direction which improves idling. Readjust the idle speed adjusting screws ((3) below). Stop the engine.

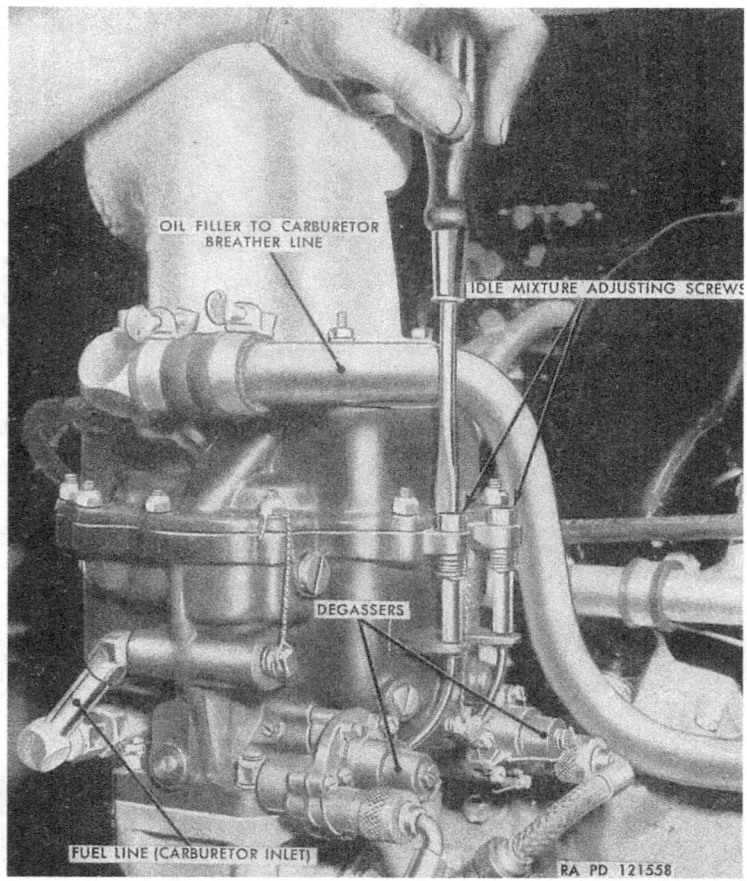

Figure 142. Adjusting idle mixture—right carburetor.

(3) *Engine idle speed.* Start and run the engine until it is warmed up to normal operating temperature. Return the hand throttle lever to the fully closed position and observe the tachometer reading, which must be 650 rpm. If the speed varies more than 50 rpm from the recommended idle speed, adjust the speed by means of the idle speed adjusting screws on each carburetor (fig. 141). Turn adjusting screws on both carburetors until the engine idles at recommended idling speed of 650 rpm. To increase the idle speed, turn the screws clockwise. To decrease the idle speed, turn the screws counterclockwise. If the engine does not idle smoothly at the recommended speed, and the float level has been tested ((1) above) and found to be correct, stop the engine and adjust the carburetor idle mixture ((2) above). Then start the engine and adjust idle speed adjusting screws. When adjustment is completed, stop the engine.

b. REMOVAL.

(1) *Remove carburetor intake elbow.* Remove front and intermediate sections of top deck (par. 158c). Remove air-cleaner-to-carburetor pipes (par. 159d (13)). Disconnect transmission vent line from left carburetor intake elbow. Remove six nuts, lock washers, and plain washers which secure each carburetor intake elbow to carburetor (fig. 143). Remove elbow and strip off gasket.

(2) *Remove right carburetor* (fig. 143). Disconnect carburetor inlet fuel line at carburetor. Disconnect throttle control linkage at carburetor. Unscrew knurled nuts and pull electrical cable connections from the two degassers. Disconnect crankcase oil breather and ignition harness vent lines. Remove two clips holding ignition harness vent line to ignition harness. Remove four jam nuts, nuts, and washers which secure carburetor to hot spot manifold and remove carburetor. Strip off gasket.

(3) *Remove left carburetor.* Disconnect carburetor inlet fuel line at carburetor. Disconnect throttle control linkage at carburetor. Disconnect electrical cables to the degassers. Disconnect starter and ignition harness vent lines. Remove four jam nuts, nuts, and washers which secure carburetor to hot spot manifold and remove left carburetor. Strip off gasket.

c. INSTALLATION.

(1) *Install left carburetor.* Place new gasket on studs and install carburetor, using four jam nuts, nuts, and washers to secure. Connect carburetor inlet fuel line and throttle control linkage. Connect starter and ignition harness vent lines and electrical cables to the two degassers. Install one clip holding ignition harness vent line to ignition harness.

A—HOT SPOT VACUUM CONTROL
B—TOP SHROUD PLATE
C—HOT SPOT VACUUM CONTROL VALVE
D—HOT SPOT RETURN TUBE
E—CARBURETOR INTAKE ELBOW
F—HOT SPOT INTAKE TUBE
G—CRANKCASE OIL BREATHER VENT LINE
H—CARBURETOR
J—DEGASSER
K—CARBURETOR MOUNTING GASKET
L—CARBURETOR MOUNTING NUT
M—HOT SPOT MANIFOLD
N—IGNITION HARNESS NAME PLATE
P—CARBURETOR INLET FUEL LINE
Q—FLEXIBLE LINE
R—FLEXIBLE LINE CLIP
S—THROTTLE CONTROL LINKAGE

RA PD 147706

Figure 143. Right carburetor—installed view.

(2) *Install right carburetor* (fig. 143). Place new gasket on studs and install carburetor, using four jam nuts, nuts, and washers to secure. Connect carburetor inlet fuel line and throttle control linkage. Connect crankcase oil breather and ignition harness vent lines and install two clips holding ignition vent line to ignition harness. Connect electrical cables to the two degassers.

(3) *Install carburetor intake elbow.* Using a new gasket, install carburetor intake elbow and secure to carburetor with six nuts, lock washers, and plain washers. Connect transmission vent line to left intake elbow. Install air-cleaner-to-carburetor pipes (par. 160*b* (11)). Start engine (par. 68) and adjust carburetors (*a* above). Inspect fuel line connections and gasket for leaks. Install front and intermediate sections of top deck (par. 161*b* (2)).

173. Hot Spot Manifold, Tubes, and Valve

a. REMOVAL.

(1) *Remove air-cleaner-to-carburetor pipes.* Remove front and intermediate portions of top deck (par. 158*c* (1)). Remove air-cleaner-to-carburetor pipes (par. 159*d* (13).

(2) *Remove hot spot tubes and vacuum controls* (fig. 143). Remove hot spot intake tube by removing two cap screws securing tube flange to hot spot manifold (right bank, flywheel side of manifold; left bank, accessory side) and pulling tube out of exhaust manifold center section. Remove hot spot return tube by removing two cap screws securing tube flange to hot spot manifold, two bolts which secure vacuum control valve to top shroud plate, and two bolts securing hot spot vacuum control to top shroud plate. Disconnect flexible line to lower end.

(3) *Remove right hot spot manifold.* Remove two clips securing degasser electrical cable to hot spot manifold. Remove right carburetor (par. 172*b*(2)). Remove three remaining jam nuts, nuts, and washers, and remove hot spot manifold from intake manifold. Strip off gasket.

(4) *Remove left hot spot manifold.* Remove two clips securing degasser electrical cable to manifold. Remove left carburetor (par. 172*b*(3)). Remove three remaining jam nuts, nuts, and washers, and remove hot spot manifold from intake manifold. Strip off gasket.

b. INSTALLATION. Replace all gaskets.

(1) *Install left hot spot manifold.* Place new gasket on studs and install hot spot manifold on left intake manifold. Secure hot spot manifold with four jam nuts, nuts, and washers

(one jam nut, nut, and washer also holds degasser electrical cable clip). Install other clip holding degasser electrical cable to hot spot manifold. Install carburetor (par. 172*c*(1)).

(2) *Install right hot spot manifold.* Install the right hot spot manifold in the same manner as the left manifold. Install carburetor (par 172*c*(2)).

(3) *Install hot spot tubes and vacuum controls* (fig. 143). Install hot spot intake tube by inserting tube through hole in top shroud plate into exhaust manifold center section and securing tube flange and gasket on hot spot manifold (flywheel side for right bank, and accessory side for left bank) with two cap screws and lock washers. Install hot spot return tube and control by securing vacuum control valve to top shroud plate with two drilled bolts ($5/16$–24 x 0.84), plain washers, slotted nuts, and cotter pins; and hot spot vacuum control to top shroud plate with two drilled bolts ($5/16$–24 x 0.84), plain washers, slotted nuts, and cotter pins. Connect flexible line to fitting in hot spot manifold.

(4) *Install air-cleaner-to-carburetor pipes.* Install air-cleaner-to-carburetor pipes (par. 160*b*(11)). Install front and intermediate sections of top deck (par. 161*b*(2)).

174. Fuel Pumps

a. FUEL PUMP VACUUM AND PRESSURE TESTS.

(1) *General.* Fuel pumps must be tested to be sure that they are pumping fuel at pressures between $4\frac{1}{2}$ and 6 psi. They must be checked on both the intake and outlet sides.

(2) *Remove right air cleaner and engine inspection cover.* To gain access to fuel pump connections, it is necessary to remove right air cleaner (par. 175) and bulkhead inspection plate (figs. 148 and 163).

(3) *Testing fuel pump vacuum.*

(*a*) Close fuel tank shut-off valve (fig. 11) and remove fuel line between fuel line filter and fuel pumps (fig. 144).

(*b*) Connect combination vacuum and pressure gage to inlet side of one pump.

(*c*) Plug inlet to other pump.

(*d*) Start and run engine at 700 rpm (par. 68).

Caution: As vacuum tests of both fuel pumps must be made on fuel in carburetors, do not run engine longer than necessary to make test readings.

(*e*) Observe gage and stop engine when maximum reading is indicated. The fuel pump, under normal operating conditions, will develop a vacuum of 10 inches of mercury.

If gage reading is less than this value, or if gage indicator fluctuates, or rapidly returns to zero, replace fuel pump (*b* below).

(*f*) Repeat this test with the other fuel pump.

(*g*) Remove gage and plug. Install fuel line between fuel line filter and fuel pumps.

(4) *Testing fuel pump pressure.*

(*a*) Disconnect fuel line from fuel pump outlet to one of the carburetors.

(*b*) Connect a combination vacuum and pressure gage between the fuel pump outlet and the carburetor, making connections secure and leakproof.

(*c*) Start and run the engine at 700 rpm.

(*d*) Observe gage and stop engine when maximum reading is indicated. Correct pressure is between 4½ and 6 psi. If pressure is too low, replace fuel pump (*b* below).

(*e*) Remove gage, connect fuel pump to carburetor line, and repeat this test for the other fuel pump.

b. REMOVAL OF FUEL PUMPS.

(1) If removal is made with engine in vehicle, remove right air cleaner (par. 175) and bulkhead inspection plate (figs. 148 and 163).

(2) Disconnect the fuel lines to carburetors and line from fuel line filter to fuel pumps (fig. 144).

(3) Remove two jam nuts, nuts, and washers which secure each fuel pump to adapter and lift off pumps.

(4) Strip off gaskets and remove fittings from pumps.

c. INSTALLATION (fig. 144). Install the inlet and outlet connections into each fuel pump. Using new gaskets, positions fuel pumps over studs and secure each with washers, nuts, and jam nuts. Connect the fuel lines to right and left carburetors and line to fuel line filter.

Note. Be sure all connections are tight.

Install bulkhead air cleaner inspection plate (par. 175).

175. Air Cleaners

a. SERVICING. The procedure for servicing both air cleaners is the same.

(1) *Remove air cleaner oil reservoir* (fig. 145). Disengage oil reservoir clamp. Turn oil reservoir slightly to release from air cleaner, and remove.

(2) *Clean oil reservoir and replace oil.* Pour the old oil out of reservoir. Scrape accumulated dirt from the reservoir and clean with a cloth soaked with volatile mineral spirits or

dry-cleaning solvent. Fill the reservoir with oil to the oil level mark, as specified on lubrication order (fig. 86).

(3) *Install air cleaner oil reservoir* (fig. 145). Position oil reservoir under air cleaner. Lift into place and turn slightly. Connect oil reservoir clamp.

b. REMOVAL. The procedure for removing both air cleaners (fig. 146) is the same. Open engine compartment side grille door and loosen hose clamp (fig. 147) on the air-cleaner-to-carburetor pipe at air cleaner. In the crew compartment, remove four lock nuts from air cleaner mounting clamps and remove clamps. Pull air cleaner forward and lift out.

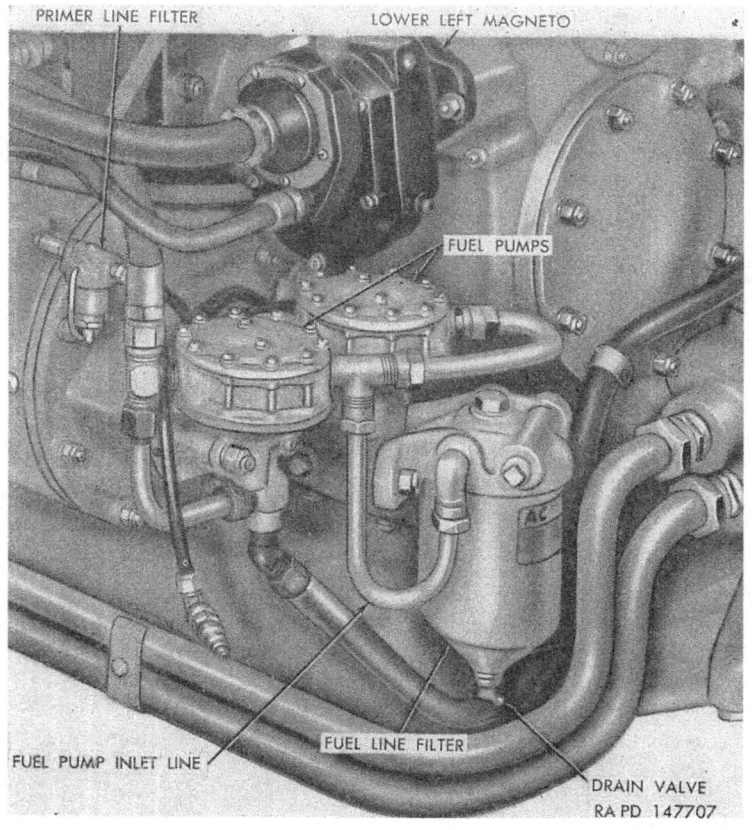

Figure 144. Fuel pumps—installed view.

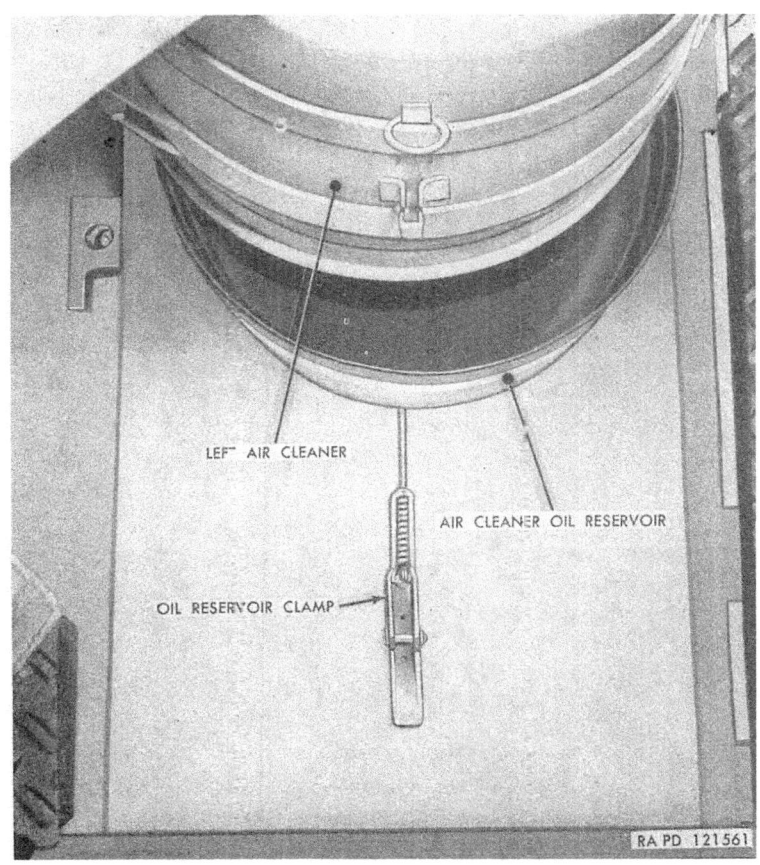

Figure 145.—Removing air cleaner oil reservoir.

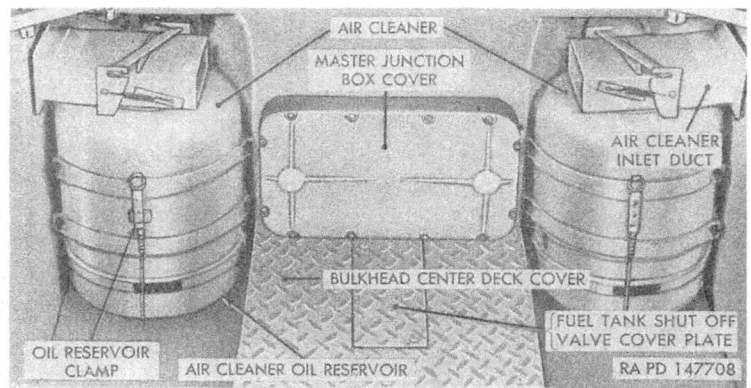

Figure 146. Air cleaners—installed view.

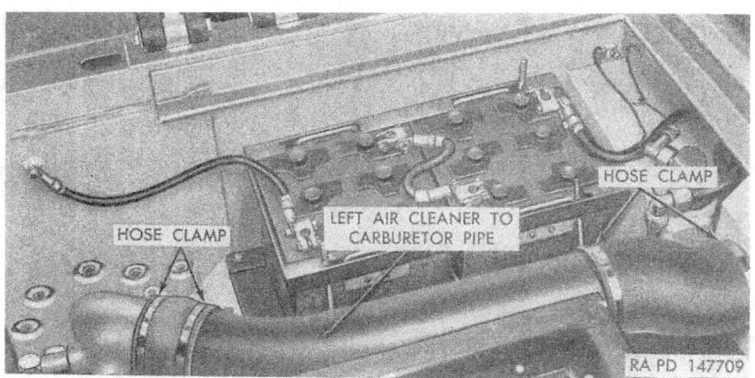

Figure 147. Air cleaner to carburetor pipe.

Figure 148. Left bulkhead—air cleaner removed.

  c. INSTALLATION. The procedure for installing both air cleaners is the same. Position air cleaner in place with outlet pipe hose through hole in bulkhead and inlet duct mating with opening in bulkhead side (figs. 146 and 147). Install air cleaner mounting clamps and secure with four lock nuts. Working through engine compartment grille doors, connect air-cleaner-to-carburetor pipe and tighten hose clamp (fig. 147).

176. Fuel Filters

  a. REMOVAL.
    (1) *Fuel tank filters* (figs. 149 and 156). Remove top deck (par. 158 *b*.) Remove power plant (par. 159). Drain fuel tanks (par. 177). Remove right-fuel-tank-shut-off valve (par. 178). To remove right-fuel-tank-fuel filter, remove six screws and lock washers which secure fuel filter cover to fuel tank. Remove filter and strip off gasket. Remove left-fuel-tank-shut-off valve (par. 178). Remove six screws and lock washers which secure fuel filter cover to left fuel tank. Remove filter and strip off gasket.

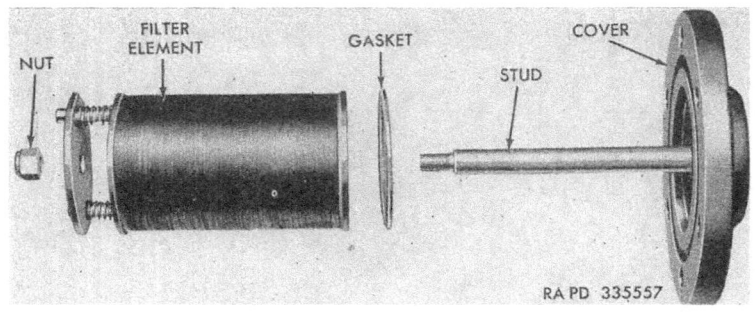

Figure 149. Fuel tank filter—exploded view.

(2) *Fuel line filter.* To gain access to the fuel line filter (fig. 130), which is on accessory end of main engine, below fuel pumps, it is necessary to remove the right air cleaner (par. 175*b*) and the bulkhead inspection plate (fig. 148). Disconnect tank-to-filter and filter-to-pumps fuel lines. Remove two jam nuts, nuts, and washers securing filter to bracket. Remove fuel line filter. Remove cover screw and cover from the shell of the fuel line filter and remove element. Discard cover screw gasket. Strip off cover gasket.

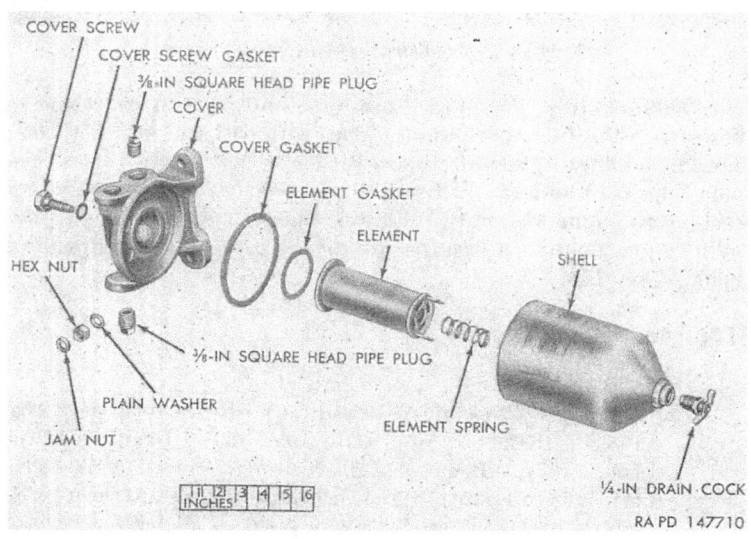

Figure 150. Fuel line filter—exploded view.

b. CLEANING (figs. 149 and 150). Remove nut from filter cover stud of fuel tank filter to relieve tension on filter element and thoroughly clean element in volatile mineral spirits or dry-cleaning solvent.

Caution: Do not use a brush to clean filter element. Carefully blow the element dry using very low pressure compressed air to prevent damage to element disks.

Clean and dry fuel-line-filter shell. Install and tighten nut on filter cover stud of fuel tank filter.

  c. INSTALLATION.

  (1) *Fuel line filter.* Replace element gasket and insert element in shell. Coat one side of a new filter cover gasket with liquid-type-gasket cement, position gasket on filter cover, and install cover on shell. Secure with cover screw and new gasket. Position filter on bracket and install washers, nuts, and jam nuts securing filter to bracket. Connect fuel lines from fuel tanks and fuel pumps to filter (fig. 130). Install bulkhead inspection plate (fig. 148) and right air cleaner (par. 175).

  (2) *Fuel tank filter.* To install left fuel tank filter, coat one side of a new filter cover gasket with liquid-type-gasket cement, position gasket on filter cover and carefully insert filter into opening on left fuel tank. Secure filter with six screws and lock washers. Tighten screws alternately to compress gasket evenly. Install left fuel tank shut-off valve (par. 178). Install right fuel tank filter in the same manner as installation of left fuel tank filter. Install right fuel tank shut-off valve (par. 178). Inspect all connections for leaks. Install power plant (par. 160). Install top deck (par. 161a).

177. Fuel Tanks

  a. FILLING. Unlock and open fuel tank filler cover and remove fuel tank filler cap (fig. 154).

  Warning: When filling fuel tanks, be sure that hose nozzle or container contacts filler neck to carry off static electricity.

Fill tanks until level is approximately 6½ inches below top of filler neck. Install cap, close and lock filler cover.

  b. DRAINING. From below the vehicle, remove the fuel tank drain access hole covers (fig. 235). Provide suitable clean containers with a total capacity of 133 gallons for draining the left tank, and 100 gallons for the right tank. Position the containers under the drain opening and remove the drain plug (fig. 151). After draining, inspect the cover gasket. Install the drain plug, gasket, and access hole cover.

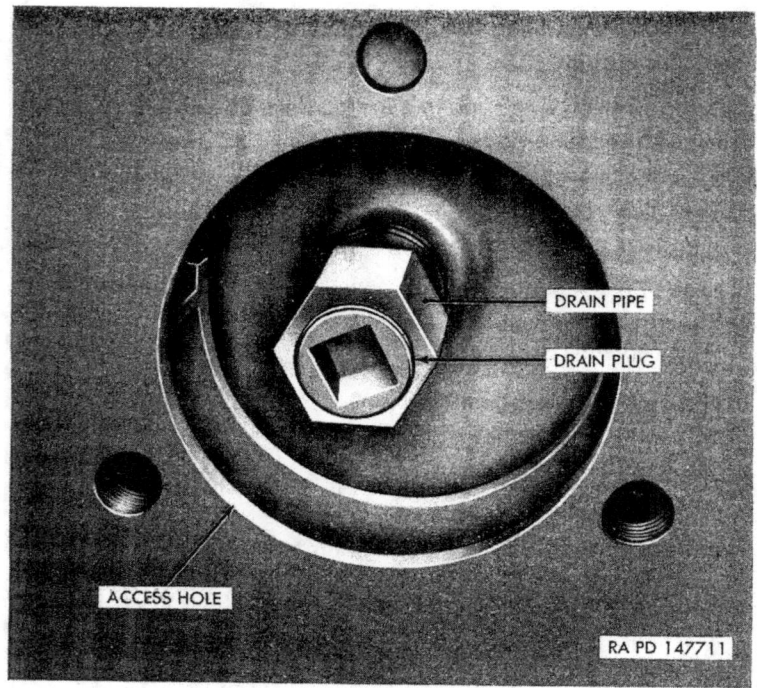

Figure 151. Fuel tank drain plug.

c. REMOVAL. Remove top deck (par. 158b). Remove power plant (par. 159). Drain fuel tanks (b above), and remove drain pipe (fig. 151). Remove fuel tank shut-off valves (par. 178).

Note. If right fuel tank is removed first, both fuel tanks can be removed without removing fuel-tank-shut-off valves. Use extreme care to prevent damage.

(1) *Remove right fuel tank* (fig. 152).

    (a) Remove auxiliary generator and engine (par. 303b).

    (b) Remove interfering electrical cables from rear of master junction box at quick-disconnects. Remove cable-holding clips and lay cables on floor. Remove electrical cables which connect to bilge pump and auxiliary generator and engine.

    (c) Remove bilge pump and pipes (par. 329).

    (d) Disconnect auxiliary engine fuel lines. Remove clips and remove auxiliary engine line. Remove main engine primer fuel line.

    (e) Disconnect electrical cables under auxiliary engine-fuel-shut-off valve.

    (f) Disconnect and remove fire extinguisher line.

(*g*) Raise the right fuel tank filler cover (fig. 154). Remove the fuel tank filler cap (fig. 154). Remove the screws from the filler neck grommet. Work the grommet up and remove. Lift out the inner grommet. Remove the screws which attach the filler neck to the adapter and remove filler neck and upper washer (fig. 155). Remove the screws which attach the adapter to the fuel tank flange. Lift out the filter, adapter, and lower washer. Cover the fuel tank opening.

(*h*) Remove fuel tank sending unit (par. 217).

(*i*) Remove three fuel tank mounting bolts (fig. 152), locking wires, washers, and springs. Two mounting bolts are located on rear of fuel tank and one on the top front. Raise the filler cover to gain access to front mounting bolt.

(*j*) Remove fuel tank. Pry fuel tank toward rear until front portion is clear and lift out of vehicle with a hoist.

Caution: Check to be certain drain pipe, all cables, lines, and clips are removed before removing tank.

(2) *Remove left fuel tank* (fig. 153).

(*a*) Remove fire extinguisher lines and nozzles.

(*b*) Remove two left side batteries, battery covers, and mounting plate (par. 224).

(*c*) Raise the left fuel tank filler cover (fig. 154) and remove filler neck. Procedure is the same as removing right filler neck ((*g*) above).

(*d*) Remove fuel tank sending unit (par. 217).

(*e*) Remove three fuel tank mounting bolts (fig. 153), locking wires, washers, and springs. Two mounting bolts are on the rear of the fuel tank. Raise the filler cover to gain access to the front mounting bolt.

(*f*) Remove the left fuel tank.

Caution: Check to make certain that drain pipe, all cables, lines, and clips are removed from the fuel tank before removing. Pry fuel tank to the rear of the vehicle until clear and lift out with a hoist.

d. INSTALLATION.

(1) *General.* If fuel tanks are to be replaced, be certain to remove any brackets, plates, etc, attached to the replaced fuel tank and install them on the new tank.

(2) *Install left fuel tank* (fig. 153).

(*a*) Position left fuel tank in vehicle and install the three fuel tank mounting bolts, locking wires, washers, and springs. Install a spring under fuel tank bracket, a spring on top of bracket, then a washer and bolt. Install the drain pipe. Install fuel tank sending unit (par. 217).

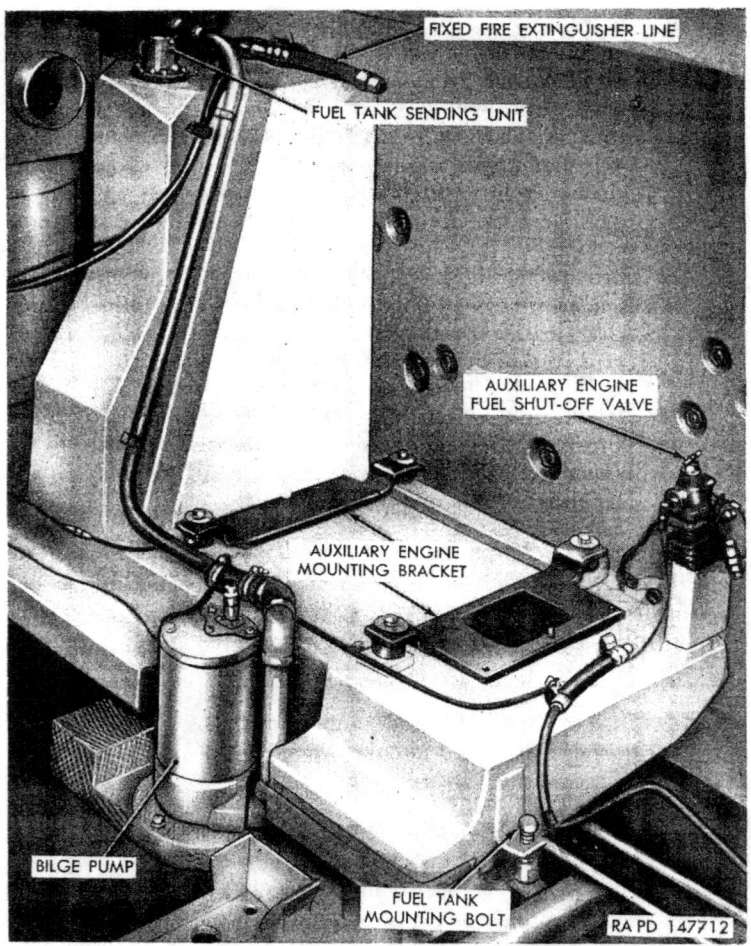

Figure 152. Right fuel tank—installed view.

    (*b*) Raise left fuel tank filler cover (fig. 154). Position the lower washer, adapter, and filter (fig. 155) on the tank and attach with screws. Insert the gaskets, upper washer, and filler neck into the fuel tank. Rotate the filler neck to aline it with the hull opening and secure filler neck to adapter with screws. (Extra holes in the filler neck are provided for this adjustment). Install the filler neck grommet (fig. 154) on the filler neck. Adjust the inner and outer grommets to fill and seal the opening between the hull and the filler neck. Tighten the grommet screws. Install the filler cap.

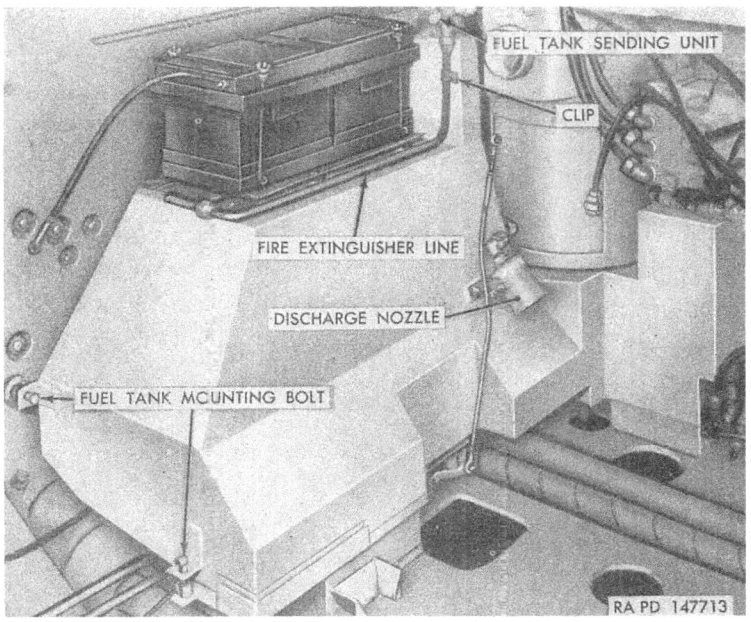

Figure 153. Left fuel tank—installed view.

(c) Install the two left side batteries, mounting plate, and covers (par. 224).

(d) Install fire extinguisher lines and nozzles.

(3) *Install right fuel tank* (fig. 152).

    (a) Position right fuel tank in vehicle and install the three mounting bolts, locking wires, washers, and springs, placing a spring on each side of fuel tank mounting bracket. Install the drain pipe. Install fuel tank sending unit (par. 217).

    (b) Install fire extinguisher line.

    (c) Install the right fuel tank filler in the same manner as installation of the left fuel tank filler ((2) (b) above).

    (d) Connect electrical cables under auxiliary-engine-fuel-shut-off valve.

    (e) Install primer fuel line and auxiliary engine fuel line. Install bilge pump and pipes (par. 329).

    (f) Install electrical cables which connect to the bilge pump and auxiliary generator and engine.

    (g) Connect the electrical cables in the rear of the master junction box. Install cable holding clips.

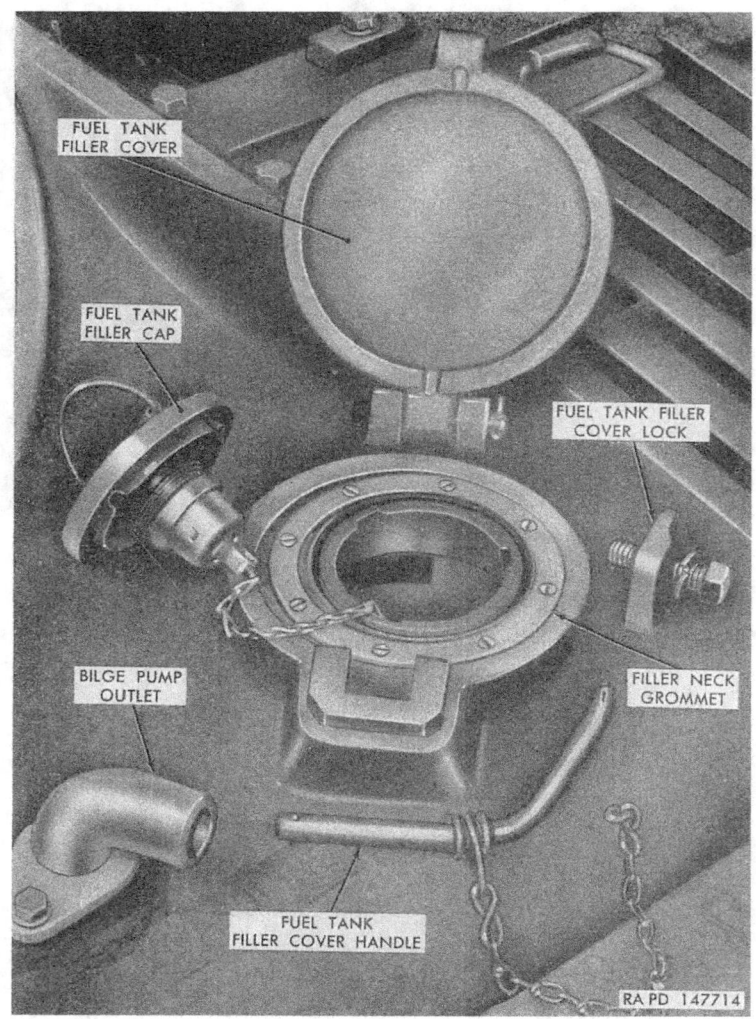

Figure 154. Fuel tank filler cover.

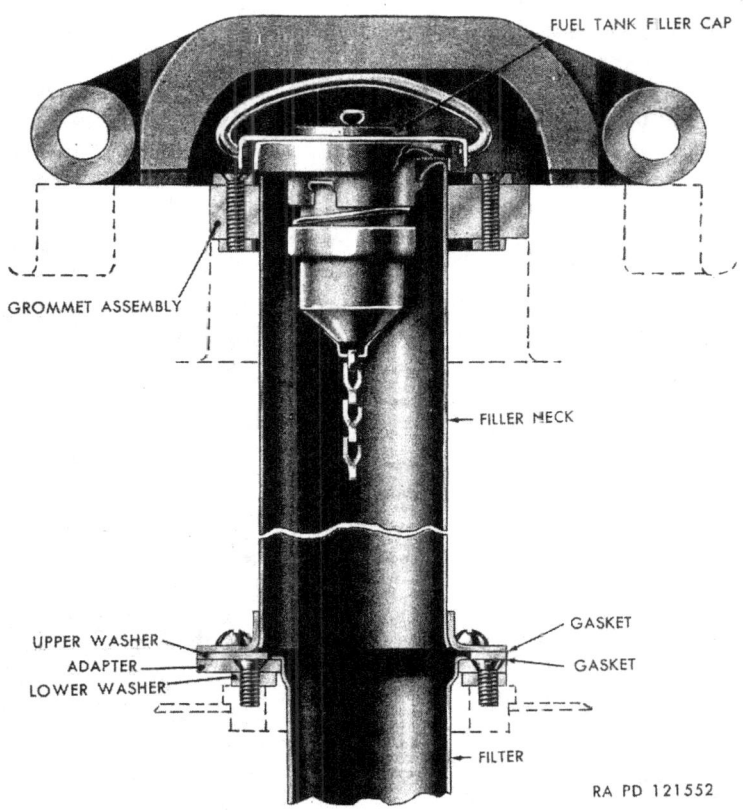

Figure 155. Fuel tank filler neck and cap—sectional view.

   (*h*) Install auxiliary generator and engine (par. 303*c*).
   (4) *Install fuel-tank-shut-off valves* (par. 178). Install power plant (par. 160). Install top deck (par. 161*a*).

178. Fuel Tank Shut-Off Valves and Controls

   a. REMOVAL (fig. 156).
   (1) *Remove power plant.* Remove top deck (par. 158*b*). Remove power plant (par. 159).
   (2) *Drain fuel tanks.* Refer to paragraph 177*b*.
   (3) *Remove right-fuel-tank-shut-off valve.* Remove fuel line to main engine. Disconnect fuel line to auxiliary engine. Disconnect and remove fuel line which connects the two fuel-tank-shut-off valves. Remove elbow and tee from right-fuel-tank-shut-off valve. Remove cotter pin and clevis pin which

connect right-fuel-tank-shut-off valve to control. Remove right-fuel-tank-shut-off valve and nipple which connects valve to right fuel tank.

(4) *Remove right-fuel-tank-shut-off-valve control.* Remove four bolts and lock washers and remove control.

(5). *Remove left-fuel-tank-shut-off valve.* Disconnect primer fuel line. Remove tee at bottom of left fuel-tank-shut-off valve. Remove cotter pin and clevis pin which connects left fuel-tank-shut-off valve to control. Remove shut-off valve and nipple which connects valve to left fuel tank.

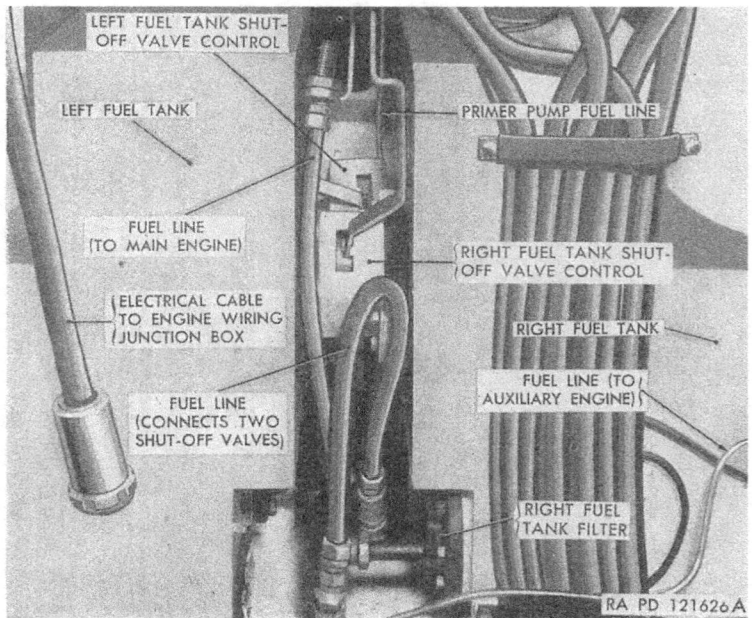

Figure 156. Fuel tank shut-off valves and lines.

(6) *Remove left-fuel-tank-shut-off-valve control.* Remove four bolts and lock washers and remove control.

Note. With engine in vehicle, the valves can be removed through the fuel-tank-shut-off-valve cover plate (fig. 146) on the crew compartment floor.

b. INSTALLATION (fig. 156).
(1) *Install left-fuel-tank-shut-off valve control.* Position control on mounting pads and secure with four bolts and lock washers.
(2) *Install left-fuel-tank-shut-off valve.* Install left shut-off valve and nipple on left fuel tank filter cover. Install clevis

pin and cotter pin to connect left-fuel-tank-shut-off valve control to the valve. Install tee on left shut-off valve. Connect primer fuel line.

(3) *Install right-fuel-tank-shut-off-valve control.* Position control on mounting pads and secure with four bolts and lock washers.

(4) *Install right-fuel-tank-shut-off valve.* Install right-fuel-tank-shut-off valve and nipple into right fuel tank cover. Install clevis pin and cotter pin which connect right-fuel-tank-shut-off valve control to the valve. Install elbow and tee on right shut-off valve. Install fuel line which connects the right- and left-fuel-tank-shut-off valves. Connect fuel line to auxiliary engine. Install fuel line to main engine.

(5) *Fill fuel tanks.* Fill both fuel tanks (par. 177a). Check all connections for leaks. Install power plant (par. 160). Install top deck (par. 161a).

179. Fuel Lines

a. Fuel lines are provided to supply fuel from either or both fuel tanks to the two carburetors on the main engine and the one on the auxiliary engine. Fuel-tank-shut-off valves are provided for selection of fuel tank to be used (par. 171a(4)). Two fuel pumps are provided for the main engine and are interconnected so either pump will furnish fuel to both carburetors. Additional fuel line supplies the crew compartment heater (par. 321) and the main engine priming system (par. 181).

b. A schematic lay-out of fuel lines is shown in figure 157. When replacing any lines or connections, be certain all connections are tight and test by opening fuel-tank-shut-off valves before placing into operation. Check to see that all lines are securely mounted with clips where provided.

180. Primer Pump

(figs. 158 and 159)

a. REMOVAL. Close primer-pump-inlet-line-shut-off valve. Disconnect the primer fuel lines at the inlet and outlet tube connections. Unscrew the collar from the pump cylinder and pull the plunger out of the pump cylinder. Remove the front mounting nut and withdraw the pump from the mounting bracket.

b. INSTALLATION. Install the rear mounting nut on the pump cylinder if it was removed. Insert the pump through the hole in the mounting bracket and install the front mounting nut. Insert the plunger into the pump cylinder and screw the collar onto cylinder. Connect the two primer fuel lines referring to figure 159 for inlet and

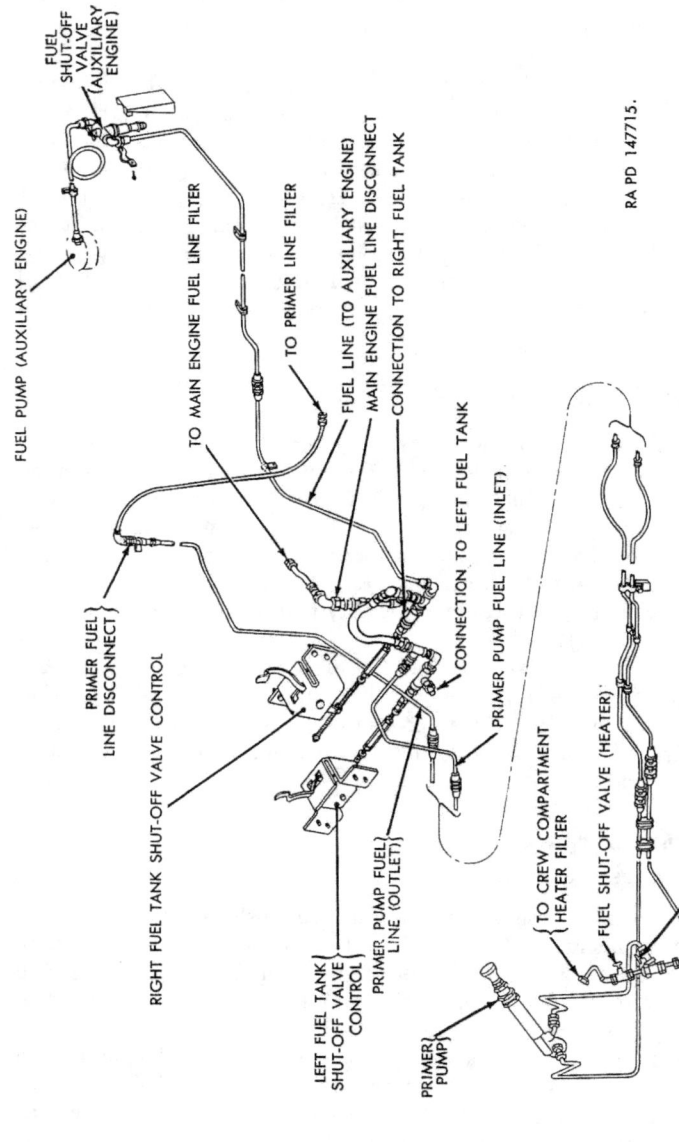

Figure 157. Fuel lines and connections schematic diagram.

Figure 158. Primer pump—installed view.

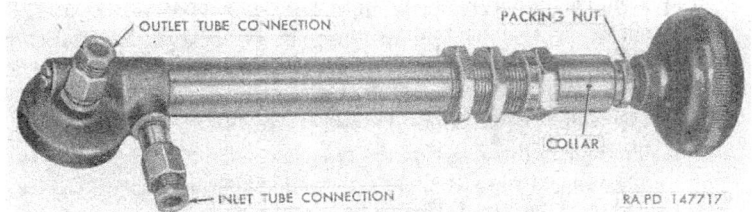

Figure 159. Primer pump.

outlet connections. Open primer-pump-inlet-line-shut-off valve. Check for leaks at connections and packing nut.

181. Primer Lines

a. GENERAL. Primer lines (fig. 157) are provided to connect the primer pump to the fuel tank and to each cylinder of the engine. When replacing lines or connections, be certain all connections are tight and test by operating primer pump to force fuel through the lines. Check to see that all lines are securely mounted with clips where provided.

b. REMOVAL.

(1) If removal is made with the power plant installed in the vehicle, remove the front and intermediate sections of the top deck (par. 158*c*(1)). Turn off master relay switch at instrument panel (fig. 18). Turn off fuel-tank-shut-off valves (par. 17).

(2) Disconnect two nut connectors on primer lines (fig. 137) at union tees at each cylinder.

Note. No. 6 cylinder on each bank has only one intercylinder primer line connection.

Lift out the intercylinder primer lines. Be careful not to kink lines.

(3) Remove connectors on primer lines to cylinders (fig. 130) at top of primer block (on accessory case). Remove two clips securing the primer line to right cylinder bank at front of engine. Remove the primer lines between the accessory case and No. 1 cylinders. Disconnect primer lines at fuel tank interconnection and at bulkhead. If necessary to remove primer line filter (fig. 130), remove primer fuel line and remove filter.

(4) Remove two center ammunition racks (par. 265) to gain access to primer pump lines under fighting compartment floor. Remove clips securing primer lines to hull. Disconnect primer lines at connections near driver's compartment. Pull out primer lines, being careful not to damage them.

(5) To remove those sections of lines in the driver's compartment, disconnect lines at primer pump and at section behind driver's seat. Make sure lines are free, then lift out.

c. INSTALLATION.

(1) Install primer nozzles to cylinders if they had been removed. Install union tees to primer nozzles if they had been removed. Install the intercylinder primer lines at the union tees. Make sure No. 6 cylinders have union tees plugged at the flywheel ends.

(2) Connect primer lines to union tees at the No. 1 cylinders and to primer block nipples. Install primer line filter (fig. 130) on primer block and install primer fuel line. Connect lines at fuel tank interconnection and at bulkhead.

(3) Position primer lines under two center ammunition racks. Connect lines near driver's compartment and install clips securing lines to hull. Install ammunition racks (par. 265).

(4) To install those sections of lines in the driver's compartment, position lines and tighten connections at primer pump and at section behind driver's seats.

(5) If removal was made with the power plant installed, turn on fuel shut-off valves (par. 17). Turn on master relay switch (fig. 18). Install front and intermediate sections of the top deck (par. 161*b*(2)).

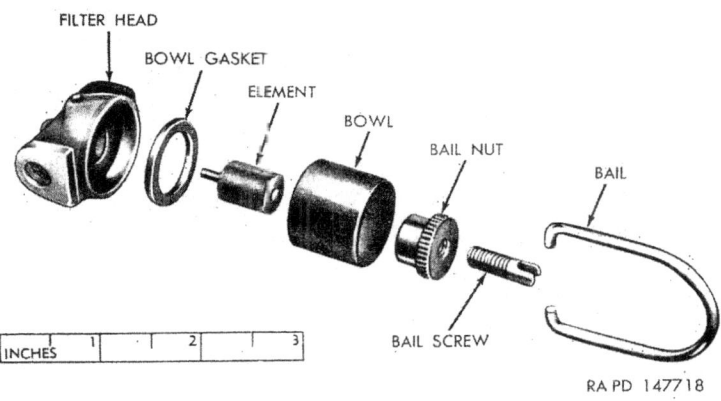

Figure 160. Primer line filter—exploded view.

d. CLEANING PRIMER LINE FILTER (fig. 160). Unscrew bail nut, located at bottom of bowl, and move bail so bowl can be removed. Remove bowl and element and thoroughly clean in volatile mineral spirits or dry-cleaning solvent. Carefully dry element using low pressure compressed air to prevent damage to element. Wipe bowl dry and install element, bowl gasket, and bowl on filter head. Position bail over bowl and tighten bail nut.

182. Exhaust Pipes and Mufflers

a. REMOVE EXHAUST PIPES. Remove cap screws securing muffler shields (fig. 3) and slide shields away from vehicle and remove. Remove two intermediate grille doors adjacent to exhaust pipe housing (fig. 122). Loosen four clamps, two on each side, and open exhaust pipe housing cover (fig. 123). Remove exhaust pipe clamps at exhaust manifold and muffler. Remove exhaust pipes.

b. REMOVE MUFFLER (fig. 123). The procedure for removing both mufflers is the same. Remove exhaust pipe clamp at muffler inlet elbow. Remove two bolts and lock washers, at front of muffler. Push muffler forward to disengage mounting flange and remove muffler.

c. INSTALL MUFFLER (fig. 123). Position the muffler on vehicle fender, engage the mounting flange and secure the muffler with two bolts and lock washers. Install exhaust pipe clamp at muffler and secure connection.

d. INSTALL EXHAUST PIPES. Position exhaust pipes and secure with clamps at exhaust manifolds and mufflers (fig. 123). Close exhaust pipe housing covers and secure with four clamps, two on each side. Position muffler shields (fig. 3) over muffler making sure clamps engage guides on fenders and secure with cap screws. Install two intermediate grille doors adjacent to exhaust pipe housing (fig. 122).

Section X. IGNITION SYSTEM

183. Description and Data

a. GENERAL.
  (1) *Sections.* The ignition system in this vehicle is composed of two similar sections, one located on each side of the engine.
  (2) *Components of each section.*
    (*a*) One magneto to provide spark for the six spark plugs on the flywheel ends of the cylinders.
    (*b*) One magneto to provide spark for the spark plugs on the accessory ends of the cylinders.
    (*c*) One rigid metallic-type ignition harness to house the cables from both magnetos to all 12 spark plugs.
    (*d*) A booster coil to facilitate starting.
    (*e*) Two filter coils for noise suppression.
  (3) *Controls.* A combination magneto, starter, and booster switch (fig. 18) is employed to ground and unground the magnetos, to energize the circuit to the starter, and to energize the booster coils. Refer to figure 161 for wiring diagram of the ignition system.

b. MAGNETOS (fig. 162). The two lower magnetos fire the spark plugs at the flywheel ends of the cylinders. The two upper magnetos fire the spark plugs at the accessory ends of the cylinders. All four magnetos are flange-mounted on the engine accessory case. They employ rotating magnets, pivotless-type breakers, and integral-type cam and distributor-cam electrodes. Rotation of the magnetos is counterclockwise as viewed from the drive end. The magnetos are gear-driven at one-half engine crankshaft speed. The magnetos are virtually weatherproof from without and electro-magnetic-radiation proof from within. Since the magnetos are sealed units, they require dry fresh air which is provided by a breather system from the engine. Forced ventilation is provided by the intake manifold vacuum. A bellows-type cam oiler (fig. 166) utilizing an oiled felt is mounted on the magneto for lubrication purposes.

c. Spark Plugs. The spark plugs are 14-mm type with double-ground electrodes. Gaskets are furnished with the plugs. The hollow head into which the spark plug cable terminal is inserted, is threaded on the outside to fit the cable nut.

d. Ignition Harnesses. Two ignition harnesses are provided, one for each cylinder bank (fig. 131). Each harness has 12 high-tension ignition cables inclosed in the conduit. The six cables for the accessory-end spark plugs, housed in an ignition harness adapter, connect to the upper magneto. The six cables for the flywheel-end spark plugs connect to the ignition harness adapter (fig. 166) on the lower magneto. On the spark plug end of each cable is a round end terminal with a spring contact. The shielding conduit over the spark plug end of the cable is furnished with a nut and ferrule at its end. This nut is screwed onto the top of the spark plug, holding cable and shielding conduit in place. A similar disconnect is provided at the ignition harness end of each spark plug cable.

e. Booster and Filter Coils. Two booster-and-filter-coils assemblies (figs. 162 and 172) are used, one on each side of the engine accessory case near the magnetos. Each booster coil, which is in series with the primary of a lower-magneto coil, is fed from the 24-volt battery when the booster switch (fig. 18) is turned on. The boosters are used to facilitate starting the engine, since the engine cranking speed will not rotate the magneto magnets fast enough to cause the magnetos to produce a satisfactory spark. Each magneto ground cable is connected through a filter coil to the main-engine magneto switch on the instrument panel. The filter coils act as radio interference suppressors, bypassing to ground any radio interference signals generated in the ignition system.

f. Data.

Magnetos _____ Bendix-Scintilla S6LN-32 (10-56600-2)
Booster and filter coils _____ Bendix-Scintilla
Spark plugs _____ Champion, 14-mm

184. Ignition Timing

Caution: Exercise special care whenever handling any of the magnetos to see that no metal articles are attracted to the magnets. These metal articles can be very damaging to the interior of the magneto when the magneto is operating. Be sure no dirt, dust, or other foreign substances enter the magneto.

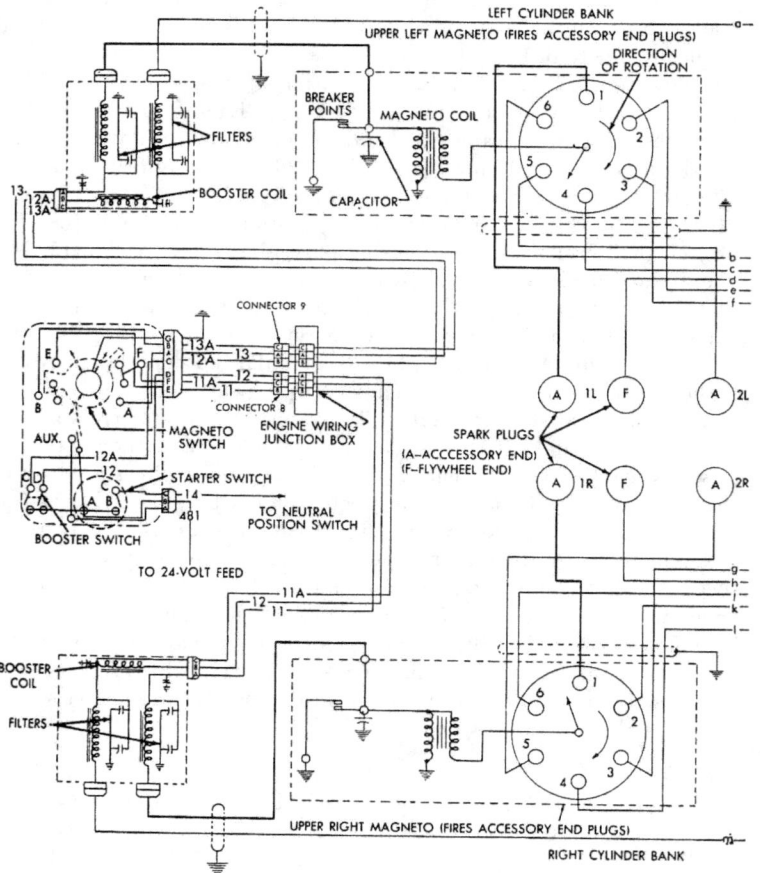

Figure 161. Ignition

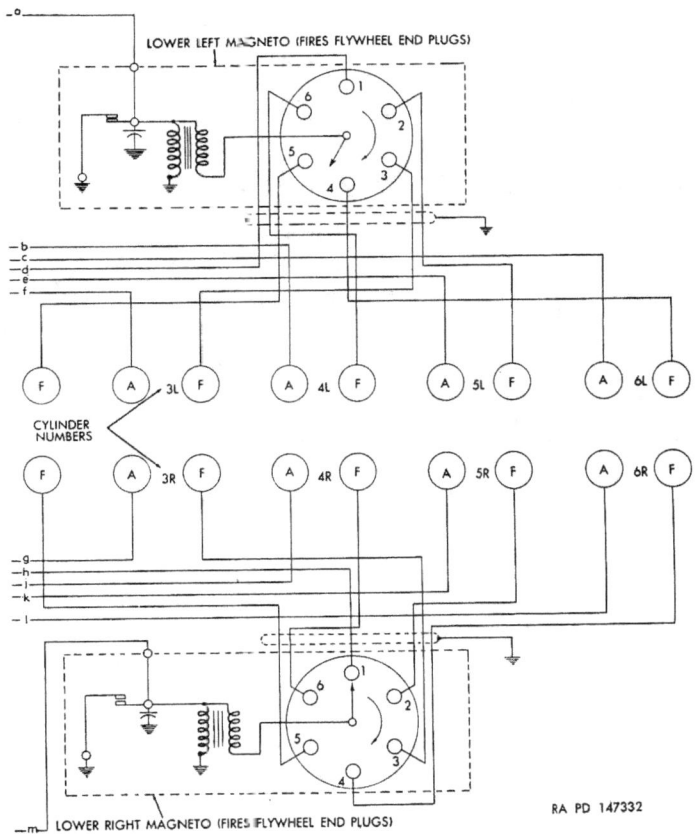

system wiring diagram.

Figure 162. Upper and lower right magnetos and booster and filter coils.

    a. Remove air cleaners (par. 175).
    b. Remove bulkhead inspection plates behind air cleaners (fig. 148).
    c. Working through opening in bulkhead (fig. 163), loosen five screws securing magneto cover and lift off cover (fig. 164). Remove cover gasket.
    d. Remove engine-timing inspection plug (fig. 229) so that timing marks on flywheel can be seen (fig. 116).
    e. Remove center access door from rear of hull (fig. 285). Remove power-take-off cover from rear center of transmission (fig. 220).
    f. Using wrench 41-W-1536-380 on shaft in rear of transmission (fig. 117), turn engine so that engine-cooling fans turn in a clockwise direction. Observe flywheel markings (fig. 116) through inspection hole and turn engine until ignition-timing mark "V-ENG, 1R, IGN" on flywheel appears for right magneto timing, or "V-ENG, 1L, IGN" for left magneto timing. Turn engine slowly until desired timing mark is adjacent to timing pointer.

    Note. Timing marks must be alined only if No. 1 cylinder on the bank being timed is on compression stroke.

To check, remove a spark plug from No. 1 cylinder (par. 186) and using an inspection light, make sure that No. 1 cylinder is on compression stroke when timing marks are alined.

g. Remove two jam nuts and loosen two mounting nuts (fig. 164) which secure magneto to mounting studs in accessory case, just enough to allow magneto to be turned. Rotate magneto until timing marks (fig. 165) on magneto are alined. Tighten two mounting nuts and install jam nuts. Recheck timing mark on flywheel.

h. Check breaker point gap with 0.018-inch feeler gage. If adjustment is correct, feeler gage will move between the points with a slight drag. If adjustment is incorrect, loosen breaker mounting screws (fig. 165) and insert feeler gage between points. Slide movable arm of

Figure 163. Left magnetos and fuel pumps—bulkhead inspection plate removed.

breaker assembly up or down until feeler gage can be moved between points with a slight drag. Tighten screws and recheck adjustment.

Note. When magneto is correctly timed, the magneto-timing marks and the flywheel ignition-timing marks, for particular magneto being timed, are in perfect alinement and the breaker points have a 0.018-inch gap.

i. Install new cover gasket and secure magneto cover by tightening five cover screws.

j. Install engine-timing inspection plug (fig. 229).

k. Remove wrench 41–W–1536–380 and install power-take-off cover in rear of transmission (fig. 220). Install center access door on rear of hull (fig. 285).

l. Install bulkhead inspection plates (fig. 148).

m. Install air cleaners (par. 175).

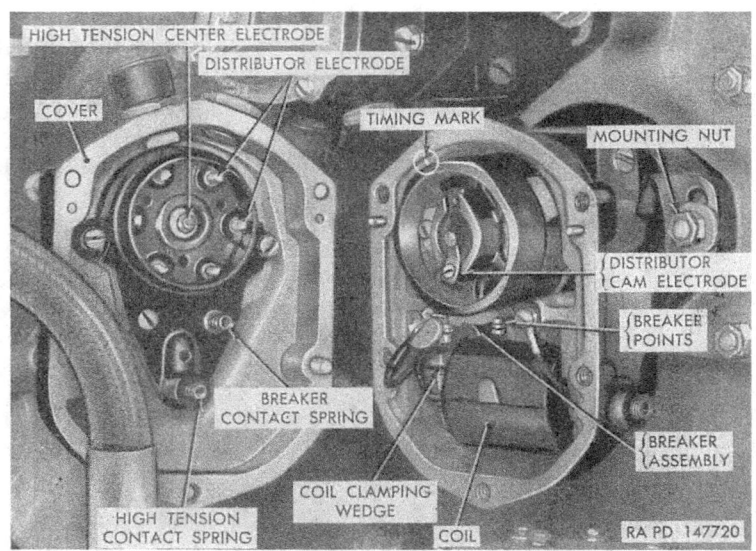

Figure 164. Lower left magneto with cover removed.

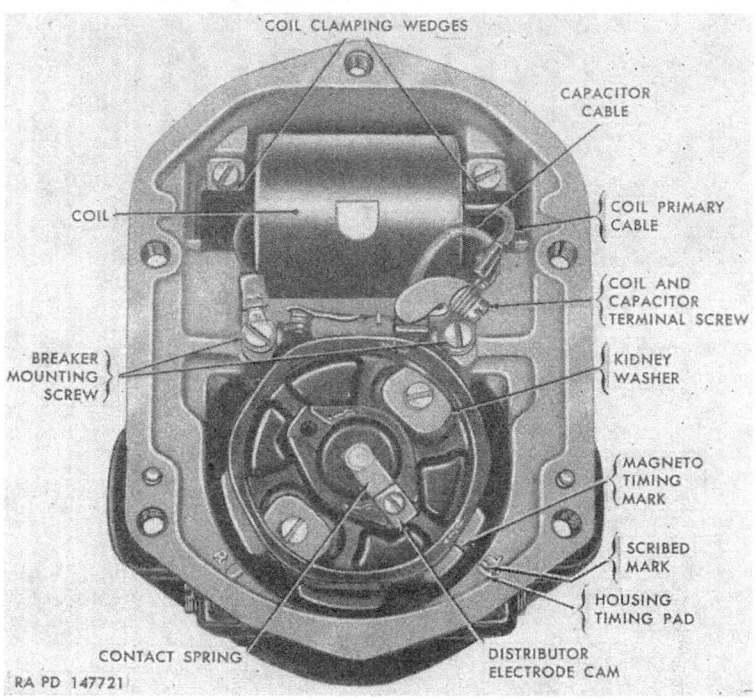

Figure 165. Magneto timing marks.

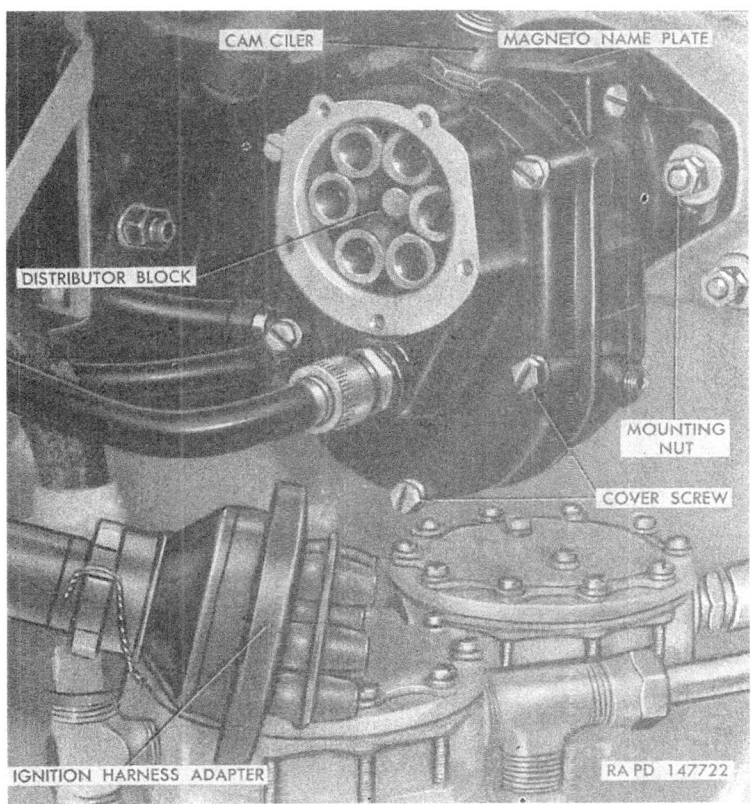

Figure 166. Lower left magneto with harness adapter removed.

185. Magnetos

a. GENERAL. This paragraph provides complete information for removal and installation of main-engine magnetos and magneto components such as magneto cover, breaker assembly, coil, and capacitor. Turn off master relay switch (fig. 18). For access to any of these components, remove air cleaner on side required (par. 175) and remove the bulkhead inspection plate (fig. 148).

b. COORDINATION WITH ORDNANCE MAINTENANCE UNIT. Repair of magnetos is normally an ordnance maintenance operation but may be performed in an emergency by the using organization, provided approval for performing this repair is obtained from the supporting ordnance officer. Tools needed for this repair, which are not carried by the using organization, may be obtained from the supporting ordnance maintenance unit.

c. MAGNETO COVER.
   (1) *Removal.* Remove five screws and remove ignition harness adapter (fig. 166). Disconnect booster cable at magneto. Loosen five magneto cover screws and remove cover (fig. 164).
   (2) *Installation.* Install new cover gasket and secure magneto cover by tightening five cover screws. Position harness adapter and secure with five screws. Connect booster cable.
d. MAGNETO BREAKER ASSEMBLY.
   (1) *Inspection.* Loosen five screws securing magneto cover to magneto and push cover to one side. Examine breaker points (fig. 164). If pitted or burned, replace breaker assembly ((2) and (3) below). If not damaged, adjust breaker point gap (par. 184*h*).
   (2) *Removal* (fig. 165). Remove coil and capacitor terminal screw and remove cables. Remove two breaker assembly mounting screws and lift off breaker assembly. Do not attempt to disassemble breaker assembly as it is removed as a unit.
   (3) *Installation* (fig. 165). Position breaker assembly to magneto frame and secure slotted end of breaker assembly and coil ground cable in place with screw, lock washer, and plain washer. Secure unslotted end with screw and lock washer. Connect capacitor (condenser) and coil primary cables to breaker assembly with screw and lock washer.
      Caution: Care must be exercised not to allow either of these terminals to contact the breaker support as this will short circuit the primary winding.
e. MAGNETO COIL.
   (1) *Removal.* Remove two screws securing coil primary and ground cables to magneto breaker assembly. Remove two screws holding coil clamping wedges in position (fig. 165) and remove coil.
      Note. If wedges will not lift out, tap housing above them gently.
   (2) *Installation.* Position coil with rounded end of spring contact pointing toward bronze bearing. Be sure coil core is seated squarely on pole shoes and secure core with two wedges, screws, and lock washers. Secure coil ground cable to slotted end of breaker assembly with screw and lock washer. Secure coil primary and capacitor cables (fig. 165) to movable arm of breaker assembly with screw and lock washer.
f. MAGNETO CAPACITOR.
   (1) *Removal.* Remove magneto coil (*e* above). Remove capacitor mounting screw and remove capacitor.
   (2) *Installation.* Position capacitor and secure with mounting screw and lock washer. Install magneto coil (*e* above).

g. REMOVAL OF MAGNETO. Disconnect booster cable at magneto. Remove five screws from harness adapter and remove adapter (fig. 166). If removing upper magneto, disconnect one vent line from top of magneto housing. If removing lower magneto, disconnect two vent lines at tee at bottom of magneto housing. Remove two jam nuts, lock washers, and plain washers securing magneto and remove magneto.

h. INSTALLATION OF MAGNETO.

(1) Loosen five cover screws on new magneto and remove cover.
(2) Turn magneto shaft clockwise until timing marks (L) on housing and on breaker cam are alined (fig. 165). Place shaft between copper-covered jaws of a vise, in a position so that timing marks are plainly visible and breaker points accessible.
(3) With timing marks alined, place 0.018-inch feeler gage between breaker points. If points are correctly adjusted, feeler gage can be moved between the points with a slight drag. If adjustment is incorrect, loosen screw securing movable arm of breaker assembly and insert feeler gage between points. Slide movable arm of breaker assembly up or down, as required, until feeler gage moves between points with a slight drag. Proper point gap is 0.018 ± 0.0006-inch when measured on high point of each cam lobe. Be sure timing marks are alined when making this adjustment.

Note. If this adjustment is properly made, no further breaker point adjustment will be required when magneto is installed on engine.

(4) Position magneto (with cover removed) to accessory case so that mounting studs are centered in adjusting slots to permit adjustment when timing. Secure with two flat washers, lock washers, nuts, and jam nuts just tight enough to permit magneto to be turned.
(5) Performs steps *a* through *f*, paragraph 184.
(6) Rotate magneto until magneto timing marks (fig. 165) are alined. Tighten magneto in place with two nuts and install two jam nuts.
(7) Install new cover gasket and secure magneto cover by tightening five cover screws.
(8) Install engine-timing inspection plug (fig. 229).
(9) Remove wrench 41–W–1536–380 and install power-take-off cover in rear of transmission (fig. 220). Install center access door on rear of hull (fig. 285).
(10) If installing upper magneto, connect vent line to top of magneto housing, using elbow from old magneto. If installing lower magneto, connect two vent lines to bottom of magneto housing, using tee from old magneto.

(11) Install ignition harness adapter with five screws and lock washers.

(12) Connect booster cable to magneto.

i. Synchronization of Magnetos. Loosen magneto cover screws (fig. 166) and remove magneto covers from the magnetos on the bank being synchronized. Secure ground cable of timing light 41–L–1439 (fig. 167) to magneto housing. Clip remaining leads to breaker posts, one to top magneto and one to bottom magneto. Rotate crankshaft (par. 184*e* and *f*) one-eighth turn in reverse direction, then bring it

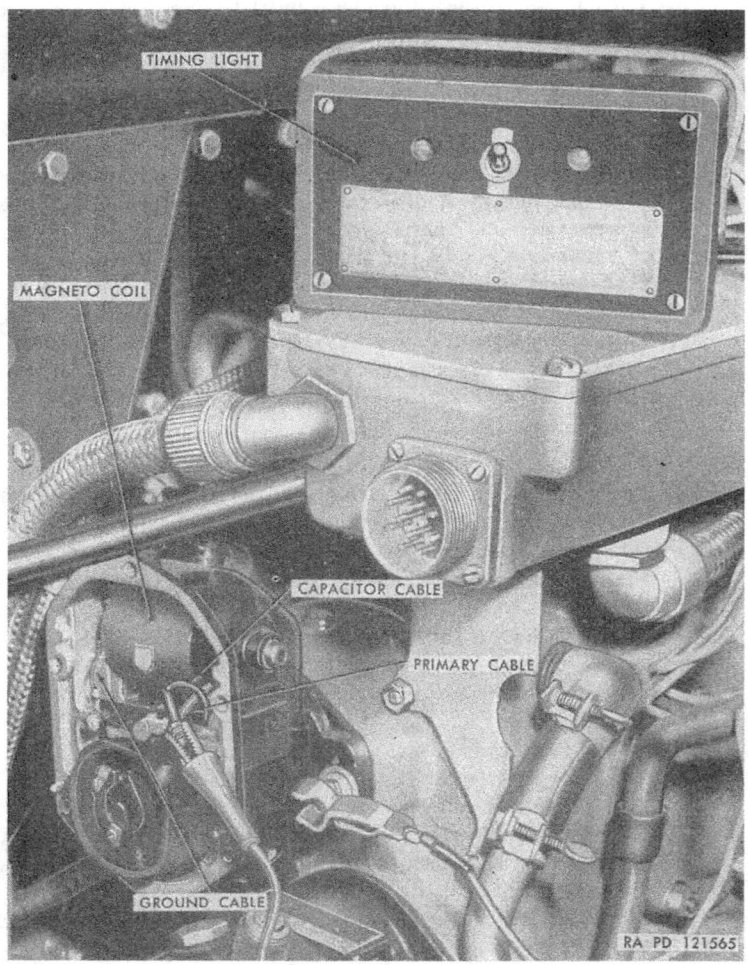

Figure 167. Synchronizing magnetos with timing light 41–L–1439.

slowly forward until timing lights go out. Both lights should go out at the same time. If they do not, loosen magneto mounting nuts and adjust magnetos until this condition is achieved.

Note. Magneto timing marks should be alined. When magnetos are synchronized and timed. If marks are not alined, breaker points must be adjusted (par 184*h*).

Replace magneto covers by tightening cover screws.

186. Spark Plugs

a. REMOVAL. Turn off master relay switch (fig. 18). Remove intermediate section of top deck (par. 158*c*). All spark plugs can now be removed except the two front and the two rear ones. To remove the two front plugs, remove the front section of the top deck (par. 158*c*). To remove the two rear spark plugs, remove the two spark plug access opening covers in engine rear shroud (fig. 221). Pull up hinged cover plates (fig. 170) over each spark plug. Detach spark plug cable from spark plug by loosening cable nut at spark plug with cable nut wrench 41–W–871–70 (fig. 168) and pulling cable free.

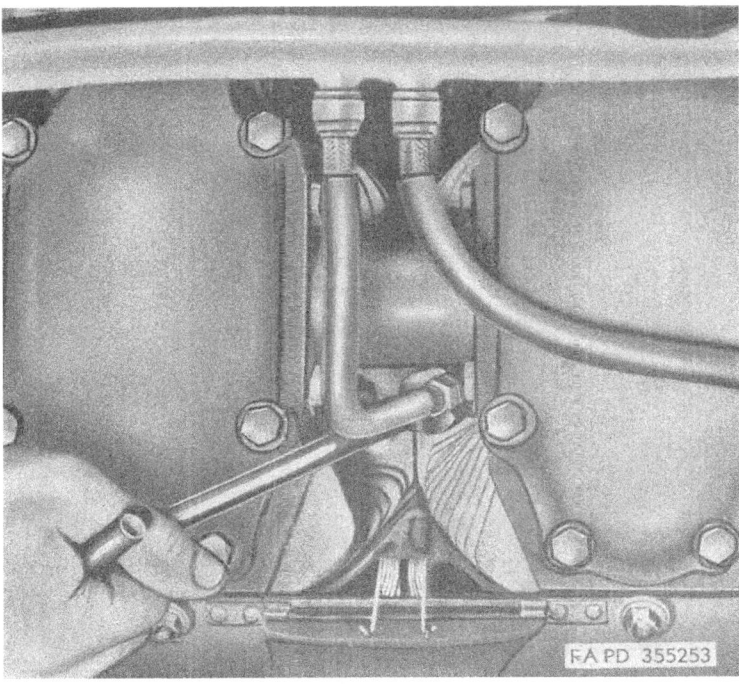

Figure 168. Removing spark plug cable nut with cable nut wrench 41–W–871–70.

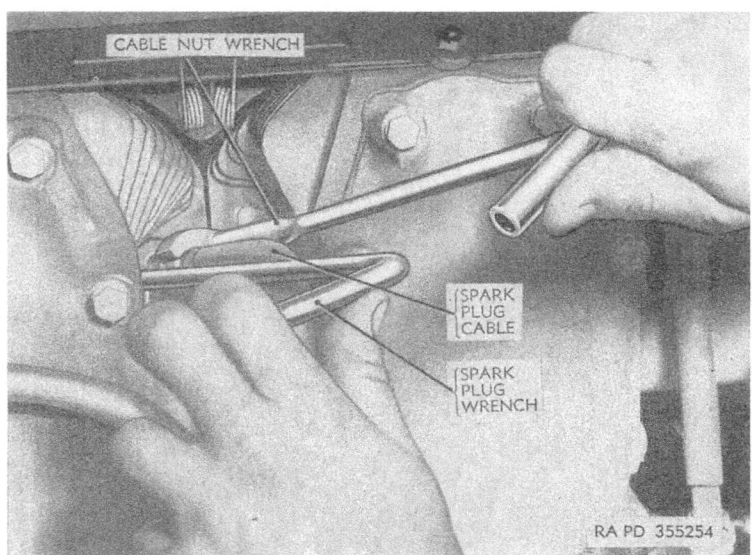

Figure 169. Removing spark pulg cable with spark plug wrench 41-W-3304-800 and cable nut wrench 41-W-871-70.

Note. If spark plug turns when removing cable nut, use spark plug wrench 41-W-3304-800 to hold plug while turning nut with cable nut wrench 41-W-871-70 (fig. 169).

Using a deep socket wrench, remove spark plug.

Note. Insert spark plug inserting and removing wrench 41-W-3306-500 (fig. 170) on top of plug to aid lifting plug out of recess.

b. INSPECTION AND ADJUSTMENT. Clean spark plugs and inspect for damage. Replace unserviceable plugs. Check plug for proper gap between electrodes. Using a round feeler gage, adjust gaps by bending outer electrode to 0.011- to 0.014-inch.

c. INSTALLATION. Install new gasket on spark plug. Screw spark plug inserting and removing wrench 41-W-3306-500 on top of spark plug (fig. 170) and insert plug into cylinder head. Tighten plug, using a deep socket wrench to 200 to 225 lb-in torque. Install spark plug cable, using cable nut wrench 41-W-871-70 (fig. 168). Close hinged cover plates (fig. 170) over each spark plug. Install spark plug access opening covers (fig. 221). Install front and intermediate sections of top deck (par. 161*b*).

187. Ignition Harness

a. REMOVAL. Turn off master relay switch (fig. 18). Remove front and intermediate sections of top deck (par. 158*c*). Remove air-

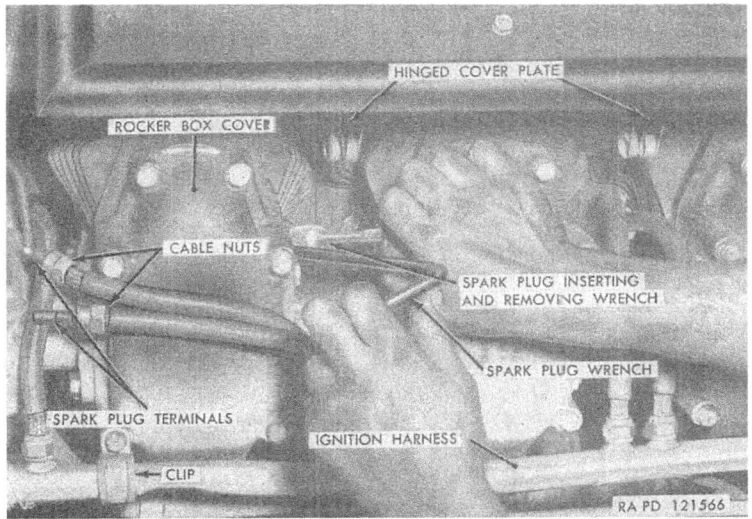

Figure 170. Using spark plug inserting and removing wrench 41-W-3306-500 and spark plug wrench 41-W-3304-800 to insert and tighten plug.

cleaner-to-carburetor pipe (fig. 147) from side on which harness is being removed. Open rear grille doors (fig. 101) over transmission. Remove spark plug access opening cover (fig. 221) on engine rear shroud by releasing two quick-disconnect-type screws attaching cover to shroud. Disconnect spark plug cables from ignition harness (fig. 171) by using cable nut wrench 41-W-871-70. Tape spark plug cable terminal springs to prevent them from catching on protrusions and being stretched out of shape. Disconnect flexible line clip (fig. 143) from ignition harness. Remove five screws from each ignition harness adapter and remove adapters (fig. 166). Disconnect ignition harness vent line at flywheel end of ignition harness. Remove four hold-down cap screws securing ignition harness main conduit to valve rocker covers (Nos. 1, 3, 4, and 6) (fig. 132). Remove ignition harness by lifting it clear of engine.

b. INSTALLATION.

Note. Make sure all spark plug cable terminal springs are taped before attempting to install ignition harness.

Position ignition harness main conduit (fig. 132) on side of cylinder bank and secure to valve rocker covers with hold-down cap screws. Remove tape from spark plug cable terminal springs and install spark plug cables, using cable nut wrench 41-W-871-70 to tighten cable nuts. Position ignition harness adapters to magnetos (fig. 166) and secure each with five screws and lock washers. Connect vent line at flywheel end of ignition harness. Connect flexible line clip (fig. 143) to ignition harness. Secure spark plug access opening cover (fig. 221) to engine rear shroud by fastening two quick-disconnect-type screws. Close rear grille doors over transmission (fig. 101). Install air-cleaner-to-carburetor pipe (fig. 147). Install front and intermediate sections of top deck (par. 161b).

188. Booster and Filter Coils

a. REMOVAL. Turn off master relay switch (fig. 18). Remove air cleaner on side on which booster and filter coils assembly is to be removed (par. 175). Remove bulkhead inspection plate (fig. 148). Remove five screws securing ignition harness adapter to each magneto and remove adapters (fig. 166) to provide working space. Remove three cables from booster and filter coils assembly (fig. 172). Remove four mounting screws at top and bottom and remove booster and filter coils assembly from bracket on accessory case.

Figure 171. Ignition harness showing spark plug cable disconnected.

b. INSTALLATION. Position booster and filter coils assembly to bracket (fig. 172) on accessory case and secure with four mounting screws and lock washers. Secure three cables to assembly. Install ignition harness adapters (fig. 166) and secure each with five screws. Install bulkhead inspection plate (fig. 148). Install air cleaner (par. 175).

Figure 172. Booster and filter coils assembly—installed view.

Section XI. STARTING SYSTEM

189. Description and Data

a. STARTING SYSTEM WIRING. The starting system wiring (fig. 173) consists of a low-current control circuit and high-current batteries-to-starter-motor circuit. The control circuit is fed from the batteries through the instrument circuit breaker to the starter switch, then through the neutral position switch into the starter current relay to ground. The main starting system circuit is fed from the batteries

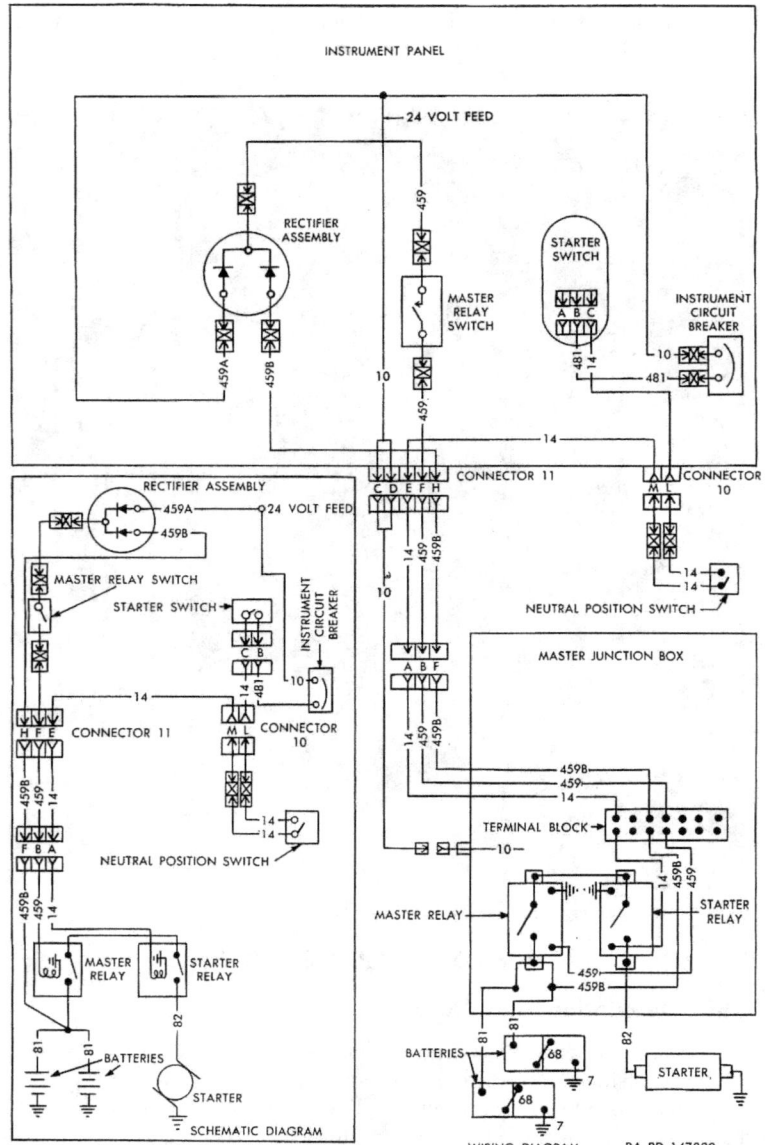

Figure 173. Starting system electrical diagram.

through the master relay, through the starter relay to the starter and then to ground.

  b. STARTER. The 24-volt, waterproof starter is mounted on the left side of the engine near the accessory end. The engine may be equipped either with an Eclipse-Pioneer starter (fig. 174) or a Jack and Heintz starter (fig. 175). These starters are essentially the same, chief differences being in the method of mounting. Each starter consists of a motor section and a gear section. The motor section comprises a commutator, field coil, and an armature-and-pinion assembly. It is a series-parallel wound, four-pole unit with four brushes. The gear section comprises reduction gears, clutch, screw shaft, and starter jaw. It is bolted to the drive end of the motor. The terminal stud extends into the back head (commutator end) and is accessible through a watertight access hole cover (figs. 176 and 177).

  c. STARTER CURRENT RELAY. The starter current relay is located in the master junction box (fig. 184). Its function is to connect the batteries to the starter motor when the starter switch (fig. 18) is turned on. If starter current relay is faulty, replace master junction box (par. 194).

  d. NEUTRAL POSITION SWITCH. The neutral position switch (fig. 180) is mounted on the rear of the manual control box. It functions as a "safety switch" in that it prevents starting the main engine while the transmission is in any shift position other than neutral. Positioning the manual control lever (fig. 10) in neutral position closes the neutral position switch, which completes the circuit from the starter switch to the starter relay.

  e. DATA.

Eclipse-Pioneer starter:
```
    Model_____ 1416-29-F
    Weight_____ 31.5 lb
    Minimum cranking speed_____ 24 rpm
    Torque_____ 400 lb-ft
    Maximum amperes_____ 175
    Rotation (viewing commutator end)_____ clockwise
```
Jack and Heintz starter:
```
    Model_____ JH6FKLW-3 (waterproof)
    Weight_____ 34 lb
    Minimum cranking speed_____ 42 rpm
    Torque_____ 800 to 900 lb-ft
    Rotation (viewing commutator end)_____ clockwise
```

190. Starter

  a. REMOVAL.

  (1) *Eclipse Pioneer.* Remove top deck (par. 158*b*) and power plant (par. 159). Remove four screws from terminal stud

Figure 174. Eclipse-pioneer starter—installed view.

Figure 175. Jack and Heintz starter—installed view.

access hole cover place and remove plate. Remove cotter pin and connector nut within starter housing. Pull terminal off terminal stud (fig. 178). Remove external cable connector nut and pull out starter cable and terminal (fig. 176). Remove starter vent line (fig. 174). Remove six jam nuts, nuts, and plain washers from mounting studs (fig. 176) and remove starter.

(2) *Jack and Heintz*. Remove top deck (par. 158*b*) and power plant (par. 159). Remove six screws from terminal stud access hole cover (fig. 177) and remove cover. Remove cotter pin and connector nut within starter housing. Pull terminal off stud. Remove external cable connector nut and pull out cable and terminal. Remove starter vent line (fig. 175). Push down on pinion lock plate (fig. 177) to release pinion gear, turn pinion gear (fig. 175) counterclockwise, and remove starter.

b. REPLACEMENT OF BRUSHES.

(1) *Remove brushes.* On Eclipse-Pioneer, remove eight safety nuts and plain washers holding starter cover to gear section flange (fig. 176) and remove starter cover. On Jack and Heintz, remove vent line connector, and three cover screws securing starter cover to flange (fig. 175), and starter cover.

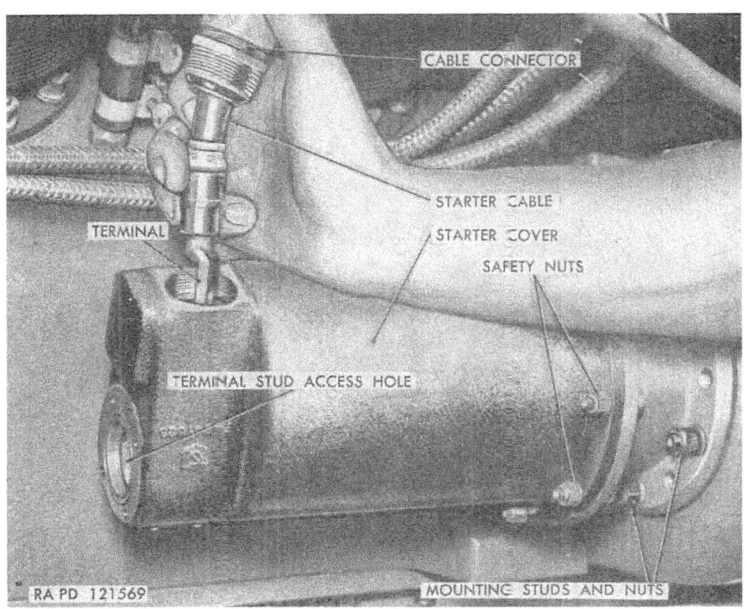

Figure 176. Removing starter cable.

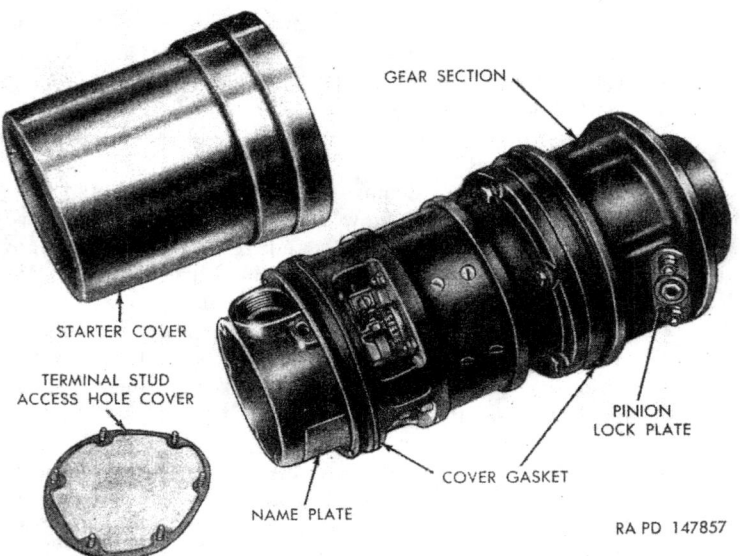

Figure 177. Jack and Heintz starter with cover removed.

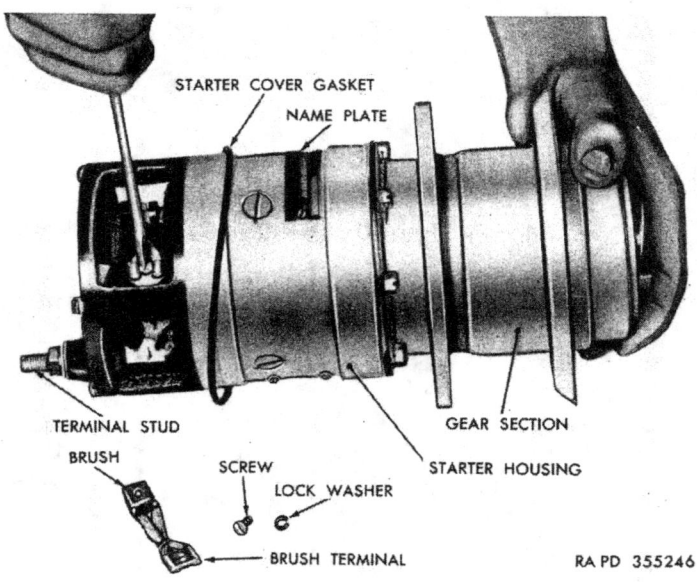

Figure 178. Replacing Eclipse-Pioneer starter brushes.

Remove brush terminal screw and lock washer (fig. 178), and pull brush out of brush holder.

(2) *Inspect brushes.* Inspect each brush for wear. If a brush is less than $11/32$-inch long, replace it.

(3) *Install brushes.* Pry brush spring up with screw driver and install brush in holder. Allow spring to snap down against brush and position itself. Fasten brush terminal to starter frame with screw and lock washer. Make sure that starter cover gasket is in position. On Eclipse-Pioneer, connect starter cover to gear section flange with eight safety nuts and plain washers. On Jack and Heintz, install vent line connector and secure starter cover to flange with three screws.

Note. Be sure terminal stud access hole is centered over stud when installing cover.

c. REPLACEMENT OF STARTER-DRIVE-GEAR-HUB OIL SEAL.

(1) *Coordination with ordnance maintenance unit.* Replacement of starter-drive-gear-hub oil seal is normally an ordnance maintenance operation, but may be performed in an emergency by the using organization, provided authority for performing this replacement is obtained from the appropriate commander. Tools required for the operation which are not carried in the using organization may be obtained from the supporting ordnance maintenance unit.

(2) *Replace oil seal* (fig. 179). If evidence of oil leakage into starter has been found, oil seal may be at fault. To remove oil seal, remove starter (*a* above), then remove oil seal from within accessory case. Install new oil seal using replacer 41–R–2388–715. Record replacement on DA AGO Form No. 478.

d. INSTALLATION.

(1) *Eclipse-Pioneer.* Position starter to mounting studs and secure with six nuts, plain washers, and jam nuts. Install cable terminal on stud and secure with nut and lock washer. Secure nut with cotter pin. Screw cable connector nut into starter housing and tighten to form a waterproof joint. Install starter vent line (fig. 174). Be sure rubber gasket is in position on terminal stud access hole cover and install cover with four screws, lock washers, and locking wire. Install power plant (par. 160) and top deck (par. 161*a*).

(2) *Jack and Heintz.* Position starter to mounting studs and turn pinion gear (fig. 175) clockwise into locking position. Release pressure on pinion gear and pinion lock plate will snap into position. Install cable terminal on stud and secure with nut and lock washer. Secure nut with cotter pin. Screw cable connector nut into starter housing and tighten

Figure 179. Replacing starter-drive-gear-hub oil seal with replacer 41–R–2388–715.

to form a waterproof joint. Install starter vent line (fig. 175). Be sure rubber gasket is in position on terminal stud access hole cover and install cover with six screws, lock washers, and locking wire. Install power plant (par. 160) and top deck (par. 161a).

191. Neutral Position Switch
(fig. 180)

a. REMOVAL. Turn off master relay switch (fig. 18). Remove four jam nuts, nuts, and washers securing switch and cover to manual control box. Remove switch and cover from the box (fig. 180). Remove two switch mounting screws and washers holding switch to switch cover. Remove two terminal screws securing two cables (14) to switch and remove switch.

b. INSTALLATION. Install two cables (14) on switch with two terminal screws. Install switch to switch cover with two screws and washers (fig. 180).

Note. Make sure that insulation between cover and switch is placed under washers and screws when installing switch.

Position cover on four mounting studs on manual control box. Secure switch with four washers, nuts, and jam nuts.

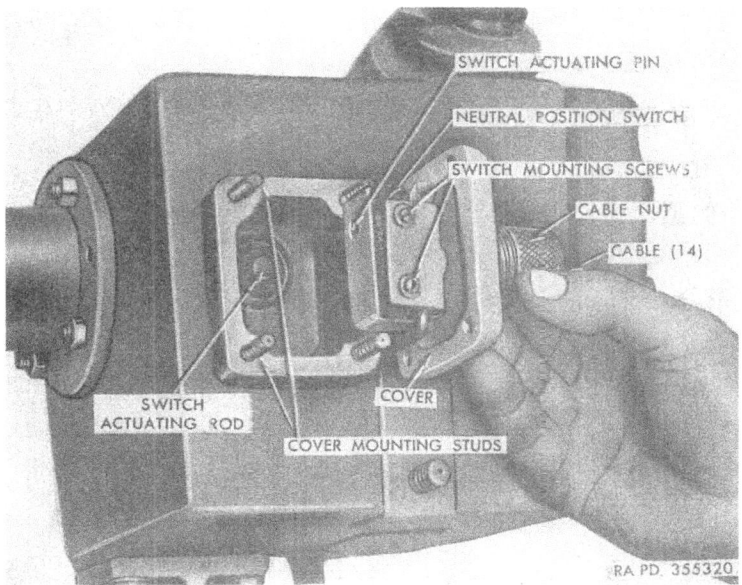

Figure 180. Neutral position switch.

Section XII. HULL ELECTRICAL SYSTEM

192. Description

a. GENERAL. The full electrical system (fig. 181) consists of several systems interconnected by means of the master junction box and engine wiring junction box through electrical cables. The electrical systems include the battery and generating system; starting system; horn and lighting system; instrument and hull accessories panel gages, circuit breakers, and switch system; ignition system; and hull accessories system, which includes bilge pumps, crew compartment heater, and electrical accessory outlet. The instrument panel (fig. 18) mounts the gages, switches, and warning-signal lights required for operation of the vehicle. The hull accessories panel (fig. 18) mounts an electrical accessory outlet socket and the controls required for operation of the hull accessories. The main bus in the master junction box feeds the main power feed cables to the turret, to the radio terminal box, and to the instrument and hull accessories panels. Electrical continuity from the instrument panel and the master junction box to the engine and transmission electrical components is accomplished by interconnecting cables within the engine wiring junction box.

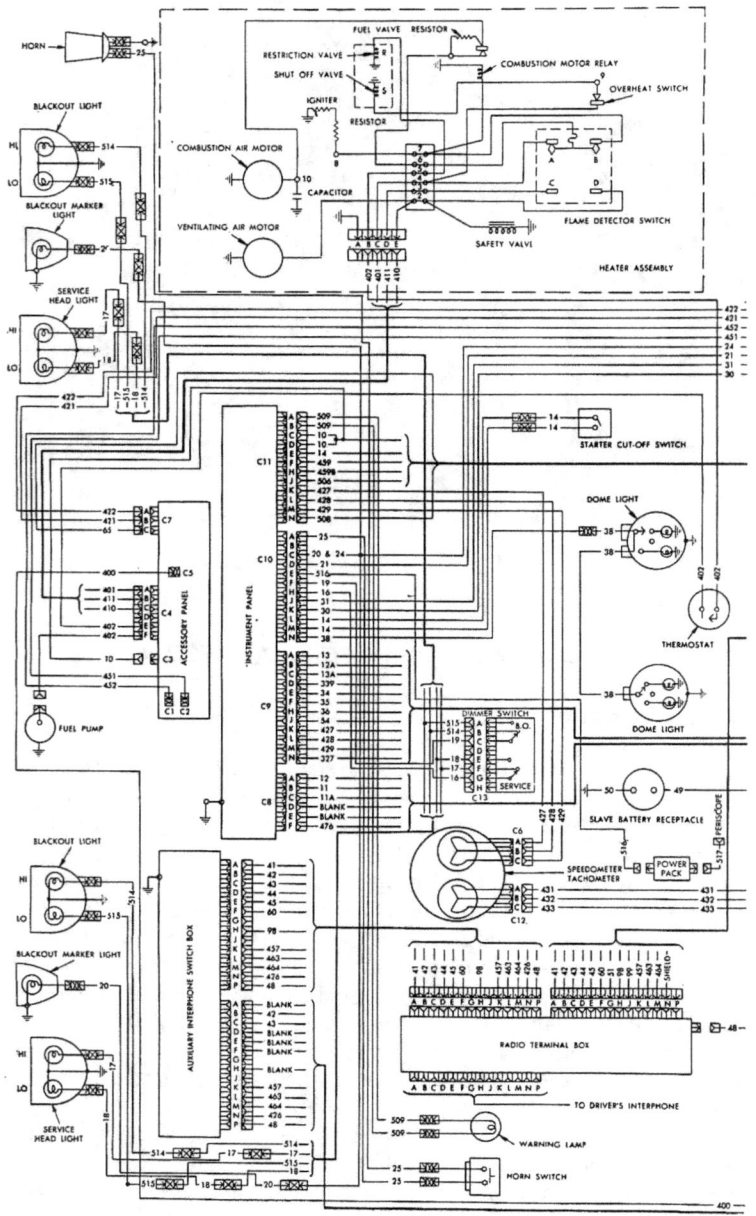

Figure 181. Hull

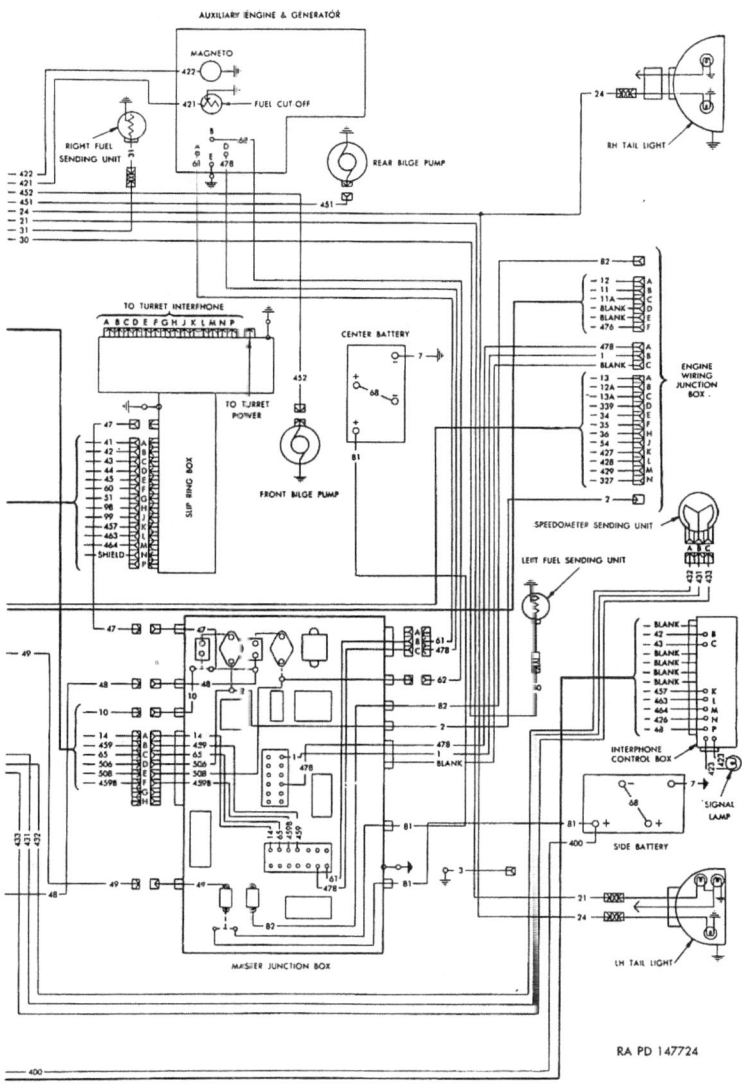

wiring diagram.

b. MASTER JUNCTION BOX (figs. 146 and 184). The master junction box provides for the control, regulation, and distribution of the outputs of the main and auxiliary generators. It is mounted on the bulkhead between the two air cleaners (fig. 146). The components in this box are made accessible by removing the box cover from within the fighting compartment (par. 194). All electrical cables connecting into the box are run through the rear of the box which is accessible from the engine compartment (fig. 182). Procedures for removal and installation of the master junction box are given in paragraph 194.

c. ENGINE WIRING JUNCTION BOX (fig. 183). The engine wiring junction box provides for the connection and disconnection of electrical circuits from the instrument panel and master junction box to the engine and transmission electrical units. It is mounted above the accessory case on the main engine (fig. 130). Two brackets (one on each side) are provided for mounting the box to studs on the accessory case. Procedures for removal and installation of this box are covered in paragraph 195.

d. CONDUITS, CABLES, and CONNECTORS.
  (1) *Conduits.*
    (*a*) *Flexible metallic conduits.* Flexible metallic conduits are used to prevent electrical radiation which interferes with operation of communication equipment and yet are flexible enough for easy bending.
    (*b*) *Solid metallic conduits.* Solid metallic conduits are used to prevent electrical radiation and yet have enough mechanical strength for protection against flattening and damage to the inclosed cables.
  (2) *Cables.* Electrical cables are rubber-covered. The ends of all cables terminate in a pin or socket on a connector plug or receptacle or at a cable terminal. The cables are always soldered to the pins, sockets, or terminals.
  (3) *Connectors.*
    (*a*) *Plug-and-receptacle type connector.* The plug-and-receptacle type of connector has two main components, a plug and a receptacle. The receptacle is secured to a box or panel by four cap screws. In connectors with watertight fittings, a coupling nut is used to compress a rubber bushing on the receptacle. The plug is secured to the receptacle by a coupling nut. This type of connector has pins and sockets either for one or for many connections.
    (*b*) *Bell-type connector.* The bell-type connector (fig. 189) are single- or dual-conductor type with two bells which inclose an insulated snap-type connector and hold rubber bushings in place to form a waterproof joint.

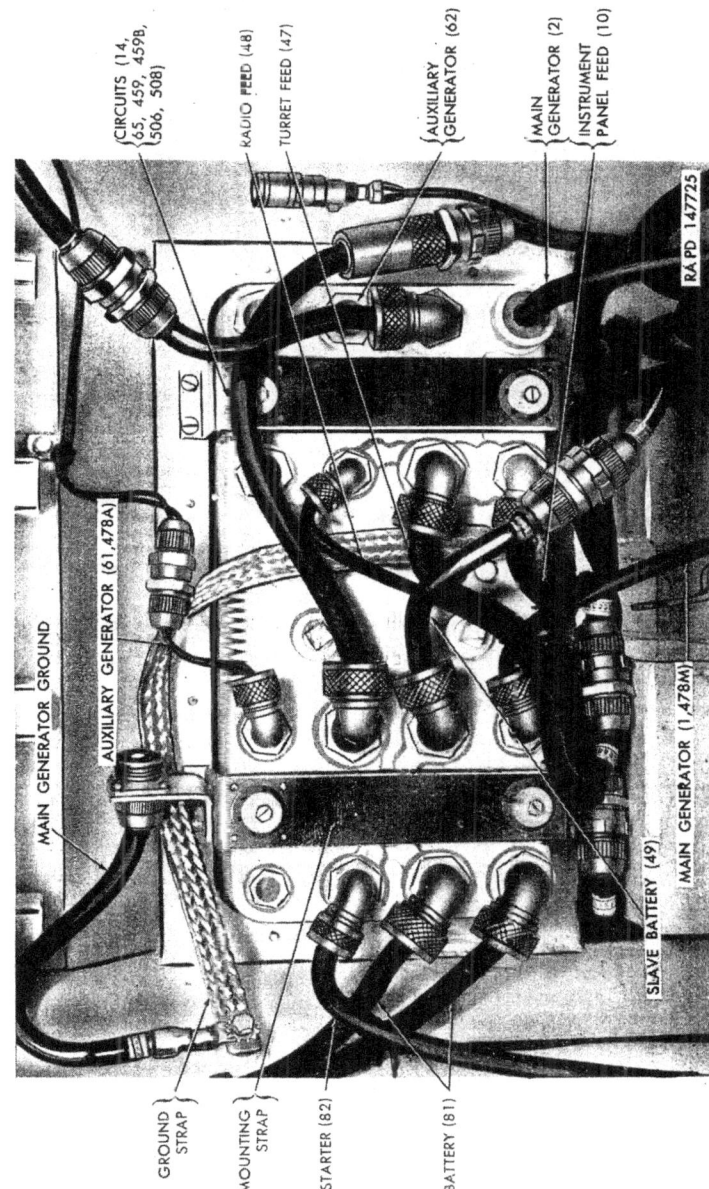

Figure 182. Identification of cables entering rear of master junction box.

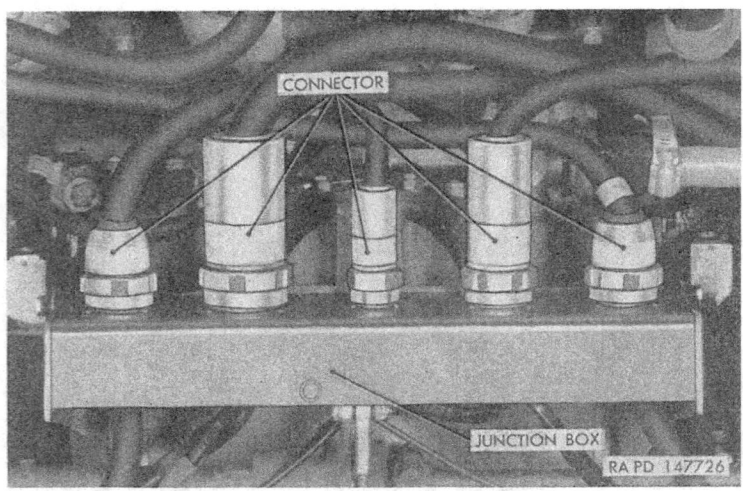

Figure 183. Engine wiring junction box—top view.

193. Electrical Circuit Numbers

Table VI "Electrical Circuit Numbers," lists the ordnance numbers assigned to the electrical circuits in this vehicle. Each number is wrapped around the proper cable and is used to identify the circuit associated with that cable. For schematic and wiring diagrams, the number is placed above a line, within a break in a line, or in a circle at the end of a line.

Table VI. Electrical Circuit Numbers

Circuit No.	Circuit name
1	Main generator field circuit.
2	Main generator armature circuit.
3	Main generator ground circuit.
7	Battery ground cable.
10	Instrument panel positive line.
11	Upper right magneto ground circuit.
11A	Lower right magneto ground circuit.
12	Right booster coil to booster switch.
12A	Left booster coil to booster switch.
13	Upper left magneto ground circuit.
13A	Lower left magneto ground circuit.
14	Starter switch to relay.
15	Main lights switch feed.
16	Lights switch to head light dimmer switch.
17	Dimmer switch to service head light, upper beam.
18	Dimmer switch to service head light, lower beam.

Table VI. Electrical Circuit Numbers—Continued

Circuit No.	Circuit name
19	Lights switch to blackout dimmer switch.
20	Lights switch to blackout marker lights.
21	Lights switch to service tail light.
24	Lights switch to blackout marker tail lights.
25	Horn circuit.
27	Instruments feed.
30	Left fuel gage circuit.
31	Right fuel gage circuit.
34	Engine low-oil-pressure-warning-signal-light circuit.
35	Engine high-oil-temperature-warning-signal-light circuit.
36	Engine oil pressure gage circuit.
37	Accessory outlet circuit.
38	Hull dome lights circuit.
41	Interphone No. 1.
42	Interphone No. 2.
43	Interphone No. 3.
44	Interphone No. 4.
45	Interphone No. 5.
47	Slip ring box feed.
48	Hull radio terminal box feed.
49	Slave battery receptacle positive line.
50	Slave battery receptacle ground cable.
51	Interphone No. 7.
54	Fuel cut-off circuit.
60	Interphone No. 6.
61	Auxiliary generator field circuit.
62	Auxiliary generator armature circuit.
63	Auxiliary generator ground circuit.
65	Auxiliary generator starter circuit.
68	Battery interconnecting cables.
81	Battery positive cables.
82	Starter relay to starter circuit.
98	Interphone No. 8.
99	Interphone No. 9.
100	Turret feed.
101	Impulse relay feed.
103	Machine gun firing solenoid to firing switch.
111	Gunner's gun firing switch feed.
111A	Commander's gun firing switch feed.
112	Firing safety switch indicator light.
113	90-mm gun firing switch to impulse relay.
137	Turret accessory outlet positive feed.
138	Turret dome lights circuit.
141	Turret interphone No. 1.
142	Turret interphone No. 2.
143	Turret interphone No. 3.
144	Turret interphone No. 4.
145	Turret interphone No. 5.
148	Turret radio terminal box positive feed.
151	Turret interphone No. 7.

Table VI. *Electrical Circuit Numbers*—Continued

Circuit No.	Circuit name
159	Ventilating blower circuit.
160	Turret interphone No. 6.
198	Turret interphone No. 8.
199	Turret interphone No. 9.
327	Transmission high-oil-temperature-warning-signal-light circuit.
339	Transmission low-lubrication-oil-pressure-warning-signal-light circuit.
400	Heater feed.
401	Heater "ON" and "OFF" circuit.
402	Heater thermostat and fuel pump circuits.
410	Heater ventilating air motor and safety valve circuits.
411	Heater ventilation fan feed.
413	Heater relay control (not used).
421	Auxiliary engine fuel cut-off valve.
422	Auxiliary engine magneto ground.
423	Auxiliary interphone signal lamp.
426	Auxiliary interphone ground circuit.
427	Tachometer, "A" terminal, send to receive.
428	Tachometer, "B" terminal, send to receive.
429	Tachometer, "C" terminal, send to receive.
431	Speedometer, "A" terminal, send to receive.
432	Speedometer, "B" terminal, send to receive.
433	Speedometer, "C" terminal, send to receive.
450	Bilge pumps feed.
451	Rear bilge pump circuit.
452	Front bilge pump circuit.
459	Master relay control circuit.
459A	Master relay rectifier feed.
459B	Master relay feed.
463	Interphone No. 11.
464	Interphone No. 12.
465	Range finder feed.
467	Oiler cooler fans controller feed.
467A	Left oil cooler fan controller circuit.
467B	Rght oil cooler fan controller circuit.
478A	Auxiliary generator equalizer circuit.
478M	Main generator equalizer circuit.
479	Heater relay control circuit (not used).
481	Magneto switch feed.
506	Main generator-warning-signal-light circuit.
508	Auxiliary generator-warning-signal-light circuit.
509	Main warning light circuit.
514	Dimmer switch to BO upper beam.
515	Dimmer switch to BO lower beam.
516	Power pack feed.
517	Periscope feed.
600	Elevation tracking motor feed.
601	Tracking motor elevating circuit.
602	Tracking motor depressing circuit.
608	Traverse tracking motor feed.
609	Tracking motor left traversing circuit.

Table VI. Electrical Circuit Numbers—Continued

Circuit No.	Circuit name
610	Tracking motor right traversing circuit.
623	Commander's override relays.
624	Commander's override switch feed.
625	Main positive feed to turret control box.
631	Safety lamp return.
632	Loader's safety switch and safety lamp feed.
633	Loader's safety relay switch control circuit.
641	Booster motor control.
642	Booster motor power.
643	Inverter circuit (not used).
645	Traverse safety relay control circuit.

194. Master Junction Box

a. DESCRIPTION.

(1) *General.* The master junction box (fig. 184) is a shock-mounted, water-proof box containing all the remote control equipment for the electrical power system of the vehicle.

(2) *Functions of master junction box.*
  (*a*) *Main generator controls.*
   (*a*) Regulation of main generator output voltage.
   (*b*) Regulation of auxiliary generator output voltage.
   (*c*) Equalizing each generator's share of load when both generators are operating at the same time.
   (*d*) Switching power to accessories.
   (*e*) Operating auxiliary generator as a starter for auxiliary engine.

(3) *Components in master junction box* (fig. 184).
  (*a*) *Main generator controls.*
   1. Carbon pile voltage regulator.
   2. Polarized relay assembly.
   3. Current relay assembly.
   4. Circuit breaker.
  (*b*) *Auxiliary generator controls.* A set of auxiliary generator controls identical to the main generator controls is provided in the master junction box.
  (*c*) *Miscellaneous units.*
   1. Master current relay assembly.
   2. Starter current relay assembly.
   3. Heater current relay assembly (not used).
   4. Ammeter shunt assembly (not used).

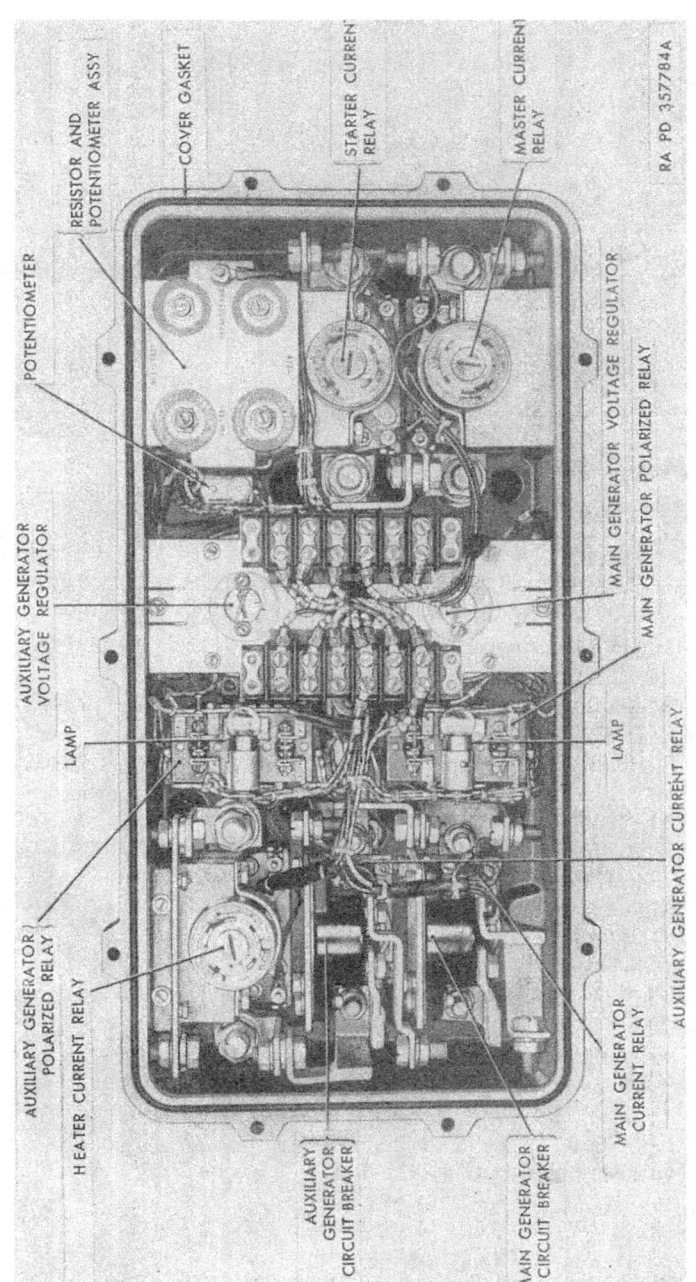

Figure 184. Master junction box with cover removed.

(4) *Cable connections.* Cable connections into the master junction box are made through the rear (fig. 182). All cables except the two battery cables (81) have quick-disconnect connectors. All cables enter the master junction box through holes threaded to receive watertight packing glands around the cables. The cables brought through these holes are secured with terminal nuts and screws to the proper terminals (fig. 187).

b. MASTER JUNCTION BOX COVER.
  (1) *Removal.* Disconnect ground cables from left side batteries and center batteries. Remove master junction box cover (fig. 146) by loosening 12 screw-type bolts holding it to master junction box until screws move freely on their undercuts.
  (2) *Installation.* Position master junction box cover and secure by tightening 12 screw-type bolts.

c. REMOVAL OF MASTER JUNCTION BOX. Traverse turret (par. 76) to obtain maximum accessibility to front section of top deck over master junction box. Turn off master relay switch (fig. 18). Remove master junction box cover (b above). Disconnect two cables (81) (fig. 187) by removing nuts from cable terminals. Remove front section of top deck (par. 158). Remove five connector plugs from engine wiring junction box (fig. 183) to provide working space at rear of master junction box. Remove two battery cables (81) by loosening mounting nuts on two angle connectors (fig. 182) and pulling connectors out from rear of master junction box. Label remaining master junction box cables (fig. 182) for proper mating and disconnect all quick-disconnect connectors at rear of master junction box.

Note. Do not remove angle connectors at entrances into rear of master junction box.

Remove master junction box ground strap (fig. 182) by removing lock washer screw securing strap to box. Remove two cap screws from each mounting strap (fig. 182) securing master junction box to bulkhead. Remove master junction box by pulling it forward through turret side of bulkhead.

d. INSTALLATION OF MASTER JUNCTION BOX. Position master junction box to mounting straps and secure to each strap with two cap screws and lock washers. Secure ground strap to master junction box (fig. 182) with lock washer screw. Connect cable connector plugs to proper mating cable connector receptacles at quick-disconnect connectors at rear of box (fig. 182). Install two cables (81) to angle connector plugs and secure each with a coupling nut. Install five engine wiring junction box connector plugs (fig. 183). Install front

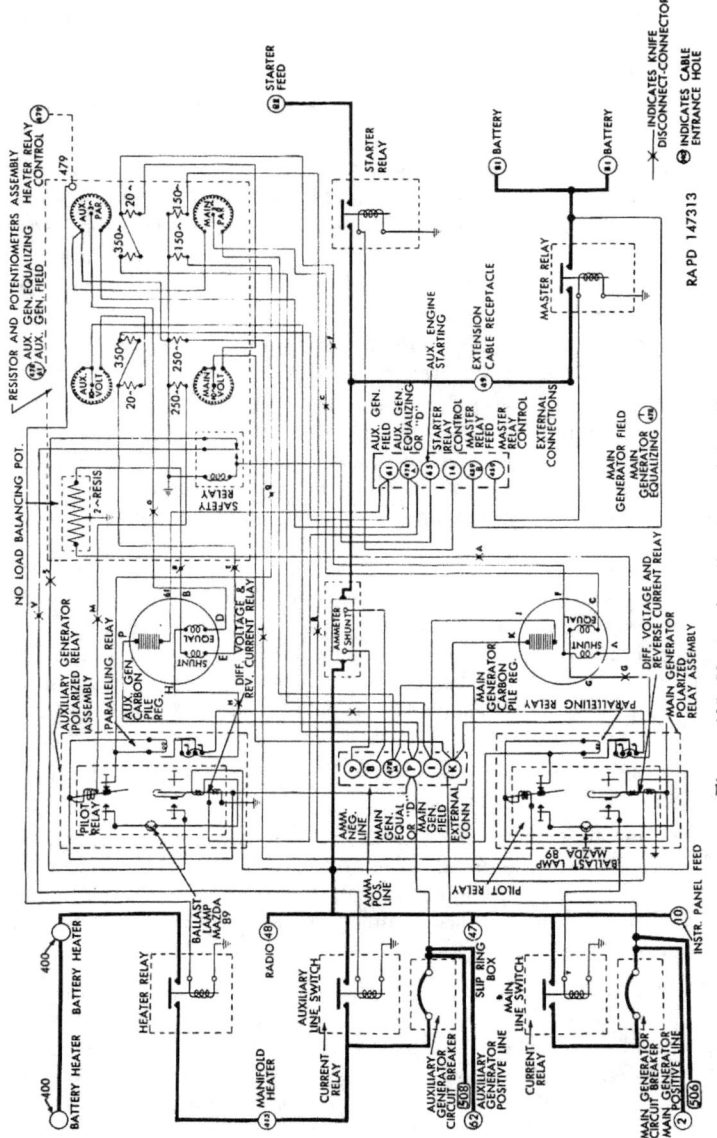

Figure 185. Master junction box wiring diagram.

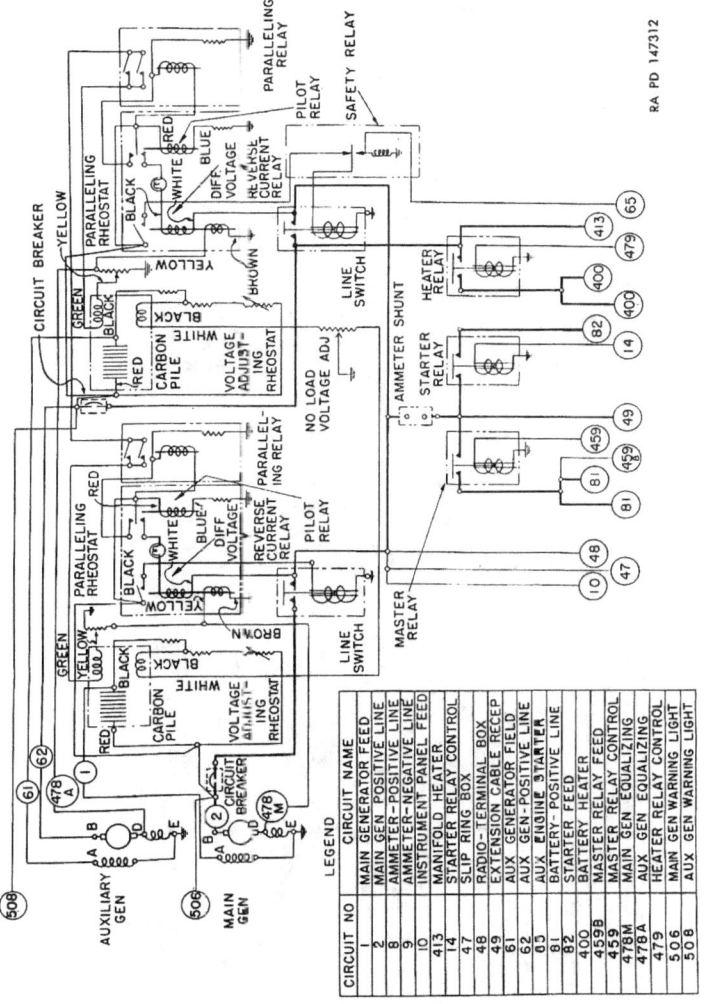

Figure 186. Master junction box and generators schematic diagram.

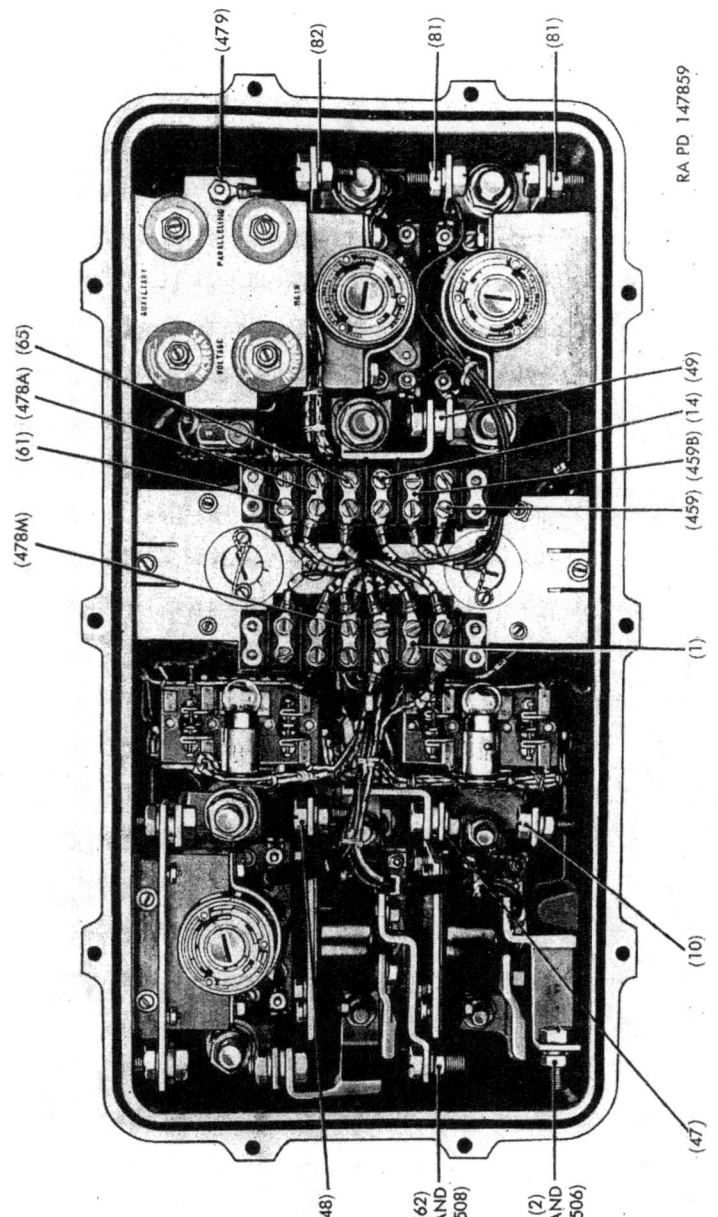

Figure 187. Location of circuit terminals in master junction box.

section of top deck (par. 161). Working from crew compartment, connect two cables (81) to cable terminals "Q" (fig. 187) inside master junction box, and fasten securely with terminal nuts. Install master junction box cover (*b* above).

195. Engine Wiring Junction Box
(fig. 183)

a. GENERAL. For description of engine wiring junction box, refer to paragraph 192. The box can be removed with the power plant installed in the vehicle.

b. REMOVAL. Turn off master relay switch (fig. 18). Remove front section of top deck (par. 158). Remove five connector plugs from engine wiring junction box (fig. 183). Loosen two nuts on each side of box that secure box to studs on engine accessory case (fig. 188). Swing engine wiring junction box to rear (fig. 188). Loosen cable clip on box (fig. 188) and free cable. Remove each cable connector receptacle from box by removing four screws and lift out engine wiring junction box.

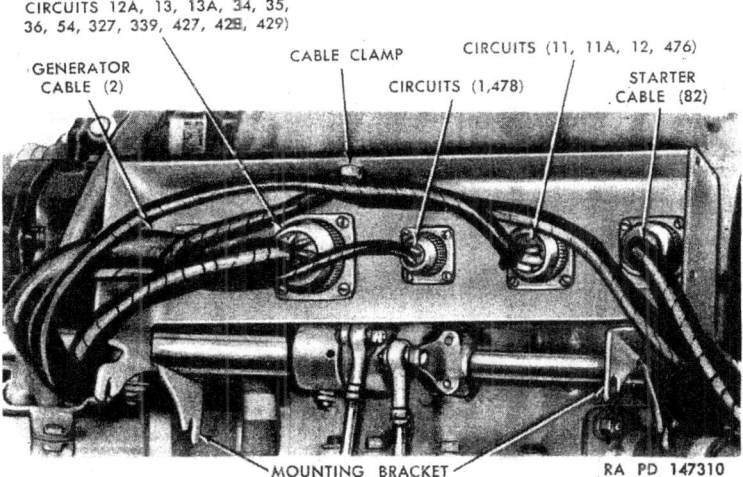

Figure 188. Engine wiring junction box—rear view.

c. INSTALLATION. Position five connector receptacles to junction box and fasten each with four screws (fig. 188). Insert cable (11, 11A, 12, 476) in clamp (fig. 188) and fasten. Position junction box to mounting studs on accessory case (fig. 188) and secure with two nuts on each side. Connect five connector plugs. Each of these plugs will connect only into proper receptacle. Install front section of top deck (par. 161).

196. Conduits, Cables, and Connectors

a. REMOVAL.
  (1) *General.* All electrical conduits and cables are removed in essentially the same manner. Disconnect the connector ((2) and (3) below) at each end and remove cable.
  (2) *Plug-and-receptacle-type connector.*
    (*a*) *Plug.* Unscrew plug coupling nut. Pull plug out of receptacle.

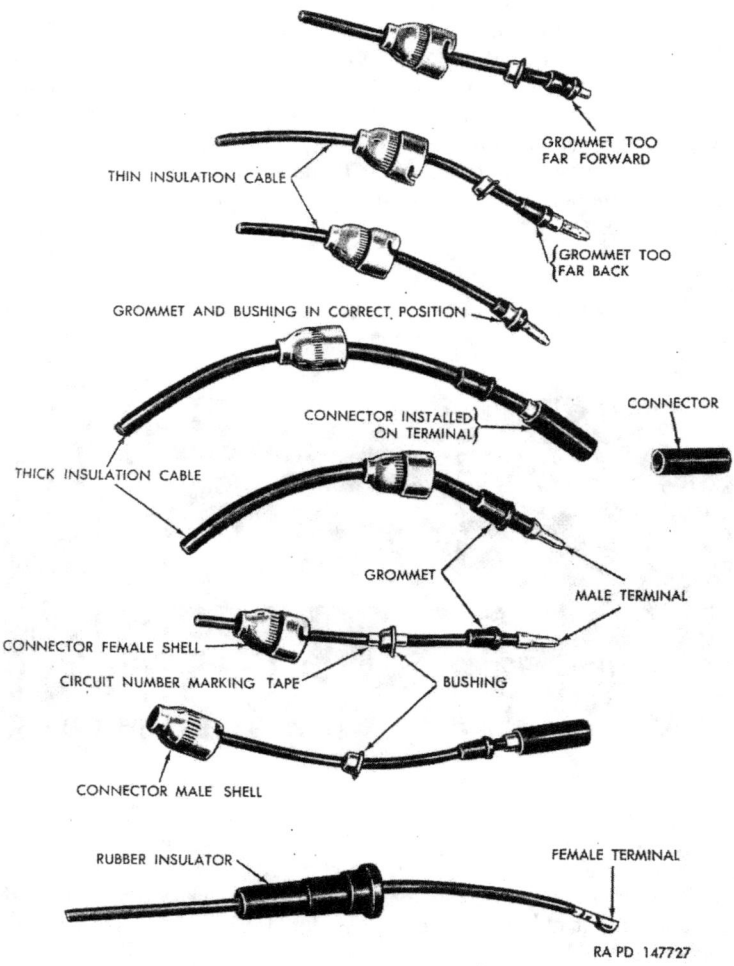

Figure 189. Bell-type connectors.

(b) *Receptacle*. Remove receptacle mounting screws and remove receptacle.

(3) *Bell-type connector* (fig. 189). Separate the two bells by a twist and pull. Pull apart the insulated, snap-type connector.

b. INSTALLATION.

(1) *General*. All electrical conduits and cables are installed in essentially the same manner, by connecting the connector ((2) and (3) below) at each end.

(2) *Plug-and-receptacle-type connector*.

(a) *Plug*. Insert plug into receptacle, making sure that positioning tongue lines up with groove.

Note. Do not force plugs. If plug is properly alined, it will enter receptacle easily.

Secure with coupling nut.

(b) *Receptacle*. Position receptacle to mounting bracket and secure with receptacle screws.

(3) *Bell-type connector* (fig. 189). Snap connector units together. Make sure rubber grommets and bushings are in proper position. For correct position see figure 189. Clip together the two bells.

197. Slave Battery Receptacle

a. GENERAL. For a description of slave battery receptacle refer to paragraph 98.

b. REMOVAL (fig. 76). Turn off master relay switch (fig. 18). Remove four lock washer screws securing slave battery receptacle and thermostat mounting bracket to vertical column to the rear of and between the driver and assistant driver. Pull cables (49) and (50) out of sockets in rear of receptacle and remove receptacle.

c. INSTALLATION (fig. 76). Insert cable (49) into positive socket at rear of slave battery receptacle and cable (50) into ground socket. Position thermostat mounting bracket and slave battery receptacle to vertical column and secure with four lock washer screws.

Note. Be sure cable slip is secured to upper left lock washer screw.

198. Power Pack

a. GENERAL. The power pack (fig. 368) is mounted on the bulkhead behind the driver's seat. It supplies 16,000 volts for operation of the periscope viewer.

Note. Early models of 90-mm gun tank M47 do not contain either power pack or periscope viewer.

Warning: Be sure to turn off master relay switch (fig. 18) before removing power pack. High voltage is dangerous.

b. Removal. Disconnect two cables at power pack. Remove four nuts and washers securing power pack and mounting brackets to shock-mounts and remove power pack and mounting brackets. Remove two bolts securing each bracket to power pack and remove power pack.

c. Installation. Secure two mounting brackets to base of power pack with two bolts. Secure power pack and mounting brackets to shock-mounts with four nuts and plain washers. Connect cables to power pack.

Section XIII. INSTRUMENT AND HULL ACCESSORIES PANELS, INSTRUMENTS, SWITCHES, AND SENDING UNITS

199. General

a. Description. This section provides descriptive information and removal and installation procedures for the instrument and hull accessories panels and their components, which are mounted in the crew compartment, and the sending units and switches, which are mounted on the engine and transmission. The instrument panel, hull accessories panel, and a speedometer-tachometer unit secured to the instrument panel are mounted in the forward portion of the crew compartment (fig. 190). On the panels are mounted electrical controls, gages, warning signal lights, rectifier units, and circuit breakers required for the operation of the vehicle and hull accessories. Warning signal lamps may be replaced from the face of the instrument panel. Removal of any other unit requires removal of instrument panel or hull accessories panel. Refer to figures 193 and 194 for wiring and schematic diagrams of the sending units and switches.

b. Cables and Connectors. Cables connecting vehicle electrical components to instrument and hull accessories panels are terminated in four connectors at rear of instrument panel and six connectors at rear of hull accessories panel (fig. 195). Speedometer and tachometer cables are terminated at two connectors at rear of the speedometer-tachometer unit (fig. 195). All connections between units within the instrument and hull accessories panels are made with bell-type connectors or with terminals over which is installed rubber covering for waterproofing. Figures 197 and 198 show the instrument panel wiring harness and hull accessories panel wiring harness respectively.

c. Procedures. It is recommended that when a unit of the instrument or hull accessories panels is to be removed, a means of good visibility be provided for facilitating operations. Do not clip or unsolder any cables for removal of any unit as each cable is equipped with a connector. Refer to wiring diagrams (figs. 191 and 192) for an understanding of functional relationships of units in the panels.

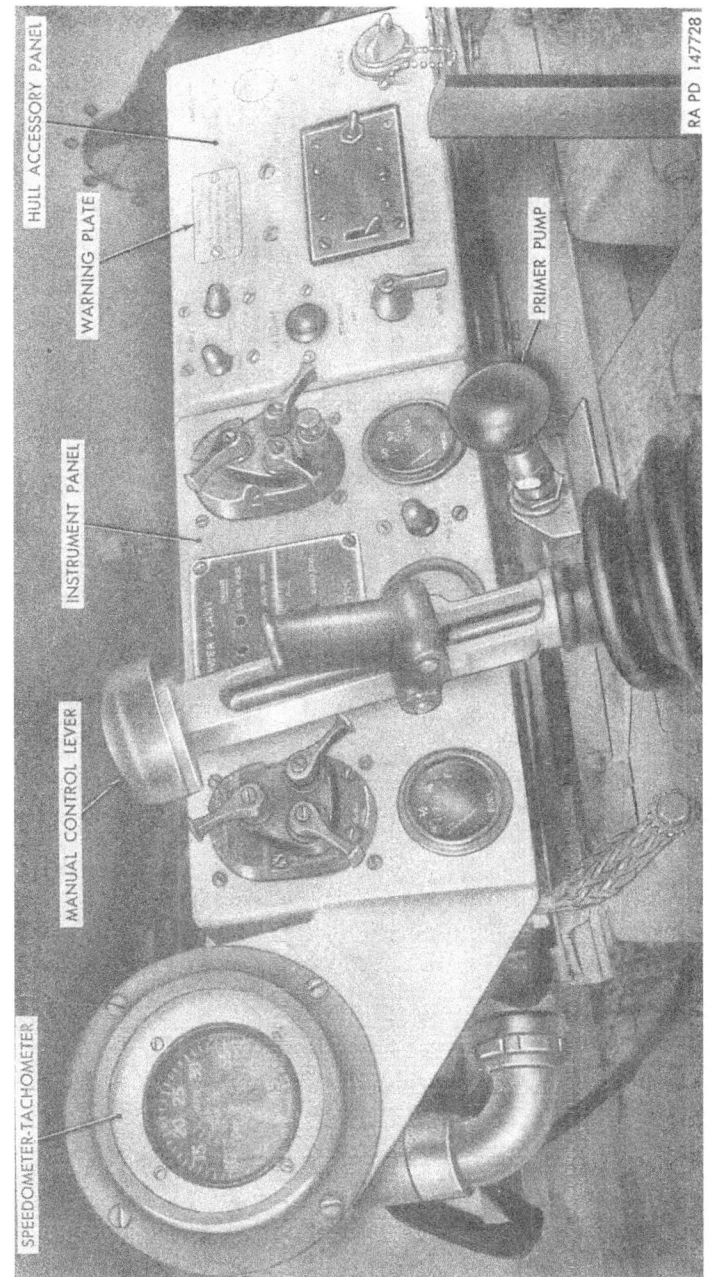

Figure 190. Instrument and hull accessories panel—installed view.

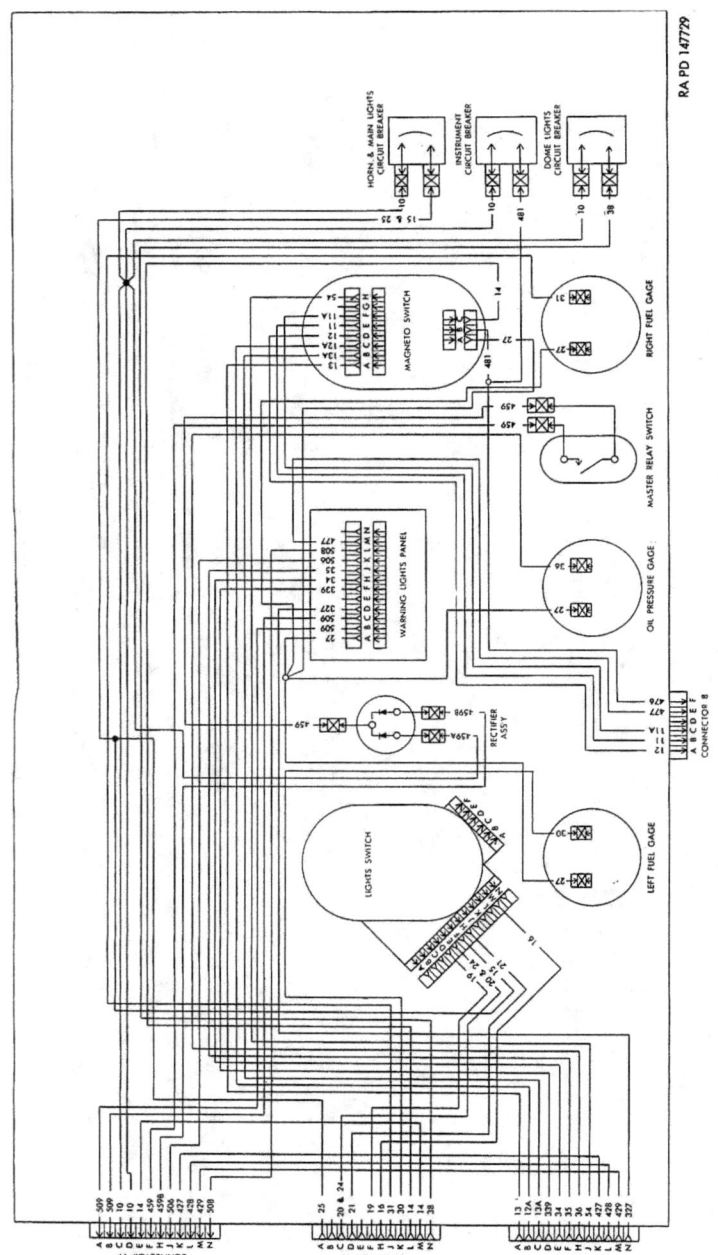

Figure 191. Instrument panel wiring diagram.

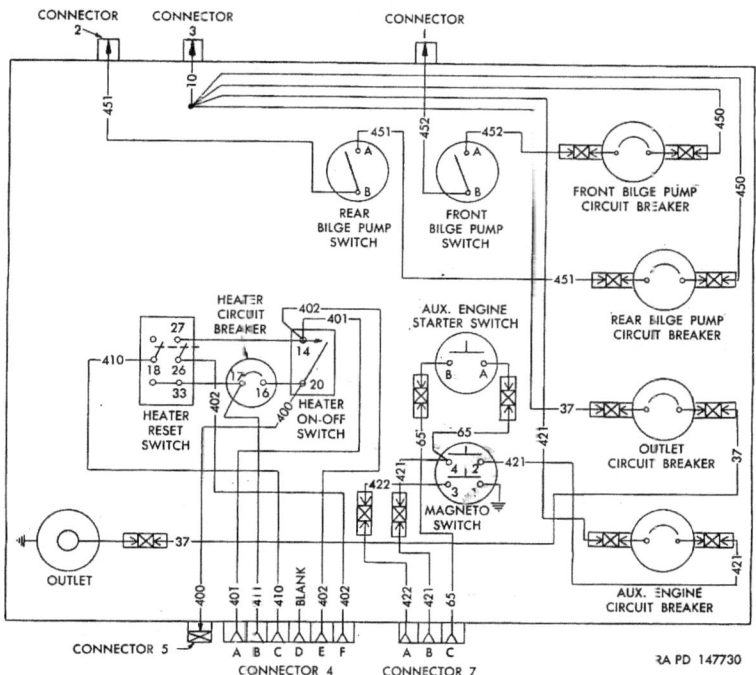

Figure 192. Hull accessories panel wiring diagram.

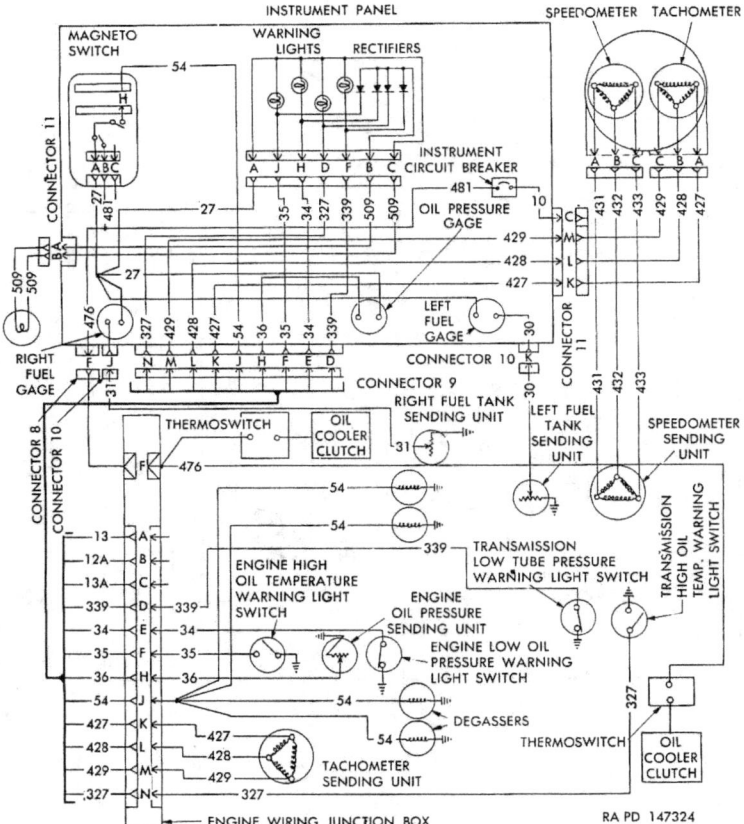

Figure 193. Engine, transmission, and fuel gages wiring diagram.

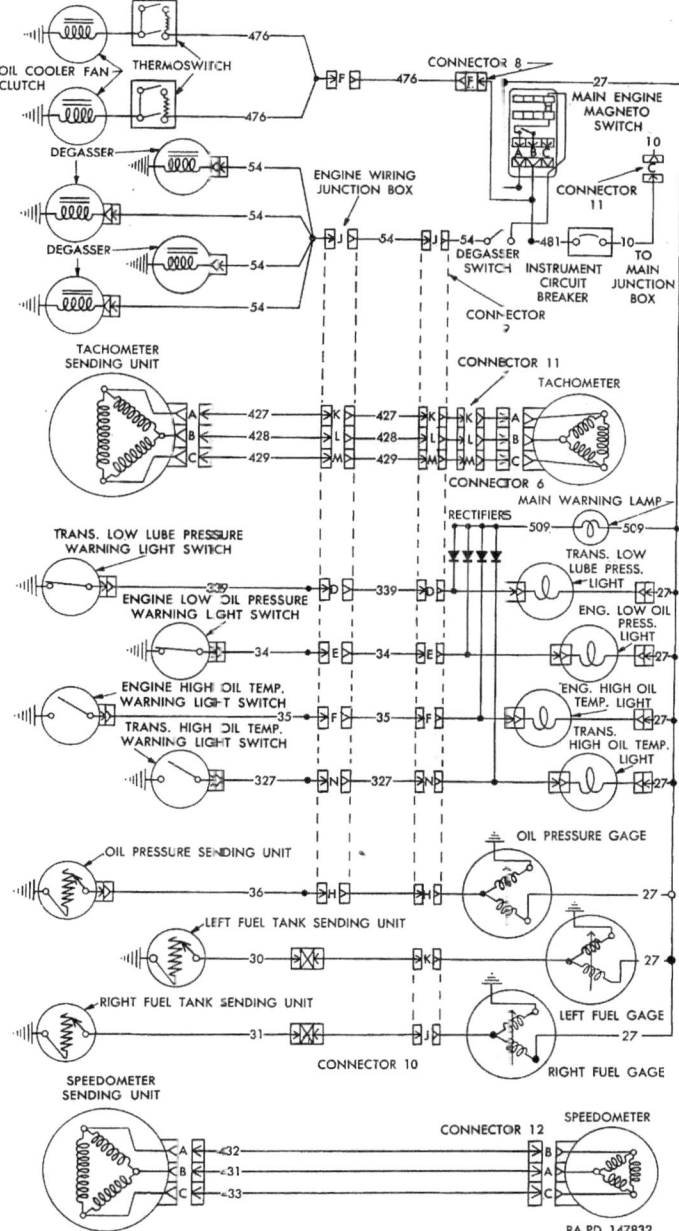

Figure 194. Engine, transmission, and fuel gages schematic diagram.

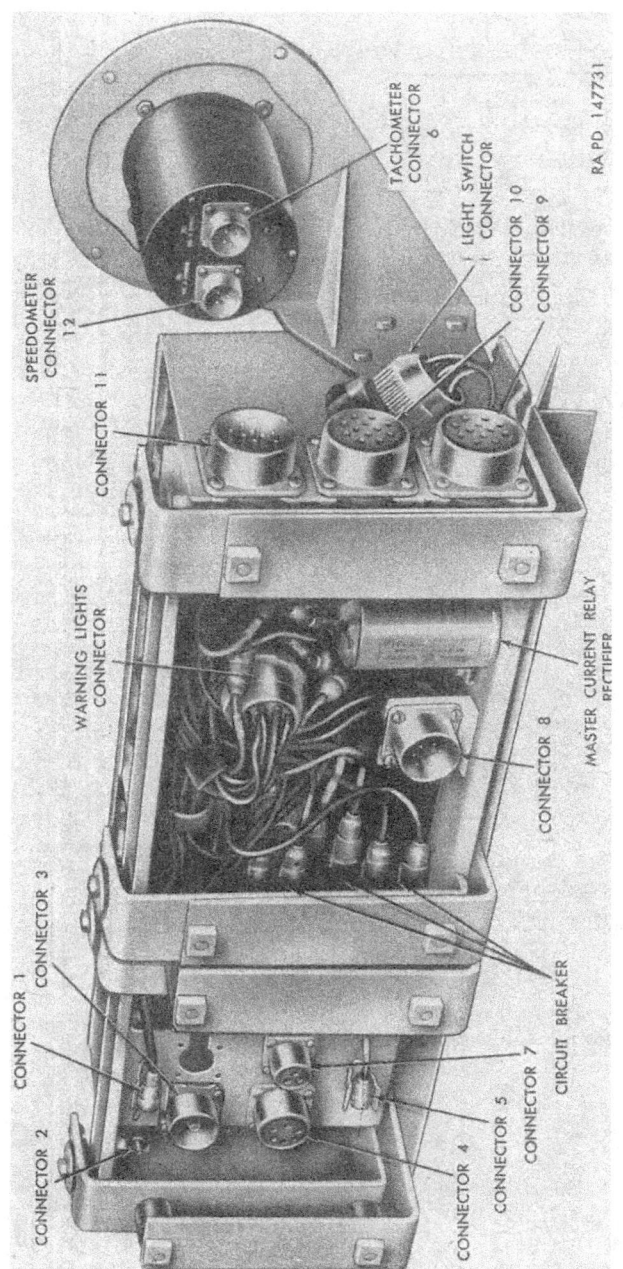

Figure 195. Instrument and hull accessories panel—rear view.

Figure 496. Instrument panel—top view.

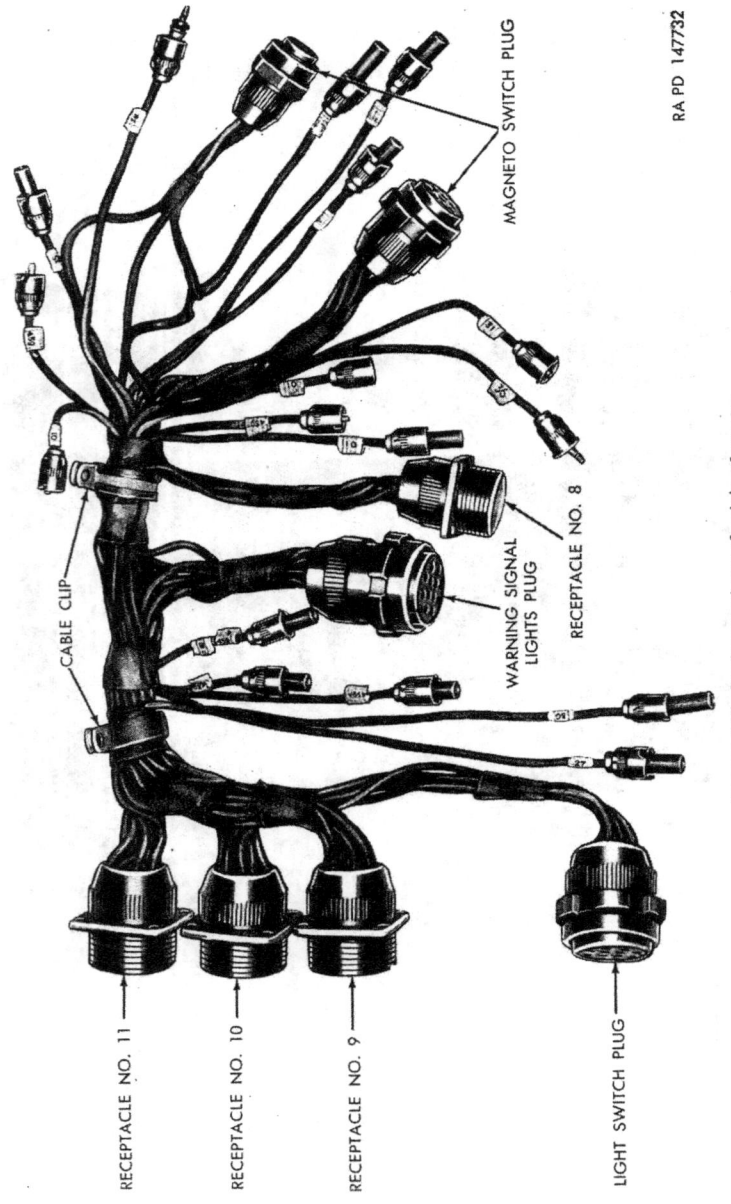

Figure 197. Instrument panel wiring harness.

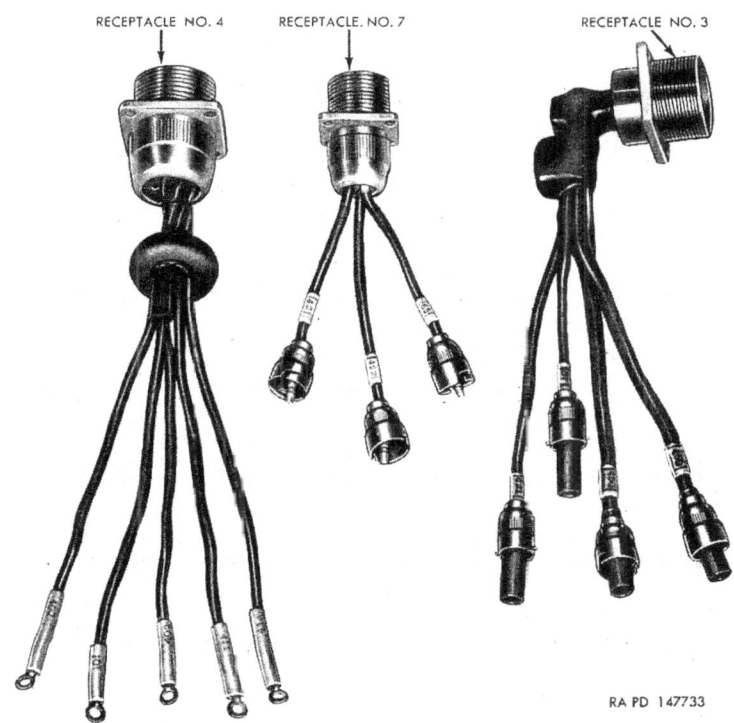

Figure 198. Hull accessories panel wiring harness.

200. Instrument Panel

a. GENERAL. The instrument panel consists of a box with no back, shock-mounted on vertical supports ahead of the driver (figs. 190, 195, and 196). The light switch, magneto-starter-booster-degasser switch, master relay switch, warning lights panel, engine oil pressure gage, and left- and right-fuel gages are secured to the face of the instrument panel (fig. 8). Three circuit breakers are mounted on the rear-lower left of the panel and a master-current-relay, rectifier unit on the rear-lower-right (fig. 195). The speedometer-tachometer unit is secured to a bracket at the left of the instrument panel (figs. 18 and 190). A main warning lamp is provided to indicate when one of the warning signal lights is on (fig. 21). Full instructions for the use of the various instruments and controls on the panel are given in chapter 2, section II.

b. REMOVAL. Turn off master relay switch (fig. 18). Remove cap screw and lock washer on each end of each instrument panel mounting bracket. Pull instrument panel forward out of brackets as far as cables will permit. Remove two connector plugs at rear of speedometer-tachometer, three connector plugs at side of instrument panel, and one connector plug at rear of panel (fig. 195).

Note. Mark all plugs and receptacles for proper assembly.

Remove instrument panel by pulling it forward out of mounting brackets.

c. INSTALLATION. Insert instrument panel into mounting brackets and connect rear connector plug (fig. 195). Position instrument panel inside brackets so that bracket holes line up with holes in panel and secure panel with one cap screw and washer at each end of bracket. Connect three connector plugs into proper receptacles on side of panel and two connector plugs into proper receptacles on rear of speedometer-tachometer (fig. 195).

201. Hull Accessories Panel

a. GENERAL. The hull accessories panel consists of a box with no back, shock-mounted on vertical supports beside the instrument panel (figs. 190 and 195). Front and rear bilge pump switches, auxiliary engine starter and magneto switches, accessory outlet, and heater control box panel with heater "ON" and "OFF" switch, manual reset circuit breaker, and safety valve reset switch are mounted on the face of the panel (fig. 18). Four circuit breakers are mounted on the rear-right side of the panel. The rear of all controls and units is latex-coated for waterproofing. Full instructions for the use of the various controls on the hull accessories panel are given in chapter 2, section IV.

b. REMOVAL. Turn off master relay switch (fig. 18). Remove cap screw and lock washer on each end of each hull accessories panel mounting bracket and also disconnect ground cable connected to lower right cap screw. Pull hull accessories panel forward as far as cables will permit. Remove three connector plugs and disconnect three bell-type connectors on rear of hull accessories panel (fig. 195).

Note. Mark all plugs, receptacles, and connectors for proper assembly.

Remove hull accessories panel by pulling it forward out of mounting brackets.

c. INSTALLATION. Insert hull accessories panel into mounting brackets and connect three connector plugs and three bell-type connectors into proper matching receptacles (fig. 195). Position hull accessories panel inside mounting brackets so that bracket holes line up with holes in panel and secure with one cap screw and washer at each end of each bracket.

Note. Be sure ground cable is securely connected to lower right cap screw.

202. Speedometer-Tachometer

a. GENERAL. The speedometer-tachometer (fig. 18) is a single unit with two dials, secured to the instrument panel by means of a bracket (fig. 195). For detailed description, refer to paragraph 31.

b. REMOVAL (fig. 195). Remove two connector plugs at rear of speedometer-tachometer unit.

Note. Mark both plugs and receptacles for proper assembly.

Remove speedometer-tachometer by removing four cap screws securing unit to mounting bracket.

c. INSTALLATION (fig. 195). Insert speedometer-tachometer into hole in mounting bracket, position on bracket, and secure with four cap screws. Connect two connector plugs into proper receptacles on rear of unit.

d. TESTING. Start main engine (par. 68). Tachometer dial should indicate 650 rpm at idle and 2,950 rpm at full throttle. Drive the vehicle and see if the speedometer operates properly. If either dial fails to register properly, replace entire unit (par. *b* above).

203. Light Switch

a. GENERAL. The light switch is located in the upper-left corner of the instrument panel (fig. 18). For a detailed description of switch, refer to paragraph 27.

b. REMOVAL. Remove instrument panel (par. 200). Remove connector receptacles Nos. 9, 10, and 11 (fig. 195) by removing four screws from each receptacle. Remove master relay rectifier (par. 213). Remove light switch connector plug (fig. 195). Remove screw from main switch handle (fig. 19) and remove handle. Remove four screws securing light switch to face of instrument panel and remove switch through rear of panel.

c. INSTALLATION. Insert switch into instrument panel from rear and secure to face of panel with four screws. Install main switch handle (fig. 19) and secure with screw and washer. Connect light switch connector plug. Install connector receptacles Nos. 9, 10, and 11 (fig. 195) on side of panel with four screws for each receptacle. Install master relay rectifier (par. 213). Install instrument panel (par. 200).

d. TESTING. Turn on master relay switch (fig. 18) and if tactical situation permits, test all positions of light switch. If tactical situation does not permit operation of lights, perform ohmmeter tests on light switch (par. 131).

204. Main-Engine Starter and Magneto Switch

a. GENERAL. The main-engine starter and magneto switch is located in the upper-right corner of the instrument panel (fig. 18) and incorporates a magneto switch, starter switch, booster switch, and fuel cut-off (degasser) switch.

b. REMOVAL. Remove instrument panel (par. 200). Remove upper two circuit breakers to provide working space (par. 214). Remove two connector plugs from rear of switch. Remove four screws securing switch to face of instrument panel and remove switch out of rear of panel box.

c. INSTALLATION. Insert switch into face of instrument panel from rear and secure with four screws. Connect two connector plugs to matching receptacles at rear of switch. Install two circuit breakers (par. 214). Install instrument panel (par. 200).

d. TESTING. Start main engine (par. 68) to test magneto, starter, and booster switches. Stop engine (par. 72) to test fuel cut-off (degasser) switch. If switch does not operate properly, refer to paragraph 131 for test procedures.

205. Gages

a. GENERAL. The following three gages are mounted on lower portion of face of instrument panel (fig. 18): left-fuel gage, right-fuel gage, and engine-oil-pressure gage. Procedure for removal and installation is the same for all three gages.

b. REMOVAL. Remove instrument panel (par. 200). Remove two nuts and lock washers from rear of gage mounting clamp. Disconnect two cables from gage and remove gage from face of instrument panel.

c. INSTALLATION. Insert gage through front of instrument panel. Position mounting clamp on gage and secure with nuts and lock washers. Connect two cables to rear of gage. Install instrument panel (par. 200).

d. TESTING. Turn on master relay switch (fig. 18). Turn main-engine magneto switch (fig. 18) to "A," "F," or "BOTH" position. Fuel gages should promptly register quantity of fuel in left- and right-fuel tanks. Oil pressure gage should register zero. Start main engine (par. 68). Oil pressure gage should now register 60 to 70 psi. If any gage does not register properly, refer to paragraph 131 for test procedure.

206. Warning Signal Lights

a. GENERAL. The warning signal lights box is located in the upper-center of the instrument panel (fig. 18). A main warning light (fig. 21) is secured to the bulkhead to the left of the driver.

b. WARNING SIGNAL LIGHTS BOX.

(1) *Warning signal lamp replacement.* Turn off master relay switch (fig. 18). Remove four screws securing warning signal lights panel to face of instrument panel and remove panel. Remove six screws and remove lens. Replace required signal lamp. Install lens and secure with six screws. Position warning signal lights panel to face of instrument panel and secure with four screws and lock washers.

(2) *Removal of warning signal lights box* (fig. 18). Turn off master relay switch (fig. 18). Remove connector plug from rear of warning signal lights box. Remove warning signal lights panel by removing four screws from face of panel. Remove four screws from front of warning signal lights box and remove box through top of instrument panel.

(3) *Installation of warning signal lights box.* Position warning lights box to face of instrument panel and secure with four screws and lock washers. Position warning signal lights panel and secure with four screws. Connect connector plug to rear of warning signal lights box.

c. MAIN WARNING LIGHT.

(1) *Main warning light replacement.* Remove two screws and remove main warning light door (fig. 21). Replace lamp. Install door and secure with two screws.

(2) *Removal of main warning light.* Disconnect two cables at quick-disconnect connectors at main warning light (fig. 21). Remove two lock washer screws securing light to brackets on hull and remove light.

(3) *Installation of main warning light.* Position light to brackets on hull wall (fig. 21) and secure with two lock washer screws. Connect two cable connectors at quick-disconnect connectors.

207. Master Relay Switch

a. GENERAL. Master relay switch is located in lower-center of instrument panel (fig. 18).

b. REMOVAL. Remove instrument panel (par. 200). Remove two cap screws which secure master relay switch to instrument panel. Disconnect cables at connectors at rear of switch and remove switch from rear of panel box.

c. INSTALLATION. Position master relay switch to face of instrument panel and secure with two screws. Connect cables at connectors at rear of switch.

Note. Make sure that cable circuit numbers are the same on both cables being connected. Install instrument panel (par. 200).

d. TESTING. Turn on master relay switch. Turn main-engine-magneto switch to "A," "F," or "BOTH" position and check to see if "LO OIL PRESS" warning signal light comes on. If light does not come on, check master relay switch (par. 131).

208. Bilge Pump Switches

a. GENERAL. Front and rear bilge pump switches are located in the upper-left corner of the hull accessories panel (fig. 18). Procedure for removal and installation is the same for both switches.

b. REMOVAL. Remove hull accessories panel (par. 201). Remove two screws securing switch to face of panel. Remove two cables from connectors at rear of switch and remove switch.

c. INSTALLATION. Position switch to hull accessories panel and secure with two screws. Connects two cables to connectors at rear of switch. Install hull accessories panel (par. 201).

d. TESTING. Turn on master relay switch (fig. 18). Operate each switch and see if proper bilge pump operates. If a switch does not operate, refer to paragraph 141 for test procedure.

209. Auxiliary-Engine-Starter Switch

a. GENERAL. The auxiliary-engine-starter switch is located on left center of hull accessories panel (fig. 18).

b. REMOVAL. Remove hull accessories panel (par. 210). Remove two screws securing auxiliary-engine-starter switch to panel. Disconnect two cables at connectors at rear of switch and remove switch.

c. INSTALLATION. Position auxiliary-engine-starter switch to hull accessories panel and secure with two screws. Connect two cables at connectors at rear of switch. Install hull accessories panel (par. 201).

d. TESTING. Start auxiliary engine (par. 93) to see if switch operates. If switch does not operate, refer to paragraph 141 for test procedure.

210. Auxiliary-Engine-Magneto Switch

a. GENERAL. The auxiliary-engine-magneto switch is located in the lower-left corner of the hull accessories panel (fig. 18).

b. REMOVAL. Remove hull accessories panel (par. 201). Disconnect four cables at rear of switch at quick-disconnect connectors. Remove switch ground cable by removing nut, lock washer, and plain washer at side of panel. Remove screw in center of switch handle and remove handle. Remove nut and lock washer from switch shaft and remove switch from rear of panel box.

c. INSTALLATION. Position auxiliary-engine-magneto switch to hull accessories panel and secure with nut and lock washer. Install handle on switch shaft with screw. Connect ground cable to panel with nut, lock washer, and plain washer. Connect four switch cables to mating cables at rear of switch. Install hull accessories panel (par. 201).

d. TESTING. Turn on master relay switch (fig. 18). Start auxiliary engine (par. 93) to test auxiliary-engine-magneto switch. If switch does not operate, refer to paragraph 141 for test procedure.

211. Heater Control Box

a. GENERAL. The heater control box is located in the lower-center portion of the hull accessories panel (fig. 18). The box contains the heater "ON" and "OFF" switch, the safety-valve-reset switch, and the heater-circuit-breaker-reset switch. The box is removable only as a unit and cannot be disassembled.

b. REMOVAL. Remove hull accessories pannel (par. 201). Remove connector receptacle No. 4 (fig. 195) by removing four screws. Pull bell-type connector No. 5 from clip. Remove two heater control box mounting screws on face of panel. Remove heater control box through rear of panel box.

c. INSTALLATION. Position heater control box to hull accessories panel and secure with two screws on face of panel. Install connector receptacle No. 4 and secure with four screws and lock washers. Install bell-type connector No. 5 into clip (fig. 195). Install hull accessories panel (par. 201).

212. Accessory Outlet Socket

a. GENERAL. The accessory outlet socket is located in the lower-right corner of the hull accessories panel (fig. 18). It is used for the connection of external electrical equipment.

b. REMOVAL. Remove hull accessories panel (par. 201). Remove two accessory outlet socket mounting screws on face of panel. Disconnect cable at connector at rear of socket and remove socket.

c. INSTALLATION. Position accessory outlet socket and secure with two mounting screws.

Note. Be sure to attach socket cover chain to left screw.

Connect cables at rear of socket. Install hull accessories panel (par. 201).

213. Master-Currency-Relay Rectifier

a. GENERAL. The master-current-relay rectifier (fig. 195) is located in the lower-right-rear corner of the instrument panel. The unit consists of two selenium rectifiers which are unidirectional current units. The rectifier unit is provided so that if the vehicle batteries are so run down that they will not close the master current relay (fig. 184) when the master relay switch is turned on, an external battery can be used to close the relay, thus eliminating the need for charging vehicle batteries from an external source. When the external battery has closed the master current relay, the main auxiliary engine can be started and the vehicle batteries will automatically be charged by the vehicle generators. Refer to figure 204 for a schematic diagram of the master relay rectifier circuit.

b. REMOVAL. Remove instrument panel (par. 200). Disconnect three cables at rectifier unit quick-disconnects (fig. 195). Remove two screws, nuts, and washers securing rectifier mounting brackets to instrument panel and remove rectifier unit.

c. INSTALLATION. Position rectifier unit at rear of instrument panel (fig. 195) and secure by tightening two mounting nuts. Connect three cables to proper mating cables at quick-disconnects at rectifier unit. Install instrument panel (par. 200).

214. Instrument Panel Circuit Breakers

a. GENERAL. Mounted on the left side of the instrument panel at the rear are three automatic-reset circuit breakers. Listed in order from top to bottom, these circuit breakers protect the following circuits: horn and main lights, instruments, and dome lights. Refer to figure 191 for wiring diagram. Procedures for removal and installation are the same for all three circuit breakers.

b. REMOVAL. Remove instrument panel (par. 200). Disconnect two cables at quick-disconnects at each circuit breaker. Remove two mounting screws, nuts, and lock washers from each circuit breaker and remove circuit breakers.

c. INSTALLATION. Position circuit breakers and secure each with two mounting screws, nuts, and lock washers. At each breaker, connect two cables to proper mating cables at quick-disconnects. Install instrument panel (par. 200).

d. TESTING. Turn on master relay switch (fig. 18). Turn main-engine-magneto switch to "A," "F," or "BOTH" position. Notice whether fuel gages operate. If they do, instrument circuit breaker is operating. If tactical situation permits, operate light switch or horn to see if horn-and-main-lights circuit breaker is operating. Turn on dome lights to see if dome lights circuit breaker is operating. If tactical situation does not permit operation of horn or lights, refer to paragraph 131 for ohmmeter test procedure.

215. Hull Accessories Panel Circuit Breakers

a. GENERAL. Mounted on the right-rear side of the hull accessories panel are four automatic-reset circuit breakers. These circuit breakers protect the following four circuits: front bilge pump, rear bilge pump, accessory outlet, and auxiliary engine starter and magneto circuits. Refer to figure 192 for wiring diagram. Procedures for removal and installation are the same for all four circuit breakers.

b. REMOVAL. Remove hull accessories panel (par. 201). Disconnect two cables at quick-disconnects at each circuit breaker. Remove two mounting screws, nuts, and lock washers from each circuit breaker and remove circuit breakers.

c. INSTALLATION. Position circuit breakers and secure each with two mounting screws, nuts, and lock washers. At each breaker, connect two cables to proper mating cables at quick-disconnects. Install hull accessories panel (par. 201).

d. TESTING. Turn on master relay switch (fig. 18). Operate front bilge pump, rear bilge pump, accessory outlet, and/or auxiliary engine to see if corresponding circuit breaker is operating. If circuit breaker does not operate, refer to paragraph 141 for ohmmeter test procedure.

216. Instrument and Hull Accessories Panels Wiring Harnesses

a. GENERAL. The instrument panel has one wiring harness assembly (fig. 197) connecting all gages, switches, and warning signal lights to the connector receptacles at the rear of the instrument panel. The hull accessories panel has three wiring harness assemblies (fig. 198). Connectors No. 3 and 7 can be individually removed. Connector No. 4 is removed only with the heater control box.

b. INSTRUMENT PANEL WIRING HARNESS.

(1) *Removal* (fig. 195). Remove instrument panel (par. 200). Remove four screws from each of connector receptacles No. 8, 9, 10, and 11 and slip each of these receptacles out of instrument panel. Disconnect warning lights connector plug, lights switch connector plug, and two connector plugs at main engine magneto switch. Disconnect all bell-type con-

nectors at rear of instrument panel. Remove two clips securing harness to top of instrument panel and remove instrument panel wiring harness.

(2) *Installation* (figs. 191 and 195). Position harness to instrument panel and secure clips in top of panel. Connect connector receptacles No. 8, 9, 10, and 11 by securing each to instrument panel with four screws. Connect warning lights connector plug, lights switch connector plug, and two connector plugs at rear of main engine magneto switch. Connect all bell-type connector plugs to proper matching receptacles. Install instrument panel (par. 200).

c. HULL ACCESSORIES PANEL WIRING HARNESS (fig. 198).

(1) *Removal*. Remove hull accessories panel (par. 201). The procedure for removing connectors No. 3 and 7 (fig. 198) is the same. Remove four screws securing receptacle to connector receptacle bracket at rear of accessories panel. Disconnect bell-type connectors and remove harness.

(2) *Installation*. Position receptacle to connector receptacle bracket and secure with four screws. Connect bell-type connectors (fig. 198). Install hull accessories panel (par. 201).

217. Sending Units

a. SPEEDOMETER SENDING UNIT.

(1) *General*. The speedometer sending unit is mounted on the transmission (fig. 221). It is an electrical unit which sends a variable voltage to the speedometer coils inclosed in the speedometer-tachometer secured to the instrument panel (fig. 18). This voltage causes rotation of the speedometer armature which moves the speedometer hand, indicating miles per hour.

(2) *Removal* (fig. 221). Turn off master relay switch (fig. 18). Open four left-rear engine compartment grille doors (fig. 101) over the transmission. Remove speedometer-sending-unit-cable-connector plug (cables 431, 432, and 433). Remove four mounting nuts and washers securing sending unit to transmission case and remove sending unit by lifting it off transmission.

(3) *Installation* (fig. 221). Position speedometer sending unit to transmission case and secure with four mounting nuts and lock washers. Install speedometer-sending-unit-cable-connector plug (cables 431, 432, and 433). Close the four left-rear engine compartment grille doors over transmission (fig. 101).

(4) *Testing.* Turn on master relay switch (fig. 18). Drive the vehicle and observe if the speedometer portion of speedometer-tachometer secured to instrument panel (fig. 18) operates.

b. TACHOMETER SENDING UNIT.
   (1) *General.* The tachometer sending unit is mounted between and to the right of the two right magnetos (fig. 162) and is coupled directly to the engine camshaft. It is an electrical unit which sends a variable voltage to the tachometer coils inclosed in the speedometer-tachometer secured to the instrument panel (fig. 18). This voltage causes rotation of the tachometer armature which moves the tachometer hand, indicating hundred of revolutions per minute.
   (2) *Removal.* Turn off master relay switch (fig. 18). Remove front section of top deck (par. 158). Remove tachometer-sending-unit-cable-connector plug (cables 427, 428, and 429). Remove four mounting nuts and lock washers securing sending unit to accessory case and remove sending unit.
   (3) *Installation.* Position tachometer sending unit to accessory case (fig. 162) and secure with four mounting nuts and lock washers. Install tachometer-sending-unit-cable-connector plug (cables 427, 428, and 429). Install front section of top deck (par. 161).
   (4) *Testing.* Turn on master relay switch (fig. 18). Start main engine (par. 58) and observe if tachometer portion of tachometer-speedometer secured to instrument panel (fig. 18) operates. Tachometer should show 650 rpm at idle.

c. ENGINE OIL-PRESSURE-GAGE SENDING UNIT.
   (1) *General.* Engine oil-pressure-gage sending unit is located on right center of engine crankcase (fig. 199). This unit is electrically connected to the engine oil pressure gage mounted on the instrument panel (fig. 18). The sending unit is an hermetically-sealed variable resistance unit, pressure operated through a diaphragm and capable of indicating a range of from 0 to 120 psi.
   (2) *Removal.* Turn off master relay switch (fig. 18). Remove main generator access hole cover (N, fig. 285) from left-bottom of hull by removing the nuts and lock washers securing it. Remove electrical cable angle connector (cable 36) plug from sending unit (fig. 199). Unscrew switch from side of engine crankcase using wrench 7950049.
   (3) *Installation.* Carefully start sending unit into engine crankcase (fig. 199) and tighten. Connect angle connector plug (cable 36) to sending unit. Position main generator access

Figure 199. Engine oil-pressure-gage sending unit and low-oil-pressure-warning-signal-light switch—installed view.

      hole cover (fig. 285) over studs on left-bottom of hull and secure with nuts and lock washers.

  (4) *Testing.* Turn on master relay switch (fig. 18). Start main engine (par. 68) and observe if oil pressure gage mounted on instrument panel (fig. 18) operates. As soon as the engine is running, the gage should indicate from 60 to 70 psi. If gage does not give correct reading, refer to paragraph 131 for test procedure.

d. FUEL TANK SENDING UNITS.

  (1) *General.* Each fuel tank has a fuel tank sending unit located on top of the tank (fig. 200). Each sending unit is electrically connected to a fuel gage on the instrument panel (fig. 18).

  (2) *Removal.* Turn off master relay switch (fig. 18). Open five engine compartment grille doors (fig. 101) over each fuel tank. Remove angle connector plug (cable 30) from left-fuel-tank-sending unit (figs. 153 and 200). Remove angle connector plug (cable 31) from right-fuel-tank-sending unit (fig. 152). Remove six screws and lock washers securing

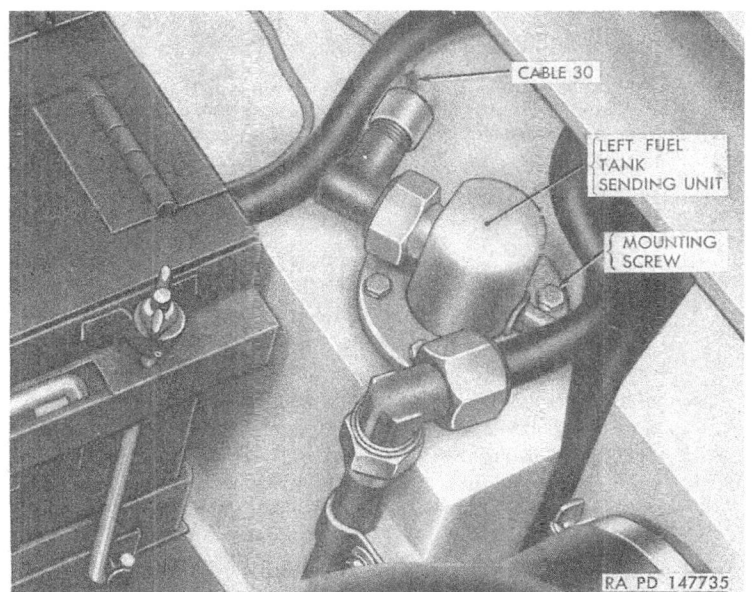

Figure 200. Left fuel tank sending unit—installed view.

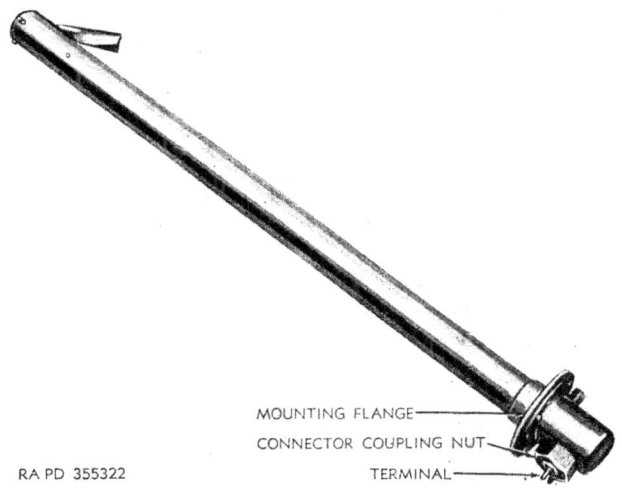

Figure 201. Fuel tank sending unit.

each sending unit to mounting flange and remove sending units and gaskets.

Note. When removing left-fuel-gage-sending-unit, remove clamp securing fire extinguisher line (fig. 153) to fuel tank and swing line free of mounting flange.

Warning: Take care not to drop anything into open fuel tank. Do not get near opening in fuel tank with an open flame.

(3) *Installation.* Apply a thin coat of liquid-type gasket cement to both sides of new gaskets. Clean off any old gasket material on tops of fuel tanks. Position gaskets on tank openings and insert sending units (figs. 152 and 153). Secure each sending unit with six screws and lock washers. Connect angle connector plug (cable 30) to left-sending-unit-terminal-connector receptacle. Connect angle connector plug (cable 31) to right-sending-unit-terminal-connector receptacle. Secure clamp securing fixed fire extinguisher line to left fuel tank (fig. 153). Close five engine compartment grille doors over each fuel tank (fig. 101).

(4) *Testing.* Turn on master relay switch (fig. 18). Check readings on both right- and left-fuel gages on instrument panel (fig. 18). If a fuel gage does not operate properly, refer to paragraph 131 for test procedure.

218. Engine Warning-Signal-Light Switches

a. ENGINE HIGH-OIL-TEMPERATURE-WARNING-SIGNAL-LIGHT SWITCH.

(1) *General.* The engine high-oil-temperature-warning-signal-light switch is located on the right side of the accessory case above the oil filter bypass valve and the oil pressure control valve (fig. 107). The switch is electrically connected to the warning signal light marked "HI OIL TEMP" on the instrument panel (fig. 18).

(2) *Removal.* Turn off master relay switch (fig. 18). Remove right air cleaner (par. 175) to gain access to bulkhead inspection plate and remove bulkhead inspection plate (fig. 148) by removing four cap screws securing it to bulkhead. Remove angle connector plug (cable 35) from switch (fig. 10). Unscrew switch from side of accessory case and remove.

(3) *Installation.* Screw switch into side of accessory case until tight (fig. 108). Connect angle connector plug (cable 35) to switch connector receptacle. Position right bulkhead inspection plate (fig. 148) and secure with four cap screws. Install right air cleaner (par. 175).

b. ENGINE-LOW-OIL-PRESSURE-WARNING-SIGNAL-LIGHT SWITCH.

(1) *General.* The engine low-oil-pressure-warning-signal-light switch is located on the right-center of the engine crankcase to the rear of the main generator (fig. 199). This switch is electrically connected to the warning signal light marked "LO OIL PRESS" on the instrument panel (fig. 18).

(2) *Removal.* Turn off master relay switch (fig. 18). Remove main generator access hole cover from left-bottom of hull (fig. 285). Remove angle connector plug (cable 34) from switch (fig. 199). Unscrew switch from side of engine crankcase and remove.

(3) *Installation.* Screw switch (fig. 199) into side of engine block until tight. Connect angle connector plug (cable 34) to switch connector receptacle. Install main generator access hole cover to bottom of hull (fig. 285).

(4) *Testing.* Turn on master relay switch (fig. 18). Engine low-oil-pressure-warning-signal light (fig. 18) should be on. Start engine (par. 68). As soon as engine starts, engine low-oil-pressure-warning-signal light should go off. If signal light does not operate properly, refer to paragraph 131 for test procedure.

219. Transmission Warning-Signal-Light Switches

a. GENERAL. There are two transmission warning-signal-light switches on the vehicle, located on top of the transmission (fig. 221). The transmission low-lubrication-pressure-warning-signal-light switch is electrically connected to warning signal light marked "LO LUBE PRESS" on instrument panel (fig. 18). The transmission high-oil-temperature-warning-signal-light switch is electrically connected to warning signal light marked "HI TEMP" on instrument panel (fig. 18).

b. REMOVAL. Turn off master relay switch (fig. 18). Open four engine compartment left rear grille doors (fig. 101). To remove low-oil-pressure-warning-signal-light switch (fig. 221), remove angle connector plug (cable 339) from switch. To remove high-oil-temperature-warning-signal-light switch (fig. 221) remove angle connector plug (cable 327) from switch. Unscrew switches from transmission and remove.

c. INSTALLATION. Screw switch into transmission housing until tight (fig. 221). Install angle connector plug (cable 339) to low-oil-pressure-warning-signal-light switch connector receptacle. Install angle connector plug (cable 327) to high-oil-temperature-warning-signal-light-connector receptacle. Close four engine compartment rear grille doors (fig. 101).

d. TESTING. Turn on master relay switch (fig. 18). Signal light marked "LO LUB PRESS" should come on and signal light marked "HI TEMP" should stay off unless transmission oil temperature rises dangerously high. If either light does not operate properly, refer to paragraph 131 for test procedure.

220. Oil Cooler Fan Controllers and Thermostats

   a. GENERAL.

   (1) *Controllers.* Two controllers are provided for the oil cooler fans. The transmission oil-cooler-fan controller (fig. 221) controls the left fan and is mounted on the left side of the transmission. The engine oil-cooler-fan controller (fig. 232) controls the right fan and is mounted on front side of right cooler. Each controller contains two thermostatic switches and a resistor (fig. 202) and is electrically connected to the respective fan clutch (fig. 237). The controllers are set to start and increase fan speed as oil temperature is increased. The right fan will start at low speed when engine oil temperature reaches 175° F., and changes to full speed when the temperature is increased to 195° F. The fan will return to low speed when temperature is reduced to 165° F. The left fan will start at low speed when transmission oil temperature reaches 210° F., and changes to full speed when temperature reaches 255° F. The fan will return to low speed when temperature is reduced to 225° F., and stop at 180° F.

   (2) *Thermostats.* Two thermostats are provided, one on each side of the transmission (fig. 220). At temperatures under 180° F., the oil is bypassed back to the transmission. At 180° F., the thermostats open, allowing the oil to flow through the oil coolers.

   b. REPLACEMENT OF THERMOSTATIC SWITCHES. Thermostatic switches in both the engine oil cooler fan controller and the transmission oil cooler fan controller are replaced in the same manner. Turn off master relay switch (fig. 18). Open engine compartment grille doors (fig. 101). Loosen four bolts securing controller cover and remove cover (fig. 202). Remove two mounting screws securing each switch in housing and lift switches from housing (fig. 203). Remove terminal screws and cables from one thermostatic switch at a time and install cables on new switch. Position switches in housing and secure each with two screws. Make sure cover gasket is in good condition and positioned in groove in top of housing (fig. 203). Secure cover with four bolts. Close engine compartment grille doors (fig. 101).

   c. REPLACEMENT OF THERMOSTATS. Open engine compartment rear grille doors (fig. 101). Unscrew thermostat (fig. 221) and remove from housing. Position new thermostat and screw into housing. Close engine compartment rear grille doors (fig. 101).

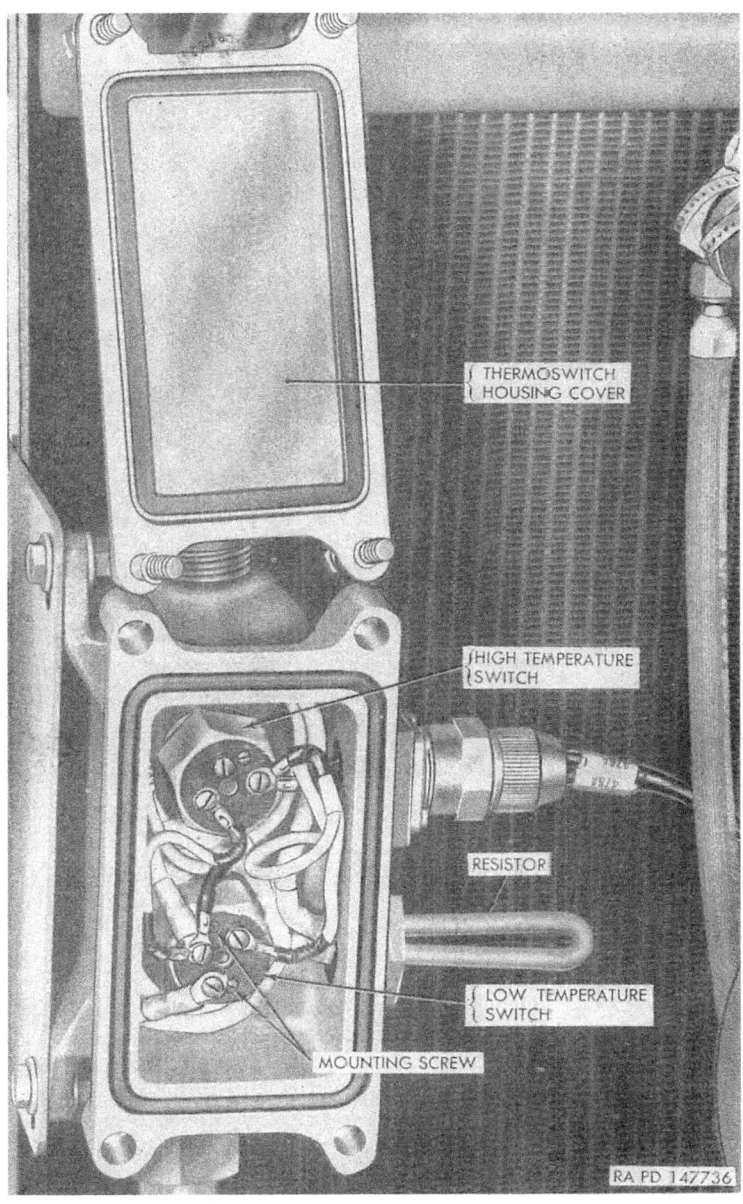

Figure 202. Engine oil cooler fan controller—cover removed.

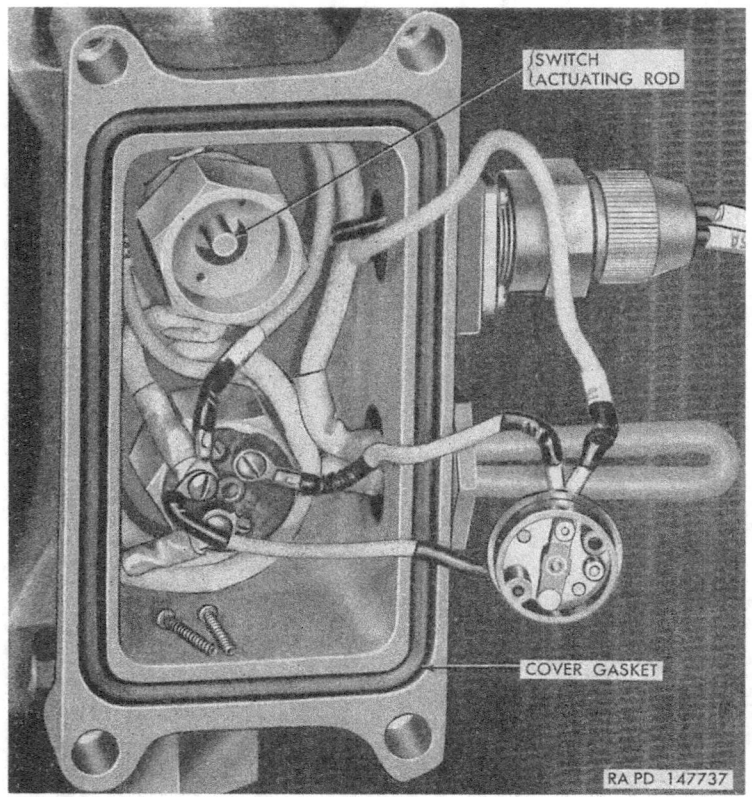

Figure 203. Engine oil cooler fan controller—low temperature switch removed.

Section XIV. BATTERIES AND GENERATING SYSTEM

221. Description and Data

a. DESCRIPTION.

(1) *System.* The batteries and generating system (figs. 204 and 205) consist of two sets of batteries, each containing two batteries; a main and an auxiliary generator circuit each including a 150-ampere, 24-volt generator with voltage and current regulators and excessive-reverse-current warning signal light; and necessary control and wiring circuits. The main generator is mounted on the right side of the accessory case (fig. 206) and is gear-driven by the engine. The auxiliary generator is mounted on the lower-left side of the auxiliary engine (fig. 308) and is gear-driven by the auxiliary engine.

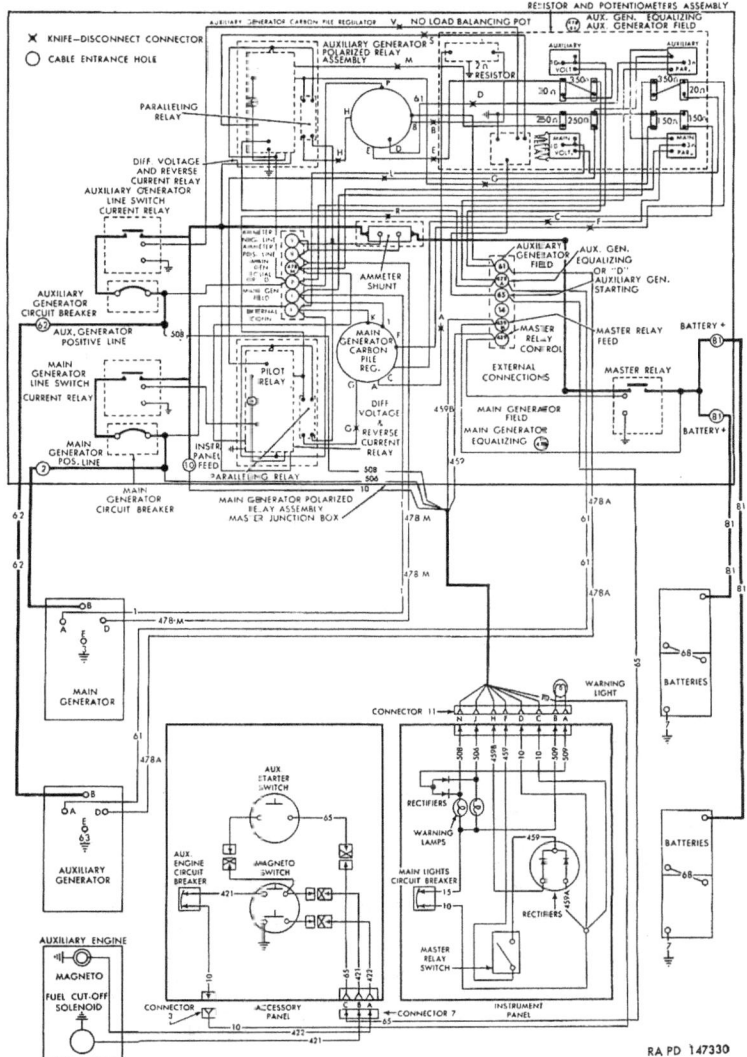

Figure 204. Generating system wiring diagram.

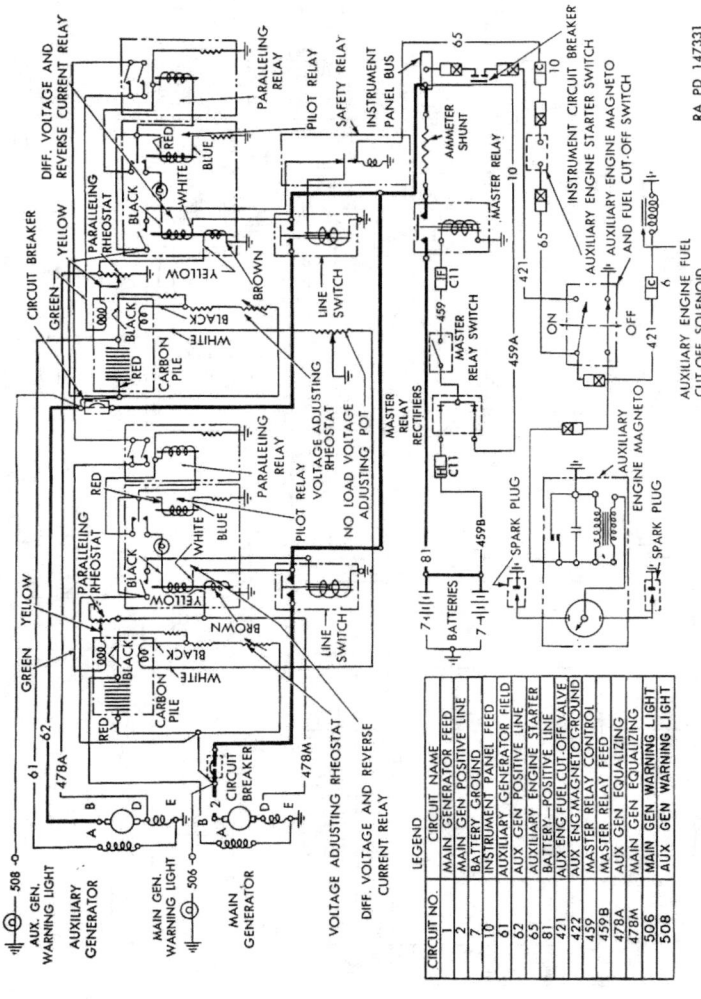

Figure 205. Generating system schematic diagram.

When necessary, both generators may be operated at the same time. A slave battery receptacle (fig. 76) is provided for connecting an external power source or a battery-charger. Refer to paragraph 98 for battery-charging procedure.

(2) *Generators* (figs. 206 and 308). Both main and auxiliary generators are 150-ampere, 24-volt, 4-pole, 4-brush, field-regulated units utilizing interpole and compensating windings. Each generator consists of five basic components: armature, commutator, fan and slip clutch, yoke and field coil, and terminal board. Positive, negative, field-circuit, and equalizing circuit cables terminate at a terminal board mounted on the fan end of the generator.

(3) *Generator regulating units.*

  (a) *General.* Two carbon pile voltage regulators (one for each generator), two polarized relays (one for each generator), and a resistor and potentiometer assembly, all mounted in the master junction box (fig. 184), comprise the generator regulating units. When both generators are operating, this system also distributes the load equally between the two generators.

  (b) *Voltage regulators* (fig. 184). Two carbon pile voltage regulators (one for each generator) function to maintain constant voltage by regulating the current in the field coils. When both generators are operating, the voltage regulators circuit combines with the paralleling relay circuit to equalize the loads on the generators.

  (c) *Polarized relays* (fig. 184). Each generator has a polarized relay assembly for the control of its operation. Each assembly consists of three relays: a differential voltage and reverse current relay, a paralleling relay, and a pilot relay. The differential voltage and reverse current relay connects its generator to the batteries if the battery voltage is low and disconnects its generator from the batteries if the generator reverse current is high. The paralleling relay permits an equalizing of the loads when both generators are operating. The pilot relay controls the action of the other relays.

  (d) *Resistor and potentiometer assembly* (fig. 184). This unit houses all the fixed and variable resistors of the master junction box.

(4) *Main- and auxiliary-generator current relays.* These current relays or line switches control the application of the output voltage to the load (fig. 184). Each relay is energized by its differential voltage and reverse current relay.

When the generator voltage exceeds the battery voltage by about one-half volt, the line switch closes, thus connecting the generator to the batteries. When a reverse current of about 20 amperes flows from the batteries to the generator, the line switch opens, disconnecting the generator from the batteries. The auxiliary-generator line switch also functions as a starting switch when the auxiliary-engine magneto switch (fig. 18) is turned on and auxiliary-engine starter button is pressed. This energizes the safety relay (figs. 185 and 186) which connects the line switch directly into the battery circuit. When the line switch is energized, the contacts close, connecting the auxiliary generator to the batteries. The generator then operates as a motor, turning the auxiliary engine for cranking purposes.

(5) *Batteries.* There are two sets of batteries supplying electrical current for the vehicle. In each set, two 12-volt, lead-acid-type batteries are connected in series by cable (68). One set is located in the top-front center of the engine compartment (fig. 101) directly in back of the turret, and feeds current to the master junction box through cable (81). The other set is located in the left side of the engine compartment (figs. 209 and 210) on top of the fuel tank and feeds current to the master junction box through another cable (81) and to the crew compartment heater control box in the hull accessories panel through cable (400). Negative terminal posts on both sets of batteries are grounded to the hull by heavy electrical cables (7) (fig. 210).

b. TABULATED DATA.

Batteries	type 6TN 12 volt
Generators	Eclipse-Pioneer (EC–30E00–3–A)
Voltage regulators	Eclipse-Pioneer (EC–104R–17–A)
Resistors and potentiometers	Eclipse-Pioneer
Polarized relays	Eclipse-Pioneer
Current relays (line switches)	Eclipse-Pioneer

222. Generator (Main Engine)

a. REMOVAL OF GENERATOR. Remove top deck (par. 158). Remove power plant (par. 159). Remove two cotter pins, nuts, and lock washers securing four cables and ground strap to generator terminal posts (fig. 206) and remove cables and ground strap. Remove locking wire from five mounting bolts (fig. 206) securing generator mounting plate to accessory case. Support generator and remove the five bolts and washers. Pull generator straight out until driving spline is clear.

Note. On some engines the generators may be installed with straight-through-type mounting bolts. In these cases, remove six jam nuts and washers securing generator to mounting studs, using wrench 41-W-1496-625.

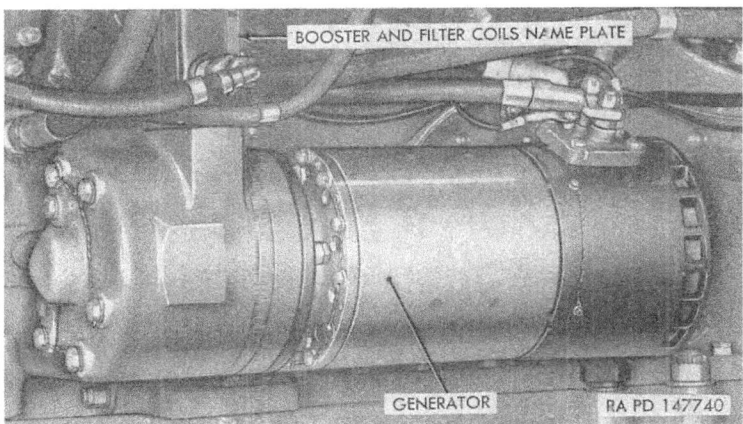

Figure 206. Generator installed on main engine.

b. BRUSH INSPECTION AND REPLACEMENT (fig. 207). Cut locking wire from screws holding commutator end cover and remove five screws and lock washers. Remove cover by tapping it gently with a soft-faced mallet. Inspect brushes for wear. If they are burned or so short that the brush springs rest almost on the brush holder, replace the brushes.

Note. The minimum length of brush is nine-sixteenths of an inch. Replace worn brushes before their wear limit is reached to assure satisfactory operation until the next brush inspection.

Inspect commutator. If badly worn or burned, replace generator. Install four brushes in brush holder and fasten brush terminals to brush holder with four screws and lock washers. Install commutator end cover and secure with five screws and lock washers. Lock with locking wire.

c. REMOVAL OF GENERATOR ADAPTER.

(1) *General.* If any oil leakage into the generator or trouble in the generator drive shaft is evidenced, the generator adapter must be removed for access to gasket at accessory case mounting and to drive shaft. Adapter is close-fitted into accessory case.

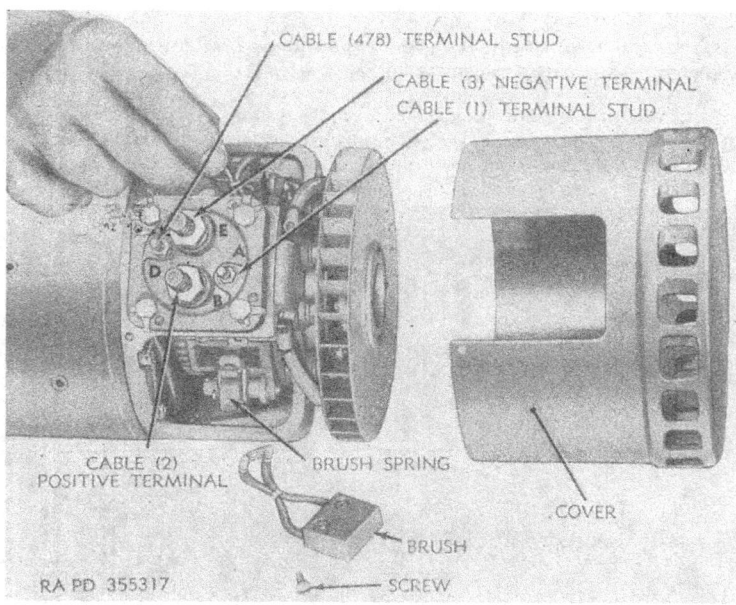

Figure 207. Installing brushes in generator.

(2) *Coordination with ordnance maintenance unit.* Replacement of generator adapter is normally an ordnance maintenance operation, but may be performed in an emergency, for purposes explained in (1) above, by using organization, provided approval for this replacement is obtained from supporting ordnance officer. Tools required for the operation which are not carried in the using organization may be obtained from supporting maintenance unit.

(3) *Removal of adapter.* Remove two screws securing adapter to accessory case. Install puller 41–P–2906–280 into the two tapped holes provided and remove adapter.

Note. Generator drive shaft will pull out with adapter.

d. INSTALLATION OF GENERATOR.

(1) If mounting plate was removed from generator, install it with six lock washers, nuts, and jam nuts. Use a new gasket between generator mounting plate and mounting head. If generator drive shaft adapter was removed, install a new gasket between accessory case and adapter, and secure adapter to accessory case with two screws. Install generator drive shaft oil seal.

(2) Record replacement on DA AGO Form No. 478.

(3) Position generator to accessory case by alining dowel pin in upper right hole in mounting plate. Install generator to accessory case by tightening five bolts and lock washers. Lock bolts with locking wire. Install four cables to terminal (fig. 207) as follows: cable (3) and ground strap on terminal "E," cable (1) on stud "A," cable (478) on stud "D," and cable (2) on terminal "B." Secure two small cables with serrated nuts, lock washers, and cotter pins. Secure two large cables with nuts, lock washers, and cotter pins. Install power plant (par. 160). Install top deck (par. 161).

223. Generator Regulating Units

Because of the complexity of the adjustments and the instruments required, the using organization is not permitted to replace any of the generator regulating units. Should one of them be suspected of failure, notify the ordnance maintenance personnel.

224. Batteries

a. SERVICE.

(1) *Specific gravity tests.* Specific gravity testing of battery fluid determines state of charge in each battery cell. Use a hydrometer and thermometer and correct hydrometer reading for temperature as indicated on temperature correction chart (fig. 208). To determine state of charge of a battery, hydrometer readings must be corrected to a temperature of 80° F. For example, if specific gravity reading is 1.280 at 0° F., the correct reading is 1.280 to 0.032 (fig. 208) or 1.248. A corrected specific gravity of 1.285 in each cell indicates a fully charged battery. A corrected specific gravity of 1.225 or less in each cell indicates that the battery must be recharged or replaced.

Table VII. Battery Freezing Temperature

Battery condition	Actual specific gravity	Freezing temperature
Fully charged	1.285	−96° F.
One-third discharged	1.255	−60° F.
One-half discharged	1.220	−31° F.
Three-fourths discharged	1.185	−8° F.
Normally discharged	1.150	+5° F.
Fully discharged	1.100	+18° F.

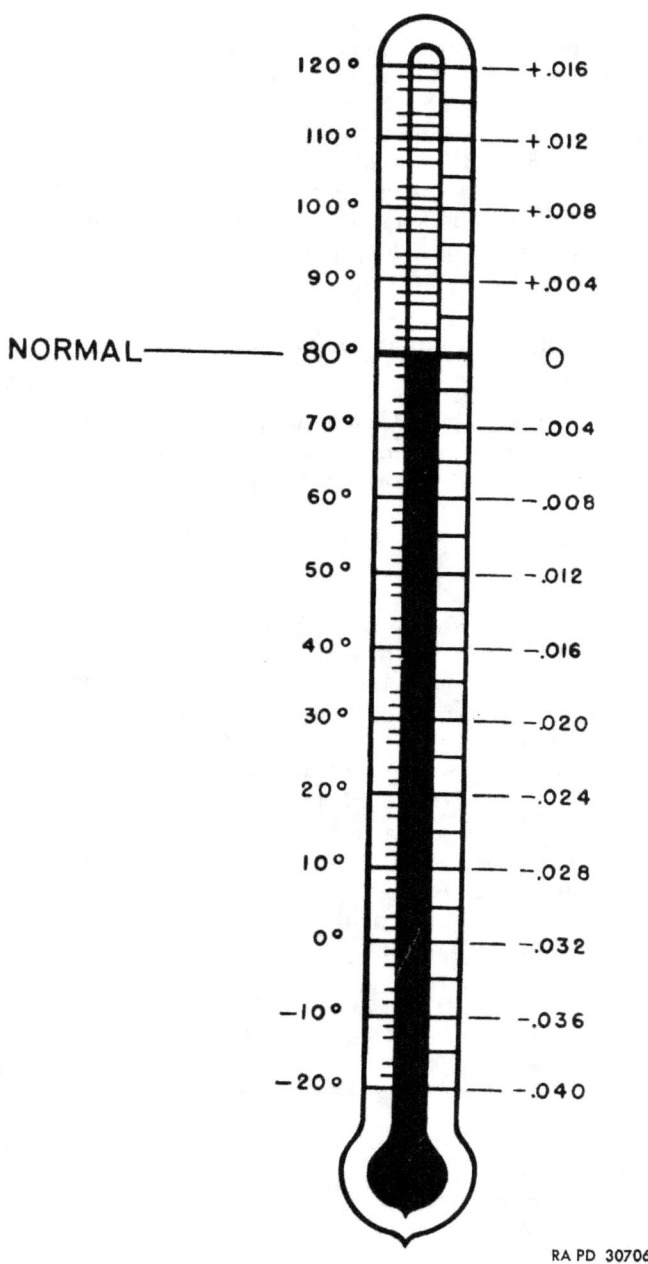

Figure 208. Specific gravity temperature correction chart.

(2) *Adding water.* Water in the battery fluid will evaporate at high temperature or with excessive charging rates. Inspect fluid level and add water when necessary to keep level about three-eighths of an inch above plates.

Caution: Do not overfill.

(3) *Cleaning battery terminals.* The top of the battery must be kept clean. Tighten vent plugs and clean battery with a brush dipped in an alkaline solution such as ammonia or a solution of bicarbonate of soda and water. After foaming stops, flush top of battery with clean water. If terminals and cable clamps are corroded, disconnect cable and clean in same manner as described above for top of battery. Corrosion can be retarded by applying a heavy coating of general-purpose grease (GAA).

b. REMOVAL.

(1) *Center batteries.* Transverse turret (par. 76) to obtain accessibility to center battery box cover (fig. 101). Turn off master relay switch (fig. 18). Loosen two hold-down clamp screws and swing clamps clear of battery cover. Lift off cover. Disconnect ground cable (7). Disconnect positive cable (81) and one side of cable (68). Remove two battery clamp hook nuts. Remove hooks and clamp. Lift batteries, one at a time, out of vehicle.

(2) *Side batteries* (fig. 209 and 210). Turn off master relay switch (fig. 18). Lift up five engine compartment grille doors (fig. 101). Unscrew four anchor hook wing nuts enough so that hooks can be turned one-quarter turn to loosen hooks from pins.

Note. Do not remove wing nuts because hooks on hull side will drop behind fuel tank.

Lift cover and four hooks off batteries. Disconnect ground cable (7). Disconnect positive cables (81) and (400) and one side of interconnecting cable (68). Remove battery hold-down frame and lift batteries, one at a time, out of vehicle.

c. INSTALLATION.

(1) *Center batteries.* Position batteries in battery box so that positive terminals are toward front of vehicle. Connect positive cable (81) to left battery positive terminal post. Connect interconnecting cable (68) to right battery positive post and left battery negative post. Connect ground cable (7) to right battery negative post. Test battery installation (*d* below). Install battery clamp, clamp hooks, and nuts. Position battery cover over dowel pins and secure by tightening hold-down clamp screws. Install center battery box cover on top deck (fig. 101).

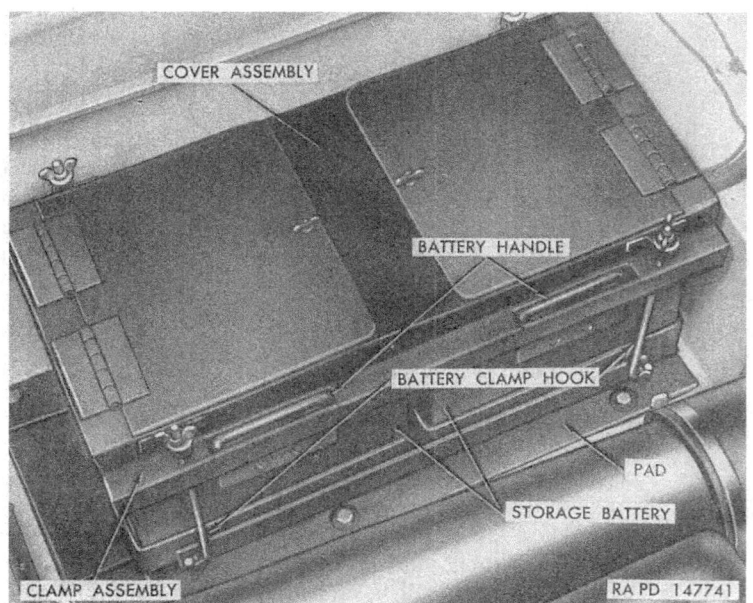

Figure 209. Left side batteries—installed view.

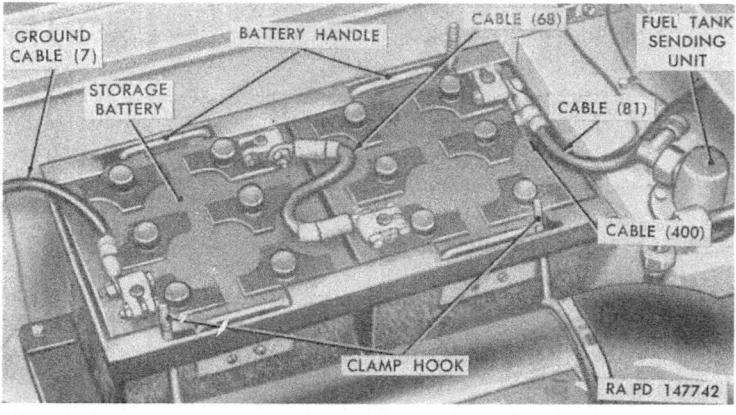

Figure 210. Left side batteries with cover removed.

(2) *Side batteries* (figs. 209 and 210). Position batteries on battery base plate so that positive terminal posts are toward left side of hull. Connect positive cables (81) and (400) to front battery positive terminal post. Connect interconnecting cable (68) to front battery negative post and rear battery positive post. Connect ground cable (7) to rear battery negative post. Test battery installation (*d* below). Close five engine compartment left grille doors (par. 101).

d. TESTING BATTERY INSTALLATION. Turn on master relay switch (fig. 18). Main warning lamp (fig. 21) should come on immediately. If it does not, either warning lamp is defective or batteries are not connected properly. Check connection in accordance with instructions given in paragraphs *c*(1) and (2) above.

Section XV. HORN AND LIGHTING SYSTEM

225. Description

a. GENERAL. The horn and lighting system (figs. 211 and 212) is comprised of the following units: two front light units, each consisting of a service headlight, blackout driving light, and blackout marker light; two taillight units, each consisting of a service taillight and a blackout marker light; two hull dome lights; and a horn and horn switch.

Note. Early models of this vehicle contain only one blackout driving light (fig. 6).

Cables from all these units enter the instrument panel through connectors (10). Current is supplied from the batteries through the master relay and two circuit breakers into the switches which control the horn and the lights. In addition, service headlight and blackout driving light cables feed through a dimmer switch (figs. 211 and 212).

b. HORN. The horn (figs. 213 and 214) is a vibrator-type unit mounted on the right fender to the right of and behind the right headlight. A bracket serves as a protective guard.

c. SERVICE HEADLIGHTS. The service headlights have sealed beam units with "HI" and "LO" beam filaments in each unit. They are mounted on the right and left sides of the hull just inside the fenders (fig. 215). A bracket serves as a protective guard for each headlight assembly.

d. BLACKOUT DRIVING LIGHTS. A sealed beam blackout driving light with "HI" and "LO" beam filaments in each lamp is mounted beside each service headlight (fig. 215).

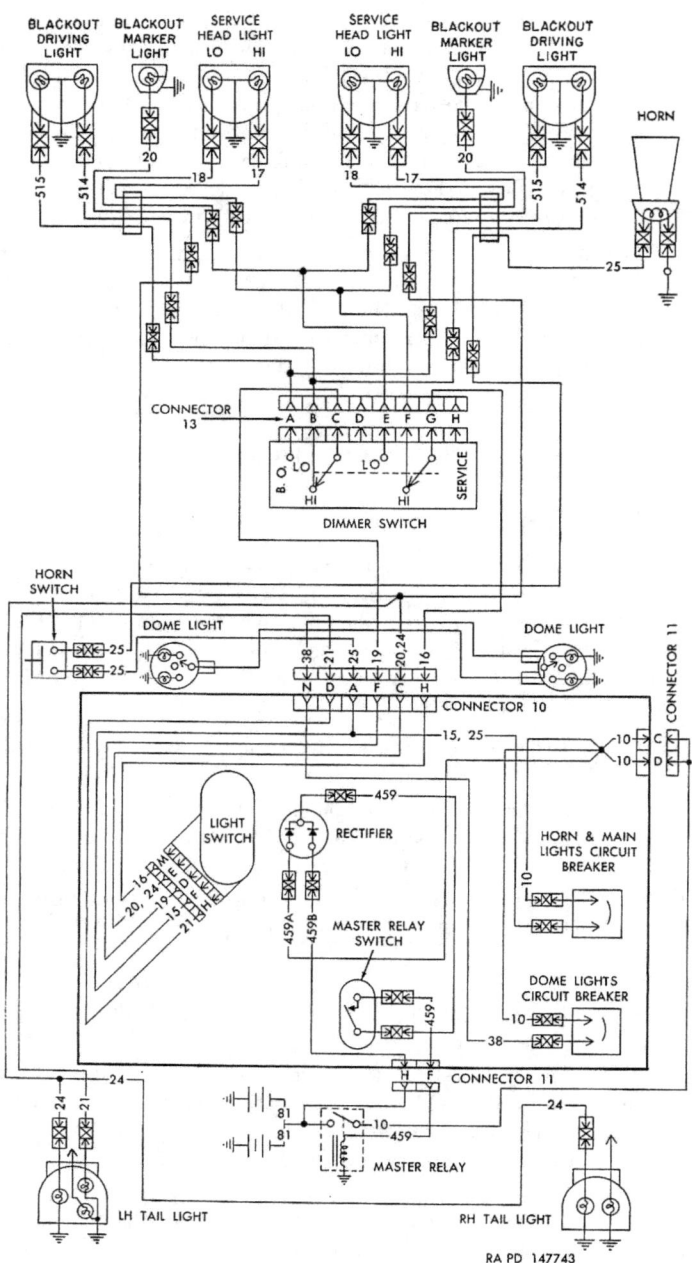

Figure 211. Horn and lighting system wiring diagram.

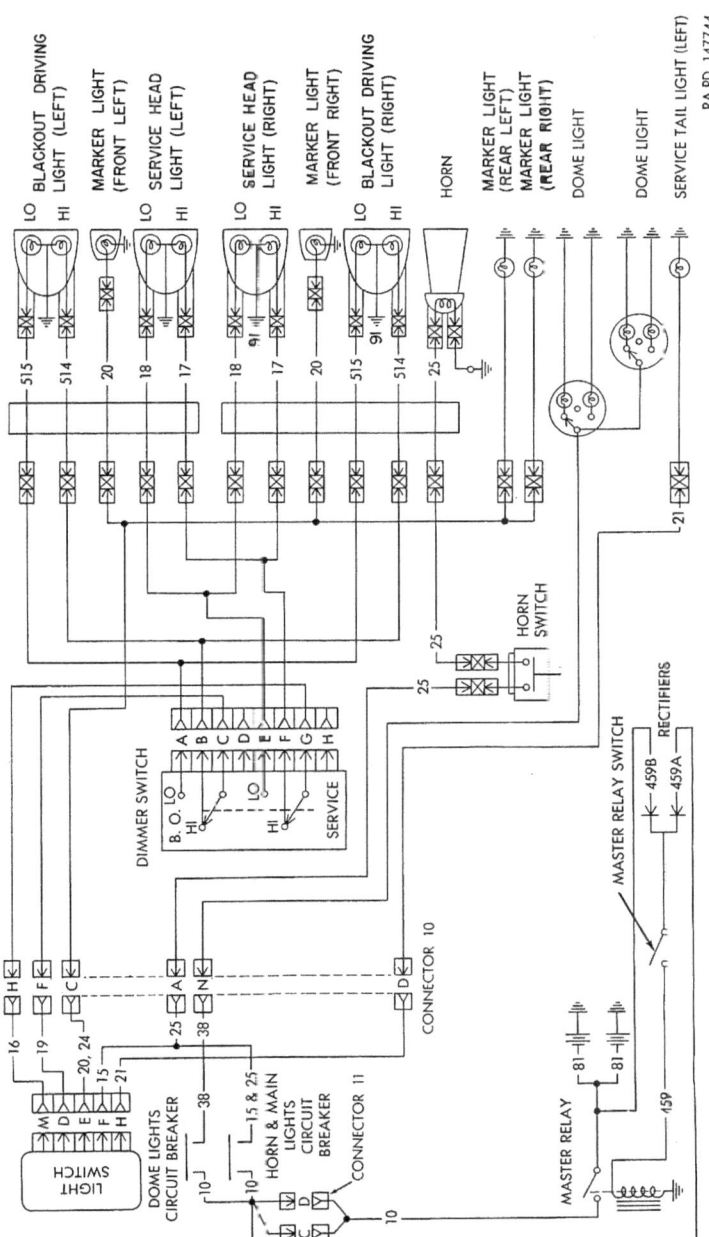

Figure 212. Horn and lighting system schematic diagram.

e. BLACKOUT MARKER LIGHTS. A blackout marker light is mounted beside and below each service headlight (fig. 215).

f. TAILLIGHTS. The two taillight assemblies are mounted on the right and left sides at the rear of the hull, just above the fenders and adjacent to the towing hooks. Each assembly contains a service taillight section and a blackout marker light section. Both sections are used in the left assembly but only the marker light section is used in the right assembly.

g. HULL DOME LIGHTS. Each dome light (fig. 219) has a partition dividing it into two sections, one section having a white lens and the other a red lens. Each section is equipped with a separate lamp. The lamp in the white lens section is turned on for regular use and the lamp in the red lens section is turned on for use in blackout areas. Each dome light has a switch handle with a safety button which must be pressed before the switch can be turned to the white lens section.

226. Horn

a. REMOVAL (fig. 213). Turn off master relay switch (fig. 18). Disconnect two cables at quick-disconnects at horn. Remove two cap screws and lock washers holding horn to mounting bracket on top of fender and remove horn.

b. INSTALLATION (fig. 213). Position horn on mounting bracket and secure with two cap screws and lock washers. Connect two cables at quick-disconnect connectors.

c. ADJUSTMENT (fig. 214). Loosen lock nut on rear cover clamp screw and loosen screw. Remove cover clamp and twist cover out of the way so that tone-adjusting nut and lock nut can be loosened. Press horn switch (fig. 21) and turn tone-adjusting nut until satisfactory tone is attained. Then release switch and tighten lock nut. Place cover in position and install cover clamp. Tighten clamp screw and lock in position with lock nut.

227. Horn Switch
  (fig. 21)

a. REMOVAL. Turn off master relay switch (fig. 18). Remove two cables (25) from switch by disconnecting two connectors. Remove switch by removing two safety nuts securing switch to mounting bracket.

b. INSTALLATION. Position switch to bracket and secure with two safety nuts. Connect two cables (25) at connectors.

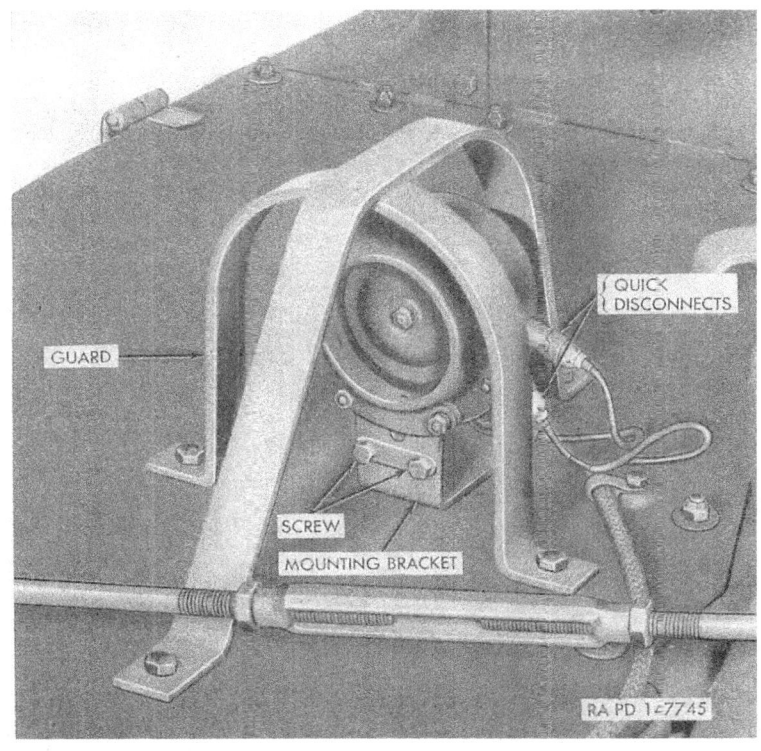

Figure 213. Horn—installed view.

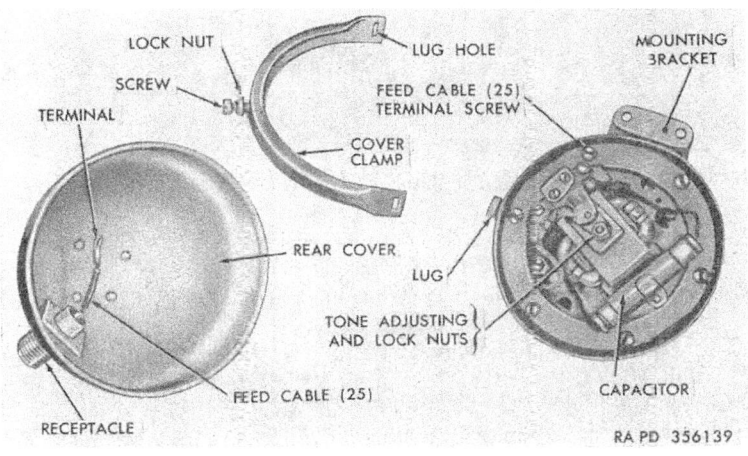

Figure 214. Horn—cover removed.

228. Service Headlights

a. REMOVAL. Turn light switch (fig. 18) to "OFF" position. Disconnect cables (17, 18, and 91) from service headlight (figs. 211, 215, and 216) at bell-type connectors. Remove mounting nut and lock washer at base of light assembly and remove headlight.

b. INSTALLATION. Position service headlight on mounting bracket (figs. 215 and 216) and secure lightly with lock washer and mounting

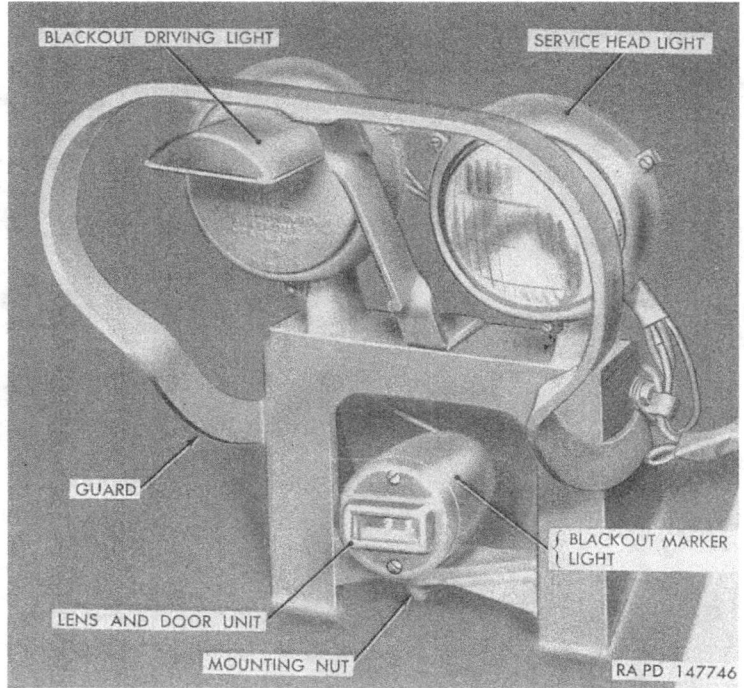

Figure 215. Left service headlight, blackout driving light, and blackout marker light.

nut. Swing headlight unit for proper beam direction and fasten in place by tightening mounting nut. Connect cables (17, 18, and 91) at connectors at rear of headlight unit.

c. REPLACEMENT OF SERVICE HEADLIGHT LAMP UNIT. Turn light switch (fig. 18) to "OFF" position. Remove three screws holding service headlight door to body and pull door and lamp-unit assembly from body. Disconnect two cables at connectors in body and remove door and lamp-unit assembly. Remove three lamp-unit retaining springs and remove lamp-unit from door. Position new lamp-unit in door, making sure lug riveted to lamp-unit engages clip inside of door. Install one lamp-unit retaining spring at top and one on each side to hold lamp-unit in door. Connect cables at connectors inside body and place connectors in clips provided inside of body. Position door and lamp-unit assembly on body, aline holes, and secure with three screws.

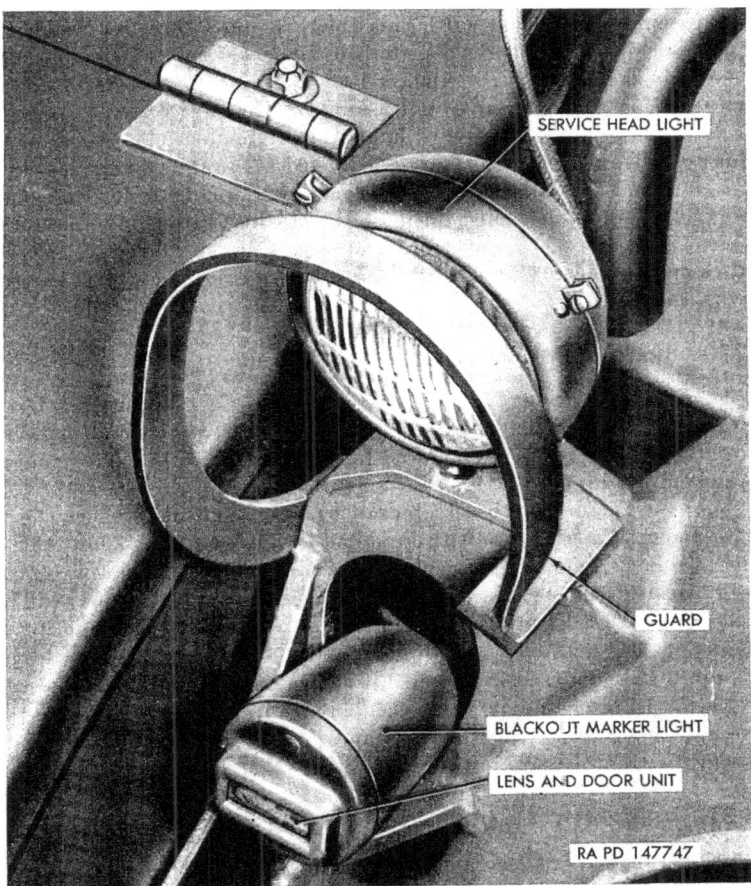

Figure 216. Right service headlight and blackout marker light.

229. Blackout Driving Lights

a. REMOVAL. Turn light switch (fig. 18) to "OFF" position. Disconnect cables (514, 515, and 91) at bell-type connectors at rear of blackout driving light. Remove mounting nut and lock washer at base of light (fig. 215) and remove blackout driving light.

387

b. INSTALLATION. Position blackout driving light on bracket (fig. 215) and secure lightly with lock washer and mounting nut. Swing light for proper beam direction and fasten in place by tightening mounting nut. Connect cables (514, 515, and 91) at bell-type connectors at rear of light.

c. REPLACEMENT OF BLACKOUT DRIVING LIGHT LAMP-UNIT. Turn light switch (fig. 18) to "OFF" position. Remove three screws holding blackout driving light door to body and pull door and lamp-unit assembly from body. Disconnect two cables at connectors in body and remove door and lamp-unit assembly. Remove three lamp-unit retaining springs and remove lamp-unit from door. Position new lamp-unit in door, making sure lug riveted to lamp-unit engages clip inside door. Install one lamp-unit retaining spring at top and one on each side to hold lamp-unit in door. Connect cables at connectors inside body and place connectors in clips provided inside body. Position door and lamp-unit assembly on body, aline holes, and secure with three screws.

230. Blackout Marker Lights

a. REMOVAL. Turn light switch (fig. 18) to "OFF" position. Disconnect cable (20) at bell-type connector at rear of blackout marker light. Remove mounting nut and lock washer (fig. 215) and remove blackout marker light.

b. INSTALLATION. Position blackout marker light on bracket (figs. 215 and 216) and secure with lock washer and mounting nut. Connect cable (20) at bell-type connector at rear of blackout marker light.

c. REPLACEMENT OF BLACKOUT MARKER LAMP. Turn light switch (fig. 18) to "OFF" position. Remove two screws (figs. 215 and 216) securing lens and door unit to body and pull off door. Press in on lamp, turn counterclockwise until it is free of socket, and pull out of socket. Insert new lamp in socket, press in, and turn clockwise to secure in position. Position lens and door unit on light body and secure with two retaining screws.

231. Taillights

a. REMOVAL (figs. 217 and 218). Turn light switch (fig. 18) to "OFF" position. Remove taillight guards by removing four cap screws attaching each guard to hull brackets. Remove two cap screws securing each taillight assembly to side of hull. Disconnect cables at bell-type connectors (one cable at right taillight assembly and two cables at left taillight assembly).

b. INSTALLATION (figs. 217 and 218). Connect cables at bell-type connectors (one-cable connector at right taillight assembly and two-cable connector at left taillight assembly). Position each taillight assembly to hull and secure with two cap screws and lock washers.

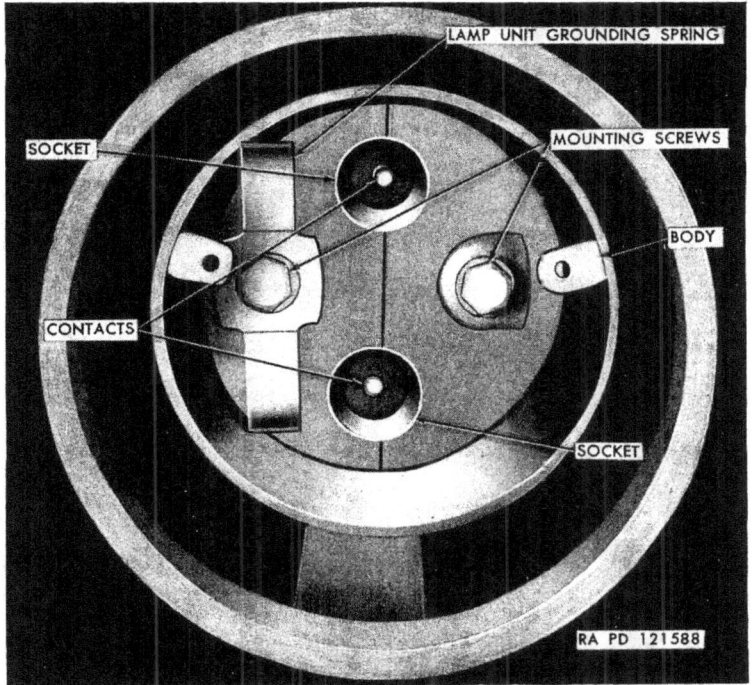

Figure 217. Right taillight body—installed view.

Make sure grounding springs are installed under correct cap screws. Position taillight guards and secure each with four cap screws.

c. REPLACEMENT OF TAIL LAMP UNITS. Remove two screws securing taillight door to light body. Remove door and pull out the two lamp-units. Insert blackout marker lamp in lower sockets of both right and left taillights. Insert service taillight unit in upper socket of left taillight and blackout stop light unit in upper socket of right taillight.

Note. The blackout stop light unit in the right taillight is not used.

Position doors over light bodies and secure each with two screws.

232. Dome Lights

a. REMOVAL (fig. 219). Turn off master relay switch (fig. 18). Remove screw in center of switch handle and remove handle. Remove six screws from door and remove door and lens. Remove terminal screw and free feed cables from terminal. Remove cable nut, washer, ferrule, and rubber bushing and pull cable out of light housing. Remove three mounting nuts and lock washers and remove dome light.

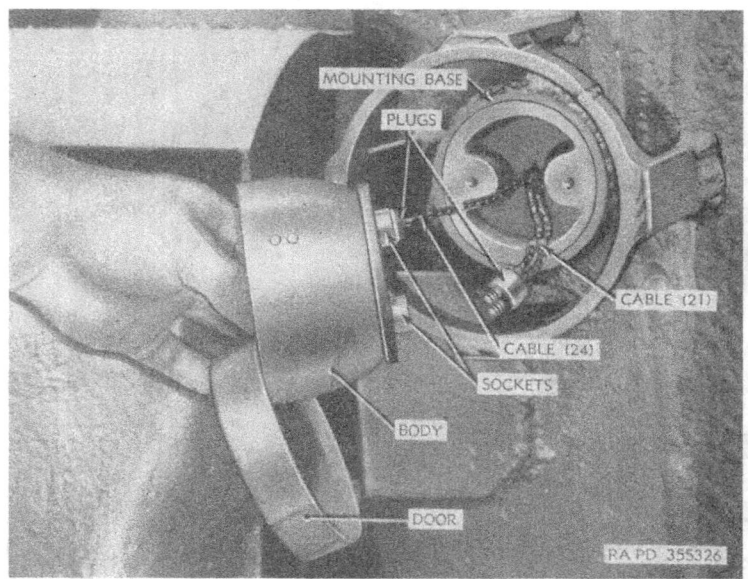

Figure 218. Removing left taillight from mounting ring.

b. INSTALLATION (fig. 219). Position dome light on mounting bracket studs and secure with three lock washers and nuts.

Note. Be sure grounding spring is attached under one of the mounting nuts.

If a new dome light is being installed, door must be removed to insert cables. To do this, remove screw in center of switch handle and remove handle. Remove six screws from door and remove door with lens. Make sure that cable nut, washer, furrule, and rubber bushing are installed on cable. Insert cable through hole in side of light body.

Note. One of the dome lights requires two cables. Install each as above.

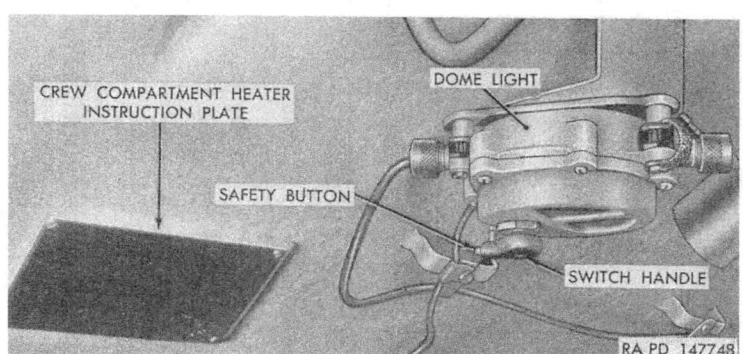

Figure 219. Hull dome light—installed view.

Connect cables to switch terminal nearest outer edge of light body. Make sure two lamps are in position in sockets. Install door with six screws. Install switch handle with screw.

c. DOME LAMP REPLACEMENT. Remove screw in center of switch handle and remove handle. Remove six screws securing door to light body and remove door. Press lamp in, turn counterclockwise, and pull lamp out of socket. To install new lamp, press lamp into socket, and turn clockwise until secure. Install door with six screws. Install switch handle with screw.

233. Headlamp Dimmer Switch

a. REMOVAL. Remove two cable connector plugs from rear of headlamp dimmer switch. Remove three lock washer screws and remove switch.

b. INSTALLATION. Position switch to mounting bracket on hull floor in front of driver and secure with three lock washer screws. Connect two cable connector plugs to matching receptacles at rear of switch.

Section XVI. TRANSMISSION

234. Description and Data

a. DESCRIPTION. The cross-drive transmission (figs. 220 and 221) used in this vehicle is a combined transmission and steering unit. It transmits power directly from the engine to the final drives and track drive sprockets. The transmission is controlled by the driver by means of a manual control lever (fig. 10), service brake pedal (fig. 9), and a hand brake lock handle (fig. 10).

(1) Characteristics of the cross-drive transmission include hydraulic torque converter, split torque drive, variable steering, and built-in disk-type service brakes. It delivers engine power to the vehicle final drive, at an output torque automatically varying according to driven-load conditions.

(2) There are two forward speed ranges and one reverse. Steering is possible in all drive ranges and in neutral. Steering in neutral causes the vehicle to spin in place, the tracks turning in opposite directions.

(3) Accessory drive gears in the transmission provide power to drive fans for cooling the oil in the oil coolers.

(4) Supplied with each transmission is a manual control box installed in the drivers' compartment and connected to the cross-drive transmission through mechanical linkage. The driver controls both steering and drive range shifting by operating the manual control lever on the manual control box.

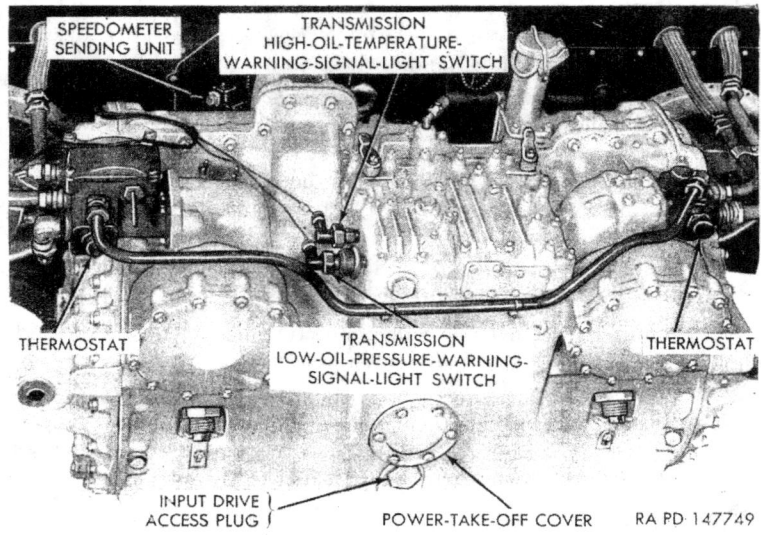

Figure 220. Transmission—rear view.

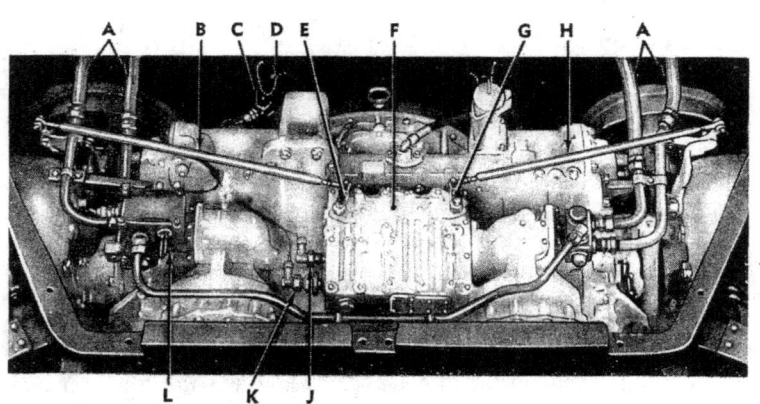

A—TRANSMISSION OIL COOLER LINES
B—STEERING CONTROL ROD
C—SPEEDOMETER SENDING UNIT
D—SPARK PLUG ACCESS OPENING COVER
E—STEERING CONTROL LEVER
F—TRANSMISSION VALVE BODY
G—SHIFTING CONTROL LEVER
H—SHIFTING CONTROL ROD
J—HIGH-OIL-TEMPERATURE-WARNING-SIGNAL-LIGHT SWITCH
K—LOW-OIL-PRESSURE-WARNING-SIGNAL-LIGHT SWITCH
L—TRANSMISSION OIL COOLER FAN CONTROLLER

Figure 221. Transmission—installed view.

b. TABULATED DATA.

Model	Allison CD-850-4
Type	cross-drive hydraulic
Weight, dry	2,946 lb
Length	30 in
Height	40⅜ in
Width, over drive flanges (aprx)	40 in
Suspension	3-point (engine and two mounting pads)
Drive ranges	low, high, and reverse
Drive range shift and steering control (external)	mechanical
Shift and steering mechanism (internal control)	hydraulic
Low range control	friction band
High range control	multidisk clutch
Reverse range control	friction band
Clutches engaged by	oil pressure
Clutches released by	spring pressure
Bands applied by	oil pressure (in servos)
Bands released by	spring pressure (in servos)
Service (and parking) brakes	2
Type	wet multiple-disk external adjustment
Application	manual
Cooled by	oil
Oil pumps	pressure, 1 scavenge
Type	gear
Driven by	3 by input shaft, 2 by output shafts
Hydraulic control system pressures	see table VII
Oil cooled by	external radiators
Oil filter type	air maze, double

235. Transmission Hydraulic and Lubricating System

a. DESCRIPTION.

(1) *Hydraulic system.*

  (*a*) The transmission hydraulic system functions in four capacities: as the applying force for control clutches, servos, and brakes; as a driving medium in the torque converter; as a lubricant; and as a cooling medium for the entire transmission, including the torque converter, clutches, and brakes.

  (*b*) Oil is pumped from the transmission pump through screened pickup tubes and forced through a filter to the valve body which is mounted on top of the rear housing.

  (*c*) The valve body contains one shifter valve and two steering clutch apply valves which are operated through external levers controlled by mechanical linkage from the driver's compartment. The valves direct the flow of oil to the clutches and servos. The two steering clutch apply valves are controlled by an internal cross lever which pulls one valve plunger into a closed position while it depresses a plunger to open the other valve.

(d) Manual movement of the range shifter valve causes main line oil pressure to be directed to the selected actuating piston or servo unit.

(2) *Lubrication system.* The cross-drive transmission is completely lubricated by the supply of oil poured into the sump through the oil filler neck under the transmission oil filler cap (fig. 229). The oil level should be maintained close to the "FULL" mark on the oil level gage. Check the oil level as indicated in *b* below.

b. CHECKING OIL LEVEL. Transmission oil level is checked by the oil level gage (bayonet type) located in the oil filler neck under the transmission oil filler cap (fig. 229).

Note. The oil level must be checked before stopping the engine, when the torque converter and oil coolers are filled with oil.

Open the engine compartment right rear grille doors (fig. 101). Remove transmission oil filler cap, withdraw oil level gage, and wipe with a clean cloth. Insert gage and withdraw to check oil level. The oil level should be maintained at, or slightly above, the "FULL" mark and in no case should fall below the "MINIMUM" mark on the oil level gage. If gage indicates oil level is low, replenish with oil as specified on the lubrication order (fig. 86). Insert oil level gage, close oil filler cap, and lock. Close three engine compartment right rear grille doors.

c. REPLACING TRANSMISSION OIL.

(1) *Drain transmission oil.*

Note. Drain only when transmission oil is at operating temperature.

Remove the transmission drain access hole covers (fig. 285) from beneath the vehicle hull, and remove oil drain plugs (fig. 223) from each side of transmission (fig. 222).

Note. Drain transmission into clean containers with at least 25 gallons total capacity.

Remove oil cooler service hole covers from beneath vehicle hull (fig. 285). Remove oil line at bottom of left oil cooler and disconnect transmission oil line at bottom of right oil cooler (fig. 222). Open engine compartment left and right rear grille doors (fig. 101). Remove three vent plugs from top of left oil cooler (fig. 233) and one vent plug from rear core of right oil cooler (fig. 234) to permit entry of air and allow oil to drain into clean containers.

(2) *Fill transmission with oil.* Install drain plugs in transmission, install oil line at bottom of left oil cooler, and connect transmission oil line at bottom of right oil cooler. Install covers on hull bottom. Install vent plugs at top of oil cooler cores. Using oil as specified by the lubrication order (fig.

86), remove transmission oil filler cap (fig. 229) and oil level gage, and pour approximately 23 gallons of oil through the oil filler neck. Start engine and run for 2 or 3 minutes with manual control lever (fig. 10) in neutral position, to fill the torque converter and oil coolers. With the engine running add or drain oil as necessary to bring the oil level to or slightly above the "FULL" mark on the oil level gage. Install oil level gage and transmission oil filler cap. Close engine left and right compartment rear grille doors (fig. 101).

d. CHECK TRANSMISSION OIL PRESSURE. Transmission oil pressure may be checked, using the transmission oil pressure gage 45–G–438–500 (fig. 224). No external adjustments can be made to correct improper oil pressures. Plugs to be removed, so that the gage adapter can be inserted into the various oil circuits, are located on top of the transmission valve body. Figure 225 indicates the locations of the plugs to be removed and table VIII indicates the normal and allowable ranges of oil pressures. To perform the tests, run engine at 1,600 rpm, with engine idling smoothly without misfire, with manual control lever in shift and steer positions as indicated.

Caution: Setting manual control lever in positions as indicated and increasing engine speed to 1,600 rpm will cause vehicle to move. Perform tests in clear space. Person reading gage must exercise great care. Stop engine each time before removing a plug and inserting gage. Install plug as soon as gage is removed. Remove only the one plug necessary to perform desired test.

236. Reverse and Low Gear Band Adjustments

a. REMOVE SIDE ACCESS DOORS. Remove three bolts and lock washers which secure each side access door on the rear of the hull (fig. 285) and remove doors.

b. ADJUST REVERSE AND LOW GEAR BANDS.

Note. The procedure for adjusting both bands is the same. The low gear band is adjusted through the right side access door and the reverse gear band through the left side access door.

Start engine (par. 68) and keep transmission running at an input speed of approximately 700 rpm. Remove bolt which secures the adjusting screw retainer, and remove retainer. Loosen lock nut, using socket wrench 41–W–3031–940, wrench extension, and torque wrench (fig. 226). Remove socket wrench 41–W–3031–940. Using common socket wrench of the proper size to fit adjusting screw, tighten band adjusting screw to 30 lb-ft (figs. 227 and 228), then back off two-thirds to five-sixths of a turn. Tighten the adjusting screw lock nut to 200 lb-ft torque. Install adjusting screw retainer and secure with a bolt and washer.

Table VIII. Transmission oil pressure tests

Test	Oil circuit	Manual control lever position	Normal range (psi)	Allowable (psi)
1	Right steering clutch	Neutral shift-full right steer	160-200	140 minimum.
2	Left steering clutch	Neutral shift-full left steer	160-200	140 minimum.
3	Torque converter regulator passage ("OUT")	Not regulated	No test required	
4	Torque converter oil feed line ("IN")	Neutral shift-neutral steer	110-120	105-140.
5	Lubricating oil	Neutral shift-neutral steer	20-40	10 minimum.
6	Main line oil pressure from pump	Neutral shift-neutral steer	170-200	150-210.
7	Reverse-range servo oil line	Reverse shift-neutral steer	170-200	150-210.
8	Low-range servo oil line	Low shift-neutral steer	170-200	150-210.
9	High-range clutch oil line	High shift-neutral steer	110-120	105-140.

Note. When operating in high range, the main line oil pressure and the full steering oil pressures are internally regulated to 105-125 psi.

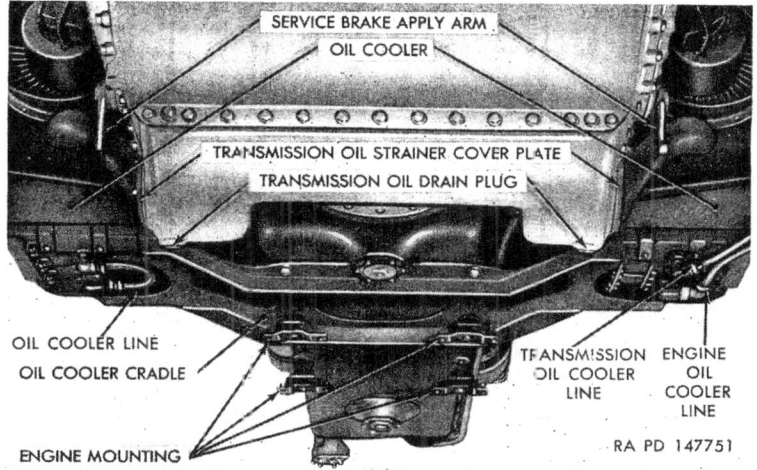

Figure 222. Power plant—rear bottom view.

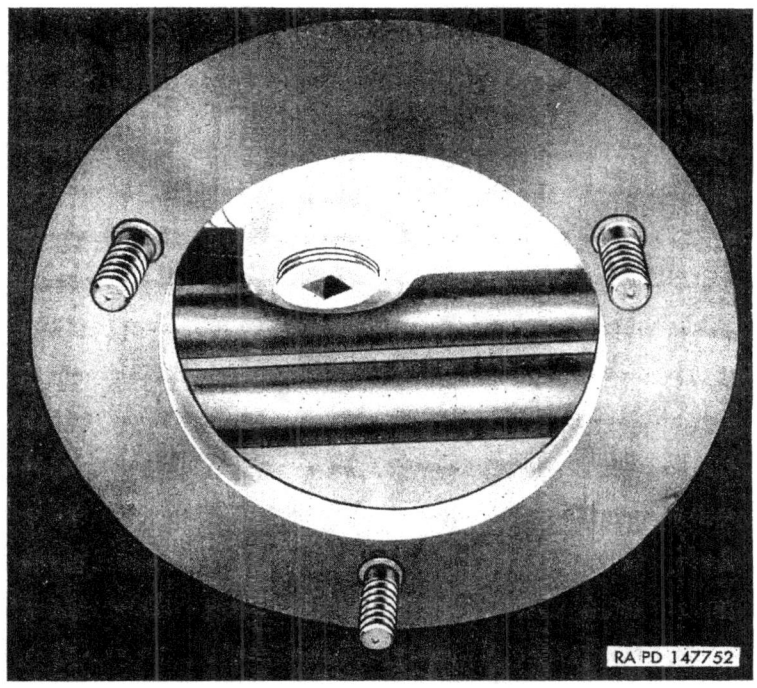

Figure 223. Transmission oil drain plug.

397

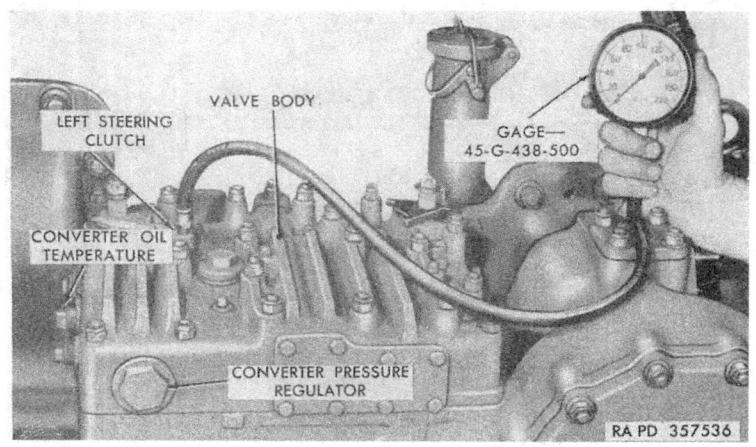

Figure 224. Checking transmission oil pressure.

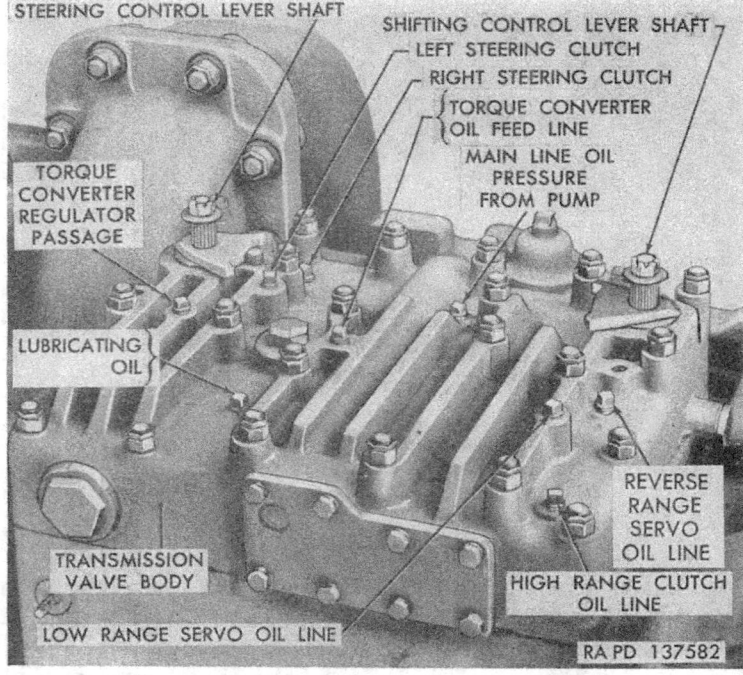

Figure 225. Transmission oil pressure and test points.

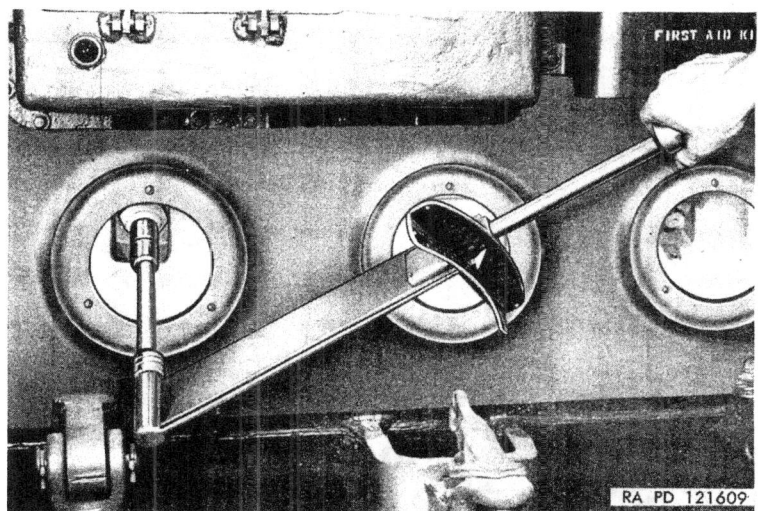

Figure 226. Loosening transmission reverse gear adjusting screw lock nut.

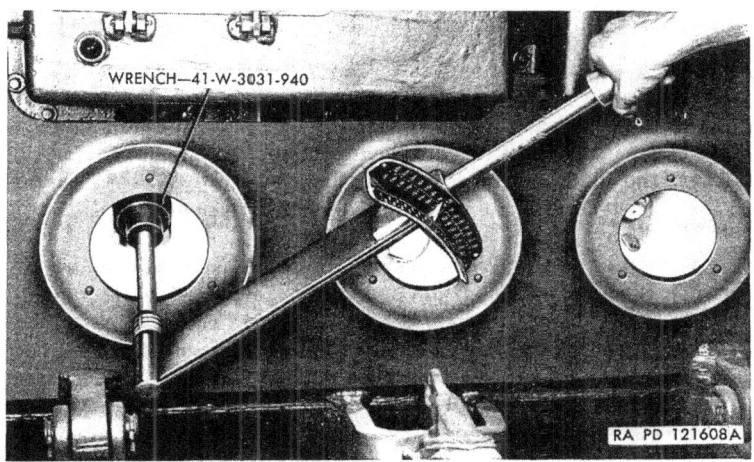

Figure 227. Adjusting transmission reverse gear band.

Figure 228. Adjusting transmission low gear band.

237. Transmission Side Oil Strainers

a. DESCRIPTION. The right and left transmission side oil strainers (fig. 132) supply oil to the right and left outlet pumps respectively. They are located in the front, lower part of the transmission case, accessible at the right and left ends of the transmission. Each consists of a cylindrical screen inside a shell. Oil passes through holes in the surface of the shell through the screen and to the outlet pump located at the inner end of the shell. The two strainers are the same both in construction and in removal and installation procedures.

b. REMOVAL.

(1) Drain oil from the transmission (par. 235*c*(1)).

(2) Remove six nuts, jam nuts, and washer securing the oil-cooler-fan-drive support (fig. 132) to transmission case. Remove locking wire, bolt, and washers at upper end of support securing oil-cooler-fan-drive housing. Remove bracket and pull strainer assembly from transmission case.

c. CLEANING. Clean thoroughly in volatile mineral spirits or dry-cleaning solvent.

d. INSTALLATION.

(1) Clean mounting surface on transmission case. Coat gasket on one side with liquid-type gasket cement and install gasket on transmission case. Install strainer assembly, pushing it

straight in so connection to outlet pump enters inner end of strainer shell.

(2) Position oil-cooler-fan-drive-housing support over cover of strainer and secure with six nuts, lock nuts, and washers.

(3) Install bolt up through hole in upper surface of support with two small washers below mounting surface and two larger washers above mounting surface. Screw bolt into oil-cooler-fan-drive housing and secure bolt with locking wire.

(4) Refill transmission (par. 235c(2)).

238. Transmission Oil Filter

a. REMOVAL. Open engine compartment left and right rear grille doors (fig. 101). Remove 10 jam nuts, nuts, and washers and remove filter (fig. 229) from transmission housing. Strip off gasket.

b. CLEANING. Flush filter in volatile mineral spirits or dry-cleaning solvent, and clean between disks with a soft, long-bristle brush.

c. INSTALLATION. Using a new gasket, position transmission oil filter on studs and secure with 10 washers, nuts, and jam nuts (fig. 229). Close engine compartment left and right rear grille doors (fig. 101).

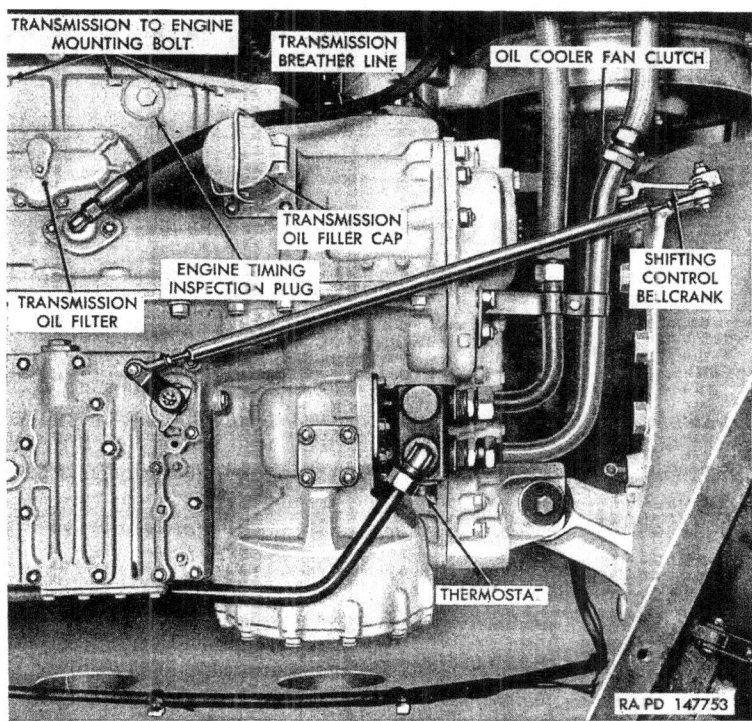

Figure 229. Right side of transmission—installed view.

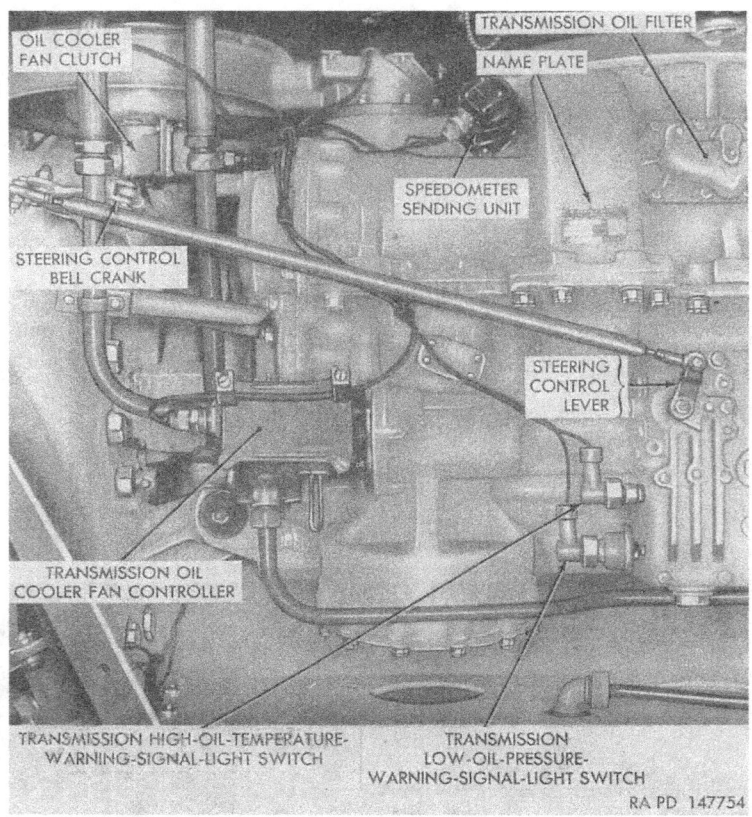

Figure 230. Left side of transmission—installed view.

239. Transmission Replacement

a. AUTHORITY. Replacement of the transmission is normally an ordnance maintenance operation but may be performed in an emergency by the using organization, provided approval for performing this replacement is obtained from the supporting ordnance officer. Tools needed for the operation which are not carried in the using organization may be obtained from the supporting ordnance maintenance unit.

b. REMOVAL.
  (1) *Remove top deck.* Refer to paragraph 158*b*.
  (2) *Remove power plant.* Refer to paragraph 159.
  (3) *Remove oil cooler fans.* Refer to paragraph 241*b*.
  (4) *Disconnect engine from transmission.* Refer to paragraph 163*e*.

c. INSTALLATION.
  (1) *General.* Remove from the transmission being replaced, all cables, connections, and parts not furnished with the replacement transmission. Install these items on the replacement transmission before installing the power plant into the vehicle.
  (2) *Connect transmission to engine.* Refer to paragraph 164*b* and *c*.
  (3) *Install oil cooler fans.* Refer to paragraph 241*b*.
  (4) *Install power plant.* Be certain that all parts removed from the vehicle with replaced transmission and complete power plant are installed on the replacement transmission ((1) above) and on the power plant. Refer to paragraph 160 for power plant installation.
  (5) *Install top deck.* Refer to paragraph 161*a*.
  (6) *Record of replacement.* Record the replacement on DA AGO Form No. 478.

Section XVII. OIL COOLERS AND LINES

240. Description

a. GENERAL. Two oil coolers are mounted in the engine compartment, just forward of the transmission, one on each side of the main engine (figs. 231, 232, 233, and 234). Each cooler has three cores arranged as follows: Three cores of the left cooler and the rear core of the right cooler are used to cool transmission oil and the remaining two forward cores of the right cooler are used to cool engine oil. Each oil cooler is provided with a fan mounted between the cooler and engine rear shroud. The fans draw air through armor louvers in the top deck, through the radiator cores, over the transmission, and out the armor louvers at rear of top deck. The engine rear shroud prevents the air from recirculating through the oil coolers. Oil lines are provided to connect the engine and transmission to their respective cores in the oil coolers (fig. 239). An oil cooler cradle, attached to the engine oil pan, supports the oil coolers (figs. 130 and 222).

b. OIL COOLER FAN, CLUTCH, AND BRUSHES. One oil cooler fan is mounted on each side of the power plant and is driven by transmission power. The fans serve to cool the oil for the engine and transmission. Power for driving the fan is transmitted from the driving gear in the transmission through a gear train to an electrically operated clutch inside the front fan shroud. The operation of the fans is controlled by their respective controllers (figs. 231 and 232), through the brushes and clutch.

c. OIL COOLER LINES. The oil cooler inlet lines (fig. 239) from transmission are connected to the rear core of the left and right

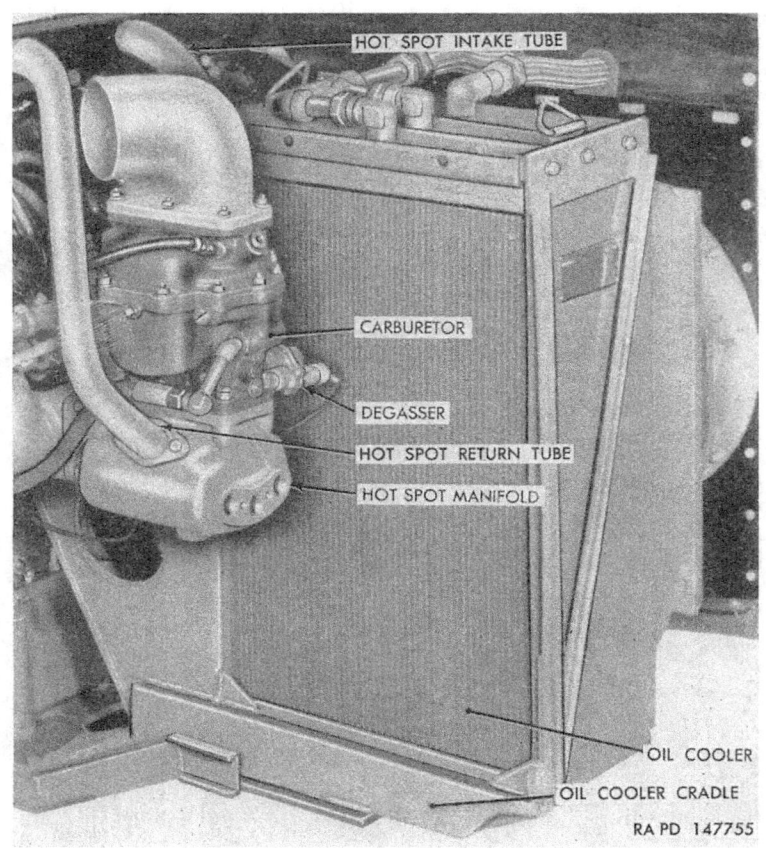

Figure 231. Left oil cooler—installed view.

Figure 232. Right oil cooler—installed view.

405

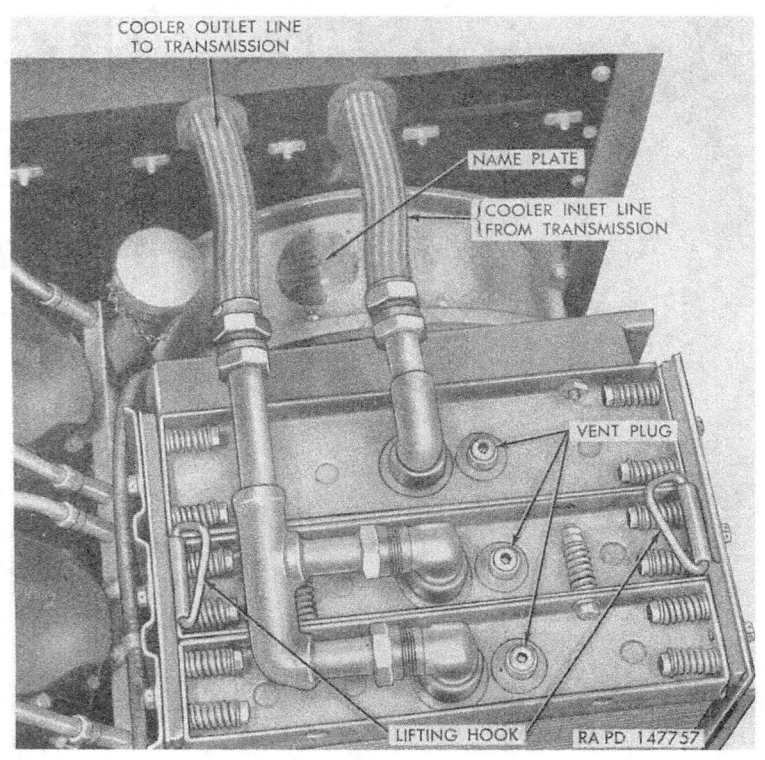

Figure 233. Left oil cooler—top view.

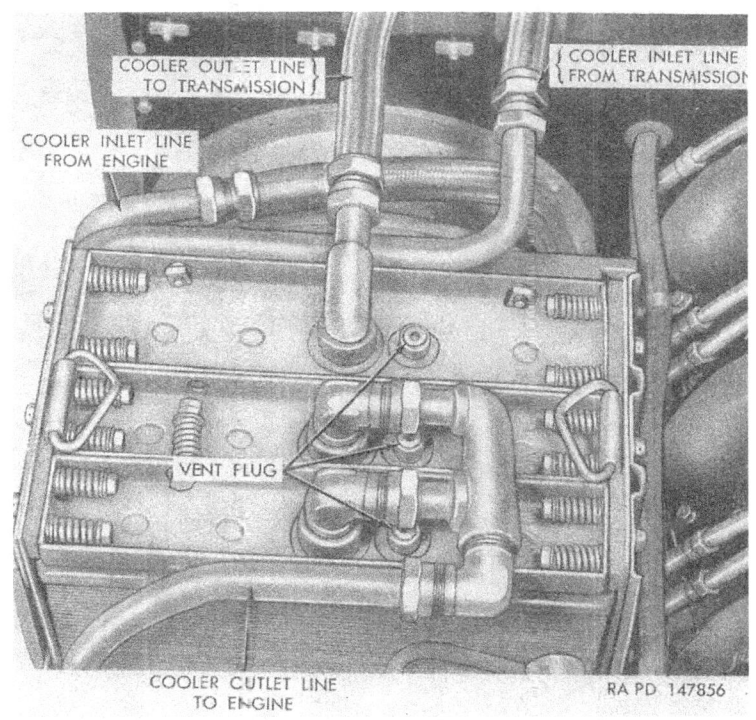

Figure 234. Right oil cooler—top view.

407

coolers. Oil cooler outlet lines to transmission are connected to two forward cores of the left cooler and rear core of the right cooler. Oil cooler inlet and outlet lines from engine (fig. 239) are connected to the two forward cores of the right oil cooler. All flexible lines are the same length to provide interchangeability.

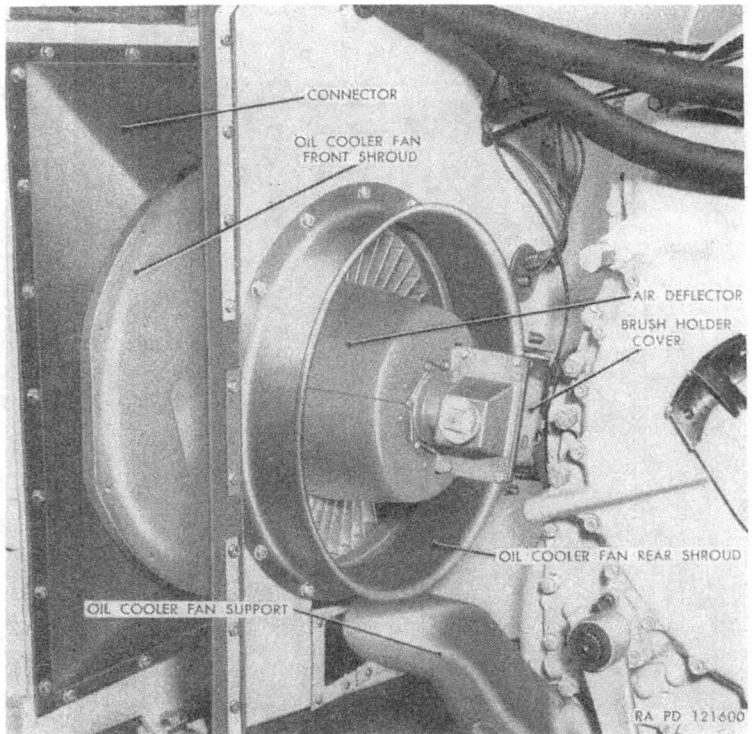

Figure 235. Oil cooler fan and fan drive—installed view.

241. Oil Cooler Fan and Clutch Brushes

a. OIL COOLER FAN CLUTCH BRUSHES.
   (1) *Removal.*
      (*a*) *Remove top deck.* Refer to paragraph 158*b*.
      (*b*) *Remove brushes* (fig. 237). Disconnect electrical cable from brush holder cover (fig. 235) and remove four bolts, nuts, and lock washers securing cover to brush holder. Remove lock nut securing cable to upper brush and remove cover. Remove slotted studs from upper and lower brushes and remove brush assemblies.

(2) *Installation.*
  (a) *Install brushes* (fig. 237). Position brush assemblies in brush holders and install slotted studs.

  Caution: Be sure to insert brush in brush holder so concave end of brush will fit shaft.

  Connect cable extending from inside brush holder cover on upper brush stud and install lock nut. Place cover on brush holder and secure with four bolts, nuts, and lock washers. Connect electrical cable to brush holder cover.

  (b) *Install top deck.* Refer to paragraph 161a.

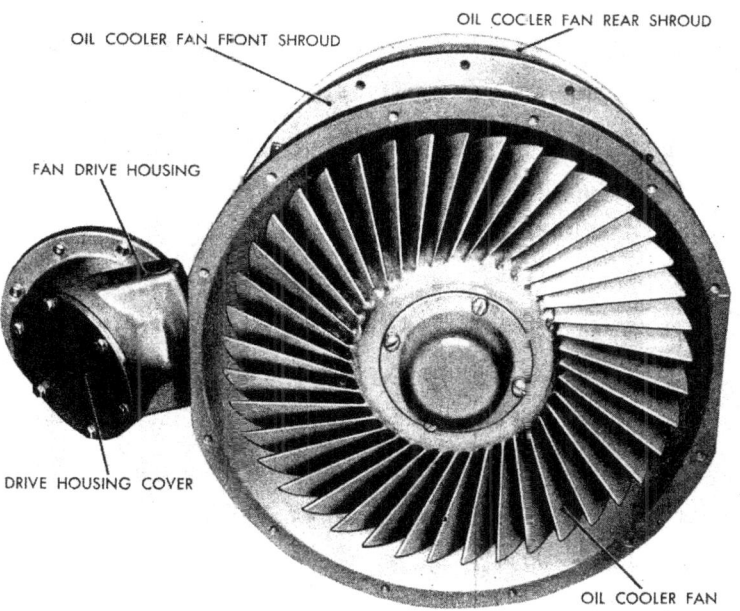

Figure 236. Oil cooler fan and fan drive.

b. OIL COOLER FAN.
  (1) *Removal.*
    (a) *Remove top deck.* Refer to paragraph 158b.
    (b) *Remove power plant.* Refer to paragraph 159.
    (c) *Remove oil coolers.* Refer to paragraph 242.
    (d) *Remove lower section of engine rear shroud.* Refer to paragraph 169.
    (e) *Remove oil cooler fan.* Disconnect fan control wires from fan terminals. Remove 12 bolts and lock washers securing connector to fan front shroud (fig. 235). Remove seven

jam nuts, and washers holding fan drive housing to transmission. Remove locking wire, nut, and washer from fan support. Remove fan and drive housing gasket.

Caution: Cover exposed fan drive bearing and gear to prevent dirt from entering. Cover opening in transmission.

Figure 237. Oil cooler fan clutch brush replacement.

(2) *Installation.*
  (a) *Install oil cooler fan.* Position oil cooler fan housing on support. Install nut, washer, and locking wire securing front shroud to fan support (fig. 236). Position fan drive housing gasket and secure fan drive housing to transmission with seven jam nuts, nuts, and washers.

  Caution: Be sure oil hole in gasket is properly alined with oil holes in fan drive housing and transmission. Install 12 bolts and lock washers securing connector to fan front shroud (fig. 235). Connect fan control wires to fan terminals.

Figure 238. Removing oil cooler.

(b) *Install lower section of engine rear shroud.* Refer to paragraph 169.
(c) *Install oil coolers.* Refer to paragraph 242.
(d) *Install power plant.* Refer to paragraph 160.
(e) *Install top deck.* Refer to paragraph 161.

242. Oil Coolers and Lines

a. OIL COOLERS. The oil coolers may be removed with the engine in the vehicle.

(1) *Removal.*
 (a) *Remove right oil cooler.* Open the three right grille doors (fig. 101) at the rear of the exhaust pipe housing.

Note. It is not necessary to drain oil coolers before removal.

Disconnect oil lines to both engine and transmission at top of oil cooler (fig. 234).

Caution: Cover all openings to prevent dirt from entering.

Remove bolts, nuts, and lock washers which secure oil cooler to top of cooler frame and cradle and remove oil cooler (fig. 238).

(b) *Remove left oil cooler.* Procedure for removal of left oil cooler is same except that the left grille doors are opened and there are only two oil lines to be disconnected.

(2) *Installation.* Procedure for installing oil coolers is the same for both oil coolers.

Note. Make sure oil coolers are installed in correct locations. Oil lines cannot be connected if coolers are reversed. Coolers may be readily identified by the oil line fittings at the top; the left oil cooler has fittings for two oil lines (fig. 233) while the right oil cooler has fittings for four oil lines (fig. 234).

Lower the oil cooler into place in the oil cooler cradle and secure to top of oil cooler frame and cradle with six bolts, nuts, and lock washers. Uncover openings in oil lines and coolers. Connect the oil cooler lines (figs. 233 and 234). Start engine (par. 68) and run for at least 3 minutes. Check oil level and replenish if necessary (pars. 148 and 235). Close left or right grille doors (fig. 101).

b. OIL COOLER CRADLE.
 (1) *Removal.*
  (a) Remove top deck (par. 158b). Remove power plant (par. 159). Remove oil coolers (a above).
  (b) Remove bolts and lock washers which secure oil cooler cradle to engine oil pan. Lift power plant from oil cooler cradle.

(2) *Installation.*
 (*a*) Lift power plant. Position oil cooler cradle under engine and secure to engine oil pan with bolts and lock washers. Lower power plant. Position oil coolers on cradle and secure to top of cooler frame and cradle with bolts, nuts, and lock washers. Connect oil lines at top of each cooler (figs. 233 and 234).
 (*b*) Install power plant (par. 160). Start engine (par. 68) and run for at least 3 minutes to allow coolers to fill with

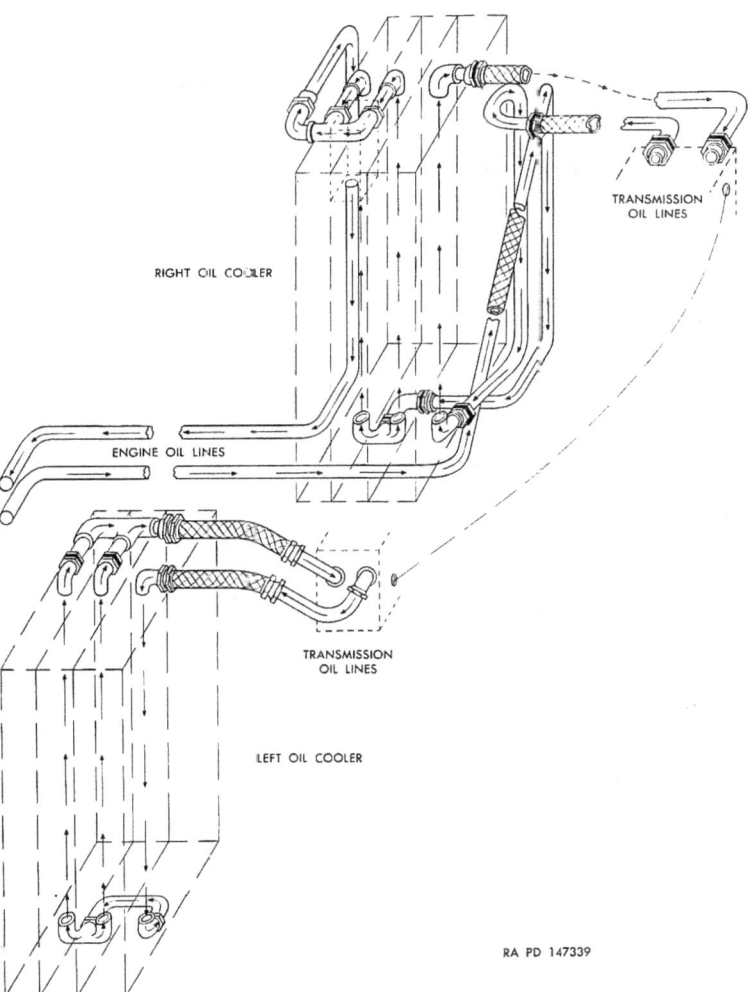

Figure 239. Oil cooler lines schematic diagram.

oil. Check and replenish oil if necessary (pars. 148 and 235). Install top deck (par. 161a).

c. CHECK OIL LINES. Check the oil lines for breaks and for leaks at the connectors. Replace the oil line sections or connectors as necessary to eliminate all leaks.

Section XVIII. FINAL DRIVES AND UNIVERSAL JOINTS

243. Description

The two universal joints and final drives transmit power from the cross-drive transmission to the track drive sprockets (figs. 240 and 244). The universal joints connect the final drives to each side of the cross-drive transmission.

244. Track Drive Sprockets

a. REMOVAL. Remove rear dust shield and fender (par. 266b). Disconnect track (par. 249b), and move it clear of the sprockets. Attach a rope sling to the outer sprocket. Remove 13 bolts and nuts, and remove outer track drive sprocket (fig. 241). To remove the inner drive sprocket remove the sprocket hub (par. 245). Stand the sprocket hub on end with the inner sprocket up. Remove 13 bolts and nuts, and remove inner sprocket.

b. CLEANING AND INSPECTION.

(1) *Cleaning.* Clean both drive sprockets thoroughly with volatile mineral spirits or dry-cleaning solvent. If available, steam may be used to remove remaining accumulations of grease and dirt. Rinse sprockets in volatile mineral spirits or dry-cleaning solvent and blow dry with compressed air.

(2) *Inspection* (fig. 241). Inspect drive sprocket for warpage, cracks, or damaged teeth. If any of these conditions are found, replace sprocket. Check sprocket teeth for excessive wear. Wear will occur on the face of each tooth in the driving direction. When one face of the tooth is excessively worn, the sprocket will be reversed on the hub at time of assembly (*c* below) so that the wear will occur on the opposite face of the tooth. An easy method of measuring to determine the extent of wear is to locate the center line of the tooth by drawing a line through a point halfway between two mounting bolt holes and the center of the opposite mounting bolt hole as shown in figure 241. Since the maximum wear occurs at the pitch diameter of the tooth, determine this point by measuring one inch down on the center line from the top of the tooth. Measure the distance from the center line to each face of the tooth at the pitch diameter. When sprocket is new this dimension will be eleven-sixteenths

inch + or − one-sixteenth inch. When this dimension is reduced to seven-sixteenths inch + or − one-sixteenth inch on one side, reverse the sprocket on the hub. When this dimension is seven-sixteenths inch + or − one-sixteenth inch on both sides of center line, replace the drive sprocket.

c. INSTALLATION. Install inner and outer sprockets on hub and secure each with 13 bolts and nuts.

Note. If vehicle is equipped with open-type-track-drive sprockets (fig. 4), be sure to tighten cap screws first and then tighten lock nuts.

Install sprockets and hub on final drive (par. 245). Connect track (par. 249*b*). Install rear fender and dust shield (par. 266*c*).

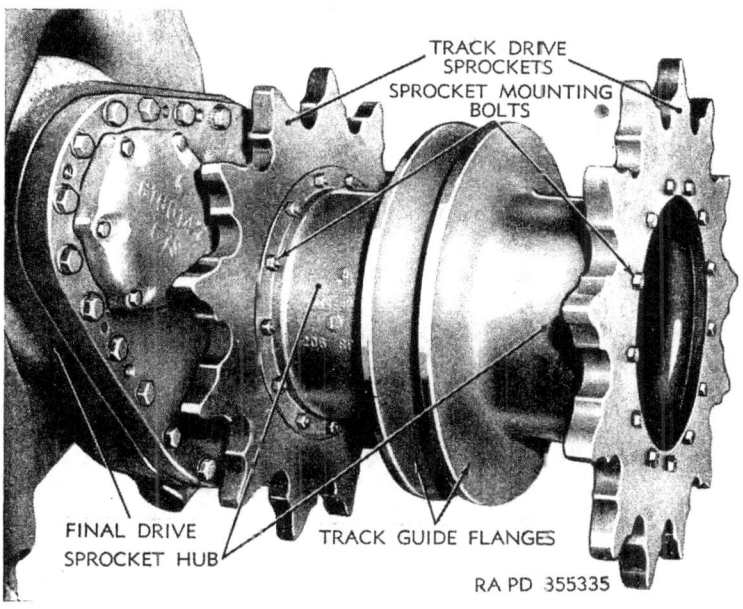

Figure 240. Final drive—installed view.

245. Sprocket Hubs

a. REMOVAL. Remove rear dust shield and fender (par. 266*b*). Disconnect track (par. 249*a*), and pull clear of sprockets. Install sling 41–S–3830–30 (fig. 242), and adjust so that sling will hold the weight of the sprockets and hub. Remove the eight safety nuts from sprocket hub mounting studs (fig. 241). Remove the sprocket hub.

b. INSTALLATION. Using sling 41–S–3830–30, lift sprocket hub onto sprocket hub mounting studs and secure with eight safety nuts (fig. 241). Remove lifting sling. Connect track. Install rear fender and dust shield (par. 266*c*).

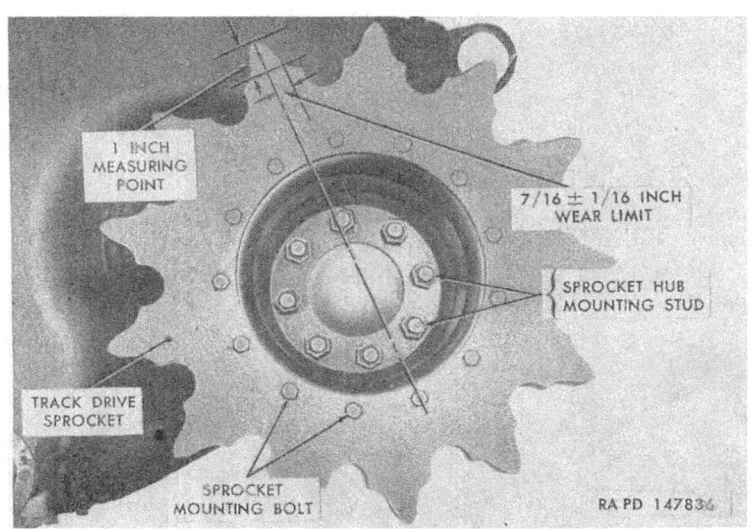

Figure 241. Track drive sprocket wear limit.

Figure 242. Removing track drive sprockets and hub.

246. Final Drives

a. Service. An oil filler and level plug and a drain plug are furnished to service the final drive (fig. 243). To check the lubricant, remove oil filler and level plug. Lubricant level should be up to lower threads of plug hole when the vehicle is on level ground. Use lubricant as specified on the lubrication order (fig. 86) when replenishing or replacing lubricant.

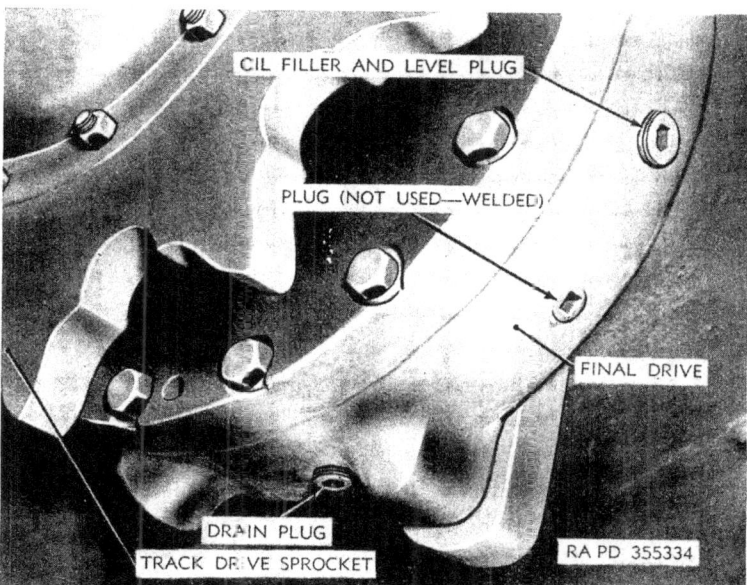

Figure 243. Final drive lubrication plugs.

b. Replacement of Complete Final Drive Assembly.

(1) *Authority.* Replacement of the final drive is normally an ordnance maintenance operation, but may be performed in an emergency by the using organization, provided authority for performing this replacement is obtained from the appropriate commander. Tools needed for the operation which are not carried in the using organization may be obtained from the supporting ordnance maintenance unit.

(2) *Removal.* Remove power plant (par. 159). Remove rear dust shields and fenders (par. 266). Disconnect track (par. 249*b*) and pull clear of sprockets. Install sling 41–S–3832–34 and sling 41–S–3830–30 (fig. 244). Lift slings sufficiently

417

to support weight of final drive. Remove two cap screws and lock washers which secure final drive to hull (outside).

Note. These two cap screws are located at the rear of hull at lower edge of final drive mating surface, and opposite the towing shackle.

Remove universal joint (par. 247*a*(3)). Remove 12 cap screws and lock washer (on circumference inside hull) which fasten final drive to hull. Remove three remaining bolts located to side of circumference bolts and in line with the two outside bolts, and lift out final drive.

Figure 244. Removing final drive, drive sprockets, and hubs.

(3) *Installation.* Install sling 41–S–3832–34 and sling 41–S–3830–30. Position final drive on hull (fig. 244). Secure with 12 cap screws and lock washers, and with five cap screws (including 12 cap screws, lock washers on circumference inside hull, 3 off to side of circumference cap screws, and 2 at outside of hull). Install universal joint (par. 247*b*(1)). Remove slings. Install power plant (par. 160). Connect track (par. 249). Install rear fenders and dust shields (par. 266).

(4) *Records of replacement.* Record the replacement on DA AGO Form No. 478.

247. Universal Joint

a. REMOVAL.
   (1) *Remove top deck.* Refer to paragraph 158*b*.
   (2) *Remove power plant.* Refer to paragraph 159.
   (3) *Remove universal joint.* Attach rope sling to the joint. Remove locking wire and four bolts which fasten the universal joint (fig. 245) to the final drive. Pry the universal joint away from the final drive shaft and lift the universal joint (fig. 246) out of the vehicle.

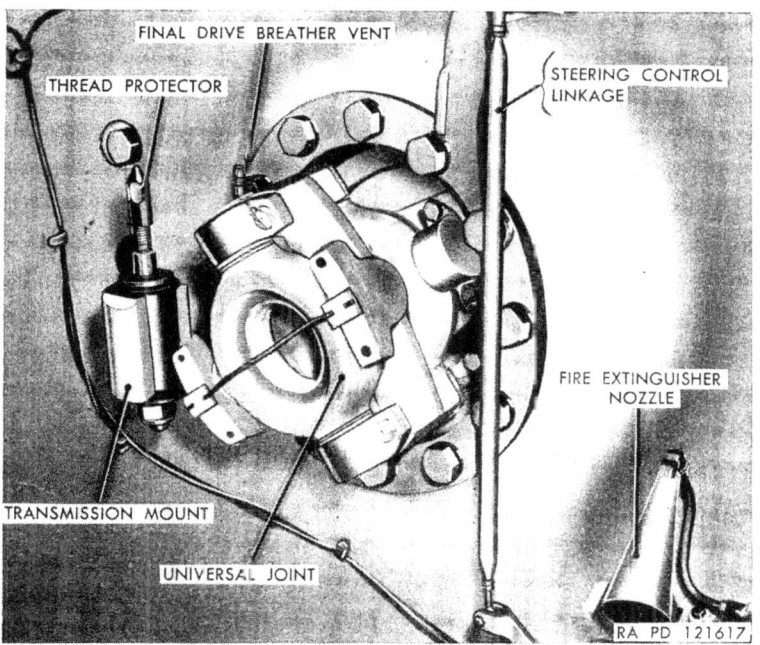

Figure 245. Universal joint—installed view.

b. INSTALLATION.
   (1) *Install universal joint.* Place sling around universal joint and lower it into position (fig. 245). Aline the flanges and install the four bolts which fasten the universal joint to the final drive. Secure bolts with locking wire.
   (2) *Install power plant.* Refer to paragraph 160.
   (3) *Install top deck.* Refer to paragraph 161*a*.

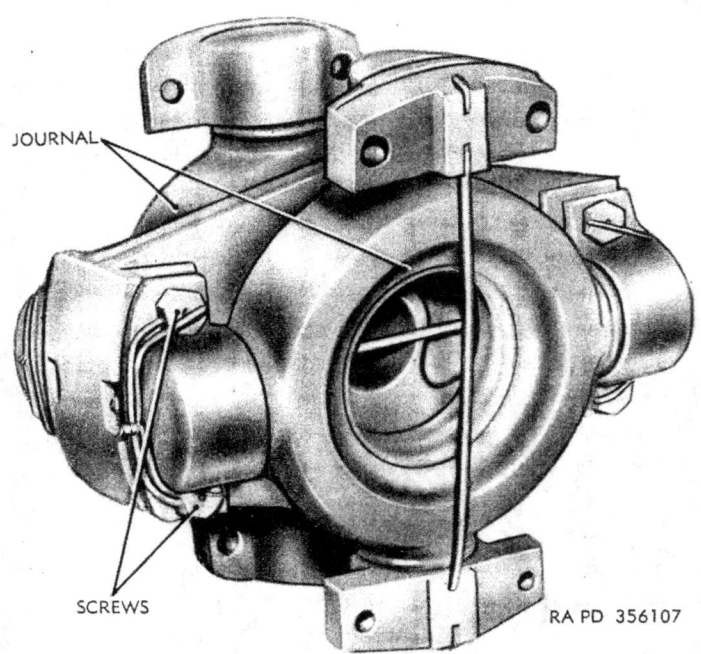

Figure 246. Universal joint.

Section XIX. TRACKS AND SUSPENSION

248. Description and Data

a. GENERAL. The tanks are supported by large-diameter, rubber-tired road wheels which roll on tracks (fig. 247). The wheels are bolted in pairs to the road wheel hubs which are mounted on inner and outer tapered roller bearings on each road wheel arm spindle. Each road wheel arm is supported in an inner and outer needle bearing in a road wheel arm support (figs. 255 and 256). The road wheel arm supports are bolted at intervals along the bottom of the hull, on each side of the vehicle. The road wheel arm supports also provide seats for the torsion bar anchor plugs which secure the ends of the torsion bars extending from the opposite side of the vehicle. The torsion bars, made of spring steel and serrated at each end, run crosswise along the hull floor, from the road wheel arms to the torsion bar anchors in the opposite road wheel arm supports. The serrations on one end of the torsion bars match with serrations in the hollow road wheel arm spindle, and at the other end with serrations in the torsion bar anchor plugs. One serration in each is eliminated to facilitate removal and installation. Each torsion bar is marked on the road wheel arm end with an arrow showing direction of rotation and a number for correct installation. The torsion bars support the weight of the vehicle and also, through torsional resistance or "twisting" action of the bars caused by the up-and-down movement of the road wheel arms and spindles, act as springs to cushion the up-and-down movement of the vehicle.

b. BUMPER SPRINGS. Volute-type bumper springs are provided for all road wheel arms (figs. 247 and 257). These bumper springs stop the travel of the road wheel arms when the track and wheels strike an obstacle of sufficient size to overcome the torsional resistance of the torsion bars and the resistance offered by the shock absorbers.

c. SHOCK ABSORBERS. Hydraulic shock absorbers (figs. 257 and 259) on each side of the vehicle control the movement of the front and rear road wheel arms. They are also provided for the front and rear intermediate road wheel arms, but not for the center wheel arms. The location of these wheels at the center of the vehicle makes shock control unnecessary. Track tension is maintained by the compensating-idler wheel (fig. 247). Pressure on the track by the compensating idler wheel is maintained through a torsion bar similar to the road wheel torsion bars. The adjusting-idler wheels are supported on an eccentric spindle in the upper end of the front road wheel arm (fig. 257). The position of the spindle in the arm is adjustable to provide correct track tension. As the lower end of the front road wheel arm is raised when the track beneath the wheel passes over an obstacle, the arm pivots on the front road wheel arm spindle. This pivoting motion moves the adjusting-idler wheel, on the upper end of the front road wheel arm, forward and downward to take up the slack in the track caused by the obstacle raising the front road wheel. At the same time, the front spring arm shackle transmits the upward thrust to the front spring arm attached to the torsion bar in the front spring arm retainer (fig. 257). The shock absorbers also assist in controlling this upward movement. The front volute bumper spring stops any movement beyond the control of these units.

d. TRACKS. Two types of tracks are used on these vehicles. The T80E6 is a steel-grouser, rubber-backed track. The T84E1 is a rubber track. The tracks are interchangeable without any alteration to the suspension system. Each track is made of 86 track shoe assemblies. Each track shoe assembly consists of a link with integral pins, track guide, and end connectors (fig. 248).

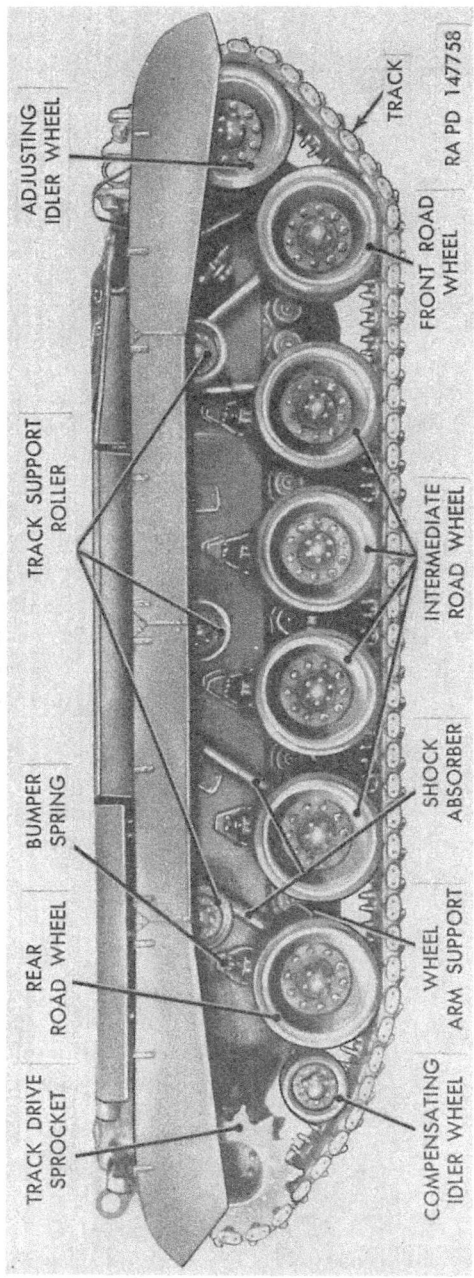

Figure 247. Tracks and suspension.

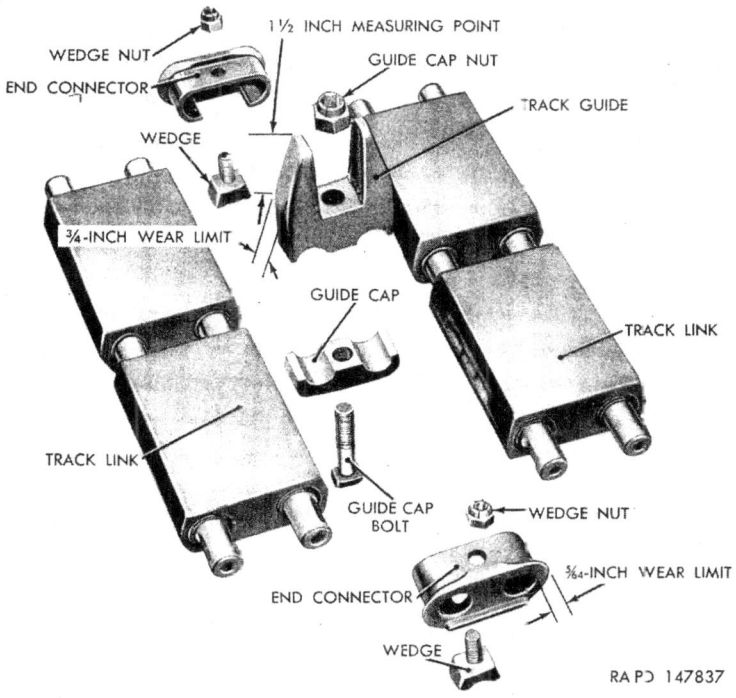

Figure 248. Rubber track shoe assembly—exploded view.

e. DATA.

Track model	Width (inches)	Weight per shoe (pounds)	Track pitch (inches)	Grouser height (inches)	
				(New)	Wear limit*
T80E6	23	57.5	6	$1\frac{1}{2}$	$1\frac{7}{32}$
T84E1	23	50	6	$1\frac{1}{2}$	$\frac{1}{4}$

*Do not ship vehicle overseas without replacing track shoes if grouser height on T80E6 track is $1\frac{1}{4}$ inches or less or grouser height on T84E1 is $1\frac{3}{8}$ inches or less.

249. Tracks

a. TRACK TENSION.

(1) *Inspection.*

(*a*) Proper track tension adjustment can be determined by any one of three methods ((*b*), (*c*), or (*d*) below). The methods discussed in (*b*) and (*d*) will result in less tension (looser tracks). However, procedures for adjusting track

423

tension ((2) below) are similar no matter which method is employed to check the adjustment.

(b) Inspect the track visually just beneath the compensating-idler wheel (fig. 247). If a slight bend in the track is observed at this point, the track is adjusted correctly.

(c) A quick and satisfactory measurement for proper track tension is to measure the distance from the top of the compensating idler wheel (fig. 247) to the underside of the track directly above the compensating-idler wheel. This distance should be between 19 and 20 inches.

Note. This is a tight adjustment and will prevent track throwing to a certain extent.

(d) A support roller cam can be used to inspect for proper track tension. Remove the three forward dust shields (par. 266) and install a support roller cam on the circumference of the center support roller. Then drive the vehicle forward (par. 70) in order to raise the track $1\frac{1}{8}$ inches over the roller by means of the cam. The track tension is properly adjusted when there is $\frac{1}{8}$-inch clearance between the track and each support roller adjacent to the center roller.

Note. A $1\frac{1}{8}$-inch wood block may be used in place of the support roller cam if a cam is not available.

Remove cam or block. If track tension is to be adjusted do not install the three forward dust shields until after adjustment ((2) below) is completed.

(2) ADJUSTMENT. Track tension is adjusted by rotation of the idler wheel adjuster (fig. 251). Allow the vehicle to coast to a stop on level ground without applying the brakes. This procedure equalizes track tension over the entire track. To adjust the track tension, proceed as follows. Remove the front fender and dust shield (par. 266).

Note. If cam method was utilized for checking track tension, the three forward dust shields have been removed.

Remove the cotter pin from the adjuster lock nut (fig. 255), and back off the lock nut approximately one-half inch using wrench 41–W–1436–25. Separate the lock ring and adjuster mating serrations by sliding the adjuster toward the lock nut with the hands. Move the adjuster until there serrations are sufficiently separated to allow the adjuster to be turned. Adjust the track tension by rotating the idler wheel adjuster with wrench 41–W–3250–875 (fig. 252). Pull up on the wrench to tighten the track and push down on the wrench to loosen the track. After the track tension has been checked ((1) above), and to see that it is properly adjusted, move the adjuster toward the lock ring until the serrations are completely in mesh. It may be necessary to rotate the adjuster slightly to match the serrations. Tighten the adjuster lock nut and install cotter pin. Either install front fender and front dust shield or the front fender and three forward dust shields (par. 266) depending upon what method of checking track adjustment was used.

b. DISCONNECTING TRACK. Remove wedge nuts and wedges from one pair of opposite end-connectors (fig. 248). Install the track connecting fixture 41-F-2995-200 (which consists of a left track fixture 41-F-2995-275 and a right track fixture 41-F-2995-375) on the track link end connector (fig. 249), and remove the connectors. Install the track connecting fixtures from below the track, on the inner and outer ends of the links, so that the opposite hooks on the fixtures fit around the exposed ends of the link pins from which the connectors were removed (fig. 250). Adjust the fixtures so that the loads on the track guide and guide cap are relieved. Remove the guide cap nut and bolt and remove the track guide and guide cap (fig. 248). Loosen and remove the track connecting fixtures.

c. CONNECTING TRACK. Install the track connecting fixture 41-F-2995-200 (fig. 250) so that the opposing hooks fit around the ends of the track link pins. Adjust the fixtures evenly until the track links are in the proper positions to install the track guide and cap. Install the track guide, guide cap, and guide cap bolt and nut (fig. 248). Tighten the nut to 300 lb-ft torque. Remove the track connecting fixtures. Install the track end connectors. Install the wedges. They may be tapped lightly with a small hammer, but do not pound them to their seats. Install the wedge nuts and tighten to 120-140 lb-ft torque. Adjust the track tension (*a* above).

d. REMOVING TRACK. Remove the two front and rear fenders and dust shields (par. 266). Disconnect the track (*b* above) at adjusting-idler wheel. Block the road wheels to prevent the vehicle from moving. Start the engine (par. 68) and place the manual control lever in reverse-shift position (par. 69). Run the engine very slowly. As the sprockets revolve, the sprocket teeth will pull the disconnected track over the support roller toward the rear of the vehicle. Guide the track links over the rollers with a crowbar. As the track comes off the sprockets, use the crowbar to carry the track away from the vehicle. This prevents the track from piling up behind the final drive and jamming the sprockets. As the removed track length and weight increase, use additional crowbars as necessary, or stop the sprockets and disconnect the track in sections; then continue removing the track as before. As the end of the track drops clear of the sprockets, place the manual control lever in neutral shift position (par. 69) and stop engine (par. 72). Carry the end of the track away from the vehicle until the track lies flat on the ground.

e. CLEANING AND INSPECTION. Cleaning and inspection of the tracks for wear limits can be accomplished without removing tracks from vehicle. A preliminary visual inspection of track for wear will determine the best place to check for excessive wear ((2) below).

(1) *Cleaning.* Clean bottom side of track link thoroughly with steam, if available, paying particular attention to the road surface of the link and grouser. If steam is not available use hot water, soap, and a stiff brush. Clean end connectors and track guide with volatile mineral spirits or dry-cleaning solvent. If available, steam may be used for cleaning after solvent has been applied. Rinse in solvent and dry with compressed air.

(2) *Inspection.*

(*a*) *Track links.* Measure height of grouser above surface of link ("A," fig. 252). When new, the grouser height on T80E6 steel track is 1½ inches. The minimum height (wear limit) is seventeen-thirty-seconds of an inch. Replace link (*j* below) when this limit is reached.

Note. The minimum grouser height on T80E6 for oversea shipment is 1¼ inches.

When new, the grouser height on T84E1 rubber track is 1½ inches. The minimum height (wear limit) is one-quarter inch. Replace link (*j* below) when this limit is reached.

Note. The minimum grouser height on T84E1 for oversea shipment is 1⅜ inches.

Inspect T80E6 steel track for cracks or other damage. Inspect T84E1 rubber track for excessive gouging, chipping, or cutting. Replace track links (*j* below) that are defective or damaged.

Caution: Replace entire track (*g* below) if grouser height between new and used track links varies sufficiently to cause an excessive "thumping" as road wheels pass over the new link. If serviceable, the old track links may be used, as required, to repair tracks on other vehicles.

(*b*) *End connectors.* Inspect end connectors for cracks and replace if damaged. Measure distance between each end of connector and edge of track-link-pin hole nearest end of connector (fig. 248). When new, the thickness of the end connector at this point is five-thirty-seconds of an inch. Replace end connector if measurement is five-sixty-fourths of an inch or less.

(*c*) *Track guides.* Inspect track guides for cracks and replace if damaged. Wear on track guides will occur on sides of prongs, reducing the thickness at this point. Measure the thickness of the prong at a point 1½ inches from the top of prong (fig. 248). Thickness at this point, when new, is 1½ inches. Replace guide if thickness is three-quarters of an inch or less.

f. INSTALLING TRACK WITH VEHICLE ON PART OF TRACK. These instructions cover track installation procedure when the vehicle is standing on part of the track to be installed, and the other part is

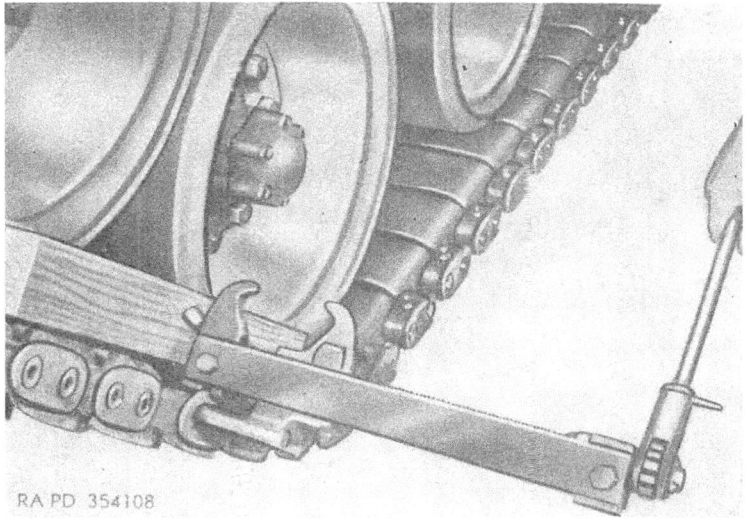

Figure 249. Removing track line end connector using fixture 41–F–2995–200.

stretched out flat on the ground behind it. For installation under other conditions refer to *g* and *h* below.

Note. The tracks are installed so that the point of the grouser "V" pattern touches the ground first when the vehicle is moving forward.

Place a crowbar under the seventh or eighth link from the free end of the track and carry that end up to the drive sprockets. Lift the free end and engage the first three links with the top drive sprocket teeth. Block the road wheels to prevent the vehicle from moving. With a crowbar, hold the track being installed in position on sprockets. Revolve the drive sprocket hub with a crowbar and force the free end of the track over the rearmost support roller. Use another crowbar to guide the track links over the roller. Be sure the manual control lever is in neutral shift position (par. 69). When three or four links have passed over the roller, remove the crowbar and start engine (par. 68). Place the manual control lever in low shift position (par. 69) and very slowly revolve the drive sprockets. As the track moves

427

forward, guide the free end over the remaining support rollers and adjusting-idler wheel with crowbar. When the track links are tight against the sprocket teeth, move manual control lever into neutral shift position (par. 69) and stop engine (par. 72). Connect track (*e* above). Adjust the track tension (*a* above). Install front and rear fenders and dust shield (par. 266). Remove the blocks from the road wheels.

g. INSTALLING NEW TRACK WITH OLD TRACK ON VEHICLE. Remove front and rear fenders and dust shields (par. 266). Remove old track

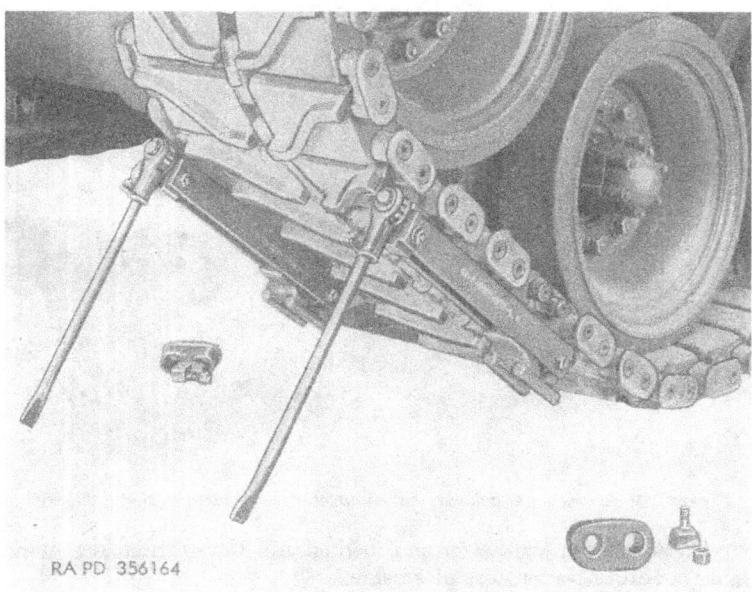

Figure 250. Disconnecting track using fixture 41–F–2995–200.

(*d* above). Place the new track on the ground in a straight line extending from the end of the old track. Be sure grousers on the track links of the new track are pointing in the same direction as those on the links of the old track. Connect the new track to the old track (*e* above). Start the engine (par. 68) and place the manual control lever in reverse shift position (par. 69). Back the vehicle slowly (pars. 69 and 70). When the front road wheel is over the seventh or eighth track link from the end of the new track connected to the old track, stop the vehicle (par. 71) and engine (par. 72), and move the manual control lever into the neutral shift position (par. 69). Disconnect the old track (*b* above). Install the new track (*f* above).

h. INSTALLING NEW TRACK WITH OLD TRACK REMOVED. With the grousers "V" pattern pointing toward the rear of the vehicle, place the new track on the ground ahead of the vehicle and directly in line with the front road wheel. For use as a ramp, place a metal plate or plank on the first track link nearest the vehicle; or dig a trench under the first track link so the link can drop down into depression and the road wheels can roll directly onto the track. Place a wood block between the third and fourth track link guides from the front end of the new track. Drive two stakes into the ground at the front end of

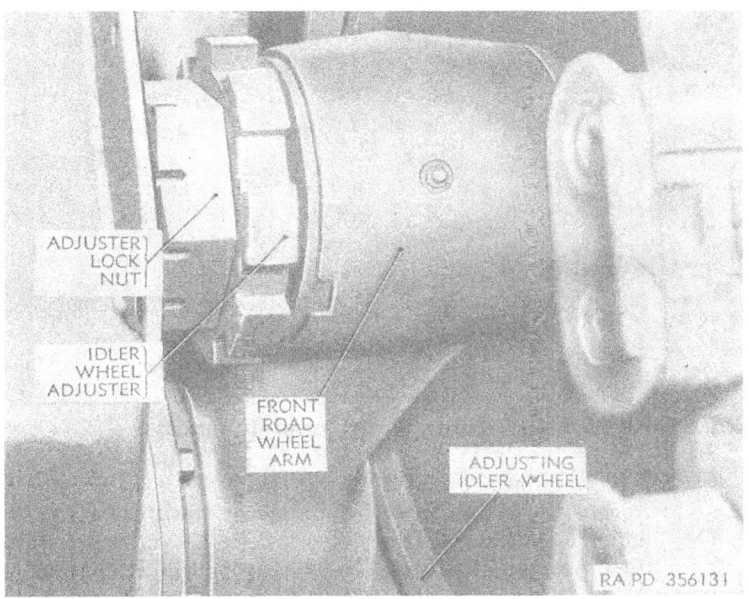

Figure 251. Adjusting-idler-wheel adjuster.

the track to prevent it from being pushed ahead when the vehicle is moving up onto the track. Remove the front and rear fenders and dust shields (par. 266). Tow or push the vehicle (par. 74) ahead onto the track until the front road wheel contacts the wood block. Continue the installation by following procedures as outlined in *f* above.

i. INSTALLING THROWN TRACK. Proceed as outlined in *h* above. Replace any damaged track links as described in *j* below.

j. INSTALLING NEW TRACK LINKS. To connect and disconnect the track, proceed as outlined in *b* or *c* above. Remove the damaged links and install the new links. Connect the track; then adjust the track tension (*a* above).

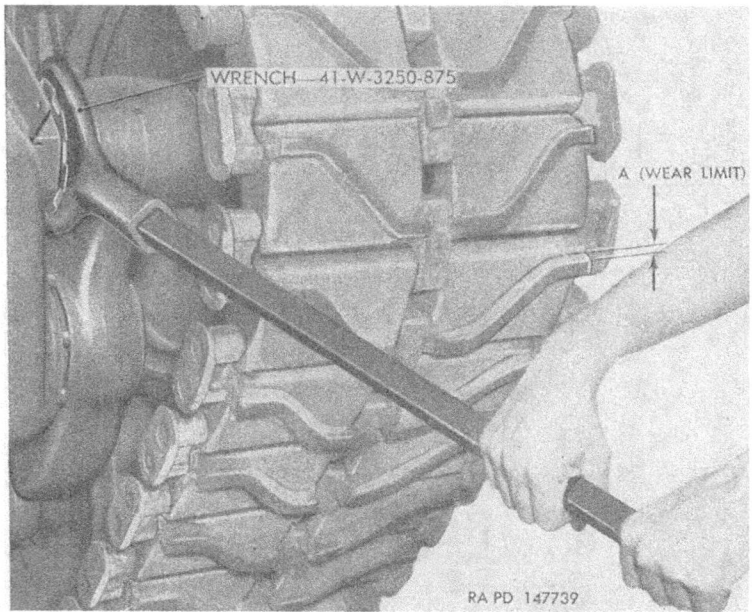

Figure 252. Adjusting track tension.

250. Road Wheels, Hubs, Arms, and Bumpers

a. INTERMEDIATE AND REAR ROAD WHEELS.

(1) *Removal.* With the road wheel on the track, loosen the ten hub nuts but do not remove them (figs. 247 and 256). Slide the upper end of lifter 41–L–1390–100 onto the inner end of the wheel spindle (fig. 253). Place the lower end of the lifter on the track, ahead of the wheel to be lifted. Drive the vehicle forward slowly, being sure the end of the lifter engages around the track link end connector. Stop the vehicle when the lifter is vertical and the wheel fully raised. Remove the hub nuts and remove the road wheel.

(2) *Cleaning and inspection.* Clean road wheel thoroughly with soap and hot water. If available, steam may be used to remove accumulations of grease and dirt. Inspect hub bolt holes for excessive wear due to loose mounting and replace wheel if wear is excessive. Check tires for gouges, chips, cuts, or separation from rim. Replace wheel if evidence of separation of tire from rim is found or if damage to tire is sufficient to cause "thumping" when installed.

(3) *Installation.* With the road wheel arm raised by means of lifter 41–L–1390–100, position the road wheels on the hub (fig. 253). Install the hub nuts. Drive the vehicle backwards slowly until the wheel is lowered onto the track and the lifter is disengaged from the track link end connector. Remove the lifter. Tighten the hub nuts to 280–300 lb.-ft. torque.

b. FRONT ROAD WHEELS.

(1) *Removal.* Loosen all hub nuts but do not remove them (fig. 256). Release the track tension (par. 249a). On the inside of the track, position lifter 41–L–1390–5 in a nearly horizontal position, with the short tip of the rounded end of the lifter on top of the eye of the front spring arm shackle which connects to the front road wheel arm spindle (figs. 254 and 258). Place the lower or forked end of the lifter on the track, with the lower fork resting around the end of the track link end connector. Drive the vehicle forward slowly until the lifter is in a vertical position and the front road wheel is raised. Remove the hub nuts and remove the wheel.

(2) *Cleaning and inspection.* Clean road wheel thoroughly with soap and hot water. If available, steam may be used to remove accumulations of grease and dirt. Inspect hub bolt holes for excessive wear due to loose mounting and replace wheel if wear is excessive. Check tires for gouges, chips, cuts, or separation from rim. Replace wheel if evidence of separation of tire from rim is found or if damage to tire is sufficient to cause "thumping" when installed.

(3) *Installation.* With the front road wheel arm raised by means of lifter 41–L–1390–5 (fig. 254), position the road wheels on the hub and install the hub nuts. Drive the vehicle backward slowly until the wheel is lowered onto the track and the lifter is disengaged from the track link end connector. Remove the lifter. Tighten the hub nuts to 280–300 lb.-ft. torque.

c. ROAD WHEEL HUBS AND BEARINGS.

(1) *Removal.* Remove the road wheel (*a* and *b* above). Remove the hub cap bolts and lock washers and remove the hub cap (figs. 256, 257, 259, and 263). Remove and discard the hub cap gasket. Remove static grounding spring. Clean out the grease. Flatten the bent portion of the lock ring lock that is pressed against the flats of the jam nuts (figs 256 and 263). Remove the jam nut, using socket wrench 41–W–3058–470 fitted on the socket wrench sliding bar 41–B–312–200 with sliding bar male head 41–H–1845–50 (fig. 260). Remove the lock ring lock, lock ring, and adjusting nut (fig. 263) with spanner wrench 41–W–3242–300 (fig. 261).

Temporarily install the hub cap to prevent the outer bearing cone from falling out when the hub is removed. Pull the hub and bearing assemblies off the spindle. Remove the hub cap and lift out the outer bearing cone. Remove the inner bearing cone. Remove the inner bearing oil seal by prying it out or driving it out with a chisel.

Note: Discard the inner bearing oil seal. Do not reuse.

(2) *Cleaning.* Clean hub assembly thoroughly in volatile mineral spirits or dry-cleaning solvent. If available, steam may be used to remove accumulations of grease and dirt after solvent has been applied. Rinse in solvent and dry with compressed air. Wash spindle with volatile mineral spirits or dry-cleaning solvent and dry with compressed air. Soak bearing cones in volatile mineral spirits or dry-cleaning solvent. After soaking, to loosen lubricant, turn rollers slowly while immersed. If lubricant is not entirely removed, strike large end of cone against wooden block to loosen remaining lubricant. Repeat soaking and striking operations until bearings are free of old lubricant.

Caution: Bearings must not be dried or spun with compressed air.

(3) *Inspection.*

    (*a*) *Hub.* Inspect hub assembly carefully for cracks or other indications of damage. Replace hub, if damaged. If hub bolt threads are damaged, repair threads or replace bolt.

    (*b*) *Bearing cups.* Inspect bearing cups for cracks, scratches, brinelling, or wear caused by contact with bearing rollers. Breaks or depressions can be felt by running the thumb nail lightly over the inside surface of the bearing cup. Replace damaged or worn cups ((4) below).

    (*c*) *Bearing cones.* Apply clean engine oil (OE) to bearing cones. Turn slowly and inspect for pitting, scoring, roughness, or looseness caused by wear. Bearing rollers must turn freely and smoothly if cones are to be used again. Replace damaged or excessively worn cones.

    (*d*) *Spindle.* Check both road wheel and road wheel arm spindles for cracks or bent condition. If either is damaged, replace road wheel arm assembly.

(4) *Repair.* If inspection ((3) above) reveals bearing cups must be replaced, remove inner bearing cup from road wheel hub, drive out with remover and replacer 41–R–2374–635. To use a bearing cup remover and replacer: first, place the two halves together and position them against rim of bearing cup; second, insert screw 41–S–1047–315 into hole made in

halves of remover and replacer, and then drive the bearing cup in or out depending upon the condition. Use remover and replacer 41–R–2374–655 to remove or install the road wheel outer bearing cup. Install the inner bearing cup in the hub, with the narrow lip outward, using remover and replacer 41–R–2374–635 and screw 41–S–1047–315. Install the outer bearing cup in hub, with narrow lip outward, using remover and replacer 41–R–2374–655 and screw 41–S–1047–315.

(5) *Installation and adjustment.* Pack the inner bearing cone with grease as specified on lubrication order (fig. 87). Install the cone in bearing cup. Install a new inner bearing oil seal, with spring side outward (fig. 256), using replacer 41–R–2392–630 and handle 41–H–1396–510 (fig. 264). Position the hub on the spindle, seating it firmly against the bearing spacer and oil seal guard mounted on the road wheel spindle (fig. 256). Pack the outer bearing cone as specified on lubrication order (fig. 87). Seat the cone in the bearing cup. Install the adjusting nut (figs. 256 and 263), and tighten with spanner wrench 41–W–3242–300 (fig. 261). At the same time, rotate the wheel, first in one direction, then in the other, until there is a slight bind, to be sure all bearing surfaces are in contact. Back off the adjusting nut one-sixth to one-quarter turn. Lock the adjusting nut by installing the lock ring. Be sure the hole indexes with the pin. Install the lock ring lock, and jam nut. Carefully tighten the jam nut, being sure not to alter the bearing adjustment. Bend the lock ring lock over the flat on the jam nut. Install the static grounding spring. Install hub cap with a new gasket and install hub cap bolts and lock washers. Install the road wheel (*a* or *b* above). Lubricate the hub as specified on lubrication order (fig. 87).

d. INTERMEDIATE AND REAR ROAD WHEEL ARMS.

(1) *Removal and disassembly.* Remove the road wheel on each side of the road wheel arm spindle to gain access to the spindle (*a* above).

Note. The front intermediate road wheel arms may be made accessible by removing only the front intermediate road wheel.

Disconnect the shock absorber from the road wheel arm (par. 255). Remove the torsion bar (par. 253). Remove the road wheel arm retainer bolts (fig. 256). Screw adapter 41–A–18–775 into the torsion-bar end-plug threads (fig. 265). Screw the threaded end of the slide-hammer puller 41–P–2957–33 (fig. 265), into the adapter and remove the road wheel arm, road wheel arm spindle, spindle inner and outer bearings, and oil seals from the road wheel arm support.

Figure 253. Raising intermediate or rear wheel using lifter 41-L-1390-100.

Figure 254. Raising front wheel using 41-L-1390-5.

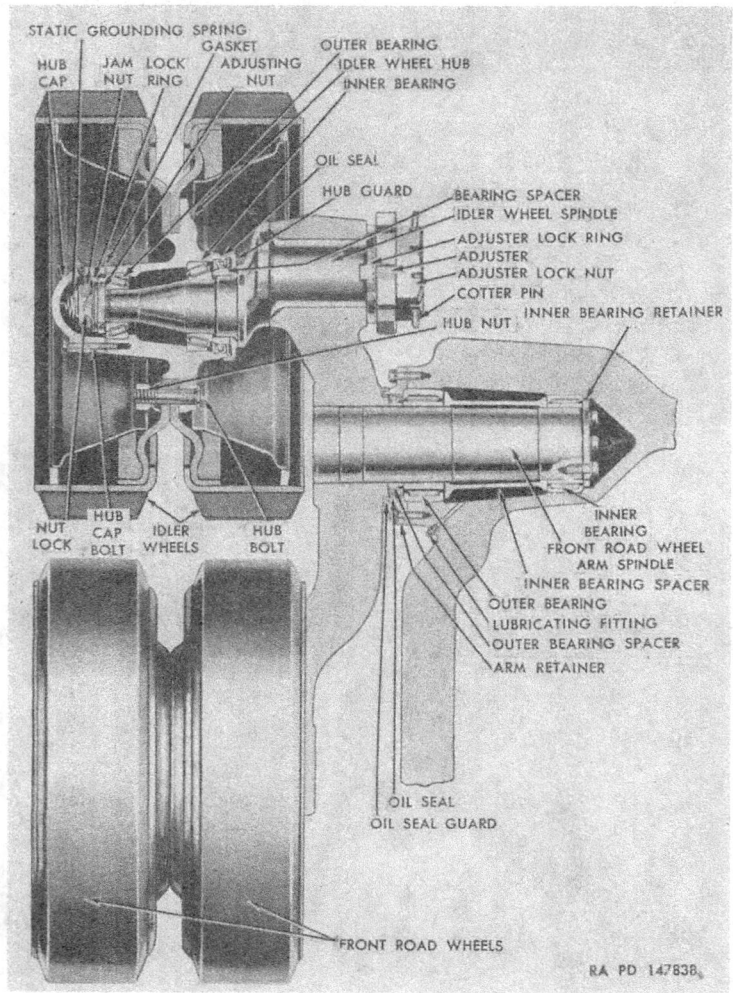

Figure 255. Adjusting-idler and front road wheels—sectionalized view.

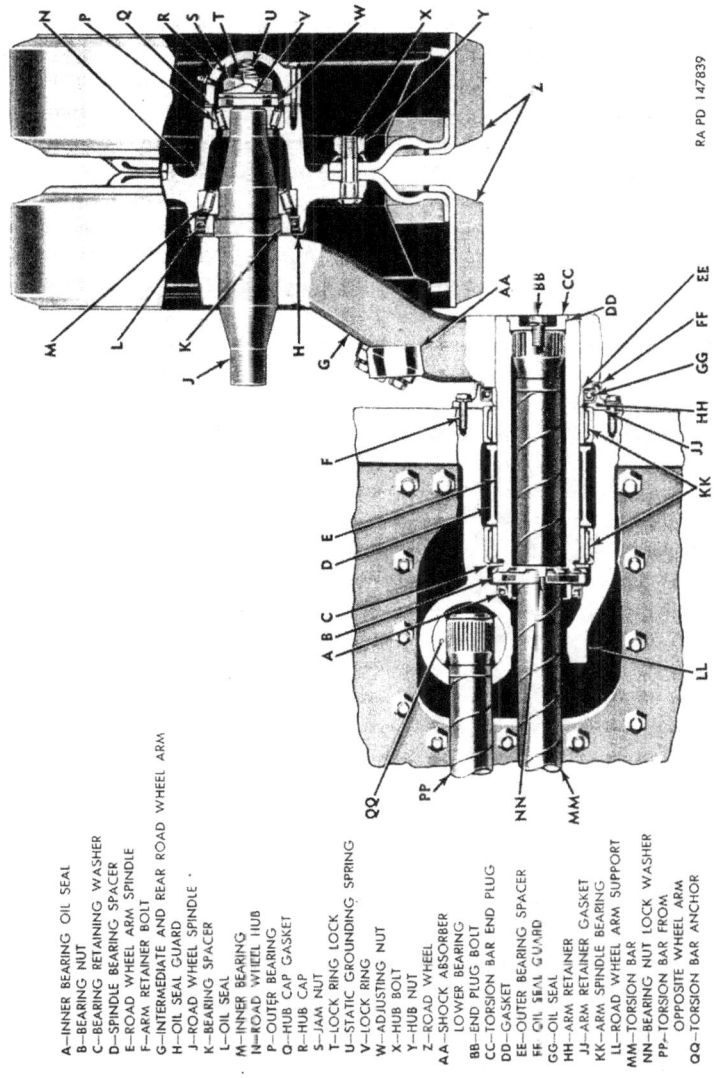

A—INNER BEARING OIL SEAL
B—BEARING NUT
C—BEARING RETAINING WASHER
D—SPINDLE BEARING SPACER
E—ROAD WHEEL ARM SPINDLE
F—ARM RETAINER BOLT
G—INTERMEDIATE AND REAR ROAD WHEEL ARM
H—OIL SEAL GUARD
J—ROAD WHEEL SPINDLE
K—BEARING SPACER
L—OIL SEAL
M—INNER BEARING
N—ROAD WHEEL HUB
P—OUTER BEARING
Q—HUB CAP GASKET
R—HUB CAP
S—JAM NUT
T—LOCK RING LOCK
U—STATIC GROUNDING SPRING
V—LOCK RING
W—ADJUSTING NUT
X—HUB BOLT
Y—HUB NUT
Z—ROAD WHEEL
AA—SHOCK ABSORBER LOWER BEARING
BB—END PLUG BOLT
CC—TORSION BAR END PLUG
DD—GASKET
EE—OUTER BEARING SPACER
FF—OIL SEAL GUARD
GG—OIL SEAL
HH—ARM RETAINER
JJ—ARM RETAINER GASKET
KK—ARM SPINDLE BEARING
LL—ROAD WHEEL ARM SUPPORT
MM—TORSION BAR
NN—BEARING NUT LOCK WASHER
PP—TORSION BAR FROM OPPOSITE WHEEL ARM
QQ—TORSION BAR ANCHOR

Figure 256. Road wheel and arm—sectionalized view.

437

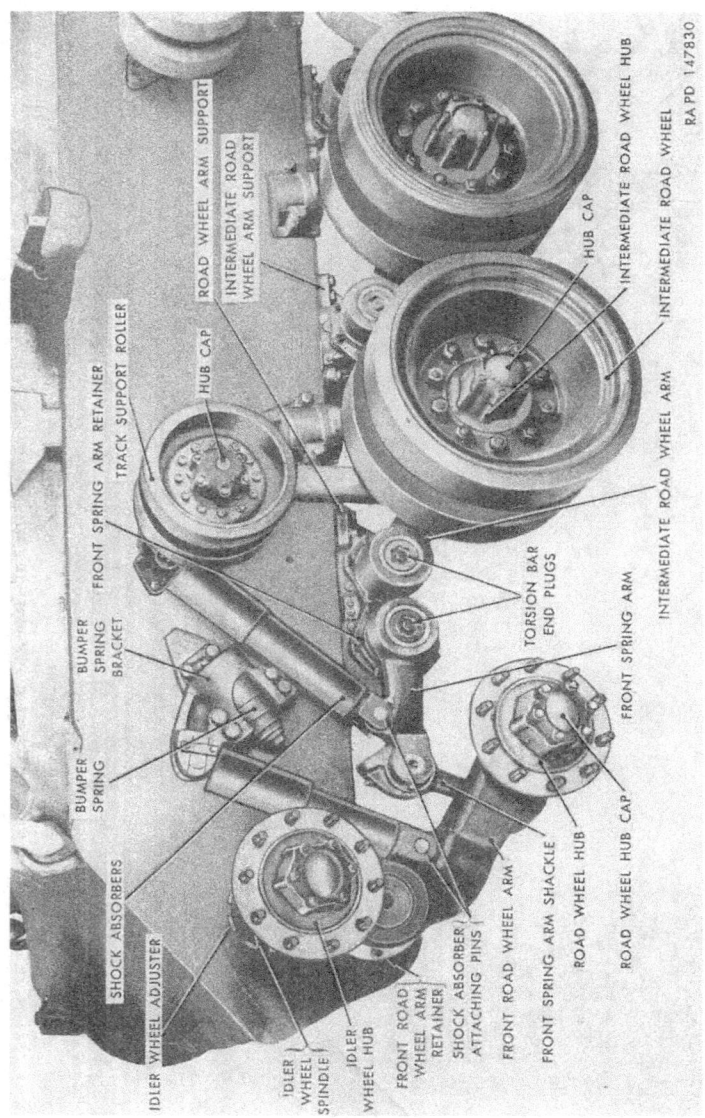

Figure 257. Left front road wheel arm.

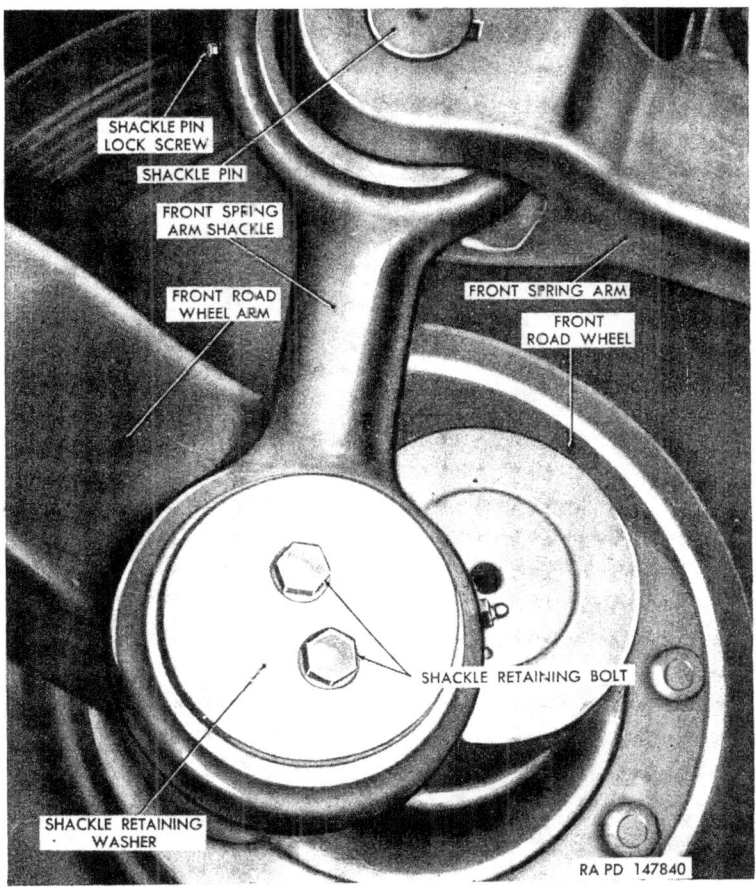

Figure 258. Front road wheel arm, front spring arm, and shackle—installed view.

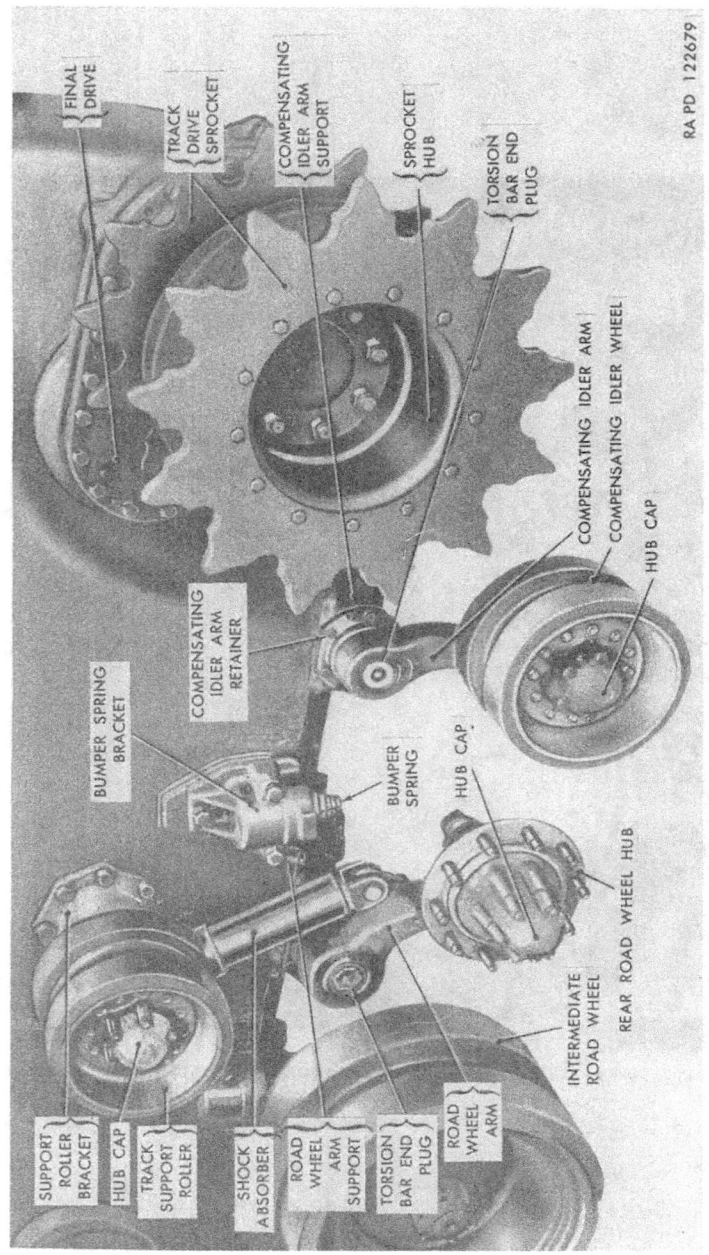

Figure 259. Left rear road wheel and compensating idler wheel.

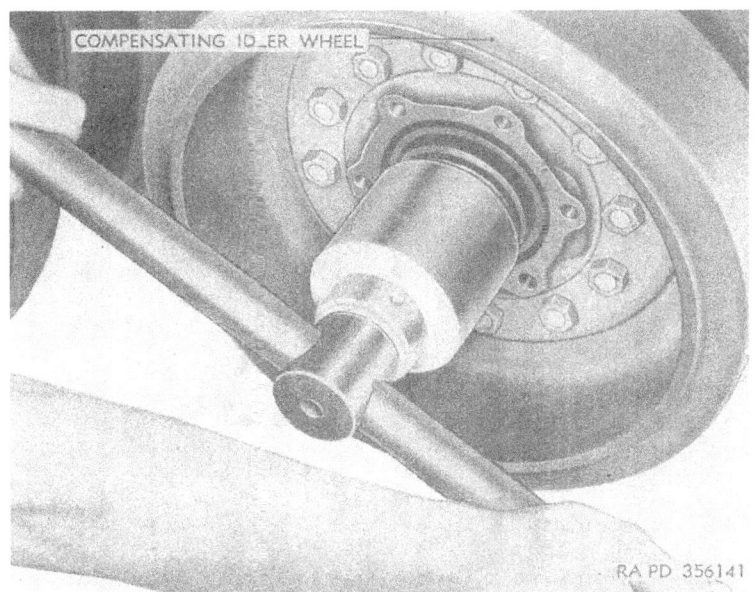

Figure 260. Removing compensating idler spindle lock nut.

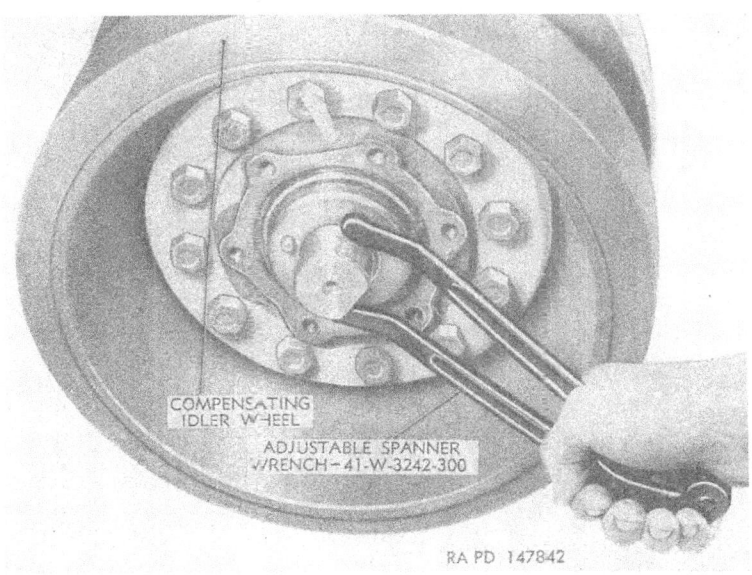

Figure 261. Removing wheel bearing adjusting nut.

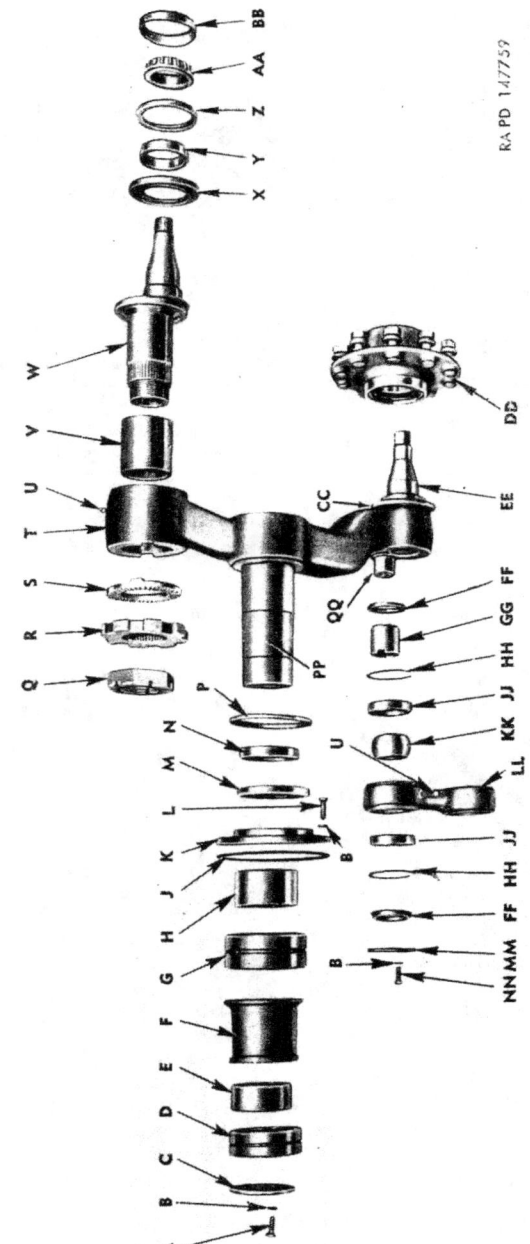

Figure 262. Front road wheel arm and spindle—exploded view.

A—BOLT
B—LOCK WASHER
C—INNER BEARING RETAINER
D—INNER BEARING OUTER RACE
E—INNER BEARING INNER RACE
F—INNER BEARING SPACER
G—OUTER BEARING OUTER RACE
H—OUTER BEARING INNER RACE
J—GASKET
K—ARM RETAINER
L—SCREW
M—OIL SEAL
N—OUTER BEARING SPACER
P—OIL SEAL GUARD
Q—ADJUSTER LOCK NUT
R—IDLER WHEEL ADJUSTER
S—ADJUSTER LOCK RING
T—FRONT ROAD WHEEL ARM
U—LUBRICATING FITTING
V—IDLER WHEEL SPINDLE BUSHING
W—IDLER WHEEL SPINDLE
X—OIL SEAL GUARD
Y—INNER BEARING SPACER
Z—INNER BEARING OIL SEAL
AA—INNER BEARING CONE
BB—INNER BEARING CUP
CC—OIL SEAL GUARD
DD—HUB ASSEMBLY
EE—FRONT ROAD WHEEL SPINDLE
FF—BEARING SEAL
GG—BEARING SLEEVE
HH—SNAP RING
JJ—BEARING SEAT
KK—BEARING BALL
LL—FRONT SPRING ARM SHACKLE
MM—SHACKLE RETAINING WASHER
NN—SHACKLE RETAINING BOLT
PP—FRONT ROAD WHEEL ARM SPINDLE
QQ—SHACKLE PIVOT

Figure 262.—Continued.

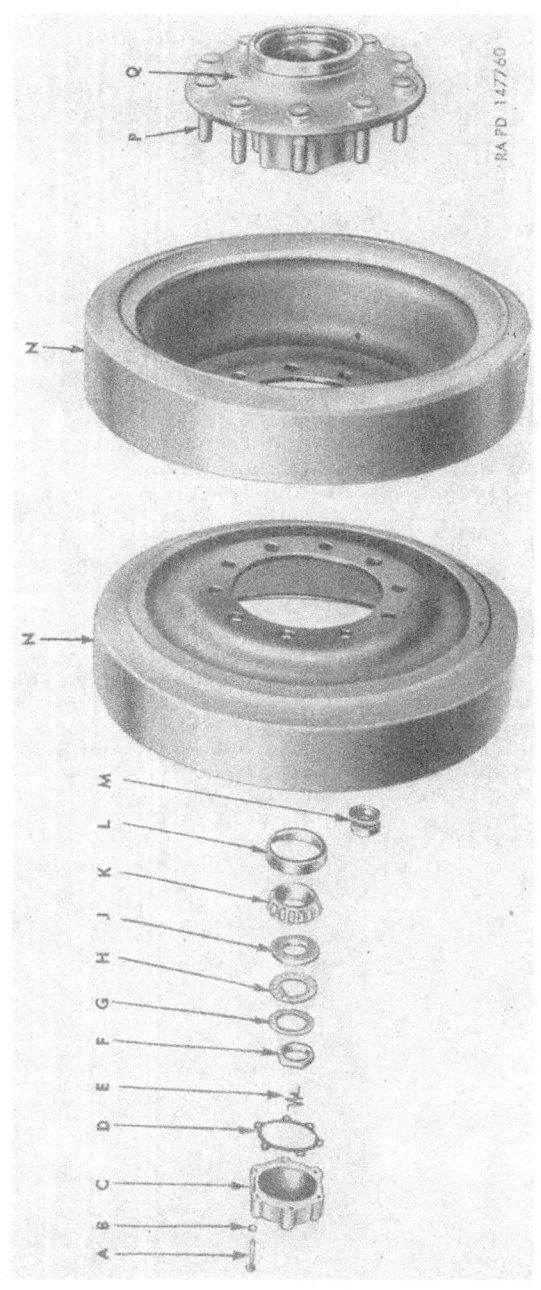

Figure 263. Idler and road wheel—exploded view.

A—HUB CAP BOLT
B—LOCK WASHER
C—HUB CAP
D—HUB CAP GASKET
E—STATIC GROUNDING SPRING
F—JAM NUT
G—LOCK RING LOCK
H—LOCK RING
J—ADJUSTING NUT
K—OUTER BEARING CONE
L—OUTER BEARING CUP
M—HUB NUT
N—WHEEL
P—HUB BOLT
Q—HUB

Figure 264. Installing road wheel bearing oil seal.

Straighten the inner bearing-nut lock washer (fig. 253) and remove the inner bearing lock nut with the spanner wrench 41-W-3252-375 (fig. 266). Remove the lock washer and bearing retaining washers. Remove the inner bearing. Remove spindle bearing spacer and outer bearing. Lift off the arm retainer and remove the oil seal from the retainer, using a drift. Discard used oil seals.

Note. Some later production vehicles are equipped with nylon-type bearings on the arm spindles in place of the needle bearings formerly used. These nylon-type bearings do not require lubrication. Care must be taken that bearing surfaces are free from oil and grease. Clean nylon-type bearings with an oily rag free from any grit. No other lubrication is required.

(2) *Cleaning and inspection.* Clean road wheel arm, bearing races, rollers, and spindles with volatile mineral spirits or dry-cleaning solvent. Inspect races and rollers for cracks, chipping, and wear and replace any defective parts. Inspect arm for distortion; the arm spindle and wheel spindle should

be parallel. Inspect spindles, especially all threads. If arm is bent or spindles damaged, replace arm assembly.

(3) *Assembly and installation.* Install a new oil seal in the road wheel arm retainer and drive into position with replacer 41-R-2392-65 and handle 41-H-1396-510 (fig. 267). Pack the bearings with grease as specified on lubrication order (fig. 87) except nylon-type bearings (*Note* above). Position the arm retainer on the spindle (figs. 256 and 266). Install the outer bearing, bearing spacer; and inner bearing on the spindle.

Note. Coat the inner race of bearings with light graphited grease before assembling except nylon-type bearings.

Install the inner bearing retaining washer, bearing nut lock washer, and bearing lock nut. Tighten the bearing lock nut with the spanner wrench 41-W-3252-375 (fig. 266). Flatten the nut lock washer tangs into the grooves in the nut. Position the arm retainer gasket on the arm retainer face. Install the road wheel arm and spindle in the arm support and install the arm retainer bolts. Install the torsion bar (par. 253). Connect the shock absorber (par. 255). Install the road wheels (*a* above). Lubricate the bearings as specified on lubrication order (fig. 87).

e. FRONT SPRING ARMS.

(1) *Removal and disassembly.* Remove the front road wheels (*b* above). Remove the torsion bar (par. 253). Disconnect the shock absorber from the front spring arm (par. 255). Disconnect the front spring arm shackle from the front spring arm (figs. 257 and 258) by loosening and backing off the shackle pin lock screw and lock washer from the front spring arm and removing the shackle pin using adapter 41-A-18-242 and slide hammer puller 41-P-2957-33. Remove the bolts from the front spring arm retainer and remove the front spring arm, spindle, and bearings by proceeding as outlined in *d* above (which describes removal of the intermediate and rear road wheel arms).

(2) *Cleaning and inspection.* Clean arm, bearing races and rollers except in case of nylon-type rollers (see *Note* above), and spindle with volatile mineral spirits or dry-cleaning solvent. Inspect arm for distortion and conditions of serrations, threads, and roughness of finished surfaces. Inspect bearing races and rollers for cracks and chipping. Replace any damaged parts. If arm is damaged in any way, replace arm assembly.

(3) *Assembly and installation.* Assemble the bearings and oil seals as outlined in *d* above. Install the front spring arm

and spindle into the arm support. Install the bolts in the front spring arm retainer (fig. 257). Install the torsion bar (par. 253). Connect the shock absorber (par. 255). Connect the front spring arm shackle by installing the shackle pin through the shackle and spring arms, and secure by installing the shackle pin lock screw and lock washer. Install the front road wheel (*b* above). Lubricate the bearings (except nylontype) as specified on lubrication order (fig. 87).

Figure 265. Removing front spring arm and spindle.

f. FRONT ROAD WHEEL ARMS AND SHACKLES.
  (1) *Removal and disassembly.*
    (*a*) Remove front road wheel (*b* above) and adjusting idler wheel (par. 251*a*(1)). Remove the shackle retaining bolts and shackle retaining washer that attach the shackle to the front road wheel arm (fig. 258). Loosen the shackle pin lock screw at the upper end of the shackle. Remove the shackle pin that holds the shackle to the front spring arm, using adapter 41–A–18–242 (fig. 79) and slide hammer puller 41–P–2957–33 (fig. 82). Remove the shackle. Remove the screws from the front road wheel arm retainer (fig. 257). Support the arm with a hoist and pull the front road wheel arm spindle (figs. 255 and 262) out of the hull.

(b) Remove the bolts and lock washers and remove the inner bearing retainer and inner bearing (figs. 255 and 262). Remove the inner bearing spacer, outer bearing, outer bearing spacer, and arm retainer. Remove the oil seal from the retainer with a drift. Discard used oil seals.

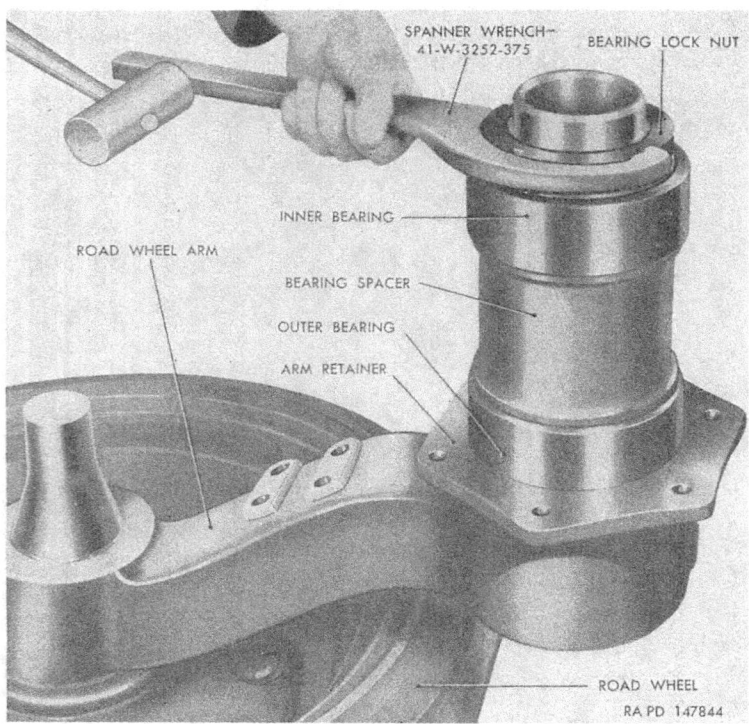

Figure 266. Removing road wheel arm spindle inner bearing lock nut.

(c) Remove bearing sleeve (fig. 262) from front spring arm shackle. Pry out bearing seals with screwdriver and discard. Remove snap rings and bearing seats. Remove bearing ball, using remover and replacer 41–R–2374–635 (fig. 79).

(d) Remove shackle upper bearing in the same manner as lower bearing ((c) above).

(2) *Cleaning and inspection.* Clean all parts (except any nylon-type rollers (see *Note*, d(1) above)) with volatile mineral spirits or dry-cleaning solvent. Inspect arm, bearing races, and rollers for cracks and chipping. Inspect spindles for roughness and damaged threads. Replace damaged parts.

(3) *Assembly and installation.*

(a) Press bearing seat (fig. 262), with the cup side in, into the opening in the upper end of the shackle, using remover and replacer 41–R–2373–460 (fig. 79). Press seat in to a point which leaves the snap ring groove clear. Install bearing ball in seat and press in the other seat in the same manner. Install snap rings on both sides of bearing to retain bearing. Install new bearing seals. Install bearing sleeve.

(b) Install lower bearing in same manner as upper bearing ((a) above).

(c) Install the oil seal in the arm retainer with replacer 41–R–2392–50 and handle 41–H–1396–510. Install the arm retainer on the front road wheel arm spindle. Coat the inside of the bearing inner races with light graphited grease except nylon-type bearings. Pack the bearings with a grease as specified on lubrication order (fig. 87) except nylon-type bearings. Install the outer bearing spacer, outer bearing, inner bearing spacer, inner bearing, and secure the inner bearing retainer with lock washers and bolts. Install a new gasket on the arm retainer.

(d) Install the spindle and arm in the hull and secure arm retainer with screws. Position the front spring arm shackle and install the shackle pin (fig. 258). Tighten the shackle pin lock screw. Connect the lower end of the shackle to the front road wheel arm and install the shackle retaining washer and bolts (fig. 258). Install the front road wheel (b above) and adjusting idler wheel (par. 251a). Connect and adjust the track (par. 249a and c). Lubricate the bearings as specified on lubrication order (fig. 87).

g. WHEEL ARM SUPPORTS.

(1) *Removal and disassembly.* Remove wheel arm from wheel arm support to be removed (d through f above). If not already accomplished, remove both torsion bars from the support (fig. 256) to be removed (one from each side of vehicle) (par. 253). Remove the support bolts and remove the wheel arm support by lowering it from hull. Remove cover and gasket and lift out the torsion bar anchor.

Note. The front support mounts both the front spring arm and the front intermediate road wheel arm.

(2) *Cleaning and inspection.* Clean all parts with volatile mineral spirits or dry-cleaning solvent. Inspect torsion bar anchor for cracks and condition of serrations. Inspect support for cracks. Replace any damaged parts.

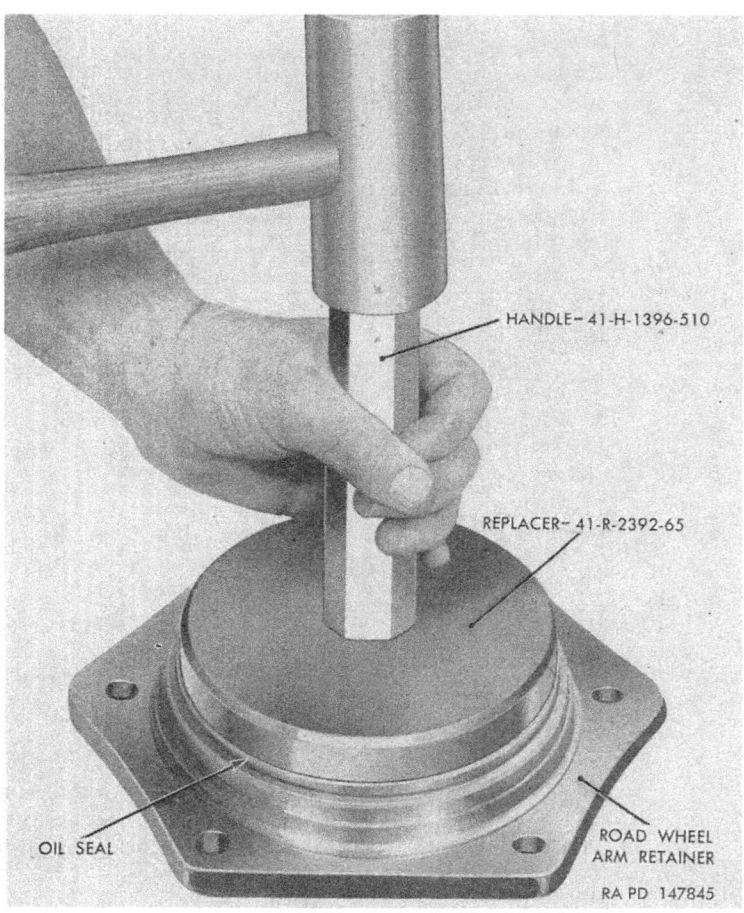

Figure 267. Installing road wheel arm spindle oil seal.

(3) *Assembly and installation.* Install torsion bar anchor in support and position anchor cover and cover gasket so anchor is held by dowels in cover in proper position to receive torsion bar. Secure cover with four bolts and lock washers. Install the wheel arm spindle inner bearing oil seal in the arm support with replacer 41-R-2395-90 for road wheel arm support and replacer 41-R-2391-425 for compensating idler arm support oil seals, and handle 41-H-1396-510 (fig. 268). Install the arm support and gasket on the bottom of the hull (fig. 256). Tighten all bolts to 150 lb-ft torque. Install the wheel arm (*d* through *f* above). Install torsion bars (par. 253). Install the wheels (*a* and *b* above).

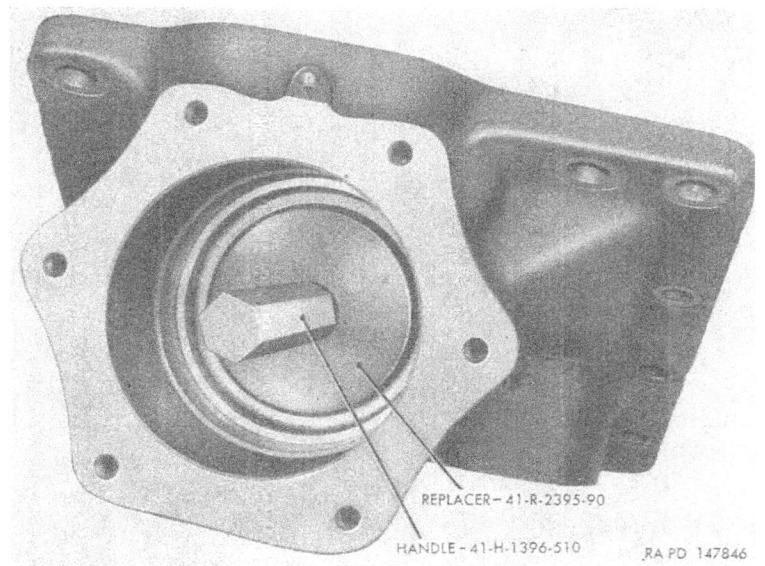

Figure 268. Installing road wheel arm spindle inner bearing oil seal in arm support.

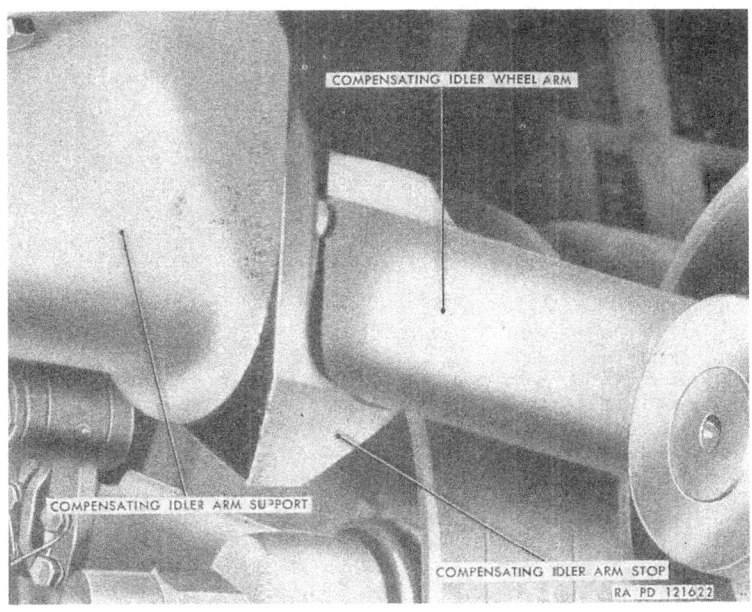

Figure 269. Compensating-idler arm and arm stop.

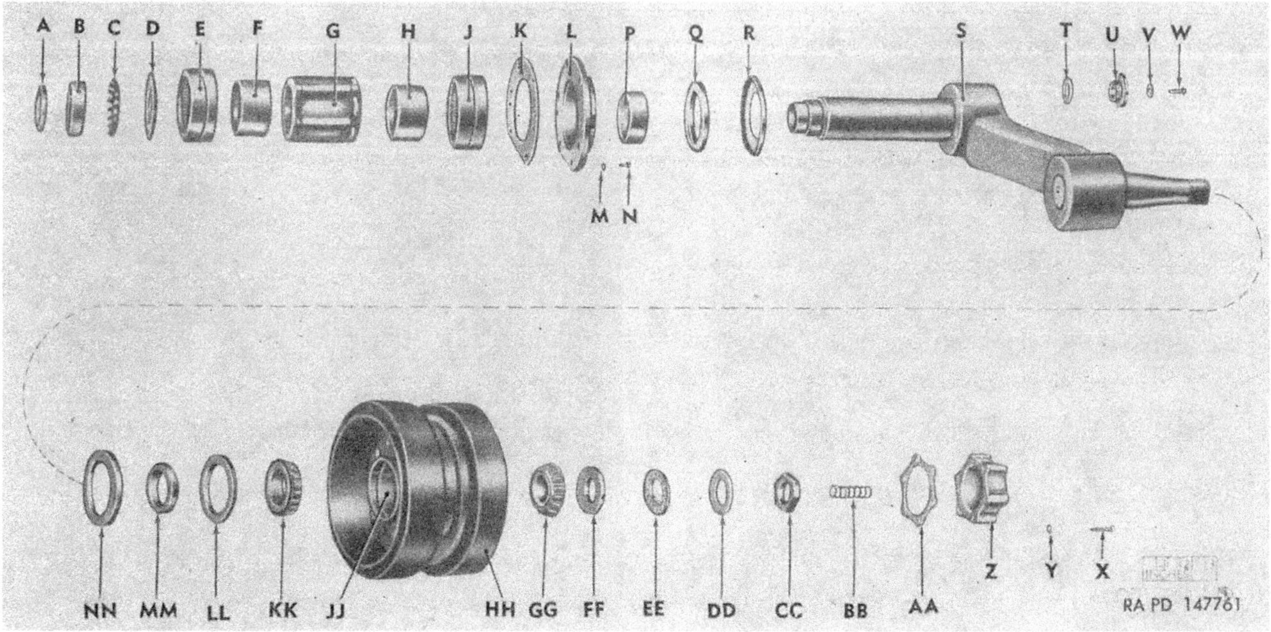

Figure 270. Compensating-idler wheel and arm—exploded view.

452

A—INNER BEARING OIL SEAL
B—LOCK NUT
C—LOCK WASHER
D—INNER BEARING RETAINER
E—INNER BEARING OUTER RACE
F—INNER BEARING INNER RACE
G—INNER BEARING SPACER
H—OUTER BEARING INNER RACE
J—OUTER BEARING OUTER RACE
K—GASKET
L—ARM RETAINER
M—BOLT
N—LOCK WASHER
P—OUTER BEARING SPACER
Q—OIL SEAL
R—OIL SEAL GUARD
S—COMPENSATING-IDLER ARM
T—GASKET
U—TORSION BAR END PLUG
V—WASHER
W—BOLT
X—HUB CAP BOLT
Y—LOCK WASHER
Z—HUB CAP
AA—GASKET
BB—STATIC GROUNDING SPRING
CC—JAM NUT
DD—LOCK RING LOCK
EE—LOCK RING
FF—ADJUSTING NUT
GG—OUTER BEARING CONE
HH—WHEEL
JJ—HUB ASSEMBLY
KK—INNER BEARING CONE
LL—OIL SEAL
MM—BEARING SPACER
NN—OIL SEAL GUARD

Figure 270—Continued.

h. BUMPER SPRINGS. To remove a bumper spring, remove the bolts and locking wires which attach the bumper spring bracket to the hull (figs. 257 and 259). Clean with volatile mineral spirits or dry-cleaning solvent and inspect visually for any damage. To install, position bracket assembly on hull and secure with bolts, tightening them to 350 lb-ft torque. Secure bolts with locking wire.

251. Adjusting Idler Wheels and Hubs

a. ADJUSTING IDLER WHEELS.
  (1) *Removal.* Remove the front fender and dust shield (par. 266). Loosen but do not remove the hub nuts (fig. 255). Disconnect the track (par. 249*c*). Move the track off the adjusting-idler wheel. Remove the hub nuts, and remove the idler wheel.
  (2) *Cleaning and inspection.* Cleaning and inspection procedure is the same as for road wheels (par. 250*a*(2)).
  (3) *Installation.* Lift the idler wheel into position on the hub. Install the hub nuts. Move the track forward over the adjusting-idler wheel. Tighten the hub nuts to 280–300 lb-ft torque. Replace track on wheel and connect the track (par. 249*c*). Adjust track tension (par. 249*a*). Install the front fender and dust shields (par. 266).

b. ADJUSTING IDLER WHEEL HUBS AND BEARINGS. Procedure for removal, cleaning, inspection, and installation is same as for road wheels (par. 250*c*).

252. Compensating Idler Wheels, Hubs, and Arms

a. COMPENSATING IDLER WHEELS.
  (1) *Removal.* Loosen hub nuts (fig. 259). Using a hydraulic jack and wood block, raise the compensating idler wheel arm. Remove the hub nuts and remove the idler wheel.
  (2) *Cleaning and inspection.* Procedure is same as for road wheels (par. 250*a*(2)).
  (3) *Installation.* With compensating idler wheel arm raised, lift the idler wheel into position on the hub. Install the hub nuts. Ease compensating idler wheel down onto track and remove hydraulic jack and wood block. Tighten the hub nuts to 280–300 lb-ft torque.

b. COMPENSATING IDLER WHEEL HUBS AND BEARINGS. Procedure for removal, cleaning and inspection, and installation is same as for road wheels (par. 250*c*) except that remover and replacer 41-R-2374-630 (fig. 79) is used for removing and installing inner bearing cup, and replacer 41-R-2391-420 (fig. 79) is used for installing inner bearing oil seal.

c. COMPENSATING IDLER ARM AND SPINDLE.

(1) *Removal and disassembly.* Remove the compensating idler wheel (*a*(1) above). Remove rear road wheel (par. 250*a*). Remove the torsion bar (par. 253). Remove six bolts and lock washers which secure compensating idler arm retainer to support (figs. 259 and 270). Screw the adapter 41–A–18–248 (fig. 79) into the torsion bar end plug threads. Attach the slide hammer puller 41–P–2957–33 (fig. 82), to the adapter and remove the compensating-idler arm and spindle. Remove bearing lock nut, using spanner wrench 41–W–3252–375 (fig. 266). Remove lock washer and inner bearing retainer (fig. 270). Remove bearings and bearing spacers. Remove oil seal and arm retainer from outer bearing spacer. Remove oil seal from retainer.

(2) *Cleaning and inspection.* Clean arm, bearing races, rollers, and spindles with volatile mineral spirits or dry-cleaning solvent. Inspect races and rollers for cracks, chipping and wear and replace any defective parts. Inspect arm for distortion; the arm spindle and wheel spindle should be parallel. Inspect spindles, especially all threads. If arm is bent or spindles damaged, replace arm assembly.

(3) *Assembly and installation.* Install oil seal into arm retainer (fig. 270) by driving it into position with replacer 41–R–2391–420 and handle (41–H–1396–510 (fig. 79). Install oil seal and retainer on outer bearing spacer by driving it into place with replacer 41–R–2391–420 and handle 41–H–1396–510. Install bearing spacer with oil seal and retainer on spindle. Pack bearings with lubricant as specified on lubrication order (fig. 87) (except nylon-type bearing (see *Note*, par. 250*d*(1)). Install outer bearing, inner bearing spacer, inner bearing, and inner bearing retainer. Install lock washer and lock nut, using spanner wrench 41–W–3252–375 (fig. 266). Flatten the washer tangs to lock the nut. Position the gasket on the arm retainer face. Install the idler arm and spindle in the compensating idler arm support (fig. 269) and install six bolts and lock washers to secure the arm retainer. Install the torsion bar (par. 253). Install the compensating idler wheel (*a*(3) above). Install rear road wheel (par. 250*a*). Lubricate as specified on the lubrication order (fig. 87).

253. Torsion Bars

a. REMOVAL OF TORSION BARS. Remove the road wheel or compensating idler wheel (pars. 250 and 252).

Note. In order to remove the torsion bar from the right compensating idler, it is first necessary to remove the right rear wheel (par. 250*a*).

Remove the road wheel lifter or jack so that the road wheel arm is down in the fully released, no-road position (figs. 257 and 259). Loosen the torsion bar anchor cover (fig. 284) in the opposite wheel arm support. Remove the bolt and lock washer from the torsion bar end plug. Install the plug wrench into the recess of the end plug and unscrew the end plug from arm.

Note. Method of using plug wrenches 41–W–1961–125 and 41–W–1961–100 is the same. The former is used to remove the road wheel torsion bar end plugs (fig. 271), while the latter is used to remove the compensating-idler-wheel-torsion-bar end plugs.

To remove the road wheel torsion bars, screw the remover and replacer 41–R–2378–950 into the tapped hole at end of torsion bar and then pull the torsion bar out from arm (fig. 272). To remove the compensating-idler-wheel-torsion bars, install adapter 41–A–18–400 to torsion bar; then attach the remover and replacer 41–R–2378–950 to the adapter, and pull out torsion bar.

Note. It may be necessary to work the torsion bar up and down before it can be loosened from the wheel arm serrations. Be sure to remove all pieces when removing a broken torsion bar.

After the torsion bar has been removed, remove the torsion bar anchor cover from the opposite wheel arm support. Remove the anchor.

Note. The anchor for the left compensating-idler-wheel-torsion bar (in right wheel compensating-idler-arm support) is accessible only from inside the engine compartment.

b. CLEANING AND INSPECTION. Clean all parts with volatile mineral spirits or dry-cleaning solvent. Inspect torsion bar and torsion bar anchor for cracks or other defects with special attention to the condition of the serrations of both torsion bar and torsion bar anchor.

c. IDENTIFICATION OF TORSION BARS. The torsion bars have designating arrows stamped on the arm end. The arrows indicate the rotation of the bars when the wheel and arm are raised, or in other words, the direction of bar "spring." The last figure of the ordnance part number which is also stamped on the arm end also can be used as a quick guide for proper identification of torsion bars. Proper installation of torsion bars which are installed from the right side of the hull is as follows: All arrows in clockwise direction with the exception of the front spring arm torsion bar which is counterclockwise. Numbers (last figures of ordnance part number) run in this order from front to rear: 0, 2, 2, 2, 2, 9, and 2 (smallest of all the bars). Identification for bars installed from the left side of the vehicle is as follows: All arrows counterclockwise with the exception of the front spring arm torsion bar which is clockwise. Numbers run in this manner from front to rear: 9, 3, 3, 3, 3, 0, and 3 (smallest of all bars on this side—for compensating idler wheel).

d. INSTALLATION OF TORSION BARS. Coat both torsion bar serrations with grease as specified in lubrication order (fig. 87). Install torsion bar anchor, cover, and gasket but do not tighten cover bolts fully until torsion bar is installed. Install the torsion bar with the remover and replacer 41–R–2378–950 (fig. 272).

Note. Use adapter 41–A–18–400 with remover and replacer—41–R–2378–950 to install compensating idler wheel torsion bar.

Caution: Be sure to install the correct torsion bar. Do not use undue force during installation.

Install the torsion bar so that the end with the identifying marks faces outward. With wheel arm at its lowest position, insert torsion bar so that the notch at the end of the bar is vertical. Push torsion bar in until serrations of torsion bar engage those of anchor. Do not engage torsion bar outer serrations. The notch on the bar should then coincide with the line on the road wheel arm spindle. Lift road wheel arm until line on torsion bar coincides with line on spindle (31 degrees clockwise from notch). Push torsion bar the rest of the way in. Fill the outer cavity with grease as specified in lubrication order (fig. 87). Install the torsion bar end plug (fig. 271).

Note. Use plug wrench 41–W–1961–125 when installing road wheel torsion bar end plugs and plug wrench 41–W–1961–100 for installing the compensating idler torsion bar end plugs.

Install the bolt and lock washer in the torsion bar end plug. Tighten the torsion-bar-anchor-cover bolts in the opposite wheel arm support. Install road wheel or compensating idler wheel (pars. 250 and 252).

e. REMOVAL OF TORSION BAR ANCHOR. Remove the torsion bar anchor cover (fig. 284) from the bottom of the opposite wheel arm support. Remove the torsion bar anchor (fig. 256).

Note. The anchor for the left compensating-idler-wheel-torsion bar (in right compensating-idler-wheel-arm support) is accessible only from inside engine compartment.

f. INSTALLATION OF TORSION BAR ANCHOR. Position the anchor in the wheel arm support. Make sure serration keyway in anchor is up.

Note. All anchors of the large bars can be interchanged. The compensating-idler anchors can be interchanged with one another also.

Install the torsion bar (*d* above). Be sure the serrations on the torsion bar are securely meshed with those in the anchor. Install the anchor cover and gasket. Tighten the anchor cover screws.

Figure 271. Removing torsion bar end plug.

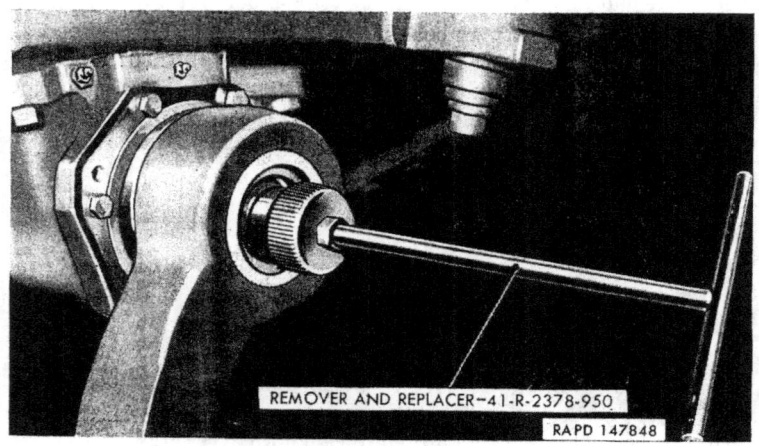

Figure 272. Removing torsion bar.

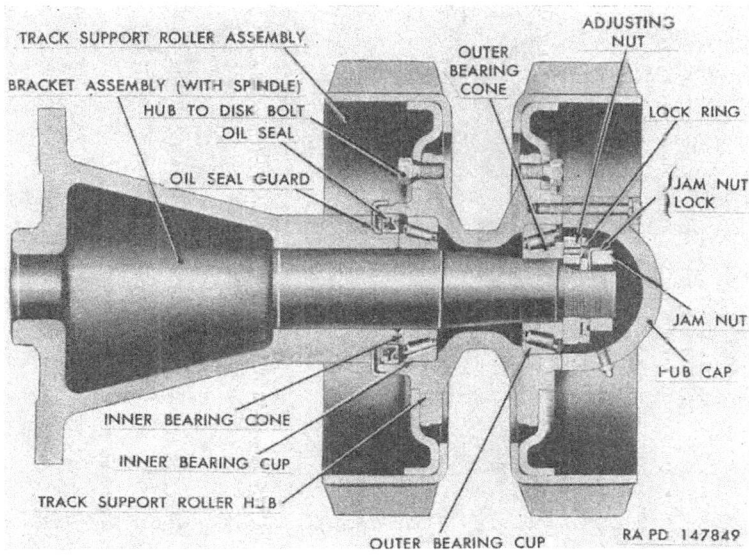

Figure 273. Track support roller—sectionalized view.

254. Track Support Rollers and Bracket Assemblies

a. REMOVAL. Remove the dust shield (par. 266). Place a board across the tops of the road wheels below the support roller to be removed. Place a jack on the board and raise the track above the support roller sufficiently to relieve the track weight on the roller. Remove the lower bolts from the support roller bracket (fig. 259). Support the roller and remove the upper bolts. Lower the roller assembly.

b. DISASSEMBLY (figs. 273 and 274). Remove the bolts and lock washers from hub cap and lift off hub cap. Remove the gasket and wipe surplus grease from hub. Straighten the ear on jam nut lock and remove jam nut, using socket wrench 41–W–3058–470, socket wrench head 41–H–1845–50, and socket wrench bar 41–B–312–200. Remove jam nut lock, lock ring, and adjusting nut. Use spanner wrench 41W–3242–300 to remove adjusting nut (fig. 261). Install hub cap temporarily to prevent outer bearing cone and rollers from dropping out when bracket assembly is removed. Pull support roller and outer bearing cone and rollers off the bracket assembly. Separate hub cap from track support roller and remove outer bearing cone and rollers. Remove inner bearing cone and rollers and pull the oil seal and oil seal guard from bracket assembly. Remove bearing cups from

459

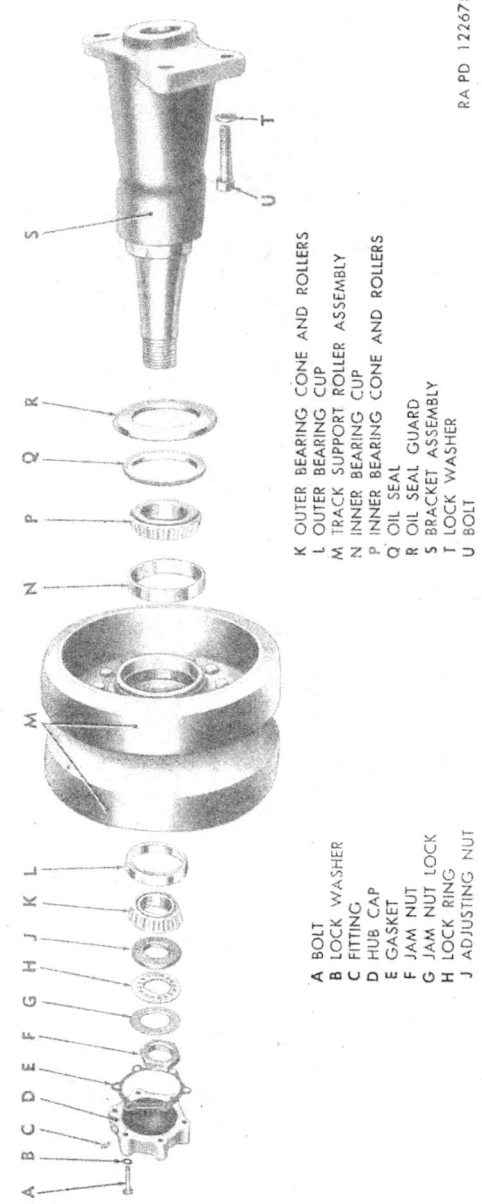

A BOLT
B LOCK WASHER
C FITTING
D HUB CAP
E GASKET
F JAM NUT
G JAM NUT LOCK
H LOCK RING
J ADJUSTING NUT
K OUTER BEARING CONE AND ROLLERS
L OUTER BEARING CUP
M TRACK SUPPORT ROLLER ASSEMBLY
N INNER BEARING CUP
P INNER BEARING CONE AND ROLLERS
Q OIL SEAL
R OIL SEAL GUARD
S BRACKET ASSEMBLY
T LOCK WASHER
U BOLT

Figure 274. Track support roller—exploded view.

hub, using screw 41-S-1047-315, remover and replacer 41-R-2374-655 for outer bearing cup, and remover and replacer 41-R-2374-630 for inner bearing cup (fig. 79).

c. CLEANING AND INSPECTION. Procedure is same as for road wheels (par. 250 *a* and *c*).

d. LUBRICATION. Pack bearings with lubricant, as specified on lubrication order (fig. 87).

e. ASSEMBLY. Install inner bearing cup, using remover and replacer 41-R-2374-630 and screw 41-S-1047-315 (fig. 275). Install outer bearing cup, using remover and replacer 41-R-2374-655 and screw 41-S-1047-315 (fig. 275). Install oil seal guard and new oil seal on bracket (fig. 273) using replacer 41-R-2391-420. Slide inner

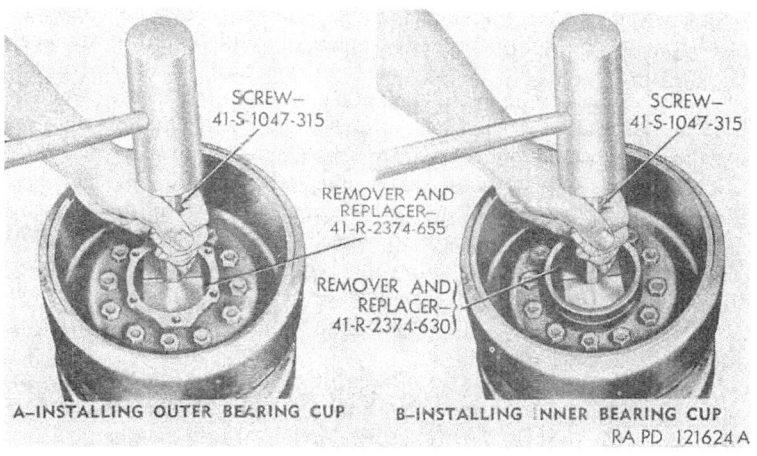

Figure 275. Installing bearing cups in track support roller.

bearing cone and rollers, track support roller, and outer bearing cone and rollers on bracket. Install adjusting nut, using spanner wrench 41-W-3242-300 (fig. 261), and tighten securely. Place lock ring on end of spindle and back off adjusting nut until dowel pin on nut and nearest hole on lock ring are in alinement. Install jam nut lock and jam nut, using socket wrench 41-W-3058-470, socket wrench head 41-H-1845-50, and socket wrench bar 41-B-312-200. Bend jam nut lock against one side of jam nut. Place a new gasket on hub, install hub cap, and secure with bolts and lock washers.

f. INSTALLATION. Cover the face of the bracket with joint-sealing compound. Position the roller bracket against the hull and install the lock washers and bolts. Tighten the bolts to 305-360 lb-ft torque. Install the dust shields (par. 266).

255. Shock Absorbers and Brackets

a. INSPECTION. Inspect the shock absorbers for evidence of oil leakage. A thin film of oil and some dust deposit is normal and does not warrant replacement of the shock absorber but if there is any indication of continuous or considerable leakage, such as drops of oil accumulating or heavy dust adhesion, replace the shock absorber. Inspect the shock absorber mounting brackets for cracks and worn mounting pin holes. Replace any damaged brackets. The most practical method of testing the shock absorber is by the temperature method. This test should be made immediately after a run of not less than 5 miles of high-speed operation or 4 miles of cross-country operation. Feel the small part of the shock absorber and then the hull near the shock absorber (figs. 257 and 259), and note the difference in temperature. If the hull and the shock absorber are the same temperature, the shock absorber is not operating and should be replaced. If the shock absorber is warmer than the hull, it is functioning properly. The difference in temperature should be clearly evident, but the small part of the shock absorber (reserve cylinder) does not need to be extremely hot to indicate a satisfactory unit.

b. REMOVAL. Remove the cotter pin and attaching pin that connect the shock absorber to the road wheel arm (or to the front spring arm if removing the front road wheel shock absorber) (fig. 257). Remove the cotter pin, nut, and washer which secure the shock absorber to the bracket on the hull (or remove the retainer that secures the front road wheel arm shock absorber to bumper spring bracket). Remove the shock absorber.

c. CLEANING. Wipe excess oil and dust off the shock absorber and clean shock absorber, pins, bolts and brackets if they have been removed, with volatile mineral spirits or dry-cleaning solvent.

d. INSTALLATION. Install brackets if they have been removed. Position the shock absorber on the hull bracket and install the washer, nut, and cotter pin (or the retainer and screws for the front road wheel arm shock absorber). Extend the shock absorber plunger to aline the shock absorber and road wheel arm holes, and insert the attaching pin and cotten pin (fig. 257).

Section XX. DRIVER'S CONTROLS AND LINKAGE

256. Description

a. MANUAL CONTROL BOX (fig. 10). The manual control box, located in the driver's compartment, is used to control shifting and steering of the vehicle. A control lever is mechanically connected within the unit. The lever controls both steering and shifting through linkage connected to levers on the transmission valve body (fig. 221). Hand grips and finger grips on the lever maintain the desired gear range and steering position. The manual control mechanism is fully inclosed and is lubricated by a supply of oil within the right and left cases.

   b. THROTTLE CONTROLS. The throttle control linkage (fig. 278) connects the hand throttle lever and the accelerator pedal to the throttle control lever on each carburetor. A governor is connected into the linkage to control engine speed.

   c. BRAKE CAM ASSEMBLY (figs. 10 and 279). The brake cam assembly is a coupling device between the service brake pedal and the transmission service brakes. The primary purpose of the brake cam assembly is to provide a cam to furnish the required mechanical effect in applying the service brakes. The assembly is connected to the service brake pedal by a cable and to the transmission service brakes by brake control rods. Movement of the service brake pedal is reflected in movement of the cable yoke on the brake cam assembly. This causes movement of the brake control rods which actuate the transmission service brake control levers. The brake cam can be locked in any position by pulling the hand brake lock handle. This pressure on the brake cam can be released by depressing the service brake pedal, which allows the yoke to fall free and thus unlock.

257. Manual Control Box and Linkage

   a. SERVICE OF MANUAL CONTROL BOX (fig. 10). Both right and left units of the manual control box contained oil for lubrication. Oil level in the units is checked by an oil level plug on the rear of each unit. Oil is poured in at the filler plugs on top of each unit and drained through an opening in the bottom by removing the drain plug. Use oil as specified in lubrication order (fig. 86).

   b. MANUAL CONTROL LEVER CONTROL ROD ADJUSTMENT. To insure proper operation of the manual control box (fig. 10), the control rod, which is actuated by the hand grip handle, must be adjusted. If this rod is out of adjustment, it will not be possible to shift in or out of neutral or reverse positions. Adjust as follows:

   (1) Place manual control lever in neutral. Remove manual control lever grip at top of lever and remove spring. Remove cotter pin from control rod adjusting nut. Back off adjusting nut several turns but do not remove from rod.

   (2) While pressing the hand grip handle, turn control rod adjusting nut down until lever will shift from neutral position. Turn down adjusting nut two additional turns and insert cotter pin. Install spring and manual control lever grip.

   c. MANUAL CONTROL BOX REMOVAL. Disconnect shifting and steering linkage at bottom of each unit of manual control box (fig. 276). Remove primer pump (par. 180). Remove four safety nuts and plain washers securing control box to mounting bracket and remove control box.

d. MANUAL CONTROL BOX INSTALLATION. Position manual control box to mounting bracket and secure with four safety nuts and plain washers. Install primer pump (par. 180). Connect steering and shifting linkage at bottom of manual control box.

e. STEERING AND SHIFTING LINKAGE. The manual control linkage for steering and shifting will not ordinarily require replacement. The linkage should be correctly adjusted, however, to eliminate any looseness or play. Movement of the manual control lever must be reflected at the control lever shafts on the transmission valve body without binding, looseness, or play.

 (1) *Steering control linkage adjustment.* Place the manual control lever in neutral steer position (par. 11). With the linkage connected, the pointer under the steering control lever (fig. 230) on the transmission valve body should point to the punch mark on the valve body. If it does not, disconnect the control rod from the valve body. Turn the pointer to the punch mark. Loosen the lock nut on the control rod and turn the eyebolt in or out as necessary to adjust the linkage. If sufficient adjustment cannot be made at this point, loosen the lock nut at the other end of the rod and follow the same procedure. Continue adjusting the linkage in this manner until the hole in the eyebolt lines up directly with the holes in the lever. When the holes are alined, connect the linkage to the control lever.

 (2) *Shifting control linkage adjustment.* Place the manual control lever in neutral shift position. With the linkage connected, the pointer under the shifting control lever (fig. 221) on the transmission valve body should point to the forward punch mark (closest to the engine) on the valve body. If it does not, disconnect the control linkage from the lever and move the pointer to the punch mark. Loosen the lock nut on the valve body end of the transmission rod and turn the eyebolt in or out as necessary to adjust the linkage. If sufficient adjustment cannot be made at this point, loosen the lock nut at the other end of the rod and follow the same procedure. Continue adjusting the linkage in this manner until the hole in the eyebolt lines up directly with the holes in the lever. When the holes are alined, connect the linkage to the control lever.

 (3) *Excessive play in control linkage.* Excessive play in the linkage usually can be traced to loose bellcranks (fig. 276). Two steering control bellcranks are mounted on brackets on the left side of the transmission (fig. 230) and two on posts under the ammunition stowage boxes (fig. 277). In each

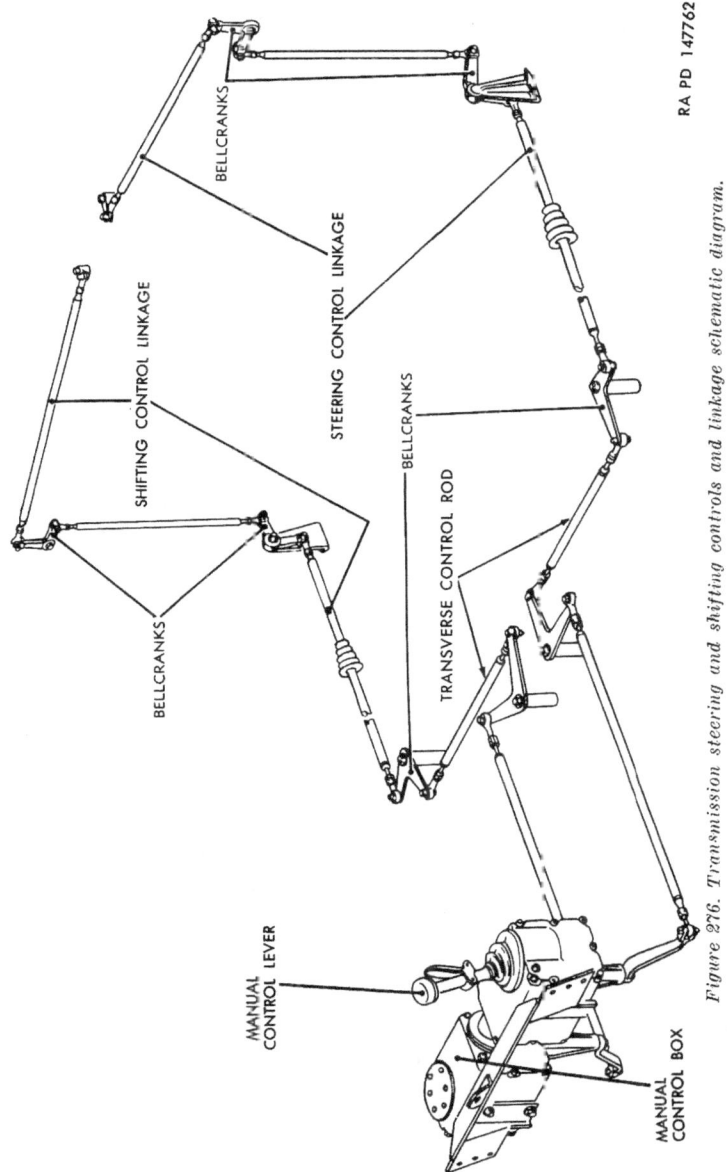

Figure 276. Transmission steering and shifting controls and linkage schematic diagram.

case the bellcrank is secured to a stud by a nut and washer. The shifting control bellcranks are similarly located on the right side of the vehicle (figs. 229 and 277). Tighten all nuts securely to eliminate looseness of the bellcranks on their mountings. The bellcrank under the transmission may be reached through the access holes in the hull under the transmission (fig. 285). To obtain access to the bellcranks under the ammunition stowage boxes open the two center boxes (fig. 288) (for the two center bellcranks) or the extreme right and left boxes (for the two side bellcranks).

(4) *Replacement of steering and shifting linkage.*

(a) *Driver's compartment* (fig. 276). Disconnect control rod at arm under manual control box by removing screw and nut. Remove screw and nut securing rod to a forward bellcrank and remove rod.

> Note. If more working space is required, remove ammunition stowage boxes (par. 265).

To install rod, position one end of rod to arm under manual control box and secure with screw and nut. Position other end of rod to forward bellcrank and secure with screw and nut. Install ammunition stowage boxes (par. 265), if removed.

(b) *Fighting compartment* (fig. 277). Remove screw and nut securing transverse control rod to bellcrank nearest bulkhead.

> Note. If more working space is required, remove ammunition stowage boxes (par. 265).

Remove nut and washer securing bellcrank to post on bottom of hull. Lift bellcrank and transverse control rod from fighting compartment. To install bellcrank and transverse control rod, position bellcrank on post and secure with nut and washer. Position control rod and secure to bellcrank nearest bulkhead with screw and nut. Install ammunition stowage boxes (par. 265), if removed.

(c) *Engine compartment.* To remove the long control rods extending from the fighting compartment to the transmission, it is necessary to remove the fuel tanks (par. 177). Remove the screws and nuts securing the control rods to the bellcranks at the rear of the fighting compartment (fig. 277). Loosen clamps securing rod to boots on bulkhead in engine compartment. Disconnect the rods at the bellcranks under the transmission. Move the rods until they are clear of the bulkhead and rear shroud channel, then lift out of the vehicle. Remove the vertical rods at the transmission by disconnecting at the upper and

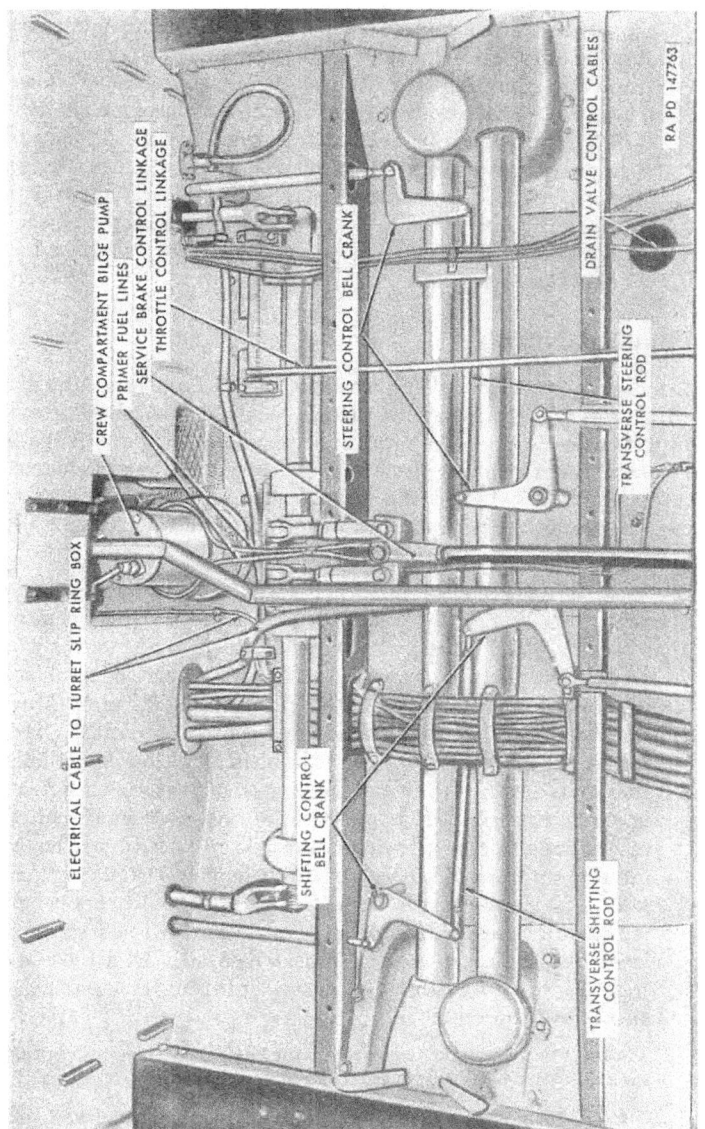

Figure 277. Hull floor with ammunition stowage boxes removed.

lower bellcranks. Remove the nuts and washers securing the four rear bellcranks to brackets mounted on the hull. Remove bellcranks. To install long control rods, push each rod from engine compartment through opening in bulkhead until rear end of rod clears shroud channel. Pull rod back, guiding rear end through hole in channel. Connect rod to bellcrank at rear of fighting compartment (fig. 277). Position bellcranks at transmission on brackets and secure each bellcrank with a nut and washer. Connect long control rod to lower bellcrank. Connect vertical rod to upper and lower bellcranks. Install screw and nut at each point of connection. Tighten boot clamp on bulkhead in engine compartment.

258. Throttle Controls

a. LINKAGE ADJUSTMENT. The throttle control linkage is adjusted at the throttle control linkage turnbuckle located under the left ammunition stowage box in the fighting compartment. To adjust the linkage, open the left stowage box floor plate and remove the bottom shell rack (fig. 277). Turn the linkage coupling (turnbuckle) to lengthen or shorten the control rod in the fighting compartment.

b. LINKAGE REPLACEMENT (fig. 278).

(1) *Removal.* Remove two extreme left ammunition stowage boxes (fig. 288). Disconnect front control rod at the accelerator pedal shaft by removing cotter pin and clevis pin. Disconnect the front control rod at the front control rod support by removing two cotter pins and clevis pin. Move the rod clear of obstructions and lift out of the vehicle. Disconnect the intermediate rod at the turnbuckle and pull the forward section of the intermediate rod out through the fighting compartment. Disconnect the rear section of the intermediate rod at the lever on the throttle cross shaft by removing cotter pin and clevis pin. Remove power plant (par. 159). Disconnect the rear control rod at the other lever on the throttle cross shaft and at the rear control rod support. Loosen boot clamp securing rod at bulkhead. Pull the rod forward into the fighting compartment through the bulkhead. Lift out of the vehicle.

Note. The control rods secured to the engine are removed with the engine and are normally replaced only during engine overhaul.

(2) *Installation* (fig. 278). Position the front control rod and secure to the accelerator pedal shaft with clevis pin and cotter pin. Position the intermediate rod and connect the front and intermediate rods to the front control rod support with

clevis pin and two cotter pins. Connect the intermediate rod to the lever on the throttle cross shaft with clevis pin and cotter pin and install the turnbuckle connecting the two sections of the rod. Position the rear control rod and connect to the

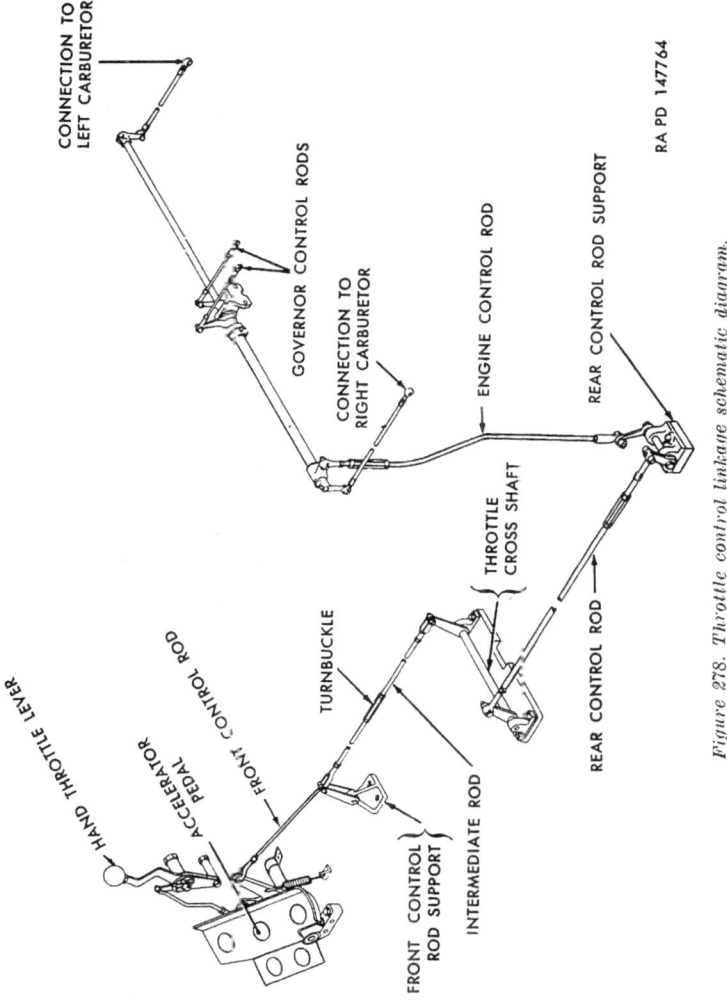

Figure 278. Throttle control linkage schematic diagram.

other lever on the throttle cross shaft and to the rear control rod support with clevis pins and cotter pins. Tighten boot clamp at bulkhead. Install the power plant (par. 160). Adjust linkage as necessary (*a* above). Install two extreme left ammunition stowage boxes (fig. 288).

469

259. Brake Controls

a. BRAKE CAM ASSEMBLY (figs. 10 and 279).
(1) *Removal.* Disconnect foot brake cable from cable yoke by removing screw. Disconnect crew compartment brake rod from brake cam by removing cotter pin and clevis pin. Re-

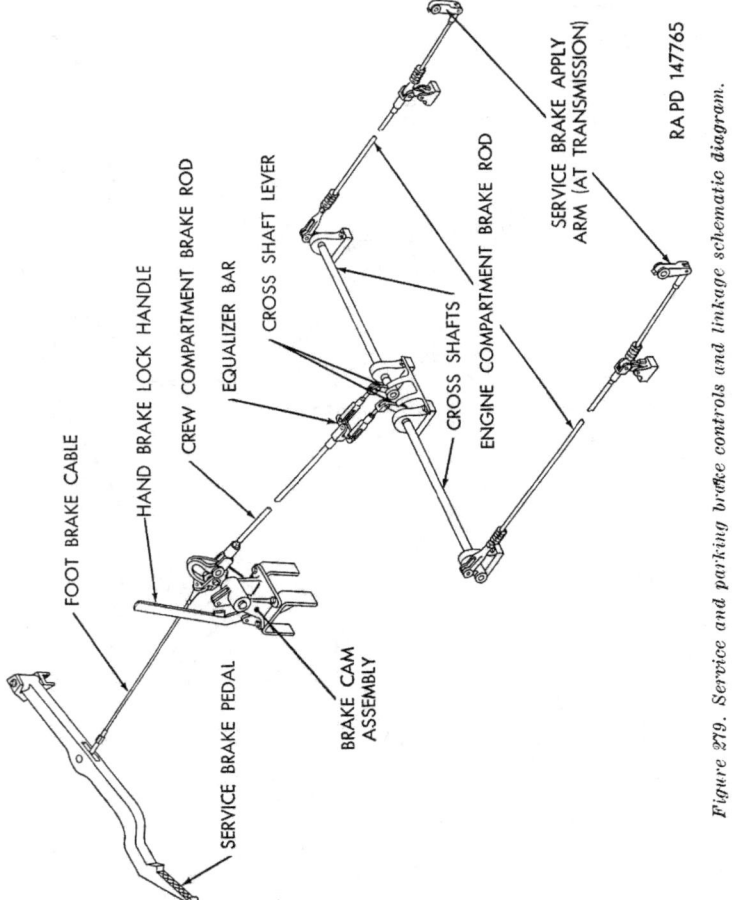

Figure 279. Service and parking brake controls and linkage schematic diagram.

move four screws securing brake cam assembly to supports on hull floor and remove brake cam assembly.
(2) *Installation.* Position brake cam assembly to supports on hull floor and secure with four screws. Connect foot brake cable to cable yoke with screw. Connect crew compartment brake rod to brake cam with clevis pin and cotter pin.

b. Adjustment of Service Brake Linkage (fig. 279). The service brake linkage must be serviced and adjusted to eliminate any play or looseness. Movement of the service brake pedal must be reflected in movement of the cable clevis on the brake cam; movement of the brake cam must be reflected in movement at the service brake apply arms on the transmission. Replace parts and adjust linkage rods or cables, as necessary, to eliminate looseness or play.

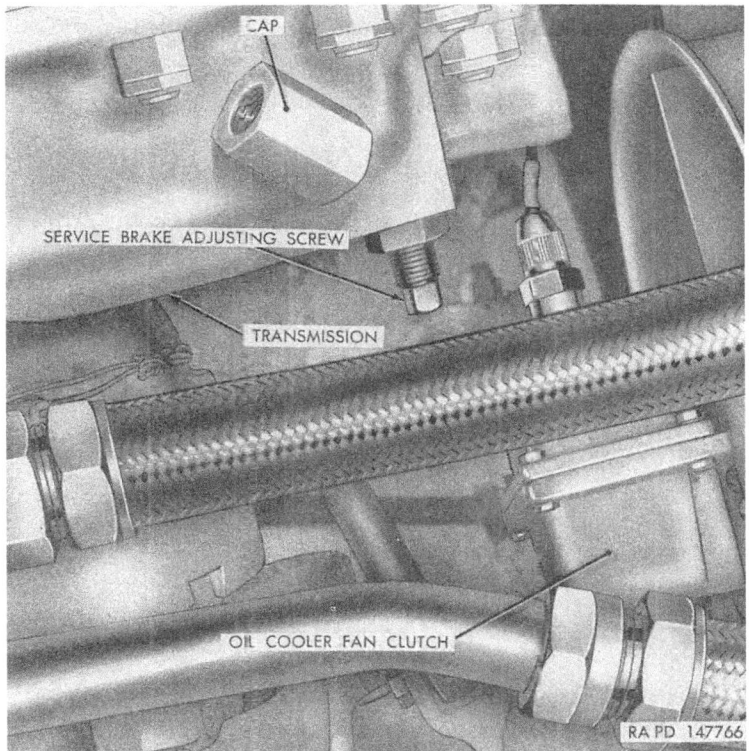

Figure 280. Service brake adjusting screw cap—removed.

c. Adjustment of Service Brakes.
  (1) *General.* A choice of two methods of brake check and adjustment are available ((2) below). Adjustment by either method is effected by shifting the position of the stationary cam ring of the disk-type brakes. A service brake adjusting screw (fig. 280) (with a worm thread on internal end) projects through each end cover of the transmission and provides an external means of adjusting brake clearance. The screw is covered by a cap (fig. 280). Turning the service brake

adjusting screw (fig. 281) actuates an internal worm, which causes rotation of the stationary cam ring.

(2) *Check service brake adjustment.*

(*a*) Brake adjustment check is made with a force of approximately 950 pounds at each of the service brake apply arms (fig. 222). This can be accomplished by fully applying the service brake pedal (fig. 9). Release the parking brake (par. 13), if applied. Remove the two brake linkage access hole covers from beneath the hull (fig. 285) to

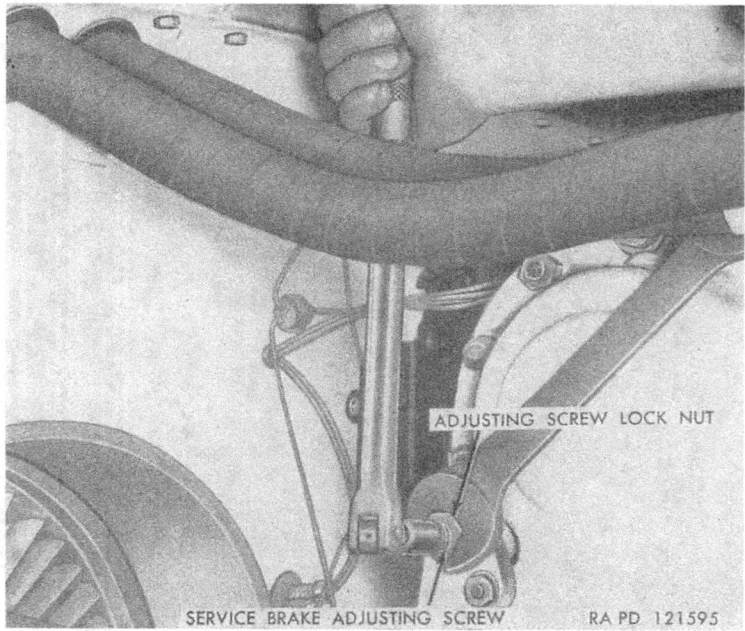

Figure 281. Adjusting service brakes.

gain access to the service brake linkage (fig. 283). Measurement of the movement of the brake rod (equivalent to movement of link pin in brake apply arms at transmission) must be made to determine travel between released and fully applied positions of the brake pedal. If the lever travel is not $1\frac{13}{16}$ inches, check to make sure that the brake linkage is correctly adjusted (*b* above). Disconnect the linkage from the brake apply arm by removing the cotter pin and connecting pin. When no force is exerted on the brake pedal, the hole in the brake apply arm should aline easily with the holes in the linkage. If it does not, turn the yoke in or out until the holes are alined. Replace the connecting pin and cotter pin. If the linkage adjustment is correct, adjust the brakes as outlined in (3) (a) below.

Warning: Operating with less than $1\tfrac{13}{16}$-inch brake apply arm travel may result in brake failure and serious damage to transmission due to increased internal friction.

(b) Alternate method of brake adjustment check is to remove the right and left side access doors on the rear of the hull (fig. 285). Remove the brake inspection hole plugs from the transmission rear housing next to the end covers (fig. 282). A circular window machined in the brake anchor

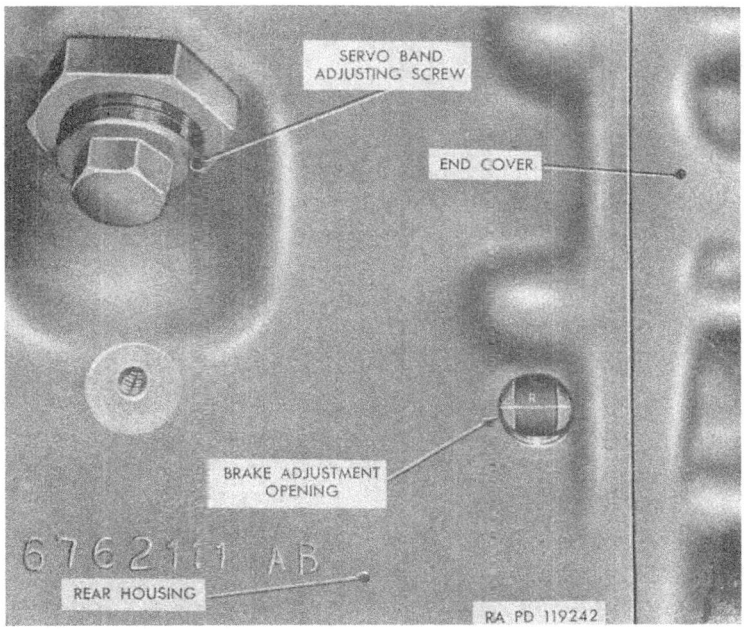

Figure 282 Opening for checking brake adjustment.

exposes the adjustment markings on the brake apply cam. One is marked "A," for apply position; the other is marked "R" for release position. Apply the brake with 120 lb-ft torque. The line marked "A" must line up with the base line chiseled on the anchor. If these lines do not line up, adjust the brakes as outlined in (3) (b) below.

(3) *Adjustment procedure.*

(a) The procedure for adjusting service brakes is the same on both right and left sides of transmission. Release the

473

parking brake (par. 13) if it is engaged. Remove engine compartment left and right rear grille doors (fig. 101). Remove the cap from the service brake adjusting screw (fig. 280) and loosen the lock nut. Turn the adjusting screw (fig. 281), so that the brake linkage rod travel is between $1^{13}/_{16}$ and $1^{27}/_{32}$ inches from the released to the fully applied positions of the service brake pedal ((2) (*a*) above). To tighten the brakes, turn the right service brake adjusting screw counterclockwise and turn the left service

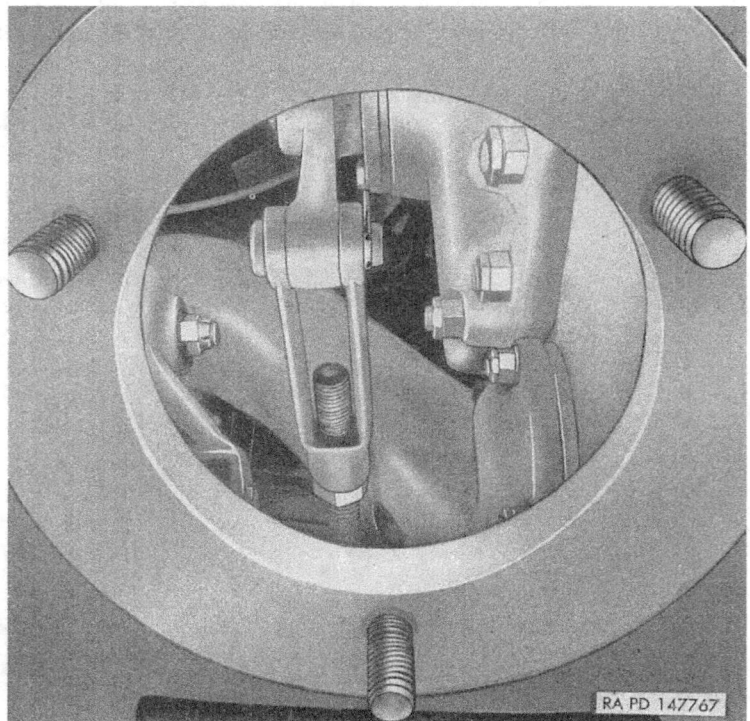

Figure 283. Service brake linkage.

brake adjusting screw clockwise. Reversing this rotation will loosen the brakes.

Caution: Correct lever travel may sometimes be obtained by turning the adjusting screw in the wrong direction. This results in no braking action.

The arrow cast in transmission housing to the rear and slightly below each adjusting screw indicates the direction of rotation to tighten brakes.

Caution: To avoid inaccurate adjustment due to internal friction and lash, it is necessary to make all final adjustments while turning the adjusting screw in the direction of tightening the brakes. If brake is too tight, back up the adjusting screw to fully loosen the brake, then proceed to tighten to proper adjustment.

After final adjustment, lock the brake adjusting screw by tightening the adjusting screw lock nut (fig. 281), and install the brake adjusting screw cap. Install the two brake linkage access hole covers (fig. 285). Install the engine compartment left and right rear grille doors (fig. 101).

(*b*) Alternate method of service brake adjustment is the same on both right and left sides of transmission. Release the parking brake (par. 13) if it is engaged. Remove engine compartment left and right rear grille doors (fig. 101). Remove the cap from the service brake adjusting screw (fig. 280) and loosen the lock nut. Adjust (fig. 281) in the direction of the arrow, adjacent to each adjusting screw, until the "A" lines up with base line. Lock brake adjusting screw with lock nut and install cap (fig. 280). Install engine compartment right and left rear grille doors (fig. 101). Install right and left side access doors on rear of hull (fig. 285).

d. REPLACEMENT OF BRAKE CONTROL LINKAGE (fig. 279).

(1) *Removal.*

(*a*) Remove ammunition stowage boxes (par. 265). To remove crew compartment brake rod, disconnect rod at yoke on brake cam assembly by removing nut. Remove cotter pin and clevis pin securing rod to equalizer bar. Pull rod toward rear of vehicle until rod clears forward section of stowage boxes and remove rod.

(*b*) At equalizer bar, remove cotter pins and clevis pins securing rods from equalizer bar to cross shafts. Remove cotter pins and clevis pins securing rods to cross shaft levers and remove rods.

(*c*) Remove cotter pins and clevis pins securing each of the extreme left and right cross shaft levers to engine compartment brake rods. Remove retaining rings from ends of cross shafts and remove levers. Remove eight bolts securing cross shaft pillow blocks and brake stop and remove cross shafts, pillow blocks, and brake stop.

(*d*) Remove cotter pins and clevis pins on yokes just forward of bulkhead. Loosen front boot clamps and remove yokes. Remove fuel tanks (par. 177). Remove cotter pins and clevis pins connecting front and rear sections of engine compartment brake rods. Pull front sections toward transmission until clear of bulkhead. Loosen rear boot clamps and remove rear sections of engine compartment brake rods.

(2) *Installation.*
 (a) Install cross shafts, pillow blocks, and brake stop and secure with eight bolts. Position four cross shaft levers and secure with retaining rings.
 (b) Position rods from equalizer bar to cross shaft levers and secure rods with a clevis pin and cotter pin at each end.
 (c) Position screw compartment brake rod and connect one end to yoke on brake cam assembly. Secure other end to equalizer bar with clevis pin and cotter pin.
 (d) From the engine compartment, push the threaded end of each of the front sections of the engine compartment brake rods far enough through the opening in the bulkhead to reach the cross shaft levers. Install a yoke on the threaded end of each rod and tighten the lock nut. Tighten front boot clamps. Secure yoke to cross shaft lever with clevis pin and cotter pin. Connect other end of rod to rear section of engine compartment brake rod and secure with clevis pin and cotter pin. Tighten rear boot clamps.
 (e) Install fuel tanks (par. 177). Install ammunition stowage boxes (par. 265). Check brake adjustment and adjust if necessary (*c* above).

Section XXI. HULL

260. Description

a. GENERAL. The hull is a welded unit made up of armor steel castings, plates, and sections welded together. The cast, V-shaped front section is welded to the hull floor at the bottom of the "V." The top of the "V" extends up and back to form the front of the vehicle and the roof of the driver's compartment. The hull rear casting forms the rear of the engine compartment. Both front and rear castings are welded to the hull upper center plates and hull side plates. The hull rear casting has two large holes machined on either side to receive the final drives and three holes for transmission inspection covers. The wheel arm supports extend through openings in the floor side plates and are bolted in position. Inspection plates and access covers are located in the bottom of the hull (fig. 285).

b. BULKHEAD. The crew compartment is separated from the engine compartment by a bulkhead welded in place. This bulkhead is provided with inspection plates for accessibility to the accessory case end of the engine. The bulkhead helps support the hull top deck and the turret ring at the rear of the turret opening, and also serves as a fire wall between the engine and crew compartment.

c. Escape Hatch Doors. Two escape hatch doors in the front floor section, one in front of each driver's seat, provide emergency exits (figs. 12, 13, and 284). The hatches are held in place by a quick-release toggle-type bar, which allows it to drop to the ground when the release handle is pulled upward.

d. Drainage System. Manually operated drain valves (figs. 16, 17, 286, and 287) and two bilge pumps (figs. 334 and 335) are provided in the crew and engine compartment to permit the hull to be drained quickly of accumulated water, oil, or fuel.

Figure 284. Bottom of hull—front view.

e. Doors. Heavy cast armor steel doors cover the openings in the section above the driver's and assistant driver's seats. These doors, which are hinged at the outer edge, can be locked in the closed position from the inside and in the open position by a catch on the outside of the door. A periscope mount is provided for each door.

f. Fenders, Dust Shields, and Stowage Boxes. The tracks are covered at the rear, front, and sides by heavy metal fenders and dust shields. These fenders are supported by outriggers welded to the hull side plates. Stowage boxes, mounted on top of the fenders, provide space for equipment.

g. Towing Shackles, Lifting Eyes, Pintle, and Gun Traveling Lock. Two towing shackles at the front (fig. 6) and two at the rear

A—SIDE ACCESS DOOR
B—CENTER ACCESS DOOR
C—BRAKE LINKAGE ACCESS HOLE COVER
D—TRANSMISSION DRAIN ACCESS HOLE COVER
E—OIL COOLER SERVICE HOLE COVER
F—ENGINE MOUNTING BOLT HOLE

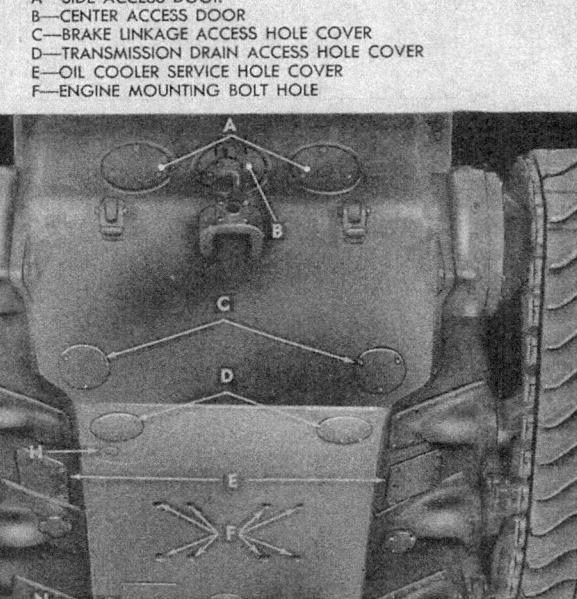

G—MAIN ENGINE DRAIN ACCESS HOLE COVER
H—DRAIN VALVE
J—FUEL TANK DRAIN ACCESS HOLE COVER
K—HULL DRAIN HOLE COVER
L—ESCAPE HATCH DOOR
M—MAIN ENGINE OIL FILTER ACCESS HOLE COVER
N—MAIN GENERATOR ACCESS HOLE COVER

RA PD 147769

Figure 285. Bottom of hull—rear view.

(fig. 7) provide means of attaching a tow bar or cable. The hull is equipped with two lifting eyes (figs. 4 and 5) at each end for use when hoisting the vehicle. A demountable towing pintle at the rear is used for towing light vehicles (fig. 290). The gun traveling lock (fig. 289) supports the gun while in traveling position thereby relieving the gun mechanism of shock-load arising with vehicle motion.

261. Doors and Cover Plates

a. DRIVER'S DOORS.

(1) *Spring adjustment* (fig. 14). Test the action of the torsion bar by opening and closing the door several times. In the closed position, an opening of approximately one inch between the door and hull must be maintained in order to have the correct adjustment. If the adjustment is not satisfactory, raise the door to release the torque on the torsion bar. Scribe a line on the torsion bar and the retainer. Remove the two screws which attach the retainer to the hull. Raise the torsion bar up enough so that the retainer can slide off the torsion bar. Tap the retainer off the torsion bar and position the retainer on the torsion bar one or two serrations from the scribe marks, depending on the amount of torque desired. Position the retainer on the hull and install the two screws securing the retainer to the hull. Test the action of the door. Repeat the adjustment if necessary.

(2) *Removal.* Raise the door into the neutral position to relieve the torque in the torsion bar. Hold or block the door securely in this position. Remove the socket head screws which attach the torsion bar rear retainer to the hull. Remove the screws which attach both of the hinge sleeve blocks to the hull (fig. 14). Lift the door off the hull. If the same torsion bar is to be installed, first scribe a reference mark on both ends of the torsion bar and the retainers so that the retainers can be installed in their original positions. Slide the rear retainer off the bar. Remove the torsion bar from the door by driving it toward the rear, out of the front retainer. Remove the screws which attach the torsion bar front retainer to the door. Remove the hinge sleeve by driving it out of the door and through the hinge sleeve blocks toward the front. Remove the locking lever plunger screw and remove the plunger. Remove the pivot pin clip and remove the locking lever. Loosen the lock nut at the door catch handle. Unscrew the handle and lock nut from the catch rod handle. Remove the cotter pin from the catch handle rod hinge pin. Drive out the hinge pin. Remove the cotter pin from the

catch handle trigger hinge pin, and drive out the hinge pin, trigger, and catch spring.

(3) *Installation.* Position the locking handle on the inside of the door. Install the pivot pin and clip. Install the plunger and plunger screw. Position the door catch spring between the catch and door. Aline the holes in the door latch trigger (fig. 14) and the bosses on the door. Install the trigger hinge pin and cotter pin. Insert the catch handle rod through the door from the inside. Aline the hole in the trigger with the hole in the end of the rod, and insert the hinge pin and cotter pin. Install the lock nut on the catch handle rod. Screw the handle on the rod until it is nearly tight and at right angles to hinge pin through the front door hinge. Position the hinge blocks on the hinge sleeve as it is driven into position against the snap ring in the rear hinge. Place the torsion bar front retainer in position, and install the two screws. From the rear, insert the torsion bar through the hinge sleeve. Aline the serrations and reference marks, if previously made. Drive the bar into the front retainer until it is flush with the face of the retainer. Aline the serrations and the reference marks (if any), and position the rear retainer on the torsion bar. Place the door on the hull in the closed position. Aline the hinge sleeve blocks, and install the attaching screws. Adjust the driver's door torsion bar (*a* above).

b. COVER PLATES.

(1) *Hull bottom covers.* There are 12 covers on the bottom of the hull, accessible from below the tank (figs. 284 and 285).

Note. All these covers are provided with gaskets to provide a watertight closure. Upon removal of cover plates, these gaskets should be stripped off and new ones used in installation of the covers.

(*a*) *Drain covers.* Two round covers, each secured with three nuts and lock washers, give access to the transmission drain plugs. Another round cover, secured with three cap screws and lock washers, gives access to the main engine drain plug. Two round covers, each secured with three bolts and lock washers, give access to the fuel tank drain plugs. A hull drain hole cover, near the front of the hull at the right side, is secured with three nuts and lock washers.

(*b*) *Service access covers.* Two round covers give access to the service brake linkage. Each is secured with three nuts and lock washers. Two rectangular covers give access to the right and left oil coolers, each secured with six nuts and lock washers. A rectangular cover gives access to the main generator. It is secured with four bolts and lock washers. A round access hole cover gives access to the main engine oil filter. It is secured with three bolts and lock washers.

(2) *Hull rear access doors.* Three access doors are located on the rear of the hull (fig. 285). Each is fitted with a gasket and is secured with three bolts and lock washers. The center access door gives access to the power-take-off and the two outer doors give access to the transmission-reverse-and-low-gear-band-adjusting screws (figs. 227 and 228) and brake adjustment opening plugs (fig. 282).

(3) *Fuel tank filler covers.* Two fuel tank filler covers on top of the hull (fig. 154) give access to the fuel tank fillers. These covers are hinged to the hull and are provided with locks.

(4) *Fuel tank shut-off valve cover plate.* The fuel tank shut-off valve cover plate (fig. 146) provides access to the fuel tank shut-off valves. It is fitted with a gasket and secured by two quick-fastening studs.

Caution: In installing cover plates, see that mounting surface is clean. Use new gaskets, coated on one side with liquid-type gasket cement.

262. Hull Seat Assemblies

a. REMOVAL. Remove seat back. Raise the seat (par. 20) and remove four bolts and lock washers which attach the seat. Lift the seat out of the vehicle through the turret.

b. INSTALLATION. Position the seat in the crew compartment. Raise the seat and install four lock washers and bolts. Install seat back.

263. Crash Pads and Head Rests

a. GENERAL. Crash pads are provided for the safety of the crew. Head rests are also provided with the periscope housings. The crash pads are secured either by welding or by adhesive.

b. REMOVAL.

(1) To remove the driver's and assistant driver's crash pads, strip off the head crash pads (inside of doors) and the shoulder crash pads on sides of hull. To remove the crash pads mounted on the door openings grind or break welds and remove pads.

(2) To remove the commander's crash pads in the cupola hatch, grind or break welds securing the circular crash pad. Remove pad assembly.

(3) Remove the crash pad on the loader's hatch door by stripping it off.

(4) Remove the crash pad on the turret ring directly behind the 90-mm gun breech by removing two cap screws, flat washers, and two lock washers.

(5) Remove the crash pad over the portable-fire extinguisher stowage door by stripping it off.

(6) Remove the head rests by removing two cap screws and lock washers.

c. INSTALLATION.

Note. Grind surfaces clear of old weldings. Clean surfaces of old adhesive by scraping clear.

(1) Use adhesive to install new driver's and assistant driver's head and shoulder pads, loader's hatch door pad, and the portable-fire extinguisher stowage door pad.

(2) Weld new pad assemblies on the driver's and assistant driver's door openings and the commander's cupola hatch circular pad.

Note. Use a piece of wet asbestos when welding new crash pads to hull and turret to prevent scorching or burning the pads.

(3) Install the turret ring pad and secure with two cap screws, flat washers, and lock washers.

(4) Install the periscope head rests and secure each with two cap screws and lock washers.

264. Drain Valves

a. GENERAL. The crew compartment drain valve (figs. 16 and 286) and the three engine compartment drain valves (figs. 286 and 287) are all replaced in the same manner.

b. REMOVAL (fig. 286). To remove drain valve, knock out tapered pin and the drain valve will separate from the knob and spring and fall from the hull bottom.

c. INSTALLATION (fig. 286). Push drain valve up through hull bottom and position spring and knob over valve. Line up the holes in valve and knob and insert the tapered pin.

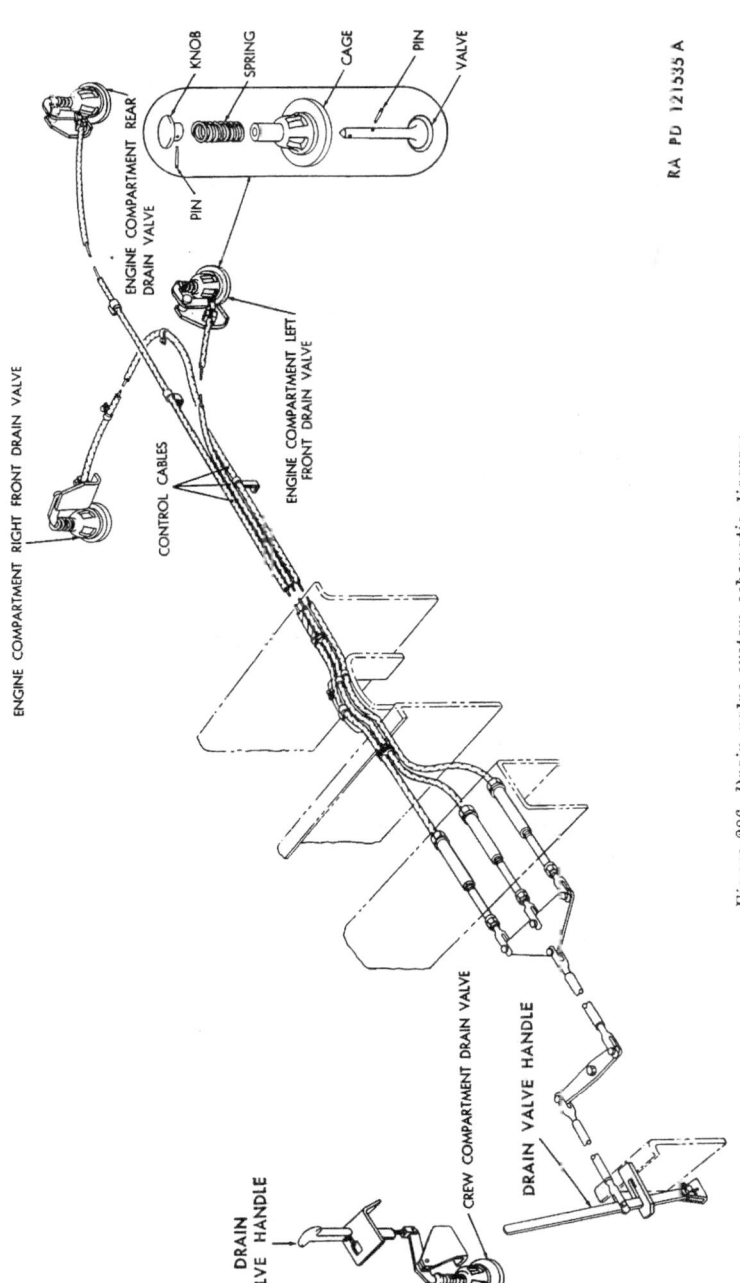

Figure 286. Drain valve system schematic diagram.

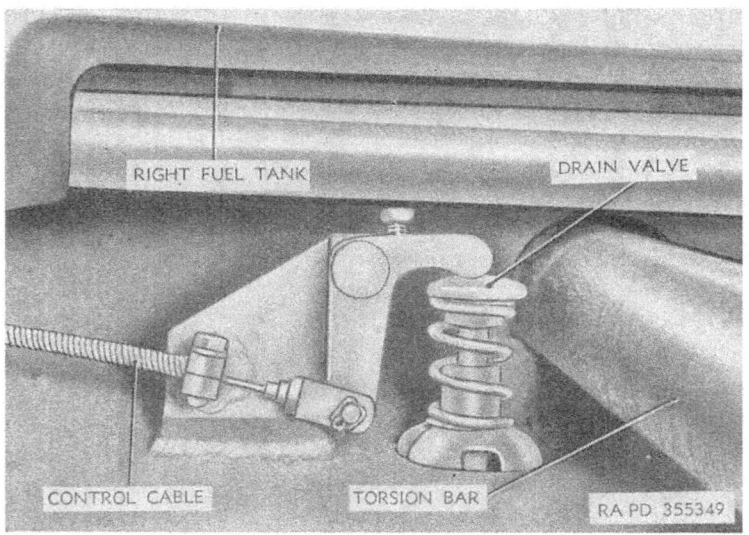

Figure 287. Engine compartment right front drain valve—installed view.

265. Ration Box and Ammunition Stowage Boxes

a. REMOVAL.

(1) *Depress 90-mm gun.* Depress the 90-mm gun to its lowest position (par. 76).

(2) *Remove the slip ring box.* Refer to paragraph 296.

(3) *Remove ration tray.* Remove the rear center compartment cover plate assembly (ration tray cover) by removing two mounting screws. Remove ration tray, secured by four screws (fig. 288).

(4) *Remove cal. .50 ammunition tray.* Remove the front center compartment cover plate assembly (cal. .50 ammunition tray cover) by removing two mounting screws. Remove ammunition tray by removing eight mounting screws (fig. 288).

(5) *Remove accessible ammunition racks.* Remove the bolts and lock washers securing the accessible ammunition racks to hull floor, and remove the racks.

(6) *Remove the two center stowage box panels.* Remove the clips resuring electrical cables to the sides of the center stowage box panels. Remove the bolts and lock washers holding the center panels at the ends and at the bottom. Remove the two center panels by lifting them out through the loader's door.

Note. The removal thus far of the ammunition stowage boxes is all that is necessary in order to gain access to the front bilge pump.

(7) Remove remaining ammunition racks. Remove the bolts and lock washers securing the remaining ammunition racks to hull floor, and remove the racks by lifting them out through the loader's door.

(8) *Remove the side stowage box panels.* Remove the stowage box panels located on both sides of the center panels by sliding them out of the bulkhead guides. Grasp the handle of the plate which is attached to the panel and slide out the plate and panel as a unit. Lift these units from inside the vehicle through the loader's door.

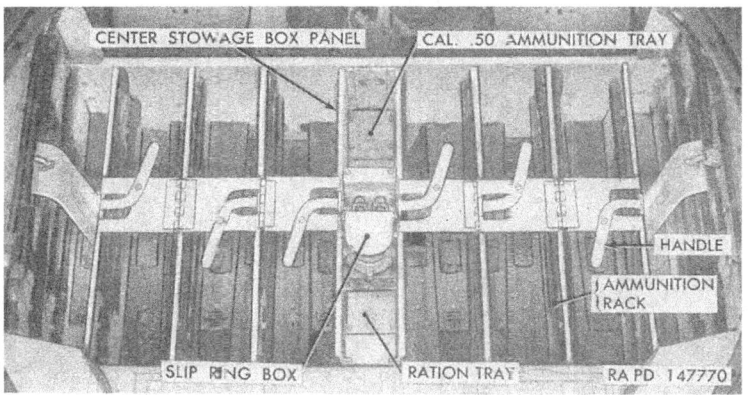

Figure 288. Ammunition stowage boxes.

b. INSTALLATION (fig. 288).

(1) *Install stowage box panels and plates* (fig 288). Lower the panels and plates through the loader's door. Slide panels and plates into the guides welded to bulkheads.

(2) *Install ammunition racks.* Lower the racks through the loader's door. Position the ammunition racks on hull floor and secure them by installing bolts and lock washers (fig. 288).

(3) *Install center panels.* Place the two center stowage box panels into position and secure them to the bulkheads and hull floor by installing bolts and lock washers at their ends and bottom. Secure the clips holding electrical cables to the center panels.

485

(4) *Install cal. .50 ammunition tray.* Install cal. .50 ammunition tray, securing with eight screws. Install the front center compartment cover plate assembly (ammunition tray cover) by securing with two screws.

(5) *Install ration tray.* Install ration tray, securing it with four screws. Install the rear center compartment cover plate assembly (ration tray cover), securing with two screws.

(6) *Install slip ring box.* Refer to paragraph 296.

(7) *Close stowage box plates.* Close the stowage box plates by means of the handles provided for closing and opening (fig. 288).

(8) *Elevate the 90-mm gun.* Refer to paragraph 76.

266. Fenders and Dust Shields

a. GENERAL. The fenders (fig. 3) are bolted to the hull along their inside edges, and to the stowage box brackets or outriggers. The dust shields (figs. 4 and 5) are attached to the side of the stowage box brackets or outriggers by hinges and quick-release fasteners. They can be removed in sections.

b. REMOVAL. To remove a dust shield, loosen the quick-release fasteners and slide dust shield off its hinges. To remove a fender, remove bolts and lock washers holding fender to stowage box brackets or outriggers, and to outer portion of hull flanges.

c. INSTALLATION. Place the fender in position on hull and install lock washers and bolts. To install dust shield, slide shield into hinges and secure to brackets by turning quick-release fasteners.

267. Gun Traveling Lock

a. REMOVAL. Remove 90-mm gun from lock. Remove retainer ring with snap ring pliers 41–P–1992–36 and knock out pin that secures bracket to engine compartment rear crossbeam bracket (fig. 289). Remove lock.

b. INSTALLATION. Position lock on engine compartment rear cross member and insert pin. Install retainer ring. Traverse turret and elevate or depress 90-mm gun, as necessary, to lock gun in gun traveling lock.

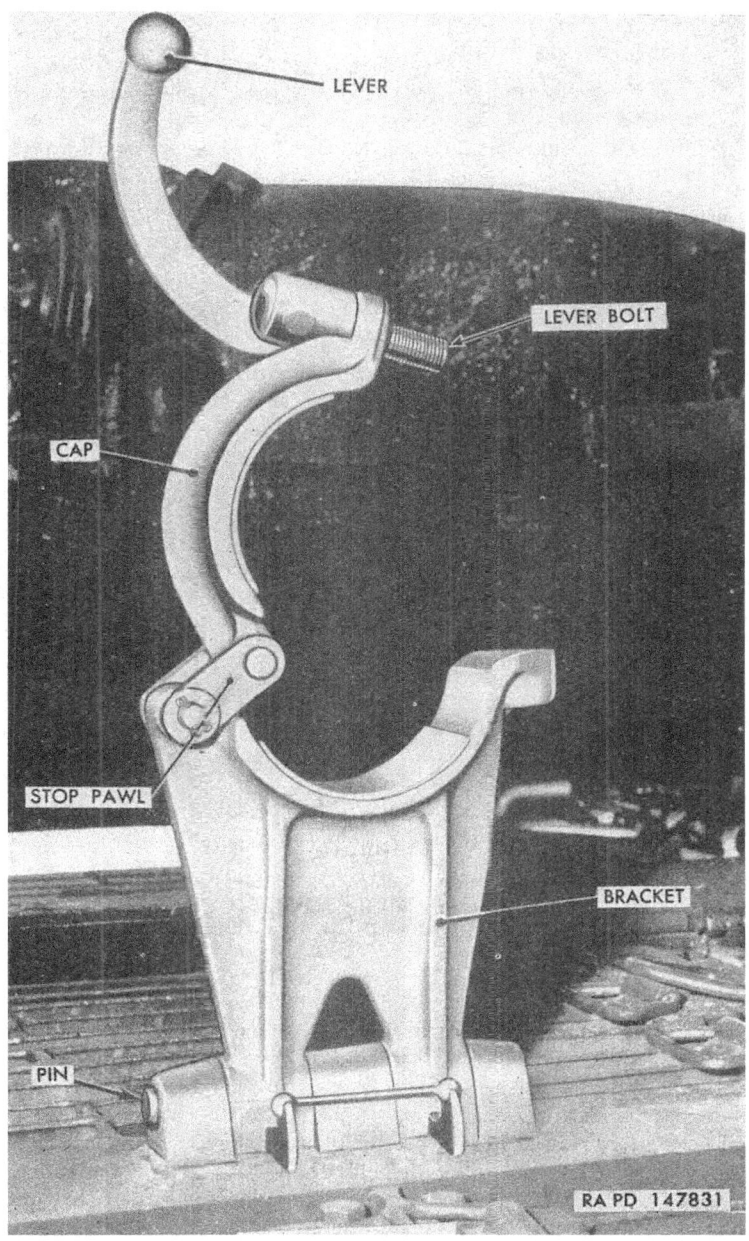

Figure 289. Gun traveling lock—installed view.

268. Towing Pintle

a. INSTALLATION FOR USE. Remove towing pintle from its stowage bracket on transmission center access door (fig. 285). Place pintle in position on mounting bracket (fig. 290) and install nut and cotter pin.

b. REMOVAL FOR STOWAGE. Remove cotter pin and nut and pull pintle from mounting bracket (fig. 290). Place towing pintle in position on stowage bracket (fig. 290) and install nut and cotter pin.

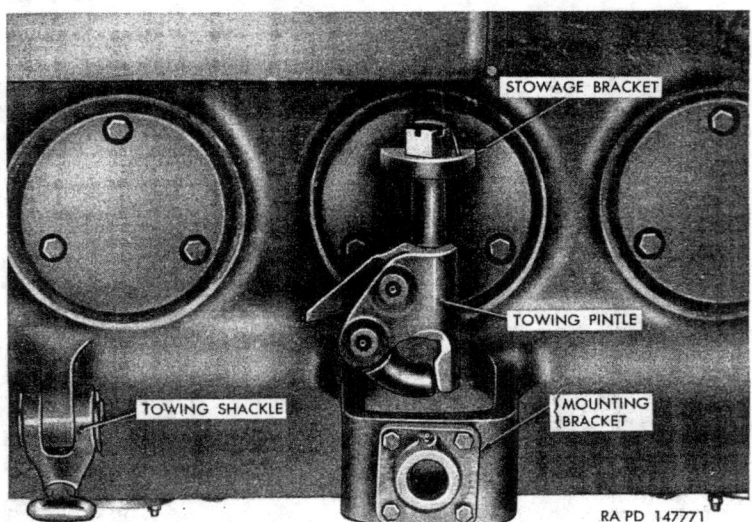

Figure 290. Towing pintle and shackles—installed view.

Section XXII. TURRET

269. Description

a. GENERAL (figs. 291, 292, 293, 294, and 295). The cast armor steel turret rotates on ball bearings in a turret race ring attached to the top of the hull. The weight of the gun is counterbalanced by a large hollow extension or bulge at the rear of the turret (fig. 292). The radio equipment, ventilating blower, and various other equipment are located or stowed in the interior of this bulge. The commander and gunner enter the turret through the door in the commander's cupola (fig. 37). The loader's escape hatch is to the left of the commander's cupola (fig. 291). A machine gun pintle stand (fig. 291) is located between the commander's cupola and loader's escape hatch. The gunner's seat (fig. 35), the commander's seat (fig. 36), and the

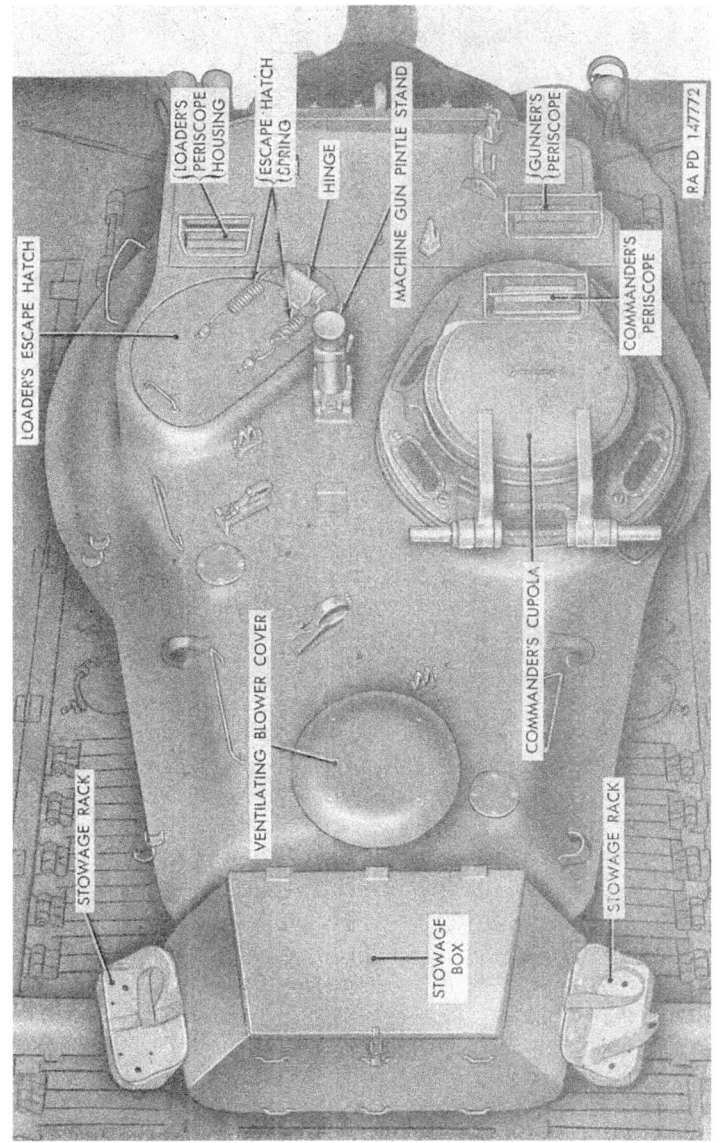

Figure 291. Turret—top view.

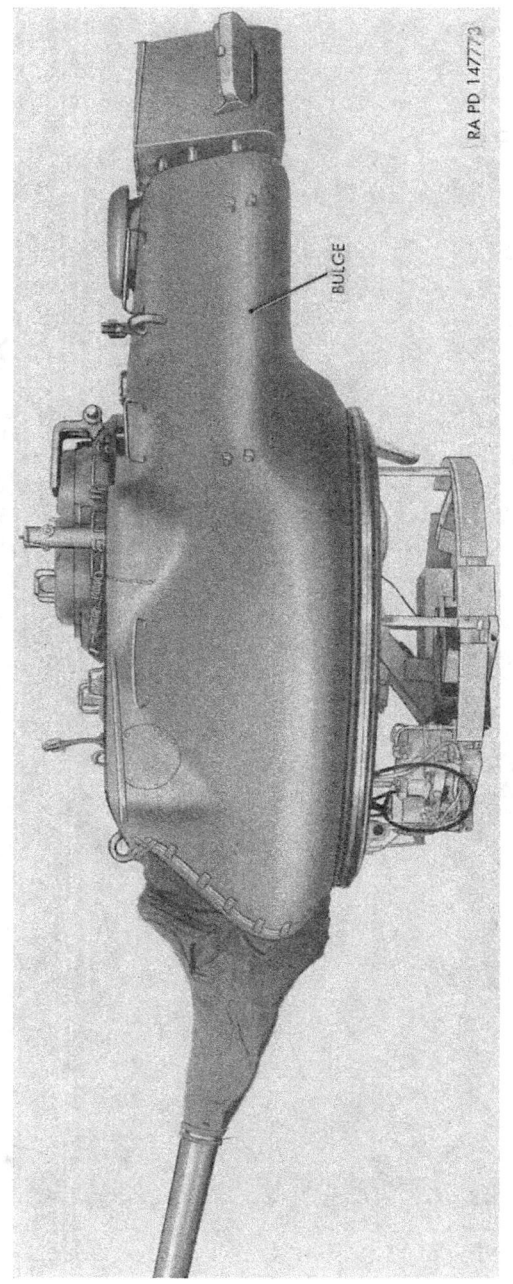

Figure 292. Turret—left-side view.

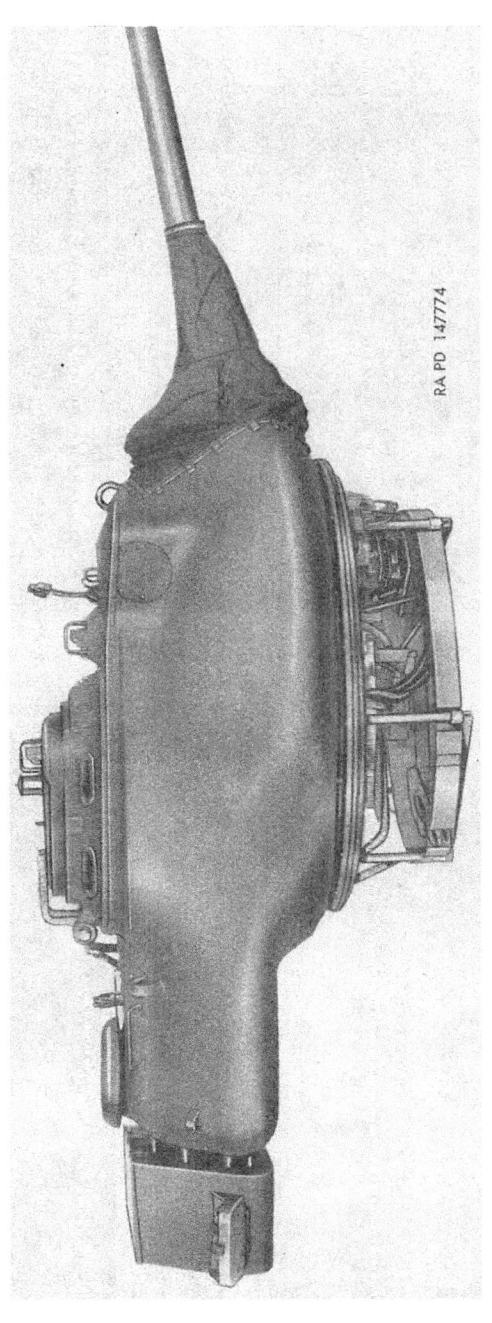

Figure 294. Turret—left-front view.

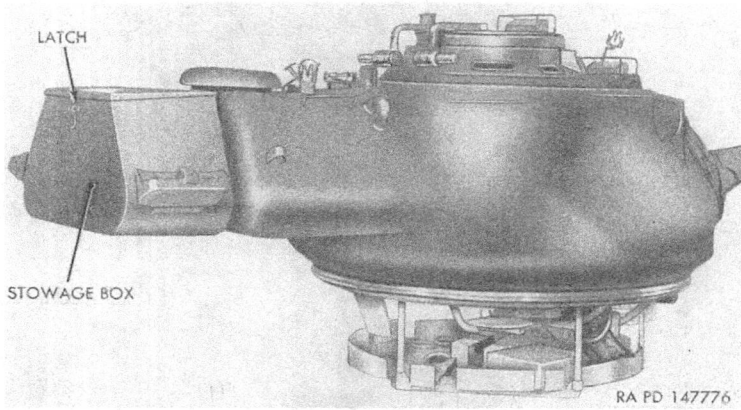

Figure 295. Turret—right-rear view.

loader's seat (fig. 34) are all attached to the turret rim. Three periscope housings are provided (fig. 291). The commander's periscope housing is located on the turret roof in front of the commander's cupola. The loader's periscope housing is located in front of the loader's escape hatch. Ammunition ready racks are provided around the rim of the turret (fig. 34).

b. TURRET TRAVERSING SYSTEMS.

   (1) *General.* The turret contains both a power and a manual traversing system. In either case, traversing is accomplished by the turret traversing mechanism (fig. 297) which contains a set of gears within a gear box. These gears mesh with the turret ring gears to traverse the turret. A turret traversing lock (fig. 31) serves to lock the turret securely in any position.

   (2) *Power traversing system* (fig. 296). In the power traversing system, electrical impulses from the commander's control handle (fig. 26) or the gunner's control handle (fig. 22) operate an electric motor which controls the turret traverse pump. This pump feeds oil from an oil reservoir to a hydraulic motor which drives the gear train to traverse the turret.

   (3) *Manual traversing system* (fig. 297). In the manual traversing system, gunner's manual traversing control handle operates the gear train directly to traverse the turret.

270. Turret Power Traversing System

a. CHECK OIL LEVEL IN OIL RESERVE (fig. 296). Remove cap from oil filler tube and check oil level in oil reservoir. Oil level must be maintained at "FULL" mark on indicator.

b. DRAIN OIL RESERVOIR. Wipe off outside of oil reservoir (fig. 296). Place a clear container of at least 1½-gallon capacity under the reservoir. Remove four drain plugs and allow reservoir to drain. Install four gaskets and drain plugs.

c. FILL OIL RESERVOIR (fig. 296). Clean top of reservoir. Remove cap from oil filler tube and check oil level as outlined in *a* above. Add oil as specified on lubrication order (fig. 87) to keep oil level at "FULL" mark.

d. CLEAN GEAR PUMP RELIEF VALVE. Lock the turret (par. 42). Remove gear pump relief valve cap at the front of the oil reservoir. Remove "O" ring, shims (if used), spring, guide, plunger, and bushing. Drain only a small portion of oil from the reservoir. Clean all parts thoroughly in volatile mineral spirits or dry-cleaning solvent. Examine valve plunger seat for scores and the spring for weakness. Turn gunner's power traversing control handle (fig. 22) fully clockwise and hold in this position. Turn on turret control switch (fig. 28) only long enough to force sufficient oil from valve port to thoroughly flush out any foreign matter. Turn off the switch and release the control handle. Assemble the valve parts by inserting the plunger and bushing with two new "O" rings into the opening in the reservoir and the spring guide into the bushing retainer, with the convex side of guide down against the bottom of the retainer. Place the spring on top of the guide. If shims were used, place them in the cap. Install

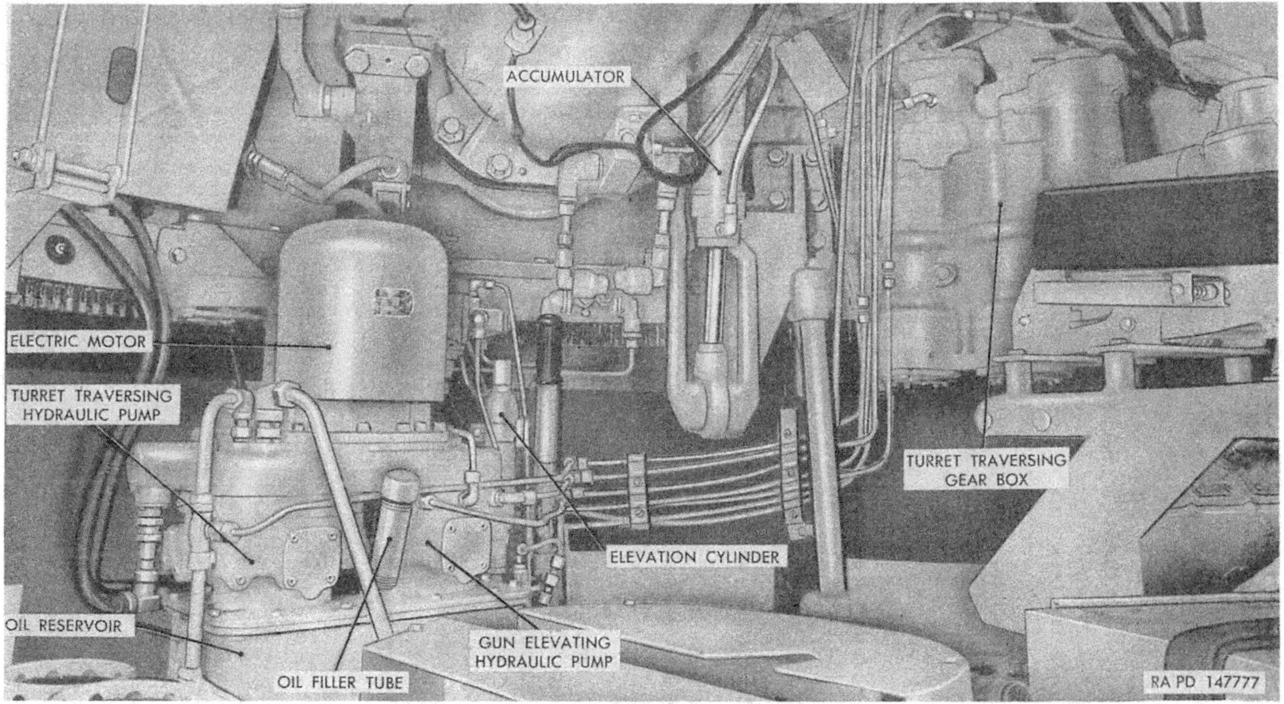

Figure 296. Turret traversing and gun elevating system.

a new "O" ring and tighten the cap securely. Fill the reservoir (*c* above). Release the turret lock (par. 42).

e. CLEAN TRAVERSE PUMP HIGH PRESSURE RELIEF VALVE. The traverse pump high pressure relief valve is a ball-type valve located in the end head of the traverse pump. Lock turret (par. 42). Remove cap, shims (if used), spring, guide, and ball. Clean valve and valve port as in *d* above. To install valve, insert ball into opening in end head of traverse pump. Insert guide and spring. Position shims, if used, in valve cap and tighten cap securely.

f. CLEAN HYDRAULIC MOTOR HIGH PRESSURE RELIEF VALVE. The hydraulic motor high pressure relief valve is located on top of the hydraulic motor. This valve is the same type as the gear pump relief valve. To clean, follow the procedure outlined in *d* above.

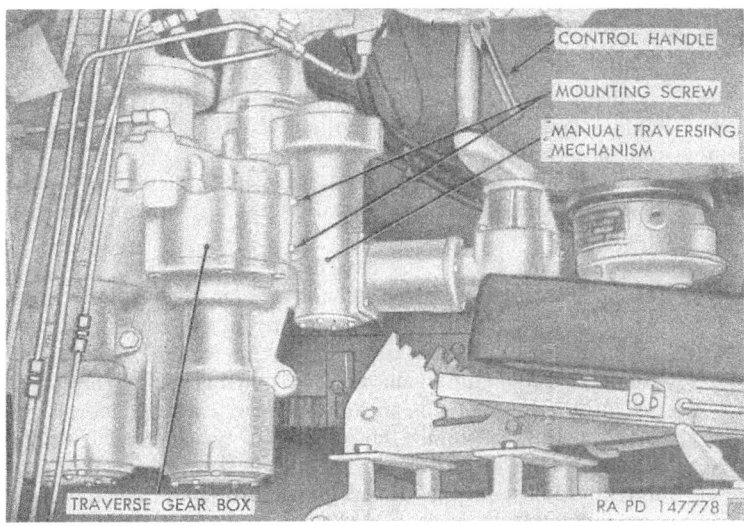

Figure 297. Turret traversing mechanism.

271. Turret Manual Traversing Mechanism

a. REMOVAL (fig. 297). Remove four mounting screws and lock washers securing manual traversing mechanism to side of traverse gear box and seven screws and lock washers securing mechanism to top of gear box. Carefully remove mechanism from gear box, being careful not to lose mounting shims.

b. INSTALLATION (fig. 297). Position manual traversing mechanism to traverse gear box. Make sure shims are in place and secure mechanism with four mounting screws and lock washers to side of gear box and seven screws and lock washers to top of gear box.

272. Turret Traversing Lock

a. REMOVAL. Lock gun in gun traveling lock (par. 55). Remove two mounting bolts securing turret traversing lock to turret (fig. 31) and remove lock.

Note. If same lock is to be installed and shims were used, the same number and size shims must be reinstalled. If new lock is to be installed, proceed as below.

b. INSTALLATION AND ADJUSTMENT. Position the turret traversing lock and secure with two mounting bolts (fig. 31). Test operation of lock (par. 42). If it binds, either loosen bolts and shift position of lock or remove lock and adjust number of shims to allow lock bolt to fully enter ring gear without binding. After the lock is properly installed, tighten bolts.

273. Commander's Cupola Assembly

a. REMOVAL. Working inside the turret, remove three bolts and washers securing cupola assembly to turret. Working outside the turret, remove eight bolts securing cupola. Using a suitable sling, remove cupola assembly.

b. INSTALLATION. Position cupola to turret and secure with eight bolts. Working inside the turret, install three bolts and washers to secure cupola.

c. REPLACEMENT OF VISION BLOCKS (fig. 37).
  (1) *Removal.* Remove bezel assembly from vision block to be removed (par. 275). Working inside the turret, remove locking wire and three screws securing vision block wedge. Remove wedge and vision block by pulling inside the turret.
  (2) *Installation.* Working inside the turret, insert vision block into opening in cupola and secure in position with wedge. Secure wedge with three screws and locking wire. Working outside the turret, install bezel assembly (par. 275).

274. Commander's Cupola Door Assembly

a. REMOVAL. (fig. 37). Open commander's cupola door far enough for door plunger to lock door in nearly vertical position. Remove three screws and washers securing each torsion spring housing to hinge and pull out torsion spring housings and spacers. Remove door assemblies.

b. INSTALLATION (fig. 37). Position door assembly to cupola in a nearly vertical position so that door plunger lines up with opening in door hinge. Insert a spacer and torsion spring housing inside each door hinge and rotate each torsion spring housing until it locks in position. Secure each torsion spring housing with three screws and washers.

275. Bezel Assembly

a. REMOVAL (fig. 37). Remove 14 screws securing bezel assembly to cupola and remove bezel.

b. INSTALLATION (fig. 37). Position assembly to cupola and secure with 14 screws.

276. Loader's Escape Hatch Assembly

a. REMOVAL. Release tension on escape hatch springs (fig. 291) by loosening two nuts and remove springs. Remove escape hatch hinge taper pin from center of hinge. Pull escape hatch hinge pin out of hinge and remove loader's escape hatch assembly.

b. INSTALLATION. Position escape hatch assembly to turret roof (figs. 38 and 291). Insert escape hatch hinge pin into hinge and secure in position by inserting escape hatch hinge taper pin. Position escape hatch springs and tighten with two nuts.

277. Loader's Periscope Housing

a. REMOVAL. Working inside the turret, remove six screws and washers securing bottom segment of periscope housing (fig. 66) to top segment of periscope housing and remove bottom segment. Working on top of the turret, remove top segment of periscope housing (fig. 291).

b. INSTALLATION. Position top segment of periscope housing to turret roof (fig. 291). Position bottom segment of periscope housing (fig. 66) to top segment, making sure dowel pins engage properly. Secure bottom segment to top segment with six screws and washers.

278. Turret Seat Assemblies

    a. COMMANDER'S SEAT (fig. 36).

        (1) *Removal.* Remove pin securing commander's seat adjusting tube at lower support. Remove four screws, nuts, lock washers, and shims (if used) securing commander's seat upper bracket and remove seat assembly.

        (2) *Installation.* Position commander's seat assembly to turret wall. Secure adjusting tube to lower support with pin. Secure upper bracket with four screws, nuts, lock washers, and shims (if required).

    b. GUNNER'S SEAT (fig. 35).

        (1) *Removal.* Remove four bolts and nuts securing gunner's seat assembly to seat support and remove seat assembly.

        (2) *Installation.* Position gunner's seat assembly to seat support and secure with four bolts and nuts.

c. LOADER'S SEAT (fig. 34).
   (1) *Removal.* Remove two bolts securing loader's seat assembly to turret wall and remove seat.
   (2) *Installation.* Position loader's seat to turret wall and secure with two bolts.

279. Ammunition Ready Racks
(fig. 34)

a. REMOVAL. Remove ready racks by removing one screw and washer securing each single-round ready rack to turret wall and two screws and washers securing each double-round ready rack to turret wall.

b. INSTALLATION. Position ready racks to turret wall. Secure each single-round ready rack with one screw and washer and each double-round ready rack with two screws and washers.

280. Turret Stowage Boxes

   a. OUTER TURRET STOWAGE BOX (figs. 291 and 295).
   (1) *Removal.* Release latch on rear of stowage box and raise cover. Remove 10 nuts and washers securing stowage box to rear of turret and remove stowage box.
   (2) *Installation.* Position stowage box to rear of turret and secure with 10 nuts and washers. Close stowage box cover and secure latch.
   b. CAL. .45 AMMUNITION BOX.
   (1) *Removal.* Remove seven screws and washers securing cal. .45 ammunition box to turret bulge floor and remove box.
   (2) *Installation.* Position cal. .45 ammunition box to turret bulge floor and secure with seven screws and washers.
   c. HAND GRENADE (SMOKE) BOX (fig. 304).
   (1) *Removal.* Remove four screws and washers securing hand grenade (smoke) box on upper left side of turret bulge and remove box.
   (2) *Installation.* Position hand grenade (smoke) box on upper left side of turret bulge and secure with four screws and washers.
   d. FRAGMENTATION GRENADE BOX (fig. 304).
   (1) *Removal.* Remove four screws and washers securing fragmentation grenade box to upper right side of turret bulge and remove box.
   (2) *Installation.* Position fragmentation grenade box to upper right side of turret bulge and secure with four screws and washers.

e. FIRST AID, CAL. .30 AMMUNITION, AND RATION BOX.
   (1) *Removal.* Remove seven screws and washers securing first aid, cal. .30 ammunition, and ration box to right side of turret bulge and remove box.
   (2) *Installation.* Position box to right side of turret bulge and secure with seven screws and washers.

f. CAL. .50 AMMUNITION BOX (fig. 303).
   (1) *Removal.* Turn off master relay switch (fig. 18). Disconnect ammunition chute and remove from ammunition box. Disconnect cables to booster motor and micro switch. Release four latches and remove booster motor housing assembly. Release latch and raise ammunition box cover. Remove eight screws securing ammunition box to rear shock mounts and six screws securing ammunition box to bottom shock mounts Remove ammunition box from vehicle.
   (2) *Installation.* If a new ammunition box is being installed, release latch and raise ammunition box cover to permit working inside box. Position box to shock mounts on left side of turret. Secure box to rear shock mounts with eight screws and to bottom shock mounts with six screws. Close ammunition box cover. Position booster motor housing assembly and secure with four latches. Connect cables to booster motor and micro switch.

Section XXIII. TURRET ELECTRICAL SYSTEM

281. Description

a. GENERAL. The turret electrical system (fig. 298) consists of a turret-traversing and gun elevating control circuit, firing control and safety circuits, cal. .50 ammunition chute booster motor circuit, radio feed and interphone circuits, and turret accessories circuits. Description and location of the components involved in these circuits are given in *b* through *t* below. Procedures for removal, installation, and testing are given in paragraphs 282 to 299.

b. OVERRIDE MOTOR CONTROL BOX. The override motor control box (fig. 300) is located on the turret floor to the rear of the gunner's control mechanism. The box connects the commander's and gunner's control mechanisms to the turret traversing and gun elevating electric motor and contains two relays to enable the commander's control circuit to override the gunner's control circuit. Refer to figure 299 for a wiring diagram of the override motor control box.

c. COMMANDER'S CONTROL HANDLE ASSEMBLY. The commander's control handle assembly (fig. 26) is secured to the turret bulkhead at the right of the commander's seat. The assembly consists of a

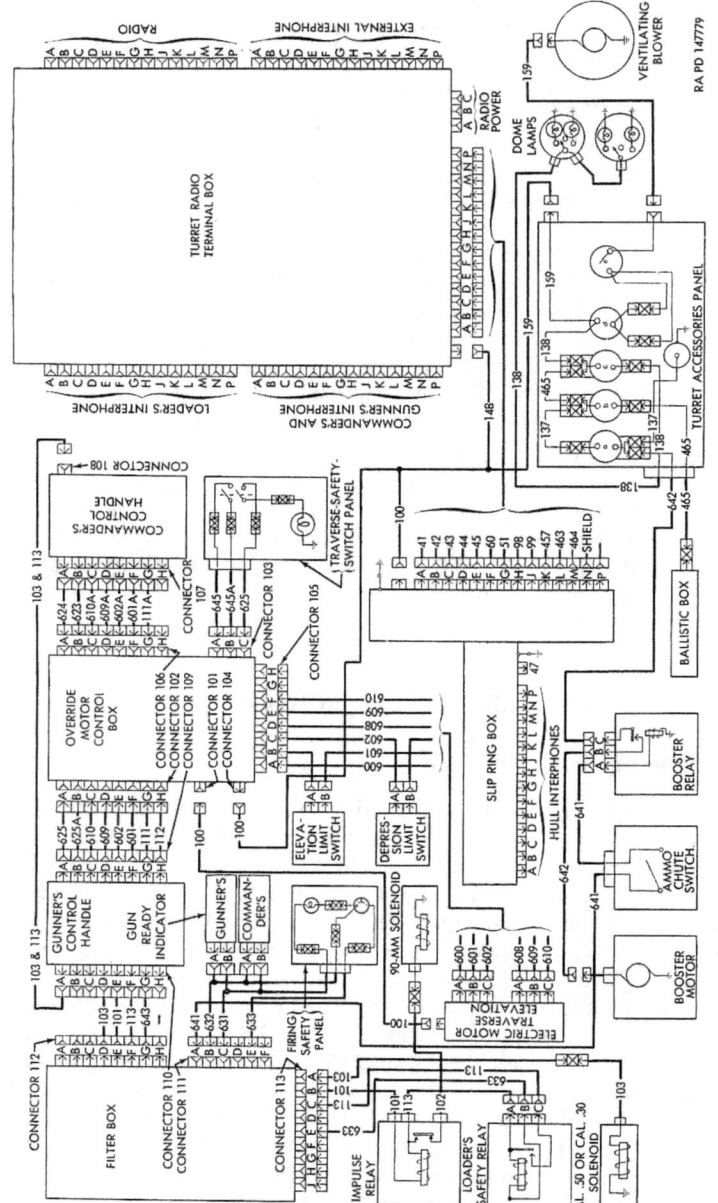

Figure 298. Turret electrical system schematic diagram.

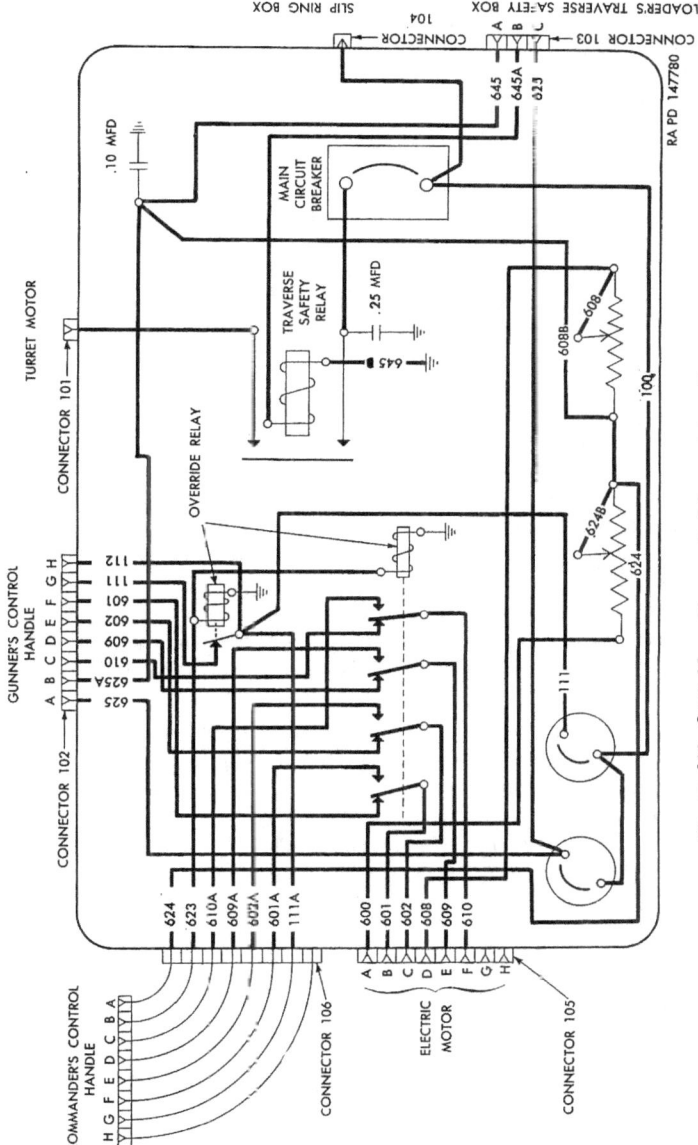

Figure 299. Override motor control box wiring diagram.

firing trigger, control handle, override lever, and traversing and elevating control box.

d. GUNNER'S CONTROL HANDLE ASSEMBLY. The gunner's control handle assembly is mounted above the turret traversing mechanism (fig. 23). The electrical portion of this assembly consists of two control handles, firing buttons, and a control box containing cal. .50 and 90-mm gun firing selector switches, a turret control switch, and three indicator lights (fig. 28).

e. LOADER'S TRAVERSE SAFETY SWITCH BOX. The loader's traverse safety switch box is secured to the turret bulkhead above the cal. .50 ammunition box (fig. 29). The control box contains a switch which turns on the positive feed to the electric motor. An indicator light is provided to indicate when the power unit acts.

f. LIMIT SWITCHES. A limit switch unit, containing an elevation limit switch and a depression limit switch, is located in the forward portion of the turret (fig. 301). If the 90-mm gun is elevated too high or depressed too low, the corresponding limit switch is opened, breaking the control circuit to the electric motor.

g. LOADER'S FIRING RESET SAFETY PANEL (fig. 30). The loader's firing reset safety panel is located on the turret bulkhead above the turret accessories panel. The panel contains a ready-to-fire indicator light, and a manual reset circuit breaker which can be reset, after it is opened by an overload, with the loader's-firing-reset-safety-switch button.

h. LOADER'S SAFETY RELAY. The loader's safety relay is located on the breech operating mechanism (fig. 40). It opens the firing safety circuit (*g* above) when the breech is open so that the gun is rendered safe.

i. GUN READY LIGHTS. Two gun ready lights are provided. One is located in the gunner's periscope mount (fig. 48), and one in the commander's periscope mount (fig. 47). They are wired in parallel with the ready-to-fire indicator light (*g* above).

j. 90-mm GUN FIRING MECHANISM. The 90-mm gun firing mechanism is secured to a bracket mounted on the 90-mm gun cradle (fig. 302). The mechanism consists of a firing solenoid and an impulse relay.

k. MACHINE GUN FIRING SOLENOIDS. The cal. .50 machine gun firing solenoid is secured to the trigger mechanism of the cal. .50 machine gun (fig. 39). The cal. .30 machine gun firing solenoid is secured to the machine gun mount (fig. 40). Each solenoid, when actuated by the firing control, moves a plunger which trips the machine gun trigger.

l. BOOSTER MOTOR. The booster motor is located within the booster motor housing located on top of the ammunition box (fig. 303). It serves to feed ammunition to the cal. .50 machine gun.

m. BOOSTER MOTOR RELAY. The booster motor relay is secured to the side of the firing safety panel (fig. 30). The relay is connected to the cal. .50 machine gun firing control circuit so that when the machine gun is fired, the booster motor is actuated.

Note. On some vehicles, the booster motor relay is attached to the turret filter box (fig. 336).

n. AMMUNITION CHUTE MICRO SWITCH. The ammunition chute micro switch is located on top of the booster motor housing (fig. 303). It serves to open the machine gun firing circuit when the ammunition chute jams.

o. TURRET RADIO TERMINAL BOX. The turret radio terminal box is mounted on the turret bulkhead near the ventilating blower (fig. 304). It contains radio feed and interphone circuit connections.

p. TURRET SLIP RING BOX. The turret slip ring box is located in the center of the turret floor (fig. 305). It provides positive feed and interphone connections from the hull to the turret. The box contains an upper portion which rotates with the turret and a lower portion which is secured to the hull. A system of rings and brushes makes possible the transfer of electrical current from the stationary portion through the rotating portion to the various electrical components of the turret.

q. TURRET ACCESSORIES PANEL. The turret accessories panel is secured to the turret bulkhead below the loader's firing reset safety switch panel (fig. 30). On the face of the panel are mounted a ventilator "ON" and "OFF" switch and an accessory outlet socket. Four circuit breakers are secured to the rear of the panel.

r. VENTILATING BLOWER. The ventilating blower is mounted in the rear of the turret (fig. 304). A handle is provided for damper control.

s. BALLISTIC BOX. An electric cable is brought into the ballistic box to provide illumination for the ballistic scales (fig. 54).

t. TURRET DOME LIGHTS. The two turret dome lights are mounted in the turret ceiling (fig. 304). They are of the same type as the hull dome lights (par. 29).

282. Override Motor Control Box
(fig. 300)

a. REMOVAL. Turn off the master relay switch (fig. 18). Disconnect six cable connector plugs on override motor control box and remove plugs and cables. Remove four lock washer screws and remove override motor control box.

b. INSTALLATION. Position override motor control box and secure with four lock washer screws. Connect six cable connector plugs to proper matching receptacles on sides of override motor control box.

c. TESTING. Turn on master relay switch (fig. 18). Traverse turret with gunner's control handle (par. 35). Operate commander's control handle (par. 38) and see if commander's handle overrides gunner's handle. If it does not, refer to paragraph 139 for trouble shooting procedure.

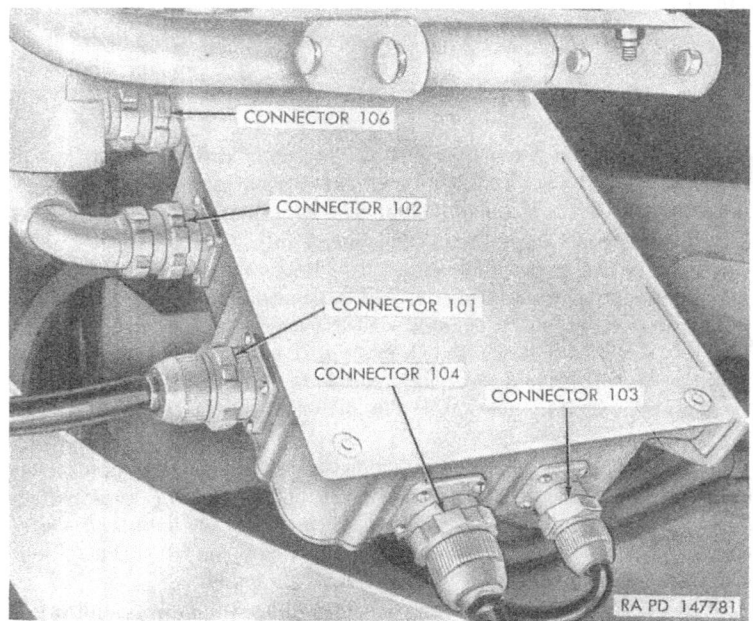

Figure 300. Override motor control box.

283. Commander's Control Handle Assembly

a. REMOVAL. Turn off master relay switch (fig. 18). Disconnect two electrical connectors at the rear of the commander's control handle assembly (fig. 26) and remove cables. Remove four screws and nuts securing assembly to mounting bracket on turret bulkhead and remove assembly.

b. INSTALLATION. Position commander's control handle assembly to mounting brackets (fig. 26) and secure with four screws, washers, and nuts. Connect two cable connector plugs into proper receptacles at rear of handle assembly.

c. TESTING. Turn on master relay switch (fig. 18). Squeeze commander's control handle and listen for a click in the override motor control box to indicate that the gunner's control has been automatically overridden. If tactical situation permits, test control handle by elevating and depressing the gun and traversing the turret (par. 38). Make sure the gun is cleared of ammunition and test the firing trigger. If commander's control handle assembly is not functioning properly, refer to paragraph 139 for trouble shooting procedure.

284. Gunner's Control Handle Assembly

a. REMOVAL. Turn off master relay switch (fig. 18). Disconnect two electrical connectors at rear of gunner's control handle assembly and remove cables. Remove four screws securing handle assembly to each of two mounting brackets and remove.

b. INSTALLATION. Position gunner's control handle assembly to mounting brackets and secure with four screws and washers to each bracket. Connect two cable connector plugs into proper receptacles on rear of control handle assembly.

c. TESTING. Turn on master relay switch (fig. 18). Set turret control switch (fig. 28) to "ON" position. If tactical situation permits, operate gunner's control handle to elevate and depress the gun and to traverse the turret (par. 35). Make sure 90-mm gun is clear of ammunition. Set 90-mm gun firing selector switch (fig. 28) to "ON" position and test both firing buttons.

Warning: Make sure cal. .50 machine gun is cleared of ammunition. Set cal. .50 machine gun firing selector switch (fig. 28) to "ON" position and press firing buttons to check buttons and control circuits. If gunner's control handle assembly does not function properly, refer to paragraph 139 for trouble shooting procedure.

285. Loader's Traverse Safety Switch Box

a. REMOVAL. Turn off master relay switch (fig. 18). Remove four screws securing loader's traverse safety switch box (fig. 29) to brackets on turret bulkhead and remove control box from brackets. Working from rear of control box, disconnect three bell-type connectors connecting control switch to external cable and remove control box.

b. INSTALLATION. Connect three bell-type connector plugs into receptacles at rear of control box. Secure control box to mounting brackets on turret bulkhead with four screws and washers.

Note. Be sure ground cable is connected to upper left screw.

c. LOADER'S TRAVERSE SAFETY SWITCH (fig. 29).

    (1) *Removal.* Remove control box (*a* above). Disconnect remaining connector at rear of control box. Remove screw in switch handle and remove handle. Remove nut from switch shaft and remove switch from rear of control box.

    (2) *Installation.* Position switch to control box and secure with mounting nut and lock washer. Connect four bell-type con-

nector plugs to receptacles at rear of control box. Secure control box to mounting brackets on turret bulkhead with four screws and washers. Be sure ground cable is connected to upper left screw. Secure switch handle to switch with screw.

d. INDICATOR LIGHT (fig. 29).

(1) *Removal.* Remove control box (*a* above). Disconnect connector at rear of light. Remove indicator light cap by removing mounting nut securing cap to face of control box. Remove two screws securing light to face of control box and remove light from rear of box.

(2) *Installation.* Position indicator light and secure to face of control box with two screws. Install indicator light cap. Be sure cap gasket is in position. Connect cable connector plugs to rear of control box. Secure control box to mounting brackets on turret bulkhead with four screws and washers. Be sure ground strap is secured to upper left screw.

e. TRAVERSE SAFETY CONTROL LAMP REPLACEMENT (fig. 29). It is not necessary to remove the traverse safety control box to replace the indicator lamp. Remove the indicator lamp cap. Replace lamp. Replace cap, making sure gasket is in position.

286. Elevation and Depression Limit Switches

a. REMOVAL. The elevation limit switch and the depression limit switch (fig. 301) are removed in the same manner. To move either switch, turn off master relay switch (fig. 18). Disconnect cable connector at switch and remove cable. Remove two socket head screws and remove limit switch.

b. INSTALLATION. Position limit switch to mounting bracket (fig. 301) and secure with two socket head screws, two lock washers, and four spacing washers. Connect limit switch cable.

c. TESTING AND ADJUSTMENT. Turn on master relay switch (fig. 18). Slowly and carefully depress or elevate the gun (par. 76), as required, to test the switch. If travel limit is not correct, adjust the adjusting screw on limit switch rod (fig. 301).

Caution: Exercise extreme care in elevating or depressing gun after installation of limit switch or serious damage may result to turret if switch is not properly adjusted.

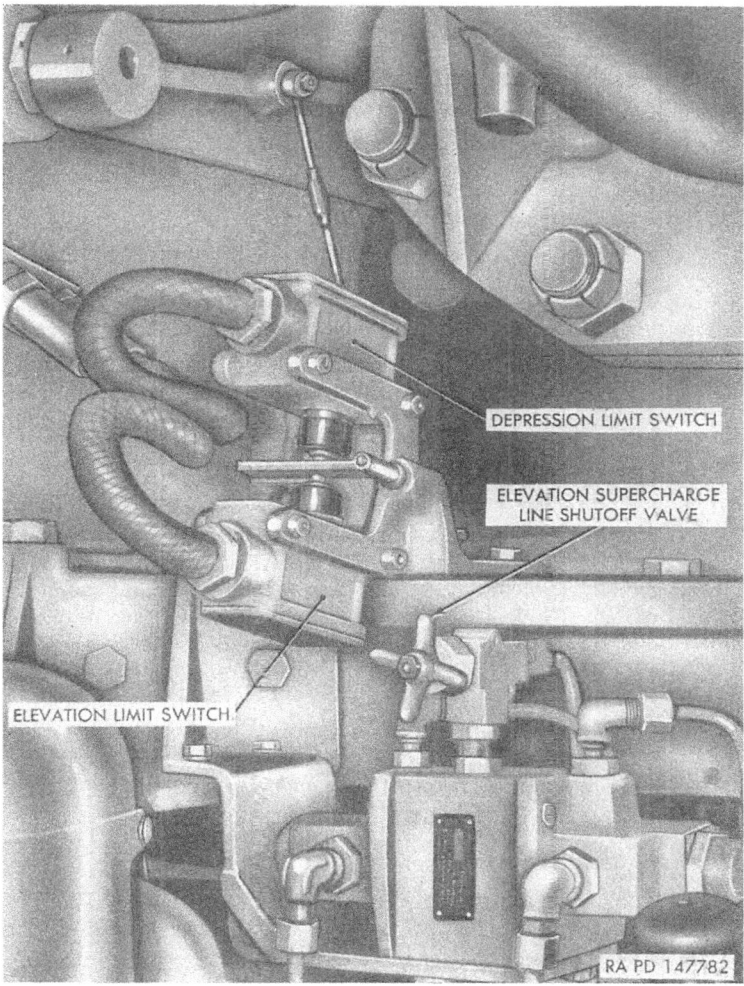

Figure 301. Elevation and Depression Limit Switches.

287. Loader's Firing Reset Safety Panel

a. REMOVAL (fig. 30). Turn off master relay switch (fig. 18). Remove booster motor relay from firing safety panel by removing two screws securing relay to right side of panel. Remove four screws securing panel to rear cover. Working inside the safety panel box, remove from the clips the connectors on the three cables that pass out the bottom of the box. Disconnect these three connectors and remove cables from box by prying loose the cable grommet in the bottom of the box. Remove the firing safety panel.

b. INSTALLATION. Insert cable grommet into hole in bottom of panel box. Connect three connector plugs into receptacles and insert connectors into clips in top of panel. Secure panel to rear cover with four screws and lock washers. Secure booster motor relay to right side of panel with two screws and lock washers.

c. LOADER'S FIRING RESET SAFETY SWITCH (fig. 30).

(1) *Removal.* Remove firing safety panel (*a* above). Remove switch cable connectors from clip. Disconnect switch-to-indicator-light connector. Remove two screws securing switch to face of panel and remove switch.

(2) *Installation.* Insert switch through hole in panel and secure with two screws and lock washers. Connect switch-to-indicator-light-cable-connector plug into proper receptacle and secure connectors in clip in top of panel. Install firing safety panel (*b* above).

d. LOADER'S FIRING RESET SAFETY SWITCH INDICATOR LIGHT (fig. 30).

(1) *Removal.* Remove firing safety panel (*a* above). Disconnect switch-to-indicator-light-cable connector at rear of panel. Remove screw and nut securing ground cable to panel. Remove two screws securing light to face of panel and remove light through rear of panel.

(2) *Installation.* Position light to face of panel and secure with two screws and lock washers. Connect switch-to-indicator-light-cable-connector plug to proper receptacle and insert into clip. Secure ground cable with screw, nut, and two washers. Install firing safety panel (*b* above).

e. FIRING SAFETY PANEL LAMP REPLACEMENT (fig. 30). It is not necessary to remove the firing safety panel to replace the lamp. Working from the front of the panel, remove the lamp cap assembly. Replace lamp. Install lamp cap, making sure gasket and washer are in position.

f. TESTING. Turn on master relay switch (fig. 18). Push loader's firing reset-safety switch (fig. 30). Indicator light should come on and firing circuits should now be energized.

Warning: Make sure guns are cleared of ammunition before testing firing controls.

If firing reset safety switch does not operate properly, refer to paragraph 139 for trouble shooting procedure.

288. Loader's Safety Relay

a. REMOVAL (fig. 40). Unscrew cable connector nut and remove cable. Remove two screws and remove loader's safety relay.

b. INSTALLATION (fig 40). Position loader's safety relay and secure with two screws and lock washers. Install cable connector plug and secure mounting nut.

289. Gun Ready Indicators

For replacement of gun ready indicators, refer to paragraph 368.

290. 90-mm Gun Firing Mechanism

a. FIRING SOLENOID (fig. 302).

(1) *Removal.* Disconnect firing solenoid cable at connector. Remove three screws securing solenoid to firing mechanism bracket and remove solenoid.

(2) *Installation.* Position firing solenoid to firing mechanism bracket and secure with three screws. Connect firing solenoid cable at connector.

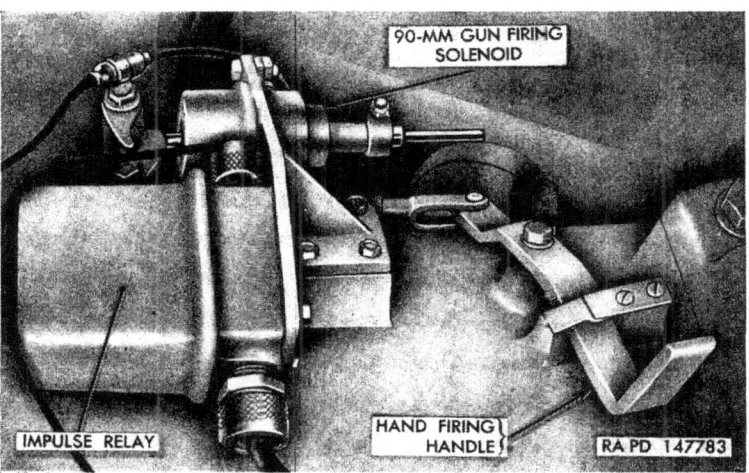

Figure 302. 90-mm gun firing mechanism.

b. IMPULSE RELAY (fig. 302).

(1) *Removal.* Using a screwdriver as a lever, unscrew and remove cap from impulse relay. Remove three cables from terminals inside impulse relay housing. Loosen two cable connector nuts at sides of impulse relay housing and pull out cables. Remove four screws and remove impulse relay from firing mechanism bracket.

(2) *Installation*. Position impulse relay to firing mechanism bracket and secure with four screws. Insert three cables through connector openings on sides of impulse relay housing and secure to cable terminals. Install impulse relay cap and screw into impulse relay housing, using a screwdriver as a lever.

291. Machine Gun Firing Solenoids

  a. CAL. .50 FIRING SOLENOID (fig. 39).

  (1) *Removal*. Disconnect cable connector at firing solenoid and remove cable. Loosen three screws securing trigger and firing solenoid to machine gun shaft and pull trigger and solenoid to the rear until free of the machine gun.

  (2) *Installation*. Install solenoid and trigger on shaft and secure with three screws.

  > *Note*. Make sure percussion mechanism, solenoid, and firing trigger are alined.

  Connect cable to firing solenoid.

  b. CAL. .30 FIRING SOLENOID (fig. 40).

  (1) *Removal*. Remove chain securing firing solenoid to machine gun cradle. Disconnect firing solenoid cable. Remove screw securing solenoid to machine gun cradle and remove solenoid.

  (2) *Installation*. Position firing solenoid and secure with screw.

  > *Note*. Be sure chain hook is attached to screw.

  Attach chain to hook. Connect firing solenoid cable.

292. Booster Motor

a. REMOVAL (fig. 303). Turn off master relay switch (fig. 18). Disconnect booster motor positive cable (642) and two booster motor ground cables. Release two booster cover latches and lift up booster cover. Remove four screws and remove booster guide assembly. Remove eight screws and lift out booster motor.

b. INSTALLATION (fig. 303). Position booster motor inside motor housing and secure with eight screws. Install booster guide and secure with four screws and lock washers. Close booster cover and secure latches. Connect booster motor positive cable (642) and two booster motor ground cables.

c. TESTING. Make sure machine gun is cleared of ammunition. Turn on master relay switch (fig. 18). Operate machine gun firing controls (par. 39) to see if booster motor operates. If it does not, refer to paragraph 139 for trouble shooting procedure.

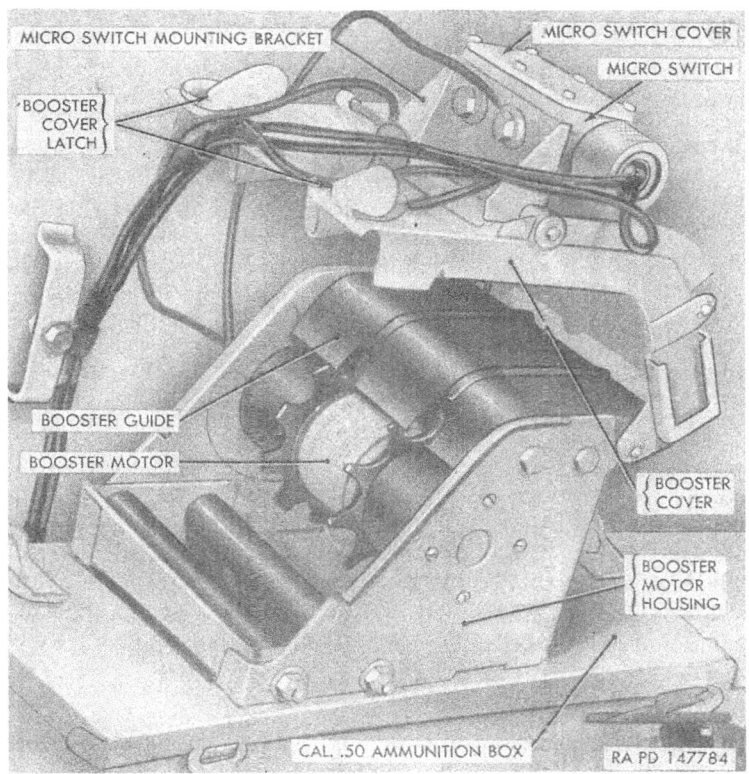

Figure 303. Booster motor and ammunition chute micro switch—booster cover removed.

293. Booster Motor Relay

a. REMOVAL. Turn off master relay switch (fig. 18). Disconnect cable connector plug from booster motor relay (fig. 30) and remove cable. Remove two screws securing relay to firing safety panel and remove relay.

b. INSTALLATION. Position relay to side of firing safety panel (fig. 30) and secure with two lock washer screws. Install cable connector plug to receptacle on relay.

c. TESTING. Make sure machine gun is cleared of ammunition. Turn on master relay switch (fig. 18). Fire the machine gun (par. 79) and note whether the relay allows current to flow to actuate the booster motor. If relay does not function properly, refer to paragraph 139 for trouble shooting procedure.

294. Ammunition Chute Micro Switch

a. REMOVAL (fig. 303). Turn off master relay switch (fig. 18). Remove six screws and remove ammunition chute micro switch cover. Disconnect two cables inside switch unit. Loosen cable connector nut and remove cables by pulling out of switch unit. Remove two screws securing switch to mounting bracket and remove switch.

b. INSTALLATION (fig. 303). Position switch to mounting bracket and secure with two screws and washers. Insert cables into opening in switch and secure to cable terminals with two screws. Tighten connector nut. Make sure cover gasket is in position. Position switch cover to top of switch and secure with six screws and lock washers.

295. Turret Radio Terminal Box

a. REMOVAL. Turn off master relay switch (fig. 18). Disconnect cable connector plugs at turret radio terminal box (fig. 304) and remove cables. Remove four lock washer screws securing box and remove box.

b. INSTALLATION. Position turret radio terminal box (fig. 304) to mounting bracket and secure with four lock washer screws. Connect cable connector plugs to terminal box receptacles.

c. TESTING. Turn on master relay switch (fig. 18) and operate radios and interphones to test turret radio terminal box.

296. Slip Ring Box

a. REMOVAL. Turn off master relay switch (fig. 18). Remove ground cable from side of slip ring box (fig. 306). Remove two jam nuts and two nuts securing box turning arm to turret bracket. Remove ground strap which is secured to top portion of slip ring box by removing two cap screws. Remove yoke and turning arm. Disconnect four cable connector plugs at slip ring box (fig. 307) and remove cables. Lift up rear stowage box door (fig. 306) and remove four cap screws securing slip ring box into bottom of recess in fighting compartment floor. Remove slip ring box.

b. INSTALLATION. Position slip ring box to bottom of floor recess (fig. 305) and secure with four cap screws and lock washers. Install yoke and turning arm. Secure arm with two jam nuts and two nuts. Connect ground strap to top of box with two cap screws and lock washers. Secure ground cable to side of box (fig. 306) with screw and lock washer. Connect four cable connector plugs to receptacles on turret slip ring box (fig. 307).

c. TESTING. Turn on master relay switch (fig. 18) and operate turret traversing system, interphones, and radio equipment to test slip ring box. If slip ring box does not function properly, check installation *b* above. Replace slip ring box if necessary.

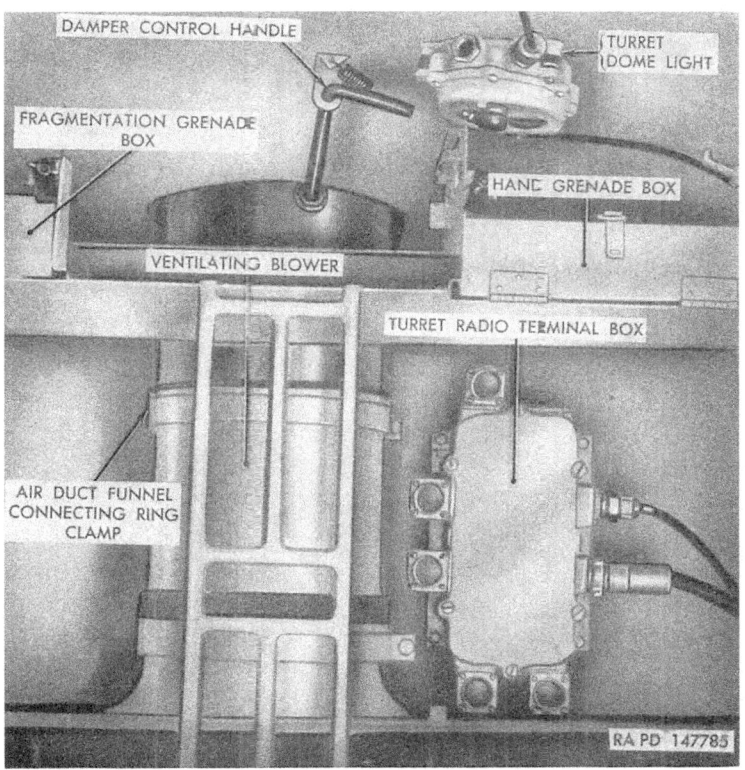

Figure 304. Ventilating blower and turret radio terminal box.

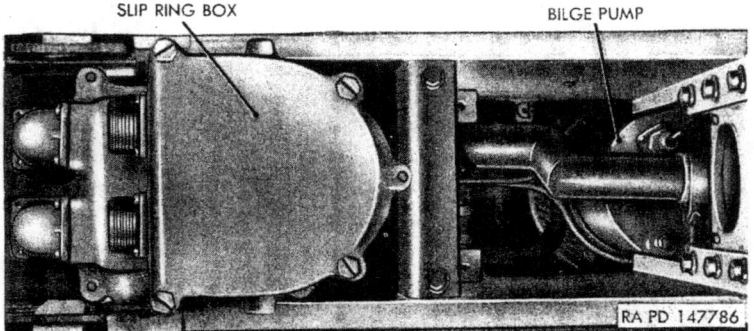

Figure 305. Slip ring box and front bilge pump—installed view.

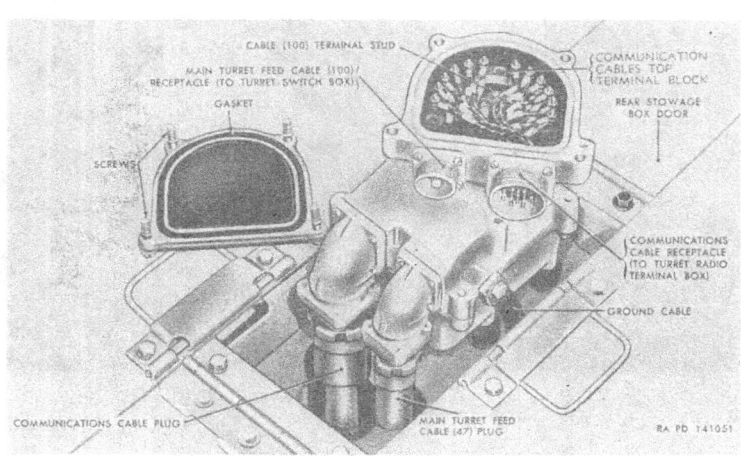

Figure 306. Slip ring box with top cover removed.

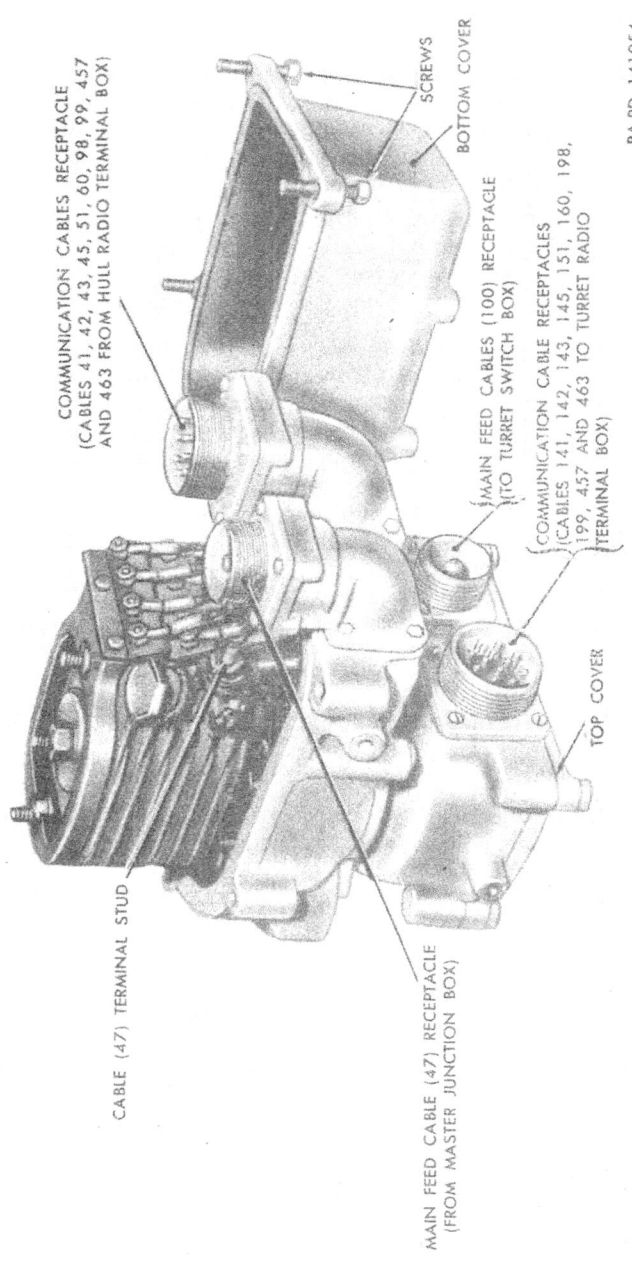

Figure 307. Slip ring box with bottom cover removed.

297. Turret Accessories Panel

a. REMOVAL (fig. 30). Turn off master relay switch (fig. 18). Disconnect two cable connector plugs at side of turret accessories panel and remove cables. Remove eight lock washer screws securing panel box bulkhead brackets and pull panel box forward. Disconnect two external cables at circuit breakers and remove panel box.

b. INSTALLATION. Connect two external cable connector plugs to receptacles at circuit breakers. Refer to figure 298 for wiring diagram. Position panel box to panel brackets and secure with eight lock washer screws and nuts. Connect two cables to side of panel box.

c. VENTILATOR SWITCH (fig. 30).

  (1) *Removal.* Remove turret accessories panel (*a* above). Disconnect two bell-type connectors at rear of ventilator switch. Remove four screws and remove female receptacle at side of panel. Working on face of instrument panel, remove two cap nuts and two lock nuts securing switch assembly to panel and remove switch through rear of panel.

  (2) *Installation.* Position switch to panel and secure with two lock washers, lock nuts, and cap nuts. Connect two bell-type connectors at rear of switch. Install female receptacle and secure with four screws, washers, and nuts. Install turret accessories panel (*b* above).

d. CIRCUIT BREAKERS (fig. 298).

  (1) *Removal.* All four circuit breakers in the turret accessories panel are removed in essentially the same manner. Remove the accessories panel (*a* above). Disconnect two cables at rear of required circuit breaker. Remove two screws securing circuit breaker to face of panel and remove circuit breaker.

  (2) *Installation.* All four circuit breakers are installed in essentially the same manner. Position circuit breaker to panel and secure with two screws, washers, and nuts. Connect cables to circuit breaker (fig. 298). Install accessory panel (*b* above).

e. ACCESSORY OUTLET SOCKET (fig. 30).

  (1) *Removal.* Remove accessories panel (*a* above). Unscrew cap from socket and remove cap. Remove cable from rear of socket. Remove two screws securing socket to face of panel and remove socket from rear of panel.

  (2) *Installation.* Position socket to panel and secure with two screws and washers. Connect cable to rear of socket. Install cap on socket. Install accessories panel (*b* above).

298. Ventilating Blower

a. REMOVAL. Unscrew damper control handle (fig. 304) and remove handle from ventilating blower fan and case assembly. Disconnect electrical cable from blower motor at base of motor. Loosen air duct funnel connecting ring clamp (fig. 304). Working outside the turret, remove turret ventilating blower cover (fig. 291) by removing four bolts. Remove ventilating blower ground strap. Remove ventilating blower and fan case assembly by removing six bolts securing assembly to turret structure.

b. INSTALLATION. Lower ventilating blower fan and case assembly through opening in turret structure and secure with six bolts and washers. Connect ventilating blower ground strap. Install blower cover (fig. 291) and secure with four bolts. Working inside the turret (fig. 304), tighten air duct funnel connecting ring clamp. Insert damper control handle through ceiling bracket and screw handle into blower fan and case assembly. Connect electrical cable at base of blower motor.

299. Turret Dome Lights

The turret dome lights are the same type as the hull dome lights (par. 29). For removal and installation procedures and replacement of bulbs, refer to paragraph 232.

Section XXIV. AUXILIARY GENERATOR AND ENGINE

300. Description and Data

a. DESCRIPTION.

(1) *General* (figs. 308, 309, and 310). The auxiliary engine is a constant-speed, air-cooled, two-cylinder, gasoline engine. Its speed is controlled at 2,800 rpm by a governor. It is provided with a manually controlled choke, magneto ignition, and both a manual starter and an electrical starting system. Fuel is supplied from either fuel tank. Lubrication is by pressure and spray from oil carried in its oil pan. The heat deflector, intake and exhaust manifolds, carburetor, choke linkage, governor, magneto, fuel pump, fuel filter, and air cleaner are mounted on the right side of the engine (fig. 310). The oil filler pipe and manual starter are mounted on the front (fig. 308). Oil drain valve (figs. 309 and 311) is at the rear of the engine, operated by an extended drain valve handle fitted with a dial indicator showing open and closed positions. Bayonet type oil gage is located on the left side

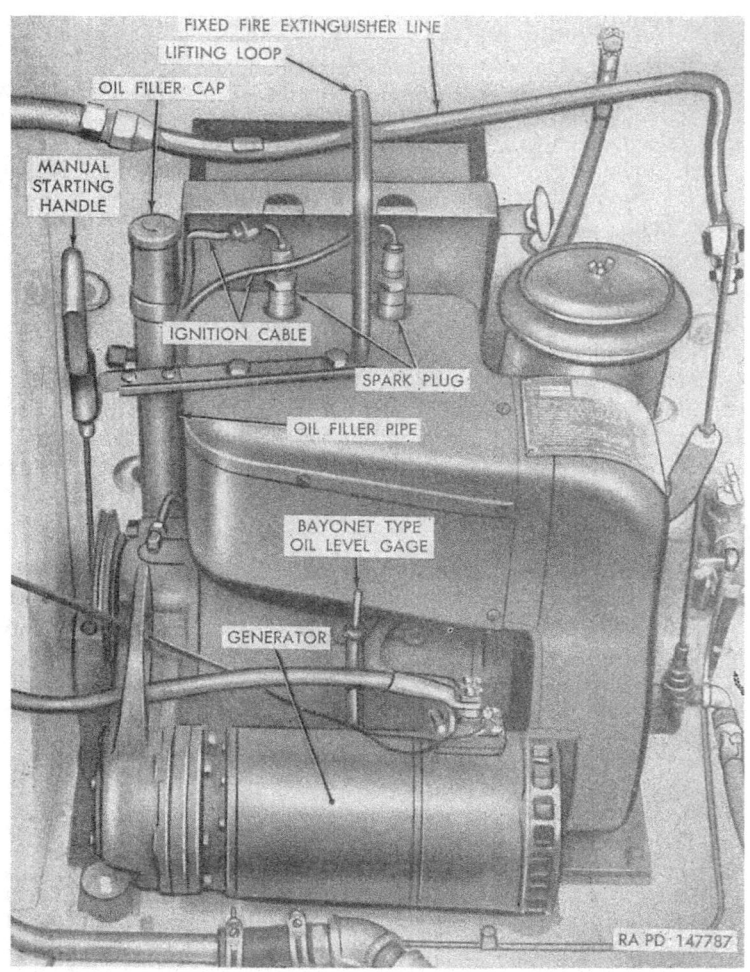

Figure 308. Auxiliary generator and engine—left-side view.

of engine. Gear box is lubricated through an oil inlet line from the engine oil pan to the top of the gear box. A return line from the bottom of the gear box permits oil to flow back to the engine oil pan. The combination flywheel and fan is located at the rear of the engine. A fuel shut-off valve is mounted on a bracket at the right rear of the engine (fig. 309). The generator, mounted on the left side of the engine (fig. 308), also acts as a motor for cranking and starting the engine. When the engine has started, the generator resumes

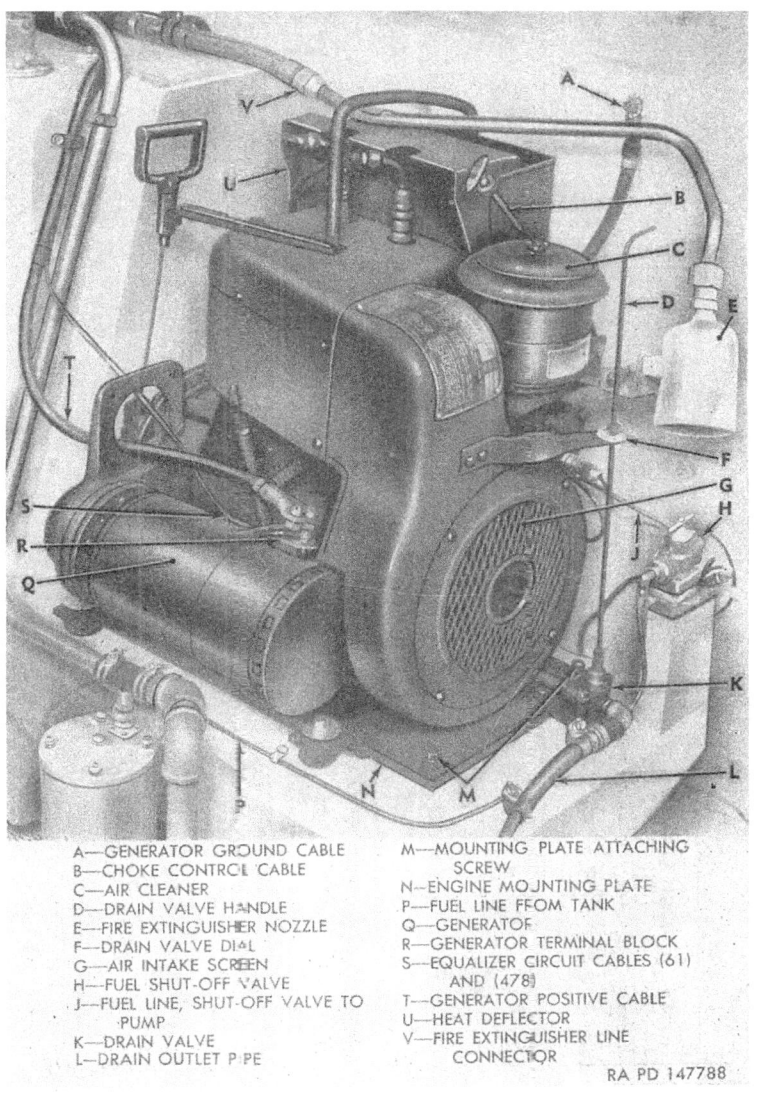

A—GENERATOR GROUND CABLE
B—CHOKE CONTROL CABLE
C—AIR CLEANER
D—DRAIN VALVE HANDLE
E—FIRE EXTINGUISHER NOZZLE
F—DRAIN VALVE DIAL
G—AIR INTAKE SCREEN
H—FUEL SHUT-OFF VALVE
J—FUEL LINE, SHUT-OFF VALVE TO PUMP
K—DRAIN VALVE
L—DRAIN OUTLET PIPE
M—MOUNTING PLATE ATTACHING SCREW
N—ENGINE MOUNTING PLATE
P—FUEL LINE FROM TANK
Q—GENERATOR
R—GENERATOR TERMINAL BLOCK
S—EQUALIZER CIRCUIT CABLES (61) AND (478)
T—GENERATOR POSITIVE CABLE
U—HEAT DEFLECTOR
V—FIRE EXTINGUISHER LINE CONNECTOR

RA PD 147788

Figure 309. Auxiliary generator and engine—left-rear view.

its normal function of supplying current to vehicular electrical equipment and to the batteries.

(2) *Carburetor* (figs. 310 and 313). The proper mixture of gasoline and air is furnished by a balanced carburetor. The main metering jet in the carburetor is of the fixed type; that is,

519

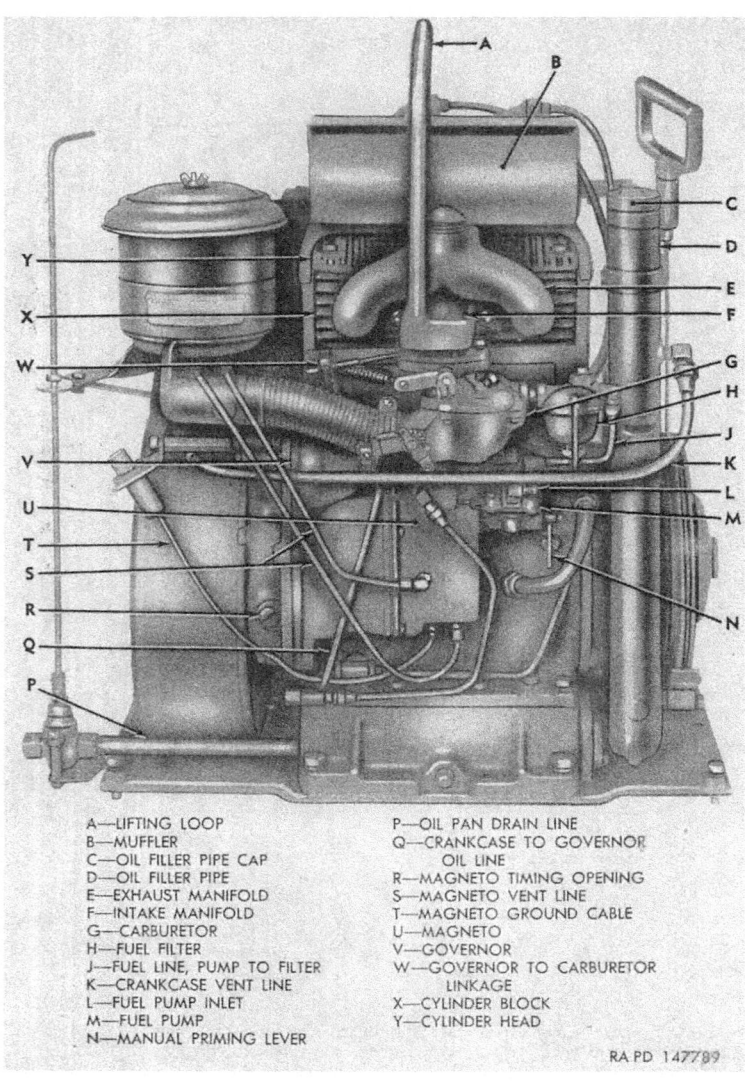

A—LIFTING LOOP	P—OIL PAN DRAIN LINE
B—MUFFLER	Q—CRANKCASE TO GOVERNOR OIL LINE
C—OIL FILLER PIPE CAP	
D—OIL FILLER PIPE	R—MAGNETO TIMING OPENING
E—EXHAUST MANIFOLD	S—MAGNETO VENT LINE
F—INTAKE MANIFOLD	T—MAGNETO GROUND CABLE
G—CARBURETOR	U—MAGNETO
H—FUEL FILTER	V—GOVERNOR
J—FUEL LINE, PUMP TO FILTER	W—GOVERNOR TO CARBURETOR LINKAGE
K—CRANKCASE VENT LINE	
L—FUEL PUMP INLET	X—CYLINDER BLOCK
M—FUEL PUMP	Y—CYLINDER HEAD
N—MANUAL PRIMING LEVER	

RA PD 147789

Figure 310. Auxiliary generator and engine—right-side view.

no adjustment is provided. The idle speed adjusting screw should be turned to adjust the idler needle for best low-speed operation, while carburetor throttle is held closed by hand.

(3) *Magneto* (figs. 310 and 315). Spark for ignition is furnished by a high-tension magneto driven off the timing gear. The magneto is fitted with an impulse coupling which makes pos-

sible a powerful spark for easy starting. In addition, the impulse coupling automatically retards the timing of the spark for starting, thus eliminating danger of a kick-back from the engine when starting.

(4) *Oil pump.* A plunger-type oil pump on the underside of the cylinder block supplies oil to a nozzle which directs oil streams against holes in the connecting rods. Part of the oil enters the rod bearing through holes in the rods, and the balance of the oil forms a spray or mist which lubricates the cylinders and other parts of the engine.

(5) *Governor* (fig. 310). A centrifugal flyball-type governor controls the engine speed by varying the throttle opening to suit the load imposed upon the engine. The governor rotates on a stationary spindle driven into the upper part of the governor and magneto adapter. The governor is driven off the camshaft gear. No maintenance on governor is authorized the using organization.

(6) *Fuel pump* (fig. 310). The fuel pump is actuated by a cam on the camshaft. A manual priming lever is provided to pump fuel into a dry carburetor.

(7) *Generator* (fig. 308). The generator on the auxiliary engine is identical to the main engine generator (par. 222).

b. DATA.

Auxiliary engine.

Make	Wisconsin
Model	TFT
Ordnance number	7722313
Type	4 cycle, gasoline
Cooling system	air cooled
Bore and stroke	3¼ x 3¼ in.
Number of cylinders	2
No. 1 cylinder	nearest flywheel
Horsepower	13.6
Firing	360° apart
Valve clearance (set when cold)	intake: 0.010 to 0.012 in.
	exhaust: 0.016 to 0.018 in.

Carburetor.

Make	**Zenith**
Model	0-10595
Ordnance number	7744753

Magneto.

Make	Fairbanks-Morse
Model	FMXFE2B7A
Ordnance number	7767295
Type	radio-shielded
Breaker-point gap	0.012 in.

Governor.
　Manufacturer's number_____ WVT965S1
　Full-load speed_____ 2,800 rpm
　No-load speed_____ 2,950 rpm
　Spring setting_____ hook governor spring in hole No. 10 (third from top).

Fuel pump.
　Make_____ AC Spark Plug
　Manufacturer's number_____ AC 1539495
　Ordnance number_____ 7767902

Spark plug.
　Make_____ Champion
　Model_____ C-26-S
　Thread_____ 18-mm
　Spark plug gap_____ 0.025 in. (nonadjustable)

Air cleaner.
　Make_____ United Specialties
　Model_____ HX55-6580-A
　Ordnance number_____ 7744744

Generator (same as main engine generator (par. 211).

301. Maintenance Operations

　a. MAINTENANCE IN VEHICLE.
　　(1) *Oil level.* Check and add oil if necessary (par. 302).
　　(2) *Drain oil.* Refer to paragraph 302.
　　(3) *Air cleaner.* Service or replace (par. 305).
　　(4) *Spark plugs.* Replace (par. 311).
　　(5) *Fuel shut-off valve.* Replace (par. 309).
　　(6) *Fuel lines.* Replace (par. 309).

　b. MAINTENANCE OUT OF VEHICLE.
　　(1) *Fuel filter.* Service or replace (par. 307).
　　(2) *Exhaust and intake manifolds and gaskets.* Replace (par. 306).
　　(3) *Muffler.* Replace (par. 306).
　　(4) *Carburetor.* Adjust or replace (par. 307).
　　(5) *Fuel pump.* Replace (par. 308).
　　(6) *Magneto breaker points.* Adjust gap or replace (par. 310).
　　(7) *Magneto.* Replace or retime (par. 310).
　　(8) *Generator brushes.* Replace (par. 304).
　　(9) *Manual starter.* Replace (par. 312).
　　(10) *Ignition cables.* Replace (par. 311).
　　(11) *Vent lines.* Replace (par. 314).
　　(12) *Mounting brackets.* Replace (par. 315).
　　(13) *External oil lines.* Replace (par. 313).

302. Servicing

a. CHECK OIL LEVEL IN AUXILIARY OIL PAN. Remove bayonet-type oil level gage (fig. 308). Wipe dry with clean rag and insert gage. Remove gage and check reading. Add oil if necessary (*c* below), to bring level up to "FULL" mark.

b. DRAIN AUXILIARY ENGINE OIL PAN. The oil pan can be drained with engine in or out of vehicle. Open drain valve at rear of engine (figs. 309 and 311). Allow the oil to drain. Close drain valve.

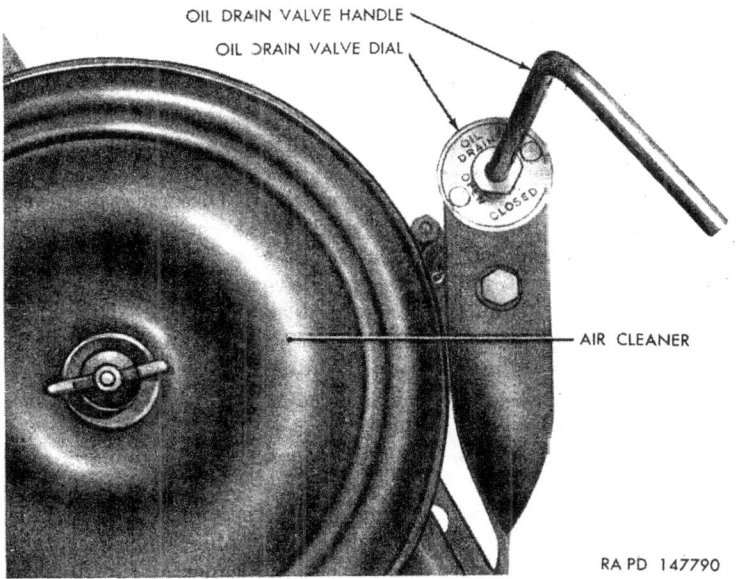

Figure 311. Auxiliary engine oil drain valve.

c. FILL AUXILIARY ENGINE OIL PAN. Remove the oil filter cap (fig. 308). Pour oil as specified on lubrication order (fig. 86) through the oil filler pipe. Check oil level with gage (*a* above).

d. CLEAN AUXILIARY ENGINE AIR CLEANER. Remove the wing nut and remove the cover from air cleaner (fig. 309). Lift out the element and clean in volatile mineral spirits or dry-cleaning solvent. Dry with compressed air. Dip the element in oil as specified on the lubrication order (fig. 86), and fill the air cleaner base with oil to oil level mark. Install the element, cover, and wing nut.

e. CLEAN FUEL FILTER BOWL. Loosen the knurled nut and swing the wire bail (fig. 313) to one side. Remove the glass bowl and clean thoroughly. Fill the bowl with gasoline and install.

303. Auxiliary Generator and Engine Replacement

a. COORDINATION WITH ORDNANCE MAINTENANCE UNIT. Replacement of the auxiliary generator and engine is normally an ordnance maintenance operation, but may be performed in an emergency by the using organization, provided authority for performing this replacement is obtained from the appropriate commander. Tools needed for the operation which are not carried in the using organization may be obtained from the supporting ordnance maintenance unit.

b. REMOVAL.

(1) *Remove main-engine air cleaner pipe.* Open the five engine compartment right grille doors (fig. 101). Remove the main-engine air-cleaner-to-carburetor pipe on the right side of the vehicle (par. 159*d*(13)).

(2) *Disconnect fire extinguisher line.* Disconnect fire extinguisher line at the connector located above the right fuel tank (fig. 309).

(3) *Disconnect two small generator cables.* Disconnect generator equalizer circuit cables (61) and (478) from generator terminals "A" and "D" (fig. 309).

(4) *Disconnect two large generator cables.* Disconnect positive cable (62) from the generator terminal block. Disconnect the ground cable by removing the cap screw securing it to the inside of the right-side of hull (fig. 309).

Note. For ease of maintenance, do not remove the negative cable from the generator at terminal block.

(5) *Remove auxiliary engine air cleaner.* Refer to paragraph 305.

(6) *Disconnect magneto ground cable.* Disconnect the auxiliary engine magneto ground cable (422) plug from receptacle mounted on bracket under the air cleaner at the flywheel end of the auxiliary engine (fig. 310).

(7) *Disconnect fuel line.* Disconnect the fuel line to auxiliary engine at the shut-off valve (fig. 309).

(8) *Disconnect oil drain outlet pipe.* Disconnect auxiliary engine oil drain outlet pipe at the drain valve (fig. 309).

(9) *Disconnect main-engine throttle control linkage.* Disconnect the main-engine throttle control linkage (fig. 96) at main-engine left carburetor and at cross shaft lever at accessory end of main engine, and remove from main engine.

(10) *Remove main-engine ignition harness.* Remove the left ignition harness from the main engine (par. 187).

(11) *Move main-engine degasser cables out of the way.* Disconnect main-engine carburetor degasser cables from degasser (fig. 130) and move them carefully out of way so as not to offer interference when removing auxiliary engine.

(12) *Remove generator and engine from vehicle.* Unscrew the four mounting plate attaching screws (fig. 309) securing the engine mounting plate to the mounting brackets on hull floor. Pass a sling through the lifting loop (fig. 310) and attach to hoist. Lift the auxiliary engine and generator out of engine compartment carefully making sure it does not strike any portion of the main engine and hull.

c. INSTALLATION.
  (1) *Install auxiliary engine in vehicle.* Pass sling through the lifting loop (fig. 310) and attach to hoist. Carefully position the auxiliary engine and generator into the vehicle, taking care not to strike any units while lowering into position. Secure the auxiliary engine by tightening the four mounting plate attaching screws (fig. 309) which secure the engine mounting plate to the mounting brackets.
  (2) *Connect auxiliary engine fuel line.* Connect the fuel line to the auxiliary engine fuel shut-off valve at the outlet side (fig. 309).
  (3) *Connect magneto ground cable.* Connect the auxiliary engine magneto ground cable (422) plug at receptacle on the bracket mounted below air cleaner, at flywheel end of engine (fig. 310).
  (4) *Connect auxiliary engine oil drain outlet pipe.* Connect the auxiliary engine oil drain outlet pipe at the drain valve (fig. 309).
  (5) *Connect the positive generator cable.* Connect the auxiliary generator positive cable (62) (fig. 309) to the generrator terminal block and secure with nut, lock washer, and cotter pin.
  (6) *Install main-engine degasser cables.* Connect the main-engine carburetor degasser cables to degassers and secure by tightening the knurled nuts (fig. 130).
  (7) *Install the main-engine ignition harness.* Refer to paragraph 187.
  (8) *Install main-engine throttle control linkage.* Connect the main-engine throttle control linkage (fig. 96) at the left carburetor and at the cross shaft lever at the accessory end of engine.
  (9) *Connect two small generator cables.* Connect the two generator equalizer circuit cables at generator terminal block, cable (61) to terminal "A" and cable (478) to "D."
  (10) *Connect the generator ground cable.* Connect the auxiliary generator ground cable to the right-side of hull (fig. 309) by installing cap screw and lock washer.
  (11) *Install auxiliary engine air cleaner.* Refer to paragraph 305.

12. *Connect fire extinguisher line and discharge nozzle.* Connect the fire extinguisher line leading to the discharge nozzle from the main line, at the connector on top of the right fuel tank (fig. 309).
13. *Install main-engine air cleaner pipe.* Install the main-engine left air-cleaner-to-carburetor pipe (par. 160*b* (11)). Close the five engine compartment right grille doors (fig. 101).
14. *Record of replacement.* Record the replacement on DA AGO Form No. 478.

304. Auxiliary Generator

a. REMOVAL. Remove auxiliary generator and engine from vehicle (par. 303*b*). Remove electrical cables from generator (fig. 309).

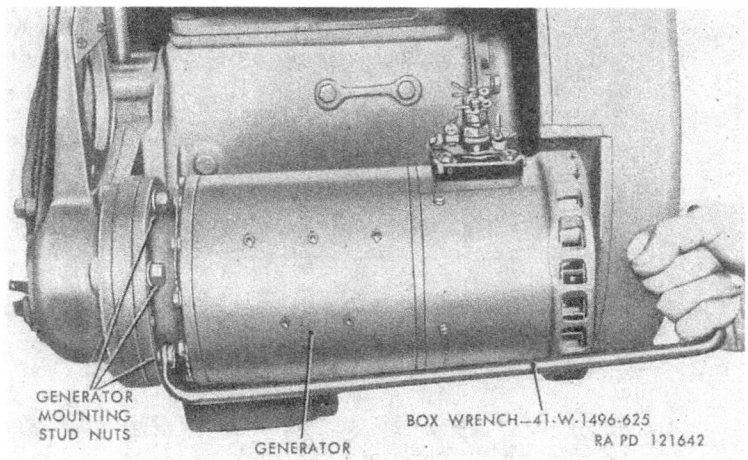

Figure 312. Removing auxiliary generator.

Using box wrench 41–W–1496–625, remove six generator mounting stud nuts and lock washers (fig. 312). Pull generator from mounting studs.

b. GENERATOR BRUSH REPLACEMENT. Remove generator (*a* above) and proceed as for main generator (par. 222). Install generator (*c* below).

c. INSTALLATION. Position generator on mounting studs. Install six lock washers and mounting stud nuts, using wrench 41–W–1496–625 (fig. 312). Insert a jumper cable between terminals "A" and "B" on terminal block (fig. 309). Start engine and check generator voltage with a voltmeter across two large terminals "B" and "E". If voltage is not less than 24 volts, remove jumper cable. If voltage is less than 24 volts, refer to paragraph 132. Install auxiliary generator and engine (par. 303c). Start auxiliary engine and test generator operation (par. 132).

305. Auxiliary Engine Air Cleaner

a. SERVICE. Refer to paragraph 302*d*.

b. REMOVAL. Open five engine compartment right grille doors (fig. 101). Remove wing nut holding air cleaner cover to stud and remove air cleaner (fig. 309). Remove air cleaner cover and element. Drain base of dirty oil and clean base and element with volatile mineral spirits or dry-cleaning solvent.

c. INSTALLATION. Position air cleaner (fig. 309) on stud and fill with oil as specified on lubrication order (fig. 86) to oil level mark. Install element, cover, and wing nut. Close five engine compartment right grille doors (fig. 101).

306. Auxiliary Engine Intake and Exhaust Manifolds and Muffler

a. REMOVAL OF MUFFLER AND EXHAUST MANIFOLDS. Remove auxiliary generator and engine (par. 303*b*). Remove fixed fire extinguisher line (fig. 308). Remove air cleaner (par. 305). Disconnect choke control cable at heat deflector bracket (fig. 309) and at carburetor and remove choke control cable and handle. Remove eight screws securing heat deflector (fig. 309), and remove heat deflector. Remove two nuts and washers securing lifting loop to engine shroud, and two nuts, washers, and spacer, holding lifting loop and exhaust manifolds in place (figs. 308 and 310). Move lifting loop out of the way and remove manifold with muffler attached. Remove gasket. Separate muffler from exhaust manifold by unscrewing.

b. INSTALLATION OF MUFFLER AND EXHAUST MANIFOLDS. Screw muffler into exhaust manifold. Position exhaust manifold and gasket, and lifting loop and secure with two spacers, washers, and nuts (fig. 310). Install two nuts and washers securing lifting loop (fig. 308) to engine shroud. Install heat deflector (fig. 309) and choke control handle mounting bracket. Secure with eight screws.

Note. One screw attaches spark plug cable clip and oil filler pipe clip to heat deflector.

Connect choke control cable at heat deflector and at carburetor. Install fire extinguisher line (fig. 308). Install auxiliary generator and engine (par. 303*c*).

c. REMOVE INTAKE MANIFOLD (fig. 310). Remove exhaust manifold and muffler (*a* above). Remove two screws and lock washers holding intake manifold to carburetor flange, and remove intake manifold.

d. INSTALL INTAKE MANIFOLD (fig. 310). Position intake manifold and gasket on carburetor flange and secure with two lock washers and screws. Install exhaust manifold and muffler (*b* above).

307. Auxiliary Engine Carburetor and Fuel Filter

a. SERVICE FUEL FILTER. Refer to paragraph 302*e*.

b. REMOVAL (figs. 310 and 313). Remove auxiliary generator and engine (par. 303*b*). Loosen clamp and screw holding choke control cable and remove cable. Loosen hose clamps on carburetor intake hose and remove hose. Remove governor-to-carburetor linkage. Disconnect fuel-pump-to-fuel-filter line at filter. Remove two screws and washers holding carburetor to intake manifold. Lift off carburetor and fuel filter. Separate fuel filter from carburetor by unscrewing filter.

c. INSTALLATION (figs. 310 and 313). Screw fuel filter into carburetor inlet. Position carburetor and fuel filter and secure to intake

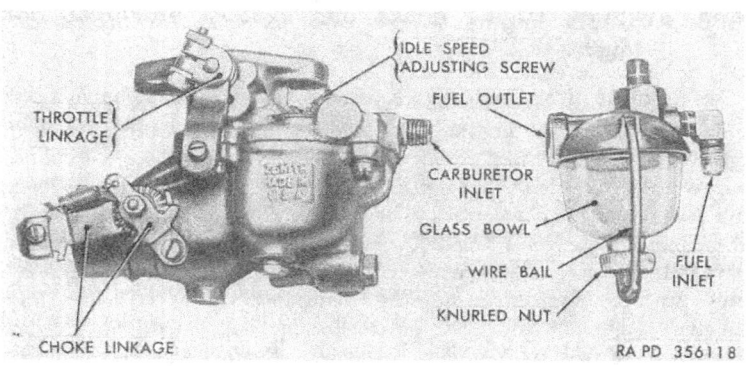

Figure 313. Auxiliary engine carburetor and fuel filter.

manifold with two washers and screws. Connect fuel-pump-to-fuel-filter line at filter. Connect governor-to-carburetor linkage at third hole from the top. Position intake hose on carburetor and secure hose clamps. Connect choke control cable to linkage and tighten screw and clamp. Install auxiliary generator and engine (par. 303*c*). Adjust carburetor (par. 300*a*(2)).

308. Auxiliary Engine Fuel Pump

a. REMOVAL (fig. 310). Open five engine compartment right grille doors (fig. 101). Remove auxiliary generator and engine (par. 303*b*). Disconnect fuel pump inlet line at pump. Disconnect fuel-pump-to-fuel-filter line. Remove two screws and lock washers and lift off fuel pump.

b. INSTALLATION (fig. 310). Position fuel pump and secure with two lock washers and screws. Connect fuel-pump-to-fuel-filter line. Connect fuel pump inlet line to the pump. Install auxiliary generator and engine (par. 303c). Close five engine compartment right grille doors (fig. 101).

309. Auxiliary Engine Fuel Shut-Off Valve and Fuel Lines

a. REMOVAL OF FUEL SHUT-OFF VALVE (fig. 314). Close fuel tank shut-off valves (par. 17). Open five engine compartment right grille doors (fig. 101). Disconnect fuel shut-off cable at valve. Disconnect

Figure 314. Auxiliary engine fuel shut-off valve—installed view.

fuel valve inlet and outlet fuel lines at valve. Remove two mounting bolts and lock washers holding valve clamp to mounting bracket and remove valve.

b. INSTALLATION OF FUEL SHUT-OFF VALVE (fig. 314). Position fuel shut-off valve and clamp on mounting bracket and install two lock washers and mounting bolts. Connect cable connector and valve inlet and outlet fuel lines. Close five engine compartment right grille doors (fig. 101). Open fuel tank shut-off valves (par. 17).

c. REMOVAL OF FUEL LINES. Open five engine compartment right grille doors (fig. 101). Close auxiliary engine fuel shut-off valve (fig. 314). Disconnect fuel line at shut-off valve and fuel pump (figs. 309 and 310). Disconnect pump-to-filter line at pump outlet and at fuel filter (fig. 310).

Caution: Take care in handling fuel lines that they do not become pinched and that no dirt gets into them.

d. INSTALLATION OF FUEL LINES. Install fuel-pump-to-filter fuel line at filter inlet and fuel pump outlet connections (fig. 310). Connect fuel line at fuel pump and fuel shut-off valve (figs. 309 and 310). Open fuel shut-off valve. Close five engine compartment right grille doors (fig. 101).

310. Auxiliary Engine Magneto

a. REMOVAL (figs. 310, 315, and 316). Remove auxiliary generator and engine (par. 303*b*). Mark ignition cables for installation pur-

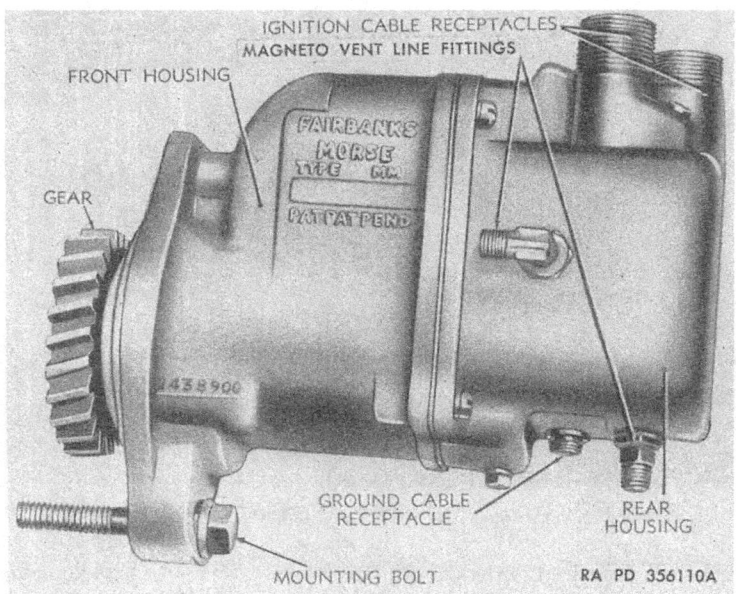

Figure 315. Auxiliary engine magneto.

poses. Remove knurled nuts holding ignition cables at spark plugs and disconnect cables. Disconnect vent line from side and vent line and ground cable from bottom of magneto. Remove fuel line, remove crankcase-to-governor oil line. Plug crankcase opening to prevent oil from escaping. Remove two bolts holding magneto to the magneto-and-governor adapter and remove magneto.

Note. Magneto ground cable receptacle below air cleaner may have to be loosened to permit removal of top bolt.

b. MAGNETO BREAKER POINTS REPLACEMENT (fig. 316).
   (1) *Removal.* Remove magneto (*a* above). Remove four screws and lock washers holding rear housing to front housing. Disconnect capacitor cable and coil cable from breaker point plate. Remove clip from breaker arm pin. Remove eccentric adjusting screw and plate lock screw and remove breaker point plate.
   (2) *Installation.* Position breaker point plate on front housing and install plate lock screw and eccentric adjusting screw. Connect capacitor and coil cable to plate. Position breaker arm on pin and install clip to secure. Adjust breaker point gap (*c* below). Time magneto and install (*d* below).

c. MAGNETO BREAKER POINT GAP ADJUSTMENT (fig. 316). Remove magneto (*a* above). Remove four screws holding rear housing to front housing. Rotate cam until points are at their maximum opening. Check gap with a feeler gage. The breaker points should have an opening of 0.012 inch. If the gap is incorrect, loosen the plate lock screw on the breaker point plate, then turn the eccentric adjusting screw, right or left, until proper gap is attained; then tighten the plate lock screw. Recheck the breaker point gap. Time magneto and install (*d* below).

d. MAGNETO TIMING.
   (1) *Preliminary procedures.* Magneto timing can only be done with auxiliary generator and engine removed from vehicle (par. 303*b*). Remove magneto (*a* above). Remove four screws and lock washers holding rear housing to front housing (fig. 316).
   (2) *Procedure.*
      (*a*) Remove the four screws which hold the air intake screen in place and remove screen (fig. 309). This will expose the timing marks on the flywheel shroud (fig. 317). Remove the spark plug from the No. 1 cylinder (nearest flywheel). Turn engine over slowly with the manual starter cable until the compression in this cylinder blows the air out of the spark plug hole.
      (*b*) The flywheel is marked with the letters "DC" near one of the air circulating vanes (fig. 317). This vane is identified further by an "X" cast on the face. When the air blows out of No. 1 spark plug hole (nearest flywheel), continue turning the crankshaft until the forward edge of the marked vane on the flywheel registers with the upper line stamped on the shroud. This is the line on the vertical center of the engine. Leave the flywheel in this position. There is another line stamped on the shroud 27° or about 2½ inches, to the left of the upper line. This

531

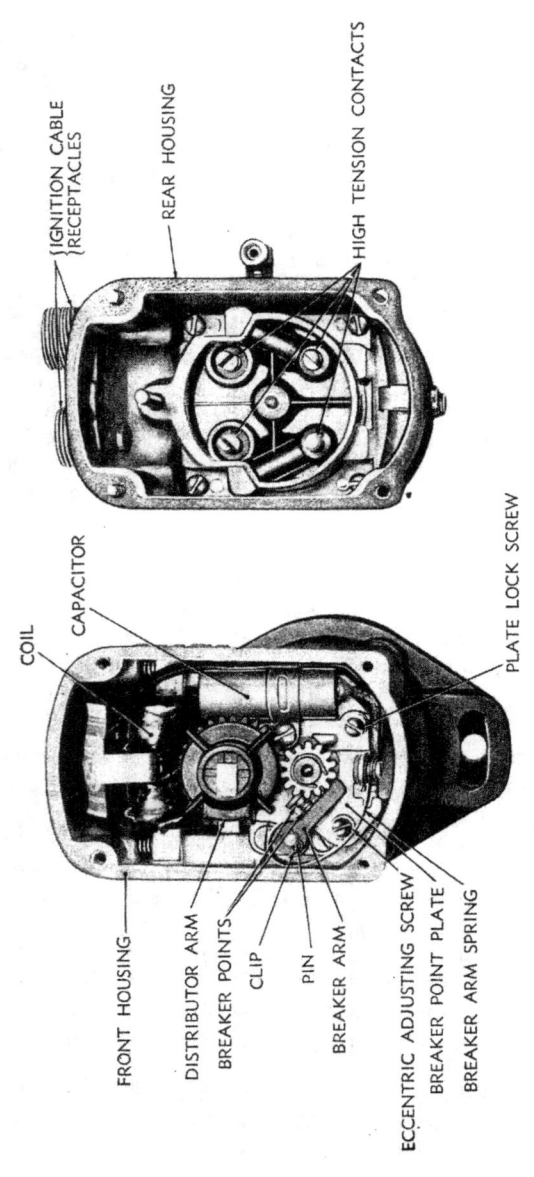
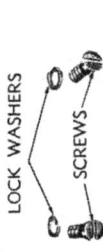

Figure 316. Auxiliary engine magneto—front and rear housings separated.

second line is the position of the marked vane on the flywheel when the spark plug fires while the engine is running.

(c) Remove the plug from the magneto timing opening (fig. 310) in the magneto and governor adapter.

(d) Turn the magneto shaft by means of the driving gear until the "X" mark on the driving gear tooth is on that side of the magneto housing which is away from engine when magneto is installed. Check to see whether distributor arm (fig. 316) is in line with the high-tension contact for the No 1 spark plug. If not, rotate gear 360° until "X" is again on outer side of housing. Hold shaft in this position and place magneto rear housing on magneto front housing and secure with four screws.

Note. Do not rotate gear after alining "X."

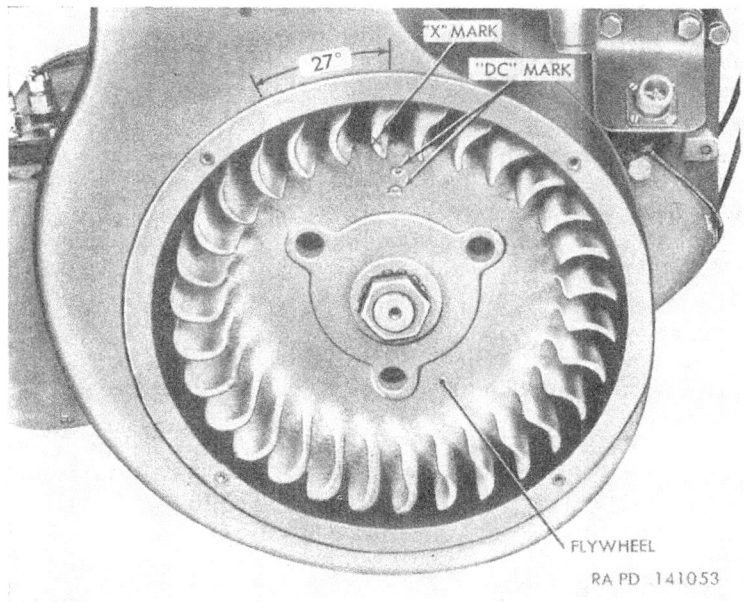

Figure 317. Auxiliary engine flywheel timing marks.

e. INSTALLATION (fig. 310). Place magneto in position on engine meshing gears so that, when the magneto is in place, the gear tooth stamped with the "X" will be visible in the center of the magneto timing opening. Install the two bolts which hold magneto to magneto-and-governor adapter. Connect magneto ground cable and vent lines. Install ignition cables into their proper receptacles as marked on disassembly and tighten knurled nuts to secure cables. Install crankcase-to-governor oil line. Install fuel line. Install auxiliary generator and engine (par. 303c). Position air intake screen (fig. 309) and secure with four screws.

311. Auxiliary Engine Spark Plugs and Ignition Cables

a. REMOVAL OF IGNITION CABLES (fig. 308). Remove auxiliary generator and engine from vehicle (par. 303b). Unscrew ignition cable knurled nuts at magneto (fig. 310), and pull out cables. Unscrew ignition cable knurled nuts at spark plugs, and pull out cables (fig. 308). Remove screw and lock washer holding ignition cable clip to heat deflector, and lift out cables.

b. INSTALLATION OF IGNITION CABLES (fig. 308). Insert ignition cables into magneto receptacles and tighten knurled nuts.

Note. Make sure No. 1 ignition cable is connected to No. 1 magneto receptacle, and No. 2 ignition cable to No. 2 magneto receptacle.

Pull ignition cables between oil filler pipe and engine shroud and connect the spark plugs.

Note. Make sure No. 1. ignition cable is connected to No. 1 spark plug and No. 2 ignition cable to No. 2 spark plug.

Position ignition cable clip against heat deflector and install lock washer and screw. Install auxiliary generator and engine in vehicle (par. 303c).

c. REMOVAL OF SPARK PLUGS (fig. 308). Open five engine compartment right grille doors (fig. 101). Unscrew ignition cable knurled nuts from spark plugs and pull out cables. Remove spark plugs from engine.

d. CHECK SPARK PLUG GAP. Gap should not be more than 0.025 inch for these nonadjustable plugs. Replace plugs if gaps are not correct.

e. INSTALLATION OF SPARK PLUGS (fig. 308). Install plugs and gaskets. Insert ignition cables into proper spark plug and tighten ignition cable knurled nuts. Close engine compartment right grille doors (fig. 101).

312. Auxiliary Engine Manual Starter

a. REMOVAL. Remove auxiliary generator and engine from vehicle (par. 303b). Remove two pulley guard stud nuts and remove pulley guard (fig. 321). Loosen set screw on pulley retaining hub (fig. 319). Install puller 41-P-2905-60 and remove pulley retaining hub (fig. 318). Remove retainer ring, spring washer, and pulley. (figs. 319 and 320). To remove manual starter cable from pulley, loosen set screw on pulley and unwind cable.

Note. Puller 41-P-2905-60 must be obtained from the supporting ordnance maintenance unit.

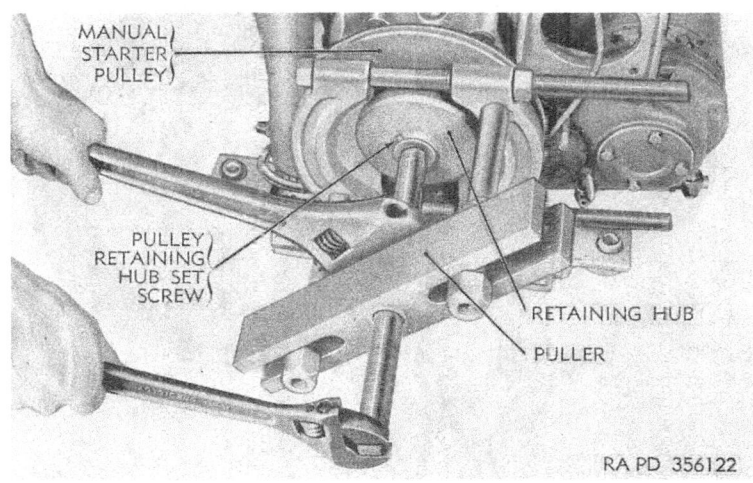

Figure 318. Removing auxiliary engine manual starter pulley retaining hub.

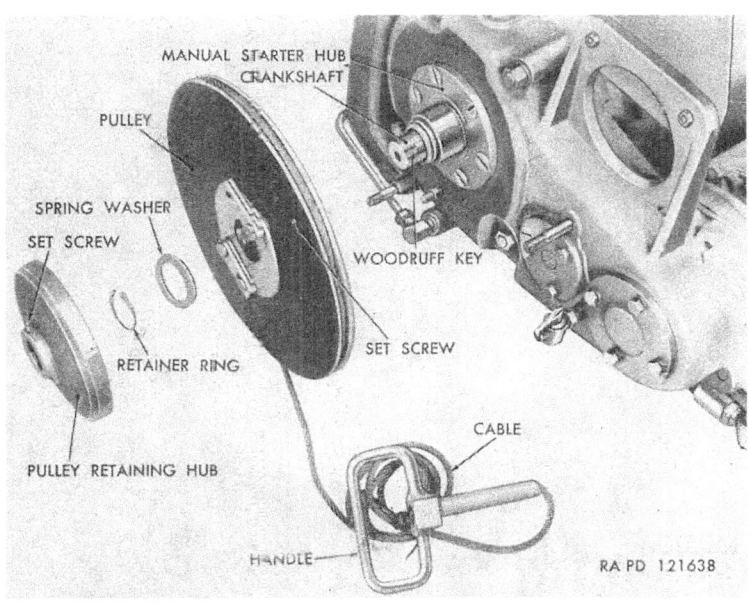

Figure 319. Auxiliary engine manual starter—disassembled view.

535

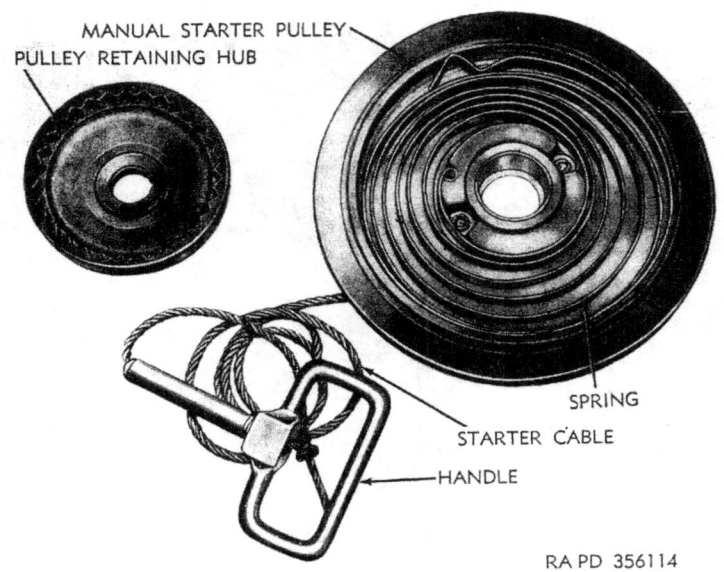

Figure 320. Auxiliary engine pulley and retaining hub.

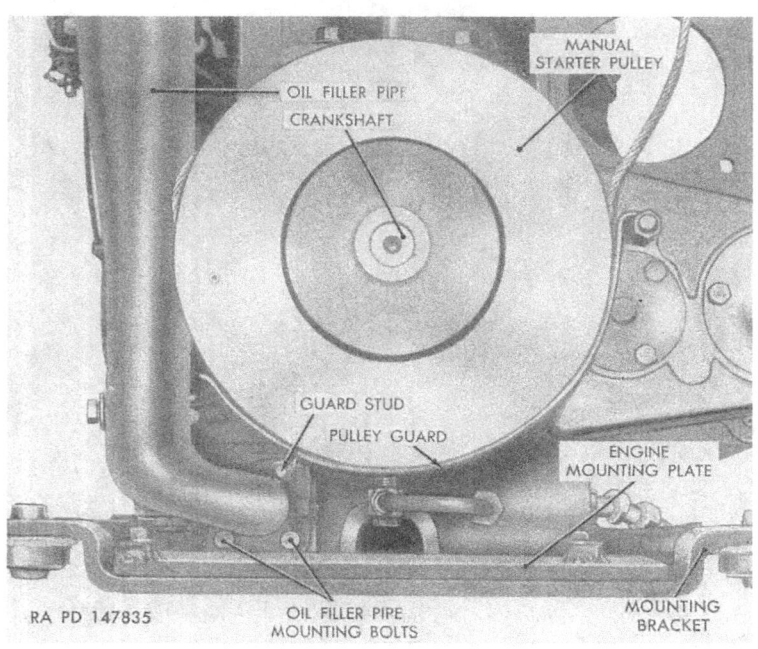

Figure 321. Auxiliary engine removed—front view.

b. INSTALLATION. Wind manual starter cable or pulley and tighten set screw (fig. 319). Place pulley on crankshaft so that keyway is in proper position and push pulley onto crankshaft. Install spring washer and retainer ring (fig. 319). Position pulley retaining hub so that keyway is in proper position and push hub onto crankshaft. Tighten set screw on pulley retaining hub. Position pulley guard (fig. 321) on studs and install two nuts. Install auxiliary generator and engine (par. 303*c*).

Figure 322. Auxiliary engine external oil lines.

313. Auxiliary Engine External Oil Lines

a. REMOVAL (fig. 322). External oil lines are accessible only with auxiliary generator and engine removed from vehicle (par. 303*b*). Disconect gear case oil inlet line from top of gear case and right side

537

of crankcase and remove line. Disconnect gear case oil return line from bottom of gear case and front of oil pan and remove line.

b. INSTALLATION (fig. 322). Connect one end of oil return line to bottom of gear case and opposite end of engine oil pan. Connect one end of oil inlet to top of gear case and opposite end of crankcase.

314. Auxiliary Engine Vent Lines

a. REMOVAL (fig. 310). All vent lines are removed in the same manner. Remove auxiliary generator and engine (par. 303*b*). Disconnect connectors at both ends of vent line and remove vent line.

Caution: Handle vent lines with care to prevent damage.

b. INSTALLATION (fig. 310). All vent lines are installed in the same manner. Position vent line and secure connectors at both ends. Install auxiliary generator and engine (par. 303*c*).

315. Auxiliary Engine Mounting Brackets

a. REMOVAL.
  (1) Remove auxiliary generator and engine (par. 303*b*).
  (2) Remove four cap screws, plain washers and lock washers (two each bracket) securing mounting brackets to hull (fig. 323).

b. INSTALLATION.
  (1) Position mounting brackets on hull (fig. 323) and secure with four cap screws, lock washers, and plain washers (two to each bracket).
  (2) Install auxiliary generator and engine (par. 303*c*).

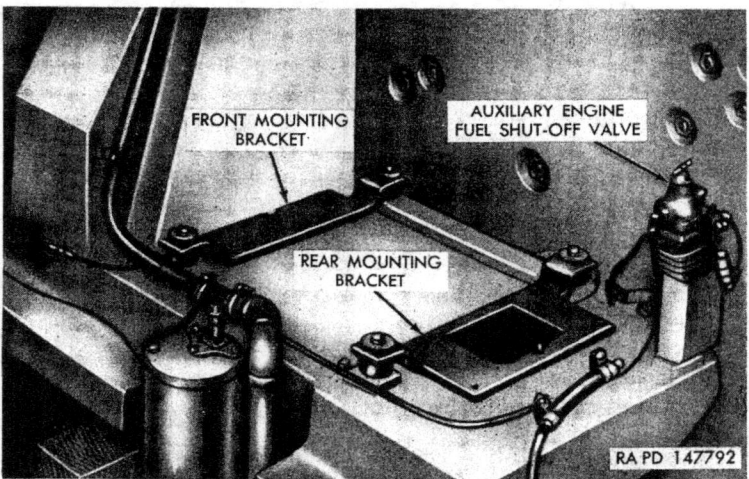

Figure 323. Auxiliary engine mounting brackets.

Section XXV. FIRE EXTINGUISHER SYSTEMS

316. Description

a. FIXED SYSTEM. A carbon dioxide (CO_2) fixed-fire-extinguisher system is provided for smothering fires in the engine compartment. The system consists of three 10-pound cylinders, control head, extinguisher lines, discharge nozzles, and an exterior control handle and cable. The 10-pound carbon dioxide cylinders are mounted in brackets between and to the rear of the driver's seats (fig. 324). All cylinders are discharged simultaneously by operating the control head lever (fig. 325) on the left cylinder or by pulling the exterior remote control handle (fig. 75). The cylinders are connected by lines to three discharge nozzles in the engine compartment (fig. 326). Each cylinder is equipped with a safety valve (fig. 324) which automatically will discharge the gas when the pressure becomes excessive due to heat.

b. PORTABLE SYSTEM. Two portable fire extinguishers are provided for fires which may occur in the fighting compartment. Each is a 5-pound cylinder filled with carbon dioxide (CO_2) and equipped with discharge horn and actuating trigger. One is located in the driver's compartment (fig. 327). The other is mounted on the turret bulge below the periscope storage.

317. Fixed Fire Extinguisher Cylinders

a. REMOVAL (fig. 324). Unscrew lower connector which secures control head to left cylinder and remove control head. Unscrew and remove connecting heads from all three cylinders. Pull cylinders out of support brackets and remove from vehicle.

Warning: Handle all charged cylinders with care. Do not bump, jar, or drop.

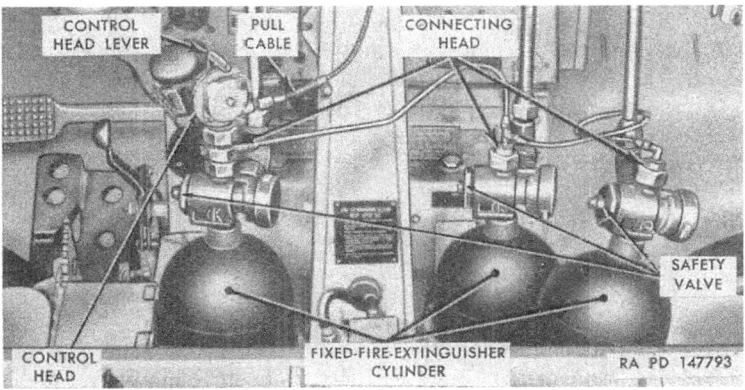

Figure 324. Fixed fire extinguisher cylinders.

539

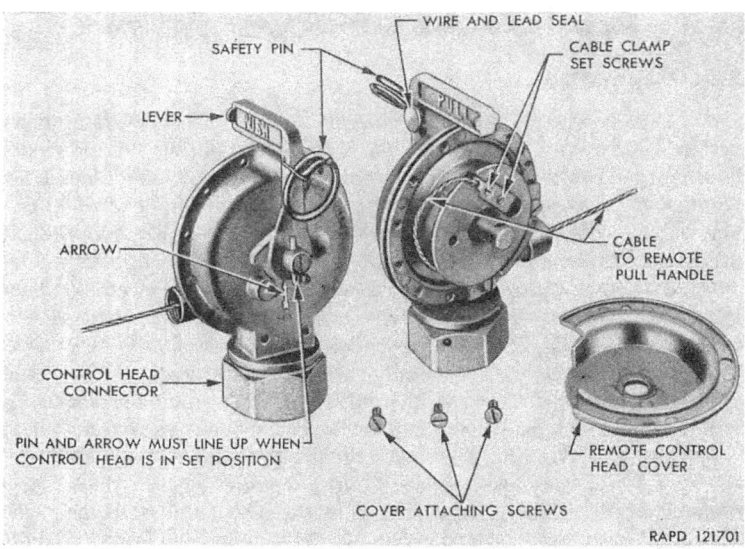

Figure 325. Fixed fire extinguisher cylinders control head.

b. INSTALLATION (fig. 324). Position cylinders into support brackets and screw connecting head into each cylinder. Screw control head into position on connecting head on left cylinder.

318. Fixed Fire Extinguisher Pull Cable

a. REMOVAL. Unscrew pull cable from control head on left cylinder (fig. 324). Remove pull cable from clips on column to the rear of and between the driver and assistant driver and on the hull ceiling (fig. 9). Disconnect the pull cable at the remote control handle (fig. 326). Remove pull cable.

b. INSTALLATION. Connect one end of pull cable to remote control handle (fig. 326). Connect other end of pull cable to control head on left fixed-fire-extinguisher cylinder (fig. 324). Secure pull cable into clips on column to the rear of and between the driver and assistant driver and to hull ceiling (fig. 9).

319. Fixed-Fire-Extinguisher Lines and Discharge Nozzles

a. LINES (fig. 326).
  (1) *Removal.* To remove a line disconnect connector at both ends, holding the elbow or fitting from which the line is being disconnected. Disconnect line clamps. Take care not to bend or dent lines as they are being removed from the vehicle. Protect them from damage after removal.

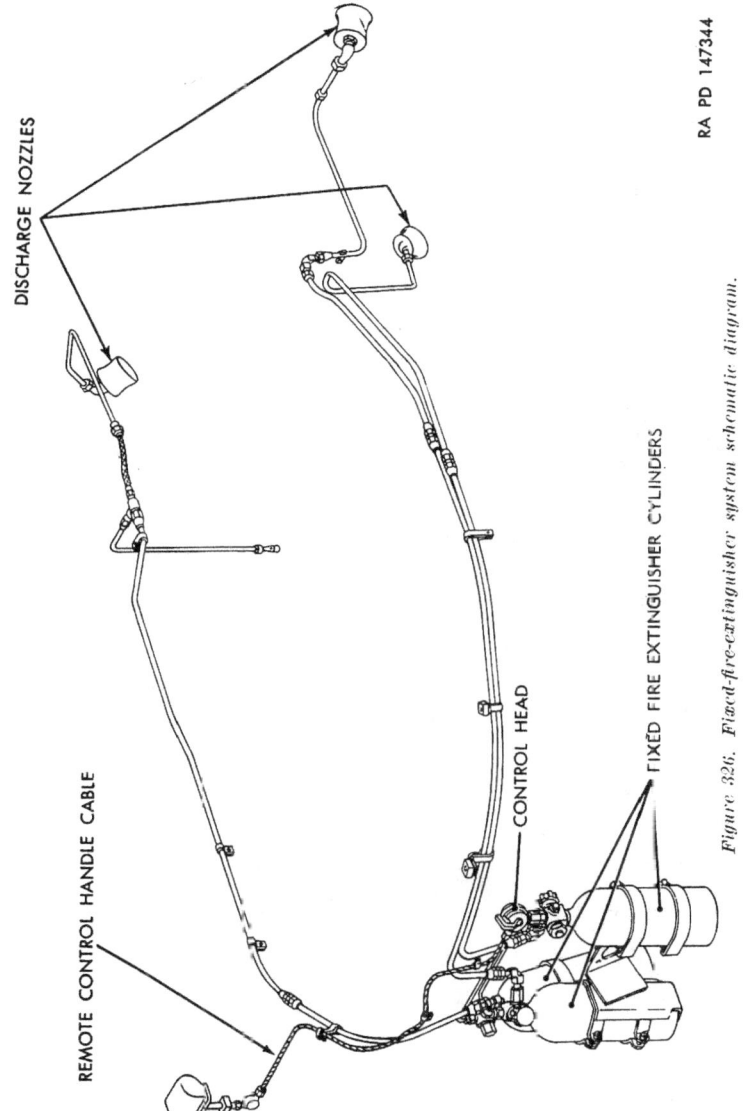

Figure 326. Fixed-fire-extinguisher system schematic diagram.

(2) *Installation.* Use preformed flared lines. Work line into position, taking care not to bend or dent lines. Make sure the flare on the line and the seat in the connection are perfectly clean. Connect both ends of line by screwing connectors on fingertight. Install attaching clips. Tighten the connector without excessive force while holding the elbow or fitting to prevent turning.

b. DISCHARGE NOZZLES (figs. 153 and 326).

(1) *Removal.* Remove fixed-fire-extinguisher line at connector on discharge nozzle, taking care not to damage line. Remove two screws and lock washers securing discharge nozzle to bracket and remove nozzle.

(2) *Installation.* Make sure discharge nozzle is clean and discharge port is not clogged. Position nozzle to bracket and secure with two screws and lock washers. Connect fixed-fire-extinguisher line at connector on discharge nozzle, taking care not to damage line.

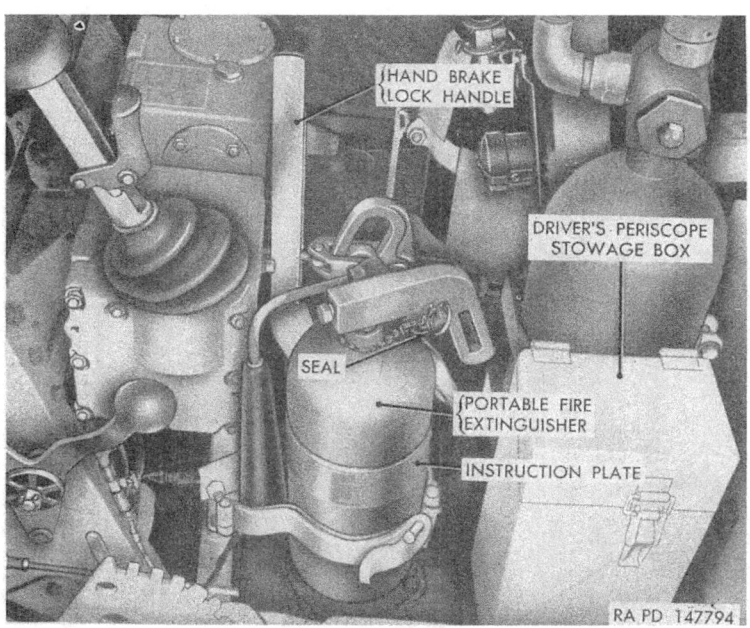

Figure 327. Portable fire extinguisher.

320. Portable Fire Extinguishers

a. REMOVAL (fig. 327). Release cylinder by opening mounting clamp and lift cylinder from mounting bracket.

b. MAINTENANCE. Information on recharging is generally included on an instruction plate, located on the side of the fire-extinguisher cylinder. Replace immediately after using. Determine contents by weight every 6 months. Replace if weight is 6 ounces below total weight stamped on valve body.

Warning: Handle all charged cylinders with care. Do not bump, jar, or drop.

c. INSTALLATION (fig. 327). Install cylinder in mounting bracket and secure by closing mounting clamp.

Section XXVI. AUXILIARY EQUIPMENT

321. Description and Data

a. CREW COMPARTMENT HEATER.

(1) *Description.* The crew compartment heater (fig. 328) which produces heat by burning a gasoline-and-air mixture inside a sealed combustion chamber, is installed in the driver's compartment under the instrument panel. The heater switch, heater safety valve reset switch, and circuit breaker reset button are grouped together on a separate personnel heater panel mounted on the accessory panel (fig. 18). The heater thermostat (fig. 76) is installed on the column to the rear of and between the driver and assistant driver. A separate fuel filter, fuel pump, safety valve, and fuel shut-off valve are mounted on a bracket (fig. 332) to the left of the heater.

(2) *Data.*
```
Make_____ Stewart-Warner
Model_____ 978M-R24
Weight_____ 23 lb
Starting load_____ 11 amp
Operating load_____ 2.5 amp
Fuel consumption, high heat_____ 1 gal 4 hr
Fuel consumption, low heat_____ 1 gal 8 hr
```

b. BILGE PUMPS.

(1) *Description.* Two centrifugal-type pumps are provided in the hull to pump out any accumulated water, oil, or fuel. The front bilge pump is mounted in the crew compartment under the ration tray box (figs. 288 and 334) and the rear bilge pump is in the engine compartment, to the left of the auxiliary generator (fig. 335). The bilge pumps are driven by motors which are controlled by switches on the accessory panel (fig. 18). Strainers are provided to prevent debris from entering the pump base and impeller area. A tubing

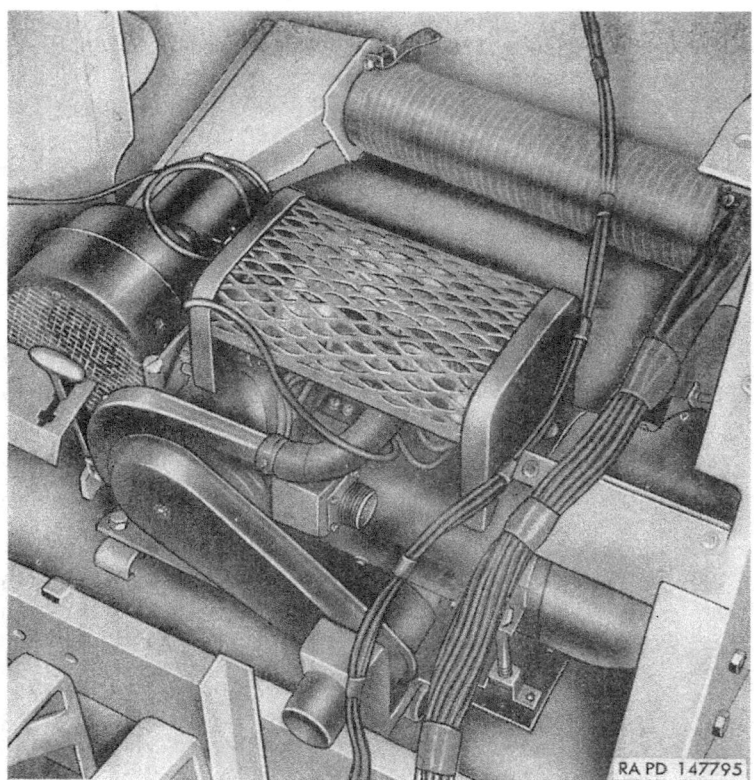

Figure 328. Crew compartment heater—installed view.

arrangement is provided from the side of the impeller housing to the outlets (figs. 3, 154, and 333) located on the top of the hull. The pumps, employing compound motors, are waterproof and capable of operating while immersed in water. Rotation, as viewed from the housing cover end, is counterclockwise.

(2) *Data.*

Volts	24
Rate of flow	50 gpm
Revolutions per minute	2,600 to 3,200
Average amperage	20

322. Crew Compartment Heater

a. REMOVAL. Turn off master relay switch (fig. 18). Remove manual control box (par. 257). Close fuel-tank-shut-off valves

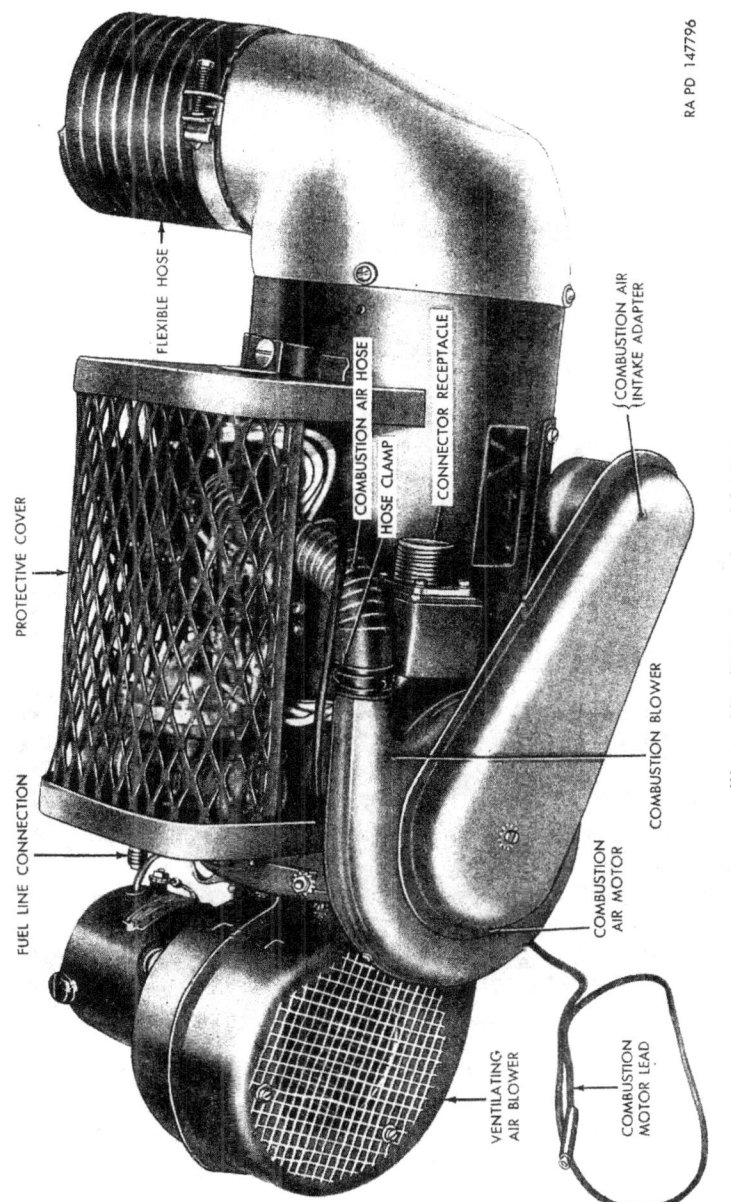

Figure 329. Crew compartment heater.

(par. 17). Disconnect line from fuel pump at fuel line connection on heater (fig. 329). Disconnect electrical cable from heater to safety valve at terminal block on safety valve (fig. 332). Disconnect flexible hose from heater. Disconnect heater exhaust line at heater. Remove electrical connector plug from connector receptacle (fig. 329) on front of heater. Remove mounting bolts from heater and remove heater.

b. INSTALLATION. Position heater in driver's compartment (fig. 328) and install mounting bolts. Connect flexible hose to heater. Connect heater exhaust line. Install electrical connector plug in connector receptacle (fig. 329) on front of heater. Connect electrical cable from heater to top terminal screw in terminal block on safety valve (fig. 332). Connect fuel line to heater at fuel line connection (fig. 329). Open fuel shut-off valves (par. 17). Install manual control box (par. 257).

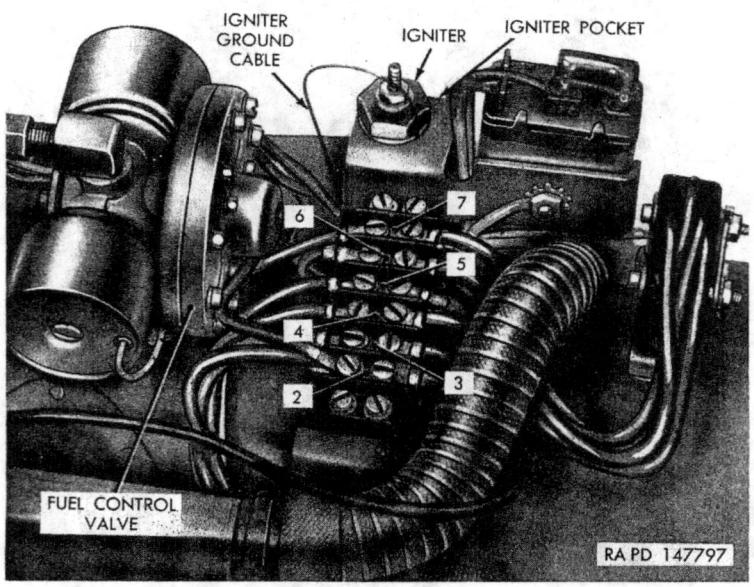

Figure 330. Crew compartment heater terminal block.

c. IGNITER REPLACEMENT AND MAINTENANCE (fig. 330).

    (1) *Removal.* Turn off master relay switch (fig. 18). Turn screw at each end of heater protective cover (fig. 329) one-quarter turn and lift off cover. Remove igniter lead No. 7 from top of igniter. Disconnect igniter ground cable from side of igniter pocket. Unscrew igniter from pocket and remove igniter.

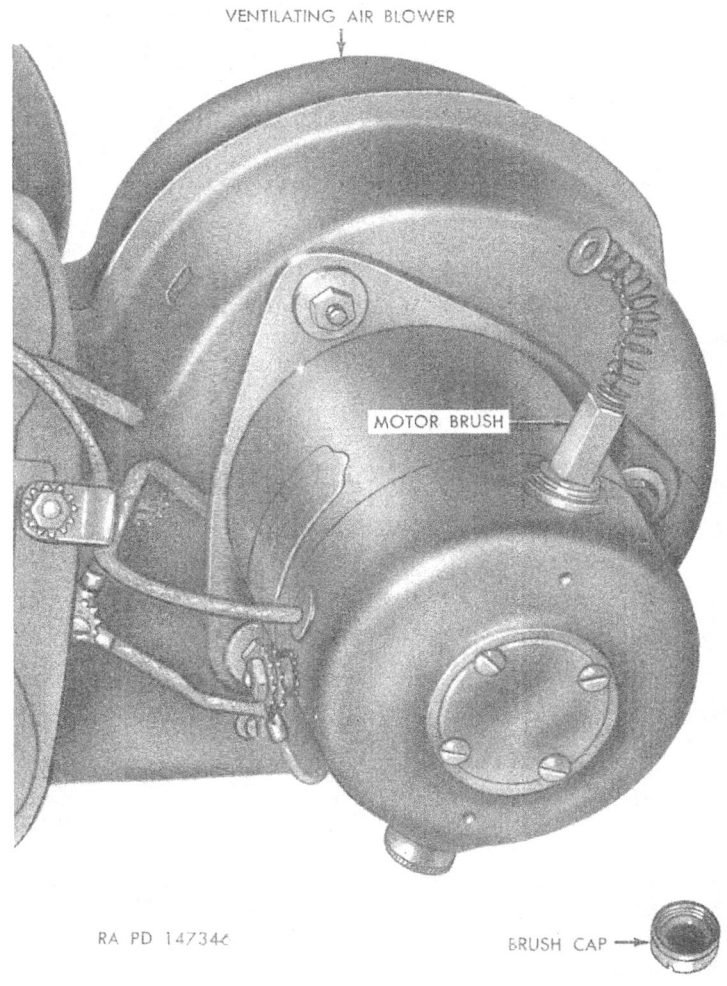

Figure 331. Removing crew compartment heater motor brushes.

(2) *Cleaning.* Using a suitable scraper, clean out igniter pocket and igniter thoroughly.

(3) *Installation.* Screw igniter into igniter pocket until secure. Connect ground cable to side of pocket. Connect igniter lead No. 7 to top of igniter. Position protective cover on heater (fig. 329) and turn screws at each end one-quarter turn to secure in place.

323. Heater Ventilating Air Blower Assembly

a. REMOVAL. Remove crew compartment heater (par. 322). Turn screw at each end of heater protective cover (fig. 329) one-quarter turn and lift off cover. Disconnect motor cable from terminal No. 2 on heater terminal block (fig. 330). Disconnect motor ground cable from heater case. Loosen four nuts holding ventilating air blower (fig. 329) to heater case. Turn blower counterclockwise until studs in heater case line up with slots in blower housing and pull blower from heater.

b. INSTALLATION. Position ventilating air blower to heater (fig. 329) and turn clockwise. Tighten four nuts holding blower to heater case. Connect motor ground cable to heater case. Connect motor cable to terminal No. 2 on heater terminal block (fig. 330). Position protective cover on heater and turn screws at each end one-quarter turn to secure in place. Install crew compartment heater (par. 322).

c. BRUSH REPLACEMENT. Remove crew compartment heater (par. 322). Remove slotted brush cap from rear of blower motor (fig. 331). Remove brush assembly. Measure length of brush. If length is five-sixteenths of an inch or less, replace brush. Insert new brush assembly in motor housing. Install brush cap and tighten securely. Install crew compartment heater (par. 322).

324. Heater Combustion Air Blower Assembly

a. REMOVAL (fig. 329). Remove crew compartment heater (par. 322). Turn screw at each end of heater protective cover one-quarter turn and remove cover. Remove combustion air hose and clamps. The clamps can be loosened by inserting screwdriver under the trigger of the clamp and lifting up. Remove the screw from the combustion air intake adapter and remove the adapter. Disconnect combustion air motor lead from terminal block screw No. 6 (fig. 330). Loosen three hex nuts which hold combustion blower to heater case flange. Remove blower by turning counterclockwise and pulling directly away from heater.

b. INSTALLATION (fig. 329). Position combustion blower to heater case flange and turn clockwise until secure. Install three hex nuts to secure blower to flange. Connect combustion air motor lead to terminal block screw No. 6 (fig. 330). Install combustion air intake adapter (fig. 329) and secure with screw and two lock washers. Install combustion air hose and clamps. Install heater protective cover and secure with two screws. Install crew compartment heater (par. 322).

325. Heater Fuel Control Valve

a. REMOVAL (fig. 330). Remove crew compartment heater (par. 322). Turn screw at each end of heater protective cover (fig. 329) one-quarter turn and remove cover. Remove shut-off solenoid lead No. 9 from overheat switch terminal. Disconnect restriction solenoid lead No. 5 from terminal block screw No. 5. Loosen compression nut on standpipe until fuel control valve (fig. 330) is free and lift valve off.

b. INSTALLATION (fig. 330). Position fuel control valve to heater case and tighten compression nut on standpipe. Connect restriction solenoid lead No. 5 to terminal block screw No. 5. Connect shut-off

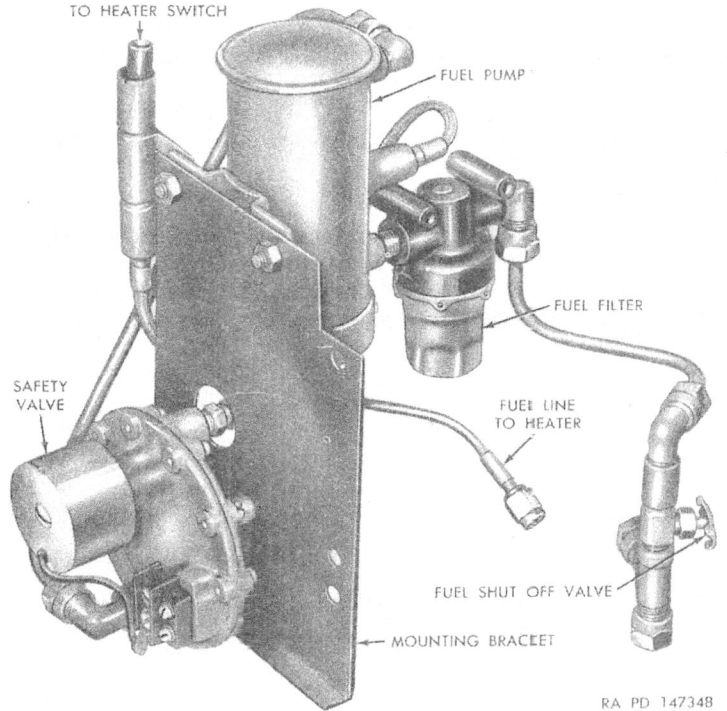

Figure 332. Crew compartment heater fuel pump, safety valve, and filter.

solenoid lead No. 9 to overheat switch terminal. Position protective cover to heater (fig. 329) and secure with two screws. Install crew compartment heater (par. 322).

326. Heater Fuel Pump

a. REMOVAL (fig. 332). Turn off master relay switch (fig. 18). Close heater-fuel-shut-off valve. Disconnect electrical cable at con-

549

nector on fuel pump. Disconnect fuel line at top of fuel pump. Remove fuel filter from fuel pump. Remove two nuts holding fuel pump to mounting bracket and remove pump.

b. INSTALLATION (fig. 332). Position fuel pump to mounting bracket and secure with two nuts. Install fuel filter on pump. Connect fuel line at top of fuel pump. Connect electrical cable at connector on fuel pump. Open heater-fuel-shut-off valve.

327. Heater Fuel Filter and Fuel Lines

The fuel filter and the sections of fuel line from fuel shut-off valve to fuel filter, from fuel pump to safety valve, and from safety valve to heater are all removed in the same manner. Shut off the heater-fuel-shut-off valve (fig. 332). Disconnect fuel line connector nuts at both ends of the section to be removed, and remove filter or fuel line section. To install, position filter or fuel line section and secure connecting nuts. Open the fuel shut-off valve to permit operation.

328. Heater Safety Valve

a. REMOVAL (fig. 332). Turn off master relay switch (fig. 18). Close heater fuel shut-off valve. Disconnect electrical cable from terminal block on safety valve. Disconnect both fuel lines at safety valve. Remove screws holding safety valve on mounting bracket and remove safety valve.

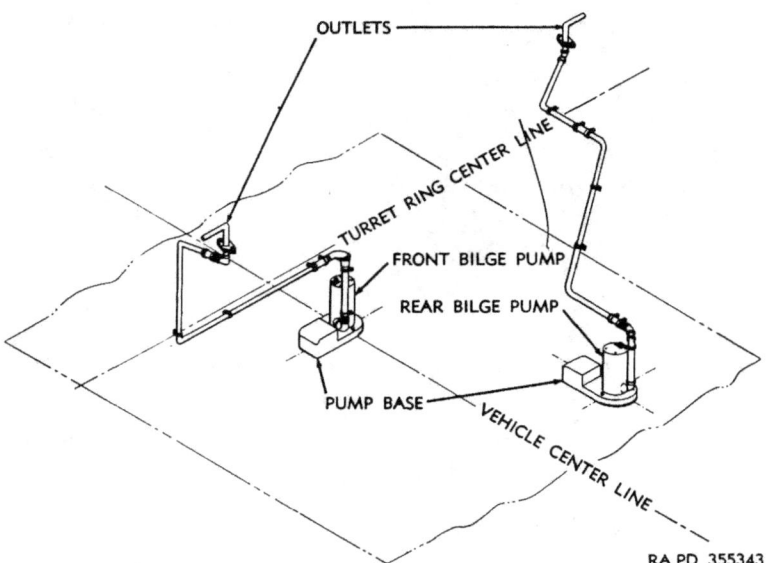

Figure 333. Bilge pumps and lines schematic diagram.

b. INSTALLATION (fig. 332). Position safety valve to mounting bracket and secure with screws. Connect both fuel lines to safety valve. Connect electrical cable from heater to top terminal screw in terminal block on safety valve. Open heater fuel shut-off valve.

Figure 334. Front bilge pump—installed view.

329. Bilge Pumps

a. FRONT BILGE PUMP (fig. 334).

(1) *Removal.* Remove ammunition stowage boxes and ration box (par. 265). Remove primer line clips and spread lines apart. Unscrew cable connector nut from bilge pump. Loosen hose clamp and pry hose off elbow. Remove three mounting screws from bilge pump base and lift bilge pump out of vehicle. Disconnect fittings from bilge pump.

(2) *Service.* To clean the bilge pump strainer, remove the four screws securing strainer to pump base and remove any foreign material which may be clogging the pump.

(3) *Installation* (fig. 334). Install fittings on bilge pump outlet. Install three mounting screws which secure bilge pump base to hull floor. Connect hose clamp to secure outlet line to pump fittings. Connect electrical cable to pump. Bend primer lines back into shape and secure them with the clip. Install ration box and ammunition stowage boxes (par. 265).

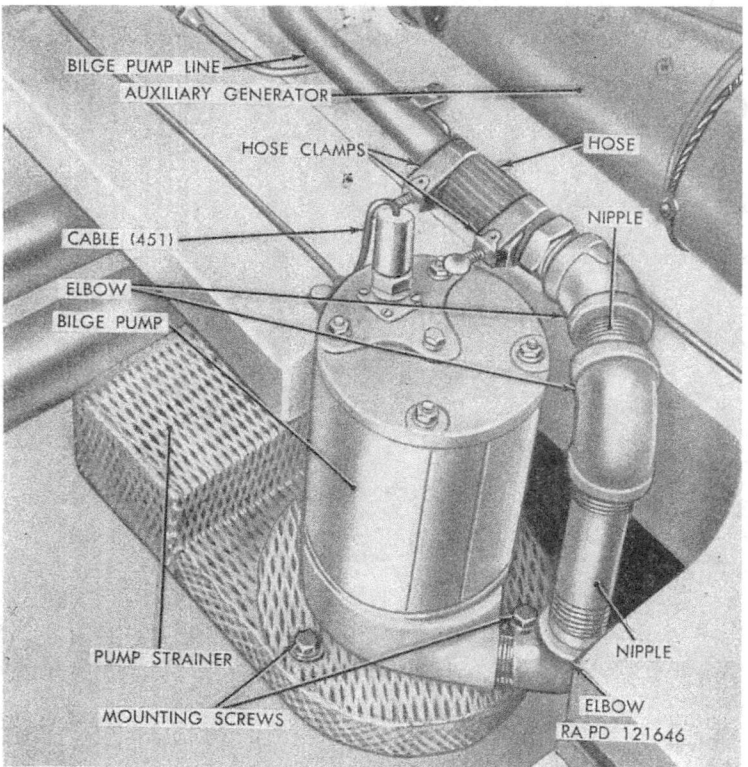

Figure 335. Rear bilge pump—installed view.

b. Rear Bilge Pump (fig. 335).

(1) *Removal.* Removal of the bilge pump can be made either with the power plant out of the vehicle (par. 159) or with the power plant in the vehicle and the auxiliary generator and engine removed (par. 303). Unscrew the cable connector nut from the bilge pump. Loosen hose clamps and remove hose from the outlet line. Remove three mounting screws from bilge pump base and lift the bilge pump out of the vehicle. Disconnect fittings from pump.

(2) *Service.* To clean the bilge pump strainer, remove the four screws which secure the strainer to the pump base and remove any foreign material which may be clogging the pump.

(3) *Installation.* Connect fittings to bilge pump outlet. Position pump and install three mounting screws which secure bilge pump base to the hull floor. Connect hose clamp to secure outlet line to pump fittings. Connect electrical cable to the pump. Depending on what method was used to gain access to the bilge pump, either install the power plant (par. 160) or the auxiliary generator and engine (par. 303).

Note. Do not tighten mounting bolts too tightly or the bilge pump impellers will bind. Tighten sufficiently to hold bilge pumps securely. Mounting bolts should be drawn down evenly. Operate bilge pumps to determine whether the impellers are binding.

Section XXVII. RADIO INTERFERENCE SUPPRESSION

330. Purpose

Radio interference suppression is the elimination or minimizing of electrical disturbances which interfere with radio reception or disclose the location of the vehicle to sensitive electrical detectors. It is important, therefore, that vehicles with as well as vehicles without radios be suppressed properly to prevent interference with radio reception of neighboring vehicles.

331. Description

Suppression in this vehicle is accomplished by the use of braided bond or ground straps and toothed washers, resistor suppressors, capacitors, and filter assemblies in all circuits where shielding of wires and units does not completely confine or dissipate electrical disturbances which otherwise would cause radio interference.

332. Ignition System

a. MAIN ENGINE.
  (1) *High-tension cables* (fig. 168). High-tension cables to the spark plugs are shielded individually by metallic hose terminating in appropriate fittings.
  (2) *Spark plugs* (fig. 168). Spark plugs are shielded integrally. If radio interference is emanating from a spark plug, replace the spark plug (par. 186).
  (3) *Magnetos* (fig. 162). Magnetos are shielded by metal caps. If radio interference is emanating from a magneto, replace the magneto (par. 185).

(4) *Magneto filter coils* (fig. 162). A radio interference filter assembly is incorporated in each low-tension ground cable between the magnetos and magneto switch. To replace filter assembly, refer to paragraph 188.

b. AUXILIARY ENGINE.

(1) *High-tension cables* (fig. 308). High-tension cables to the spark plugs are individually shielded by tinned copper braid over loom.

(2) *Spark plugs* (fig. 308). Each spark plug is equipped with a 10,000-ohm insert-type resistor-suppressor. If resistor is faulty, replace spark plug (par. 311).

(3) *Magneto ground cable* (fig. 310). The magneto ground cable is shielded by means of tinned copper braid from the magneto to the auxiliary engine magneto switch (fig. 18).

333. Generating System

a. MAIN GENERATOR (fig. 206). Low-tension cables to the main-generator are individually shielded by metallic hose.

b. AUXILIARY GENERATOR (fig. 308). Armature terminal "B" of the auxiliary generator is bypassed to the generator frame by means of a 0.1-mfd, 100-volt capacitor.

334. Hull and Turret Accessories Systems

a. BILGE PUMPS (figs. 334 and 335). The positive feed cable of each bilge pump motor is bypassed to the pump motor frame by means of a 0.01-mfd, 100-volt capacitor.

b. CREW COMPARTMENT HEATER (fig. 329). The positive terminal of the crew compartment heater combustion air motor is bypassed to the motor frame by means of a 0.01-mfd, 100-volt capacitor.

c. VENTILATING BLOWER (fig. 304). The ungrounded brush of the ventilating blower motor is bypassed to the motor frame by a 0.1-mfd, 100-volt capacitor mounted within the motor frame.

335. Turret Electrical System

a. TURRET FILTER BOX (fig. 336).

(1) *Description.* A radio interference suppression filter box assembly is secured to the turret bulkhead in front of the gunner's control assembly. It contains capacitors for suppressing any radio interference generated in the firing circuits.

(2) *Removal.* Turn off master relay switch (fig. 18). Disconnect cable connector plugs from the sides of the turret filter box and remove cables. Remove four screws securing filter box body to cover and remove filter box body. To remove cover, remove four lock washer screws securing cover to mounting brackets and remove cover.

(3) *Installation.* Position cover, if removed, to mounting brackets and secure with four lock washer screws. Position filter box body to cover (fig. 336) and secure with four screws and lock washers. Install cable connector plugs into filter box receptacles.

b. Override Motor Control Box Capacitors (fig. 299). Installed in the override motor control box (fig. 300) are a 0.10- and a 0.25-mfd capacitor to provide radio interference suppression while the turret is being traversed. If either of these capacitors is faulty, replace the override motor control box (par. 282).

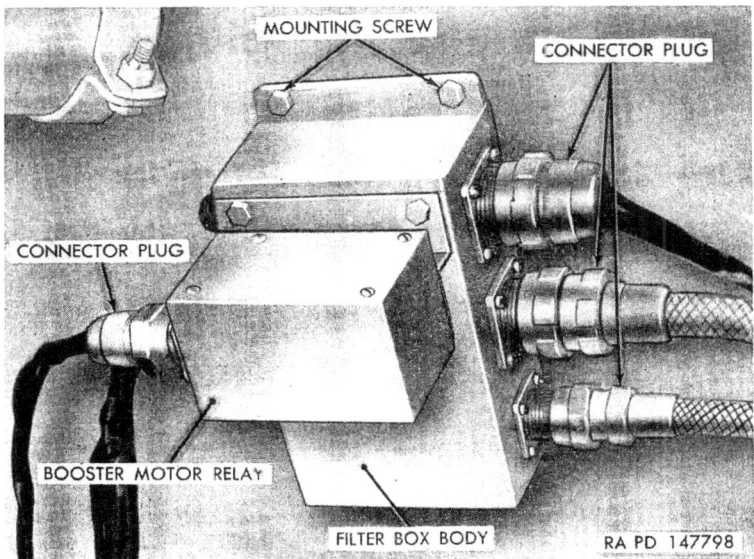

Figure 336. Turret filter box.

336. Fasteners and Bond Straps

Dome lights (fig. 20), ventilating blower (fig 304), and master junction box (fig. 182) are grounded by means of braided bond straps to suppress radio interference from these units. Toothed washers are employed at bond strap connections to provide good contact to ground.

SECTION XXVIII. 90-MM GUN T119E1

337. Description and Data

a. Description.
  (1) The 90-mm gun T119E1 (fig. 337) is a light artillery weapon. It uses fixed ammunition which is loaded into the gun man-

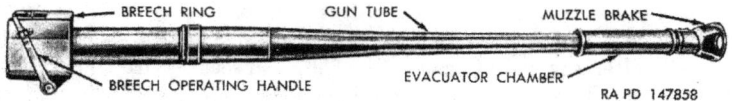

Figure 337. 90-mm gun T119E1—right side view.

ually. The loading of a round permits a closing spring mechanism to close the breech. The breech is opened automatically during counterrecoil of the gun, extracting the empty shell case. The ammunition is fired by an inertia-type percussion mechanism housed in a vertical sliding breechblock.

(2) *Muzzle brake.* The gun tube is equipped with a muzzle brake (fig. 338). The high velocity gases following the projectile

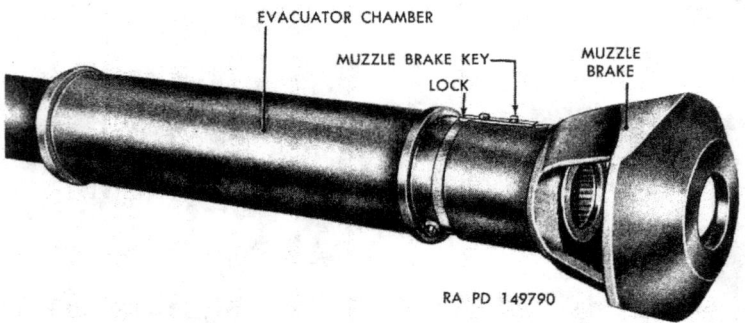

Figure 338. Muzzle end of 90-mm gun T119E1.

through the tube impinge upon the baffles of the brake and are deflected rearward and sideways into the outside atmosphere. This creates forward forces on the gun which counteract partially the force of recoil and also reduces obscuration of the target by reducing blast cloud eect and dust disturbance.

(3) *Evacuator chamber.* The 90-mm gun is equipped with an evacuator chamber (fig. 338) mounted on the gun tube to the rear of the muzzle brake. The function of the evacuator chamber is to offset the objectionable powder gases from accumulating in the fighting compartment. When a round is fired and after the projectile has passed by the evacuator some of the powder gases following the projectile flow into the evacuator chamber creating a high pressure. After the projectile leaves the tube the pressure in the tube drops. Pressure in the evacuator chamber forces the tube gases to

flow in the direction of the muzzle, and as the breechblock opens a sucking action is created in the direction of the muzzle which cleans the bore of residual powder gases.

   b. TABULATED DATA. See paragraph 5c.

338. Disassembly

   a. MUZZLE BRAKE AND EVACUATOR CHAMBER.
   (1) To disassemble the muzzle brake and evacuator chamber, unscrew the socket head cap screw which fastens the lock and key at the rear end of the muzzle brake (figs. 338 and 339). Remove the lock and key. Pry the outside lug of the key with a screwdriver, if necessary, to free the key.
   (2) Unscrew the muzzle brake from the tube. A bar may be inserted through the ports of the muzzle brake and used as a lever in unscrewing the brake.
   (3) With muzzle brake removed, unscrew and remove the evacuator chamber. If necessary, spanner wrench 7237270 (figs. 85 and 340) may be applied to the lugs in the front head of the chamber to unscrew it.

   Warning: Extreme care must be taken in assembly, disassembly, and handling of the evacuator chamber to avoid damage to the bore and inner lips of the front and rear head of the evacuator. The lips are formed to a thin edge to provide a seal against the tube and any deformation of these lips will destroy the effectiveness of the sealing action and prevent proper functioning of the evacuator. When removing or replacing the evacuator, keep the evacuator in line with the gun axis until completely free, and use the minimum force necessary to slide the evacuator on or off the tube. After removal from the tube the evacuator should be placed in a protected location.

   (4) If it is necessary to replace the inserts (fig. 339), they may be unscrewed from their tapped holes in the tube.

   Note. The newer tubes will not be tapped to receive inserts; the holes in the tube wall, near the muzzle, will be drilled and tapered to act as orifices, thus eliminating inserts.

   b. BREECH RING. Refer to section XXIX.

339. Maintenance

   a. GENERAL.
   (1) It is of vital importance to keep all parts of the matériel in proper condition for immediate service. The intended use of tools and accessories, and lubricating, cleaning, and preserving materials is to keep the matériel in operating condition.

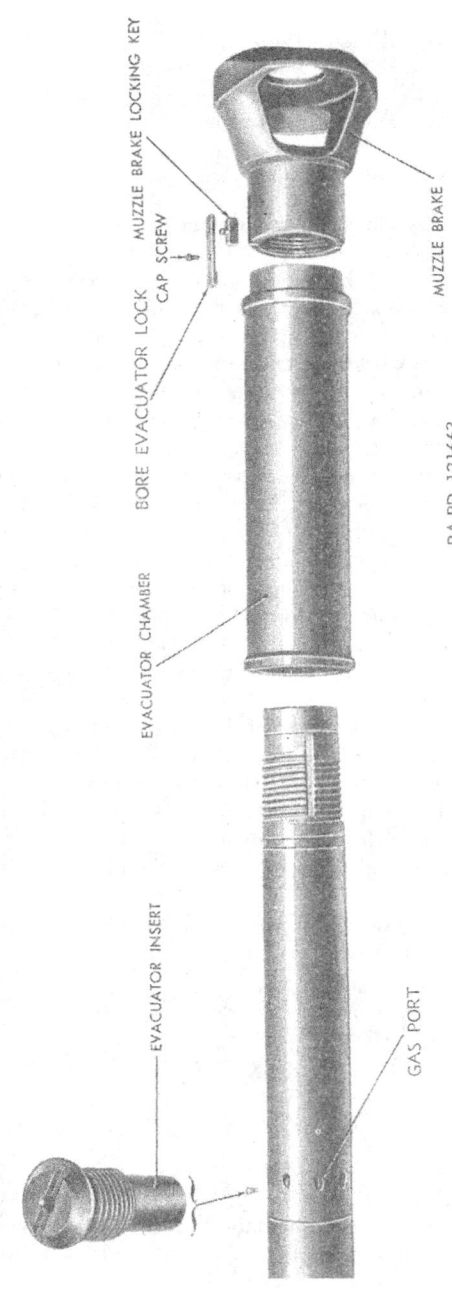

Figure 339. Evacuator chamber—exploded view.

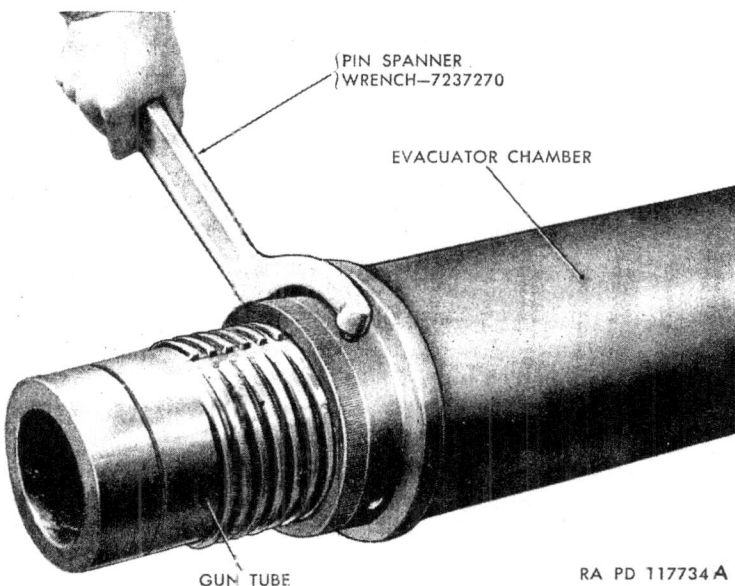

Figure 340 Removing bore evacuator chamber.

(2) Proper lubrication, using the designated lubricants at the intervals specified in lubrication order (fig. 87), is essential to the care and preservation of the matériel. Examination should be made periodically to insure that lubricants reach the parts for which they are intended.

(3) All protective covers for the gun and mount should be installed when in traveling position or when the gun is not in service. If the matériel is not to be used for a considerable length of time, all exposed, bright, unpainted surfaces should be cleaned with rifle bore cleaner volatile mineral spirits or dry-cleaning solvent, thoroughly dried, and protected with a coat of rust-preventive compound.

(4) When disassembling and assembling, thoroughly clean dirt and grit from parts to be removed or assembled Clean parts thoroughly before oiling and assembling. Do not use a steel hammer directly on any part when assembling. If a copper or lead hammer is not available, use a block of wood as a buffer. Always use the tool intended for the part to be assembled or disassembled. Tools and accessories should be cleaned frequently and the metal surfaces protected with preservative lubricating oil (special).

(5) Water from a high-pressure hose must never be played into any part of the weapon. Washing should be done with a sponge. Carefully dry parts which have become wet from washing, then oil in the manner prescribed.

(6) The leveling plates should be protected. Tools or other articles must not be placed upon them.

b. EVACUATOR CHAMBER. If the objectional effects of residual gases are not entirely or largely offset by the bore evacuator, proceed as follows:

(1) Remove the muzzle brake and evacuator (par. 338).

(2) Clean the gas ports in the gun tube (fig. 339). If any of the evacuator inserts have been removed from the gun tube, screw them into their tapped holes to the rear of the muzzle threads on the tube (fig. 339). Normal maintenance does not require removal of these inserts.

(3) Install evacuator and muzzle brake (par. 340).

c. GUN TUBE. Organizational maintenance of the gun tube and breech ring is limited to operations herein; for all others, notify ordnance maintenance personnel.

(1) Wear of bore does not depend entirely upon the number of rounds fired; it also depends on the care given the bore in cleaning.

(2) Note general appearance of gun bore for wear and deformation of lands and grooves and for pitting and pastilles (indentations in bore which resemble gas pockets in a casting). Examine for evidence of powder fouling and rust. Do not confuse coppering of bore with powder fouling. The removal of copper fouling is prohibited. A clean bore is not necessarily a shiny bore and may frequently have a dull gray appearance. A shiny, polished bore might indicate that abrasives have been used in cleaning operations. If lands or grooves are deformed or pitted or if there is indication of cracking (inside or outside tube), notify ordnance maintenance personnel. Remove any burs on breech face of tube and in extractor pits with crocus cloth.

(3) Before firing, clean bore as follows:

(*a*) Assemble staff sections 6197240 and install brush 6181980 (fig. 85) to the end staff.

(*b*) Apply rifle bore cleaner to the brush, insert it into bore and thoroughly scrub with a push-pull action from one end of the tube to the other.

(*c*) Replace the bore brush with wiper ring 6181986 (fig. 85), insert dry jute through loop of wiper ring and run it through bore until all surfaces are dry.

(4) After firing and on two consecutive days thereafter thoroughly clean the bore with rifle bore cleaner ((3) above) but do not wipe dry. On third day after firing, clean bore with rifle bore cleaner and wipe dry. Saturate wiping cloth with preservative lubricating oil (special) and wring out; attach cloth to wiper ring and run through bore several times. Weekly thereafter, clean with rifle bore cleaner, wipe dry and reoil with preservative lubricating oil.

340. Assembly

a. BREECH RING. Refer to section XXIX.

b. EVACUATOR CHAMBER.
  (1) If the evacuator inserts have been removed from the tube, insert them, slotted end outward, into the tapped holes to the rear of the muzzle threads on the tube, and screw them in to a firm seat (fig. 339).
  (2) With the internally threaded head of the evacuator chamber forward, slide the evacuator chamber rearward over the muzzle end of the tube until the threads engage; then screw the chamber rearward as far as it will go without jamming (fig. 340).
  (3) Screw the muzzle brake rearward onto the muzzle until the muzzle face of the tube is flush with the adjacent rear wall of of the muzzle brake chamber (fig. 339). Then rotate the brake sufficiently to the right or left to aline its keyslot with the nearest keyway in the tube.
  (4) Insert and seat the key to retain the brake in position and screw the evacuator forward with spanner wrench 7237270 firmly against the muzzle brake.

   Note. Tight contact between evacuator chamber and muzzle brake is essential, as the evacuator serves as a lock nut on the brake.

  Then place the lock over the stud of the key, with the serrated face of the lock against the serrated surface on the evacuator chamber head; insert the cap screw and tighten to secure the evacuator chamber and muzzle brake in position.

Section XXIX. BREECH MECHANISM

341. Description

a. GENERAL. The breech mechanism is of the vertical sliding type consisting principally of a breech ring, breechblock, firing parts, extractors, and breech operating parts.

b. BREECH RING. The breech ring body (fig. 345) houses the breechblock, firing parts, extractors, and breech operating parts.

Leveling plates, used when boresighting the gun, are inlaid in the top of the breech ring.

c. BREECHBLOCK. The breechblock and related parts (fig. 346) slide vertically in the breech ring. The top of the breechblock is U shaped to guide the cartridge into the chamber, and the upper front edge is beveled to drive the round home as the breech is closed. The two vertical holes in the bottom of the breechblock are for lightening the breechblock.

d. FIRING PARTS. The firing parts include the percussion mechanism, firing spring, firing spring retainer, sear, and trigger (fig. 346). Firing is accomplished by actuating the trigger plunger (through the firing linkage (sec. XXXI)), which protrudes from the upper, left, front face of the breech ring (fig. 349).

e. EXTRACTORS. The function of the extractors is to extract the empty shell case and to lock the breech in open position until the next round is loaded into the chamber.

f. BREECH OPERATING PARTS. The breech mechanism is designed for automatic opening during counterrecoil and automatic closing by spring action upon the insertion of a round. At the instant the breechblock reaches its maximum down position (open) it is locked by the extractors. The compression induced in the closing spring when the

Figure 341. Breech closed—left rear view.

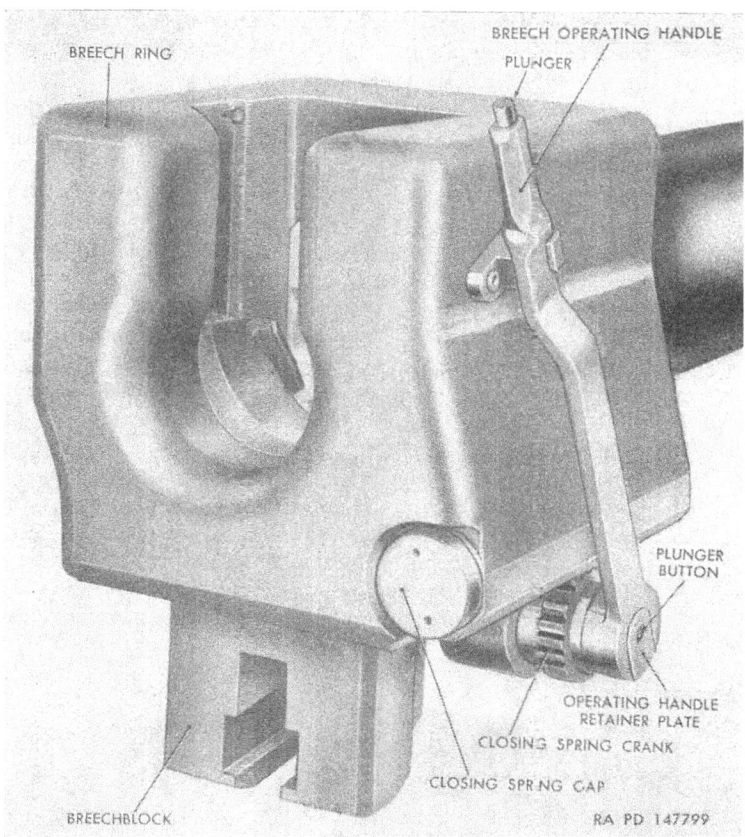

Figure 342. Breech open—right rear view.

breechblock moves down serves to cushion the effect of breech opening. Upon the insertion of a round into the chamber, the extractors are tripped (released) by the rim of the cartridge case and the breech closes.

342. Removal

a. Breechblock and Related Parts.

(1) Install the breechblock removing eyebolt 41–E–3150 (fig. 85) in the threaded hole in the top of the breechblock. Insert the breechblock removing tool 7237831 in the central breechblock bore, rotating the long flange of the tool so that it rests on the U-shaped part of the breech ring. The breechblock removing tool holds the breechblock in this position while loosening the closing spring and disengaging the breechblock

crank. With a small screwdriver, push up on the breechblock crank stop plunger (fig. 345) through the vertical hole in the breechblock crank stop, sliding the stop forward enough to free it from the plunger. Withdraw screwdriver and push stop forward to clear the breechblock. Remove the cap lock screw and loosen the closing spring cap with face-type spanner wrench 41-W-3247-720 until the spring pressure is released. Support the eyebolt; take out the breechblock removing tool, and lower the breechblock until the breechblock crank is disengaged, and the extractors can be removed. Remove the extractors from the their breech ring pivots and then lift the breechblock out of the breech ring recess, being certain that the trigger plunger is not depressed.

(2) Using the face-type spanner wrench 41-W-3247-720, unscrew the closing spring cap in the rear face of the breech ring and remove the closing spring (fig. 345). Depress the exposed plunger button in the right end of the breech operating shaft and remove the operating shaft plate (fig. 345) from the operating handle hub. Depress the operating handle plunger and remove the handle off the right end of the shaft. Push on the handle end of the shaft and slide the shaft out of the left bottom breech ring lug and remove the

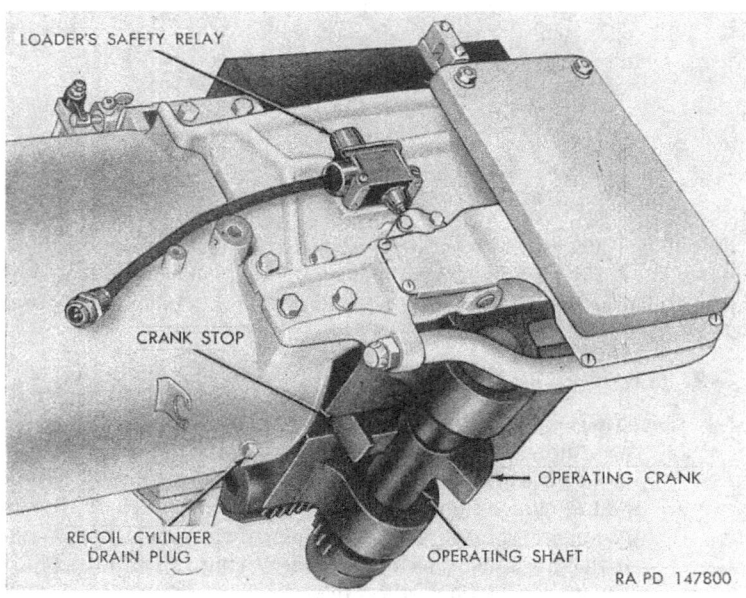

Figure 343. Breech operating mechanism—bottom view.

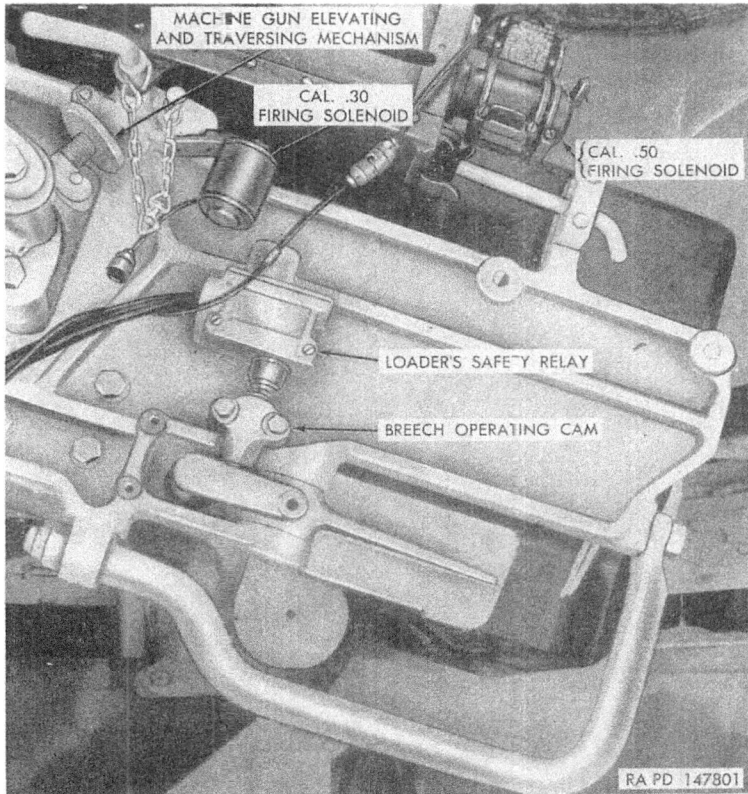

Figure 344. Breech operating mechanism—left-side view.

spacers as they are freed by the shaft removal. If the shaft is stuck, use the shaft removing push rod 5206983 (fig. 85). Place the rounded end of the rod against the right end of the shaft and tap the flat end of the rod with a hammer to ease removal of the shaft.

Caution: Be sure the shaft plunger or the edge of the plunger hole is not tapped with the rod.

After the operating shaft has been removed, unscrew the two detents in the rear of each breech ring lug. Remove the operating crank from the left lug of the breech ring and the closing spring crank from the right breech ring lug (fig. 345). Push the closing spring piston out of the front of the closing spring bore. Slide the crossheads from the breechblock crank pivot if necessary.

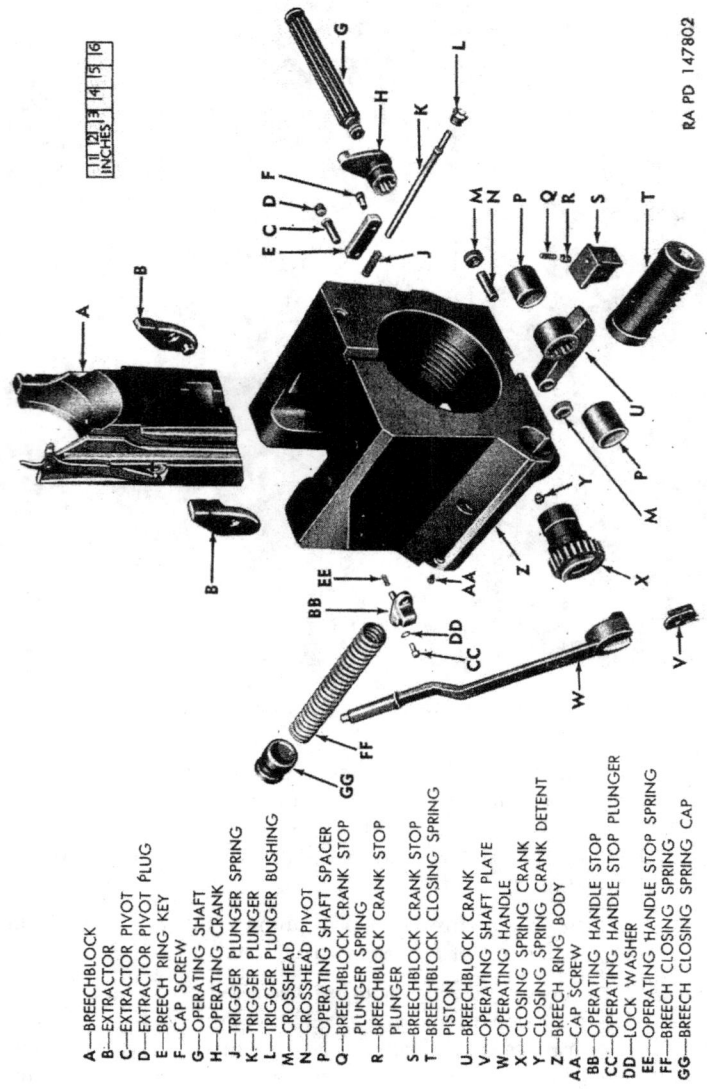

A—BREECHBLOCK
B—EXTRACTOR
C—EXTRACTOR PIVOT
D—EXTRACTOR PIVOT PLUG
E—BREECH RING KEY
F—CAP SCREW
G—OPERATING SHAFT
H—OPERATING CRANK
J—TRIGGER PLUNGER SPRING
K—TRIGGER PLUNGER
L—TRIGGER PLUNGER BUSHING
M—CROSSHEAD
N—CROSSHEAD PIVOT
P—OPERATING SHAFT SPACER
Q—BREECHBLOCK CRANK STOP PLUNGER SPRING
R—BREECHBLOCK CRANK STOP PLUNGER
S—BREECHBLOCK CRANK STOP
T—BREECHBLOCK CLOSING SPRING PISTON
U—BREECHBLOCK CRANK
V—OPERATING SHAFT PLATE
W—OPERATING HANDLE
X—CLOSING SPRING CRANK
Y—CLOSING SPRING CRANK DETENT
Z—BREECH RING BODY
AA—CAP SCREW
BB—OPERATING HANDLE STOP
CC—OPERATING HANDLE STOP PLUNGER
DD—LOCK WASHER
EE—OPERATING HANDLE STOP SPRING
FF—BREECH CLOSING SPRING
GG—BREECH CLOSING SPRING CAP

Figure 345. *Breech ring—exploded view.*

b. FIRING MECHANISM.

(1) With the chamber empty and the breech closed, operate the firing mechanism (par. 349) on the mount to actuate the trigger plunger and trigger and trip the sear. Press the firing spring retainer (fig. 346) inward against the firing spring and rotate the retainer 90° until the slot in the retainer is horizontal. Remove the retainer and firing spring. Hold one hand over the exposed hole and pull rearward on the cocking lever (fig. 341) to move the percussion mechanism to the rear and beyond the sear. Grasp the rear end of the firing pin guide and withdraw the percussion mechanism from the central breechblock bore. To disassemble the percussion mechanism, see paragraph 343*b*.

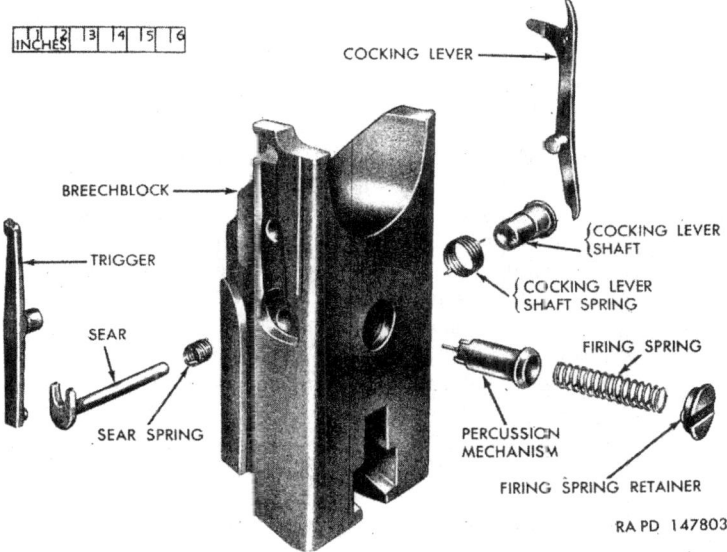

Figure 346. Breechblock and attaching parts—exploded view.

(2) Removal of the sear, trigger, and cocking mechanism requires removal of the breechblock (*a* above). With the breechblock removed and standing upright, pull the cocking lever (fig. 346) out of its seat in the right side of the breechblock. Remove the cocking lever shaft and the cocking lever shaft spring from the bore in the right side of the breechblock. Remove the trigger from its seat in the left side of the breechblock. Remove the sear and sear spring from the bore on the left side of the breechblock.

567

343. Disassembly

a. BREECH OPERATING PARTS.

(1) *Trigger plunger.* The trigger plunger group, which activates the trigger, need not be removed in ordinary maintenance but only to replace unserviceable parts or correct a malfunction. Unscrew the trigger plunger bushing (fig. 345), using wrench 41–W–3735–510 (fig. 85), from the front face of the breech ring. Remove the plunger and spring from their bore in the breech ring.

(2) *Breech ring key.* To remove the breech ring key (fig. 345), unscrew the two socket head cap screws which fasten the key to the left side of the breech ring and remove the key.

(3) *Operating handle stop.* To remove the operating handle stop (fig. 345), unscrew the two socket head cap screws which fasten the stop to the right side of the breech ring. Remove the screws, washers, and stop, being careful to catch the spring and plunger seated in the stop as the stop is being removed.

(4) *Extractor pivots.* To remove the extractor pivots (fig. 345), unscrew the two plugs, one in the right, and one in the left side of the breech ring, and push the pivots out of their holes.

(5) *Operating handle plunger.* To remove the operating handle plunger, punch out the transverse pin in the handgrip of the operating handle. Remove plunger and spring from the end of the operating handle.

(6) *Extractor plunger.* To remove the extractor plunger, unscrew the small pin set screw in the flat face of the extractor allowing the plunger and spring to be lifted out of the extractor trunnion.

(7) *Operating shaft plunger.* To remove the operating shaft plunger, depress the plunger into the shaft and smooth the four stake marks at the outside shoulder of the hole. Remove plunger and spring from the shaft.

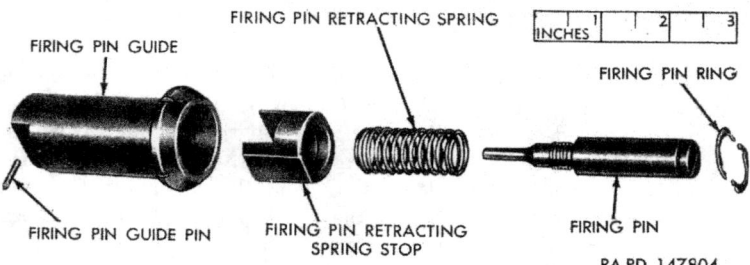

Figure 347. Percussion mechanism—exploded view.

b. Percussion Mechanism. The percussion mechanism (fig. 347) should not be disassembled except to replace unserviceable parts or to correct a malfunction. When necessary it may be disassembled by pressing the prongs of the stop rearward into the guide and punching out the pin which fixes the firing pin to the guide (fig. 348). Return the stop to normal position and unscrew the firing pin from the guide. Remove the stop from the guide, and slide the retracting spring off the forward end of the firing pin. The snap ring should not be removed from the firing pin unless damaged.

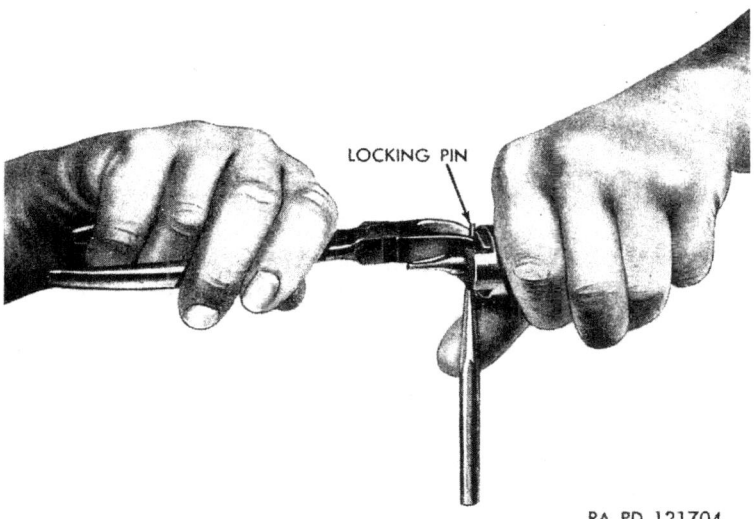

Figure 348. Removing locking pin from percussion mechanism.

344. Maintenance

Examine breech and firing mechanism for operation, adjustment and lubrication. Clean and lubricate for proper functioning. Examine bearing surfaces of breech block and breech ring for scores or burs and if necessary, smooth with crocus cloth. Check firing pin for proper protrusion, any misalinement, pitting, or wear. Clean and oil all parts and exposed surfaces daily and after firing. Check all springs for breakage or weakness. For a list of the spare parts authorized the using arm, see ORD 7 SNL G-262.

345. Assembly

a. Percussion Mechanism (fig. 347). Insert the firing spring stop into the guide, pronged end first. The prongs project through the forward openings of the guide. Assemble the snap ring in its groove on the rear end of the firing pin. Put the retracting spring over the firing pin ahead of the snap ring. Screw the firing pin into the guide. While holding the prongs of the stop inward, aline the pin hole in the firing pin and guide and insert the pin, allowing the stop prongs to go forward into normal position.

 b. BREECH OPERATING PARTS.

  (1) *Operating shaft plunger.* To assemble the plunger in the operating shaft, insert the spring and plunger, bored end first, into the shaft. Depress the plunger and stake the outside of the plunger hole in four places (stake marks 90° apart).

  (2) *Extractor plunger.* To assemble the extractor plungers in the extractors, insert the spring and plunger, flat end first, into the trunnion of each extractor. Depress the plunger and screw the pin set screw into the flat face of each extractor with its inside end retaining the plunger in the extractor trunnion.

  (3) *Operating handle plunger.* To assemble the operating handle plunger mechanism, insert the spring and plunger, tapered end first, into the handle counterbore. Rotate the plunger and aline the slot in the plunger with the transverse hole in the handle. Insert the pin into the hole and stake in place.

  (4) *Extractor pivots.* To assemble the extractor pivots in the breech ring insert the pivots, small ends first, one in the left and one in the right side of the breech ring. Screw in the two pivot plugs which hold the pivots in the breech ring.

  (5) *Operating handle stop.* To fasten the operating handle stop to the breech ring, insert the plunger, small end first, and spring, into the bottom of the stop, hold in place, and insert the lug of the stop into its breech ring hole. Aline stop holes with those of the breech ring, the jaw of the stop forward, and fasten the stop to the breech with the two washers and socket head cap screws.

  (6) *Breech ring key.* Fasten the breech ring key to the breech ring by inserting it in the slot in the left face of the breech ring and screwing the two socket head screws in place which fasten the key to the breech ring.

  (7) *Trigger plunger.* Install the trigger plunger spring over the rear end of the trigger plunger and insert the spring and plunger into the bore in the left front face of the breech ring. The rounded end of the plunger protrudes. Slide the trigger plunger bushing over this protruding end and screw the bushing into the breech ring.

346. Installation

a. FIRING MECHANISM.
  (1) Assembly of the sear, trigger, and cocking mechanism is accomplished with the breechblock outside of the breech ring. To assemble the sear in the breechblock, first insert the sear spring into the counterbore in the left side of the breechblock, being sure that the inner end of the spring fits into the small hole in the bottom of the breechblock counterbore. Insert the sear, small end first into the same bore through the sear spring until the outside end of the spring seats in its small hole in the "horseshoe" of the sear. Insert the trigger into its seat in the left side of the breechblock. The central pivot lug of the trigger goes into a mating hole in the breechblock and the bottom round lug fits into the "horseshoe" of the sear. Next assemble the cocking mechanism by inserting the cocking lever shaft spring into the counterbore in the right side of the breechblock, seating the inside end of the spring in its small hole in the bottom of the counterbore. Insert the cocking lever shaft, small end first, and seat the outside end of the spring in the small hole through the outside cap of the shaft. Insert the cocking lever into the right side of the breechblock, its central pivot fitting into the bore in the right side of the breechblock. The lower lip of the cocking lever should cam the lug of the cocking lever shaft as the top of the cocking lever is pulled rearward.
  (2) Replacement of the percussion mechanism, firing spring, and firing spring retainer is normally performed with the breechblock assembled in the breech ring and in closed position. To install, rotate the sear by depressing the trigger plunger at the front upper face of the breech ring and push the percussion mechanism, firing pin first, past the sear as far as it will go into the central rear breechblock bore. Release the trigger. Insert the firing spring into the percussion bore and seat the forward end of the spring in the rear end of the guide. Seat the rear end of the spring in the forward face of the retainer. With its rear slot horizontal, push the retainer inward into the rear of the bore. When the "ears" of the retainer are inside their corresponding cut-outs of the breech block, rotate the retainer until the slot is vertical (90°). The retainer "ears" should snap into undercuts on the inside of the breechblock bore. Release pressure on the retainer, leaving it in locked position in the breechblock.

b. BREECHBLOCK AND RELATED PARTS.
  (1) Place the extractors over the pivots in the breech ring recess, with the lips of the extractors in the recesses of the tube.

Screw the eyebolt into the top of the breechblock, and before the breechblock is lowered into the breech recess, the trigger and cocking mechanism should be installed in the breechblock ($a(1)$ above). Lower the breechblock into the breech recess, being sure the firing plunger in the breech ring is not depressed and that the trunnions of the extractors line up with their breechblock grooves. Insert the breechblock removing tool 7237831 into the rear breechblock bore, long flange down, and let the tool support the weight of the breechblock. Assemble the breechblock crank, pivot, and crosshead. Insert the closing spring piston into the front end of the bore in the right side of the breech ring, open end first, rack teeth fitting into the lower breech ring slot. Push the piston back until the front face is flush with the front wall of the breech ring. Insert the closing spring crank into the right bottom breech ring lug. Aline the arrow on the piston rack with the arrow on the gear segment of the crank before meshing the gear teeth of the crank with the rack. Insert the operating crank in the left bottom breech ring lug. Screw the detents, one into the rear of each breech ring lug to prevent the cranks from coming out. The inside of each detent seats itself in the annular slot in the cylindrical part of each crank. Support the breechblock weight and remove the breechblock tool from the rear of the breechblock bore. Lower the breechblock enough to insert the crossheads and breechblock crank into the bottom of the breechblock, the flat end of the crank stop facing to the front. Push the crossheads into the breechblock "T" slot and lift the breechblock up, supporting its weight again with the breechblock tool. The round lug of the tool goes into the breechblock rear bore and the rounded flange rests on the "U" of the breech ring. Insert the operating shaft, small end first, in the left breech ring lug, through the hub of the operating crank, left spacer, breechblock crank, right spacer, and closing spring crank. Splines of components are alined as the shaft is pushed into position. Push the shaft all the way in, its left collar seating in the outside counterbore of the operating crank. Put the hub of the operating handle over the right end of the operating shaft. Make sure the shaft is all the way home, the inside lug of the operating handle hub should butt up against the protruding sector of the closing spring crank with the handle locked to its stop on the right side of the breech ring. Slip the operating shaft plate into the outside of the handle hub, flat side out, depress the plunger in the shaft and slide the plate into its handle slot. The shaft plunger snaps into the hole in the operating shaft plate and locks the latter in position after the inside jaw of the plate has locked into the right peripheral groove of the shaft. Insert the closing spring in the rear of its bore in the breech ring and screw the closing spring cap into the rear of the closing spring hole, using face spanner wrench 41-W-3247-720, to put the initial compression on the spring. Be sure the cap is screwed in until the correct cap hole becomes alined with the housing screw hole. Remove the breechblock tool from the rear bore of the breechblock and unscrew the eyebolt 41-E-3150 from the top of the breechblock. The closing spring supports the weight of the breechblock when the breech is closed.

(2) Install the percussion mechanism ($a(2)$ above).

Section XXX. COMBINATION GUN MOUNT M78

347. Description and Data

a. GENERAL. The combination gun mount M78 (figs. 350 and 351) consists principally of the gun shield and adapter assembly, the hydrospring type recoil mechanism concentric with the 90-mm gun, the breech guard, and the cradle assembly which mounts, coaxially with the 90-mm gun, either a cal. .50 HB Browning machine gun M2 or a cal. .30 M1919A4 Browning machine gun. On the combination mount are mounted the impulse relay and firing solenoid for firing the 90-mm gun electrically and the mechanism for hand-firing it, the firing linkage (fig. 349) to the gun trigger plunger in the breech ring, and the cam which operates the breech operating mechanism. The solenoid for electrical firing of a cal. .30 machine gun mounted in the coaxial machine gun mount is on the combination gun mount. The firing solenoid for the cal. .50 machine gun mounts on the machine gun and is actuated by the same firing cable and control. Provision is made for elevating, depressing, and traversing the coaxial machine gun mount to aline it exactly with the 90-mm gun.

b. TABULATED DATA.

Type of recoil mechanism_____ concentric, hydrospring
Elevating mechanism_____ hydraulic
Oil capacity_____ 5½ gal
Maximum elevation_____ 19° (338 mils)
Maximum depression _____ 5° (89 mils)
Transverse by turret (hydraulic)_____ 360° (6,400 mils)
Length of recoil (maximum)_____ 14 in
Length of recoil (normal)_____ 12 in
Minimum time required for power traverse of 360°_____ 10 sec
Maximum rate of power elevation_____ 4° per sec

348. Maintenance

No maintenance of the combination gun mount is required except that authorized under recoil mechanism (par. 358), firing linkage (par. 352), and elevating mechanism (sec. XXXIII).

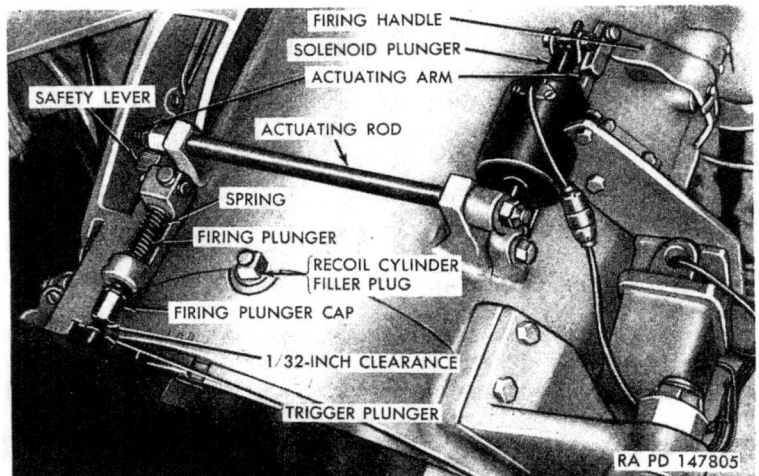

Figure 349. Firing linkage.

Section XXXI. FIRING LINKAGE

349. Description

a. The firing linkage of the 90-mm gun T119E1 (fig. 349) consists of a firing plunger, spring, firing plunger cap, actuating rod and arms, solenoid plunger, and firing handle, and actuating mechanism. An electrical relay and firing solenoid are provided to enable the gun to be fired electrically as well as manually (par. 79). The firing mechanism is rendered inoperative by a safety lever on the firing plunger (figs. 41 and 349).

b. When the manual firing control is used (par. 54), the firing handle arm presses back on the right actuating rod arm, thus operating the firing linkage. The firing linkage on the mount then strikes the trigger plunger in the breech ring which in turn releases the percussion mechanism which fires the round.

c. When the gun is fired electrically (par. 79), the circuit is closed by the relay and current flows through the solenoid. The solenoid is then energized setting up a magnetic field which acts upon the solenoid plunger to pull it backward to operate the firing linkage.

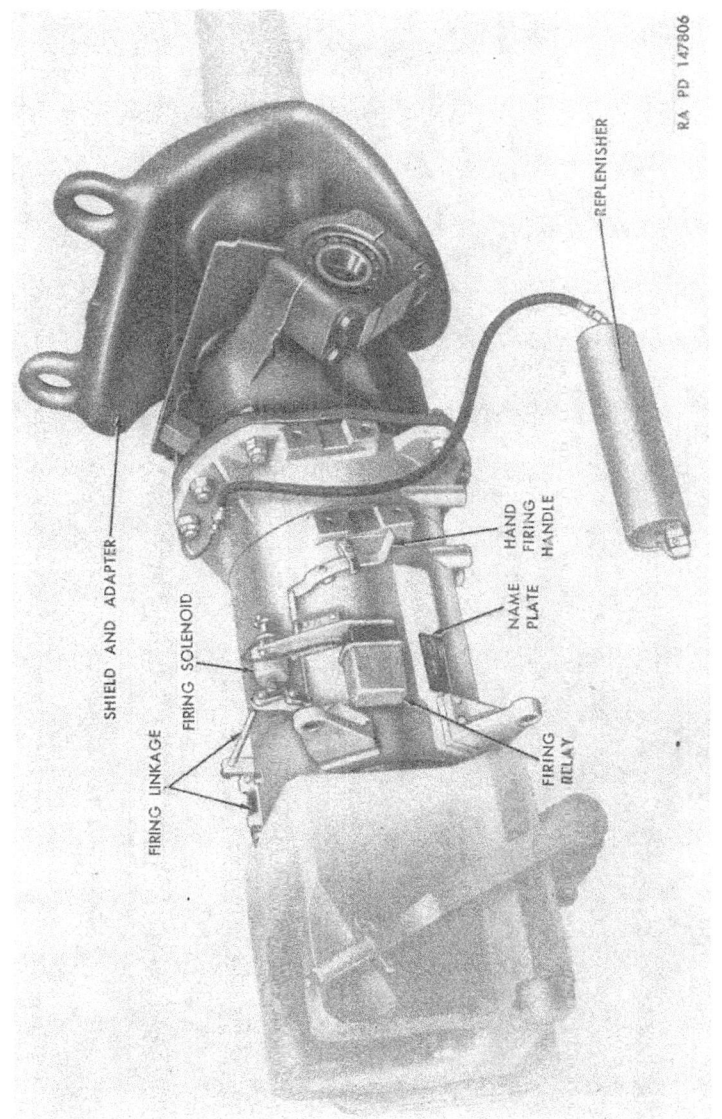

Figure 350. Combination gun mount M78—right rear view.

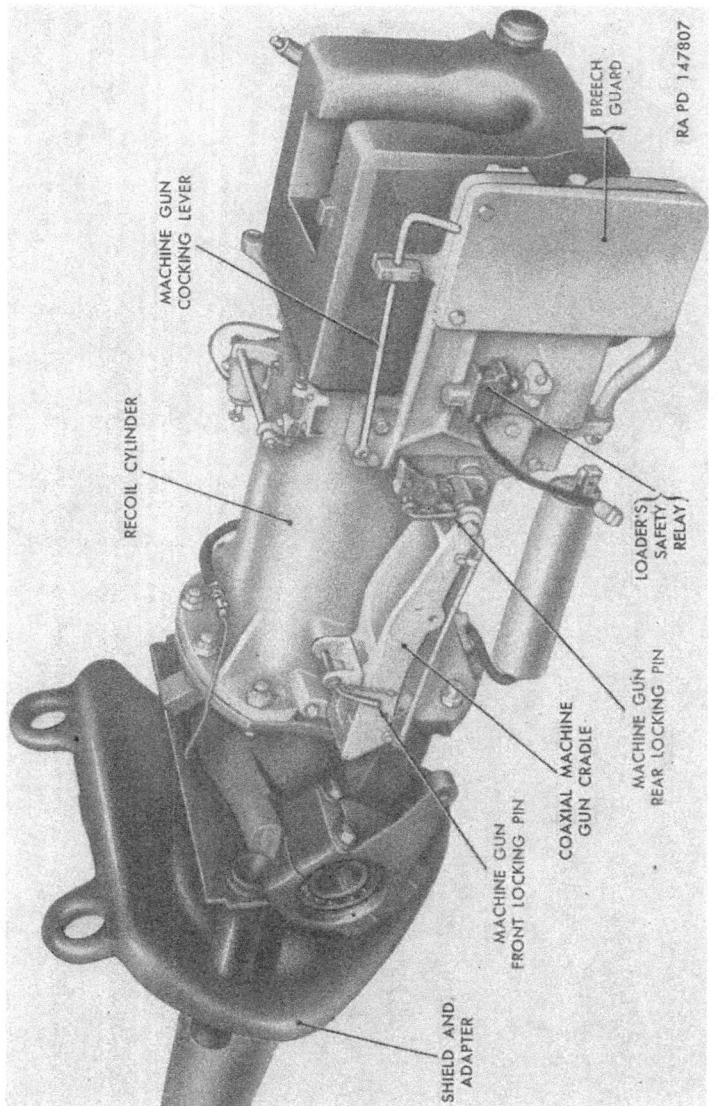

Figure 351. Combination gun mount M78—left rear view.

d. When the firing controls are released, the firing plunger is returned to its original position by the action of the firing plunger return spring (fig. 349).

350. Removal

a. FIRING PLUNGER GROUP. To remove the firing plunger group (fig. 349), remove two cap screws and washers which secure the supporting bracket to the gun cradle and tilt the forward end upward to free the pluger from the left actuating arm.

b. FIRING SOLENOID. Refer to paragraph 290*a*.

c. IMPULSE RELAY. Refer to paragraph 290*b*.

d. HAND-FIRING MECHANISM. Remove hand-firing handle screw which secures the hand-firing handle (fig. 350) to the gun cradle. Remove shoulder screw securing hand-firing arm. Remove hand-firing mechanism.

351. Disassembly

a. FIRING PLUNGER GROUP. Remove firing plunger group (par. 350*a*). Compress spring and remove retaining ring. Loosen lock nut securing firing plunger cap and remove cap and lock nut. Place safety lever in open position and pull plunger out of bracket.

b. FIRING PLUNGER ACTUATING MECHANISM. Remove nut and washer from both ends of actuating rod. Remove actuating arms from actuating rod. Remove rod.

352. Maintenance

a. Check the clearance between the firing plunger and the trigger plunger. This should be $\frac{1}{32}$ inch when the percussion mechanism is cocked. If clearance is greater or smaller than $\frac{1}{32}$ inch, push the gun safety level in "SAFE" position, loosen the jam nut, adjust the firing plunger cap until the required clearance is obtained, and tighten the jam nut.

Caution: Never perform this adjustment with a round in the chamber.

b. With firing mechanism safety lever in "SAFE" position, operate both electrical and hand-firing controls. Linkage should remain inoperative; if not, notify ordnance maintenance personnel. Repeat test with safety lever in "FIRE" position. If percussion mechanism is not released or firing linkage fails to return to its original position, notify ordnance maintenance personnel.

353. Assembly

a. FIRING PLUNGER ACTUATING MECHANISM. Insert actuating rod in place through bosses on gun cradle. Place the forked arm on the squared left end of the actuating rod and the other arm on the right of the actuating rod so the two arms are alined. Secure with nuts and lock washers.

b. FIRING PLUNGER GROUP. With the gun safety lever in "SAFE" position, insert the plunger into bracket and through the retaining ring, washer, and spring. Install firing plunger cap and jam nut. Compress spring and install retainer ring in position on plunger.

354. Installation

a. HAND-FIRING MECHANISM. Position hand-firing mechanism and secure hand-firing arm with shoulder bolt and hand-firing handle with shoulder screw.

b. IMPULSE RELAY. Refer to paragraph 290*b*.

c. FIRING SOLENOID. Refer to paragraph 290*a*.

d. FIRING PLUNGER GROUP. Position firing plunger, making sure that front end of plunger properly engaged left plunger actuating arm. Secure bracket to gun cradle with two screws. Adjust firing plunger cap (par. 352*a*).

Section XXXII. RECOIL MECHANISM

355. Description

a. CONCENTRIC RECOIL MECHANISM. The concentric recoil mechanism which incloses the tube of the 90-mm gun, is of the short recoil, hydrospring type and is provided with a replenisher which maintains a constant quantity of oil in the recoil cylinder, a breech operating mechanism, and a firing mechanism.

b. RECOIL CYLINDER ASSEMBLY. The recoil cylinder is composed principally of the cradle, the piston, and the recoil spring. The recoil cylinder consists of the tubular space between the inside of the cradle and the outside of the gun tube.

c. REPLENISHER ASSEMBLY. The replenisher assembly is composed principally of a cylinder, piston, indicator tape with serrated edges, a helical spring, a spring loaded ball valve, and openings for filling, draining, and hose connection to recoil cylinder (fig. 42). The replenisher assembly is connected to the recoil cylinder by a flexible hose. As the gun tube forms one surface of the recoil cylinder the recoil oil becomes heated very rapidly during firing. This causes the oil to expand, forcing the excess oil through the hose into the replenisher cylinder applying pressure on the replenisher piston. As the temperature of the oil increases, the increased pressure forces the replenisher piston back to provide the needed extra oil capacity. This travel of the piston compresses the spring and causes the indicator tape to wind itself around the screw in the indicator assembly. As the indicator assembly (fig. 42) is constructed so the edges of the tape are accessible to the fingers, this allows direct readings of the quantity of oil in the system (par. 358c). As the gun cools, the oil contracts and the spring elongates forcing the piston back to its original position, returning oil to the recoil cylinder, thus keeping the oil reservoir filled at all times. The replenisher also serves to compensate for any oil lost through seepage.

356. Functioning

a. RECOIL. As the gun recoils, the piston, which is securely keyed to the gun tube, compresses the helical spring and forces the recoil oil from the rear of the piston to the front of the piston. The inside diameter of the cradle tapers toward the rear so that opening between the outside diameter of the piston and the inside diameter of the cradle becomes less as the recoil progresses, practically shutting off the flow of oil at the end of recoil. This throttling of the recoil oil and the high compression of the spring bring the gun to rest.

b. COUNTERRECOIL. The compressed spring immediately starts the counterrecoil action, forcing the oil past the piston from front to rear. Near the end of counterrecoil an enlarged portion of the gun tube enters the oil filled buffer chamber in the muzzle end of the recoil cylinder, sharply constricting the opening through which the oil must travel in escaping from the buffer chamber, thus cushioning the counterrecoil and allowing the gun to return to battery without severe shock. To compensate for the greater viscosity of the recoil oil in extremely cold weather when the gun returns to battery too slowly or not at all, there is provided a buffer regulator which provides additional passage for the escape of oil from the buffer chamber.

Note. This buffer regulator must be fully closed unless actual firing in cold weather shows greatly increased recoil cycle time.

If this regulator is left open during normal temperature operation the gun will slam back to battery.

c. OPERATING FUNCTIONS. In addition to bringing the gun back to battery, the counterrecoil opens the breech, thereby extracting the empty case, cocks the gun for the next firing, and compresses the closing spring which returns the breechblock to closed position upon insertion of a new round.

357. Disassembly

Disassembly is not authorized the using arm.

358. Maintenance

a. GENERAL. Organizational maintenance of the recoil mechanism is limited to the inspections and operations described below. For any other operations, notify ordnance maintenance personnel.

b. OIL LEAKAGE. Check for oil leakage around all the plugs in cradle and replenisher and replace and defective plugs or gaskets. If there is evidence of undue leakage around the gun tube, notify ordnance maintenance personnel.

c. OIL LEVEL. The replenisher indicator assembly (fig. 42) is used to indicate the amount of oil in the system. To determine whether there is too little oil reserve in the replenisher, using the fingers, grasp the indicator tape on both edges through the accessible part of the indicator assembly. If serrated edges are felt on both edges, the replenisher cylinder is empty and needs refilling; if one edge is smooth and the other serrated, the replenisher has the normal amount of oil; if both edges are smooth, the replenisher has too much oil and should be bled (*e* below).

d. FILLING REPLENISHER. Remove the replenisher valve plug (fig. 42) and using oil filler gun 41-G-1362-500, fill the replenisher through the ball valve with the oil prescribed by the lubrication order.

Note. Do not fill past the operating range or normal reserve of the replenisher (*c* above).

e. BLEEDING EXCESS OIL. If the replenisher indicator (fig. 42) shows too much oil (*c* above), loosen one of the pipe plugs in replenisher cylinder head and allow excess oil to seep out into a container.

Note. Do not remove the plug or oil will gush out.

Tighten the pipe plug as soon as indicator tape shows oil to be at normal level.

f. DRAINING RECOIL MECHANISM. Notify ordnance maintenance personnel.

g. FILLING RECOIL MECHANISM. Notify ordnance maintenance personnel.

h. CARE OF RECOIL OIL.

(1) Care must be exercised not to let moisture get into the recoil mechanism through open plug holes. Exposure of recoil oil to the atmosphere in an open can is to be avoided in order to keep out moisture and dirt. Use of cans without covers or of cans with condensed water on inner walls is prohibited. Condensation in a container partly filled with oil, or pouring from one container to another which has moisture on its inner walls, results in moisture being carried along with the oil into the recoil mechanism.

(2) Recoil oil should be tested for presence of water. Fill a clean glass bottle with recoil oil and allow to settle in a warm place. If water is present, it will settle on the bottom. Also, if the bottle is slightly tilted, drops or bubbles will form in the lower portion. Invert the bottle and hold it to the light; drops or bubbles of water, if present, may be seen slowly settling in the oil. A cloudy appearance may be due to particles of water.

(3) If this test shows the presence of water, the oil on hand should not be used.

(4) The transfer of recoil oil to a container not marked with the name of the oil may result in the wrong oil getting into the recoil mechanism or in the use of recoil oil for lubricating purposes. Recoil oil must not be put into any container not marked with the name of the oil. Dirty recoil oil should be replaced with new oil except in an emergency, when it may be strained through clean cloth before reuse. Do not mix recoil oils with any other type of oil.

359. Assembly

No assembly is authorized the using arm.

Note. On some M47 tanks, a tee replaces the elbow connecting the replenisher hose to the recoil cylinder (fig. 351), and a tube connects this tee with a fitting in the recoil cylinder front follower. If this tube or connectors become damaged, remove tube and discard it. The recoil mechanism will function properly without this tube, therefore, it will not be maintained. Close openings at tee and follower with pipe plugs.

Section XXXIII. ELEVATING MECHANISM

360. Description

Elevation or depression of the combination gun mount M78 supporting the 90-mm gun and the coaxial machine gun, is accomplished by a hydraulic elevation cylinder (fig. 296) controlled manually or by power. It elevates the guns to a maximum of 19° above horizontal and depresses them to 5° below horizontal. The elevating mechanism can be operated while the turret is being traversed as well as while it is at rest.

361. Power Elevating Mechanism

a. DESCRIPTION. The power elevating mechanism consists principally of a hydraulic pump, a hydraulic elevating cylinder, and gunner's (and commander's) controls. Hydraulic pressure for activating the power elevating mechanism is supplied from the elevating-pump side of the double hydraulic pump (fig. 296). Control is by either the gunner's power traversing and elevating control handle (fig. 22) or the commander's power traversing and elevating control handle (fig. 26). The power elevating mechanism can elevate or depress the guns at any speed up to 4° per second, governed by the rate at which the control handle is rotated.

b. MAINTENANCE. Check for oil leaks at all connections and tighten or replace gaskets if necessary. In case of breakage or malfunctioning, notify ordnance maintenance personnel.

362. Manual Elevating Mechanism

a. DESCRIPTION. The manual elevating mechanism consists principally of the elevation cylinder, a hand pump, an accumulator, an oil return valve, an elevation cylinder supercharging valve, gunner's manual elevating control handle, and necessary tubing and fittings. When shifting from power to manual operation, open the elevation supercharge line shut-off valve (fig. 32) for 1 second to allow passage of a few drops of oil; then close. To elevate the guns, rotate the gunner's manual elevating control handle (fig. 25) clockwise; to depress the guns, rotate handle counterclockwise. The rate at which the handle is rotated regulates the rate at which the gun is elevated or depressed. The hand pump is used to charge the manual elevation system and is used only if the vehicle has been inactive for a period of time or the system has been serviced. The pump is operated by means of a hand pump handle located to left of the gunner's seat (fig. 32). To operate the pump, move the pump handle back and forth several times. An accumulator, which has a chamber filled with an inert gas, is located on the front of the turret floor to the right of the hydraulic pumps (fig. 296). The function of the accumulator is to keep the manual elevating system under hydraulic pressure. A small amount of oil from the power pack reservoir should be occasionally pumped into the accumulator by the hand pump where it is stored under pressure until required during normal operation. The accumulator is also charged by oil from the traverse pump when the turret is accelerated and decelerated in power traverse.

b. MAINTENANCE. Check for oil leaks in fittings and tubing and tighten or replace leaking parts as required. Check to see that oil return valve (fig. 33) and elevation supercharge line shut-off valve (fig. 32) are closed. In case of breakage or malfunction, notify ordnance maintenance personnel.

Section XXXIV. MACHINE GUN MOUNTS

363. Description

There are three machine gun mounts. One on the combination gun mount M78 (fig. 351) will mount either a cal. .50 or a cal. .30 machine gun coaxially with the 90-mm gun. A flexible ball mount in the driver's compartment mounts a cal. .30 machine gun. A pintle mount on the turret roof mounts a cal. .50 machine gun (figs. 5, 6, and 352). The coaxial machine gun is fired by either the gunner or the commander with the same firing controls that fire the main gun or it can be fired manually. The bow gun is operated by the assistant driver and is capable of 24° elevation and 10° depression. The turret roof gun, which is primarily for anti-aircraft service, is controlled and fired manually. Its mount locks in the pintle stand assembly and gun and assembly can be removed as a unit if desired. The pintle stand is located adjacent to the loader's hatch. A tripod mount M2 is provided for the cal. .30 machine gun M1919A4. This light weight (14 pounds) portable folding mount is used for ground fire. The tripod mount M2 is stored in the left front fender box when not in use. For information pertaining to the cal. .30 and cal. .50 machine guns, see FM 23–55 and FM 23–65 respectively.

364. Installation and Removal of Coaxial Cal. .50 HB Browning Machine Gun in M2 Machine Gun Cradle (in Combination Gun Mount M78)

a. INSTALLATION. Place the cal. .50 HB Browning machine gun M2 in position in the machine gun cradle (fig. 351) at left of the combination gun mount (fig. 39). Aline the front mounting holes in gun and mount and insert the machine gun front locking pin (fig. 351). Swing the elevating and traversing mechanism up to aline its mounting hole with the rear mounting hole of the machine gun and insert the machine gun rear locking pin. Install the empty cartridge bag on its support below the gun mount and connect the ammunition chute. Connect the firing cable to the firing solenoid mounted on the machine gun.

b. MAINTENANCE. No maintenance is authorized the using arm.

c. REMOVAL. Check that gun is not loaded. Disconnect ammunition chute and firing cable. Remove the coaxial machine gun by removing the machine gun front and rear locking pins and lift gun out.

365. Installation and Removal of Cal. .30 Browning Machine Gun M1919A4 in Cal. .30 Machine Gun Mount (BOW)

a. INSTALLATION. Place the cal. .30 Browning machine gun M1919A4 in position on the bow mount so the front mounting holes in the gun and mount aline. Secure the gun by inserting the locking pin. Swing the bow gun locking rod out from side of hull and insert into lock assembly mounted in machine gun rear mounting hole.

b. MAINTENANCE. No maintenance is authorized the using arm.

c. REMOVAL. Check that gun is not loaded. Remove locking rod from lock assembly and secure under clip to side of hull. Remove fastening pin and lift gun out.

366. Installation and Removal of Cal. .50 HB Browning Machine Gun M2 in Cal. .50 Machine Gun Mount (Turret Roof)

a. INSTALLATION. Remove cover plug assembly from pintle stand (fig. 352). Place pintle lock in open position (handle down). Insert pintle of gun mount in pintle stand and lock (handle up). Place the cal. .50 HB Browning machine gun M2 in the cradle assembly of the mount and aline the front and rear mounting holes in the machine gun and mount (fig. 353). Insert the front and rear fastening pins (fig. 353).

b. MAINTENANCE. No maintenance is authorized the using arm.

c. REMOVAL. Check that gun is not loaded. Remove the locking pins and lift out the gun. Unlock pintle and lift out mount. Install cover plug assembly in pintle stand.

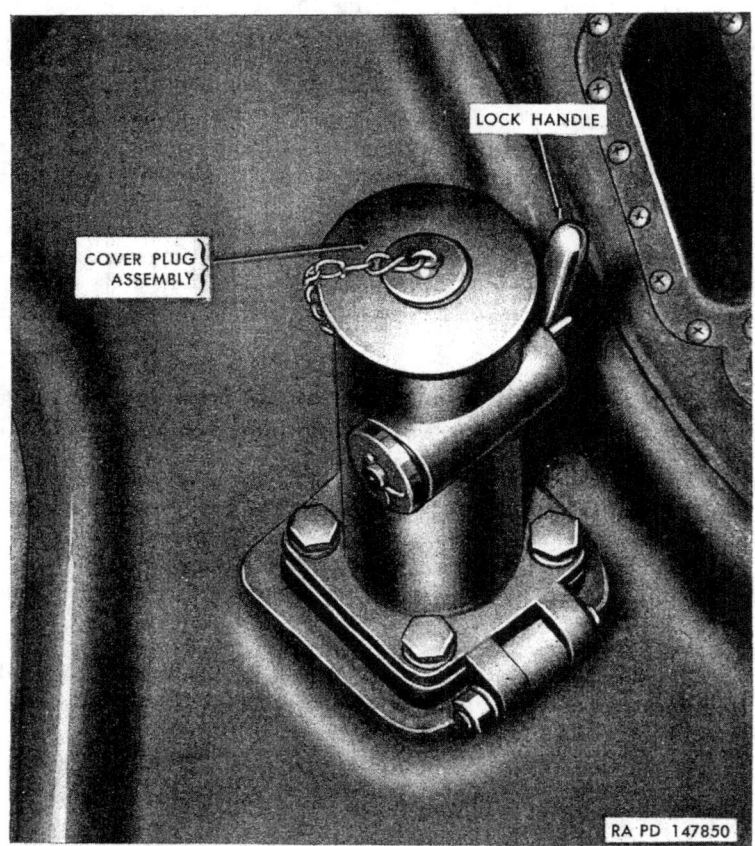

Figure 352. Turret machine gun pintle stand.

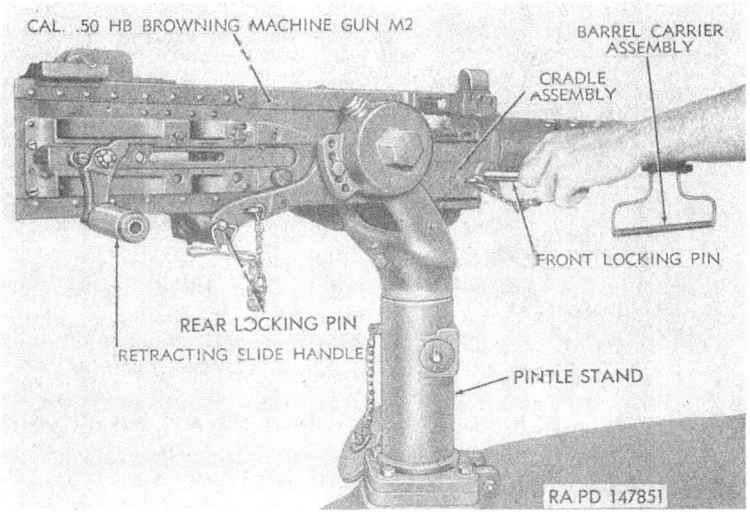

Figure 353. Installing cal. .50 HB Browning machine gun M2 on mount.

Section XXXV. SIGHTING AND FIRE CONTROL EQUIPMENT

367. General

a. CARE IN HANDLING SIGHTING AND FIRE CONTROL EQUIPMENT.
  (1) Sighting and fire control instruments are, in general, rugged and suited for the designed purpose. They will not, however, stand rough handling or abuse. Inaccuracy or malfunctioning will result from mistreatment.
  (2) Unnecessary turning of screws or other parts not required in the use of the instruments is forbidden.
  (3) Stops are provided on instruments to limit the travel of the moving parts. Do not attempt to force the rotation of any knob beyond the stop limit.
  (4) Keep the instruments as dry as possible. If instrument is wet, dry it carefully before placing it in its carrying case.
  (5) When not in use, instruments must be kept in the carrying cases provided, or covered and protected from dust and moisture.
  (6) Any instruments which indicate incorrectly or fail to function properly after the authorized tests and adjustments have been made, are to be turned in for repair by ordnance personnel. Adjustments other than those expressly authorized are not the responsibility of the using arm personnel.
  (7) Painting of sighting or fire control equipment by the using personnel is not permitted.

(8) Do not keep a telescope pointed directly at the sun unless a filter is used, as the heat of the focused rays may melt the cement used in compound lenses.

b. OPTICAL PARTS.
  (1) To obtain satisfactory vision, it is necessary that the exposed surfaces of the lenses and other parts are kept clean and dry. Corrosion and etching of the surface of the glass, which interfere with vision, can be prevented or greatly retarded by keeping the glass clean and dry.
  (2) Under no circumstances should polishing liquids, pastes, or abrasives be used for polishing lenses and windows.
  (3) For wiping optical parts use only lens tissue paper specially intended for cleaning optical glass. Use of cleaning cloths is not permitted. To remove dust, brush the glass lightly with a clean artist's camel's-hair brush, and rap the brush against a hard body in order to knock out the small particles of dust that cling to the hairs. Repeat this operation until all dust is removed.
  (4) Exercise particular care to keep optical parts free from oil and grease. Do not touch the lenses or windows with the bare fingers. To remove oil or grease from optical surfaces, apply liquid lens cleaning soap with a tuft of lens tissue paper, and wipe gently with clean lens tissue paper. If liquid lens cleaning soap is not available, breathe heavily on the glass and wipe off with clean lens tissue paper. Repeat this operation until clean.
  (5) In cold weather, optical surfaces should be cleaned with lens tissue paper moistened with alcohol. If alcohol is not available, use dry lens tissue paper. Alcohol should never be applied directly to the lens surfaces, as any excess may injure the sealing compound. Do not breathe on the oculars.
  (6) Because of condensation, moisture may collect on the optical parts of the instrument when the temperature of the parts is lower than that of the surrounding air. This moisture, if not excessive, can be removed by placing the instruments in a warm place. Heat from strongly concentrated sources should not be applied directly, as it may cause unequal expansion of parts, thereby resulting in damage to optical parts and inaccuracies of observation.

c. BATTERIES.
  (1) Dry-cell-type batteries are used in the instrument lights and should habitually be removed whenever the lights are not in use. Chemical reaction set up in an exhausted battery will damage the battery tube.
  (2) To replace batteries, remove the cap on the battery tube. The cap is secured by bayonet pins and is removed by press-

ing the cap inward and then turning slightly until free. When replacing the batteries, be sure they go back into the tube in the same position as when removed. See that the bayonet pins in the cap engage the slots in the tube to insure a tight contact between the battery terminals.

d. LUBRICATION. Only ordnance maintenance personnel will disassemble to clean and lubricate sighting and fire control instruments.

368. Replacement of Authorized Spare Parts

a. GENERAL. The only spare parts authorized for replacement by the using arm on the sighting and fire control equipment are replacement of lamps (*b* through *e* below). If any other parts are required, notify ordnance maintenance personnel.

b. REPLACEMENT OF LAMPS IN RANGE FINDER T41 (fig. 44). Spare lamps are carried in the spare lamp case on range finder T41. To replace the lamps provided for the scales and the left stereoscopic reticle, open the lamp housing cover on the range finder by means of the reticle lamp replacement knob. Replace the bayonet-type lamps and close the cover. The lamp housing for right stereoscopic reticle is located on the right underside of the range finder. To replace the lamp, proceed as above.

c. REPLACEMENT OF GUN READY INDICATOR LAMP ON PERISCOPE MOUNTS T176 AND T177 (fig. 354). Unscrew lamp assembly from periscope mount. Replace lamp. Install lamp assembly.

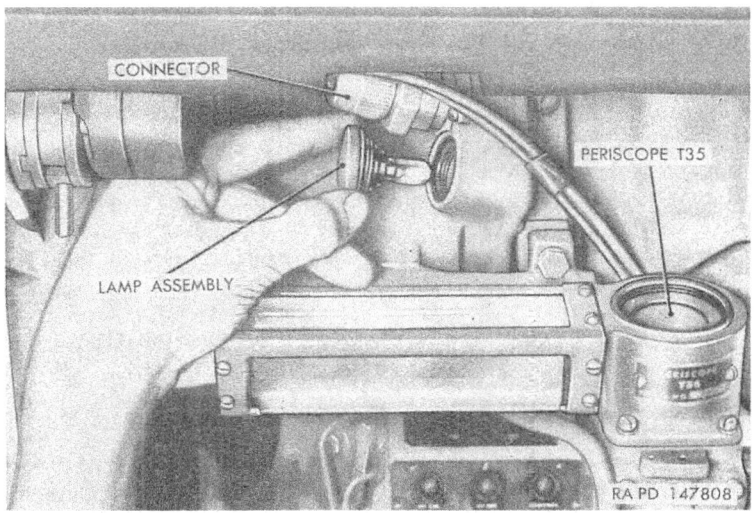

Figure 354. Replacing gun ready indicator lamp on periscope mount T176 and T177.

587

d. REPLACEMENT OF LAMP IN BALLISTIC DRIVE T23E1 (fig. 54). Unscrew lamp assembly from left side of ballistic drive T23E1 and replace lamp assembly. Install lamp assembly.

e. REPLACEMENT OF LAMP ASSEMBLY IN AZIMUTH INDICATOR T24 (fig. 355). Unscrew lamp assembly from side of azimuth indicator. Install new lamp assembly.

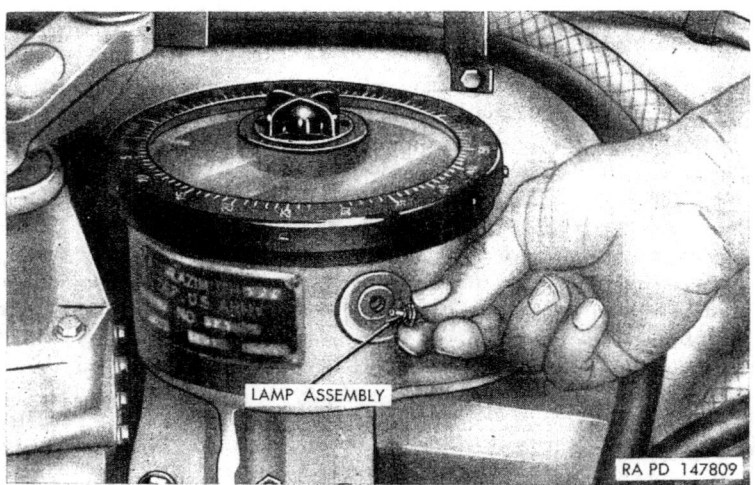

Figure 355. Replacing lamp in azimuth indicator T24.

369. Adjustment of Elevation Quadrant Micrometer

Place gunner's quadrant M1 on breech ring of 90-mm gun T119E1 so that it is parallel to the axis of the gun bore (fig. 356). Adjust elevation knob on gunner's quadrant M1 to 0. Elevate or depress the gun (par. 76) until the bubble in the level vial of the gunner's quadrant M1 is centered. Grasp the elevation micrometer of elevation quadrant T21 with the fingers of the right hand and the elevation knob with the fingers of the left hand (fig. 357). Turn the knob until the elevation micrometer index indicates zero on the elevation micrometer.

Section XXXVI. MAINTENANCE UNDER UNUSUAL CONDITIONS

370. Extreme-Cold Weather Maintenance Problems

a. The importance of maintenance must be impressed on all concerned, with special emphasis on organizational (preventive) maintenance. Maintenance of mechanical equipment in extreme cold is exceptionally difficult in the field. Even shop maintenance cannot be completed with normal speed, because the equipment must be allowed

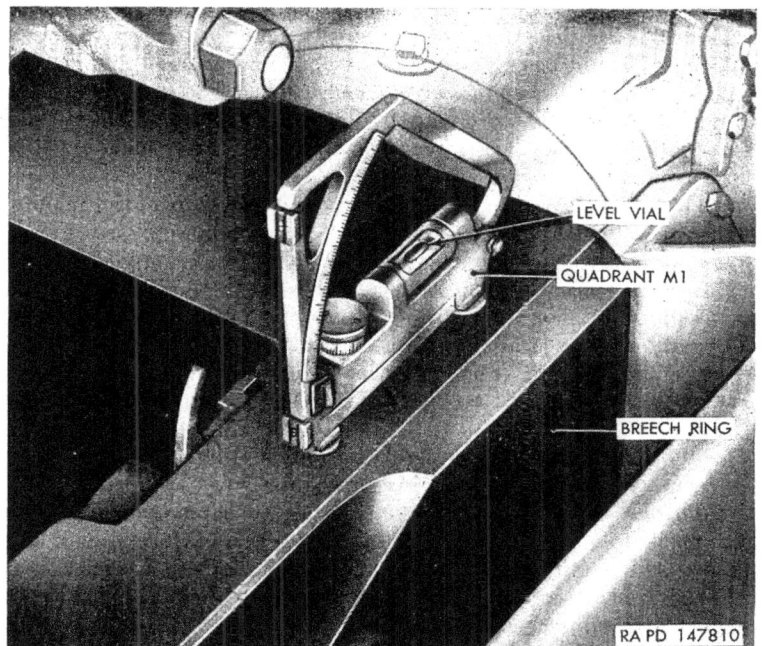

Figure 356. Gunner's quadrant M1 on breech ring of 90-mm gun T119E1.

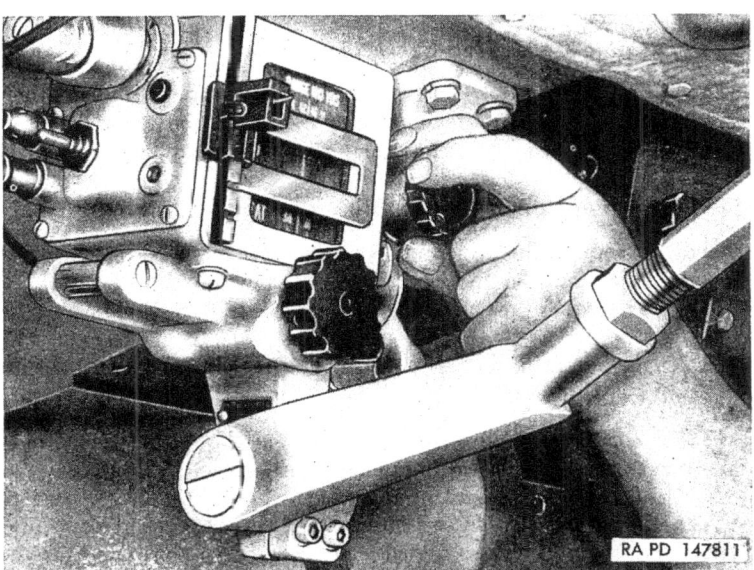

Figure 357. Adjusting elevation micrometer on elevation quadrant T21.

to thaw out and warm up before the mechanic can make satisfactory repairs. In the field, maintenance must be undertaken under the most difficult of conditions. Bare hands stick to cold metal. Fuel in contact with the hands results in super-cooling due to evaporation, and the hands can be painfully frozen in the matter of minutes. Engine oils, except subzero grade, are unpourable at temperatures below $-40°$ F. Ordinary greases become as solid as cold butter.

 b. These difficulties increase the time required to perform maintenance. At temperatures below $-40°$ F., maintenance requires up to five times the normal amount of time. The time required to warm up a vehicle so that it is operable at temperatures as low as $-50°$ F. may approach 2 hours. Vehicles in poor mechanical condition probably will not start at all, or only after many hours of laborious maintenance and heating. Complete winterization, diligent maintenance, and well-trained crews are the key to efficient Arctic-winter operations.

 c. Refer to TM 9–2855 and TB ORD 193 for specific information on extreme-cold weather maintenance procedures.

 Caution: It is imperative that the approved maintenance procedures be followed. TM 9–2855 contains general information which is specifically applicable to this vehicle as well as all other vehicles. It must be considered an essential part of this manual, not merely an explanatory supplement to it.

 d. Refer to SB 9–16 for information on winterization kit for this vehicle.

371. Extreme-Cold Weather Maintenance—Vehicle

Carefully drain hull of all accumulated water during cold weather. Do not attempt to start bilge pumps when they are frozen. Thaw them out first, otherwise serious damage to the pumps may result.

372. Extreme-Cold Weather Maintenance—Armament

 a. Keep bore of gun and firing mechanism covered when not in use in order to prevent the entrance of snow.

 b. Clean bore of gun while the weapon is still hot, if possible.

373. Extreme-Cold Weather Maintenance—Sighting and Fire Control

 a. The sighting and fire control equipment should be prepared to operate at the lowest prevailing temperature expected.

 b. Sighting and fire control equipment should not be exposed to sudden changes in temperature because of the dangers of condensation, and the effects of sudden changes of temperatures on the accuracy of the equipment.

374. Extreme-Hot Weather Maintenance

   a. BATTERIES.
   (1) *Electrolyte level.* In torrid zones, check level of electrolyte in cells daily and replenish, if necessary, with pure distilled water. If this is not available, rain or drinking water may be used. However, continuous use of water with high mineral content will eventually cause damage to batteries and should be avoided.
   (2) *Specific gravity.* Batteries operating in torrid climates should have a weaker electrolyte than for temperate climates. Instead of 1.280 specific gravity as issued, the electrolyte (sulphuric acid, sp gr 1.280) should be diluted to 1.200 to 1.240 specific gravity (TM 9–2857). This is the correct reading for a fully-charged battery. This procedure will prolong the life of the negative plates and separators. Under this condition, a discharge battery should be recharged at about 1.160 specific gravity.
   (3) *Self-discharge.* A battery will self-discharge at a greater rate at high temperatures if standing for long periods. This must be considered when operating in torrid zones. If necessary to park for several days, remove batteries and store in a cool place.

   Note. Do not store acid-type storage batteries near stacks of tires, as the acid fumes have a harmful effect on rubber.

   b. HULL AND TURRET.
   (1) In hot damp climates, corrosive action will occur on all parts of the vehicle and will be accelerated during the rainy season. Evidences will appear in the form of rust and pain blisters on metal surfaces and mildew, mold, or fungus growth on fabrics, leather, and glass.
   (2) Protect exterior surfaces from corrosion by touch-up painting and keep a film of engine lubricating oil (OE–10) on unfinished exposed metal surfaces. Cables and terminals should be protected by ignition insulation compound.
   (3) Make frequent inspections of idle, inactive vehicles. Remove corrosion from exterior metal surfaces with abrasive paper or cloth and apply a protective coating of paint, oil, or suitable rust preventive.

   c. SIGHTING AND FIRE CONTROL. In hot, humid areas, inspect parts frequently for moisture corrosion or fungus growth. In dry, dusty or sandy areas, keep oculars protected from etching by sand.

375. Maintenance After Fording

   a. GENERAL. Although the vehicle unit housings are sealed to prevent the free flow of water into the housings, it must be realized that, due to the necessary design of these assemblies, some water may enter, especially during submersion. The following services should be accomplished on all vehicles which have been exposed to some depth of water or completely submerged, especially in salt water. Precautions should be taken as soon as practicable to halt deterioration and avoid damage before the vehicle is driven extensively in regular service.

b. HULL AND TURRET. Drain and clean out hull. Clean all exposed surfaces and touch up paint where necessary. Coat unpainted metal parts with preservative lubricating oil.

c. ENGINE, TRANSMISSION, AND FINAL DRIVES. Check the lubricant in the engine, transmission, and final drives. Should there be evidence that water has entered, drain, flush, and refill with the correct lubricant. Remove and clean engine and transmission oil filter.

d. SUSPENSION. Clean and lubricate all parts as specified on the lubrication order. Remove road wheels, idlers, and support rollers, clean and repack bearings. Make sure that lubricant is generously forced into each lubrication fitting to force out any water present.

e. BATTERIES. Check the batteries for quantity and specific gravity of electrolyte to be sure no water entered through the vent caps. This is of special importance should the vehicle have been submerged in salt water.

f. ELECTRICAL CONNECTIONS. Check all electrical connections for corrosion, particularly the bayonet-type connectors.

g. FUEL SYSTEM. Drain fuel tanks of any accumulated water; clean fuel filter and lines as necessary. If water is found in the air cleaner, clean and refill with oil.

h. CONDENSATION. Although most units are sealed, the sudden cooling of the warm interior air upon submersion may cause condensation of moisture within the cases or instruments. A period of exposure to warm air after fording should eliminate this condition. Cases which can be opened may be uncovered and dried.

i. ALUMINUM OR MAGNESIUM PARTS. If vehicle remains in salt water for any appreciable length of time, aluminum or magnesium parts which were exposed to the water will probably be unfit for further use and must be replaced.

j. ARMAMENT. Armament subjected to submersion should be disassembled, dried, and lubricated.

k. SIGHTING AND FIRE CONTROL INSTRUMENTS. If moisture has entered optical instruments turn in to ordnance maintenance for repair at earliest opportunity.

l. DEEP-WATER FORDING. Refer to TM 9-2853 for deep-water fording kit information.

376. Maintenance After Operation on Unusual Terrain

a. MUD. Thorough cleaning and lubrication of all parts affected must be accomplished as soon as possible after operation in mud, particularly when a sea of liquid mud has been traversed. Clean all suspension components and lubricate as specified on the lubrication order.

b. SAND OR DUST. Clean engine and engine compartment. Touch up all painted surfaces damaged by sandblasting. Lubricate completely to force out lubricants contaminated by sand or dust. Air cleaners, fuel and oil filters must be cleaned at least daily. Engine grilles and other exposed vents should be covered with cloth when vehicle is not in use.

CHAPTER 4

MATERIEL USED IN CONJUNCTION WITH MAJOR ITEM

Section I. AMMUNITION FOR 90-MM GUN T119E1

377. General

Complete rounds of ammunition authorized for use in the 90-mm gun M3A1 are also authorized for use in the 90-mm gun T119E1. These rounds are assembled with the M19, M19B1, or T27 cartridge cases. Other rounds, intended for use only in the T119E1 gun, are assembled with the T24 cartridge case. Some rounds, assembled with newly designed projectiles, are assembled with the M19, M19B1, or T24 cartridge case. All such rounds are authorized for use in the T119E1 gun. The T24 cartridge case has a slightly longer shoulder (has less taper) than the M19, M19B1, or T27 and therefore will not normally chamber in the M3A1 gun (refer to par. 382). The ammunition is issued in the form of fuzed and unfuzed complete rounds of fixed ammunition. The round consists of a primer and propelling charge contained in a cartridge case which is crimped rigidly to the projectile. The term "fixed," used in connection with ammunition, signifies that the propelling charge is fixed (not adjustable) and that the round is loaded into the gun as a unit. Generally rounds used against ground targets are fitted with impact or VT type fuzes whereas rounds used for antiaircraft fire are fuzed with combination, mechanical, or VT types. Rounds containing deep-cavity shell are issued with supplementary charge and fuze or closing plug assembled; certain lots of unfuzed deep-cavity shell of earlier manufacture have been issued without supplementary charge. Certain fuzes, including proximity (VT) and concrete-piercing (CP), are issued separately to be assembled to the shell in the field. The fuze MTSQ M500 (75 sec.) is also issued separately for high burst ranging (for adjustment of terrestrial fire) purposes.

378. Firing Tables

Firing data published in FT 90–F–2 (abridged) and FT 90–L–1 (abridged) are applicable for use with the 90-mm T119E1 gun. Firing tables for rounds intended for use only in the T119E1 gun, when available, will be indexed in SR 310–20–3. Graphical firing tables, when available, will be listed in ORD 7 SNL's F–237 and

F-331. Firing data for cal. .50 machine gun used for subcaliber purposes are given in FT 0.50-H-1.

379. Classification

Dependent upon type of projectile, ammunition for the 90-mm gun T119E1 is classified as armor-piercing with tracer (AP-T), armor-piercing-capped with tracer (APC-T), hyper velocity armor-piercing with tracer (HVAP-T), hyper velocity armor-piercing-discarding sabot with tracer (HVAP-DS-T), high-explosive (HE or HE-T), high-explosive, antitank (HE, AT), HE, Marker (green, red, or yellow), smoke (WP or WP-T), target practice (TP), hyper velocity target-practice with tracer (HVTP-T), blank, or drill.

  a. The armor-piercing with tracer projectile is a capped or uncapped solid shot. It has a tracer element incorporated in its base which provides a luminous trace for observation purposes during the first stages of projectile's flight.

  b. The armor-piercing-capped projectile is a heavy walled projectile with a small base cavity which is loaded with a bursting charge of high explosive. The projectile is fitted with an armor-piercing cap and a light weight windshield. These projectiles are intended primarily for penetration of armor plate and explosive effect behind the plate. This projectile is fitted with a base detonating fuze which functions with delay action. A tracer element is incorporated in the base of the fuze and functions independently of the fuze.

  c. The hyper velocity armor-piercing with tracer shot consists of a lightweight body which carries a dense core. The dense core is the armor-piercing element. The projectile, because of its light weight, has a high muzzle velocity. The projectile is inert except for the tracer element incorporated in its base. Later models of this shot have two sintered iron rotating bands.

  d. The hyper velocity armor-piercing-discarding sabot with tracer shot consists of a steel sheath, a dense core, and a sabot or cup with two sintered iron rotating bands. The sabot fits over the steel sheath and upon firing separates from the sheath a short distance from the gun. The dense core, which is the armor-piercing element, is carried within the sheath. The projectile, because of its light weight, has a high muzzle velocity. The projectile is inert except for the tracer in its base.

  e. The high-explosive (HE) projectile (M71) is a conventionally designed nose fuzed shell containing a high-explosive bursting charge. The HE-T, T91 projectile is also a conventionally designed high-explosive projectile but it has a hemispherical base which also contains a tracer. They are intended principally for fragmentation, blast, or mining effect. The projectile may be fitted with point detonating fuze M51A5 or M51A4, concrete-piercing fuze M78A1 or M78, mechanical time fuzes M67A3 or MTSQ M500, TSQ, fuze M55A3, or VT fuze M97. Assembled with a concrete-piercing fuze, this high-explosive shell is effective against concrete and will destroy reinforcing bars and remove debris from within the impact area.

f. The high-explosive with tracer (HE–T) projectile T91 is a conventionally designed nose fuzed high-explosive shell, but it has a hemispherical base and two sintered iron rotating bands. A projection on the base contains the tracer element. The projectile contains a normal cavity and is not designed for use with VT fuzes. The projectile is fuzed with the point detonating fuze M51A5.

g. High-explosive-antitank projectiles are a special type which contain a shaped high-explosive bursting charge. It is designed primarily for penetrating armored targets.

h. The HE, Marker (green, red, or yellow) shell, consists of the standard 90-mm M71 projectile metal parts assembly loaded with an axial cylindrical bursting charge of baratol (barium nitrate—TNT mixture), surrounded by a pressed mixture of a dye. The shell is fuzed with the MTSQ M500 fuze. Functioning of the fuze and booster initiates the bursting charge. The detonation of the bursting charge fragments the shell and disperses the smoke charge, forming a colored smoke burst. The burst is discernible from aircraft several thousand feet above the burst. This shell is used for marking bomb release lines and marking of targets by means of ground impact and for marking and signalling by means of high altitude air burst.

i. The smoke projectile (M313) is a phosphorus filled bursting type shell which has a central tube or burster containing a high explosive. It is intended primarily to produce a screening smoke. This projectile is fitted with the point detonating fuze M48A3 or M48A2. The M57 fuze, which is limited standard, is assembled to some rounds.

j. The smoke with tracer (WP–T) projectile T92 is a conventionally designed, nose fuzed, phosphorus filled bursting type shell which has a central tube or burster. The projectile has a hemispherical base and two sintered iron rotating bands. A projection on the base contains the tracer element. The projectile is fuzed with the point detonating fuze M48A3.

k. Target practice with tracer and hyper velocity target practice with tracer projectiles are respectively of the same size, shape, and weight as the high-explosive and hyper velocity armor-piercing projectiles. Both target practice projectiles are inert except for the tracer on the hyper velocity shot.

l. Blank ammunition contains no projectile and is provided for use in combat vehicles primarily for simulating fire and saluting.

m. Drill ammunition is completely inert and is intended for practice in service of the piece and handling.

380. Identification

a. GENERAL. Ammunition and ammunition components are completely identified by the painting and marking (including an ammunition lot number) on the ammunition items and on all original packing containers. The components of various types of rounds may be identified by the marking thereon and by the color scheme (when employed). Refer to figures 358 through 367.

b. MODEL. To identify a particular design, a model designation is assigned at the time the item is classified as an adopted type. This model designation becomes an essential part of the standard nomenclature and is included in the marking on the item. The present system of model designation consists of the letter "M" followed by an Arabic numeral, for example, "M1." Modifications are indicated by adding the letter "A" and appropriate Arabic numeral. Thus, "M1A1" indicates the first modification of an item for which the original model designation was "M1." Similarly, a system applied to development items involves the use of a "T" designation to indicate the basic design and an "E" to indicate modifications thereof. Thus, T108E11 would indicate the eleventh modification of a development item originally designated T108.

c. AMMUNITION LOT NUMBER. When ammunition is manufactured, an ammunition lot number, which becomes an essential part of the marking, is assigned in accordance with pertinent specifications. The lot number, which consists of the loader's initials or symbol and the number of the lot, is stamped or marked on every loaded item as issued, size permitting, and on all packing containers. It is required for all purposes of record including reports on condition, functioning, or accidents in which the ammunition may be involved. To provide for the most uniform functioning, all of the components in any one lot are manufactured under as nearly identical conditions as practicable. Complete rounds of fixed ammunition of any one lot are similarly manufactured and consist of components, each type of which is of one lot, for example: all of the rounds in any one ammunition lot consist of projectiles of one lot number (one type and weight zone), fuzes of one lot number, propellent powder of one lot number, and primers of one lot number. Hence, insofar as the ammunition is concerned, to obtain the greatest accurracy when firing, successive rounds should be of the same lot number whenever practicable.

d. WEIGHT-ZONE MARKING. When it is not practicable to manufacture projectiles within the narrow weight limits necessary for the required accuracy of fire, they are grouped into weight zones in order that appropriate ballistic corrections, as shown on firing tables, may be applied. The weight zones are indicated on the projectile by means of squares of the same color as the marking. One, two, or three squares are marked on the projectile, dependent upon its weight as loaded. Two squares indicate "standard" or "normal" for which no weight corrections are necessary when computing firing data. An exception to the above is the HE, Marker shell T49 which has four zones. Armor-piercing projectiles (all types) and HE, AT projectiles do not have weight-zone markings.

e. PAINTING. Artillery projectiles are painted primarily to prevent rust and to provide, by the color, a ready means of identification as to type. Lusterless paint is used to meet requirements for camouflage. The color scheme is as follows:

Armor-piercing shot	Black, marking in white.
High-explosive; HE, Marker; armor-piercing with explosive filler.	Olive drab, marking in yellow.
Smoke	Gray; one yellow band indicates smoke. Marking on the ammunition is in the same color as the band.
Practice	Blue, marking in white.
Drill (dummy)	Unpainted when made of bronze; otherwise black, marking in white.

f. MARKING.
(1) *On the projectile (stenciled):*
Caliber and type of cannon in which fired.
Kind of filler, for example, "COMP B," "TNT," "HE, MARKER (Green, Red, or Yellow)," "EXP D," "WP SMOKE," as applicable.
Model of projectile.
When the projectile contains a tracer or supplementary charge, it is marked: "_____ T" or "W/SUPPL CHG _____" whichever is applicable.
Weight-zone marking, when required.
Lot number of the filled projectile. Ordinarily, the projectile lot number is not required after the round is assembled. Hence, it is stenciled below the rotating band, in which position it is hidden from view by the neck of the cartridge case.
Shell loaded with WP smoke, in addition to the stenciling to indicate the filler, will have a yellow band.

(2) *On the cartridge case:*
(*a*) *Stamped in the metal.*
Caliber and model of cartridge case.
Cartridge case lot number, including initials or symbol of the cartridge case manufacturer.
Year of manufacture.

(*b*) *Stenciled.*

Ammunition lot number and loader's initials.
This number is marked on HVAP-T, HVTP-T, and HVAP-DS-T rounds in red; black is used on all others.

Burning characteristics of the propelling charge such as "FLASHLESS," "SMOKELESS," or "FLASHLESS-SMOKELESS" are stenciled on the base of the cartridge case. When space is limited, the abbreviations "FLHLS," "SMKLS," or "FLHLS-SMKLS" are used as applicable. These are marked in red on HVAP-T, HVTP-T, and HVAP-DS-T rounds; black is used on all others.

HVAP-T, HVTP-T, and HVAP-DS-T rounds have the word "HYPERVELOCITY" and the muzzle velocity stenciled in red on the side of the cartridge case.

(3) *On the fuze* (*stamped in the metal unless otherwise indicated*):

Type and model of fuze.
Loader's lot number, including loader's initials.
Year of manufacture.
Action—this includes: "SQ," "DELAY," the length of delay as ".05 sec" or time limit of time fuzes by a graduated scale (time ring). The tip of the *nondelay* CP fuze M78A1 and M78 is painted white.

381. Care, Handling, and Preservation

Warning: Explosive ammunition or components containing explosives must be handled with appropriate care at all times. The explosive elements in primers and fuzes are particularly sensitive to undue shock and high temperature. Explosive ammunition and components should not be dropped, thrown, tumbled, or dragged.

a. Ammunition is packed to withstand conditions ordinarily encountered in the field. Care must be observed to keep packings from becoming broken or damaged. All broken packings must be repaired immediately and careful attention given to the transfer of all markings to the new parts. Complete rounds are shipped in individual moisture-resistant fiber containers inclosed in a wooden packing box, or in individual metal containers when the ammunition is to be shipped to theaters where excessively humid conditions prevail.

b. When it is necessary to leave ammunition in the open, raise it on dunnage at least 6 inches from the ground and cover it with a double thickness of paulin, leaving enough space for the circulation of air. Where practicable, dunnage strips should be placed under each layer of ammunition boxes and other ammunition components. Suitable trenches should be dug to prevent water from running under the pile.

c. Since explosives are adversely affected by moisture and high temperature, due consideration should be given to the following:
  (1) Do not break the moisture-resistant seals until ammunition is to be used. Ammunition removed from airtight containers, particularly in damp climates, is apt to corrode, thereby rendering the ammunition unserviceable.
  (2) Protect ammunition particularly fuzes, from high temperature and direct rays of the sun. More uniform firing is obtained if the rounds are at the same temperature.
  (3) Whenever practicable, WP loaded shell should be stored at temperatures below the melting point (105° F.) of the WP filler. If this is not practicable, WP rounds should be stored on their bases so that, should the WP filler melt, it will resolidify with the void space in its normal position in the nose of the shell, when the temperature falls below the melting point of the filler. Prematures have been caused by voids in the base end of WP shell and erratic performance may result from voids in the side.

d. Do not attempt to disassemble the complete round or any of its components.

e. Do not remove protective or safety devices from fuzes until just before use.

Caution: Do not attempt to disassemble any fuze.

f. Before loading the complete round into the weapon, each of the components should be free of foreign matter, sand, mud, moisture, frost, snow, ice, oil, and grease.

g. Blank ammunition with loose or broken closing cup or plug will not be used but will be reported to the ordnance officer for disposition.

h. Brass cartridge cases are easily dented and should be protected from hard knocks and blows.

i. Each round to bel oaded, which contains a time or point detonating fuze, should be kept well out of the path of recoil of the gun. If the fuze on a round is accidentally hit by recoil, it is not to be fired under any circumstances but will be placed immediately in a segregated location and reported to the local ordnance officer for examination and necessary action.

j. Do not handle duds. Because their fuzes may be armed, duds are extremely dangerous. They will not be moved or touched but will be destroyed in a place in accordance with TM 9–1900. Unlike other fuzes, VT fuzes may be considered safe for handling 24 hours after the projectile has been fired, but they should be handled with care since they contain an unignited powder train and booster charge.

k. The following pertains to the storage and handling of VT fuzes:
  (1) In addition to the usual care in handling explosive ammunition, VT fuzes and VT-fuzed rounds should be protected from long exposure to high humidity, temperature below

−20° and above +130° F., and excessive jolting and shocks, which may result in improper fuze action.

(2) In temperate climates, the fuzes may be expected to remain serviceable for 2 months after removal from their original sealed containers. The fuzes should not be removed from their original sealed containers, particularly in tropical climates, until just before use. Exposure to rain or immersion in water will result in accelerated deterioration.

(3) VT fuzes will withstand normal handling without danger of detonation or damage when in their original packing containers or when assembled to projectiles. However, care should be taken not to strike or drop fuzes or fuzed rounds as these actions may increase the number of duds. A drop of 4 feet may cause the electrolyte vial in the fuze battery to break, thus creating a dud. Excessive rough handling will not decrease fuze safety but may increase the number of duds.

(4) Supplementary charges which have been removed from shell, prior to assembling VT fuzes, will be packed in the containers from which the VT fuzes have been removed. The containers should be properly marked and returned to ordnance personnel for disposition.

(5) VT-fuzed ammunition may be safely transported short distances, with normal care and handling. However, when such ammunition is to be transported considerable distances, it is advisable to remove the fuze from the shell and return the fuze to its original marked container. The supplementary charge and original fuze or closing plug (with gasket and spacer) should be reassembled to the shell, making certain that the supplementary charge is properly inserted (felt-pad end innermost).

(6) If projectiles on which fuzes have been changed are returned to their original packing, care must be taken to change markings on the containers and boxes to show the fuze actually on the projectile.

(7) Rounds fuzed with VT fuzes must be adequately padded if temporarily placed in fiber containers. The U-shaped support which engages the wrench slots of nonproximity point fuzes will not fit the VT fuze wrench slots and should, therefore, be omitted. The play that results is taken up by placing extra corrugated cardboard pads under the base end of the projectile before closing the container.

(8) VT fuzes of certain lots, as issued, have a wax coating on the plastic ogive This wax coating is necessary for the proper functioning of the fuzes of such lots. Removal of this coating will usually result in malfunctioning of the fuze. VT

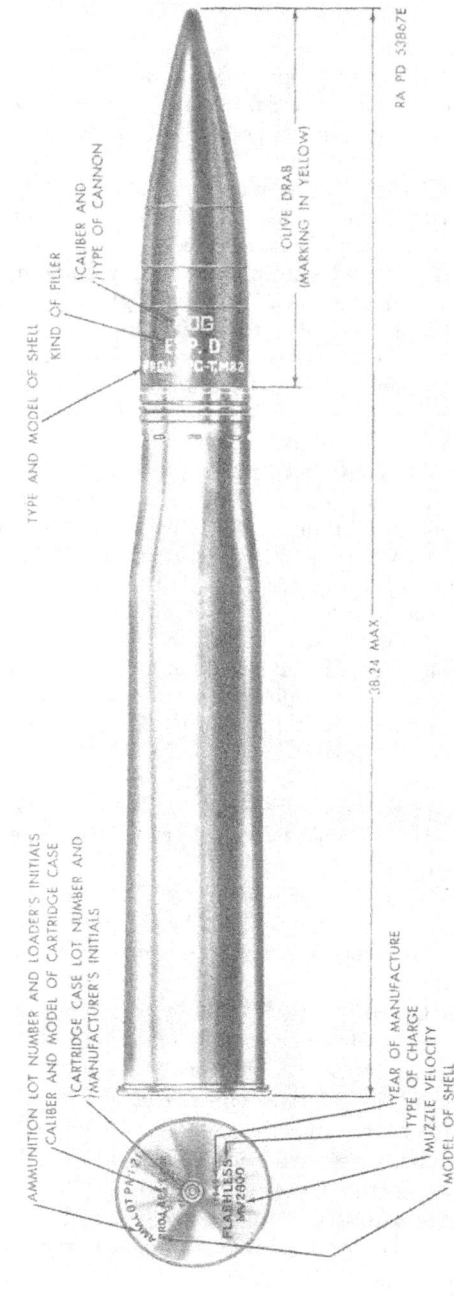

Figure 358. Projectile, fixed, APC-T, M82, MV 2,800, flashless, for 90-mm guns M1, M2, M3, and T8.

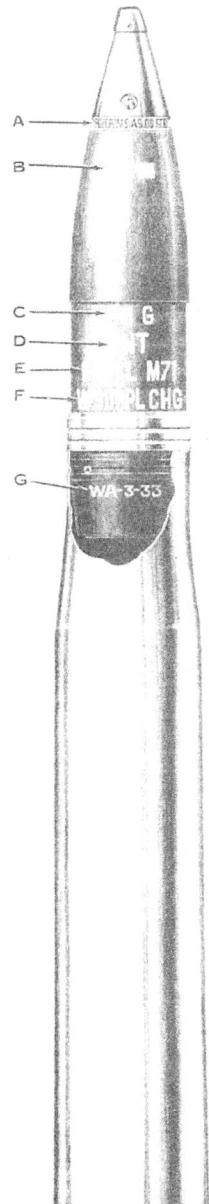

- A — TYPE, MODEL, AND ACTION OF FUZE
- B — WEIGHT ZONE MARKING.
- C — CALIBER AND DESIGNATION OF WEAPON
- D — KIND OF FILLER.
- E — TYPE AND MODEL OF SHELL.
- F — FOR DEEP-CAVITY SHELL CONTAINING SUPPLEMENTARY CHARGE. FOR DEEP CAVITY SHELL W/O FUZE AND W/O SUPPLEMENTARY CHARGE, "FOR VT FUZE" OR "FOR FUZE M___".
- G — LOT NUMBER OF FILLED SHELL.
- H — AMMUNITION LOT NUMBER. AN X AFTER THE SERIAL NUMBER INDICATES STEEL CARTRIDGE CASE.
- I — CALIBER AND MODEL OF CASE.
- J — LOT NUMBER OF CASE
- K — YEAR OF MANUFACTURE.
- L — PERFORMANCE OF ROUND UPON FIRING. AS FLASHLESS, OR FLHLS SMOKELESS, OR SMKLS FLASHLESS-SMOKELESS, OR FLHLS-SMKLS.
- M — TYPE AND MODEL OF SHELL.

Figure 359. Shell, fixed, HE, M71, smokeless, w/suppl CHG and fuze, PD, M51A5, 0.05-sec delay, for 90-mm guns M1, M2, M3, and T8.

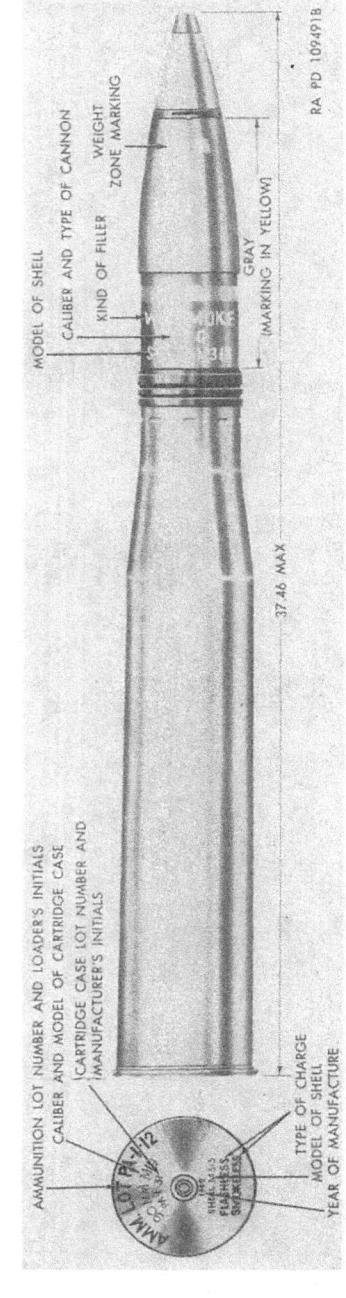

Figure 360. Shell, fixed, smoke, WP, M313, smokeless, w/fuze, PD, M48A3, 0.05-sec delay, for 90-mm guns M1, M2, M3, and T8.

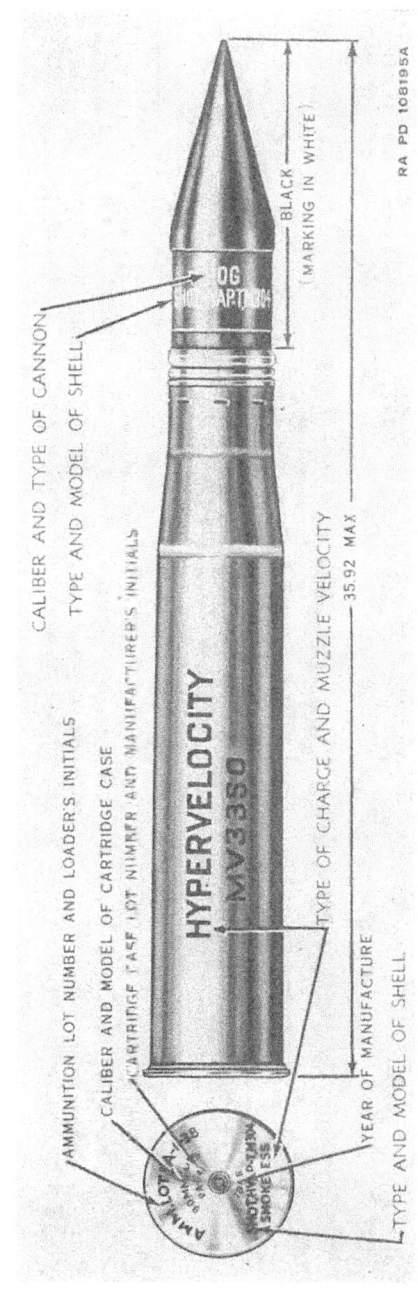

Figure 361. Shot, fixed, HVAP–T, M304 (T30E16), smokeless, for 90-mm guns M1, M2, M3, and T8.

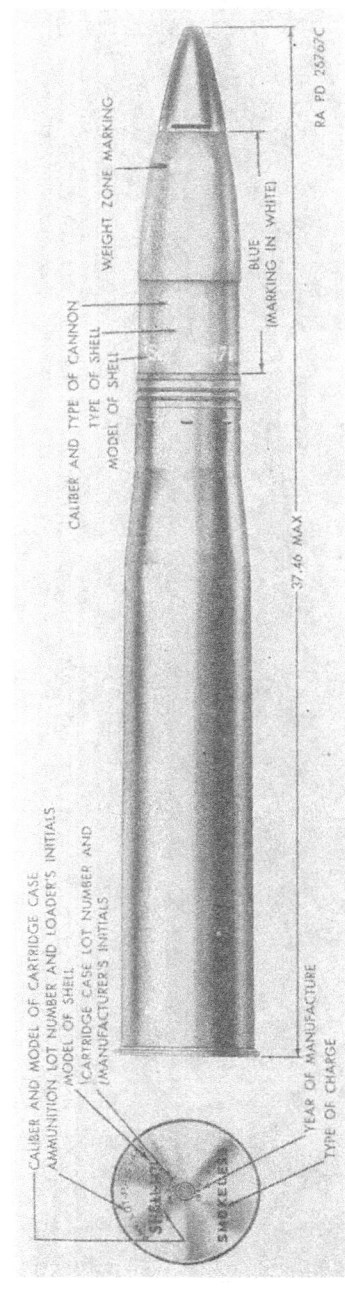

Figure 362. Shot, fixed, TP, M71, smokeless, w/fuze, dummy, M73, for 90-mm guns M1, M2, M3, and T8.

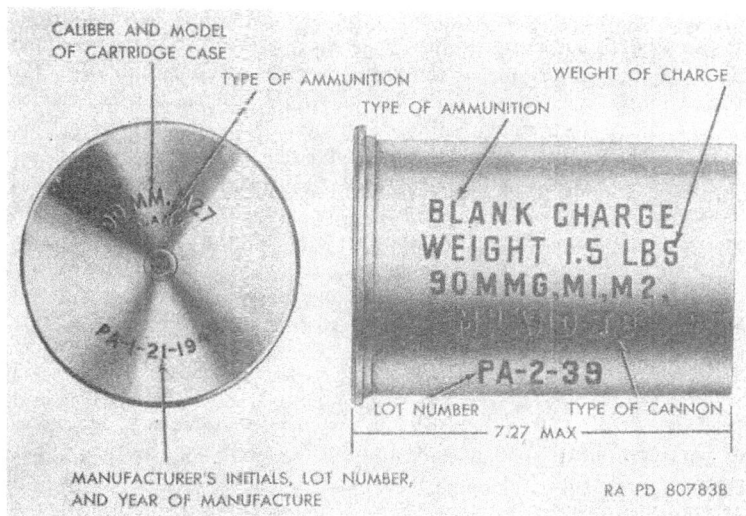

Figure 363. Ammunition, blank, for 90-mm guns M1, M2, M3, and T8.

fuzes should be used as issued; that is, with the wax coating on the plastic ogive if so issued, or without a wax coating if issued without the coating.

(9) In view of the security classification of VT fuzes, special precautions will be observed in connection with their storage and use to prevent any possible breach of security.

382. Authorized Rounds

a. Ammunition authorized for use in the 90-mm gun M3A1 is also authorized for use in the T119E1 gun (refer to table IX). However, rounds authorized for use only in the T119E1 gun must not be used in the M3A1 gun or other models of 90-mm guns having the same chamber as the M3A1 gun. The table of authorized rounds (table IX) is given in two parts; the first part consists of items authorized for use only in the T119E1 gun; the second part consists of items authorized for use in the M3A1 and T119E1 guns.

b. Ordinarily, issue of this ammunition is in proportion by types to meet tactical requirements so that substitution of fuzes in the field is not required. For special purposes, certain substitutions are authorized as indicated in table IX. To prepare shell for firing with substitute fuzes, refer to paragraph 383.

c. Some of the HE shells M71 are loaded with a deep-fuze cavity provided for insertion of VT fuzes. These shells may also be used with fuzes fitted with M21A4 type boosters if a supplementary charge of TNT is first inserted into the fuze cavity. Rounds assembled with this type shell are now shipped completely assembled with a supplementary charge and one of the point fuzes as indicated in table IX. Such rounds are marked "W/SUPPL CHG" on the shell. Earlier shipments of deep-cavity shell were made with or without supplementary charge but without fuze, a closing plug being used in lieu of the fuze. Such shells are marked "FOR VT FUZE." The CP fuzes M78A1 and M78, and booster M25, or the MTSQ fuze M500, or the mechanical time fuze may be used in the 90-mm HE shell M71; these fuzes are issued separately for assembly to projectiles in the field. Some of these HE shells are issued assembled with the MTSQ M500 or mechanical time M67A3 fuzes.

383. Preparation for Firing

a. GENERAL. After removal from packing materials, fuzed rounds of fixed ammunition for the 90-mm gun T119E1 are ready for firing except for setting the point fuzes (par. 384) to obtain the desired action, or for substitution of fuzes (*b*(1) below). Unfuzed rounds require the addition of a fuze (*b*(1) *below*). Armor-piercing rounds require no preparation for firing after removal from packing materials. Rounds and their components prepared for firing but not fired will be returned to their original condition and packings and be appropriately marked. Such rounds will be used first in subsequent firing in order that stocks of opened packings may be kept to a minimum.

Note. Upon removing a point-fuzed round from its container, withdraw the U-shaped packing stop from the wrench slots in the fuze. Serious damage to the fuze setter or gun tube may result if this stop is not removed.

b. PREPARATION OF HE SHELL.

(1) *HE shell M71*. HE rounds to be fired with their original fuzes require only the adjustment of fuzes described in paragraph 384. Rounds to be fuzed or to be fitted with substitute fuzes are prepared for firing as indicated in table X.

(2) *HE–T shell T91*. This round requires the adjustment of its fuze to be prepared for firing. Substitute fuzes at the present time are not authorized.

384. Fuzes

a. GENERAL. A fuze is a device used with a projectile to explode it at the time and under the circumstances required.

b. CLASSIFICATION. Fuzes are classified according to their manner of functioning as "time and impact," or "time," or "impact." Impact fuzes function upon striking a resistant object. They are classified, according to the time of functioning after impact, as *superquick*, *nondelay*, or *delay*. Point detonating (PD) and base detonating (BD) fuzes, which are impact fuzes, and are so called because of their location on a projectile. Time fuzes are designed to function primarily

Table IV.

PART I. AUTHORIZED ROUNDS FOR 90-MM GUN T119E1
(Not to be Used in M1, M2, M3, or T8 Guns)

Standard nomenclature	Complete round Weight (lb)	Complete round Length (in)	Projectile weight as fired (lb)	Action of fuze	Special purpose of substitute fuzes
Service Ammunition					
CARTRIDGE, AP-T, T33E4, smokeless, MV 3,050, for 90-mm gun T119.	44.75	37.43	23.82	None	None.
CARTRIDGE, HE-T, COMP B, T91, flashless-smokeless, w/fuze, PD, M51A5, 0.05-sec delay, for 90-mm gun T119.	------	37.47	18.05	SQ and 0.05 sec delay.	None.
CARTRIDGE, HVAP-T, M332 (T67), smokeless, MV 4,100, for 90-mm gun T119.	32.9	35.93	12.2	None	None.
CARTRIDGE, HVAP-DS-T, T65, MV 4,100, for 90-mm gun T119	------	36.89	18.03	SQ and 0.05 sec delay.	None.
CARTRIDGE, smoke, WP-T, T02, flashless-smokeless, w/fuze, PD, M48A3, 0.05-sec delay, for 90-mm gun T119.	32.21				
Target Practice					
CARTRIDGE, HVTP-T, M333 (T83), smokeless, MV 4,100, for 90-mm gun T119.	32.9	35.93	12.2	None	None.
CARTRIDGE, TP-T, T225, smokeless, MV 3,050, for 90-mm gun T119.	44.86	37.43	24.21	None	None.

Table IX.—Continued

PART II. AUTHORIZED ROUNDS FOR 90-MM GUNS M1, M2, M3, T8, AND T119E1

Standard nomenclature	Complete round		Projectile weight as fired (lb)	Action of fuze	Special purpose or substitute fuzes
	Weight (lb)	Length (in)			
Service Ammunition					
CARTRIDGE, HE, AT, T108 (All Mods) w/fuze, PI, T209, for 90-mm guns M1, M2, M3, and T8.	34.5	36.97	14.3	Nondelay	None.
PROJECTILE, fixed, APC–T, M82, MV 2,600, flashless, w/fuze, BD, M68, for 90-mm guns M1, M2, M3, and T8.	43.87	38.24	24.11	Delay	None.
PROJECTILE, fixed, APC–T, M82, MV 2,800, flashless, w/fuze, BD, M68, for 90-mm guns M1, M2, M3, and T8.	43.87	38.24	24.11	Delay	None.
PROJECTILE, fixed, APC–T, M82, MV 2,600, smokeless, w/fuze, BD, M68, for 90-mm guns M1, M2, M3, and T8.	43.87	38.24	24.11	Delay	None.
PROJECTILE, fixed, APC–T, M82, MV 2,800, smokeless, w/fuze, BD, M68, for 90-mm guns M1, M2, M3, and T8 (double base powder used).	43.87	38.24	24.11	Delay	None.
SHELL, fixed, HE, COMP B, M71, flashless, w/suppl chg, and fuze, PD, M51A5, 0.05-sec delay, for 90-mm guns M1, M2, M3, and T8.	42.17	37.44	23.24	SQ and 0.05 sec. delay.	CP, M78A1 or M78; time, mechanical, M67A3; TSQ, M55A3; MTSQ, M500.

Ammunition				Fuzes
SHELL, fixed, HE, COMP B, M71, smokeless, w/suppl chg, and fuze, PD, M51A5, 0.05-sec delay, for 90-mm guns M1, M2, M3, and T8.	42.17	37.44	23.29	SQ and 0.05 sec. delay.	CP M78A1 or M78; time, mechanical, M67A3; TSQ, M55A3; MTSQ, M500.
SHELL, fixed, HE, Green Marker, M71, flashless, w/fuze, MTSQ, M500, for 90-mm guns M1, M2, M3, and T8.[a]	41.8	37.44	23.5	Time (to 75 sec.) and SQ.	Time mechanical, M67A3.
SHELL, fixed, HE, Red Marker, M71, flashless, w/fuze, MTSQ, M500, for 90-mm guns M1, M2, M3, and T8.[a]	41.8	37.44	23.5	Time (to 75 sec.) and SQ.	Time mechanical, M67A3.
SHELL, fixed, HE, Yellow Marker, M71, flashless, w/fuze, MTSQ, M500, for 90-mm guns M1, M2, M3, and T8.[a]	42.0	37.44	23.7	Time (to 75 sec.) and SQ.	Time mechanical, M67A3.
SHELL, fixed, HE, M71, flashless, w/fuze, PD, M48A2, 0.15-sec delay, for 90-mm guns M1, M2, M3, and T8.	41.84	37.46	23.40	SQ and 0.15 sec. delay.	CP M78 or M78A1; time, mechanical, M67A3; TSQ M55A3.
SHELL, fixed, HE, M71, flashless, w/fuze, PD, M51A4 (M48A2), 0.05-sec delay, for 90-mm guns M1, M2, M3, and T8.	41.93	37.46	23.40	SQ and 0.05 sec. delay.	CP M78 or M78A1; time, mechanical, M67A3; TSQ M55A3.
SHELL, fixed, HE, M71, flashless, w/fuze, PD, M51A4 (M48A2), 0.15-sec delay, for 90-mm guns M1, M2, M3, and T8.	41.93	37.46	23.40	SQ and 0.15 sec. delay.	CP M78 or M78A1; time, mechanical, M67A3; TSQ M55A3.
SHELL, fixed, HE, M71, flashless, w/fuze, PD, M51A5, 0.05-sec delay, for 90-mm guns M1, M2, M3, and T8.	41.84	37.46	23.40	SQ and 0.05 sec. delay.	CP M78 or M78A1; time, mechanical, M67A3; TSQ M55A3.
SHELL, fixed, HE, M71, flashless, w/suppl chg, and fuze, PD, M51A4 (M48A2), 0.05-sec delay, for 90-mm guns M1, M2, M3, and T8.	41.84	37.46	23.40	SQ and 0.05 sec. delay.	CP M78 or M78A1; time, mechanical, M67A3; TSQ M55A3; VT M92, M93, M97.
SHELL, fixed, HE, M71, flashless, w/suppl chg, and fuze, PD, M51A4 (M48A2), 0.15-sec delay, for 90-mm guns M1, M2, M3, and T8.	41.84	37.46	23.40	SQ and 0.15 sec. delay.	CP M78 or M78A1; time, mechanical, M67A3; TSQ M55A3; VT M92, M93, M97.

See footnotes at end of table.

Table IX.—Continued

PART II. AUTHORIZED ROUNDS FOR 90-MM GUNS M1, M2, M3, T8, AND T119E1—Continued

Standard nomenclature	Complete round		Projectile weight as fired (lb)	Action of fuze	Special purpose or substitute fuzes
	Weight (lb)	Length (in)			
Service Ammunition—Continued					
SHELL, fixed, HE, M71, flashless, w/suppl chg, and fuze, PD, M51A5, 0.05-sec delay, for 90-mm guns M1, M2, M3, and T8.	41.84	37.46	23.40	SQ and 0.05 sec. delay.	CP M78 or M78A1; time, mechanical, M67A3; TSQ M55A3; VT M92, M93, M97.
SHELL, fixed, HE, M71, flashless, w/suppl chg, w/o fuze, for 90-mm guns M1, M2, M3, and T8.	40.08	b 35.00	See footnote b.	See footnote c.	See footnote c.
SHELL, fixed, HE, M71, flashless, w/o fuze, for VT fuze, for 90-mm guns M1, M2, M3, and T8.	40.08	b 35.00	23.40	VT	See footnote c.
SHELL, fixed, HE, M71, flashless-smokeless, w/suppl chg, and fuze, MTSQ, M500, for 90-mm guns M1, M2, M3, and T8.	d 41.84	d 37.44	d 23.20	Time (to 75 sec.) and SQ.	CP M78 or M78A1; time, mechanical, M67A3; TSQ M55A3; VT M92, M93, M97.
SHELL, fixed, HE, M71, flashless-smokeless, w/suppl chg, and fuze, MTSQ, M502, for 90-mm guns M1, M2, M3, and T8.	41.84	37.44	23.20	Time (to 30 sec.) and SQ.	CP M78 or M78A1; time, mechanical, M67A3; VT M92, M93, M97.
SHELL, fixed, HE, M71, flashless-smokeless, w/fuze, PD, M51A4, 0.05-sec delay, for 90-mm guns M1, M2, M3, and T8.	41.93	37.46	23.20	SQ and 0.05 sec. delay.	CP M78 or M78A1; time, mechanical, M67A3; TSQ M55A3.
SHELL, fixed, HE, M71, flashless-smokeless, w/fuze, PD, M51A4, 0.15-sec. delay, for 90-mm guns M1, M2, M3, and T8.	41.93	37.46	23.20	SQ and 0.15 sec. delay.	CP M78 or M78A1; time mechanical, M67A3; TSQ M55A3.

612

Ammunition				
SHELL, fixed, HE, M71, flashless-smokeless, w/fuze, PD, M51A5, 0.05-sec delay, for 90-mm guns M1, M2, M3, and T8.	41.93	37.46	23.20	SQ and 0.05 sec. delay.	CP M78 or M78A1; time, mechanical, M76A3; TSQ M55A3.
SHELL, fixed, HE, M71, flashless-smokeless, w/fuze, TSQ M55A3, for 90-mm guns M1, M2, M3, and T8.	41.93	37.44	23.40	Time (to 25 sec.) and SQ.	CP M78 or M78A1; PD M51A4 (M48A2) or M51A5; time, mechanical M67A3.
SHELL, fixed, HE, M71, flashless-smokeless, w/suppl chg, and fuze, PD, M51A4, 0.05-sec delay, for 90-mm guns M1, M2, M3, and T8.	41.84	37.46	23.20	SQ and 0.05 sec. delay.	CP M78 or M78A1; time, mechanical, M67A3; TSQ M55A3; VT M92, M93, M97.
SHELL, fixed, HE, M71, flashless-smokeless, w/suppl chg, and fuze, PD, M51A4, 0.15-sec delay, for 90-mm guns M1, M2, M3, and T8.	41.84	37.46	23.20	SQ and 0.15 sec. delay.	CP M78 or M78A1; time, mechanical, M67A3; TSQ M55A3; VT M92, M93, M97.
SHELL, fixed, HE, M71, flashless-smokeless, w/suppl chg, and fuze, PD, M51A5, 0.05-sec delay, for 90-mm guns M1, M2, M3, and T8.	41.84	37.46	23.20	SQ and 0.15 sec. delay.	CP M78 or M78A1; time, mechanical, M76A3; TSQ M55A3; VT M92, M93, M97.
SHELL, fixed, HE, M71, flashless-smokeless, w/suppl chg, w/o fuze, for 90-mm guns M1, M2, M3, and T8.	41.42	b 35.00	See footnote.	See footnote.	See footnote e.
SHELL, fixed, HE, M71, smokeless, w/fuze, MTSQ M500, for 90-mm guns M1, M2, M3, and T8.	d 41.84	d 37.44	d 23.20	Time (to 75 sec.) and SQ.	CP M78 or M78A1; PD M51A4 (M48A2) or M51A5; time, mechanical, M67A3; TSQ M55A3.
SHELL, fixed, HE, M71, smokeless, w/fuze, PD, M51A4, 0.05-sec delay, for 90-mm guns M1, M2, M3, and T8.	41.93	37.46	23.20	SQ and 0.05 sec. delay.	CP M78 or M78A1; time, mechanical, M67A3; TSQ M55A3.
SHELL, fixed, HE, M71, smokeless, w/fuze, PD, M51A4, 0.15-sec delay, for 90-mm guns M1, M2, M3, and T8.	41.93	37.46	23.20	SQ and 0.15 sec. delay.	CP M78 or M78A1; time, mechanical, M67A3; TSQ M55A3.
SHELL, fixed, HE, M71, smokeless, w/fuze, PD, M51A5, 0.05-sec delay, for 90-mm guns M1, M2, M3, and T8.	41.93	37.46	23.20	SQ and 0.05 sec. delay.	CP M78 or M78A1; time, mechanical, M67A3; TSQ M55A3.

See footnotes at end of table.

Table IX.—Continued

PART II. AUTHORIZED ROUNDS FOR 90-MM GUNS M1, M2, M3, T8, AND T119E1—Continued

Standard nomenclature	Complete round		Projectile weight as fired (lb)	Action of fuze	Special purpose or substitute fuzes
	Weight (lb)	Length (in)			
Service Ammunition—Continued					
SHELL, fixed, HE, M71, smokeless, w/fuze, TSQ M55A3, for 90-mm guns M1, M2, M3, and T8.	41.93	37.44	23.40	Time (to 25 sec) and SQ.	CP M78 or M78A1; PD M51A4 (M48A2) or M51A5; time, mechanical, M67A3.
SHELL, fixed, HE, M71, smokeless, w/suppl chg, and fuze, PD, M51A4, 0.05-sec delay, for 90-mm guns M1, M2, M3, and T8.	41.84	37.46	23.20	SQ and 0.05 sec delay.	CP M78 or M78A1; time, mechanical, M67A3; TSQ M55A3; VT M92, M93, M97.
SHELL, fixed, HE, M71, smokeless, w/suppl chg, and fuze, PD, M51A4, 0.15-sec delay, for 90-mm guns M1, M2, M3, and T8.	41.93	37.46	23.20	SQ and 0.15 sec delay.	CP M78 or M78A1; time, mechanical, M67A3; TSQ M55A3; VT M92, M93, M97.
SHELL, fixed, HE, M71, smokeless, w/suppl chg, and fuze, PD, M51A5, 0.05-sec delay, for 90-mm guns M1, M2, M3, and T8.	41.93	37.46	23.20	SQ and 0.05 sec delay.	CP M78 or M78A1; time, mechanical, M67A3; TSQ M55A3; VT M92, M93, M97.
SHELL, fixed, smoke, WP, M313, flashless-smokeless, w/fuze, PD, M48A2, 0.05-sec delay, for 90-mm guns M1, M2, M3, and T8.	42.04	37.46	23.40	SQ and 0.05 sec delay.	None.
SHELL, fixed, smoke, WP, M313, flashless-smokeless, w/fuze, PD, M48A3, 0.05-sec delay, for 90-mm guns M1, M2, M3, and T8.	42.04	37.46	23.40	SQ and 0.05 sec delay.	None.
SHELL, fixed, smoke, WP, M313, flashless-smokeless, w/fuze, PD, M57, for 90-mm guns M1, M2, M3, and T8.	42.04	37.46	23.40	SQ	None.

SHELL, fixed, smoke, WP, M313, smokeless, w/fuze, PD, M48A2, 0.05-sec delay, for 90-mm guns M1, M2, M3, and T8.	42.04	37.46	23.40	SQ and 0.05 sec delay.	None.
SHELL, fixed, smoke, WP, M313, smokeless, w/fuze, PD, M48A3, 0.05-sec delay, for 90-mm guns M1, M2, M3, and T8.	42.04	37.46	23.40	SQ and 0.05 sec delay.	None.
SHELL, fixed, smoke, WP, M313, smokeless, w/fuze, PD, M57, for 90-mm guns M1, M2, M3, and T8.	42.04	37.46	23.40	SQ	None.
SHELL, fixed, smoke-HE (See under SHELL, fixed, HE, Green (Red or Yellow) Marker).[a]					
SHOT, fixed, AP–T, M77, flashless, for 90-mm guns M1, M2, M3, and T8.[d][f]	42.04	32.75	23.40	None	None.
SHOT, fixed, AP–T, M77, smokeless, for 90-mm guns M1, M2, M3, and T8.[d][f]	42.04	32.75	23.40	None	None.
SHOT, fixed, AP–T, M318 (T33), flashless, for 90-mm guns M1, M2, M3, and T8.	43.98	37.43	24.15	None	None.
SHOT, fixed, HVAP–T, M304 (T30E16), smokeless, for 90-mm guns M1, M2, M3, and T8.	37.13	35.92	16.80	None	None.
SHOT, fixed, HVAP–T, M332 (T67), smokeless, MV 3,900, for 90-mm guns M1, M2, M3, and T8.	32.3	35.92	12.2	None	None.
SHOT, fixed, HVAP–T, T30E15, smokeless, for 90-mm guns M1, M2, M3, and T8.	37.13	35.92	16.80	None	None.

See footnotes at end of table.

Table IX.—Continued

PART II. AUTHORIZED ROUNDS FOR 90-MM GUNS M1, M2, M3, T8, AND T119E1—Continued

Standard nomenclature	Complete round		Projectile weight as fired (lb)	Action of fuze	Special purpose or substitute fuzes
	Weight (lb)	Length (in)			
Practice Ammunition					
SHELL, fixed, TP, M71, flashless, w/fuze, dummy, M73, for 90-mm guns M1, M2, M3, and T8.	42.04	37.44	23.40	Inert	See footnote g.
SHELL, fixed, TP, M71, smokeless, w/fuze, dummy, M73, for 90-mm guns M1, M2, M3, and T8.	42.04	37.44	23.40	Inert	See footnote g.
SHOT, fixed, HVTP-T, M317A1, smokeless, for 90-mm guns M1, M2, M3, and T8.	37.49	35.92	16.62	None	None.
SHOT, fixed, HVTP-T, M333 (T83), smokeless, MV 3,900, for 90-mm guns M1, M2, M3, and T8.	32.3	35.92	12.2	None	None.
Blank Ammunition					
AMMUNITION, blank, for 90-mm guns M1, M2, M3, and T8.	8.23	7.27	None	None	None.

Drill Ammunition

CARTRIDGE, drill, M12B2, w/fuze, time, mechanical, M43A2, inert, for 90-mm guns M1, M2, M3, and TS.	42.04	37.44	See footnote h.	Inert	None.

APC-T	—armor-piercing with cap and tracer.
AP-T	—armor-piercing with tracer.
DD	—base detonating.
COMP B	—a high explosive.
CP	—concrete-piercing.
HE	—high-explosive.
HE, AT	—high-explosive, antitank.
HE-T	—high-explosive with tracer.
HVAP-DS-T	—hyper velocity armor-piercing-discarding sabot with tracer.
HVAP-T	—hyper velocity armor-piercing with tracer.
HVTP-T	—hyper velocity target practice with tracer.
MTSQ	—mechanical time superquick.
MV	—muzzle velocity.
PD	—point detonating.
sec	—second.
SQ	—superquick.
suppl chg	—supplementary charge.
TP	—target practice.
TP-T	—target practice with tracer.
TSQ	—time and super-quick.

VT	—proximity.
w/	—with.
w/o	—without.
WP	—white phosphorus.
WP-T	—white phosphorus with tracer.

ª Round formerly designated SHELL, fixed, smoke—HE, green (red or yellow).

ᵇ If alternate type closing plug is used, length becomes 36.50 inches.

ᶜ With supplementary charge in place, the round may be fuzed with standard type fuzes CP M78A1, M78, PD M51A5, M51A4 (M48A2), TSQ M55A3, time, mechanical, M67A3 or MTSQ M500, supplementary charge removed, the round may be fitted with VT fuze M97. Weight of projectile as fired with VT fuze is 23.40 pounds—with standard type fuze, 23.17 pounds.

ᵈ Estimated.

ᵉ Basically a smoke round but contains an HE burster.

ᶠ For training purposes only.

ᵍ May be fitted, as shipped, with FUZE, dummy, M73, with an inert fuze of the M51 series, or with FUZE, dummy, M44A2 assembled with an inert booster of the M20 or M21 series.

ʰ Round is not fired.

617

Table X. *Preparation of 90-mm HE Shell M71 for Firing With Special Purpose or Substitute Fuzes*

	90-MM HE SHELL M71 AS ISSUED									
	Fuzed [1]				Unfuzed [2]					
	With standard fuze cavity		With deep cavity and supplementary charge			With deep cavity and supplementary charge			With deep cavity but without supplementary charge		
	CP M78A1 or M78	MTSQ, M500, TIME, mechanical M67A3	CP M78A1 or M78	MTSQ, M500, TIME, mechanical M67A3	VT M97	Fuzes with M21A4 boosters [3]	CP M78A1 or M78	VT M97	Fuzes with M21A4 boosters [3]	CP M78A1 or M78	VT M97
	MODEL OF FUZE TO BE ASSEMBLED TO SHELL										
1. Place round to be refused on its side—*protect primer and cartridge case from being struck or damaged.* Loosen booster set screw, if present	R	R	R	R	R	R	R	R	R	R	R
2. Remove fuze or nose plug from shell as issued with appropriate fuze wrench.[4]	R	R	R	R	R	R	R	R	R	R	R
3. If booster remains in shell, remove with fuze wrench M16	R	R	R	R	R	R	NR	NR	NR	NR	NR
4. Remove supplementary charge	NR	NR	NR	NR	R	R	NR	R	NR	NR	NR
5. Inspect the cavity for damage. Remove any loose material from the cavity. If the HE filler forming the cavity is broken, reject the shell. Remove any explosive adhering to the threads—*use a non-ferrous or wooden instrument.* Do not use shell or fuzes having damaged threads	R	R	R	R	R	R	R	R	R	R	R
6. Insert supplementary charge—felt pad end innermost	NR	NR	NR	NR	NR	NR	NR	NR	R	R	R

618

7. Assemble booster to shell	[5] R	NR	[5] R	NR	NR	NR	[6] R	[5] R	NR	[6] R	[5] R	NR
8. Assemble fuze to shell by hand and tighten securely with appropriate fuze wrench.[4] Do not hammer on the wrench or use an extension handle. If the fuze cannot be tightened to form a good seat between fuze and shell, reject the component at fault. Do not stake the fuze to the shell	R	[7] R	R	[7] R	R	R	[6][7] R	R	R	[6][7] R	R	R
9. Although booster set screw is not required, if originally present, it should be tightened sufficiently so that the screw does not project above the ogival surface of the shell	R	R	R	R	R	R	R	R	R	R	R	R

R—operation required.
NR—operation not required.

[1] Shell as shipped may be fitted with one of the following fuzes: M48A3, M51A5, M55A3, or M500.
[2] Shell as shipped is fitted with closing plug in place of fuze.
[3] Fuzes with M21A4, booster: M48A3, M51A5, M55A3, M67A3, and M500.
[4] Use fuze wrench M18 for alternative fuzes, fuze wrench M16 for CP fuze M78A1 or M78 and Booster, M25, and fuze wrench M18 for VT fuzes.
[5] Remove safety pin from booster M25, if one is present, and assemble boster to shell, tightening it in place with fuze wrench M16.
[6] Remove safety pin from booster M21A4, if one is present, before assembling to shell. Discard boosters issued without safety pins.
[6] Remove safety pin from booster M21A4, if one is present, before assembling to shell. This applies only in the cases of fuze M48A2. For these fuzes, the booster may first be assembled to the fuze in which case see operation 8. The booster M20A1 may be used as a substitute for the M21A4. All other fuzes in this group are fitted with booster M21A4 as issued.
[7] For fuzes fitted with the booster M21A4, remove safety pin from booster, if one is present, prior to assembling fuze to shell.

619

A — SUPPLEMENTARY CHARGE &
PD FUZE, W/BOOSTER

B — SUPPLEMENTARY CHARGE,
SPACER, AND PLUG

C — PLUG, ALONE

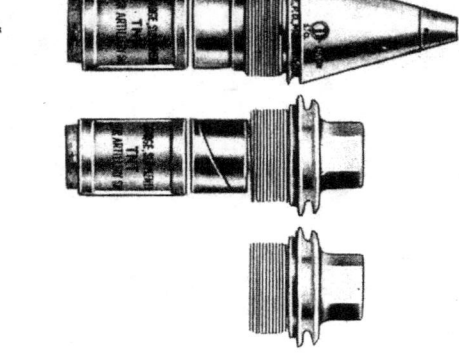

COMPONENTS ISSUED WITH DEEP-CAVITY SHELL

A — SUPPLEMENTARY CHARGE &
PD FUSE, W/BOOSTER

B — SUPPLEMENTARY CHARGE &
CP FUZE, W/BOOSTER

C — VT FUZE

COMPONENT ASSEMBLIES FOR SERVICE USE WITH DEEP-CAVITY SHELL

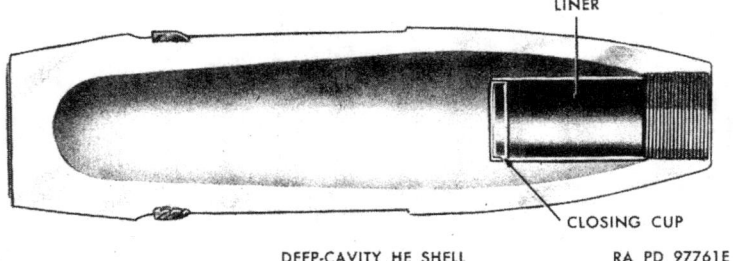

DEEP-CAVITY HE SHELL   RA PD 97761E

Figure 364. Components used with deep-cavity HE shell M71.

M51A5 PD FUZE .05 SEC DELAY

M55A3 TSQ FUZE

M67A3 MECHANICAL TIME FUZE   RA PD 89322B

Figure 365. Fuzes, (M51A5, M55A3, and M67A3).

while the projectile is still in flight; certain time fuzes are also provided with an impact element. These time or airburst fuzes are of three types, mechanical time, powder train time, and VT. Powder train time fuzes differ essentially from mechanical time fuzes in that the former uses a compressed black powder time train to delay functioning for a preset length of time whereas the mechanical time fuze uses a clockwork mechanism to achieve the same result. VT fuzes are radio-actuated point fuzes which function on approach to the target.

c. BORESAFE AND NONBORESAFE. A boresafe (detonator-safe) fuze is one in which the explosive train is so interrupted that, while the projectile is still in the bore of the weapon, premature action of the bursting charge is prevented should any of the more sensitive elements (primer or detonator) function. The fuzes used with the ammunition described herein, except the BD fuze M68, are considered boresafe. However, the M68 fuze may be used under the same conditions as boresafe fuzes.

Caution: Fuzes will not be disassembled. Any attempt to disassembly fuzes in the field is dangerous and is prohibited except under specific directions from the Chief of Ordnance.

d. FUZE PD, M51A5, OR M48A3.

(1) *Description.* The FUZE, PD, M51A5, 0.05-second delay (fig. 365), is an impact type which may be adjusted prior to firing to function with superquick action or with a delay action of 0.05 second. The M51A5 differs from the M48A3 by having a booster (M21A4) assembled to it. The fuzes M48A2 and M51A4, fuzes of earlier manufacture, were made with 0.05- or 0.15-second delay. The delay is marked on the fuze. It should be noted that if the fuze is set for superquick action, and this action fails, the projectile will be detonated with delay action rather than become a dud. The fuze is fitted with the booster M21A4, which is a component of the fuze, and is assembled to it at the time of manufacture. As shipped, the fuze is set for superquick action; that is, the slot in the head of the setting sleeve is parallel to the vertical axis of the fuze and is alined with the index mark for superquick (SQ) action as shown in figure 365. For precautions in firing this fuze, refer to paragraph 385.

(2) *Setting.* To set the fuze for delay action, it is only necessary to turn the slot 90° to aline with the index mark for delay (DELAY) action. The setting may be changed at will (with the screwdriver end of fuze wrench M18 or M7A1 or with a similar tool) at any time before firing; this can be done even in the dark by feeling the position of the slot.

e. FUZE, TSQ, M55A3.

(1) *Description.* This fuze (fig. 365), which is classified as limited standard, is a combination time and superquick type. It is a unit assembly consisting of an M54 fuze and an M21A4 booster. A safety pull wire extends through the fuze to secure the plunger during shipment and handling. The fuze contains two actions, time and superquick. The superquick action is always operative and will function on impact unless prior functioning has been caused by time action; therefore, to set the fuze for superquick action, it is required that the time action be set either at "S" (safe) or for a time longer than the expected time of flight. The time ring is graduated for 25 seconds and contains a compressed black powder time train. To offset extremely rapid action, an internal safety feature prevents the time action from functioning should the fuze be set for less than 0.4 second; therefore, when setting for time action, the setting should be greater than 0.4 second. The fuze, as shipped, is set safe (S).

(2) *Setting.* Remove the safety wire from the fuze and set the fuze to the desired time, using the fuze setter M27, M26, or M14.

> *Note.* If, after setting the fuze preparatory to firing, the round is not fired, the fuze will be reset "SAFE" and the safety wire replaced in its proper position before the round is unfuzed and the components returned to their packing containers.

f. FUZE, MECHANICAL TIME SUPERQUICK, M500.
(1) *Description.* The M500 (fig. 366) is a combination mechanical time and impact fuze with settings for time action (3 to 75 sec.) and an impact element for superquick action. The time action of the fuze is based on a clockwork principle.

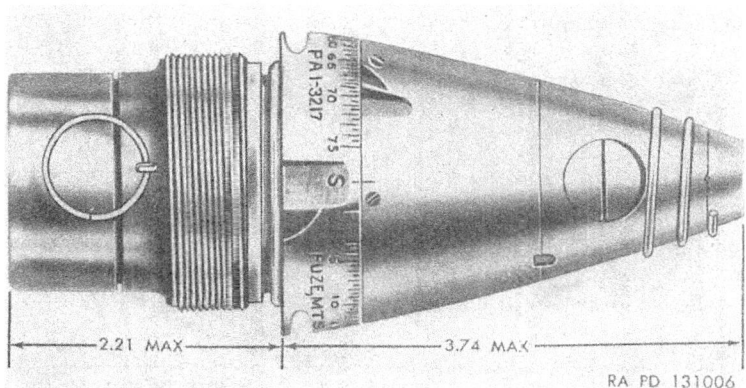

Figure 366. Fuze, mechanical time superquick, M500.

623

The time ring is calibrated in ½-second intervals. A safety leaf prevents functioning of the fuze at less than 1.5 seconds time of flight. The fuze is set safe as shipped. A safety wire, which must be removed before firing, extends through the fuze body and the firing pin, and provides positive safety during shipment and handling. When preparing the fuze for time action, the wire is removed before setting the fuze.

(2) *Setting.* Remove the safety wire from the fuze and set the fuze to the desired time, using the fuze setter M27, M26, M14, or M23.

Note. If, after setting the fuze preparatory to firing, the round is not fired, the fuze will be reset "SAFE" and the safety wire replaced in its proper position before the round is unfuzed and the components returned to their packing containers.

g. FUZE, BD, M68. This fuze, which is assembled in the base of armor-piercing projectiles (APC–T) used in this gun, is known as a base detonating (BD) fuze. It functions with delay action, upon impact. A tracer composition is incorporated in the fuze body and functions independently of the fuze mechanism. Due to its location in the projectile, the fuze is not visible in assembled rounds of APC–T ammunition.

h. FUZE, CP, M78A1, WITH BOOSTER, M25.

(1) *Description.* The concrete-piercing fuze M78A1 with booster M25 (fig. 367) are used to convert HE shell M71 into a projectile capable of penetrating or severely damaging concrete or other heavy targets. Both the fuze and booster are shipped as separate components in the same container. The fuze is a solid hardened steel nose plug which contains a detonator and delay plunger assembly in its base. It is shorter and heavier than the standard type impact fuzes.

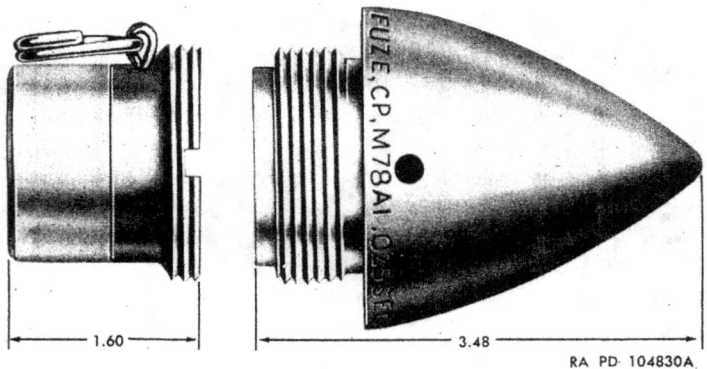

Figure 367. Fuze, CP, M78A1, with booster, M25.

The CP fuze M78A1 is fitted with either a nonadjustable 0.025-second delay or a nondelay plunger assembly; the amount of delay is stamped on the fuze. As an aid in identification, 1 inch of the nose of the nondelay fuze is painted white. Nondelay fuzes are used primarily for spotting purposes. Fuzes with 0.025-second delay plunger assemblies are used when firing for effect. Booster M25 is a modified M21A4 booster having three rather than six external threads. A cotter pin with pull ring is located in the booster body and must be removed prior to assembly of the booster to the shell. This booster is intended for use only with CP fuzes M78A1 and M78. FUZE, CP, M78 is limited standard.

(2) *Setting.* This fuze requires no setting

i. FUZE, VT, M97.

(1) *Description.* The M97 fuze (fig. 364) is a proximity fuze and is provided for use in terrestrial fire with deep-cavity high-explosive shell (suppl chg removed). It is essentially a self-powered radio transmitting receiving unit. In flight, the armed fuze transmits radio waves. When any part of the radio wave front is reflected back from the target, it interacts with the transmitted wave. The ripple or beat caused by this interaction trips a switch which closes an electric circuit and initiates detonation of the fuze explosive train when at optimum distance from the target. Bore-safety is provided by an arming switch which delays arming of the fuze for approximately 5 seconds. When armed, the fuze will function on close approach to any object capable of reflecting the transmitted waves.

(2) *Setting.* No setting is required. Any attempt to set the fuze may result in a malfunction.

385. Precautions in Firing

The following precautions should be closely observed in order to prevent injury to personnel or damage to matériel.

a. If the PD fuzes M48A3 or M51A5, the TSQ fuze M55A3, or the MTSQ fuze M500 are fired during extremely heavy rainfall, premature functioning may occur. The rainfall necessary to cause malfunctioning is comparable to the exceedingly heavy down pours which occur during summer thunderstorms. In the case of the PD M48A2 and M51A4 fuzes, such prematures may be prevented by setting the fuze for delay action, thus making the "SQ" action inoperative. However, no corresponding change can be made in the case of the PD M55A3, or MTSQ M500 fuzes since their "SQ" action is always operative. The VT M97 fuze will function properly at temperatures within 0° and 120° F. and should not be used outside these limits. Also, if the VT-fuze round is loaded into the chamber of a hot gun and not fired before 30 seconds, the fuze probably will cause either an early burst or a dud. Darkness has no effect on the functioning of VT fuzes.

   b. Do not remove safety devices from fuzes until just before use.

   c. Make sure the safety wire has been removed from the TSQ M55A3 and MTSQ M500 fuzes before firing.

   d. Make sure that the U-shaped packing stop has been removed from the wrench slots in any point fuze before loading the round into the weapon.

   e. Exercise care when loading to avoid striking the fuze. Keep the round out of the path of recoil.

   f. Before loading into the weapon, the ammunition should be free of foreign matter, sand, mud, moisture, frost, snow, ice, or grease.

   g. If, at the time firing may be interrupted, a round is in the chamber of a hot weapon, the round should be removed promptly to prevent the possibility of a cook-off.

   h. The HVAP–DS–T round will not be fired over the heads of friendly troops, unless they are protected by adequate cover, due to the danger of being struck by the discarded sabot (cup). The danger area extends up to *800 yards* from the gun along the trajectory and spreads out to 50 yards on each side of the trajectory at that range (refer to par. 379*d*).

   i. Misfires will be handled in accordance with paragraph 143 and SR 385–310–1.

386. Packing and Marking

   a. PACKING.

   (1) Service and drill rounds authorized for use in the 90-mm gun T119E1 are packed one per CONTAINER, fiber, M53A1 two containers per wooden box. The volume of the box is 2.43 cubic feet; weight with two fuzed rounds, 130 pounds; dimensions (in.) of the box are 43⅞ x 13 x 7⅜.

   (2) Armor-piercing rounds are packed similarly in the CONTAINER, fiber, M96A1. The volume of the box is 2.48 cubic feet; weight with two rounds, 131 pounds; dimensions (in.) of the box are 44⅝ x 13 x 7⅜. For shipment to certain areas, service rounds are also packed one per metal container M159A2. The volume of the container is 0.94 cubic foot; weight with one fuzed round, 69 pounds; dimensions (in.) of the container are 43⅜ x 6⅛ x 6⅛.

   (3) Practice rounds authorized for use in the 90-mm gun T119E1 are packed one per CONTAINER, fiber, M212, two containers per wooden box. The box dimensions are the same as those for service rounds ((1) above). Blank ammunition for 90-mm guns is packed one per CONTAINER, fiber, M125,

eight containers per box. The volume of the box is 2.10 cubic feet; weight with eight containers and ammunition, 99 pounds; dimensions (in.) of the box are 25⅞ x 13 x 10⅞.

(4) Complete data are published in Department of the Army Supply Catalogs ORD 3 SNL's P-5 and P-8.

b. MARKING.
(1) The following information is marked in black on packing boxes of 90-mm ammunition:
  (a) Interstate Commerce Commission (ICC) shipping designation.
  (b) Ammunition Identification Code (AIC) symbol.
  (c) Ammunition lot number.
  (d) Gross weight of package and contents.
  (e) Cubical displacement of package.
  (f) Bursting charge in projectile (if applicable).
  (g) Zone marking on projectile (if applicable).
  (h) Date loaded.
  (i) Descriptive nomenclature of packed item.
  (j) Caliber and weapon designation.
  (k) Ordnance insignia.
  (l) Name and address of box manufacturer and date manufactured.
  (m) Inspector's stamp.
(2) Yellow adhesive tape is used to seal fiber containers containing HE ammunition; blue, for practice ammunition; gray, for WP smoke ammunition; black, for AP ammunition.

387. Subcaliber Ammunition

a. AMMUNITION CARTRIDGE, ball, cal. .50, M2 is authorized as subcaliber ammunition when firing the coaxial, HB, cal. .50 Browning machine gun, M2, as a subcaliber weapon. The muzzle velocity of the 709.5 grain bullets is 2,930 fps in the 45-inch barrel and the maximum ground impact horizontal range is 7,400 yards at 35° elevation. Additional data are given in TM 9-1990.

b. PACKING. The cartridges are packed 10 rounds per carton, 35 cartons (350 rounds) per metal-lined box; 10 rounds per carton, 6 cartons per metal can, 2 cans (120 rounds) per wooden box; or 265 rounds in metallic link belt, 1 belt per metal-lined box. Representative data for these packings are given as follows:

Packing	Dimensions (in)	Weight (lb)	Volume (cu ft)
120-round	14¼ x 10½ x 7⅞	44	0.7
265-round	18½ x 9½ x 14⅞	99	1.5
350-round	18½ x 9½ x 14⅞	112	1.5

Complete packing data are published in Department of the Army Supply Catalog ORD 3 SNL T-1.

Section II. COMMUNICATION SYSTEM

388. General

90-mm gun tank M47 is equipped with radio sets AN/GRC-3, 4, 5, 6, 7, or 8 with appropriate antennas and auxiliary interphone equipment AN/VIA-1.

389. Radio Sets AN/GRC-3, 4, 5, 6, 7, and 8 Description

a. Table XI charts the principal components of these radio sets.

Table XI. Components of Radio Sets AN/GRC-3, 4, 5, 6, 7, and 8

	AN/GRC-3	AN/GRC-4	AN/GRC-5	AN/GRC-6	AN/GRC-7	AN/GRC-8
Receiver-Transmitter RT-66/GRC ("A" Set)	X	X				
Receiver-Transmitter RT-67/GRC ("A" Set)			X	X		
Receiver-Transmitter RT-68/GRC ("A" Set)					X	X
Receiver-Transmitter RT-70/GRC ("B" Set)	X	X	X	X	X	X
AF Amplifier AM-65/GRC (Interphone Amplifier)	X	X	X	X	X	X
Power Supply PP-112/GR	X	X	X	X	X	X
Control C-435/GRC	X	X	X	X	X	X
Receiver R-108/GRC	X					
Receiver R-109/GRC			X			
Receiver R-110/GRC					X	

b. For each radio set, the components noted in table XI are installed on mounting MT-297/GR (fig. 368), except control C-435/GRC, which is plugged into the underside of the U-shaped, terminal box portion of the mounting. Mounting MT-297/GR is a steel-framed unit with equally spaced, recessed channels, running from front to rear. Latching springs, which lock the feet of the components in the channels, are activated by levers along the front edge of the mounting. Shock mounts attach the mounting surface to hold-down plates. A U-shaped terminal box for connection of control and power wiring is provided under the mountings. Power feed and control cables are connected to the turret radio terminal box by pin and socket connectors.

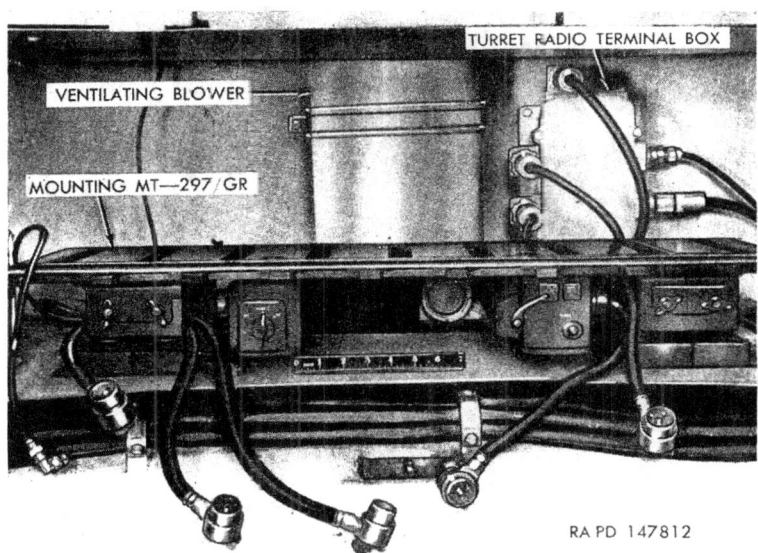

Figure 368. Mounting MT-297/GR, component of radio sets AN/GRC-3 and 4—installed view.

c. Receiver-Transmitters RT-66/GRC, RT-67/GRC, and RT-68/GRC (fig. 369) (referred to as the "A" set), which provide intertank radio communication, are identical in shape and size and differ only in the frequency band in which they operate. Receiver-Transmitter RT-70/GRC (referred to as the "B" set), which provides radio communication between tank and infantry units, is a small, light-weight unit intended to supplement the "A" set. Power supply PP-112/GR is the 24-volt vibrator power supply which powers the "A" set. AF amplifier AM-65/GRC contains the audio amplifier and electronic mixer circuits necessary to provide interphone communication and radio monitoring at control boxes C-375/VRC. Power supply circuits required for operation of the "B" set are provided in the AM-65/GRC and it also serves as a junction box for all "B" set connections.

d. Radio Receivers R-108/GRC, R-109/GRC, and R-110/GRC (fig. 369), are auxiliary receivers and are identical except for the frequency band in which they operate. Operating frequencies of the receivers correspond to the frequencies of the "A" set receiver-transmitters. Control C-435/GRC is a retransmission unit for providing automatic or manual retransmission between the "A" set and "B" set. Control boxes C-375/VRC (figs. 370, 371, and 372) are used for control of the "A" set, "B" set, and interphone amplifier. This

Figure 369. Radio set AN/GRC-3.

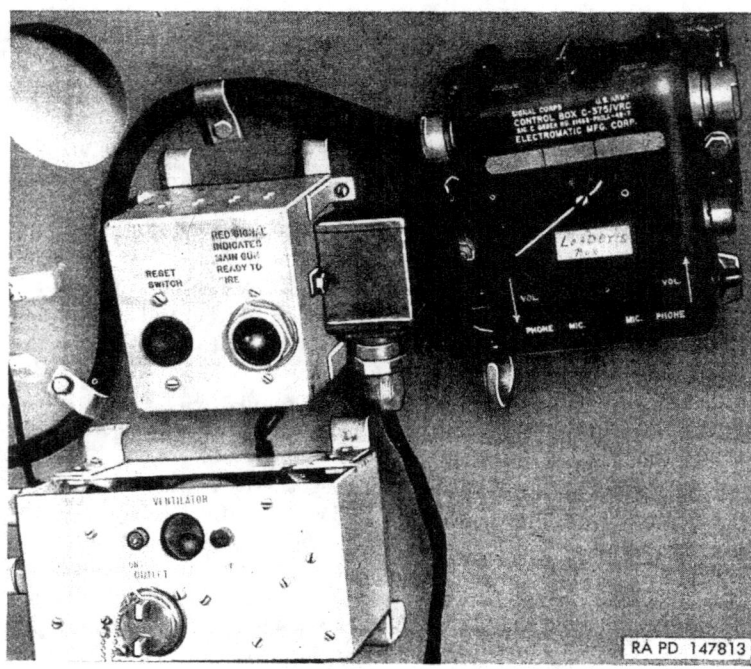

Figure 370. Loader's control box C-375/VRC.

box has a main control switch in the front to select the desired radio and interphone facilities. A spring-loaded, radio-transmit switch is located on the top of the box. Three control boxes C–375/VRC are secured directly to brackets in the vehicle by two screws. Cables from the control boxes are connected to the radio terminal boxes in the turret (fig. 373) and hull (fig. 374), by means of pin-and-socket connectors.

Figure 371. Commander's and gunner's control box C–375/VRC.

390. Radio Sets ANGRC–3, 4, 5, 6, 7, or 8 Removal and Installation

a. REMOVAL.

(1) To remove the principal components located on top of mounting MT–297/GR, first disconnect the antenna lead-in cables at the radio sets (figs. 369 and 376). Disconnect the cables leading to the U-shaped box under mounting MT–297/GR by disconnecting the connector plugs from receptacles on the front panels of the components. Also disconnect the cables interconnecting the auxiliary receiver, "A" set power supply, amplifier, and "B" set. Release switch levers on the top front of mounting MT–297/GR, and remove the principal components. To remove mounting MT–297/GR, first shut off the master relay switch (fig. 18) then disconnect

Figure 372. Driver's and assistant driver's control box C-375/VRC.

the pin-and-socket connectors of the two cables from mounting MT-297/GR at the turret radio terminal box (fig. 373). Remove cable clips securing the cables to the vehicle. Loosen wing nuts at front of mounting which engage the base plate to the shock mounts and lift the hinged base plate, making accessible the eight screws which secure the mounting holddown plates to the vehicle. Remove screws, coil the interphone cable and power feed cable, and withdraw the equipment from the vehicle.

(2) To remove control boxes C-375/VRC, first disconnect the interphone cables from the turret and hull radio terminal boxes (figs. 373 and 374). Remove control boxes C-375/VRC by removing the two screws which secure the box to the brackets provided in the vehicle.

b. INSTALLATION. For installation, refer to Signal Corps installation instructions, Signal Corps Stock No. 6D12843-V110.1.

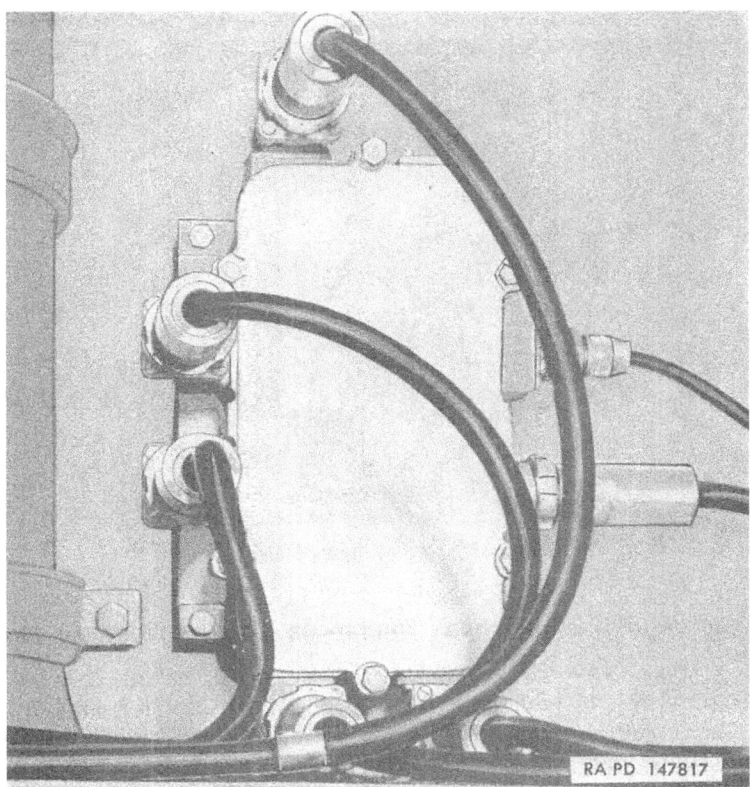

Figure 373. Turret radio terminal box showing entry of cordage from components of radio sets AN/GRC-3 and 4.

391. Antennas

a. DESCRIPTION. Two whip-type antennas are used. The location and components of each antenna are given in table XII.

b. MAST BASE AB-15/GR. Mast base AB-15/GR (fig. 375) consists of a flexible stem, a feed-through porcelain insulator, and provisions on the bottom for connecting the antenna lead-in cable (fig. 376). The mast sections are of flexible steel tubing.

c. REMOVAL. Unscrew bottom mast section from mast base (fig. 375), separate sections from each other, and stow in bag CW-206/GR supplied with the radio set. Disconnect antenna lead-in cable (fig. 376). Hold bottom insulator of mast base stationary (fig. 375) and turn flexible stem counterclockwise until upper and lower sections of mast base can be separated and removed.

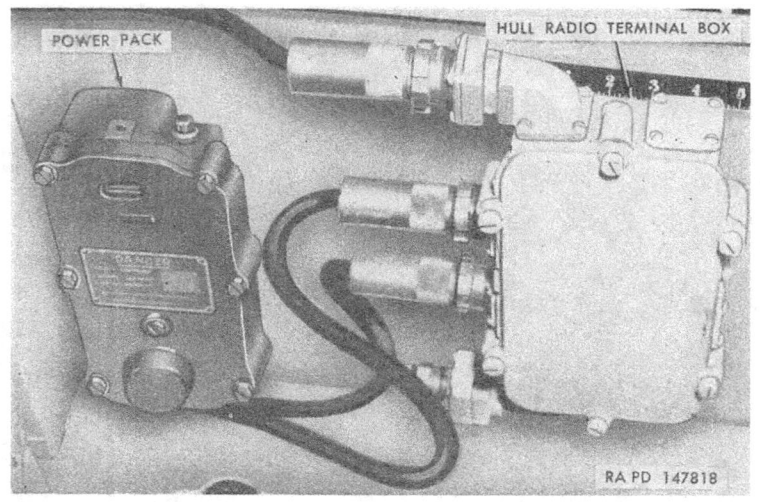

Figure 374. Hull radio terminal box.

392. Auxiliary Interphone Equipment AN/VIA-1 Description

a. CABLE REEL RL-149/VIA-1. This unit consists of a hand-set, with 40 feet of cable on a take-up reel, located at the rear of the vehicle (figs. 377 and 378). Using this hand-set, a supporting infantryman may talk over the vehicle interphone system to the tank-crew members. He may signal the vehicle loader by pressing his hand-set switch, thus flashing a light near the loader's position in the turret.

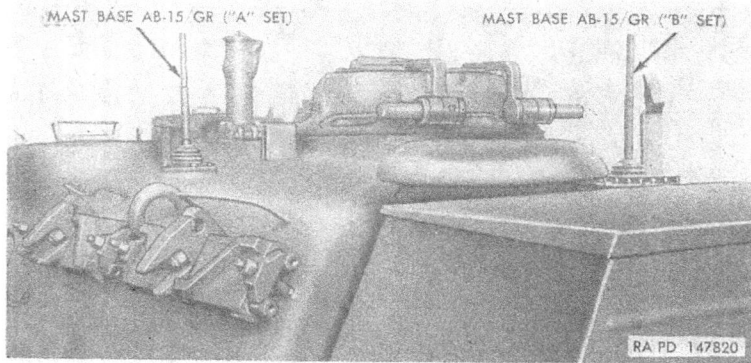

Figure 375. Mast bases AB-15/GR.

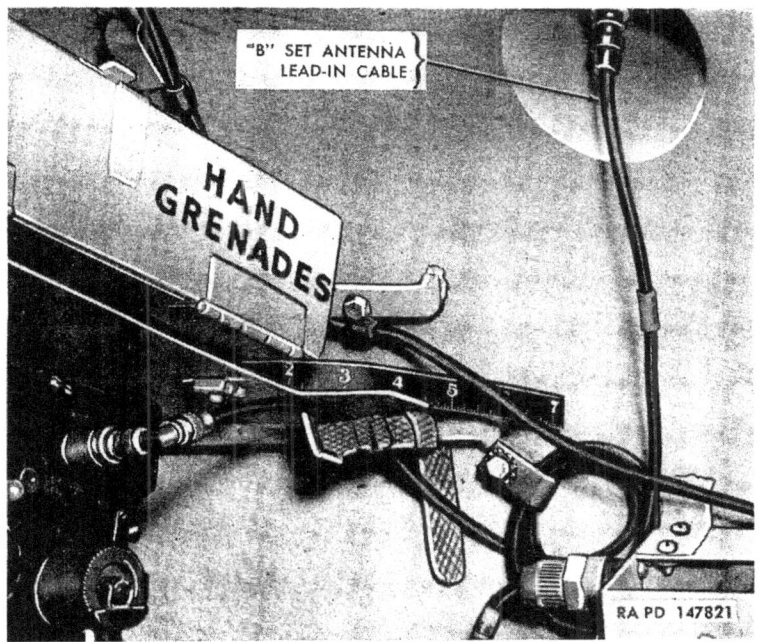

Figure 376. "B" set antenna lead-in cable.

Figure 377. Housing for auxiliary interphone equipment AN/VIA-1

Table XII. Antenna Components

Radio set	Antenna location	Mast base type No.	Mast sections type No.	TYPE OF antenna lead-in
AN/GRC-3 "A" set	On top of turret (right rear)	AB-15/GR	One MS-116-A One MS-117-A One MS-118-A One AB-22/GR One AB-24/GR	RF Cable Assy CG-568/U (4 ft, 2 in.).
"B" set	On top of turret (left side)	AB-15/GR	One AB-22/GR One AB-24/GR	RF Cable Assy CG-530/U (4 ft, 2 in.).
AN/GRC-4 "A" set	On top of turret (right rear)	AB-15/GR	One MS-116-A One MS-117-A One MS-118-A	RF Cable Assy CG-568/U (4 ft, 2 in.).
"B" set	On top of turret (left side)	AB-15/GR	One AB-22/GR One AB-24/GR	RF Cable Assy CG-530/U (4 ft, 2 in.).
AN/GRC-5 "A" set	On top of turret (right rear)	AB-15/GR	One MS-116-A One MS-117-A One MS-118-A	RF Cable Assy CG-530/U (4 ft, 2 in.).
"B" set	On top of turret (left side)	AB-15/GR	One AB-22/GR One AB-24/GR	RF Cable Assy CG-530/U (4 ft, 2 in.).
AN/GRC-6 "A" set	On top of turret (right rear)	AB-15/GR	One MS-116-A One MS-117-A One MS-118-A	RF Cable Assy CG-530/U (4 ft, 2 in.).
"B" set	On top of turret (left side)	AB-15/GR	One AB-22/GR One AB-24/GR	RF Cable Assy CG-530/U (4 ft, 2 in.).

AN/GRC-7 "A" set	On top of turret (right rear)	AB-15/GR	One MS-117-A One AB-24/GR	RF Cable Assy CG-530/U (4 ft, 2 in.)
"B" set	On top of turret (left side)	AB-15/GR	One AB-22/GR One AB-24/GR	RF Cable Assy CG-530/U (4 ft, 2 in.)
AN/GRC-8 "A" set	On top of turret (right rear)	AB-15/GR	One MS-117-A One AB-24/GR	RF Cable Assy CG-530/U (4 ft, 2 in.)
"B" set	On top of turret (left side)	AB-15/GR	One AB-22/GR One AB-24/GR	RF Cable Assy CG-530/U (4 ft, 2 in.)

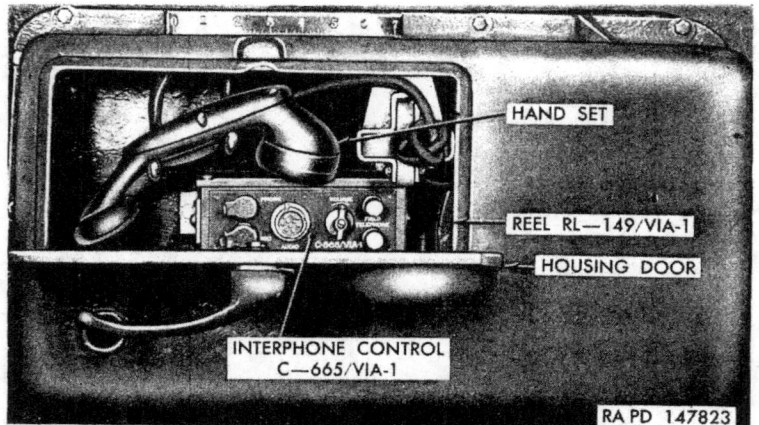

Figure 378. Housing for auxiliary interphone equipment AN/VIA-1—Door open.

b. INTERPHONE CONTROL C-665/VIA-1. This unit is a control box located at the rear of the vehicle (figs. 378 and 379), containing volume control, jacks for auxiliary hand-set or head set and microphone, and telephone binding posts. These binding posts are connected to similar binding posts located on interphone control C-664/VIA-1 adjacent to the driver. These binding posts may be used to connect a number of tanks in a party telephone system. The hand-set of cable

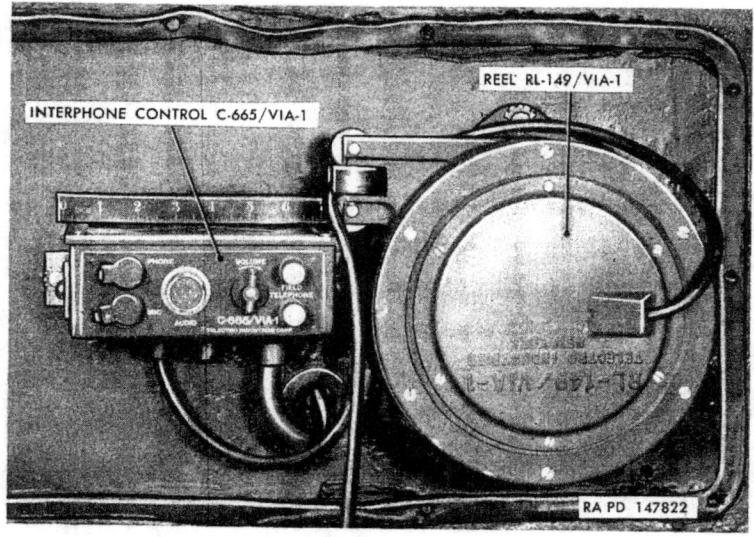

Figure 379. Auxiliary interphone equipment AN/VIA-1—housing removed.

reel RL-149/VIA-1 and interphone control C-665/VIA-1 are accessible by opening the armored door at the rear of the tank (fig. 378).

c. INTERPHONE CONTROL C-663/VIA-1. This unit is a disconnect switch box located near the driver (fig. 380), and is used to disconnect the circuits leading to the rear hand-set and control box in case of damage to the rear components by shell fire. This box also contains telephone binding posts.

Figure 380. Interphone control C-663/VIA-1.

d. INTERPHONE CONTROL C-664/VIA-1. This unit is a signal box located in the turret near the loader (fig. 381). This box has two switches; one permits the loader to flash a light at the rear of the tank (fig. 377) to signal the external personnel, and the other is a test switch to determine that the rear light is operating.

393. Auxiliary Interphone Equipment AN/VIA-1 Removal and Installation

a. REMOVAL.
  (1) Remove housing (fig. 377) which houses the hand-set, take-up reel, and control box at rear of hull (figs. 378 and 379), by removing 16 screws. Remove cover of control box and

639

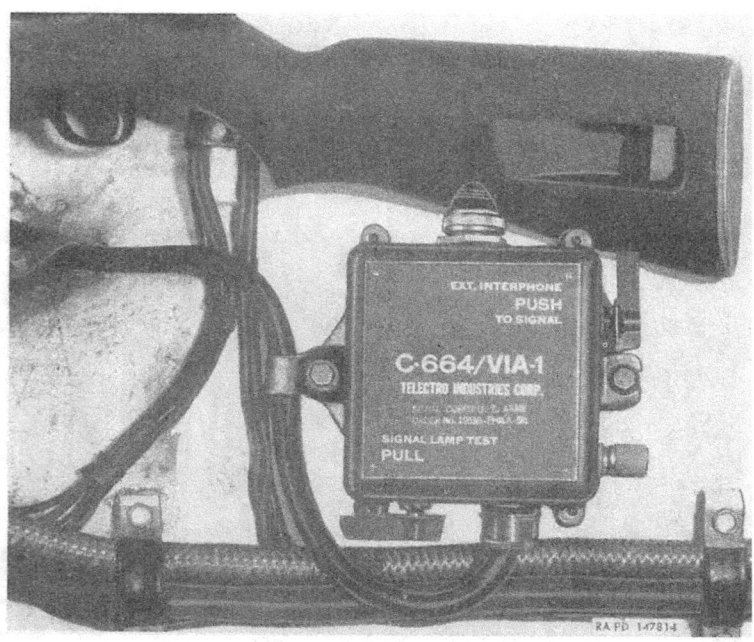

Figure 381. Interphone control C-664/VIA-1.

remove cables from terminal strip. Tag each conductor with its proper terminal number. Remove reel by removing the three screws securing reel to rear of hull. Remove control box by removing the two mounting screws.

    (2) To remove interphone control C-663/VIA-1 (fig. 380), disconnect the two plugs on back of unit. Remove screws holding box to mounting bracket and remove box.

    (3) Remove interphone control C-664/VIA-1 (fig. 381), by removing the two screws securing box to the bracket. Remove clamps from cable leading to turret radio terminal box, and disconnect plug from turret radio terminal box (fig. 373).

  b. INSTALLATION. For installation, refer to Signal Corps installation instructions, Signal Corps Stock No. 6D12843-V110.1.

394. Hull Radio Terminal Box

  a. REMOVAL (fig. 374). Disconnect four cables at hull radio terminal box. Remove four screws securing box to ammunition rack behind driver's seat and remove box.

  b. INSTALLATION. (fig. 374). Position hull radio terminal box to ammunition rack behind driver's seat and secure with four screws. Connect four cables to box.

CHAPTER 5

SHIPMENT AND LIMITED STORAGE AND DESTRUCTION OF MATÉRIEL TO PREVENT ENEMY USE

Section I. SHIPMENT AND LIMITED STORAGE

395. Domestic Shipping Instructions

a. PREPARATION FOR SHIPMENT IN ZONE OF INTERIOR. When shipping the 90-mm gun tank M47 interstate or within the zone of interior, the officer in charge of preparing the shipment *will be responsible* for furnishing tanks to the carriers for transport in a *serviceable* condition, properly cleaned, preserved, painted and lubricated as prescribed in SB 9–4.

Note. For loading and blocking instructions for these tanks on flatcars, refer to paragraphs 397 and 398.

b. PREPARATION FOR SHIPMENT TO PORTS.
   (1) *Inspection.* All used tanks destined for oversea use will be inspected prior to shipment in accordance with TB ORD 385.
   (2) *Processing for shipment to ports.* All tanks destined to ports of embarkation for oversea shipment will be further processed in accordance with SB 9–4 and MIL-P-11290 (ORD).

   Note. Ports of embarkation will supplement any necessary or previously omitted processing upon a receipt of tanks.

c. REMOVAL OF PRESERVATIVES FOR SHIPMENT. Personnel withdrawing tanks from a limited storage for domestic shipment must not remove preservatives, other than to insure that they are complete and serviceable. If it has been determined that preservatives have been removed, they must be restored prior to domestic shipment. The removal of preservatives is the responsibility of depots, ports, or field installations (posts, camps, and stations) receiving the shipments.

d. ARMY SHIPPING DOCUMENTS. Prepare all Army shipping documents accompanying freight in accordance with TM 38–705.

e. DEEP-WATER FORDING. If during the course of shipment, operations embrace deep-water fording, prepare tanks in accordance with TM 9–2853.

f. ARTILLER. GUN BOOK. During transfer of shipment, locate the artillery gun book in the gun book envelope and secure to the top of the breech mechanism with water-resistant pressure-sensitive adhesive tape. Under one of the wrappings of tape, insert one end of a tab reading "GUN BOOK HERE."

396. Limited Storage Instructions

a. GENERAL.
  (1) Tanks received for storage already processed for domestic shipment, as indicated on the Vehicle Processing Record Tag (DA AGO Form 9–3), must not be reprocessed unless the inspection performed on receipt of tanks reveals corrosion, deterioration, etc.
  (2) Completely process tanks upon receipt directly from manufacturing facilities, or if the processing data recorded on the tag indicates that they have been rendered ineffective by operation or freight shipping damage.
  (3) Tanks to be prepared for limited storage must be given a limited technical inspection and be processed as prescribed in SB 9–63. The results and classification of tanks will be entered on DA AGO Form 461–5.

b. RECEIVING INSPECTIONS.
  (1) Report of tanks received for storage in a damaged condition or improperly prepared for shipment will be reported on DD Form 6 in accordance with SR 745–45–5. Report of tanks received in a unsatisfactory condition (chronic failure or mulfunction of the tank or equipment) will be reported on the Unsatisfactory Equipment Report DA AGO Form 468 in accordance with SR 700–45–5.
  (2) When tanks are inactivated, they are to be stored in a limited storage status for periods not to exceed 90 days. Stand-by storage for periods in excess of 90 days will normally be handled by ordnance maintenance personnel only.
  (3) Immediately upon receipt of tanks for storage, they must be inspected and serviced as prescribed in chapter 2, section I. Perform a systematic inspection and replace or repair all missing or broken parts. If repairs are beyond the scope of the unit and the tanks will be inactivated for an appreciable length of time, store them in a limited storage status and attach tags specifying the repairs needed. The reports of these conditions will be submitted by the unit commander for action by an ordnance maintenance unit.

c. INSPECTIONS DURING STORAGE. Perform a visual inspection periodically to determine general condition. If corrosion is found on any part, remove the rust spots, clean, paint, and treat with the prescribed preservatives.

Note. Touch-up painting will be in accordance with TM 9-2851.

  d. REMOVAL FROM LIMITED STORAGE.
  (1) If the tanks are not shipped or issued upon expiration of the limited storage period, they may either be processed for another limited storage period or be further treated for stand-by storage (tanks inactivated for periods in excess of 90 days up to 3 years) by ordnance maintenance personnel.
  (2) If tanks to be shipped will reach their destination within the scope of the limited storage period, they need not be reprocessed upon removal from storage unless inspection reveals it to be necessary according to anticipated in-transit weather conditions.

  Note. All tanks being reissued through the depot supply system to troops within the continental limits of the United States must meet the requirements of TB ORD 385. This is NOT required for so-called reissues, exchanges, or redistribution among troop units, where the depot supply system is not involved.

  (3) Deprocess tanks when it has been ascertained that they are to be placed into immediate service. Remove all rust preventive compounds and thoroughly lubricate as prescribed in chapter 3, section II. Inspect and service vehicles as prescribed in chapter 2, section I.
  (4) Repair and/or replace all items tagged in accordance with *b*(3) above.

  e. STORAGE SITE. The preferred type of storage for tanks is under cover in open sheds or warehouses whenever possible. Where it is found necessary to store tanks outdoors, the storage site must be selected in accordance with AR 700-105 and protected against the elements as prescribed in TB ORD 379.

397. Loading the 90-mm Gun Tank M47 on Railroad Flatcars

  a. PREPARATION.
  (1) When tanks are shipped by rail, every precaution must be taken to see that they are properly loaded and securely fastened and blocked to the floor of flatcar. All on vehicle matériel (OVM) will be thoroughly cleaned, preserved, packed (boxed or crated), and securely stowed in or on the tank or on flatcar for transit.
  (2) Prepare all tanks for rail shipment in accordance with paragraph 395*a*. In addition, take the following precautions:
    (*a*) If tank is to be shipped within the continental United States, except directly to ports of embarkation, disconnect the battery cables from battery. Clean if necessary and

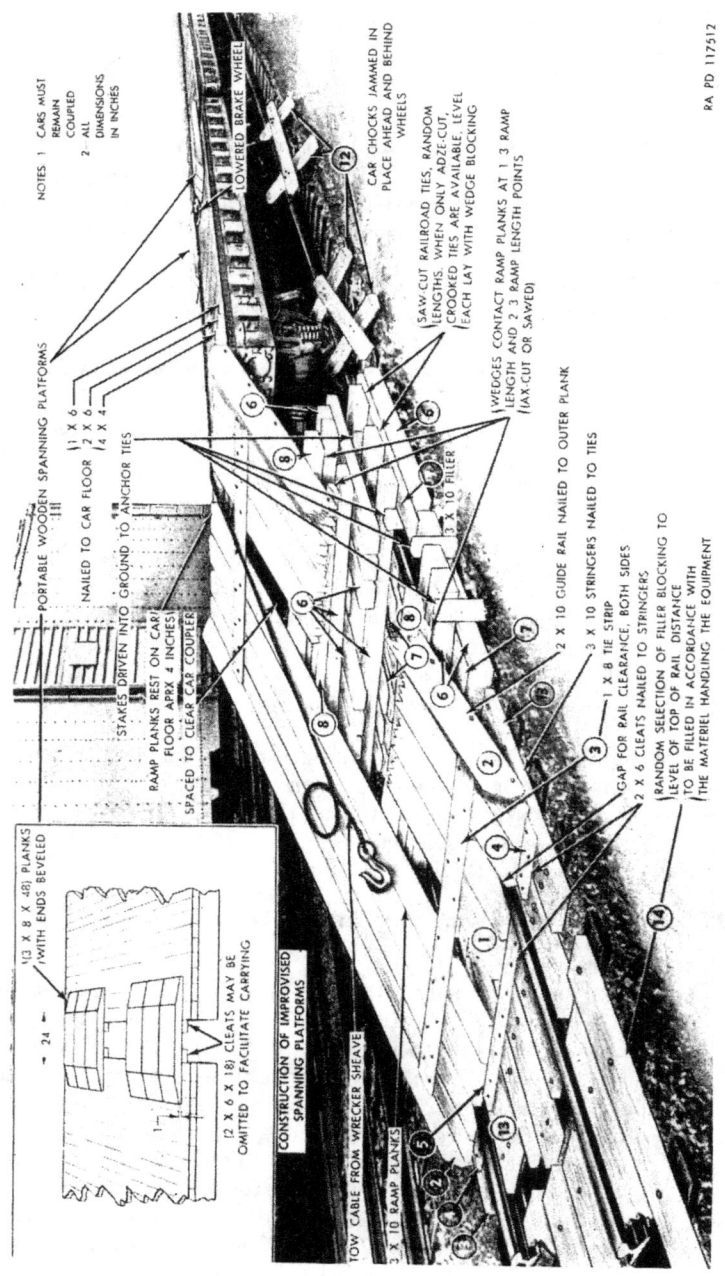

Figure 382. Construction of improvised loading ramp and spanning platforms.

NOTES:

1. RAMP SHOWN IS OF CAPACITY OF LARGEST END-LOADING FREIGHT CAR. FOR LESSER LOADS, REDUCE NUMBER OF RAMP PLANKS.
2. WIDTH DETERMINED BY TREAD OF MATERIEL BEING LOADED
3. FOR LOADING TWO WHEELED ARTILLERY TRAILERS, OR SHORT WHEELBASE MATERIEL, RAMP PLANKS MAY BE SHORTER.
   CAUTION: WHEN RAMP IS TOO SHORT, UNDERPINNING OF MATERIEL WILL STRIKE END OF RAMP (EX: 90 MM AA GUN).
4. OPENING AT CENTER MAY BE FILLED UP TO THE CAR COUPLER TO AVOID INJURY TO MANEUVERING PERSONNEL.
5. FOR LOADS OVER 40-TONS, APPROACH END OF FLATCAR MUST BE BLOCKED UP TO AVOID TIPPING OF FLATCAR
6. THIS TYPE RAMP IS ADAPTABLE TO DROP-END GONDOLA AND AUTO END-DOOR BOX CAR LOADING.
7. WHEN LOADING AN AUTO END-DOOR BOX CAR, IT MAY BE NECESSARY TO LOAD A FLATCAR COUPLED TO THE BOX CAR, TO GAIN OVERHEAD LOADING CLEARANCE.
8. WHEN LOADING BY WRECKER CABLE, WITH PULL AT 90-DEGREES TO TRAIN, USING A SHEAVE, FLATCAR AT POINT OF PULL MUST BE LASHED TO ADJACENT RAILS, CARS, OR OTHER FIXED OBJECT.

RA PD 117513

BILL OF MATERIALS FOR RAMP AS ILLUSTRATED					
PART NO	QUANT REQ'D	PART NAME	LENGTH	WIDTH	THICKNESS
1	8	RAMP PLANKS	00 ft	10 in	3 in
2	2	GUIDE RAILS	20 ft	8 in	2 in
3	2	TIE STRIPS	8 ft	8 in	1 in
4	2	CLEATS	18 in	6 in	2 in
5	1	CLEAT	56 in	6 in	2 in
6	31	RAILROAD TIES	8 ft	8 in	8 in
7	AS REQD	FILLERS	AS REQD	10 in	3 in
8	AS REQD	WEDGES (CUT TO FIT)			
9	1	STEPDOWN PIECE	8 ft	4 in	4 in
10	1	STEPDOWN PIECE	8 ft	6 in	2 in
11	1	STEPDOWN PIECE	8 ft	6 in	1 in
12	4	CHOCK BLOCKS	AS REQD	4 in	4 in
13	AS REQD	STRINGERS	AS REQD	10 in	3 in
14	AS REQD	GROUND DUNNAGE	AS REQD		

Figure 383. Bill of materials for improvised loading ramp.

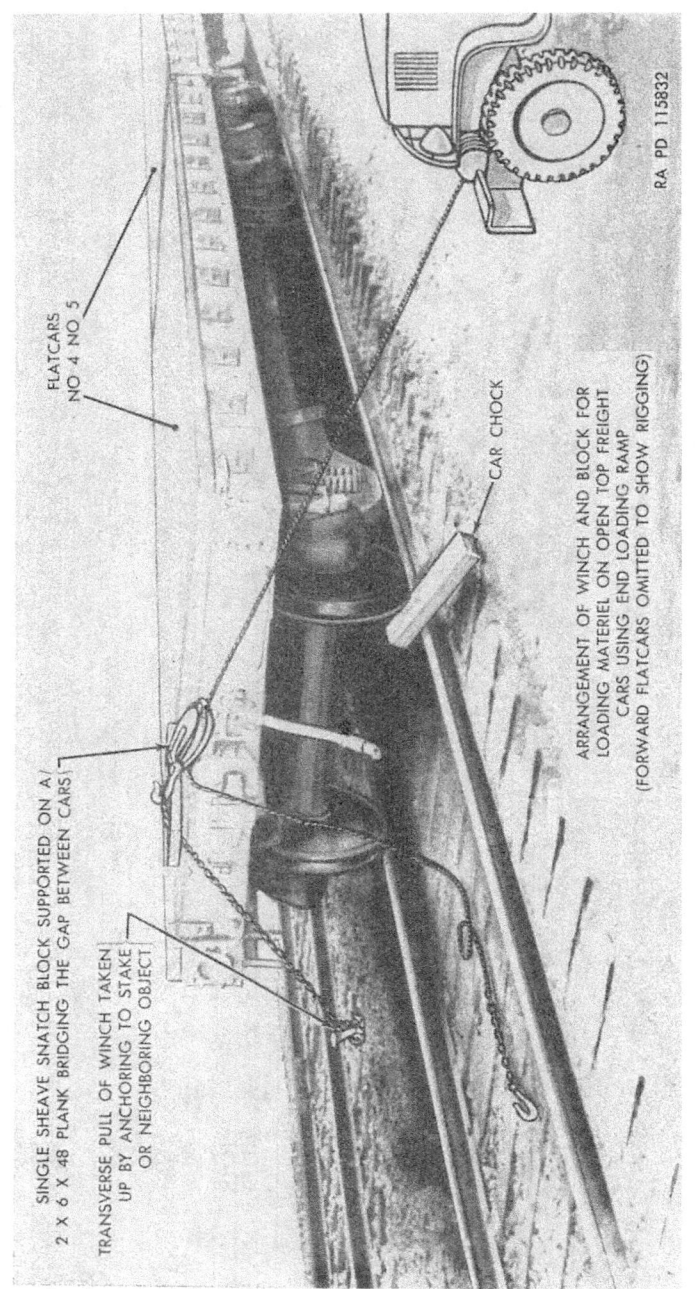

Figure 384. Method of powering the towing cable.

wrap cable terminals and battery posts with nonhygroscopic adhesive tape. Secure terminals away from battery.

Note. Not required for drive-away movement.

(*b*) If tank is to be shipped directly to ports of embarkation, *except* when tank is to be combat loaded, remove batteries, plug vents and clean with an alkali-type cleaning compound or a solution of trisodium-phosphate diluted with water. Rinse with *cool* water and remove vent plugs. Scrape or wire brush and clean cable terminals and battery box (holder) with above cleaning solution. Rinse with *cool* water. Coat cable terminals with No. 2 general purpose lubricating grease. Paint battery boxes with black acid resisting paint. Battery will be shipped wet charged and boxed in accordance with TM 9-2854 and secured in vehicle with OVM.

(*c*) Apply the parking brakes and place the transmission in neutral position after the tank has been finally spotted on the flatcar. The tank must be loaded on the flatcar in such a manner as to prevent the flatcar from carrying an unbalanced load.

b. TYPE OF CARS. Instructions contained herein pertain to the loading of tanks on flatcars (cars with wooden floors laid over sills and without sides and ends but equipped with stake pockets).

c. METHOD OF LOADING 90-MM GUN TANK M47 ON FLATCARS.

(1) Tanks will be loaded and unloaded with the use of hoisting equipment when available. When suitable hoisting equipment is not available for loading on or for subsequent unloading from a flatcar, an end ramp must be used in cases where the tank is not on a level with the flatcar deck. Tanks on a warehouse platform or loading dock can be pivoted over spanning platforms aboard a flatcar adjacent to the platform, then again pivoted into lateral position on the flatcar.

(2) When tanks must be loaded from ground level, a ramp may be improvised (*d* below) by borrowing railroad ties normally found stacked in railroad yards and by procuring necessary planking. An efficient end ramp is shown in place in figure 382. The bill of materials for constructing this ramp is shown in figure 383.

Note. Railroad ties, alone, stacked without deck planking and not securely anchored, provide a very unstable ramp and should not be attempted except under conditions of extreme emergency.

(3) Tanks which can be loaded under their own power will be driven onto the improvised apron at base of ramp and then be carefully guided up the ramp to their positions on flatcar.

(4) To load tanks which cannot be operated due to processing, tow onto the improvised apron at base of ramp and unhitch. Using a cable laid along the center line of the flatcar, attached to tank, the tank is pivoted to point towards the ramp.

Caution: Follow-up forward movement of the tank by chocking behind tracks on the ramp.

(5) After the first tank is loaded on the flatcar, additional tanks or other vehicles may be similarly hauled aboard by passing the towing cable beneath the loaded tank. When a train of flatcars is being loaded, steel or wooden spanning platforms or bridges are used to cover the gap between cars. Flatcar brake wheels must first be lowered to floor level to permit passage. A pair of improvised spanning platforms are shown in the insert in figure 382. These spanning platforms are moved along the train by hand as the tank advances.

(6) The above method of train loading requires careful advance planning as to the order of loading, so that tanks are arranged on each flatcar under prescribed methods and combinations.

(7) For powering the towing cable, a vehicle with winch is spotted at *right angles* to the train. It is located at about the third or fourth flatcar to facilitate signaling and because of cable length limits. A single-sheave snatch block located between cars on the train center line will provide the necessary *lateral* pull. A tank passing this point can be towed by a vehicle on the ground with personnel guiding its passage. A long, tow cable from the towing vehicle will lessen the tendency of the towed tank to stray from the center line of the train.

Note. The snatch block fastening chain must be lashed to an adjacent solidly fixed object or stake to offset the cross pull of the powered winch (fig. 384).

d. LOADING RAMP.

(1) A ramp for end-loading of tanks on open-top freight cars may be improvised when no permanent ramps or hoisting facilities are available. A ramp suitable for the loading of most ordnance items is shown in figure 382. For loading the 90-mm gun tank M47, the width of the ramp may be reduced to two double-plank runways, each cleated together. Length of planking must be determined with consideration to underhull clearance, in order to clear the hump at the upper end of ramp.

Caution: Personnel guiding the tank up the ramp must exercise care when working close to the edges of the ramp planking.

(2) The flatcar bearing the ramp must be securely blocked against rolling, particularly when the car brakes are not applied as in train loading. Successive cars must remain

coupled and be additionally chocked at several points along the train when ground towing of tanks aboard the train is being effected.

(3) Whenever the flatcars are not on an isolated track or blocked siding, each end approach to the train must be posted with a blue flag or light to advise that men are at work and that the siding may not be entered beyond those points.

(4) Upon completion of the loading operation, the ramp planks and bridging devices should be loaded on the train for use in unloading operations. Random sizes of timbers used in building the approach apron up to rail level should be included. All materials should be securely fastened to the car floors, after tanks are blocked in place, and entered upon the bill of lading (B/L). Railroad ties borrowed for the operation should not be forwarded to the unloading point unless specifically required and only with the consent of the owner.

e. LOADING RULES. For general loading rules pertaining to rail shipment of ordnance vehicles, refer to TB 9–OSSC–G.

Warning: The height and width of tanks when prepared for rail transportation must not exceed the limitations indicated by the loading table as prescribed in AR 700–105, section II. Whenever possible, local transportation officers must be consulted about the limitations of the particular railroad lines to be used for the movement to avoid delays, danger, or damage to equipment.

398. Blocking the 90-mm Gun Tank M47 on Railroad Flatcar

a. GENERAL.

(1) All blocking instructions specified herein are minimum and are in accordance with the Association of American Railroads "Rules Governing the Loading of Commodities on Open Top Cars." Additional blocking may be added as required at the discretion of the officer in charge. Double-headed nails may be used if available, except in the lower piece of two-piece cleats. All item reference letters given below refer to the details and locations as shown in figures 385, 386, 387, 388, and 389.

Note. Any loading methods or instructions developed by any source which appear in conflict with this publication or existing loading rules of the carriers, must be submitted to the Chief of Ordnance, Washington 25, D. C., for approval.

(2) The method of blocking the 90-mm gun tank M47 described herein provides two methods of securing tanks on flatcars. Field blocking method (*c* below) is for using organizations where the necessary materials and machine shop equipment

for manufacture are not available. Depot or arsenal blocking method (*d* below) is for use by higher echelons where such materials and equipment are available. Any of the methods described herein or a combination of the two may be effectively used for securing tanks on flatcars.

b. BRAKE WHEEL CLEARANCE "A." Load tanks on flatcars with a minimum clearance of at least 4 inches below and 6 inches above, behind, and to each side of the brake wheel (figs. 385 and 387). Increase clearance as much as is consistent with proper location of load.

Note. Prior to loading the 90-mm gun tank M47 on flatcars, locate the track inside cleats "C" and spacers "D" (fig. 385) and angle cleats "C" (fig. 387) and nail to car floor. This is accomplished by measuring the width between inside of tracks and applying chalk marks on car floor to establish exact location of cleats (exception is noted in *c*(1) and *d*(1) below).

c. FIELD BLOCKING (fig. 385).

(1) *Track inside cleats "C" (2 x 6 inches and 2 x 4 inches, length to suit, two of each required).* (See note in *b* above.) Locate the 2 x 6-inch cleats on previously measured chalk marks and nail to car floor with thirtypenny nails, staggered along their lengths. Locate the 2 x 4-inch cleats on top of the 2 x 6-inch cleats with one edge flush with the edge of the lower cleat and butting against the inside track. Nail the upper cleat to the lower cleat and car floor with fortypenny nails staggered along their length.

Note. Cleats "C" may be located against the outside surfaces of each track if flatcars are received with ample width to allow nailing these cleats to floor of car. If cleats "C" are located against the outer surfaces of each track, spacers "D" ((2) below) will not be required.

(2) *Track inside spacers "D" (2 x 4 inches, length to suit, 10 required).* (See note in *b* above.) Locate five spacers equidistant along "C" crosswise of car floor with ends butting against the sides of lower cleats "C." Nail to car floor with thirtypenny nails. Locate upper spacers on top of lower spacers with ends butting against the sides of upper cleats "C." Nail to lower spacers and car floor with fortypenny nails staggered along their length.

Note. Spacers "D" are not required if cleats "C" ((1) above) are located against the outside surfaces of tracks.

(3) *Chock blocks "B" (6 x 8 x 24 inches, eight required).* After positioning tank on flatcar, locate the 45-degree surface of two chock blocks against the rear of each track and the 33-degree surface of two chock blocks against the front of each track. Nail heel of chock blocks to car floor with three forty-

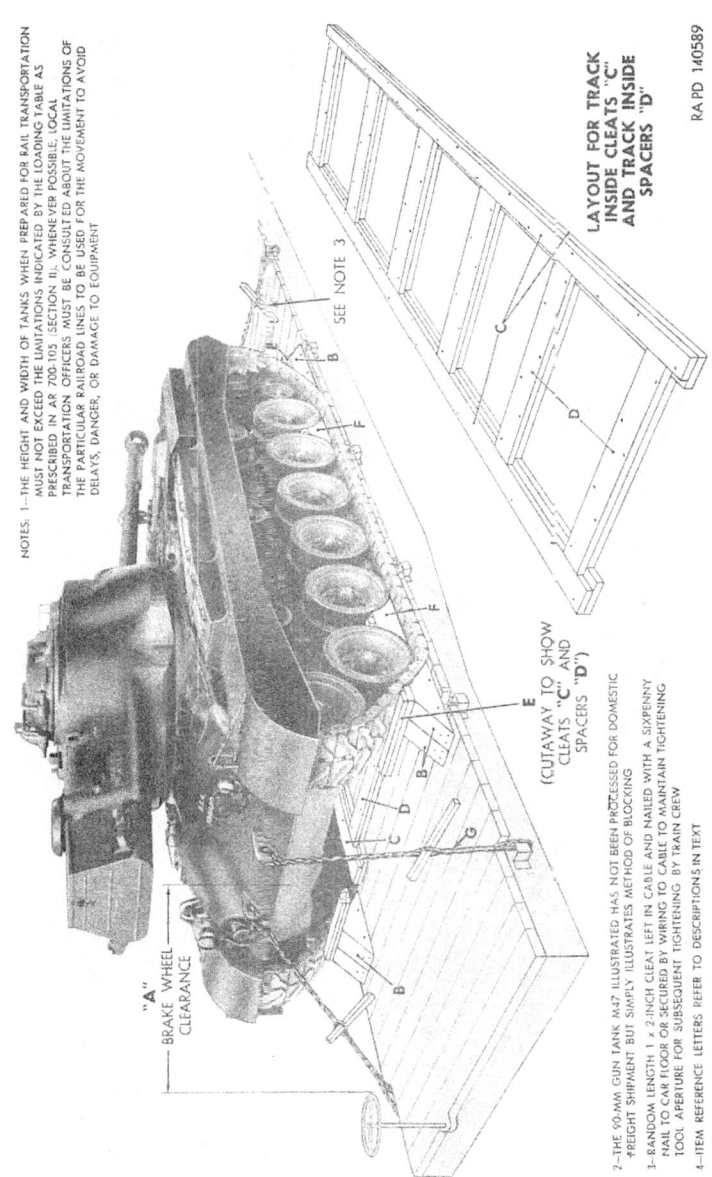

Figure 385. Method of blocking 90-mm gun tank M47 for rail shipment—field blocking.

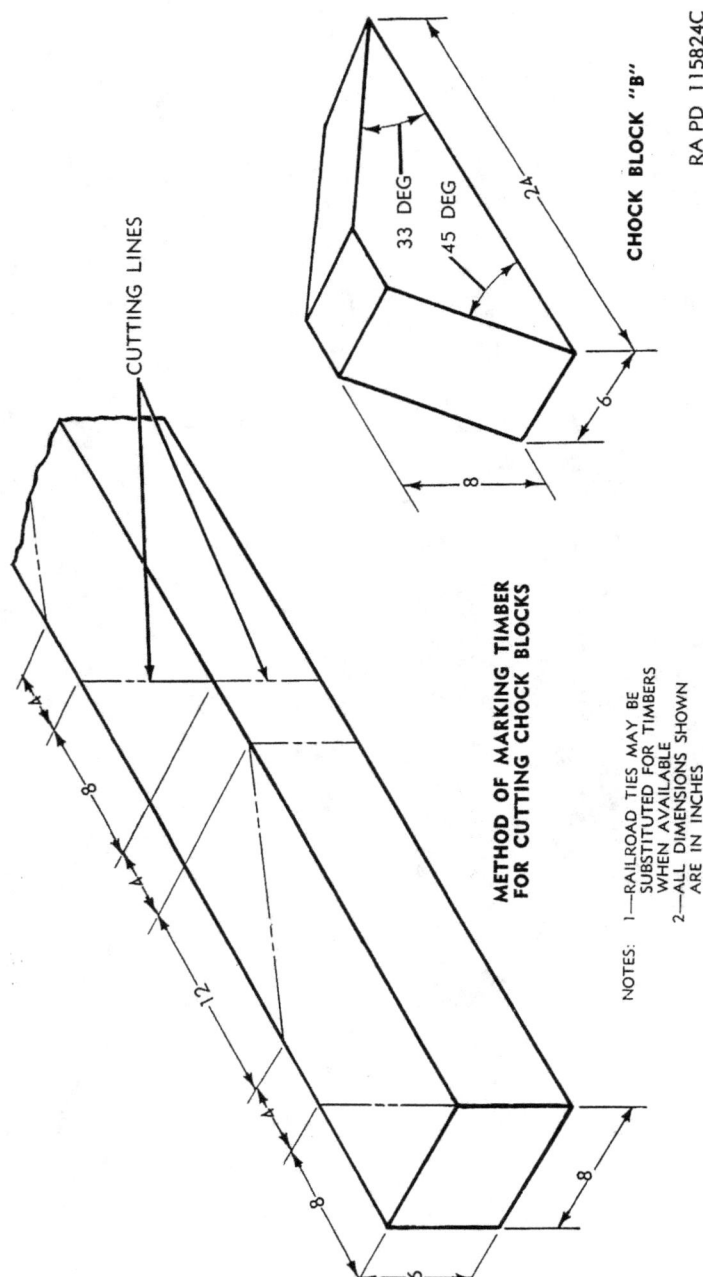

Figure 386. *Cutting chock blocks from timbers.*

penny nails. Toenail the outer surfaces of each block to car floor with two fortypenny nails.

Note. Chock blocks may be cut from timbers or railroad ties (when available) as shown in figure 386.

(4) *Cross cleats "E"* (*2 x 4 inches, length to suit, four required*). Locate lower cleat on top of chock blocks against tracks across the front and rear of tank. Nail to top of each chock block with thirtypenny nails. Locate the upper cleats on top of the lower cleats against tracks. Nail to lower cleats and chock blocks with fortypenny nails in each block.

(5) *Bogie blocks "F"* (*see fig. 388, detail 6, eight required*). Cut blocks "F" to conform with the radius of bogie wheels as shown in figure 388, detail 6. Locate blocks between bogie wheels on both sides of the tracks as shown in figure 385. Locate wedges under blocks to insure a snug fit against wheels. Nail a 1 x 4 x 15-inch cleat across the top of blocks to prevent bogie blocks from sliding out of position.

(6) *Tie-down "G"* (*No. 8 gage black annealed wire*). Twist-tie six strands of wire together to form a single cable. Pass one end of cable through the towing lug on front right side of hull. Extend end beyond half the distance to a stake pocket forward of tank. Pass other end of cable through the stake pocket and form a 6-inch loop in end, twisting each strand of wire *tightly* around cable. Make certain that the loop is positioned above the span of the cable. Insert the free end of cable through the loop hand tight and form another 6-inch loop in the free end. Insert a tightening tool in one of the loops. Locate a random length 1 x 2-inch block between cables. Twist-tie cables with tightening tool just enough to take up slack. Keep random length block intact for subsequent tightening by train crew when required. Repeat operation for other side of hull on opposite side of flatcar. Secure rear end of tank as described above.

d. DEPOT OR ARSENAL BLOCKING (fig. 387).

(1) *Angle cleats "C"* (*see fig. 388, detail 3, four required*). (See note in *b* above.) Locate angle cleats "C" approximately two feet from both ends of each track so that the 90-degree angle portion rests flush against inside of track at the points indicated. Hammer the spiked portion of angle cleats into the car floor. Nail cleats to car floor with forty penny nails through the three $5/16$-inch diameter holes and spikes through the two $3/8$-inch square holes.

Note. Angle cleats "C" may be located against the outside surfaces of each track if flatcars are received with ample width to allow nailing these cleats to floor of car.

Figure 387. Method of blocking 90-mm gun tank M47 for rail shipment—depot or arsenal blocking.

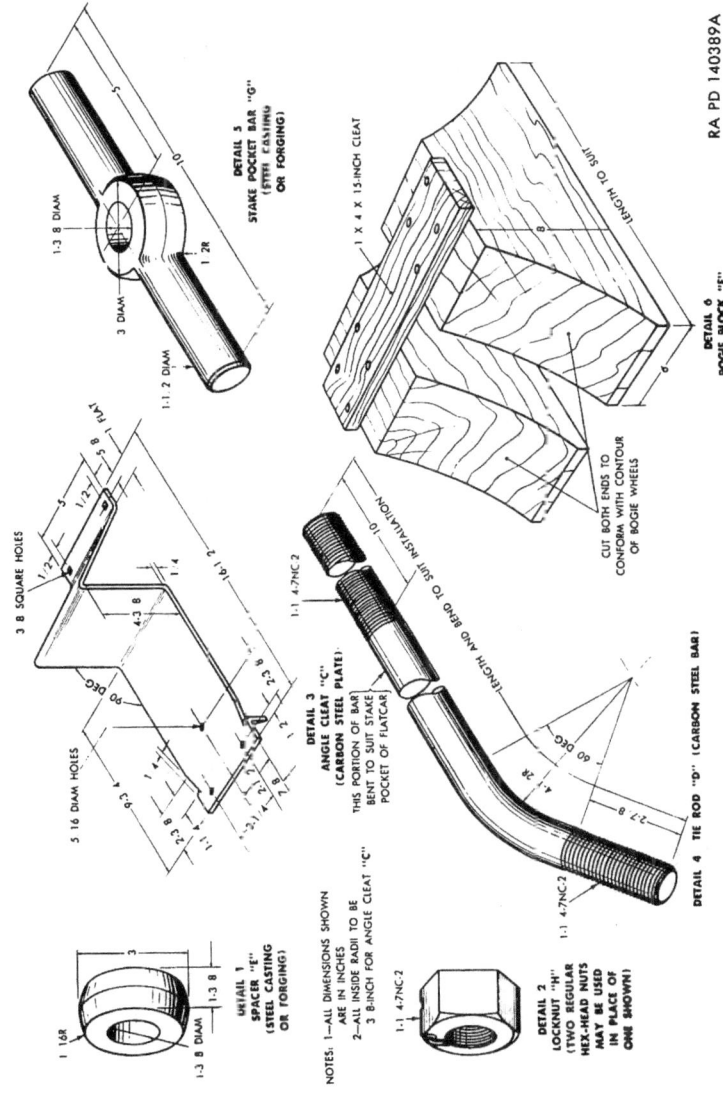

Figure 388. Method of blocking 90-mm gun tank M47 for rail shipment—blocking details.

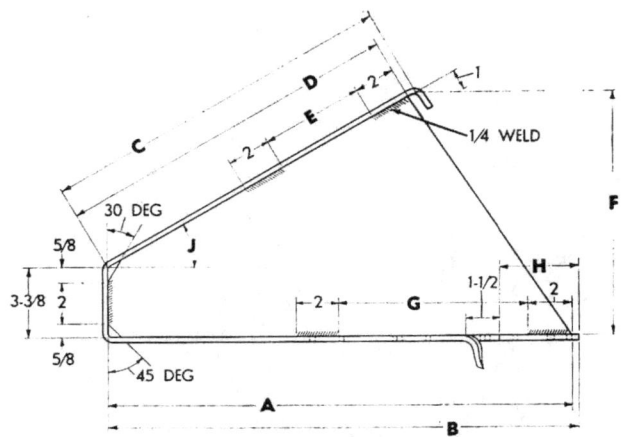

KEY	CHOCK BLOCK DIMENSIONS—INCHES	
	FRONT ASSY B	REAR ASSY B-1
A	18-5/8	22
B	19	22-3/8
C	13-3/4	17
D	13-1/4	16-1/2
E	3	5
F	11-1/8	11-5/8
G	7	9
H	4	3-3/4
J	34-1/2 DEG	29 DEG

NOTES: 1—ALL DIMENSIONS SHOWN IN INCHES
2—ALL INSIDE RADII 3/8-INCH UNLESS OTHERWISE DIMENSIONED
3—ALL TOLERANCES ON ANGLES ARE ± 1 DEGREE AND FRACTIONS ± 1/16 INCH

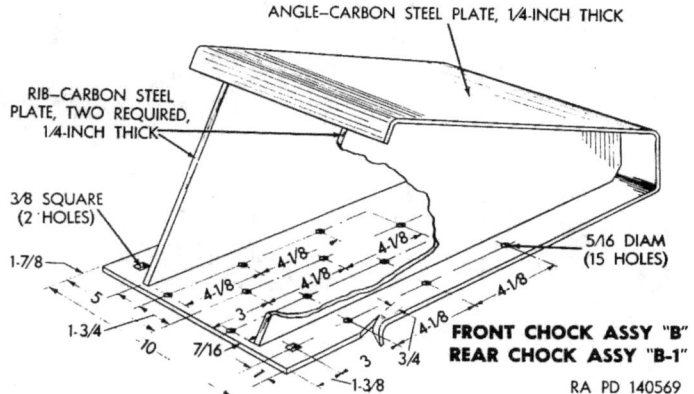

Figure 389. Method of blocking 90-mm gun tank M47 for rail shipment—blocking details.

(2) *Chock blocks "B" and "B–1" (see fig. 389, two of each required).* After positioning tank on flatcar, locate the front chock block assembly "B" and the rear chock block assembly "B–1" against tracks as shown in figure 587. Hammer the spiked portion of chock blocks into the car floor. Nail chock blocks to car floor with fortypenny nails through the fifteen $5/16$-inch diameter holes and spikes through the two $3/8$-inch square holes.

(3) *Bogie block "F" (see fig. 388, detail 6, eight required).* Refer to $c(5)$ above.

(4) *Tie rod "D" (see fig. 388, detail 4, four required).* Insert one end of rod through towing lug on front of tank and secure with spacers "E" and locknuts "H" (see fig. 388, details 1 and 2). Insert other end of tie rod through stake pockets forward of tank and secure with stake pocket bar "G", spacers "E" and locknut "H" (see fig. 388, details 1, 2, and 5).

Section II. DESTRUCTION OF MATÉRIEL TO PREVENT ENEMY USE

399. General

a. Destruction of the tank, armament, and equipment, when subject to capture or abandonment in the combat zone, will be undertaken by the using arm only when, in the judgment of the unit commander, such action is necessary in accordance with orders of, or policy established by, the army commander.

b. The information which follows is for guidance only. Certain of the procedures outlined require the use of explosives and incendiary grenades which normally may not be authorized items for the tank. The issue of these and related materials and the conditions under which destruction will be effected are command decisions in each case, according to the tactical situation. Of the several means of destruction, those most generally applicable are—

Mechanical	Requires ax, pick mattock, sledge, crowbar or similar implement.
Burning	Requires gasoline, oil, incendiary grenades, or other inflammables.
Demolition	Requires suitable explosives or ammunition.
Gunfire	Includes artillery, machine guns, rifles using rifle grenades, and launchers using antitank rockets. Under some circumstances hand grenades may be used.

In general, destruction of essential parts, followed by burning, will usually be sufficient to render the tank, armament, and equipment useless. However, selection of the particular method of destruction requires imagination and resourcefulness in the utilization of the facilities at hand under the existing conditions. Time is usually critical.

c. If destruction to prevent enemy use is resorted to, the tank, armament, and equipment must be so badly damaged that they cannot be restored to a usable condition in the combat zone either by repair or cannibalization. Adequate destruction requires that all parts essential to the operation of the tank, armament, and equipment, including essential spare parts, be destroyed or damaged beyond repair. However, when lack of time and personnel prevents destruction of all parts, priority is given to the destruction of those parts most difficult to replace. Equally important, the same essential parts must be destroyed on all like matériel so that the enemy cannot construct one complete operating unit from several damaged ones.

d. If destruction is directed, due consideration should be given to—
  (1) Selection of a point of destruction that will cause greatest obstruction to enemy movement and also prevent hazard to friendly troops from fragments or ricocheting projectiles which may occur incidental to the destruction.
  (2) Observance of appropriate safety precautions.

400. Evacuation of Sighting and Fire Control Instruments and Equipment

All items of sighting and fire control instruments and equipment, including such items as periscopes, ballistic drives, and elevation quadrants, are costly, difficult to replace, yet relatively light, hence, whenever practicable, they should be conserved and evacuated rather than destroyed. In the event of subsequent abandonment, the instruments and equipment will be completely destroyed and all optical elements and mountings smashed; firing tables, trajectory charts, and inflammable items, burned.

401. Destruction of Armament

a. GENERAL. Ordinarily the armament should be destroyed in conjunction with the destruction of the tank (par. 403); however, if limitation of time, personnel, and materials preclude simultaneous destruction of the tank, armament, and equipment, priority will be given to destruction of the armament.

b. 90-MM GUN T119 AND COMBINATION GUN MOUNT M78.
  (1) *Method No. 1—by demolition.*
    (a) *With HE ammunition.* With the gun at 0-degree elevation, open drain plugs on the recoil mechanism and allow recoil fluid to drain. It is not necessary to wait for the

recoil fluid to drain completely before proceeding as follows:
1. Remove the safety pin from an HE antitank rifle grenade or an HE antitank rocket. Then insert the grenade or rocket into the bore through the breech end of the weapon with the nose pointing toward the breech. The nose of the grenade or rocket should be about 20 inches forward of the origin of rifling. Next, load an HE round in the chamber.

Caution: Exercise extreme caution in handling the armed grenade or rocket.

2. As an alternative to *1* above, jam a fuzed HE round into the muzzle end of the gun and load the gun with a fuzed HE round, fuze set "SQ." An armor-piercing round, if used in lieu of the HE round, will not be as satisfactory.
3. Fire the weapon from cover using a lanyard about 100 feet long. The danger area is approximately 200 yards. Elapsed time: about 3 minutes.

(*b*) *With demolition materials.* Planning for simultaneous detonation, prepare and place the charges of EXPLOSIVE, TNT (using 1-pound blocks or equivalent together with the necessary detonating cord to make up each charge) as indicated below:

Charge	Location of charge
3-lb	Insert the charge in the muzzle end of the tube. Plug the muzzle tightly with any available material, such as rags or mud, to a distance of about 1 foot.
5-lb	Insert the charge into the chamber. Before closing the breechblock, insert an object, such as a hammer handle, in the breech opening to prevent damage to the detonating cord due to full closing of the breechblock.
2-lb	Place the charge on the power traversing and elevating control.

Connect these charges for simultaneous detonation with detonating cord. For methods of detonating these charges, refer to paragraph 403*c* (7), (8), and (9). Elapsed time: about 10 minutes.

(2) *Method No. 2—by burning (incendiary grenades).* With the gun at approximately zero-degree elevation, insert four unfuzed incendiary grenades end to end, midway in the tube. Ignite the four grenades by a fifth one fitted with a length of safety fuse to give a 15-second delay (safe fuze burns at the

rate of 1 foot in 30 to 45 seconds; test before using). The metal from the grenades will fuse with the tube and fill the grooves. Place four unfuzed incendiary grenades on the power traversing and elevating control. Ignite the four grenades by a fifth one fitted with a length of safety fuse to give a 15-second delay. After placing the grenades, first ignite those in the tube and then those on the power traversing and elevating control. Elapsed time: about 3 minutes.

c. MACHINE GUNS—BY MECHANICAL MEANS. The machine guns mounted on this vehicle, unless required for protection of troops during withdrawal, may be destroyed simultaneously with the vehicle as follows: Using an ax, sledge, or other heavy implement, destroy the machine gun by smashing the retracting slide handle assembly, spade grip assembly, cover assembly, feedway, receiver, barrel jacket, and barrel. If removed to cover withdrawal of troops, this weapon, if subsequently abandoned, should be destroyed as indicated above. Elapsed time: about 2 minutes.

d. MACHINE GUN MOUNTS BY MECHANICAL MEANS. Using an ax, sledge, machine gun barrel or similar implement, smash the essential parts of the machine gun mounts. Elapsed time: about 1 minute.

402. Destruction of Communication Equipment

Radios and intercommunication equipment will be damaged mechanically in conjunction with the destruction of the tank (par. 403); circuit and wiring diagrams and performance data will be burned.

403. Destruction of the Tank

a. GENERAL. Although varying degrees of damage to the armament and equipment of the tank may be expected incidental to the destruction of the tank by any one of the methods outlined below, complete destruction of the combat vehicle (tank, armament, and equipment) requires that applicable parts of the procedure for destruction of the armament and equipment (pars. 400, 401, and 402) be coordinated with the method employed for destruction of the tank.

b. METHOD NO. 1—BY BURNING.
  (1) Remove and empty portable fire extinguishers and discharge the fixed fire extinguisher system.
  (2) Drain or puncture the fuel tanks, collecting the gasoline in containers for use as outlined in (9) below.
  (3) If quantities of combustibles are limited, smash all vital elements such as magnetos, carburetors, air cleaners, generators, spark plugs, lights, instruments, and control levers.
  (4) Destroy the interphone amplifier by shearing off all panel knobs, dials, and switches.

(5) Break sockets, smash tubes, coils, microphones, earphones, and batteries in communication equipment.
(6) Smash the power traversing and elevating control box.
(7) Smash engine blocks, crankcases, and transmission.
(8) Explosive ammunition should be removed from packing or other protective material. Place ammunition in and about the tank so that greatest damage will result from its detonation. Remove any safety devices from the ammunition.
(9) With all doors and hatches open to admit air for combustion, pour gasoline and oil in and over the entire vehicle; ignite and take cover.

Caution: When igniting the gasoline, due consideration should be given to the highly inflammable nature of gasoline and its vapor. Carelessness in its use may result in painful burns.

Elapsed time: about 6 minutes.

c. METHOD NO. 2—BY DEMOLITION.
(1) Remove and empty portable fire extinguishers and discharge the fixed fire extinguisher system.
(2) Drain the fuel tanks or puncture them as near the bottom as possible.
(3) Smash all vital elements as outlined in *b* (3) and (6) above.
(4) For armament, prepare and set charges as outlined in paragraph 401.
(5) For the engine compartment and tracks, prepare six 2-pound charges of EXPLOSIVE, TNT (using two 1-pound blocks or equivalent together with necessary detonating cord to make up each charge). Place the charges as follows:
  (*a*) Set the *first* charge on the accessory drive housing at the forward end of the engine.
  (*b*) Set the *second* and *third* charges on the engine; one on the *left* side and the other on the *right* side.
  (*c*) Set the *fourth* charge between the engine and the cross drive transmission, as close to the flywheel housing as possible.
  (*d*) Set the *fifth* and *sixth* charges on the left and right track driving sprockets.
(6) Connect these six charges for simultaneous detonation with detonating cord.
(7) Provide for dual priming to minimize the possibility of a misfire. For priming, either a nonelectric blasting cap crimped to at least 5 feet of safety fuse (safety fuse burns at the rate of 1 foot in 30 to 45 seconds; test before using) or an electric blasting cap and firing wire may be used. If a nonelectric blasting cap and safety fuse are used, the fuse should be sufficiently long and so positioned that it may be

ignited from the outside of the tank since gasoline which is draining from the fuel tanks may be exploded prematurely by the burning fuse. Safety fuse, which contains black powder, and nonelectric blasting caps must be protected from moisture at all times. The safety fuse may be ignited by a fuse lighter or a match; the electric blasting cap requires a blasting machine or equivalent source of electricity.

(8) Connect all charges, the charges for the armament and the charges for the engine compartment and tracks, for simultaneous detonation with detonating cord. Detonate these charges as indicated in (8) below.

Caution: Keep the blasting caps, detonating cord, and safety fuse separated from the charge until required for use.

Note. For the successful execution of methods of destruction involving the use of demolition materials, all personnel concerned will be thoroughly familiar with the provisions of FM 5-25. Training and careful planning are essential.

(9) Detonate the charges. If primed with nonelectric blasting cap and safety fuse, ignite and take cover. If primed with electric blasting cap, take cover before firing. Elapsed time: about 10 minutes.

d. METHOD NO. 3—BY GUNFIRE.

(1) Remove and empty portable fire extinguishers and discharge the fixed fire extinguisher system.

(2) Drain or puncture the fuel tanks.

(3) Smash all vital elements as outlined in *b* above.

(4) Destroy the tank by gunfire using adjacent tanks, artillery, machine guns, rifles using rifle grenades, or launchers using antitank rockets. Fire on the tank aiming at the track driving sprockets, bogies, idlers, tracks, engine compartment, turret, and armament. Although one well-placed direct hit may destroy the tank, several hits are usually required for complete destruction unless an intense fire is started, in which case the tank may be considered destroyed.

Caution: Firing at ranges of 500 yards or less should be from cover.

Elapsed time: about 6 minutes.

(5) Unless evacuated, destroy the last remaining tank or weapon by the best means possible.

APPENDIX
REFERENCES

1. Publication Indexes

The following publication indexes and lists of current issue should be consulted frequently for latest changes or revisions of references given in this appendix and for any new publications relating to matériel covered in this manual:

Index of Administrative Publications (Army Regulations, Special Regulations, Joint Army-Air Force Adjustment Regulations, General Orders, Bulletins, Circulars, Commercial Traffic Bulletins, Army Procurement Circulars, Department of the Army Pamphlets, and ASF Manuals)	SR 310-20-5
Index of Army Motion Pictures, Kinescope Recordings, and Film Strips	SR 110-1-1
Index of Training Publications (Field Manuals, Training Circulars, Firing Tables and Charts, Army Training Programs, Mobilization Training Programs, Army Training Tests, Graphic Training Aids, Joint Army-Navy Air Force Publications, Combined Communications Board Publications, and Army Communications Publications)	SR 310-20-3
Index of Blank Forms and Army Personnel Classification Tests	SR 310-20-6
Index of Technical Manuals, Technical Regulations, Technical Bulletins, Supply Bulletins, Lubrication Orders, Modification Work Orders, Tables of Organization and Equipment, Reduction Tables, Tables of Allowances, Tables of Organization, and Tables of Equipment	SR 310-20-4
Introduction and Index (supply catalog)	ORD 1
Military Training Aids	FM 21-8

2. Supply Catalogs

a. AMMUNITION.

The following catalogs of the Department of the Army Supply Catalog pertain to this matériel:

Ammunition for Antiaircraft Artillery _____ ORD 3 SNL P-5
Ammunition Instruction Material for Antiaircraft, Harbor Defense, Heavy Field Artillery, Including Complete Round Data _____ ORD 3 SNL P-8
Fuzes, Primers, Blank Ammunition, and Miscellaneous Items for Antiaircraft, Harbor Defense, Heavy Field, and Railway Artillery_ ORD 11 SNL P-7
Service Fuzes and Primers for Pack, Light, and Medium Field, Aircraft, Tank, and Antitank Artillery _____ ORD 3 SNL R-3

b. ARMAMENT.
  Gun, Machine, Cal. .30, Browning, M1919A4, Fixed and Flexible; M1919A5, Fixed; M1919-A6; and Ground mounts _____ ORD (*) SNL A-6.
  Gun, Machine, Cal. .50, Browning, M2, Heavy Barrel, Fixed and Flexible; and Ground Mounts _____ ORD (*) SNL A-39
  Mount, Machine Gun, AA, Cal. .50 (6580030) _____ ORD (*) SNL A-55, Sec. 38

c. DESTRUCTION TO PREVENT ENEMY USE.
Ammunition:
  Land Mines and Fuzes, Demolition Material, and Ammunition for Simulated Artillery and Grenade Fire _____ ORD 11 SNL R-7

d. MAINTENANCE AND REPAIR.
  Cleaners, Preservatives, Lubricants, Recoil Fluids, Special Oils, and Related Maintenance Materials _____ ORD 3 SNL K-1
  Items of Soldering, Metallizing, Brazing, and Welding Materials; Gases and Related Items _____ ORD 3 SNL K-2
  Lubricating Equipment, Accessories, and Related Dispensers _____ ORD (*) SNL K-3
  Tool Set, Organizational Maintenance (2d echelon), Set No. 1, Common _____ ORD (*) SNL J-7, Sec. 1

*See ORD 1, Introduction and Index, for published catalogs of the ordnance section of the Department of the Army Supply Catalog.

Tool Set, Organizational Maintenance (2d echelon), Set No. 1, Supplemental _____ ORD (*) SNL J-7, Sec. 2
Tool Set, Organizational Maintenance (2d echelon), Set No. 2, Common _____ ORD (*) SNL J-7, Sec. 3
Tool Set, Organizational Maintenance (2d echelon), Set No. 2, Supplemental _____ ORD (*) SNL J-7, Sec. 4
Tool Set, Organizational Maintenance (2d echelon), Set No. 4, Block and Tackle _____ ORD (*) SNL J-7, Sec. 5
Tool Set, Organizational Maintenance (2d echelon), Set No. 5, Oxy-Acetylene _____ ORD (*) SNL J-7, Sec. 6

e. SIGHTING AND FIRE CONTROL EQUIPMENT.

Light, Aiming Post, M14, M41, M43; Light, Instrument, M1, M2, M10, M12, M13, M18, M19, M20, M22, M28, M30, M31, M32, M33, M34, M35, M36, M37, M38, M39C, M42, M45, M46, M47, M48 _____ ORD (*) SNL F-205
Periscope M6 _____ ORD (*) SNL F-235, Vol. 2
Periscope M12, M13, M13B1, M14, M14A1, M17, and T22 _____ ORD (*) SNL F-235, Vol. 5
Quadrant, Gunner's, M1 _____ ORD (*) SNL F-140
Setter, Fuze M22, M23 _____ ORD (*) SNL F-293
Slides M1; Table, Graphical Firing, M22, M23, M24, M25, M26, M26A1, M27, M28, M30, M31, M32, M33, M34, M35, M37 and M38 _____ ORD (*) SNL F-237

f. VEHICLE.

Tank, 90-mm, Gun M47 _____ ORD (*) SNL G-262

3. Forms

The following forms are applicable to this matériel:
Standard Form 91, Operator's Report of Motor Vehicle Accident.
Standard Form 91A, Transcript of Operator's Report of Motor Vehicle Accident.
Standard Form 93, Report of Investigating Officer.
Standard Form 94, Statement of Witness.
DA Form 30b, Report of Claims Officer.

---

*See ORD 1, Introduction and Index, for published catalogs of the ordnance section of the Department of the Army Supply Catalog.

DA AGO Form 9-3, Processing Record for Storage and Shipment (Tag).
DA AGO Form 9-4, Vehicular Storage and Servicing Record (Card).
DA AGO Form 9-69, Spot Check Inspection Report for all Full-Track and Tank-Like Wheeled Vehicles.
DA AGO Form 9-74, Motor Vehicle Operator's Permit.
DA AGO Form 9-75, Daily Dispatching Record of Motor Vehicle.
DA AGO Form 348, Driver's Qualification Record.
DA AGO Form 460, Preventive Maintenance Roster.
DA AGO Form 461-5, Limited Technical Inspection.
DA AGO Form 462, Preventive Maintenance Service and Inspection for Full-Track Vehicles.
DA AGO Form 468, Unsatisfactory Equipment Report.
DA AGO Form 478, MWO and Major Unit Assembly Replacement Record and Organizational Equipment File.
DA AGO Form 811, Work Request and Job Order.
DA AGO Form 811-1, Work Request and Hand Receipt.
DD Form 6, Report of Damaged or Improper Shipment.
DD Form 317, Preventive Maintenance Service Due (Sticker).
OO Form 5825, Artillery Gun Book.

4. Other Publications

The following explanatory publications contain information pertinent to this matériel and associated equipment:

a. AMMUNITION.

Distribution of Ammunition and Explosives for Training Purposes	SR 710-60-50
Ammunition, General	TM 9-1900
Ammunition Inspection Guide	TM 9-1904
Artillery Ammunition	TM 9-1901
Ballistic Data, Performance of Ammunition	TM 9-1907
Qualification in Arms and Ammunition Training Allowances	AR 775-10
Regulations for Firing Ammunition for Training, Target Practice, and Combat	SR 385-310-1
Small Arms Ammunition	TM 9-1990

b. ARMAMENT.

Browning Machine Guns, Cal. .30, M1917A1, M1919A4, and M1919A6	FM 23-55
Browning Machine Gun, Cal. .50, HB, M2	FM 23-65
Fundamentals of Artillery Weapons	TM 9-2305

c. CAMOUFLAGE.

Camouflage	TM 5-267
Camouflage, Basic Principles	FM 5-20
Camouflage of Vehicles	FM 5-20B

d. COMMUNICATIONS.

Electrical Communication Systems Equipment	FM 11-487
Field Radio Techniques	FM 24-18
Radio Direction Finding	TM 11-476
Radio Operator's Manual, Army Ground Forces	FM 24-6
Signal Communications in Armored Division	FM 17-70
Suppression of Radio Noises	TM 11-483

e. DECONTAMINATION.

Decontamination	TM 3-220
Decontamination of Armored Force Vehicles	FM 17-59
Defense Against Chemical Attack	FM 21-40

f. DESTRUCTION TO PREVENT ENEMY USE.

Explosives and Demolitions	FM 5-25

g. GENERAL.

Driver Selection, Training and Supervision, Half-Track and Full-Track Vehicle	TM 21-301
Instruction Guide: Operation and Maintenance of Ordnance Matériel in Extreme Cold (0° to −65° F.)	TM 9-2855
Manual for the Full-Track Driver	TM 21-306
Motor Vehicles	AR 700-105
Mountain Operations	FM 70-10
Operations in Snow and Extreme Cold	FM 70-15
Precautions In Handling Gasoline	AR 850-20
Preparation of Ordnance Matériel for Deep-Water Fording	TM 9-2853
Principles of Automotive Vehicles	TM 9-2700
Report of Accident Experience	SR 385-10-40
Spark Plugs	TB ORD 313
Storage Batteries: Lead-Acid Type	TM 9-2857
Supplies and Equipment: Unsatisfactory Equipment Report	SR 700-45-5

h. MAINTENANCE AND REPAIR.

Cleaning, Preserving, Sealing, and Related Materials Issued for Ordnance Matériel	TM 9-850
Hand, Measuring, and Power Tools	TM 10-590
Maintenance and Care of Hand Tools	TM 9-867
Maintenance and Care of Pneumatic Tires and Rubber Treads	TM 31-200
Maintenance of Supplies and Equipment: Maintenance Responsibilities and Shop Operation	AR 750-5
Motor Vehicle Inspection and Preventive Maintenance Services	TM 37-2810

Painting Instructions for Field Use.......... TM 9-2851
Tracklaying Vehicles: Tracks Currently Applicable.......... TB ORD 391

i. SHIPMENT AND LIMITED STORAGE.
Army Shipping Document.......... TM 38-705
Catalog of Approve Packaging Instructions for Major Items and Spare Parts for Ordnance General Supplies.......... [1] PS 1000
Instruction Guide: Ordnance Packaging and Shipping (Post, Camps, and Stations)...... TM 9-2854
Marking and Packing of Supplies and Equipment: Marking of Oversea Supply.......... SR 746-30-5
Ordnance Storage and Shipment Chart—Group G.......... TB 9-0SSC-G
Packaging of Tank, Medium, M46; Tank, Medium, M46A1; Tank, 90-mm Gun, M47; Tank, 90-mm Gun, T48 (For Oversea Shipment or Storage).......... [2] MIL-P-11290 (ORD)
Preparation of Unboxed Matériel for Shipment.......... SB 9-4
Protection of Ordnance General Supplies in Open Storage.......... TB ORD 379
Shipment of Supplies and Equipment: Report of Damaged or Improper Shipment.......... SR 745-45-5
Standards for Oversea Shipment and Domestic Issue of Ordnance Matériel Other Than Ammunition and Army Aircraft.......... TB ORD 385
Storage, Inspection, and Issue of Unboxed Serviceable Motor Vehicles; Preparation of Unserviceable Vehicles for Storage; and Deprocessing of Matériel Prior to Operation. SB 9-63

j. SIGHTING AND FIRE CONTROL EQUIPMENT.
12-Inch Graphical Firing Tables.......... TM 9-524
Auxiliary Sighting and Fire Control Equipment.......... TM 9-575
Cold Weather Lubrication: Operation and Maintenance of Artillery and Sighting and Fire Control Matériel.......... TB ORD 193
General Supply: Winterization Equipment for Automotive Matériel.......... SB 9-16
Inspection Guide: Elementary Optics and Applications to Fire Control Instruments... TM 9-2601
Lubrication of Ordnance Matériel for Operation in Extreme Cold.......... TB 9-2855-1

[1] Copies may be obtained from the Raritan Arsenal Publications Division, Metuchen, New Jersey.
[2] This specification may be requisitioned from the Commanding General, Ordnance Depot, Aberdeen Proving Ground, Maryland.

INDEX

	Paragraph	Page
90-mm gun T119E1:		
Ammunition	381	599
Assembly:		
Breech ring	340	561
Evacuator chamber	340	561
Data	337	555
Description	337	555
Disassembly:		
Breech ring	338	557
Muzzle brake and evacuator	338	557
Maintenance:		
Evacuator chamber	339	560
General	339	557
Gun tube	339	560
Tabulated data	337	557
90-mm gun tank M47:		
Auxiliary generator and engine	3	14
Bilge pump	3	14
Crew compartment heater	3	14
Data	3	13
Description	3	13
Fire extinguishers	3	14
General	3	13
Hull	3	13
Power train and suspension	3	13
Turret	3	13
Ventilating blower	3	14
Abnormal noise in hydraulic pump or motor	138	194
Accelerator pedal:		
Location	14	23
Accessory outlet	141	207
Accessories panel (turret):		
Installation	297	516
Location	281	503
Removal	297	516
Accessory drive ratios (main engine)	145	222
Adjusting—idler wheels and hubs:		
Cleaning	251	454
Inspections	251	454
Installation	251	454
Removal	251	454
Adjustment:		
Bands	236	395
Brake control linkage	259	470
Carburetors (main engine)	172	278
Commander's seat control handle	49	50

	Paragraph	Page
Adjustment—Continued		
Doors and cover plates	261	479
Driver's controls and linkage	137	192
Driver's seats	20	26
Elevation and depression limit switches	286	506
Elevation quadrant micrometer	369	588
Horn	226	384
Manual control box and linkage	257	463
Reverse and low gear band	236	395
Rocker replacement	156	243
Shifting control linkage	257	464
Spark plugs	186	315
Steering control linkage	257	464
Throttle control linkage	258	468
Tracks	249	423
Traversing lock (turret)	272	496
Valve timing	155	237
Valves:		
Oil cooler bypass	151	231
Oil filter bypass	151	231
Aiming point method of bore sighting	91	99
Air cleaners:		
Auxiliary engine:		
Installation	305	527
Removal	305	527
Main engine:		
Installation	175	289
Removal	175	286
Servicing	175	285
Air-intake system:		
Description	171	275
Tabulated data	171	278
Ammunition for 90-mm gun T119E1:		
Authorized rounds (table IX)	382	607
Boresafe	384	608
Care, handling, and preservation	381	599
Classification	379	595
Firing tables	378	594
Fuzes:		
Description	384	622
Setting	384	622
General	377	594
Identification:		
General	380	597
Lot number	380	597
Model	380	597
Weight zone marking	380	597
Marking	380	598
Nonboresafe	384	622
Packing and marking	386	626
Painting	380	598
Precautions in firing	385	625
Preparation for firing (table X)	383	608
Subcaliber ammunition	387	627

	Paragraph	Page
Ammunition chute micro switch:		
Installation	294	512
Location	281	503
Removal	294	512
Ammunition ready racks:		
Description	47	50
Installation	279	498
Removal	279	498
Antennas. (See communications system.)		
Armament:		
Cal. .50 coaxial machine gun	53	55
Cal. .50 HB Browning machine gun M2	366	584
Destruction	401	658
Extracting fired cartridge case	80	88
Failure of breech block to open	143	215
Failure to fire:		
90-mm gun	81, 143	88, 212
Coaxial machine gun	81	88
Firing:		
90-mm gun	79	87
Coaxial machine gun	79	87
Gun not elevated or depressed	139	198
Machine gun, manual controls	54	55
Premature firing	143	215
Preparation:		
For traveling	83	88
Under unusual conditions	101	112
Procedure (table V)	123	153
Tabulated data:		
90-mm gun T119E1	5	18
Gun mount M78	5	18
Unloading a stuck round	143	216
Assembly:		
90-mm gun T119E1	340	561
Breech operating mechanism	341	561
Compensating—idler arm and spindle	252	454
Firing linkage	353	577
Front road wheel arms and shackles	250	447
Front spring arms	250	446
Intermediate and rear road wheel arms	250	433
Percussion mechanism	345	569
Recoil mechanism	359	581
Track support rollers and bracket assemblies	254	459
Valves:		
Oil cooler bypass	151	231
Oil filter bypass	151	231
Wheel arm supports	250	449
Authorized forms:		
Artillery gun book	2	2
Auxiliary engine. (See auxiliary generator and engine.)		
Auxiliary equipment (see also hull accessories panel components and auxiliary equipment)	141	207

	Paragraph	Page
Auxiliary equipment:		
Bilge pumps:		
Data	321	544
Description	321	543
Installation	329	552
Removal	329	551
Service	329	551
Crew compartment heater:		
Data	321	544
Description	321	543
Installation	322	546
Removal	322	544
Heater combustion air blower:		
Installation	324	548
Removal	324	548
Heater fuel control valve:		
Installation	325	549
Removal	325	548
Heater fuel filter and fuel lines:		
Installation	327	550
Removal	327	550
Heater fuel pump:		
Installation	326	550
Removal	326	549
Heater safety valve:		
Installation	328	551
Removal	328	550
Heating ventilator air blower assembly:		
Brush replacement	323	548
Installation	323	548
Removal	323	548
Ignition replacement and maintenance:		
Cleaning	321	543
Installation	321	543
Removal	321	543
Auxiliary generator and engine:		
Air cleaner:		
Data	300	521
Installation	305	527
Removal	305	527
Carburetor:		
Data	300	521
Installation	307	528
Removal	307	528
Data	3	13
Description	3, 93, 300	13, 105, 517
External oil lines:		
Installation	313	539
Removal	313	537
Fails to run at full speed	140	206
Fails to start	140	206
Fuel pump:		
Data	300	521

Auxiliary generator and engine—Continued

	Paragraph	Page
Fuel pump—Continued		
Installation	308	529
Removal	308	528
Fuel shut-off valve and fuel lines:		
Installation	309	529
Removal	309	529
General	140, 300	206, 517
Generator:		
Data	221	370
Governor:		
Data	300	522
Inspection	93	105
Installation	303	525
Intake and exhaust manifolds and mufflers:		
Installation	306	527
Removal	306	527
Magneto:		
Installation	310	533
Procedure	310	531
Removal	310	530
Timing	310	531
Maintenance operations:		
In vehicle	301	522
Out of vehicle	301	522
Manual starter:		
Installation	312	537
Removal	312	534
Mounting Brackets:		
Installation	315	538
Removal	315	538
Oil pump	300	521
Procedure (table V).		
Replacement	303	524
Servicing:		
Check oil level in oil pan	302	523
Clean engine air cleaner	302	523
Clean fuel filter bowl	302	523
Drain engine oil pan	302	523
Fill engine oil pan	302	523
Spark plugs and cables:		
Data	300	522
Installation	311	534
Removal	311	534
Starting:		
Electric	93	107
Manual	93	107
Starts, fails to run	140	206
Stopping	93	107
Testing	132	182
Trouble shooting	132, 140	182, 206
Vent lines		
Installation	314	538
Removal	314	538

Auxiliary interphone equipment. (*See* communications system.)

	Paragraph	Page
Azimuth indicator T24:		
Data	5	17
General	64	73
Illumination	64	73
Ballistic box (turret)	139, 281	204, 503
Ballistic drive T23E1:		
Controls	61	68
Data	5	17
Description	61	68
Illumination	61	68
Batteries:		
Installation	224	379
Removal	224	379
Service	224	377
Testing installation	224	381
Bezel assembly:		
Installation	275	497
Removal	275	497
Bilge pump:		
Description	3	13
Operation	95	108
Blackout driving lights:		
Installation	229	388
Removal	229	387
Replacement of lamp unit	229	388
Blackout marker lights:		
Installation	230	388
Removal	230	388
Replacement	230	388
Booster and filter coils:		
Installation	188	319
Removal	188	318
Booster and filter coils (ignition system)	183	305
Booster motor:		
Installation	292	510
Location	281	502
Removal	292	510
Testing	292	510
Booster motor relay:		
Installation	293	511
Removal	293	511
Testing	293	511
Booster switch (main engine)	25	34
Bore sighting:		
Aiming point method	91	99
Emplacement	91	99
Installation	91	99
Procedure	91	99
Purpose	91	99
Testing target method	91	103
Brake control linkage:		
Adjustment	259	471
Installation	259	470
Removal	259	470
Replacement	259	475

	Paragraph	Page
Breechblock:		
Description	341	561
Installation	346	571
Maintenance	344	569
Removal	342	567
Breech mechanism:		
General	77, 341	86, 561
Operation:		
Open	77	86
Closing	77	86
Breech operating mechanism:		
Assembly	341	561
Description	341	561
Disassembly	343	568
Operation	341	362
Breech ring:		
Description	341	561
Disassembly	343	568
Maintenance	344	569
Cables and connectors	199	344
Cal. .30 Browning machine gun M1919A:		
Installation	365	583
Maintenance	365	583
Removal	365	583
Cal. .50 ammunition feed control (turret)	139	196
Cal. .50 HB Browning machine gun M2:		
Installation	366	584
Maintenance	366	584
Removal	366	584
Carburetors:		
Auxiliary engine:		
Installation	307	528
Removal	307	528
Main engine:		
Adjustment	172	278
Floods	128	171
Installation	172	281
Removal	172	281
Charging system	132	182
Check:		
Driver's controls and linkage	137	192
Engine tune up:		
Battery connections	146	222
Carburetor	146	222
Compression	146	222
Ignition	146	222
Valves	146	222
Oil level on transmission	235	394
Oil pressure on transmission	235	395
Starting main engine	68	77
Turret traversing and elevating systems	138	193
Circuit breaker tests:		
Auxiliary engine starter and magneto circuit	141	207
Bulge pump, front and rear	141	207

675

	Paragraph	Page
Cleaning:		
Adjusting idler wheels and hubs	251	454
Compensating—idler arm and spindle	252	455
Compensating—idler wheels	252	454
Exhaust manifold	153	234
Front road wheel arms and shackles	250	447
Front road wheels	250	431
Front spring arms	250	446
Fuel filters	176	289
Igniter replacement and maintenance	321	543
Intake manifold	167	268
Intermediate and rear road wheels and arms	250	433
Oil filter (main engine)	149	227
Oil pressure control valve	150	228
Primer lines	181	302
Road wheel hubs and bearings	250	431
Road wheels, hubs, arms, and bumpers	250	430
Rocker replacement	156	243
Shock absorbers and brackets	255	462
Torsion bars	253	455
Track drive sprockets	244	414
Tracks	249	423
Tracks and suspension	248	420
Track support rollers and bracket assemblies	254	459
Transmission:		
Oil filter	238	401
Side oil strainers	237	400
Valves:		
Oil cooler bypass	151	231
Oil filter bypass	151	231
Wheel arm supports	250	449
Coaxial cal. .50 HB Browning machine gun M2:		
Installation	364	583
Maintenance	364	583
Removal	364	583
Combination gun mount M78:		
Data	347	573
Description	347	573
Erratic recoil action	144	217
General	347	573
Gun:		
Does not return to battery	144	217
Overrecoils	144	217
Returns to battery, too great a shock	144	217
Underrecoils	144	217
Maintenance	347	573
Tabulated data	347	573
Communications system:		
Antennas:		
Components (table XII)	391	633
Description	391	633
Mast base AB-15/GR	391	633

	Paragraph	Page
Communications system—Continued		
Auxiliary interphone equipment:		
Description	392	634
Installation	393	640
Removal	393	639
Description	5	17
Destruction of equipment	402	660
General	388	628
Hull radio terminal box:		
Installation	394	640
Removal	394	640
Radio sets:		
Description (table XI)	389	628
Installation	390	632
Removal	390	631
Compensating—idler arm and spindle:		
Assembly	252	455
Cleaning	252	455
Disassembly	252	455
Inspection	252	455
Installation	252	455
Removal	252	455
Compensating—idler wheels:		
Cleaning	252	454
Inspection	252	454
Installation	252	454
Removal	252	454
Commander's control handle assembly:		
Description	281	499
Installation	283	504
Removal	283	504
Testing	283	504
Commander's control inoperative	138	194
Commander's coupola assembly:		
Installation	273	496
Removal	273	496
Replacement	273	496
Commander's coupola controls:		
Description	51	51
Commander's cupola door assembly:		
Installation	274	496
Removal	274	496
Commander's direct view prisms (*see also* observation of terrain)	89	98
Commander's-power-traversing-and-elevating-control handle:		
Location	38	42
Operation	38	41
Override of gunner	38	41
Commander's seat:		
Control handle, adjustment	49	50
Installation	278	497
Removal	278	497
Common tools and equipment	111	117
Conduits, cables, and connectors:		
General	196	343
Installation	196	343

677

	Paragraph	Page
Conduits, cables, and connectors—Continued		
Removal	196	342
Conduits, cables, and connectors (*see also* hull electrical system)	192	330
Connecting:		
Tracks and suspension	249	423
Continuity tests:		
Ohmmeter method of electrical testing	125	161
Controls:		
Ballistic drive T23E1	61	68
Periscope	59	64
Range finder T41	58	60
Controls and instruments:		
Arrangement and use of sighting and fire control equipment	57	57
Description:		
90-mm gun manual controls	54	55
Cal. .50 coaxial machine gun	53	55
Gun traveling lock	55	57
Location:		
Ammunition ready racks	47	50
Commander's-power-traversing-and-elevating control handle	38	43
Elevation-cylinder-control valve	44	46
Elevation-superchare-line-shut-off valve	46	50
Firing selector switches	39	44
Gunner's:		
Manual-elevating-control handle	37	41
Manual-traversing-control handle	36	40
Power-traversing-and-elevating-control handle	35	39
Seat control handle	48	50
Hand pump	43	46
Indicator lights	39	44
Loader's:		
Firing-reset-safety switch with indicator light	41	46
Traverse-safety switch and indicator light	40	44
Oil-return-valve handle	45	49
Turret lock	42	46
Correction of deficiencies	7	21
Crash pads and head rests:		
General	263	481
Installation	263	482
Removal	263	481
Crew compartment heater:		
Description	3	13
Installation	322	546
Location	97	108
Removal	322	544
Cylinder air deflectors:		
Installation	166	267
Removal	166	267
Data:		
90-mm gun T119E1	337	555
90-mm gun tank M47	3	13
Air-intake system	171	275
Armament	5	18

	Paragraph	Page
Data—Continued		
Armor:		
Combination gun mount	347	573
Hull	5	17
Turret	5	17
Auxiliary generator and engine:		
Air cleaner	300	522
Carburetor	300	521
Fuel pump	300	522
Generator	221	373
Governor	300	522
Spark plugs and cables	300	522
Auxiliary equipment:		
Bilge pumps	321	543
Crew compartment heater	321	543
Batteries and generating system	221	543
Communications system	5	19
Exhaust system	171	275
Fuel system	171	275
Ignition system	171	275
Main engine	145	217
Performance	5	19
Periscope mounts:		
Commander's T177	5	19
Gunner's T176	5	19
Range finder T41	5	19
Sighting and fire control:		
Azimuth indicator T24	5	18
Ballistic drive T23E1	5	18
Elevation quadrant T21	5	18
Periscopes	5	18
Starting system	189	319
Telescope	5	19
Tracks and suspension	248	420
Transmission	234	391
Description:		
90-mm firing mechanism	281	502
90-mm gun T119E1	337	555
90-mm gun tank M47	3	13
90-mm gun manual controls	54	55
Accessory panel	281	503
Air-intake system	171	275
Ammunition chute micro switch	281	503
Ammunition ready racks	47	50
Antennas	391	633
Auxiliary:		
Equipment	321	543
Generator and engine	171, 300	275, 522
Interphone equipment	392	634
Ballistic drive T23E1	61	68
Batteries and generating system	221	373
Bilge pumps	3, 321	14, 543
Booster motor	281	502
Booster motor relay	281	503
Breechblock	341	562

	Paragraph	Page
Description—Continued		
Breech operating mechanism	341	562
Breech ring	341	561
Cables and connectors	199	344
Cal. .50 coaxial machine gun manual controls	53	55
Combination gun mount	347	573
Commander's control handle assembly	281	499
Commander's cupola controls	51	50
Communications system:		
Antennas	391	633
Radio sets	389	628
Crew compartment heater	321	543
Dimmer switch	28	35
Drain valves:		
Crew compartment	21	30
Engine compartment	21	30
Driver's controls and linkage	256	462
Driver's doors	19	26
Elevation quadrant	63	72
Elevation mechanism	360	581
Elevating gun	145	217
Engine lubrication	127	167
Engine lubrication system	148	223
Exhaust system	171	275
Extractors	341	562
Extracting fired cartridge case	80	88
Failure to fire	81	88
Final drives	243	414
Fire extinguishers:		
Fixed	316	539
Portable	316	539
Firing:		
90-mm gun and coaxial machine gun	79	87
Linkage	349	574
Mechanism	341	561
Safety panel	281	502
Fuel lines	179	289
Fuel system	171	275
Fuze setter M14	66	75
Generator:		
Auxiliary engine	300	517
Main engine	221	373
Governor:		
Auxiliary engine	300	517
Main engine	171	275
Gunner's periscope mount T176	60	66
Horn and lighting system	133, 225	185, 381
Hull	260	476
Hull electrical system	192	327
Indicator lights	39, 40, 41	44, 46
Instrument lights	65	75
Loader's hatch door controls	52	55
Loader's traverse safety box	281	502
Loading 90-mm gun	78	87
Machine gun mounts	363	582

Description—Continued	Paragraph	Page
Main engine	145	217
Manual control lever	11	23
Manual elevating mechanism	362	582
Master junction box	192, 194	327, 335
Oil coolers and lines	240	403
Periscope M13, M13B1, and M6	62	71
Periscope T35:		
Commander's	59	64
Gunner's	59	64
Placing vehicle in motion	69	79
Power elevating mechanism	361	581
Radio interference suppression system	331	553
Radio terminal box	281	503
Radio sets (table XI)	389	628
Range finder T41	58	60
Recoil mechanism	355	578
Replenisher indicator	56	57
Slave battery receptacle	98	109
Slip ring box	281	503
Starting system	189	319
Superelevation transmitter T13	58	60
Tracks and suspension	248	420
Towing the vehicle	74	83
Transmission:		
Hydraulic and lubrication system	235	393
Oil pressure tests (table VIII)	236	396
Side oil strainers	237	400
Turret	269	488
Turret electrical system	139, 335	195, 554
Universal joints	243	414
Ventilating blower	281	503
Destruction of material to prevent enemy use:		
Armament	401	658
Communication equipment	402	660
General	399	657
Sighting and fire control instruments and equipment	400	658
Tank	403	660
Dimmer switch:		
Description	28	35
Direct fire:		
Fire control equipment and sighting	57	57
Operation	86	95
Disassembly:		
90-mm gun T119E1	338	557
Breech operating mechanism	343	568
Breech ring	343	568
Compensating—idler arm and spindle	252	
Firing linkage	351	577
Front road wheel arms and shackles	250	447
Front spring arms	250	446
Intermediate and rear road wheel arms	250	433
Oil pressure control valve	150	228
Percussion mechanism	343	569
Recoil mechanism	357	580

	Paragraph	Page
Disassembly—Continued		
Track support rollers and bracket assemblies	254	459
Valves:		
Oil cooler bypass	151	232
Oil filter bypass	151	231
Wheel arm supports	250	430
Dome light controls:		
Installation	29	35
Dome lights:		
Do not operate	139	195
Installation	232	390
Removal	232	389
Replacement	232	391
Turret	281	503
Doors and cover plates:		
Adjustment	261	479
Installation	261	480
Removal	261	479
Drain engine lubrication system	148	225
Draining fuel tanks	177	291
Drain valves:		
Description	21	30
General	264	482
Installation	264	482
Removal	264	482
Driver's controls and linkage:		
Adjustment	137	192
Check	137	192
Description	256	462
Driver's doors	19	26
Driver's or operators preventive maintenance services:		
Armament (table IV)	122	145
Vehicle (table III)	122	145
Driver's seats:		
Adjustment	20	26
Installation	262	481
Removal	262	481
Driving precautions	73	82
Driving vehicle:		
Instructions	70	81
Steering	70	81
Electrical circuit numbers (table VI)	193	332
Elevate or depress gun	38	41
Elevating mechanism:		
Description	360	581
Elevating systems (*see also* turret traversing and elevating systems)	138	193
Elevation and depression limit switches:		
Adjustment	286	506
Installation	286	506
Removal	286	506
Testing	286	506
Elevation and traversing mechanism:		
Run-in procedure	9	22

	Paragraph	Page
Elevation-cylinder-control valve:		
Location	44	46
Elevation quadrant T21:		
Adjustment	369	588
Data	5	17
Description	63	72
Elevation-supercharge-line-shut-off valve:		
Location	46	50
Purpose	46	50
Engine. (*See* Auxiliary generator and engine; main engine.)		
Engine wiring junction box (*see also* hull electrical system)	192	327
Erratic recoil action (gun mount M78)	144	217
Escape hatch doors	18	26
Excessive oil consumption	126	166
Excessive oil pressure	127	168
Excessive radio interference	142	207
Exhaust manifold:		
Cleaning	153	234
Inspection	153	234
Installation	153	235
Removal	153	234
Exhaust pipes and mufflers: Installation and removal	182	
Exhaust system:		
Data	171	275
Description	171	275
Tabulated data	171	278
External oil lines:		
Installation	168	271
Removal	168	271
Extracting	341	562
Extractors. (*See* Breech mechanism and breech operating mechanism.)		
Failure of breechblock to close (90-mm gun T119E1)	143	215
Failure of breechblock to open (90-mm gun T119E1)	143	215
Failure of round to chamber (90-mm gun T119E1)	143	213
Failure to extract cartridge case (90-mm gun T119E1)	143	216
Failure to fire (90-mm gun T119E1)	143	213
Fasteners and bond straps	336	555
Fenders and dust shields		
General	266	486
Installation	266	486
Removal	266	486
Field report of accidents	2	2
Final drives:		
Description	243	414
Installation	246	418
Removal	246	417
Replacement	246	417
Service	246	417
Final drives (*see also* transmission and final drives)	134	187
Fire control equipment and sighting:		
Arrangement	57	57
Direct fire	57	57
General	57	57

	Paragraph	Page
Fire control equipment and sighting—Continued		
Indirect fire	57	60
Miscellaneous equipment	57	60
Use	57	57
Fire control instruments for traveling:		
Preparation:		
Fuze setter M14 and M27	90	99
Periscope head T35	90	98
Periscope M13 or M6	90	99
Fire extinguishers:		
Description	3	13
Fixed:		
Cylinders:		
Installation	317	540
Removal	317	539
Description	316	539
Lines and discharge nozzle:		
Installation	319	542
Removal	319	540
Pull cable:		
Installation	318	540
Removal	318	540
Location	94	107
Portable:		
Description	316	539
Installation	320	543
Removal	320	542
Firing 90-mm gun and coaxial machine gun	79	87
Firing control (turret)	139	196
Firing linkage:		
Assembly	353	577
Description	349	574
Disassembly	351	577
Installation	354	578
Maintenance	352	577
Removal	350	577
Firing mechanism:		
Description	341	561
Installation	346	571
Maintenance	344	569
Removal	342	563
Firing safety panel:		
Installation	287	508
Location	281	502
Removal	287	507
Firing selector switches:		
Location	39	44
Forms:		
Authorized	2	2
General	2	2
Front road wheel arms and shackles:		
Assembly	250	446
Cleaning	250	445
Disassembly	250	447

	Paragraph	Page
Front road wheel arms and shackles—Continued		
Inspection	250	448
Installation	250	449
Removal	250	447
Front road wheels:		
Cleaning	250	431
Inspection	250	431
Installation	250	431
Removal	250	431
Front spring arms:		
Assembly	250	446
Cleaning	250	446
Disassembly	250	446
Inspection	250	446
Installation	250	446
Removal	250	446
Fuel, air-intake, and exhaust systems	171	275
Fuel-cut-off switch	26	34
Fuel filters:		
Cleaning	176	290
Installation	176	291
Removal	176	289
Fuel lines:		
Description	179	299
Fuel mixture too lean:		
Fuel system	128	170
Fuel not reaching carburetor	128	169
Fuel pumps:		
General	174	284
Installation	174	285
Removal	174	285
Fuel system:		
Carburetor floods	128	171
Data	171	275
Description	171	275
Engine slows down but does not stop	128	171
Fuel not reaching carburetor	128	169
Fuel mixture too lean	128	170
General	128	169
Operation of fuel cut-off switch	128	171
Tabulated data	171	278
Fuel tanks:		
Draining	177	291
Installation	177	293
Removal	177	292
Fuel-tank-shut-off-valve handles:		
Location	17	25
Fuel valves and controls:		
Installation	178	298
Removal	178	297
Functioning of recoil mechanism	356	579
Fuze setter M14:		
Description	66	75
Operation	85	93

	Paragraph	Page
Gages:		
Run-in procedure	9	22
General:		
90-mm gun tank M47	3	13
Accessory outlet socket	212	359
Auxiliary-engine:		
Megneto switch	210	359
Starter switch	209	358
Auxiliary generator and engine	140, 300	206, 517
Azimuth indicator T24	64	73
Bilge pump switches	208	358
Breech mechanism	77, 341	86, 561
Combination gun mount	347	573
Conduits, cables, and connectors	196	343
Communications system	388	628
Crash pads and head rests	263	481
Drain valves	264	482
Destruction of matériel to prevent enemy use	399	657
Driving precautions	73	82
Engine-warning-signal-light switches	218	366
Fenders and dust shields	266	486
Fire control equipment and sighting	57	60
Forms	2	2
Fuel pumps	174	284
Fuel system	128	169
Gages	205	356
Heater box control	205	356
Horn and lighting system	133	185
Hull	260	476
Hull accessories panel	201	354
Hull accessories panel circuit breaker	215	361
Hull electrical system	192	327
Ignition system	129, 183	171, 304
Instrument panel	185	311
Instrument panel circuit breaker	214	360
Light switch	203	355
Magnetos	213	360
Main engine:		
Starter and magneto switch	204	356
Wiring junction box	195	341
Master current relay rectifier	213	360
Master junction box	194	335
Master relay switch	207	358
Ohmmeter method of electrically testing	125	161
Oil cooler fan controllers and thermostats	220	368
Oil coolers and lines	240	403
Operation under unusual conditions	99	109
Power pack	198	343
Primer lines	181	302
Replacement of authorized parts	368	587
Scope	1	1
Sending units	217	362
Sighting and fire control equipment	367	585
Speedometer-tachometer	202	355
Starting main engine	68	77

	Paragraph	Page
General—Continued		
Top deck	158	245
Towing the vehicle	74	83
Tracks and suspension	248	420
Transmission-warning-signal-light switches	219	367
Traversing turret and elevating gun	76	85
Turret	269	488
Valves:		
Oil cooler bypass	151	231
Oil filter bypass	151	231
Warning signal lights	206	357
Wiring harnesses	216	361
Generator:		
Auxiliary engine:		
Description	300	517
Installation	304	526
Removal	304	526
Main engine:		
Description	221	370
Installation	222	376
Removal	222	374
Regulating units	223	377
Generating system. (*See* Radio interference suppression system.)		
Governor:		
Auxiliary engine	300	521
Main engine	171	277
Gun does not return to battery (combination gun mount M78)	144	217
Gun elevating (turret)	139	195
Gun mount M78 (*see also* combination gun mount M78)	144	217
Gunner's control handle assembly:		
Installation	284	505
Location	281	502
Removal	284	505
Testing	284	505
Gunner's control inoperative	138	194
Gunner's-manual-elevating-control handle:		
Location	37	41
Gunner's-manual-traversing-control handle:		
Location	36	39
Gunner's periscope mount T176:		
Description	60	66
Gun ready light	60	66
Gunner's-power-traversing-and-elevating-control handles:		
Location	35	39
Operation	35	39
Gunner's seat:		
Control handle:		
Installation	278	497
Removal	278	497
Gun overrecoils (combination gun mount M78)	144	217
Gun ready indicators:		
Location	281	502
Replacement	289	509
Gun ready light:		
Gunner's periscope mount T176	60	66

	Paragraph	Page
Gun returns to battery, too great a shock (combination gun mount M78)	144	217
Gun traveling lock:		
Installation	267	486
Operation	55	57
Removal	267	486
Hand brake lock handle	13	23
Hand elevating mechanism inoperative or weak	138	194
Hand pump	43	46
Hand throttle lever	15	23
Hard riding	136	191
Head lamp dimmer switch:		
Installation	233	391
Removal	233	391
High oil temperature	127	167
Horn:		
Adjustment	226	384
Installation	226	384
Removal	226	384
Horn and lighting system:		
Description	133, 225	185, 381
General	133	185
Testing	133	187
Trouble shooting	133	186
Horn switch:		
Installation	227	384
Location	30	35
Removal	227	384
Hot spot manifold, tubes and valve:		
Installation and removal	173	283
Hubs and bearings (see also road wheels, hubs, arms, and bumpers)	250	430
Hull:		
90-mm gun tank M47	3	13
Description	260	476
General	260	476
Hull accessories panel components and auxiliary equipment:		
Tests	141	207
Hull and turret accessories system. (See Radio interference suppression system.)		
Hull armor:		
Tabulated data	5	17
Hull electrical system:		
Conduits, cables, and connectors	192	330
Description	192	327
Engine wiring junction box	192	330
General	192	327
Master junction box	192, 194	330, 335
Hull seat assemblies:		
Installation	262	481
Removal	262	481
Hydraulic pump operates but turret cannot turn	138	193
Hydraulic pump or electric motor fails	138	193

	Paragraph	Page
Ignition system:		
Booster and filter coils	133	305
Data	133	305
General	129, 133	171, 304
Ignition wiring harness	133	308
Magnetos	133	304
No spark or unsatisfactory spark	129	172
Resistance and continuity tests	129	175
Spark plugs	133	305
Timing	134	305
Trouble shooting	129	174
Ignition system (*see also* radio interference suppression system)	332	553
Ignition wiring harness:		
Installation	187	317
Removal	187	316
Illumination:		
Azimuth indicator T24	64	73
Ballistic drive T23E1	61	68
Range finder T41	58	61
Indicator lights:		
Description	39, 40, 41	44, 46
Installation	285	505
Removal	285	505
Indirect fire:		
Fire control equipment and sighting	57	57
Operation	87	96
Inspection:		
Adjusting-idler wheels and hubs	251	454
Auxiliary generator and engine	93	105
Compensating-idler arm and spindle	252	454
Compensating-idler wheels	252	454
Exhaust manifold	153	234
Front road wheel arms and shackles	250	447
Front road wheels	250	431
Front spring arms	250	446
Intake manifold	167	268
Intermediate and rear road wheels and arms	250	430
Magnetos	185	312
Oil pressure control valve	150	228
Preparation of matériel for firing	75	84
Rear shroud (main engine)	169	271
Road wheel hubs and bearings	250	430
Road wheels, hubs, arms, and bumpers	250	430
Rocker replacement	156	243
Shock absorbers and brackets	255	462
Spark plugs	186	316
Torsion bars	253	455
Track drive sprockets	244	414
Tracks and suspension	249	423
Track support rollers and bracket assemblies	254	459
Transmission and final drivers	124	188
Valves:		
Oil cooler bypass	151	232
Oil filter bypass	151	231
Wheel arm supports	250	449

689

	Paragraph	Page
Instruction plates	4	14
Instructions:		
Driving vehicle	70	81
Painting	119	143
Radio interference suppression system	142	207
Stopping main engine	72	82
Stopping vehicle	71	82
Instructions, operating:		
Correction of deficiencies	7	21
Preliminary service	8	22
Procedure:		
Air accessories	8	22
Air cleaners	8	22
Batteries	8	22
Electrical wiring	8	22
Engine controls	8	22
Engine-high-oil-temperature-warning light	8	22
Engine-oil-pressure gage	8	22
Fuel gages	8	22
Gun traveling lock	8	22
Horn, blower and heater	8	22
Hull and paulin	8	22
Leaks	8	22
Lubrication	8	22
Main-generator-warning light	8	22
Oil	8	22
Primer pump	8	22
Suspension	8	22
Tachometer	8	22
Test auxiliary engine generator	8	22
Towing connections	8	22
Tracks	8	22
Transmission-high-oil-temperature-warning-signal light	8	22
Transmission-low-lubrication-pressure-warning-signal light	8	22
Transmission steering and shifting linkage	8	22
Universal joints	8	22
Vision devices	8	22
Wheel and hub nuts	8	22
Instrument and hull accessories panels, instruments, switches, and sending units:		
Accessory outlet socket:		
General	212	359
Installation	212	360
Removal	212	359
Auxiliary-engine magneto switch:		
General	210	359
Installation	210	359
Removal	210	359
Testing	210	359
Auxiliary-engine starter switch:		
General	209	358
Installation	209	358

Instrument and hull accessories panels, instruments, switches, and sending units—Continued

	Paragraph	Page
Auxiliary-engine starter switch—Continued		
Removal	209	358
Testing	209	358
Batteries and generating system:		
Description and data	221	370
Tabulated data	221	374
Bilge pump switches:		
General	208	358
Installation	208	358
Removal	208	358
Testing	208	358
Cables and connectors:		
Description	199	334
Procedure	199	334
Engine warning-signal-light switches:		
General	218	367
Installation	218	367
Removal	218	367
Testing	218	367
Gages:		
General	205	356
Installation	205	356
Removal	205	356
Testing	205	357
Heater box control:		
General	211	359
Installation	211	359
Removal	211	359
Hull accessories panel:		
General	201	354
Installation	201	355
Removal	201	355
Hull accessories panel circuit breakers:		
General	215	361
Installation	215	361
Removal	215	361
Testing	215	361
Instrument panel:		
General	200	354
Installation	200	354
Removal	200	354
Instrument panel circuit breakers:		
General	214	360
Installation	214	360
Removal	214	360
Testing	214	361
Light switch:		
General	203	355
Installation	203	356
Removal	203	355
Testing	203	356

Instrument and hull accessories panels, instruments, switches, and sending units—Continued

	Paragraph	Page
Main engine starter and magneto switch:		
General	204	356
Installation	204	356
Removal	204	356
Testing	204	356
Master current relay rectifier:		
General	213	360
Installation	213	360
Removal	213	360
Master relay switch:		
General	207	358
Installation	207	358
Removal	207	358
Testing	207	358
Oil cooler fan controllers and thermostats:		
General	220	358
Replacement of thermostats	220	358
Sending units:		
General	217	362
Installation	217	362
Removal	217	362
Testing	217	363
Speedometer-tachometer:		
General	202	355
Installation	202	355
Removal	202	355
Testing	202	355
Transmission-warning-signal-light switch:		
General	219	367
Installation	219	367
Removal	219	367
Testing	219	367
Warning signal lights:		
General	206	357
Main warning light	206	357
Wiring harnesses:		
General	216	361
Installation	216	362
Removal	216	361
Instrument lights	65	75
Instrument panel components:		
Testing	131	177
Instruments:		
Run-in procedures	9	22
Instruments. (*See* Instrument and hull accessories panels.)		
Instrument and controls:		
General	10	22
Intake manifold:		
Cleaning	167	270
Inspection	167	270
Installation	167	270
Removal	167	268

	Paragraph	Page
Intermediate and rear road wheel arms:		
Assembly	250	446
Cleaning	250	445
Disassembly	250	433
Inspection	250	445
Installation	250	446
Removal	250	433
Intermediate and rear road wheels:		
Cleaning	250	430
Inspection	250	430
Installation	250	431
Removal	250	430
Leaks	9	22
Light switch:		
Location	27	34
Operation:		
Blackout driving	27	35
Blackout marker	27	35
Service head	27	35
Lighting system (see also horn and lighting system)	133	185
Limit switches	281	502
Lines (see also oil coolers and lines)	135	190
Linkage (see also driver's controls and linkage)	137	192
Loader's escape hatch assembly:		
Installation	276	497
Removal	276	497
Loader's firing reset safety switch:		
Description	41	46
Installation	287	508
Removal	287	508
Testing	287	508
Loader's firing reset safety switch indicator lamp:		
Installation	287	508
Removal	287	508
Loader's hatch door controls	52	55
Loader's periscope housing		
Installation	277	497
Removal	277	497
Loader's safety relay (turret):		
Installation	288	509
Location	281	502
Removal	288	508
Loader's seat:		
Installation	278	497
Removal	278	497
Loader's traverse safety box (turret):		
Description	281	502
Installation	285	505
Removal	285	505
Loader's-traverse-safety switch and indicator light	40	44
Loading (90-mm gun)	78	87
Location:		
Accelerator pedal	14	23
Commander's-power-traversing-and-elevating-control handle	38	41

693

Location—Continued	Paragraph	Page
Crew compartment heater	97	108
Elevation-cylinder-control valve	44	46
Elevation-supercharge-line-shut-off valve	46	50
Escape hatch doors	18	26
Fire extinguishers	94	107
Firing selector switches	39	44
Firing safety panel	281	502
Fuel-cut-off switch	26	34
Fuel-tank-shut-off valve handles	17	25
Hand brake lock handle	13	23
Hand pump	43	46
Hand throttle lever	15	23
Horn switch	30	35
Gunner's-manual-elevating control handle	37	41
Gunner's-manual-traversing control handle	36	39
Gunner's-power-traversing-and-elevating-control handles	35	39
Indicator lights	39	44
Light switch:		
Main switch	27	30
Mechanical lock	27	30
Loader's-firing-reset-safety switch and indicator light	41	46
Loader's-traverse-safety switch and indicator light	40	44
Machine gun firing mechanism	281	502
Main engine:		
Booster switch	25	34
Magneto switch	23	32
Starter switch	24	34
Master relay switch	22	32
Oil-return valve handle	45	49
Primer pump	16	25
Service brake pedal	12	23
Speedomer-tachometer	31	38
Turret control switch	39	44
Turret lock	42	46
Ventilating blower	96	108
Low oil pressure	127	167
Low-oil-pressure-warning-signal light comes on	134	190
Lubrication and painting	113	132
Lubrication:		
Track support rollers and bracket assemblies	254	459
Lubrication instructions:		
After dusty and sandy conditions	118	143
After fording	117	133
Application	114	132
Equipment	114	132
Instructions	114	132
Operation below 0° F	116	133
Order	113	132
Reports and records	114	132
Unusual conditions	115	132
Lubrication system (main engine)	127, 148	167, 223
Machine gun firing mechanism	281	502
Machine gun mounts	363	582

	Paragraph	Page
Machine guns:		
Cal. .30 Browning:		
Installation	365	583
Removal	365	583
Cal. .50 HB Browning M2:		
Installation	366	584
Removal	366	584
Coaxial mount	53	55
Firing solenoids:		
Installation	291	510
Removal	291	510
Magnetos:		
General	185	311
Inspection	185	316
Installation	185	313
Removal	185	313
Magnetos (ignition system)	183	304
Magneto switch (main engine)	23	32
Main engine:		
Accessory drive ratios	145	222
Booster switch	25	34
Cooling fan rotors:		
Installation	154	237
Removal	154	237
Cranks slowly	130	176
Data	145	217
Description	145	218
Does not develop full power	126	165
Excessive oil consumption	126	166
Fails:		
To crank	126	163
To keep running	126	165
To start	126	163
General	126	163
Installation	164	264
In vehicle (operation)	147	222
Low oil pressure	126	166
Lubrication system:		
Description	127, 148	167, 223
Drain engine	148	225
Excessive oil consumption and pressure	127	168
Fill engine lubrication	148	225
High oil temperature	127	169
Low oil pressure	127	167
No oil pressure	127	169
Oil level check	148	225
Trouble shooting	127	169
Magneto switch:		
Location and operation	23	32
Mounting:		
Installation and removal	170	275
Oil filter:		
Cleaning	149	228
Installation and removal	149	227
Oil pressure gage (mounted)	33	38

	Paragraph	Page
Main engine—Continued		
Overheats	126	163
Rear shroud:		
Inspection	169	274
Installation	169	274
Removal	169	271
Removal	164	264
Removal from vehicle (operation)	165	267
Run-in procedure	9	22
Slow down but does not stop	128	171
Starter switch	24	34
Tabulated data	145	218
Temperature high, oil pressure low	135	190
Top shroud:		
Installation and removal	152	232
Transmission temperature high, oil pressure normal	135	191
Trouble shooting	126	165
Tune up	146	222
Will not start	130	175
Wiring junction box:		
General	195	341
Installation and removal	195	341
Main generator:		
Testing	132	182
Trouble shooting	132	184
Maintenance:		
90-mm gun T119E1	339	557
Breechblock	344	569
Breech ring	344	569
Coaxial cal. .50 HB Browning machine gun M2 in machine gun cradle	364	583
Cal. .30 Browning machine gun M1919A4 in cal. .30 ball gun mount 7351999	365	583
Cal. .50 HB Browning machine gun M2 in cal. .50 machine gun mount 6580030	366	584
Combination gun mount	347	573
Firing linkage	352	577
Firing mechanism	344	569
Operations	301	522
Power elevating mechanism	361	581
Preventive services:		
Cleaning	121	145
General:		
After-operation	120	143
At-the-halt	120	143
Before-operation	120	143
During-operation	120	143
Weekly services	120	143
Preventive maintenance (tables III, IV, V)		
Procedures (organizational maintenance mechanics (table V))	123	153
Maintenance operations, procedure (table V)	123	154
Maintenance under unusual conditions:		
After fording	375	591
After operation on unusual terrain	376	592

Maintenance under unusual conditions—Continued

	Paragraph	Page
Extreme-cold weather:		
Armament	372	590
Sighting and fire control	373	590
Vehicle	371	590
Extreme-hot weather	374	591
Manual control box and linkage:		
Adjustment	257	463
Installation	257	464
Removal	257	463
Service	257	463
Manual control lever:		
Description	11	23
Placing vehicle in motion	69	79
Manual control lever and brakes:		
Run-in procedure	9	22
Manual elevating mechanism:		
Description	362	582
Maintenance	362	582
Manual traversing mechanism (turret):		
Installation	271	495
Removal	271	495
Master junction box:		
General	194	337
Installation	194	337
Removal	194	337
Trouble shooting	132	182
Master relay switch:		
Location	22	32
Miscellaneous equipment:		
Fire control equipment and sighting	57	57
Mufflers (see also exhaust pipes and mufflers)	182	303
Name plate	4	14
Neutral position switch:		
Installation and removal	191	326
No oil pressure:		
Engine lubrication	127	169
Normal driving of vehicle	69	81
Observation of fire	88	97
Observation of terrain:		
Commander's direct-view prisms	89	98
Periscope M13, M13B1, or M6	89	98
Periscope T35	89	98
Ohmmeter method of electrical testing:		
Continuity tests	125	161
General	125	161
Resistance reading	125	161
Terminal and ground designations	125	161
Oil cooler bypass valves (see also oil filter bypass and oil cooler bypass valves)	151	231
Oil cooler fan and clutch brushes:		
Installation	241	409
Removal	241	408
Oil cooler fan controllers and thermostats	220	368

	Paragraph	Page
Oil cooler lines	240, 242	403, 412
Oil coolers and lines:		
Description	240	403
Engine and transmission temperature high, oil pressure:		
High	135	190
Normal	135	190
General	240	403
Installation	242	412
Oil cooler fan, clutch, and brushes	240	403
Oil cooler lines	240, 242	403, 412
Removal	242	412
Transmission temperature high, oil pressure low	135	190
Trouble shooting	135	190
Oil filter bypass and oil cooler by pass valves:		
Adjustment	151	231
Assembly	151	232
Cleaning	151	232
Disassembly	151	231
General	151	231
Inspection	151	232
Installation	151	232
Removal	151	231
Oil level check (engine lubrication)	148	225
Oil pressure control valve:		
Assembly	150	230
Cleaning	150	230
Disassembly	150	228
Inspection	150	230
Installation	150	230
Removal	150	228
Oil pressure gage (main engine)	33	38
Oil pump	300	521
Oil-return-valve handle:		
Location	45	49
Oil-temperature-warning-signal light comes on	134	190
Operating instructions (*see also* instructions)	6	21
Operating to elevate or depress gun	38	41
Operating to traverse the turret	38	41
Operation:		
Bilge pump	95	108
Blackout driving and marker lights	27	35
Breech mechanism	77	86
Breech operating mechanism	341	561
Commander's - power - traversing - and - elevating - control handle	38	43
Direct fire:		
Primary sighting	87	96
Secondary sighting	87	96
Engine removed	165	267
Fuel cut-off switch	128	171
Fuze setter M14	85	93
Gun traveling lock	55	57
Gunner's-power-traversing-and-elevating-control handles	35	39

	Paragraph	Page
Operation—Continued		
Indirect fire:		
Deflection correction	87	96
Laying gun	87	96
Light switch	27	34
Main engine-magneto switch	23	32
Performance with engine in vehicle	147	222
Range finder T41	58	60
Sighting and fire control instruments:		
Azimuth indicator T24	84	93
Elevation quadrant T21	84	93
Periscope M13, M13B1, or M6	84	93
Periscope T35	84	89
Superelevation transmitter T13	58	60
Turret and elevating gun	76	85
Unsteady turret	138	194
Usual conditions	67	77
Operation under unusual conditions:		
Armament	101	113
Extreme—cold	100, 101	111, 113
Extreme—hot	102	114
Fording	104	115
General	99	109
Ice	103	115
Lubrication	100, 105, 106, 107, 108	111, 116
Mud	103	114
Sand	103	115
Sighting and firing control matériel	101	113
Snow	103	114
Winterization	100	112
Outlet, accessory:		
Tests	141	207
Override of gunner:		
Commander's - power - traversing - and - elevating - control handle	38	43
Override motor control box:		
Location	281	499
Painting instructions	119	143
Percussion mechanism:		
Assembly	345	569
Disassembly	343	568
Periscope M13, M13B1, or M6:		
Description	62	71
Observation of terrain	89	98
Tabulated data	5	17
Periscope T35:		
Controls	59	66
Description	59	64
Observation of terrain	89	98
Reticle pattern	59	66
Periscope mounts:		
Tabulated data:		
Commander's T177	5	17
Gunner's T176	5	17

699

	Paragraph	Page
Placing vehicle in motion:		
Description	69	79
Manual control lever	69	80
Normal driving	69	81
Reverse vehicle	69	81
Power elevating mechanism:		
Description	361	581
Inoperative or slow	138	194
Maintenance	361	581
Power pack:		
General	198	343
Installation	198	344
Removal	198	344
Power plant:		
Installation:		
Instructions	160	254
Procedure	160	254
Removal:		
Disconnect points	159	252
General	159	249
Instructions	159	249
Procedures	159	252
Replacement	157	245
Power train and suspension (90-mm gun tank M47)	3	13
Power traversing system (turret):		
Check, clean, drain, and fill oil reservoir	270	493
Precautions:		
Traversing turret and elevating guns	76	85
Preliminary service	8	22
Premature firing (90-mm gun T119E1)	143	215
Preparation:		
Armament for traveling	83	88
Fire control instruments for traveling	90	98
Matériel for firing:		
Inspection:		
After	75	84
Before	75	84
During	75	84
Procedure	75	84
Preventive maintenance by driver or operator (table IV):		
Armament:		
Unusual conditions	122	151
Usual conditions	122	150
Vehicle:		
Unusual conditions	122	147
Usual conditions	122	146
Preventive maintenance services by organizational mechanic or maintenance crew (table V):		
Intervals	123	152
Procedure	123	153
Primer lines:		
Cleaning	181	303
General	181	302
Installation	181	302
Removal	181	302

	Paragraph	Page
Primer pump:		
Installation	180	299
Location	16	25
Removal	180	299
Procedure:		
Air accessories	8	22
Air cleaners	8	22
Batteries	8	22
Bore sighting:		
90-mm gun	91	99
Coaxial machine gun	91	100
Periscope T35	91	104
Range finder T41	91	104
Cables and connectors	199	344
Electrical wiring	8	22
Engine:		
High-oil-temperature-warning light	8	22
Oil-pressure gage	8	22
Warm up	8	22
Fenders, dust shield, and guards	8	22
Fire extinguisher	8	22
Fuel:		
Filters	8	22
Gages	8	22
Gun traveling lock	8	22
Horn, blower, and heater	8	22
Hull and paulin	8	22
Leaks	8	22
Lights	8	22
Lubrication	8	22
Magneto	310	531
Main-generator-warning light	8	22
Oil	8	22
Preparation of matériel for firing	75	84
Primer pump	8	22
Suspension	8	22
Tachometer	8	22
Test auxiliary engine generator	8	22
Towing connections	8	22
Tracks	8	22
Transmission:		
High-oil-temperature-warning-signal light	8	22
Low-lubrication-pressure-warning-signal light	8	22
Steering and shifting linkage	8	22
Universal joints	8	22
Vision devices	8	22
Wheel and hub nuts	8	22
Procedure (table V):		
Armament	123	156
Auxiliary-generator and engine	123	156
Maintenance operations	123	157
Road test	123	155
Tools and equipment	123	153
Unusual conditions	123	160
Pump, bilge (front and rear)	141	207

	Paragraph	Page
Radio interference suppression system:		
Description	331	553
Excessive radio interference, vehicle in motion	142	211
Fasteners and bond straps	336	555
Generating system:		
Auxiliary generator	333	554
Main generator	333	554
Hull and turret accessories system:		
Bilge pumps	334	554
Crew compartment heater	334	554
Ventilating blower	334	554
Ignition system:		
Auxiliary engine	332	554
Main engine	332	553
Instructions	142	207
Purpose	330	553
Radio interference:		
Auxiliary generator operating	142	210
Crew compartment heater operating	142	211
Turret traversed	142	211
Vehicle not in motion, engine running	142	207
Ventilating blower operating	142	211
Turret electrical system:		
Description	335	554
Installation	335	555
Removal	335	554
Radio sets. (*See* Communications system.)		
Radio terminal box (turret):		
Installation	295	512
Location	281	503
Removal	295	512
Testing	295	512
Range finder T41:		
Controls	58	63
Description	58	60
Illumination	58	64
Operation	58	60
Reticle pattern	58	61
Tabulated data	5	17
Ration box and ammunition boxes:		
Installation	265	485
Removal	265	484
Recoil mechanism:		
Assembly	359	581
Description	355	578
Disassembly	357	579
Functioning	356	579
Maintenance	358	579
Replacement:		
Authorized parts	368	587
Auxiliary generator and engine	303	524
Blackout marker lights	230	388
Brake control linkage	259	471
Brushes (starter)	190	323
Commander's cupola assembly	273	496

	Paragraph	Page
Replacement—Continued		
Dome lights	232	389
Final drives	246	417
Gun ready indicator	289	509
Lamp unit	228, 229	386, 388
Power plant	157	245
Shifting control linkage	257	464
Starter-drive-gear-hub oil seal (starter)	190	325
Steering control linkage	257	464
Switches and thermostats	220	368
Transmission	239	402
Replacement of switches and thermostats. (*See* Oil cooler fan controllers and thermostats.)		
Replacing transmission oil	235	394
Replenisher indicator:		
Description	56	57
Resistance and continuity tests:		
Ignition system	129	171
Starting system	130	175
Turret	139	195
Resistance reading:		
Ohmmeter method of electrical testing	125	161
Reticle pattern:		
Periscope T35	59	64
Range finder T41	58	60
Reverse and low gear band adjustments:		
Adjust bands	236	395
Remove inspection covers	236	395
Reverse vehicle	69	79
Road test, procedure (table V)	123	155
Road wheel hubs and bearings:		
Cleaning	250	432
Inspection	250	432
Installation	250	433
Removal	250	431
Road wheels, hubs, arms, and bumpers:		
Cleaning	250	430
Inspection	250	430
Installation	250	431
Removal	250	430
Rocker replacement:		
Adjustment	156	243
Cleaning	156	244
Inspection	156	244
Installation	156	244
Removal	156	243
Run-in procedure:		
Eelevation and traversing mechanism	9	22
Engine	9	22
Instruments and gages	9	22
Leaks	9	22
Manual control lever	9	22
Temperature	9	22
Tracks and suspension	9	22
Unusual noise	9	22

	Para-graph	Page
Scope:		
General	1	1
Trouble shooting	124	160
Sending units:		
Testing	131	177
Sending units. (*See* Instruments and hull accessories panels.)		
Service:		
Auxiliary generator and engine	302	523
Bilge pumps	329	551
Final drives	246	417
Manual control box and linkage	257	463
Service brake pedal:		
Location	12	23
Service headlights:		
Installation	228	386
Removal	228	386
Replacement of lamp unit	228	386
Servicing:		
Air cleaners	175	285
Batteries	224	377
Shifting control linkage:		
Adjustment	257	463
Replacement	257	466
Shipment and limited storage:		
Blocking gun tank on flatcar	398	649
Domestic shipping	395	641
Limited storage	396	642
Loading on flatcars	397	643
Shock absorbers and brackets:		
Cleaning	255	462
Inspection	255	462
Installation	255	462
Removal	255	462
Sighting and fire control equipment:		
Arrangement	57	57
Destruction	400	658
Direct fire	57	57
General	57, 367	57, 585
Indirect fire	57	60
Miscellaneous equipment	57	60
Use	57	57
Sighting and fire control instruments:		
Operation:		
Azimuth indicator T24	84	89
Elevation quadrant T21	84	89
Periscope M13, M3B1, or M6	84	89
Periscope T35	84	89
Tabulated data	5	17
Slave battery receptable:		
Description	98	109
Installation	197	343
Removal	197	343
Slip ring box:		
Installation	296	512
Location	281	503

	Paragraph	Page
Slip ring box—Continued		
Removal	296	512
Testing	296	512
Spark plugs:		
Adjustment	186	316
Inspection	186	316
Installation	186	316
Removal	186	315
Spark plugs (ignition system)	183	315
Special tools and equipment (table II)	112	118
Speedometer-tachometer	31	38
Sprocket hubs:		
Installation	245	415
Removal	245	415
Starter:		
Installation	190	325
Removal	190	321
Replacement of brushes	190	323
Replacement of starter-drive-fear-hub oil seal	190	325
Starter switch (main engine)	24	34
Starting:		
Auxiliary generator and engine	93	105
Main engine	68	77
Starting system:		
Data	189	319
Description	189	319
Engine cranks slowly	130	176
Engine will not start	130	175
Resistance and continuity tests	130	176
Starter spins but does not crank engine	130	176
Trouble shooting	130	176
Steering control linkage:		
Adjustment	257	463
Replacement	257	466
Steering vehicle	70	81
Stopping:		
Auxiliary generator and engine	93	105
Main engine	72	82
Vehicle	71	82
Stowage boxes (turret):		
Installation	280	499
Removal	280	498
Superelevation transmitter T13:		
Description	58	60
Operation	58	60
Suspension (see also tracks and suspension)	136	191
Switches:		
Tests:		
Auxiliary engine magneto and starter	141	207
Bilge pump front and rear	141	207
Heater "ON"—"OFF"	141	207
Heater safety valve seat	141	207
Switches. (See Instrument and hull accessories panels.)		

	Paragraph	Page
Tabulated data:		
90-mm gun T119E1	337	555
Air-intake system	171	275
Batteries and generating system	221	374
Combination gun mount M78	347	573
Exhaust system	171	275
Fuel system	171	275
General	5	13
Main engine	145	218
Transmission	234	391
Tail lights:		
Installation	231	388
Removal	231	388
Tank, destruction	403	660
Telescope:		
Tabulated data	5	17
Temperature:		
Run-in procedure	9	22
Terminal and ground designation:		
Ohmmeter method of electrical testing	125	161
Testing:		
Auxiliary generator and engine	132	182
Batteries	224	381
Booster motor	292	510
Booster motor relay	293	511
Bore sighting	91	99
Charging system	132	182
Commander's control handle assembly	283	504
Elevation and depression limit switches	286	506
Loader's-firing-reset-safety switch	287	507
Radio terminal box	295	512
Sending units	131	176
Slip ring box	296	512
Testing target method:		
Bore sighting	91	99
Tests:		
Auxiliary equipment	141	207
Auxiliary generator	132	182
Batteries	224	377
Bilge pump switches	208	358
Charging system	132	182
Circuit breaker:		
Auxiliary-engine-magneto switch	210	359
Auxiliary-engine starter and magneto circuit	141	207
Engine-warning-signal-light switches	218	366
Front bilge pump	141	207
Rear bilge pump	141	207
Gages	205	356
Horn and lighting system	133	185
Hull accessories panel:		
Circuit	215	361
Components	141	207
Instrument panel components	131	176
Main generator	132	182
Master relay switch	207	358

	Paragraph	Page
Tests—Continued		
Outlet, accessory	141	207
Sending units	131, 217	176, 364
Speedometer-tachometer	202	355
Switches:		
Auxiliary engine magneto	141	207
Auxiliary engine starter	141	207
Bilge pump (front)	141	207
Bilge pump (rear)	141	207
Heater "ON"—"OFF"	141	207
Heater safety valve reset	141	207
Transmission-warning-signal-light switches	219	367
Throttle control linkage:		
Adjustment	258	468
Installation	258	468
Removal	258	468
Thrown track	136	191
Timing:		
Ignition system	184	305
Magneto	310	530
Tools and equipment (table II)	112	118
Tools and equipment, procedure (table V)	123	153
Top deck:		
General	158	245
Installation	161	261
Removal:		
Engine compartment	158	247
Torsion bars:		
Cleaning	253	456
Identification	253	456
Inspection	253	456
Installation	253	456
Removal	253	455
Towing pintle:		
Installation for use	268	488
Removal for stowage	268	488
Towing the vehicle:		
Description	74	83
General	74	83
Track drive sprockets:		
Cleaning	244	414
Inspection	244	414
Installation	244	415
Removal	244	414
Tracks:		
Adjustment	249	424
Cleaning	249	426
Connecting	249	425
Disconnecting	249	425
Inspection	249	423
Installation	249	427
Removal	249	425
Tracks and suspension:		
Cleaning	249	428
Connecting	249	425

	Paragraph	Page
Tracks and suspension—Continued		
Data	248	420
Description	248	420
General	248	420
Hard riding	136	191
Inspection	249	423
Installation	249	427
Removal	249	425
Run-in procedure	9	22
Thrown track	136	191
Vehicle leads to one side	136	191
Vehicle sags to one side	136	191
Track support rollers and bracket assemblies:		
Assembly	254	461
Cleaning	254	461
Disassembly	254	459
Inspection	254	461
Installation	254	461
Lubrication	254	461
Removal	254	459
Transmission:		
Data	234	391
Description	234	391
Tabulated data	234	393
Transmission and final drives:		
Inspection	134	188
Low-oil-pressure-warning-signal light comes on	134	190
Oil-temperature-warning-signal light comes on	134	190
Trouble shooting	134	188
Vehicle will not drive	134	187
Transmission hydraulic and lubrication system:		
Check oil level	235	394
Check transmission oil pressure	235	395
Description	235	393
Oil pressure tests. (*See* table VII.)		
Replacing transmission oil	235	394
Transmission oil filter:		
Cleaning	238	401
Installation	238	401
Removal	238	401
Transmission replacement:		
Installation	239	403
Removal	239	402
Transmission side oil strainers:		
Cleaning	237	400
Description	237	400
Installation	237	400
Removal	237	400
Transmission temperature high; oil pressure low	135	190
Traversing lock (turret):		
Adjustment	272	496
Installation	272	496
Removal	272	496

	Paragraph	Page
Traversing turret and elevating gun:		
General	76	85
Precautions	76	85
Trouble shooting:		
Auxiliary generator and engine	140	206
Fuel system	128	169
Horn and lighting system	133	185
Ignition system	129	171
Main engine:		
Cranks slowly	130	176
Does not develop full power	126	165
Excessive oil consumption	126	166
Fails to crank	126	163
Fails to keep running	126	165
Fails to start	126	163
Low oil pressure	126	166
Lubrication	127	167
Overheats	126	167
Will not start	130	175
Oil coolers and lines	135	190
Reistance and continuity tests	130	176
Scope	124	160
Starter spins but does not crank engine	130	176
Starting system	130	175
Transmission and final drives	134	187
Turret:		
90-mm firing mechanism:		
Description	281	502
Installation	290	509
Removal	290	509
90-mm gun tank M47	3	13
Accessory panel:		
Description	281	499
Installation	297	516
Removal	297	516
Ammunition chute micro switch:		
Description	281	503
Installation	294	512
Removal	294	512
Armor, data	5	17
Ballistic box	281	503
Booster motor:		
Description	281	502
Installation	292	510
Removal	292	510
Testing	292	510
Booster motor relay:		
Description	281	503
Installation	297	516
Removal	297	516
Testing	297	516
Circuit breakers:		
Installation	297	516
Removal	297	516

Turret—Continued

	Paragraph	Page
Commander's control handle assembly:		
Description	281	499
Installation	283	504
Removal	283	504
Testing	283	504
Control switch, location	39	44
Description	269, 291	488
Dome lights	281	503
Elevating gun:		
Description	3	13
Operating	76	85
Unsteady	138	194
Elevation and depression limit switches:		
Adjustment	286	506
Installation	286	506
Removal	286	506
Testing	286	506
Firing safety panel:		
Description	281	502
Installation	287	508
Removal	287	507
General	269, 281	488, 499
Gun ready indicator	281	502
Gunner's control handle assembly:		
Installation	284	505
Location	281	502
Removal	284	505
Testing	284	505
Indicator light:		
Installation	285	505
Removal	285	505
Limit switches	281	502
Loader's firing reset safety switch:		
Installation	287	508
Removal	287	507
Testing	287	508
Loader's firing reset safety switch indicator lamp:		
Installation	287	508
Removal	287	508
Loader's traverse safety box:		
Description	281	502
Installation	285	505
Removal	285	505
Lock, location	42	46
Machine gun firing mechanism:		
Installation	291	510
Removal	291	510
Manual traversing mechanism:		
Installation	271	495
Removal	271	495
Override motor control box:		
Installation	282	503
Location	281	499

Turret—Continued
- Override motor control box—Continued

	Paragraph	Page
Removal	282	504
Testing	282	503
Power traversing system:		
Check	270	493
Clean	270	493
Drain	270	493
Radio terminal box:		
Description	281	503
Installation	295	512
Removal	295	512
Testing	295	512
Seat assemblies:		
Installation	278	497
Removal	278	497
Slip ring box:		
Description	281	503
Installation	296	512
Removal	296	512
Testing	296	512
Stowage boxes:		
Installation	280	498
Removal	280	498
Traversing and elevating systems:		
Abnormal noise in hydraulic pump or motor	138	194
Check	138	193
Commander's control inoperative	138	194
Gunner's control inoperative	138	194
Hand elevating mechanism inoperative or weak	138	194
Hydraulic pump operates but turret cannot turn	138	193
Hydraulic pump or electric motor	138	193
Power elevating mechanism inoperative or slow	138	195
Sluggish or unsteady turret operation	138	193
Speed low or erratic	138	194
Turret creeps excessively	138	194
Turret creeps in one direction	138	194
Turret does not traverse freely	138	193
Traversing lock:		
Adjustment	272	496
Installation	272	496
Removal	272	496
Trouble shooting:		
Accessories	139	197
Ammunition feed control	139	202
Ballistic box	139	204
Cannot be elevated or depressed	139	198
Description	139	195
Dome lights	139	204
Elevates too far or not far enough	139	198
Firing control	139	196
Gun elevating	139	195
Resistance and continuity tests	139	204
Traverses too far or not far enough	139	198
Ventilating blower does not operate	139	203

	Paragraph	Page
Turret—Continued		
Ventilating blower:		
Description	281	503
Installation	298	517
Removal	298	517
Ventilator switch:		
Installation	297	516
Removal	297	516
Universal joints:		
Description	243	414
Installation	247	419
Removal	247	419
Unloading a stuck round (90-mm gun T119E1)	143	216
Unusual conditions, procedure	123	160
Unusual noise:		
Run-in procedure	9	22
Use:		
Fire control equipment and sighting	57	57
Valve timing.		
Adjustment	155	237
Vehicle leads to one side	136	191
Vehicle sags to one side	136	191
Vehicle will not drive:		
Any gear	134	187
High gear	134	187
Low gear	134	187
Reverse gear	134	190
Ventilating blower:		
90-mm gun tank M47	3	13
Location	96	108
Turret:		
Does not operate	139	204
Installation	298	517
Location	281	503
Removal	298	517
Ventilator switch:		
Installation	297	516
Removal	297	516
Wheel arm supports:		
Assembly	250	449
Cleaning	250	449
Dissasembly	250	449
Inspection	250	449
Installation	250	450
Removal	250	449
Wiring harnesses:		
Installation	216	362
Removal	216	361

©2013 Periscope Film LLC
All Rights Reserved
ISBN#978-1-937684-55-6
www.PeriscopeFilm.com